The Dirichlet Space and Related Function Spaces

Mathematical Surveys and Monographs

Volume 239

The Dirichlet Space and Related Function Spaces

Nicola Arcozzi
Richard Rochberg
Eric T. Sawyer
Brett D. Wick

AMERICAN MATHEMATICAL SOCIETY
Providence, Rhode Island

2010 *Mathematics Subject Classification.* Primary 30-XX, 31-XX, 32-XX, 39-XX, 46-XX, 47-XX.

For additional information and updates on this book, visit
www.ams.org/bookpages/surv-239

Library of Congress Cataloging-in-Publication Data

Names: Arcozzi, N. (Nicola), 1963– author. | Rochberg, Richard, author. | Sawyer, E. T. (Eric T.), 1951– author. | Wick, Brett D., author.

Title: The Dirichlet space and related function spaces / Nicola Arcozzi, Richard Rochberg, Eric T. Sawyer, Brett D. Wick.

Description: Providence, Rhode Island : American Mathematical Society, [2019] | Series: Mathematical surveys and monographs ; volume 239 | Includes bibliographical references and index.

Identifiers: LCCN 2019011699 | ISBN 9781470450823 (alk. paper)

Subjects: LCSH: Function spaces. | Dirichlet principle. | Hilbert space. | Functional analysis. | AMS: Potential theory. msc | Several complex variables and analytic spaces. msc | Difference and functional equations. msc | Functional analysis. msc | Operator theory. msc

Classification: LCC QA323 .D55 2019 | DDC 515/.98–dc23

LC record available at https://lccn.loc.gov/2019011699

Visit the AMS home page at `https://www.ams.org/`

10 9 8 7 6 5 4 3 2 1   24 23 22 21 20 19

To Melinda, Nan, Sheryl and Elodie

Contents

Preface

Overview

The vibrant interaction between function theory and functional analysis is more than one hundred years old [**262**], [**277**], and it continues to prosper. In this volume we consider topics in that tradition, all centered around the classical Dirichlet space, $\mathcal{D}$. That space is the set of functions f which are holomorphic on the disk having finite Dirichlet integral:

$$\mathfrak{D}(f) := \frac{1}{\pi}\int_{\mathbb{D}} |f'|^2 \, dA < \infty.$$

In the first part of the book we focus on $\mathcal{D}$ as a reproducing kernel Hilbert space. From that viewpoint we investigate multipliers, Carleson measures, zero sets, and interpolating sequences for both the space and for its multiplier algebra. The fact that the reproducing kernel for $\mathcal{D}$ has the complete Pick property is often a central factor.

In the second part of the book we consider more specialized topics related to $\mathcal{D}$, including discussions of interpolation, boundary behavior, alternative norms, the local Dirichlet integral, shift invariant subspaces, and Hankel forms. These treatments are largely independent of each other and can be read in any order. In those sections the Hilbert space structure of $\mathcal{D}$ is still used, but the general theory of reproducing kernel Hilbert spaces has a smaller role and the complete Pick property is rarely mentioned. On the other hand estimates from function theory and potential theory have a larger role.

Some of the results we discuss are in a definitive form. Others are only incomplete resolutions of natural questions. In fact, for many interesting questions about the Dirichlet space, including describing the zero sets, the boundary zero sets, the onto interpolating sequences, and the shift invariant subspaces, only partial results are known.

In the final part we expand our viewpoint in two fundamental ways. We move beyond functions of a single variable and consider spaces of holomorphic functions on the ball in $\mathbb{C}^n$. Also, we go beyond Hilbert spaces and consider Banach spaces of functions. With these changes, new techniques need to be introduced. In particular, analysis of the Besov spaces on the ball requires obtaining good local estimates for higher order derivatives in the context of the anisotropic geometry of the ball, and doing so requires new machinery. Also, as we move to Banach spaces of functions we lose access to tools from the theory of reproducing kernel Hilbert spaces. As a consequence, for instance, when characterizing interpolating sequences in the ball, we obtain interpolating functions constructively, not by abstract existence theorems as done in one variable. Those chapters present basic facts about Carleson measures, multipliers, and interpolating sequences for spaces of holomorphic functions as well as for their discrete model spaces.

The final chapter has a corona type theorem for Besov spaces on the ball as well as a classical multiplier corona theorem for the Hilbert spaces in the family, including H^2, $\mathcal{D}$, and the finite dimensional Drury-Arveson spaces. Interestingly, at this point the discussion comes full circle. The theory of reproducing kernel Hilbert spaces with complete Pick kernels, a tool that had been set aside, is used for passage from "corona type theorems" to classical "corona theorems."

The final chapters contain only a few theorems, some with elaborate proofs. However we feel, and hope some readers feel, that the true heart of those sections lies beyond those specific results, and is in the techniques and tools that are developed. Those techniques include a tool for quantifying local oscillation of Besov space functions, the "almost invariant holomorphic derivative", the development of new types of function spaces on trees which model spaces of holomorphic functions on the ball, and new methods for solving $\overline{\partial}$ equations in several variables with estimates in Besov spaces.

There are two themes which run as subtexts throughout the book. The first is the background role of the Hardy space H^2. The study of H^2 is the historical and conceptual basis of much of modern function theoretic operator theory and understanding how classical Hardy space results do or do not generalize is a continuing major theme. There is also the more recent insight that both H^2 and $\mathcal{D}$ are Hilbert spaces whose reproducing kernels have the complete Pick property and, as such, they share some basic properties. We will discuss the commonalities as we go along.

The other ongoing theme is the study of discrete "tree" models of the disk and ball, as well as function spaces on those trees. The simplest of those are, in the language of physicists, toy models. It is the authors' experience that spaces on trees are very valuable tools, both conceptually and technically. In particular, they are used in essential ways in the proofs of Theorems 13.2, 15.1, and 15.2. Also, it is the authors' opinion that these models are intrinsically very interesting. As we go along we present those views in detail.

The book includes two appendices. The first is a small collection of results from functional analysis. The second concerns Schur's lemma, a method for proving boundedness of linear operators given by positive kernels. We use that basic tool in several contexts and with several variations, and we felt it would be convenient for the authors, and we hope for the readers, to collect that material in a single location.

Most material is presented at the level of a topics course, appropriate for students who have taken the introductory graduate course (e.g. complex analysis, measure theory, and functional analysis). Selections from the material in the first seven chapters could be used as a basis for a graduate course at the level just beyond the qualifying exams. Various subsets of the later chapters could be used as the basis for an advanced graduate course, reading course, or seminar. The material in those later chapters is generally more sophisticated; in particular the discussion of "holomorphic Besov spaces on trees" and the proof of the corona theorem for the Drury-Arveson space are best suited for advanced students or researchers.

Although much of our presentation is self contained, some parts are not. We try to remain clear as we go along which mode we are in, and ask the readers indulgence for the times we are unclear. In the early chapters, working with functions of one variable, we do not offer full discussions of Hardy space theory or of potential theory on function spaces. In later chapters, working with functions of several variables, we use external sources for basic facts about the geometry of the complex ball, and about the functional analysis of Besov spaces. In the final chapter, working on corona questions, we make use of relatively recent research on solution operators for $\overline{\partial}$ equations and on the Koszul complex.

Also, as we go along we will mention without proof some interesting results related to our main narrative; sometimes with a small discussion, in other cases with just a reference.

There are a number of exercises in the first part of the book. Their intention is to encourage the reader to keep the pencil and paper nearby:

> Can one learn mathematics by reading it? I am inclined to say no. Reading has an edge over listening because reading is more active – but not much. Reading with pencil and paper on the side is very much better – it is a big step in the right direction. The very best way to read a book, however, with, to be sure, pencil and paper on the side, is to keep the pencil busy on the paper and throw the book away.

Paul Halmos, "The problem of learning to teach", American Mathematical Monthly 82 (1975), 466–476.

These are rapidly evolving topics. Our hopes of providing an up-to-date and comprehensive treatment collapsed under the weight of material. Thus, this is not a systematic survey of the areas mentioned; rather we present questions, answers, and techniques that the authors hope are an invitation to the area. We begin each chapter with a brief discussion to help create a context for that chapter. We focus on the specifics we know best and which we hope move our narrative forward. We end chapters with a section of notes which include some supplemental references and comments.

We are telling a very rich story, and our telling reflects both our interests and our ignorance. Our apologies to those who are ill treated in our telling.

Sources

The theory of the Hardy space, which is the background arena for many of the early topics, is presented in many places. The sources the authors rely on heavily include [**161**] and [**245**].

The theory of the Dirichlet space is presented in [**136**] and some of that material is also presented here. That presentation gives more emphasis to the potential theoretic aspects of the theory, ours to the Hilbert space theoretic. There are also surveys on the Dirichlet space which discuss many of the results here; see the following [**340**], [**287**], [**51**], [**205**].

Our discussion of interpolating sequences is informed by the presentations in [**161**], [**245**], [**302**], [**294**], and [**4**].

The book [**4**] is a basic reference for Hilbert spaces with reproducing kernels as well as the more specialized theory of kernels with the complete Pick property. The more recent [**255**] also contains a great deal of related material.

The books [**288**], [**346**], and [**345**] are our basic references for function theory and function spaces on the unit ball in complex Euclidean space.

Background about corona problems in several variables is in [**130**], [**129**], and [**196**].

A substantial fraction of the contents here is from research papers, new and old, by the authors and others. We will mention those as we go along; in the text, in the notes at the end of each chapter, or both. However we will certainly overlook some relevant references and we apologize for that.

Some of the material we discuss has not been published before. That includes material in the widely circulated but unpublished manuscripts [**3**], [**72**], and [**220**]; as well as from the more recent [**47**]. Some results here are new; including the John-Nirenberg type theorem in Section 3.8, and the results on interpolating sequences for Besov spaces on the ball, beginning in Chapter 14, which completes the program begun in [**47**].

There are also informal presentations available of some of the topics here. Although we do not use them as references in the text, readers may find them useful. Evolved versions of some of that material is in this volume. First, in the academic year 2010-2011 Brett Wick ran an Internet Analysis Seminar on the topic "The Dirichlet Space." The lectures and some other material from that seminar are available online [**339**]. Second, in the summer of 2011 the Mathematical Sciences Research Institute sponsored a Summer Graduate School titled "The Dirichlet Space, Connections Between Operator Theory, Function Theory, and Complex Analysis." Videos of many of the lectures given there, as well as other material, is available online at [**221**]. Finally, early versions of the first part of the book [**255**] are, as of now, available online [**254**].

Finally, the general viewpoint here as well as some details are similar to those in [**294**], generally intentionally.

We have additionally some papers that are tangential to the themes of this book; while they are not immediately connected the interested reader can consult them and see the commonalities between this book and tools and techniques in the papers [**14, 61, 81, 82, 100, 137, 138, 140, 151, 153, 167, 169, 206, 211, 233, 244, 253, 286, 310, 319, 321, 322, 349**].

Topics Not Included

As often happens in presentations of mathematical material, we begin by noting some things we will not talk about. Here are a few.

For many mathematicians "Dirichlet spaces" is a topic in abstract potential theory which traces back to the classical work of Beurling and Deny [**71**]. Although $\mathcal{D}$ is their basic model, that work extends the ideas in directions that are very different from those we consider. We will mention it no more.

We will introduce trees and function spaces on trees. That is a huge topic and we will not even try give an overview of it. The history goes back, at least, to Poincaré [**162**]. We only present our small corner of that larger picture. Some samples of work that resonates with what we are doing here are [**195**], [**261**], [**216**], [**204**].

Also, we do not give a systematic development of the potential theory associated with spaces of smooth functions, or even the details of potential theory associated to the Dirichlet space. Sources for that material include [**1**] and [**266**] as well as the discussions in [**136**]. Nor do we systematically consider potential theory for function spaces on trees; see [**216**] and the references there.

Although tempted, we do not have a discussion of corona questions for spaces of functions on the disk.

Bookkeeping and Notation

We take an inconsistent policy on the normalizing constants associated with integrals. Often, when we know them we include them. However when we are unsure, or when we feel the exact value is not essential, we use a generic "C" or "c" to denote that unspecified constant. Thus, the letter is an oxymoronically named *variable constant.* Worse still, at times we will forget to include the constant.

We will generally use the notation $A(x) \lesssim B(x)$ to indicate that there is an unspecified, but finite, C such that, for all x (or all x in an obvious range) $A(x) \leq CB(x)$. This avoids introducing C as another variable constant. We write $A(x) \approx B(x)$ if both $A(x) \lesssim B(x)$ and $A(x) \gtrsim B(x)$ hold. We will write $f \sim g$ when we mean that there is a formal relationship between f and g that will be useful as a guide to the reader.

All our Hilbert spaces are assumed separable.

We will often use the notation $A := B$ to indicate that the expression on the left is being defined by the expression on the right.

Acknowledgments

Writing this book was a journey of many years. During that time we have been greatly helped by the support and assistance of many people and many organizations. We are most grateful to them all.

In particular, we appreciate the financial support from the COFIN project Analisi Armonica, and the PRIN project Verietà Real e Complesse: Geometria, Topologia e Analisi Armonica, funded by the Italian Minister for Research, European Union's Horizon 2020: Marie Skłodowska-Curie grant agreement No 777822, the National Science and Engineering Council of Canada, and the National Science Foundation of the United States (RR was partially supported by DMS 1001488; BDW was partially supported by DMS 0955432, 1001098, 1500509, 1560995, 1603246).

Our universities were very gracious hosts during the many visits we made to each other: Universita di Bologna, Bologna, Italy; McMaster University, Hamilton, Ontario, Canada; Georgia Institute of Technology, Atlanta, Georgia, USA; Washington University, St. Louis, Missouri, USA.

We would like to thank the organizers and the supporting institutions of the meetings, conferences, and programs which we attended and which gave us opportunities to work together. We especially benefited from the opportunity to work together for two weeks in June of 2009 in Trento, Italy as part of the Research in Pairs program of CIRM, Centro Internazionale Per La Ricerca Matematica, and from the program The Dirichlet Space: Connections Between Operator Theory, Function Theory, And Complex Analysis held at the Mathematical Sciences Research Institute, Berkeley, California in June of 2011.

Preliminary versions of some of this material were prepared and used as course lecture notes. Our thanks to the students in those courses, for their patience and their many helpful suggestions.

Parts of this manuscript were reviewed by students, and we thank them for their many helpful suggestions; particular thanks to Philip Benge, Ishwari Kunwar, Robert Rahm, Marie-Jose Saad, and Cody Stockdale.

Part 1

The Dirichlet Space: Foundations

The Hardy space and the Dirichlet space are both reproducing kernel Hilbert spaces of holomorphic functions on the unit disk, and we begin by looking at general properties of such spaces. We then show that for both spaces the multipliers are bounded functions which satisfy a Carleson measure condition. We also describe the Carleson measures; for the Hardy space they are characterized by a local measure theoretic condition; for the Dirichlet space we give a condition using logarithmic capacity

The kernels for both spaces have the complete Pick property and we discuss consequences of that. In particular we show that, for the Hardy space and for the Dirichlet space, the Hilbert space interpolating sequences and the multiplier interpolating sequences are the same. For both spaces those sequences are characterized by a separation condition together with a Carleson measure condition.

We also introduce a discrete model of the Dirichlet space, the "tree Dirichlet space." We show that functions in the model space and also functions in the Dirichlet space have radial boundary values off sets of capacity zero. Using the model space as a framework we develop a measure theoretic description of Dirichlet space Carleson measures which is quite different in appearance from Stegenga's classical capacity theoretic description.

CHAPTER 1

Geometry and Analysis on the Disk

In this chapter we present some notation and terminology for the unit disk, $\mathbb{D} = \mathbb{B}_1$, in the complex plane $\mathbb{C}$. Similar background material for the complex ball, $\mathbb{B}_n \subset \mathbb{C}^n$, is postponed until Chapter 14.

Many of the ideas in this chapter have simple, and for us fundamental, geometric formulations. The reader who is new to these ideas would be extremely well served by making informal sketches along the way to help understand the geometric content.

1.1. Conformal Automorphisms

The disk has a transitive group of biholomorphic automorphisms, $\text{Aut}(\mathbb{D})$. For $\alpha \in \mathbb{D}$, $\alpha \neq 0$, the function $\varphi_\alpha(z)$ defined, for $z \in \mathbb{D}$, by

$$\varphi_\alpha(z) := \frac{\alpha - z}{1 - \overline{\alpha} z} \tag{1.1}$$

is in $\text{Aut}(\mathbb{D})$. The map φ_α interchanges 0 and α, $\varphi_\alpha(0) = \alpha$, and $\varphi_\alpha(\alpha) = 0$, and φ_α is the only element of $\text{Aut}(\mathbb{D})$ with that property. It is involutive, $\varphi_\alpha(\varphi_\alpha(z)) = z$.

Straightforward computation shows that for $\alpha, z, w \in \mathbb{D}$ we have

$$\varphi_\alpha'(z) = -\frac{1 - |\alpha|^2}{(1 - \overline{\alpha} z)^2} \tag{1.2}$$

and

$$1 - \overline{\varphi_a(w)} \varphi_a(z) = \frac{(1 - \overline{a} a)(1 - \overline{w} z)}{(1 - \overline{w} a)(1 - \overline{a} z)}. \tag{1.3}$$

The rotations, given for real θ by $r_\theta(z) = e^{i\theta z}$ are also in $\text{Aut}(\mathbb{D})$ and together with the φ_α, they generate $\text{Aut}(\mathbb{D})$. In fact any $\omega \in \text{Aut}(\mathbb{D})$ is of the form $\omega = r_\theta \circ \varphi_\alpha = \varphi_{\alpha'} \circ r_{\theta'}$ for some $\theta, \theta' \in \mathbb{R}$, $\alpha, \alpha' \in \mathbb{D}$ [**161**].

1.2. Distances

The pseudohyperbolic distance (sometimes called the Mobius distance), ρ, is the function on $\mathbb{D} \times \mathbb{D}$ given by

$$\rho(x, y) := \left| \frac{x - y}{1 - \overline{x} y} \right| = |\varphi_x(y)| = |\varphi_y(x)|. \tag{1.4}$$

It satisfies $0 \leq \rho < 1$ and is a metric on the disk. It is clearly symmetric and positive. It satisfies a strengthened triangle inequality; $\forall x, y, z \in \mathbb{D}$

$$\frac{|\rho(x, z) - \rho(z, y)|}{1 - \rho(x, z)\rho(z, y)} \leq \rho(x, y) \leq \frac{\rho(x, z) + \rho(z, y)}{1 + \rho(x, z)\rho(z, y)}. \tag{1.5}$$

Setting $a = x$ and $w = z = y$ in (1.3) produces the following useful identity: for $x, y \in \mathbb{D}$

$$(1.6) \qquad 1 - \rho(x,y)^2 = 1 - \left|\frac{x-y}{1-\overline{x}y}\right|^2 = \frac{(1-|x|^2)(1-|y|^2)}{|1-\overline{x}y|^2}.$$

In terms of the automorphisms

$$1 - |\varphi_x(y)|^2 = (1-|y|^2)\,|\varphi_x'(y)|\,.$$

The pseudohyperbolic metric is not a length metric, it is not defined by lengths of curves. If we use ρ to define the length of curves and then set $\beta(x,y)$ to be twice the infimum of the length of curves connecting x and y then $\beta \neq \rho$. The function β that is obtained this way is the hyperbolic metric (also called the Poincaré or Bergman metric). It is related to ρ by

$$(1.7) \qquad \beta(x,y) := \log\left(\frac{1+\rho(x,y)}{1-\rho(x,y)}\right);\ \rho(x,y) = \tanh\left(\frac{\beta(x,y)}{2}\right).$$

In particular, for $x \in [0,1)$

$$(1.8) \qquad \rho(0,x) = x$$

$$\beta(0,x) = \log\left(\frac{1+x}{1-x}\right) = \log\left(\frac{1}{1-x}\right) + r(x).$$

$$(1.9) \qquad r(x) = O_{x\to 1}(1),\ r(x) = O_{x\to 0}(x).$$

The hyperbolic metric is a Riemannian metric with arc length element, $ds := 2(1-|z|^2)^{-1}\,|dz|$. It has Gaussian curvature -1; the volume element is

$$(1.10) \qquad d\lambda := 4(1-|z|^2)^{-2}dA.$$

(Some authors define the hyperbolic metric without the factor of 2. In that case the distance from the origin to z is $|z| + O(z^2)$, using either the Euclidean or pseudohyperbolic or hyperbolic distance.)

EXERCISE 1.1. *Prove* (1.5). *One path is to use the triangle inequality for* β, (1.7) *and the addition formula for the hyperbolic tangent.*

Both ρ and β are invariant under the action of elements of $\operatorname{Aut}(\mathbb{D})$. If $x, y \in \mathbb{D}$ and $\varphi \in \operatorname{Aut}(\mathbb{D})$ then $\rho(x,y) = \rho(\varphi(x), \varphi(y))$ and $\beta(x,y) = \beta(\varphi(x), \varphi(y))$.

The disk, together with the metric β, is the Poincare model for hyperbolic geometry. Although we will not explore that geometry systematically, the rich relation between hyperbolic geometry on the disk and function theory on the disk, discussed, for instance, in [**62**], is the stage for much of what follows. The disk, with its hyperbolic geometry is also closely related to the geometry of the rooted dyadic tree which we will introduce in Chapter 5 as a model for the unit disk. The relation between those two geometries has a rich history; an interesting discussion of its early days is given in [**162**].

1.3. Disks

We will denote the Euclidean ball, or disk, with center z and radius r by $B(z,r)$. For $z \in \mathbb{D}$ we denote the pseudohyperbolic ball of pseudohyperbolic radius r by $B_\rho(z,r)$. Similarly we will denote the hyperbolic balls by $B_\beta(z,r)$. The distances ρ and β are monotone functions of each other and hence the sets of pseudohyperbolic and hyperbolic disks are the same.

Also true, but less obvious, is that the set of pseudohyperbolic disks is the same as the set of Euclidean disks inside $\mathbb{D}$. This can be seen by direct computation. Alternatively it can be seen by combining several geometric observations. First note that $\mathrm{Aut}(\mathbb{D})$ acts transitively on the points of $\mathbb{D}$ and isometrically with respect to ρ. Hence $\mathrm{Aut}(\mathbb{D})$ acts transitively on the set of pseudohyperbolic disks. Next, note from (1.8) that the class of Euclidean disks centered at 0 is the same as the class of pseudohyperbolic disks centered there. Thus we will be finished if we show that $\mathrm{Aut}(\mathbb{D})$ acts transitively on the set of Euclidean disks inside $\mathbb{D}$. Suppose $\Delta \subset \mathbb{D}$ is a Euclidean disk and $\Lambda \in \mathrm{Aut}(\mathbb{D})$. Because Λ is a linear fractional map of the complex plane, $\Lambda(\Delta)$ will be a Euclidean disk or a halfplane; $\Lambda(\Delta)$ is contained in $\mathbb{D}$ and hence is a disk. It remains to show that if Δ is not centered at the origin then there is a $\Gamma \in \mathrm{Aut}(\mathbb{D})$ such that $\Gamma(\Delta)$ is a disk centered at the origin. Let α and β be the two points where the straight line connecting the origin and the center of Δ cuts the boundary circle of Δ. Let Γ be an element of $\mathrm{Aut}(\mathbb{D})$ which map α and β to two points on the real axis, one the negative of the other. This Γ does the job.

EXERCISE 1.2. *Fill in the details at the end of the previous argument. Why is there such a Γ? Why is the image of Δ a Euclidean disk centered at 0?*

For fixed small $\alpha > 0$ and $0 < \gamma < 1$, set $z_n = 1 - \gamma^n\alpha$. The Euclidean distances $|z_n - z_{n+1}|$ decrease geometrically. The $\rho-$distances, $\rho(z_n, z_{n+1})$, are approximately constant, hence also the β distances, $\beta(z_n, z_{n+1})$. For large n, k, $\beta(z_n, z_{n+k}) = -k \log \gamma + O(1)$.

1.4. Differential Operators

For functions defined on the disk, writing $z = x + iy$, we introduce differential operators:

$$\partial := \frac{\partial}{\partial z} = \frac{1}{2}\left(\frac{\partial}{\partial x} - i\frac{\partial}{\partial y}\right);$$

$$\overline{\partial} := \frac{\partial}{\partial \overline{z}} = \frac{1}{2}\left(\frac{\partial}{\partial x} + i\frac{\partial}{\partial y}\right),$$

$$\Delta := \frac{\partial^2}{\partial x^2} + \frac{\partial^2}{\partial y^2} = 4\partial\overline{\partial} = 4\overline{\partial}\partial.$$

Using this notation, the Cauchy-Riemann equations, which characterize analyticity for a function $f(z)$, take the form $\overline{\partial} f = 0$. The operator Δ is called the *Laplacian* and the equation $\Delta f = 0$ is referred to as *Laplace's equation.* Functions, f, which satisfy this equation are called *harmonic*.

> **Note:** We will only rarely use the notation Δ for either the Laplace operator, as above; or for a Euclidean disk, as slightly earlier. In the coming chapters we will use Δ frequently to denote a difference operator for functions on a tree. We will use it in later sections for a geometric quantity related to points in the complex ball.

These differential operators do not interact cleanly with the conformal automorphisms of the disk . However, the *invariant gradient,* $\widetilde{\nabla}$, and the *invariant*

Laplacian, $\widetilde{\Delta}$, defined for smooth functions f on the disk by

$$(1.11)\qquad \begin{aligned}(\widetilde{\nabla} f)(a) &:= (1-|a|^2)(\nabla f)(a),\\ \widetilde{\Delta} f(a) &:= (1-|a|^2)^2(\Delta f)(a),\end{aligned}$$

do transform nicely. Indeed we have

$$\begin{aligned}|(\widetilde{\nabla}\,(f\circ\varphi))\,(z)\,| &= |(\widetilde{\nabla} f)(\varphi(z))|, \text{ and}\\ (\widetilde{\Delta}\,(f\circ\varphi))(z) &= (\widetilde{\Delta} f)(\varphi(z)).\end{aligned}$$

In particular, with $\varphi = \varphi_a$, we have:

$$\begin{aligned}|(\widetilde{\nabla} f)(a)| &= |\nabla\,(f\circ\varphi_a)\,(0)|\,, \text{ and}\\ (\widetilde{\Delta} f)(a) &= \Delta\,(f\circ\varphi_a)\,(0).\end{aligned}$$

EXERCISE 1.3. *Check those formulas.*

1.5. Harmonic Functions and Fourier Series

The Dirichlet problem for Laplace's equation on the disk is the following boundary value problem. Given $U(e^{i\theta})$, a continuous function on the unit circle $\mathbb{T}$, find a function $u(z) = u(x,y)$ which is harmonic in $\mathbb{D}$, is continuous on $\overline{\mathbb{D}} = \mathbb{D}\cup\mathbb{T}$, and which agrees with U on $\mathbb{T}$.

If U is given by a finite Fourier series; that is

$$(1.12)\qquad U(e^{i\theta}) = \sum_{n=-N}^{N} \hat{U}(n)e^{in\theta},\quad e^{i\theta}\in\mathbb{T}$$

with

$$(1.13)\qquad \hat{U}(n) := \frac{1}{2\pi}\int_0^{2\pi} U(e^{i\theta})\overline{e^{in\theta}}\,d\theta,\quad n\in\mathbb{Z}$$

then $u(z) = u(re^{i\theta})$ given by

$$(1.14)\qquad u(re^{i\theta}) = \sum_{n=-N}^{N} \hat{U}(n)r^{|n|}e^{in\theta},\quad re^{i\theta}\in\mathbb{D}$$

is a solution to the Dirichlet problem with boundary data U. Because U is a finite sum it suffices to verify this for each term. For the individual terms note that the extension of the function $e^{in\theta}$ on $\mathbb{T}$ to a harmonic function on $\mathbb{D}$ is given by $r^{|n|}e^{in\theta}$.

Alternatively, with $z = re^{i\theta}$ noting that, on $\mathbb{T}$, $z = e^{i\theta}$ we find that if the boundary function is

$$U(z) = \sum_{n=-N}^{-1} \hat{U}(n)\overline{z}^n + \sum_{n=0}^{N}\hat{U}(n)z^n;\quad |z|=1$$

then

$$u(z) = \sum_{n=-N}^{-1} \hat{U}(n)\overline{z}^n + \sum_{n=0}^{N}\hat{U}(n)z^n;\quad |z|<1.$$

is the solution function.

With additional technical work this approach can be extended to more general U. Alternatively we can use the Poisson kernel and the associated Poisson integral formula. The Poisson kernel is a function $P(r, e^{i\theta})$ defined on $[0,1)\times\mathbb{T}$ such that, writing $z = re^{i\phi}$, we have, for any continuous U,

$$u(re^{i\phi}) := \frac{1}{2\pi}\int_0^{2\pi} U(e^{i\theta})P(r, e^{i(\phi-\theta)})d\theta, \tag{1.15}$$

is the *unique* solution to the Dirichlet problem with boundary data $U(e^{i\theta})$. The function P is given by the formula

$$P(r, e^{i\alpha}) := \frac{1-r^2}{1-2r\cos\alpha + r^2}, \tag{1.16}$$

or, with $z = re^{i\alpha}$,

$$P(r, e^{i\alpha}) = \frac{1-|z|^2}{|1-z|^2} = \operatorname{Re}\left(\frac{1+z}{1-z}\right). \tag{1.17}$$

The crucial facts about the Poisson kernel are: for any r, $P(r,\cdot)$ is a nonnegative function having integral one; as $r \to 1$ the measures $P(1, e^{i\theta})d\theta$ converge weakly to δ_1, the point mass at 1; and, for each ϕ the function $x + iy = re^{i\theta} \to P(|x+iy|, e^{i(\phi-\theta)})$ is a harmonic function of (x, y).

An important special case is when $U(z)$ is a given function which is harmonic in $\mathbb{D}$ and continuous on $\overline{\mathbb{D}}$. The solution to the Dirichlet problem is unique and both u and U are solutions, hence $u(z) = U(z)$ in the disk. That is, (1.15) recaptures the interior values of the harmonic function from its boundary values. In particular (1.15) can be used to recapture the values of holomorphic functions from their boundary values, as an alternative to the Cauchy integral formula. We will use this fact a number of times going forward.

Both the Fourier series approach given by (1.14) and the Poisson kernel approach using (1.15) can be extended to solve the Dirichlet problem for U in $L^2(\mathbb{T}, d\theta/2\pi)$. The function $u(z)$ produced by a Fourier series formula analogous to (1.14) and u produced by (1.15) are the same, are harmonic, and have boundary limits which satisfy $\lim_{r\to 1} u(re^{i\theta}) = U(e^{i\theta})$ for a set of $e^{i\theta}$ of full measure which includes every point of continuity of U.

The inner product on $L^2(\mathbb{T}, d\theta/2\pi)$ can be given with an integral or using Fourier series. Given $F, G \in L^2(\mathbb{T}, d\theta/2\pi)$ with Fourier coefficients $\{\hat{F}(n)\}$, and $\{\hat{G}(n)\}$ we have

$$\langle F, G\rangle_{L^2(\mathbb{T},d\theta/2\pi)} := \frac{1}{2\pi}\int_0^{2\pi} F(e^{i\theta})\overline{G(e^{i\theta})}d\theta = \sum_{n=-\infty}^{\infty} \hat{F}(n)\overline{\hat{G}(n)}$$

where the second equality expresses the fact that the map $F \mapsto \hat{F}$ is a Hilbert space isomorphism between $L^2(\mathbb{T})$ and $\ell^2(\mathbb{N})$. This fact can be used to show the equivalence of the two paths we described to solving the Dirichlet problem.

1.6. Convergence Regions and Maximal Functions

We recall some definitions and basic results; proofs can be found in [**161**].

If a function $f(z)$ is defined on the disk $\mathbb{D}$ and if $\omega \in \mathbb{T}$ is a boundary point then

$$\lim_{z\to\omega} f(z)$$

is the conventional shorthand for the more awkward but more precise

$$\lim_{\substack{z\to\omega\\ z\in\mathbb{D}}} f(z). \tag{1.18}$$

It turns out to be useful to also consider less demanding limits in which convergence is only required as z approaches w through a constrained region. Radial convergence is a basic example. In that case z is restricted to lie on the radius $R(\omega)$ connecting the origin and ω. Thus the *radial limit* at ω is $\lim_{r\nearrow 1} f(r\omega)$, or, in other notation,

$$\lim_{\substack{z\to\omega\\ z\in R(\omega)}} f(z). \tag{1.19}$$

Nontangential convergence is the convergence in which z is restricted to approach ω through a region $\Gamma(\alpha,\omega)$ which is larger than the $R(\omega)$ but smaller than the full disk. For $\omega\in\mathbb{T}$ and $\alpha>1$ we define the nontangential convergence region by

$$\Gamma_\alpha(\omega) := \{\lambda\in\mathbb{D} : |\lambda-\omega| < \alpha(1-|\lambda|)\}. \tag{1.20}$$

Thus near ω, the set $\Gamma_\alpha(\omega)$ is approximately the intersection of the disk with a cone having vertex at ω and bisected by the radial segment to ω. The angular aperture of that cone is $\theta = 2\sec^{-1}\alpha$ (and so, when $\alpha = 1+\varepsilon$, $\theta\approx\varepsilon$). The name refers to the fact that $\Gamma_\alpha(\omega)$ does not contain any curve tangent to the boundary circle. For most purposes the exact value of α does not play an essential role.

Suppose $\alpha>1$ is fixed. The *nontangential limit* of f at ω is defined as

$$f^\beta(\omega) := \lim_{\substack{z\to\omega\\ z\in\Gamma_\alpha(\omega)}} f(z).$$

Maximal functions are an important tool in studying nontangential limiting behavior of functions on the disk. For F defined on $\mathbb{D}$, F^*, the *nontangential maximal function* of F, is the function on the boundary circle $\mathbb{T}$

$$F^*(\omega) := \sup\{|F(z)| : z\in\Gamma_\alpha(\omega)\}.$$

The value $F^*(\omega)=\infty$ is allowed. The starting point for seeing the relation between this maximal function and nontangential convergence is the elementary observation that, given F, the set of $\omega\in\mathbb{T}$ for which F has a nontangential limit at ω is contained in the set of ω for which $F^*(\omega)<\infty$.

A closely related maximal function is the Hardy-Littlewood maximal function of F, MF, defined for functions on $\mathbb{T}$ by

$$MF(\omega) := \sup_{\omega\in J\in\mathcal{I}} \frac{1}{|J|}\int_J |F(\gamma)|\,|d\gamma|$$

where $\mathcal{I}$ is the collection of all subintervals of $\mathbb{T}$ and $|d\gamma|$ is normalized Lebesgue measure on the circle.

Suppose p and α are fixed, $1<p<\infty, \alpha>1$. The Hardy-Littlewood maximal theorem states that if $U\in L^p(\mathbb{T})$ and $u(z)$ is the associated harmonic function on the disk given by (1.15), or, equivalently, (1.14); then MU and u^* are roughly the same size as each other and as U.

PROPOSITION 1.4 (Hardy-Littlewood maximal theorem). *With constants of equivalence that depend on p,α but not U;*

$$\|U\|_{L^p(\mathbb{T})} \approx \|MU\|_{L^p(\mathbb{T})} \approx \|u^*\|_{L^p(\mathbb{T})}.$$

The ideas of convergence regions and associated maximal functions can be extended to more general geometries and there are delicate relationships between the shape of the convergence regions and the convergence results. If the cones $\Gamma_\alpha(\omega)$ are replaced by oricycles at $\omega \in \mathbb{T}$ (circles internally tangent to $\mathbb{T}$ at ω, also called horocycles) of radius $1/\alpha$ we have oricyclic convergence. We will encounter oricyclic convergence in Proposition 10.17 when considering boundary behavior of functions in the Dirichlet space. We discuss more general convergence regions for Dirichlet space functions briefly in Section 9.1.4.

1.7. Subdividing the Disk and the Circle

1.7.1. Whitney Covers. We now describe a *Whitney cover*, or *Whitney decomposition*, of $\mathbb{D} \subset \mathbb{C}$. Although our construction will be detailed, the details are not the central point. What is important is to have a covering with certain metric properties. Constructions of this sort are available in great generality and are a powerful tool. Very informally, a Whitney cover of an open set Ω in $\mathbb{R}^n$ is a collection of nice subsets $\{\Omega_i\}$ such that

- $\bigcup \Omega_i = \Omega$
- The Ω_i have disjoint interiors,
- The Ω_i are, roughly, all the same shape.
- The Ω_i are of different sizes.
  - The size of each Ω_i is roughly proportional to the distance from Ω_i to the complement $\mathbb{R}^n \setminus \Omega$.
  - If Ω_i and Ω_j are near each other, then they are of comparable size.
- There is a uniform upper bound on the number of Ω_j that are relatively close to any given Ω_i.

By "nice" we mean that there will be no trouble when we are casual about distinguishing between the sets, their interiors, and their closures. In fact, in many cases the sets can be taken to be cubes or other very regular sets. Details of constructing such covers in general situations, as well as applications, can be seen in, for example, [**316**], [**164**, Appendix J].

We will be particularly interested in the disk, $\mathbb{D} = \mathbb{B}_1 \subset \mathbb{C}$. In that case a simple example of a Whitney cover is given by the *dyadic decomposition of the disk*, an explicit decomposition for which the geometric properties we stated before are clear. For integers $n \geq 0$ and $1 \leq j \leq 2^n$, we define the "cubes" $Q_{n,j}$ by

$$Q_{n,j} := \left\{ z = re^{2\pi it} \in \mathbb{D} : 2^{-n-1} \leq 1 - r \leq 2^{-n} \,, \frac{j-1}{2^n} \leq t \leq \frac{j}{2^n} \right\}.$$

We write $z_{n,j}$ for the center of $Q_{n,j}$ when $n \geq 1$,

$$z_{n,j} := (1 - 2^{-n-1/2}) \exp(2\pi i(j - 1/2)/2^n).$$

We denote the projection of $Q_{n,j}$ onto the boundary $\mathbb{T}$ of $\mathbb{D}$ by $I_{n,j}$, $I_{n,j} = \{z/|z| : z \in Q_{n,j}\}$. The collection $\{I_{n,j}\}$ is called the *dyadic decomposition of the circle*.

Later we will introduce trees as a model for the disk. The dyadic decomposition offers a convenient visualization of one such tree. The centers $\{z_{n,j}\}$ can be viewed as the vertices of the tree, with an edge connecting each vertex to the two "below" it. Alternatively, but equivalently, the intervals $\{I_{n,j}\}$ can be regarded as the vertices of the tree and an edge connects two vertices if and only if one of the associated intervals is half of the other.

Here is an alternative, more generic, construction of a Whitney cover for the disk. Pick a small positive ε. Pick a set of points $\{\zeta_i\}$ in the disk such that the pseudohyperbolic distance, ρ, between any two is at least ε, and which is maximal with respect to that condition. We then have that the balls $\{B_\rho(\xi_i, \varepsilon/2)\}$ are disjoint and the balls $\{B_\rho(\xi_i, \varepsilon)\}$ form a covering. Define Ω_j inductively for $j = 1, 2, \ldots$.

$$\Omega_j := B_\rho(\xi_j, \varepsilon) \setminus \left[(\cup_{i<j}\Omega_i) \bigcup (\cup_{i>j} B_\rho(\xi_i, \varepsilon/2)) \right].$$

We then have that $B_\rho(\xi_j, \varepsilon/2) \subseteq \Omega_j \subseteq B_\rho(\xi_j, \varepsilon)$ and the other properties mentioned earlier. This construction adapts easily to other metric spaces we will encounter and we use variations on it in several places, for example in Lemma 14.1.

1.7.2. Tents and Carleson Boxes. We will need specific versions of the general idea of "the part of the disk near a specified boundary interval." There are several standard constructions of such regions; here we describe two. For our purposes these variations and several others can be used interchangeably.

The first construction is a *Carleson box* over I, $B(I)$. Fix I, an interval in $\mathbb{T}$, and let $|I|$ denote its length and c_I its center. A *Carleson box* over I, $B(I)$, is the set

$$B(I) := \{z : z \in \mathbb{D},\ z/|z| \in I,\ 1 - |I|/2\pi < |z| < 1\}.$$

The same object is sometimes called a *Carleson square over* I, and denoted $S(I)$.

The second construction is a *tent over* I, sometimes called a *Carleson tent over* I, $T(I)$. Let m_I be the midpoint of I and let c_I, the vertex of the tent, be $(1 - \frac{|I|}{2\pi})m_I$. The tent $T(I)$ is the convex hull of I and c_I. We will also write $z(T)$ or $z(I)$ for the vertex and $T(z)$ for the tent with that vertex.

Finally we will need a notion of "close to the boundary interval but not too close." A standard choice is the *top half of the Carleson box*, $TB(I)$:

$$TB(I) := \{z : z \in \mathbb{D},\ z/|z| \in I,\ 1 - |I|/2\pi < |z| < 1 - |I|/4\pi\}.$$

The top halves $TB(I)$ tile the disk as I ranges over dyadic intervals.

Note that for dyadic intervals I the relation between $I, B(I)$, and $TB(I)$ can be described in terms of the $Q_{j,k}$ of the earlier construction and their projections.

EXERCISE 1.5. *In terms of the ρ metric all of the $TB(I)$ are, roughly balls of the same size. Precisely, there are $\theta_1, \theta_2 > 0$ such that each $TB(I)$ contains a $\rho-$ball of $\rho-$radius θ_1 and is contained in a $\rho-$ball of $\rho-$radius θ_2. A similar statement holds for $\beta-$balls.*

1.8. Notes and Comments

The material in this chapter is covered in several of the texts listed in the bibliography. The authors often use [**161**] as a reference.

CHAPTER 2

Hilbert Spaces of Holomorphic Functions

In this chapter we introduce Hilbert spaces with reproducing kernels, denoted as RKHS. Our emphasis is on spaces of holomorphic functions on the disk; in particular the Bergman, Hardy, and Dirichlet space. We introduce multipliers, Carleson measures, and a distance function; and develop some of their basic properties.

2.1. The Dirichlet Space

The Dirichlet space, $\mathcal{D}$, is the Hilbert space of holomorphic functions, f, defined on the unit disk, $f \in \operatorname{Hol}(\mathbb{D})$, with power series $f(z) = \sum_{n=0}^{\infty} a_n z^n$ satisfying

$$\|f\|_{\mathcal{D}}^2 := \sum_{n=0}^{\infty} (n+1) |a_n|^2 < \infty. \tag{2.1}$$

Note that this condition ensures that the radius of convergence of the power series is at least 1.

For $g(z) = \sum_{n=0}^{\infty} b_n z^n \in \mathcal{D}$ the inner product $\langle f, g \rangle_{\mathcal{D}}$ is given by

$$\langle f, g \rangle = \langle f, g \rangle_{\mathcal{D}} := \sum_{n=0}^{\infty} (n+1) a_n \overline{b_n}. \tag{2.2}$$

2.1.1. $\mathcal{D}$ is a Reproducing Kernel Hilbert Space. The Dirichlet space is a Hilbert space with additional structure. It is a space of functions on $\mathbb{D}$, and for each $\zeta \in \mathbb{D}$ the evaluation functional e_ζ defined by $e_\zeta(f) = f(\zeta)$, is a continuous linear functional on $\mathcal{D}$. Hence the Riesz representation theorem ensures that there is a $k_\zeta \in \mathcal{D}$, the *reproducing kernel for* ζ, with the property that

$$f(\zeta) = \langle f, k_\zeta \rangle \quad \forall f \in \mathcal{D}. \tag{2.3}$$

This structure – a Hilbert space, H, consisting of functions on a set, X, and such that for all x in X the evaluation functional $f \to f(x)$ is continuous on H – is a *reproducing kernel Hilbert space*, RKHS.

For the Dirichlet space, rather than prove the evaluation functionals are continuous on $\mathcal{D}$ and as a consequence know the existence of the k_ζ, we construct $k_\zeta \in \mathcal{D}$ so that (2.3) holds, and thus conclude that evaluation is a continuous functional.

Using (2.1) one checks that the polynomials are dense in $\mathcal{D}$ and that the normalized monomials $e_n(z) = z^n/\sqrt{n+1}$, $n = 0, 1, \ldots$ are an orthonormal basis. Set

$$k_\zeta(z) := \sum_{n=0}^{\infty} e_n(z) \overline{e_n(\zeta)}. \tag{2.4}$$

Using this we get the explicit formula

$$k_\zeta(z) := \sum_{n=0}^{\infty} e_n(z)\overline{e_n(\zeta)} = \sum_{n=0}^{\infty} \frac{1}{n+1}(z\bar{\zeta})^n = \frac{1}{z\zeta}\log\frac{1}{1-z\bar{\zeta}}, \tag{2.5}$$

and then evaluating at $z=\zeta$ we find the norm of the kernel

$$\|k_\zeta\|^2 = \langle k_\zeta, k_\zeta\rangle = k_\zeta(\zeta) = \frac{1}{|\zeta|^2}\log\frac{1}{1-|\zeta|^2},$$

which is finite for ζ in the disk. Thus $k_\zeta \in \mathcal{D}$. The reproducing property is a consequence of the structure of the formula; we compute

$$\begin{aligned}
\langle f, k_\zeta\rangle &= \left\langle f(z), \sum_{n=0}^{\infty} e_n(z)\overline{e_n(\zeta)}\right\rangle \\
&= \sum_{n=0}^{\infty}\left\langle f(z), e_n(z)\overline{e_n(\zeta)}\right\rangle \\
&= \sum_{n=0}^{\infty}\langle f(z), e_n(z)\rangle\, e_n(\zeta) \\
&= \left(\sum_{n=0}^{\infty}\langle f, e_n\rangle\, e_n\right)(\zeta) \\
&= f(\zeta).
\end{aligned} \tag{2.6}$$

In the passage to the last line we used the fact that $\{e_n\}$ is an orthonormal basis.

It will be convenient to have a separate notation for the normalized reproducing kernels. Set

$$\widehat{k_\zeta} := \frac{k_\zeta}{\|k_\zeta\|}.$$

Also, note that for any $z, w \in \mathbb{D}$, $\langle k_z, k_w\rangle = \overline{\langle k_w, k_z\rangle}$ and hence

$$\overline{k_z(w)} = k_w(z). \tag{2.7}$$

2.1.2. A Geometric Definition. The area of $f(\mathbb{D})$, the image of the unit disk under a holomorphic map f, with area counted according to multiplicity, equals $\pi\mathfrak{D}(f)$, where $\mathfrak{D}(f)$ is the *Dirichlet integral* of f defined by

$$\mathfrak{D}(f) := \frac{1}{\pi}\int_{\mathbb{D}} |f'|^2\, dA = \sum_{n=1}^{\infty} n\,|a_n|^2. \tag{2.8}$$

If f is one to one this is a direct consequence of the change of variables formula for the integral measuring the area. The general case is almost as straightforward.

This gives a geometric characterization of functions in the Dirichlet space; f is in the Dirichlet space exactly if the area of $f(\mathbb{D})$ is finite. Using this characterization we can check that $\mathcal{D}$ neither contains nor is contained in $H^\infty(\mathbb{D})$, the space of bounded holomorphic functions on $\mathbb{D}$. An example of an unbounded function in $\mathcal{D}$ is the Riemann map of the unit disk onto any unbounded region of finite area. An example of a bounded function not in $\mathcal{D}$ is given by any infinite Blaschke product B (defined below in (3.7)). Alternatively, one can give an elementary example:

$$g(z) := \exp\left(i\log\frac{1+z}{1-z}\right). \tag{2.9}$$

The map of z to $(1+z)/(1-z)$ takes the disk to the right halfplane. The logarithm takes the right half plane to the horizontal strip $\{z = x+iy : |y| < \pi/2\}$. Multiplication by i then produces the vertical strip with $-\pi/2 < x < \pi/2$ which is then mapped in an infinite-valent manner to the ring

$$\{z : \exp(-\pi/2) < |z| < \exp(\pi/2)\}.$$

Thus g is bounded but, because $\pi\mathfrak{D}(g)$ computes the area of the image with multiplicity, $\mathfrak{D}(g) = \infty$, and hence g is not in the Dirichlet space.

The functional $\mathfrak{D}(f)$ is a quasinorm on $\mathcal{D}$ (it annihilates constants). An alternative norm on $\mathcal{D}$, equivalent to $\|\cdot\|_{\mathcal{D}}$, is given by $\|\cdot\|^2_{\mathcal{D},\,\mathtt{alt}}$

$$\begin{aligned}\|f\|^2_{\mathcal{D},\mathtt{alt}} &:= |a_0|^2 + \frac{1}{\pi}\int_{\mathbb{D}} |f'(z)|^2\,dA \\ &= |a_0|^2 + \sum_{n=1}^{\infty} n\,|a_n|^2 = |f(0)|^2 + \mathfrak{D}(f).\end{aligned} \tag{2.10}$$

The associated inner product is given by, for $f = \sum_{n=0}^{\infty} a_n z^n$, $g = \sum_{n=0}^{\infty} b_n z^n$,

$$\langle f, g\rangle_{\mathcal{D},\,\mathtt{alt}} = f(0)\overline{g(0)} + \frac{1}{\pi}\int_{\mathbb{D}} f'(z)\overline{g'(z)}\,dA = a_0\overline{b_0} + \sum_{n=1}^{\infty} n a_n \overline{b_n}.$$

EXERCISE 2.1. *Show that if the norm* $\|f\|^2_{\mathcal{D},\,\mathtt{alt}}$ *is used then the reproducing kernel for* $\mathcal{D}$ *is given by*

$$k_\zeta(z) = 1 + \log\frac{1}{1-z\overline{\zeta}}. \tag{2.11}$$

The space of functions with $\|f\|^2_{\mathcal{D}\ \mathtt{alt}} < \infty$ is the same as the space of functions with $\|f\|^2_{\mathcal{D}} < \infty$ and the two norms are equivalent. Both are also sometimes referred to as "the Dirichlet space" and for many purposes the choice makes no difference.

One area where the distinction matters is that when the norm $\|\cdot\|^2_{\mathcal{D}}$ is used the space $\mathcal{D}$ has a complete Pick kernel (defined later), but that fails when $\|\cdot\|^2_{\mathcal{D}\ \mathtt{alt}}$ is used. With that in mind, when we make no other comment, we will work with $\|\cdot\|^2_{\mathcal{D}}$.

On the other hand, the exponential integrability results of Beurling-Chang-Marshall in Section 9.2 have a more elegant formulation using the space $\boldsymbol{D}$ defined by

$$\boldsymbol{D} := \{f \in \mathrm{Hol}(\mathbb{D}) : f(0) = 0,\ \mathfrak{D}(f) < \infty\}$$

and normed by $\|f\|_{\boldsymbol{D}} := \mathfrak{D}(f)^{1/2}$. This space also has an interesting invariance property. For $\phi \in \mathrm{Aut}(\mathbb{D})$ we have, noting the area interpretation of $\mathfrak{D}(\cdot)$, $\mathfrak{D}(f) = \mathfrak{D}(f\circ\phi)$. Thus the map of $\boldsymbol{D}$ to itself which takes f to $f\circ\phi - f(\phi(0))$ is an isometry of $\boldsymbol{D}$. A theorem of Arazy and Fisher [**34**] states that, essentially, this property characterizes the Dirichlet space among Hilbert spaces of holomorphic functions on the disk. This isometry is basic to the analysis in [**34**].

2.1.3. Defining Spaces by Kernel Functions. We defined the Dirichlet space by giving an inner product on a space of functions. We then used the fact that the monomials were an orthogonal basis to obtain a formula for the reproducing kernel. We will do the same for the Hardy and Bergman spaces in a moment.

It is also possible to reverse the procedure and start with the kernel functions. Here is an informal description of that process. The Moore-Aronszajn theorem [**54**] states that any RKHS can be obtained this way.

Suppose we are given a set X and a complex-valued function $K(x, y)$ defined on $X \times X$ which is a *positive kernel function*. That is, we require that for any $\{x_i\}_{i=1}^n \subset X$, the matrix $(K(x_i, x_j))_{i,j=1}^n$ is positive definite; for any sequence $\{a_i\}_{i=1}^n \subset \mathbb{C}$ we have

$$\sum_{i,j=1}^{n} K(x_i, x_j) a_i \overline{a_j} \geq 0 \tag{2.12}$$

with equality holding only if all the a_i are zero. We then define, for $x \in X$, the functions $k_x(\cdot) = K(\cdot, x)$.

We define an inner product on the set of functions $\{k_x\}$ by setting (the middle equality is the definition)

$$\langle k_y, k_x \rangle = \langle K(\cdot, y), K(\cdot, x) \rangle = K(x, y) = k_y(x). \tag{2.13}$$

Next we extend the definition of the inner product to finite linear combinations of the k_x so that it is linear in the first variable and conjugate linear in the second. By virtue of (2.12) this defines a positive definite inner product and thus the set of linear combinations is a pre-Hilbert space of functions on X. Let H be the Hilbert space obtained as the completion of that pre-Hilbert space. We regard h in H as a function on X by setting $h(x) := \langle h, k_x \rangle$ and note that this is consistent with the action of the evaluation functional on the pre-Hilbert space. Thus the k's are the reproducing kernels for the RKHS H.

This type of description can be quite useful but it may hide a great deal. For instance, suppose $\mathcal{X}$ is the interval $(0, 1)$ and consider the RKHS $\mathcal{X}$ associated with the kernel function $K(x, y)$ defined on $\mathcal{X} \times \mathcal{X}$ by

$$K(x, y) = \frac{1}{xy} \log\left(\frac{1}{1 - xy}\right).$$

Tracing through the definitions and comparing this with (2.5) we see that there is a natural isometric imbedding of $\mathcal{X}$ into $\mathcal{D}$. Next, note that a function in $\mathcal{D}$ that is orthogonal to the image of $\mathcal{X}$ must vanish on $(0, 1)$ and hence must be the zero function. Thus $\mathcal{X}$ is, or is naturally isometrically isomorphic to, $\mathcal{D}$. In particular, functions in $\mathcal{X}$, originally defined only on $(0, 1)$, all extend to be holomorphic on the unit disk; a fact not clear from the definition of $\mathcal{X}$.

2.2. The Hardy Space

The Hardy space, H^2, is the Hilbert space of $f \in \operatorname{Hol}(\mathbb{D})$ with power series $f(z) = \sum_{n=0}^{\infty} a_n z^n$ satisfying

$$\|f\|_{H^2}^2 := \sum_{n=0}^{\infty} |a_n|^2 < \infty. \tag{2.14}$$

In particular $\mathcal{D} \subset H^2$.

The inner product for the Hardy space is given by, for $g(z)=\sum_{n=0}^{\infty} b_n z^n \in H^2$,

$$\langle f, g\rangle_{H^2} := \sum_{n=0}^{\infty} a_n \overline{b_n}.$$

Again, the power series of functions in this space have radius of convergence at least one, the point evaluations e_z, $z \in \mathbb{D}$, defined by $e_z(f) = f(z)$ are continuous and the values can be obtained by taking the inner product with the (Hardy space) reproducing kernel. The analysis we used to find the formula for the reproducing kernel for the Dirichlet space can be adapted for the Hardy space. We find

$$k_\zeta(z) = k_{H^2,\zeta}(z) := \sum_{n=0}^{\infty} (z\overline{\zeta})^n = \frac{1}{1-z\overline{\zeta}}.$$

EXERCISE 2.2. *Do the computation.*

Here we are using subscripts on the notation for the kernel function to denote the space being considered, H^2 in this case. We will use similar notation below but we will generally omit the subscripts when we think it is safe.

There is an alternative formula for the norm on H^2 which resembles the norm (2.10) of the Dirichlet space. Reduction to the case of monomials and then straightforward computations establish

$$\|f\|_{H^2}^2 = |f(0)|^2 + \frac{2}{\pi}\int_{\mathbb{D}} |f'(z)|^2 \log\frac{1}{|z|} dA.$$

Hence we also have the less precise but more convenient

$$\|f\|_{H^2}^2 \approx |f(0)|^2 + \int_{\mathbb{D}} |f'(z)|^2 (1-|z|^2) dA. \tag{2.15}$$

EXERCISE 2.3. *Check these formulas.*

2.2.1. Boundary Values. A powerful tool in the study of functions analytic on the closed disk, $\overline{\mathbb{D}}$, is the ability to pass information back and forth between the boundary function, $f(e^{i\theta})$, defined on $\mathbb{T}$, and the holomorphic function $f(z)$ defined on the open domain $\mathbb{D}$. We would like to have that tool for Dirichlet space and Hardy space functions but those functions are not defined on $\mathbb{T}$. An attempt to define them on $\mathbb{T}$ using limits runs into problems.

EXERCISE 2.4.

(1) *Show that the linear functional $e_1(f) = f(1)$ is well defined on a dense subspace of H^2 but cannot be extended to be a continuous functional on H^2. Do the same for $\mathcal{D}$.*
(2) *Show that the linear functional $\lim_{r\nearrow 1} f(r)$ is well defined on a dense subspace of H^2 but cannot be extended to be a continuous functional on H^2.*

There is a technically sophisticated way to define boundary values of functions in H^2. We will describe the approach and use its outcome, but we will not give the details of the development. Those details can be found in [**161**, Chapter 2].

For $f(z)=\sum_{n=0}^{\infty} a_n z^n \in H^2$ and $e^{i\theta}$ in $\mathbb{T}$ we define the boundary value function $f^\beta(e^{i\theta})$ to be the radial limit (1.19) of $f(z)$ at $e^{i\theta}$;

$$f^\beta(e^{i\theta}) := \lim_{r\to 1^-} f(re^{i\theta}).$$

It is clear how this works if f is a polynomial, but for general f this limit may not exist. However the general theory ensures that for any $f \in H^2$ a limit $f^\beta(e^{i\theta})$ does exist except perhaps for a set of $e^{i\theta}$ of Lebesgue measure zero on the circle. Furthermore, the boundary value limit function, f^β, will be in $L^2 = L^2(\mathbb{T}, d\theta/2\pi)$. As such it has a Fourier series and, exactly as formal computations suggest, the Fourier coefficients of f^β are the power series coefficients of f.

$$f^\beta(e^{i\theta}) = \sum_{n=0}^{\infty} a_n e^{in\theta}.$$

The norm of $f^\beta \in L^2$ is given by

$$\left\|f^\beta\right\|_{L^2}^2 = \frac{1}{2\pi}\int_{\mathbb{T}} \left|f^\beta(e^{i\theta})\right|^2 d\theta = \sum_{n=0}^{\infty} |a_n|^2 . \tag{2.16}$$

EXERCISE 2.5. *Show that the H^2 inner product can be given by a boundary integral,*

$$\langle f, g\rangle = \frac{1}{2\pi}\int_{\mathbb{T}} f^\beta(e^{i\theta})\overline{g^\beta(e^{i\theta})}d\theta.$$

Thus the map of f in H^2 to f^β in L^2 is an isomorphism onto the closed subspace of L^2 consisting of Fourier series of *power series type*, that is, Fourier series whose Fourier coefficients of negative index are all zero. That subspace of L^2 is also called H^2 and one generally writes $f(e^{i\theta})$ instead of $f^\beta(e^{i\theta})$.

2.3. The Bergman Space

The Bergman space A^2 (also denoted $L_a^2(\mathbb{D})$, shorthand for the analytic functions in $L^2(\mathbb{D})$) is another RKHS of holomorphic functions on $\mathbb{D}$. For $f(z) = \sum_{n=0}^{\infty} a_n z^n$ and $g(z) = \sum_{n=0}^{\infty} b_n z^n$ the Bergman space norm and inner product are given by

$$\|f\|_{A^2}^2 := \sum_{n=0}^{\infty} \frac{1}{n+1} |a_n|^2 = \frac{1}{\pi}\int_{\mathbb{D}} |f(z)|^2 \, dA \tag{2.17}$$

$$\langle f, g\rangle_{A^2} = \sum_{n=0}^{\infty} \frac{1}{n+1} a_n \overline{b_n} = \frac{1}{\pi}\int_{\mathbb{D}} f(z)\overline{g(z)}dA.$$

There is also an equivalent norm which is in the spirit of (2.10) and (2.15):

$$\|f\|_{A^2}^2 \approx |f(0)|^2 + \frac{1}{\pi}\int_{\mathbb{D}} \left|f'(z)\right|^2 (1-|z|^2)^2 dA. \tag{2.18}$$

Again, functions in this space have power series with radius of convergence at least one and again the functionals of evaluation at points $z \in \mathbb{D}$ are continuous. As before, using the fact that the monomials are an orthogonal basis we can compute the reproducing kernel, which is often called the *Bergman kernel*:

$$k_\zeta(z) = k_{A^2,\zeta}(z) := \sum_{n=0}^{\infty} (n+1)(z\overline{\zeta})^n = \frac{1}{(1-z\overline{\zeta})^2}.$$

The space A^2 is a closed subspace of the Hilbert space $L^2(\mathbb{D}) = L^2(\mathbb{D}, dA/\pi)$. The Bergman projection P, the orthogonal projection P of $L^2(\mathbb{D})$ onto A^2, maps

F in L^2 to PF given by

$$PF(\zeta) := \langle F, k_\zeta \rangle_{L^2(\mathbb{D})} = \frac{1}{\pi}\int_{\mathbb{D}} F(z)\frac{1}{(1-\bar{z}\zeta)^2}dA. \tag{2.19}$$

EXERCISE 2.6. *Verify this.*

In particular if $F \in A^2$ we have

$$F(\zeta) = \frac{1}{\pi}\int_{\mathbb{D}} F(z)\frac{1}{(1-\bar{z}\zeta)^2}dA. \tag{2.20}$$

which is sometimes called the *Bergman reproducing formula.*

We mentioned that functions in the Hardy space have radial boundary limits *a.e.*. The Dirichlet space is a subspace of the Hardy space and hence functions in the Dirichlet space also have boundary limits *a.e.*. However, functions in the Bergman space need not have radial limits on any radius. Examples of this can be built using power series with gaps, for instance $\sum_{n=0}^{\infty} z^{2^n}$ [**24**] and [**174**, Chapter 1].

General references about the Bergman space include [**127**], [**348**], [**174**], and [**133**].

2.4. Results About the Spaces

To repeat: the spaces $\mathcal{D}$, H^2, and A^2 are not only Hilbert spaces, they have additional structure. For each there is an associated set X, in these cases $X = \mathbb{D}$, and the elements of the Hilbert space are functions on X. Furthermore, in each case, for any $x \in X$, the evaluation functional e_x, defined by $e_x(f) = f(x)$, is a continuous linear functional on the Hilbert space. Hence, in each case, and for each $x \in \mathbb{D}$, there is a *reproducing kernel*, k_x, an element of the space which realizes the evaluation functional, $e_x(f) = f(x) = \langle f, k_x \rangle$.

Some properties of these spaces can be formulated and proved using the language of RKHS. Often that proof is neither longer nor shorter; but it does emphasize the generality of the approach and results. With this in mind, for now we will work with a generic RKHS H, a space of functions on some X, with, for $x \in X$, the kernel function and normalized kernel function denoted by, k_x and $\hat{k}_x$ respectively.

2.4.1. Size and Smoothness.

There is a standard estimate for the size of the values of functions in RKHS. The result for the Dirichlet space is

PROPOSITION 2.7. *Suppose* $f \in \mathcal{D}$, $\|f\| \le 1$, $z \in \mathbb{D}$, *then* $|f(z)| \le \widehat{k_z}(z) = k_z(z)^{1/2} = |z|^{-1}\sqrt{-\log(1-|z|^2)}$.

That is the specialization to $\mathcal{D}$ of the general RKHS result.

PROPOSITION 2.8. *Suppose H is a RKHS on X and that $x \in X$. We have*

$$\sup\{\operatorname{Re} f(x) : f \in H, \|f\| \le 1\} = \widehat{k_x}(x) = k_x(x)^{1/2} = \|k_x\|_H\,.$$

Furthermore, the unique f for which equality is attained is

$$f(\cdot) = \widehat{k_x}(\cdot) = k_x(x)^{-1/2}k_x(\cdot).$$

Hence the kernel function is given by $k_x(\cdot) = \overline{f(x)}f(\cdot)$.

PROOF. The proof is shorter than the statement. We know $\operatorname{Re} f(x) = \operatorname{Re}\langle f, k_x\rangle$. We know from Hilbert space basics that $\sup\{\operatorname{Re}\langle f, g\rangle : \|f\| \leq 1\}$ is attained with the unique choice $f = g/\|g\|$. □

EXERCISE 2.9. *Given $f \in H^2$, $\|f\| \leq 1$, $\zeta \in \mathbb{D}$, what is the maximum possible value of $|f'(\zeta)|$? Suggestion: find $j_\zeta \in H^2$ such that $\langle f, j_\zeta\rangle = f'(\zeta)$ and then maximize the inner product. Does the same procedure work for $\mathcal{D}$?*

The previous proposition shows that for any f in H,

$$\limsup_{|w|\to 1} \frac{|f(w)|}{\|k_w\|} \leq \|f\|.$$

But once f is fixed, more is true.

PROPOSITION 2.10. *Suppose $f \in H$ for $H = \mathcal{D}$, H^2, or A^2. Then*

$$\lim_{|w|\to 1} \frac{|f(w)|}{\|k_w\|} = 0.$$

PROOF. The restriction on H gives us easy access to a dense set of nice functions, the polynomials. Pick $\varepsilon > 0$ and a polynomial p_ε with $\|f - p_\varepsilon\| \leq \varepsilon$. We estimate

$$\begin{aligned}
\limsup_{|w|\to 1} \frac{|f(w)|}{\|k_w\|} &\leq \limsup_{|w|\to 1} \frac{|p_\varepsilon(w)|}{\|k_w\|} + \limsup_{|w|\to 1} \frac{|(f - p_\varepsilon)(w)|}{\|k_w\|} \\
&\leq \limsup_{|w|\to 1} \frac{|\langle p_\varepsilon, k_w\rangle|}{\|k_w\|} + \|f - p_\varepsilon\| \\
&\leq \limsup_{|w|\to 1} \frac{|\langle p_\varepsilon, k_w\rangle|}{\|k_w\|} + \varepsilon. \\
&\leq \varepsilon.
\end{aligned}$$

where the last estimate uses the following exercise. Letting ε tend to 0 completes the proof. □

EXERCISE 2.11. *To finish the proof of the proposition, show that for the spaces considered, and for any fixed polynomial p_ε,*

$$\lim_{|w|\to 1} \frac{|\langle p_\varepsilon, k_w\rangle|}{\|k_w\|} = 0.$$

If V is a closed subspace of H then V is also a RKHS and it has reproducing kernels $\{k_{V,x} : x \in X\}$. That is, there are functions $k_{V,x} \in V$ such that if $v \in V$ then $v(x) = \langle v, k_{V,x}\rangle$. These kernel functions can be obtained using P_V, the orthogonal projection of H to V.

PROPOSITION 2.12. $\forall x \in X$, $k_{V,x} = P_V k_x$.

PROOF. The presence of the projection P_V ensures that $k_{V,x} \in V$. For $v \in V$ we have

$$\langle v, P_V k_x\rangle = \langle P_V v, k_x\rangle = \langle v, k_x\rangle = v(x),$$

as required. □

In particular, suppose $x \in X$ and $V = V_x$, the subspace of H of functions that vanish at x. In that case $(V_x)^\perp$ is the one dimensional space spanned by k_x. Hence the orthogonal projection, $P_x^\perp$, of some $v \in H$ into $(V_x)^\perp$ is given by $P_x^\perp(v) = \left\langle v, \widehat{k_x} \right\rangle \widehat{k_x}$ The projection P_V equals $I - P_x^\perp$. Hence

$$v = (v - \left\langle v, \widehat{k_x} \right\rangle \widehat{k_x}) + \left\langle v, \widehat{k_x} \right\rangle \widehat{k_x}$$

is the decomposition of $v \in H^2$ with respect to the decomposition $H = (V_x)^\perp \oplus V_x$. Combining these observations with the previous proposition we have the following. For $w \in X, w \neq x$; the function h_w, the reproducing kernel in the space V_x for the point w, is given by

$$h_w(\cdot) = P_V k_w(\cdot) = k_w(\cdot) - \left\langle k_w, \widehat{k_x} \right\rangle \widehat{k_x}(\cdot). \tag{2.21}$$

Proposition 2.8 gave information about the size of functions in H. We now collect similar information about their oscillation. It will be convenient to have the following notation, for any $x, w \in X$,

$$\delta(x, w) := \delta_H(x, w) = \sqrt{1 - \left|\left\langle \widehat{k_x}, \widehat{k_w} \right\rangle\right|^2}. \tag{2.22}$$

PROPOSITION 2.13. *Suppose $x, w \in X$, $f \in H$, $\|f\| \leq 1$. If $f(x) = 0$ then* $\operatorname{Re} f(w) \leq \delta_H(x, w) k_w(w)^{1/2}$. *Furthermore there is a unique $f_{\mathtt{ext}}$ for which equality holds,*

$$f_{\mathtt{ext}} = \delta_H(z, w)^{-1} P_x^\perp \widehat{k_w} = \delta_H(z, w)^{-1} \left(\widehat{k_w} - \left\langle \widehat{k_w}, \widehat{k_x} \right\rangle \widehat{k_x}\right).$$

PROOF. Let V be the subspace of H consisting of functions that vanish at x. By Proposition 2.8 the extremal value we seek is $h_w(w)/\|h_w\|$ where h_w is the reproducing kernel of the space V for evaluation at w. That proposition also ensures that the unique extremal function is $h_w(\cdot)/\|h_w\|$.

The formula for h_w is given in (2.21). In order to normalize h we compute

$$\begin{aligned}
\|h_w\|^2 &= \left\| k_w - \left\langle k_w, \widehat{k_x} \right\rangle \widehat{k_x} \right\|^2 \\
&= \|k_w\|^2 - 2\operatorname{Re}\left\langle k_w, \left\langle k_w, \widehat{k_x} \right\rangle \widehat{k_x} \right\rangle + \left\| \left\langle k_w, \widehat{k_x} \right\rangle \widehat{k_x} \right\|^2 \\
&= \|k_w\|^2 - \left|\left\langle k_w, \widehat{k_x} \right\rangle\right|^2 \\
&= k_w(w) - \frac{|k_w(x)|^2}{k_x(x)} \\
&= k_w(w)\left(1 - \left|\left\langle \widehat{k_x}, \widehat{k_w} \right\rangle\right|^2\right).
\end{aligned}$$

Thus

$$\|h_w\| = k_w(w)^{1/2}\sqrt{1 - \left|\left\langle \widehat{k_x}, \widehat{k_w} \right\rangle\right|^2}$$

as required. $\square$

We can also obtain estimates for $|f(z) - f(w)|$. Here is a specific result for the Dirichlet space. Recall from (1.7) that the hyperbolic metric on the disk, β, is related to the pseudohyperbolic metric ρ, by

$$\beta(z,w) = \log\left(\frac{1+\rho(z,w)}{1-\rho(z,w)}\right). \tag{2.23}$$

PROPOSITION 2.14. *If $f \in \mathcal{D}$, $\|f\| \le 1$, $z, w \in \mathbb{D}$ then*

$$|f(z) - f(w)| \le \beta(z,w)^{1/2}. \tag{2.24}$$

PROOF. Let $\tilde{\mathcal{D}}$ be the space of functions in $\mathcal{D}$ which vanish at the origin, normed using the Dirichlet integral (2.8); $\|f\|^2_{\tilde{\mathcal{D}}} = \mathfrak{D}(f)$. The reproducing kernels for $\tilde{\mathcal{D}}$ are the functions $h_\zeta(z) = -\log\left(1 - \overline{\zeta} z\right)$.

Let k_z and k_w be reproducing kernels in $\mathcal{D}$ for z, w, and let h_z, h_w be the corresponding kernels in $\tilde{\mathcal{D}}$. For $f \in \mathcal{D}$, $\|f\| \le 1$ we have

$$\begin{aligned}
|f(z) - f(w)|^2 &\le \sup\left\{|\langle f, k_z - k_w\rangle| : f \in \mathcal{D}, \|f\|_{\mathcal{D}} \le 1\right\}^2 \\
&\le \sup\left\{|\langle f, k_z - k_w\rangle| : f \in \mathcal{D}, f(0) = 0, \|f\|_{\mathcal{D}} \le 1\right\}^2 \\
&\le \sup\left\{|\langle f, h_z - h_w\rangle_{\tilde{\mathcal{D}}}| : f \in \tilde{\mathcal{D}}, \|f\|_{\tilde{\mathcal{D}}} \le 1\right\}^2 \\
&= \|h_z - h_w\|^2_{\tilde{\mathcal{D}}} \\
&= \langle h_z - h_w, h_z - h_w\rangle_{\tilde{\mathcal{D}}} && (2.25) \\
&= -\log(1 - |z|^2) - \log(1 - |w|^2) + 2\log|1 - \overline{z}w| \\
&= \beta(z,w) - 2\log\left(1 + \rho(z,w)\right) && (2.26) \\
&\le \beta(z,w).
\end{aligned}$$

In passing from the first line to the second we replace $f(z)$ by $f(z) - f(0)$ which does not change the inner product and makes the norm of f smaller. In passing to the third line the value of the inner product is again unchanged and the set of candidate functions is expanded. The passage to (2.26) uses (1.6) and (2.23). □

Thus Dirichlet functions, as maps from the disk with its hyperbolic metric to the plane with its Euclidean metric, are Hölder continuous of order $1/2$. Although this estimate is not good for nearby pairs of points, it is optimal when the points are far from each other. We will see analogous results later for Besov spaces, for instance Lemma 14.12, but the exponent in those estimates must be modified depending on the space being considered.

2.4.2. Multipliers and Carleson Measures. Suppose H is a RKHS on the space X. The elements of H can be viewed as functions and hence we can multiply them by each other or by other functions. We say that a function m defined on X is a *multiplier* (*of* H or *for* H) if pointwise multiplication by m, denoted as an operator by M_m, maps H boundedly to itself; that is, if there is a $C = C(m)$ so that for all $h \in H$

$$\|M_m h\|_H = \|mh\|_H \lesssim \|h\|_H. \tag{2.27}$$

EXERCISE 2.15. *In fact* (2.27) *is a consequence of the superficially weaker hypothesis that for all $h \in H$, $mh \in H$. (See Theorem* A.1.*)*

Let $\text{Mult}(H)$ be the space of all multipliers of H. For $m \in \text{Mult}(H)$ let $\|m\| = \|m\|_{\text{Mult}(H)} = \|M_m\|_{\texttt{operator}}$. With this norm $\text{Mult}(H)$ is a commutative Banach algebra and understanding this algebra is an important part of understanding H.

EXERCISE 2.16 (Smooth functions are multipliers). *Show that multiplication by the monomial z^j is a bounded operator on $\mathcal{D}$, on H^2, and on A^2. In each case, estimate the norm of the operator as a function of j. Using those estimates show that if $m \in \text{Hol}(\overline{\mathbb{D}})$ then multiplication by $m(z)$ is a bounded operator on each of those spaces. Furthermore m is the limit in multiplier norm of polynomial multipliers.*

If $m \in \text{Mult}(H)$ then the action of M_m^*, the adjoint of the multiplication operator, on kernel functions takes a very simple form.

PROPOSITION 2.17. *Let H be a RKHS on X. For $\zeta \in X$; let $k_{H,\zeta}$ be the associated kernel function. For $m \in \text{Mult}(H)$ and $\zeta \in X$ we have*

$$M_m^* k_\zeta(z) = \overline{m(\zeta)} k_\zeta.$$

PROOF. We select $\zeta, z \in X$ and compute

$$\begin{aligned} M_m^* k_\zeta(z) &= \langle M_m^* k_\zeta, k_z \rangle \\ &= \langle k_\zeta, M_m k_z \rangle \\ &= \langle k_\zeta, m k_z \rangle \\ &= \overline{\langle m k_z, k_\zeta \rangle} \\ &= \overline{m(\zeta) k_z(\zeta)} \\ &= \overline{m(\zeta)} k_\zeta(z); \end{aligned}$$

the last step using (2.7). □

COROLLARY 2.18. *Multipliers are bounded functions. More precisely,*

$$\sup_X |m(\zeta)| = \|m(\cdot)\|_\infty \le \|M\|.$$

PROOF. We just saw that the kernel function k_ζ is an eigenvector of M^* with eigenvalue $\overline{m(\zeta)}$. Hence $\left|\overline{m(\zeta)}\right| \le \|M^*\| = \|M\|$. Taking the supremum over $\zeta \in \mathbb{D}$ gives the conclusion. □

PROPOSITION 2.19. *Suppose H is a RKHS and δ_H is the distance function given by (2.22). For $z, w \in X$ and $m \in \text{Mult}(H)$, $\|m\| \le 1$, if $m(z) = 0$ then $|m(w)| \le \delta_H(z, w)$.*

PROOF. Let $\hat{k}_w$ be the normalized reproducing kernel for w. Because $\|m\| \le 1$ the vector $m\hat{k}_w$ has norm at most one. Also, $m\hat{k}_w(z) = 0$. Thus $m\hat{k}_w$ is a competitor in the extremal problem described in Proposition 2.13. Hence, from the conclusion of that proposition we have $\left|m(w)\hat{k}_w(w)\right| \le k_w(w)^{1/2} \delta_H(z, w)$. This gives what we want. □

This estimate is sharp for some choices of H and not sharp for others. We will find out more about that in Chapter 6 when we consider the Pick property.

For specific H it can be easy or difficult to find a description of $\operatorname{Mult}(H)$; the details depend on the particular space. With that in mind we now narrow our focus from general RKHS H to the three specific choices $H = \mathcal{D}$, H^2, and A^2.

For those H, $\|1\|_H = 1$ and hence there is a contractive inclusion of the multipliers into H

$$\|m\|_H \leq \|m\|_{\operatorname{Mult}(H)} .$$

In particular, the multipliers are holomorphic. Thus, taking into account Corollary 2.18, the multiplier algebras are contractively contained in H^∞; for all $m \in \operatorname{Mult}(H)$, $\|m\|_{H^\infty} \leq \|m\|_{\operatorname{Mult}(H)}$.

For the Bergman space that is the full story; $\operatorname{Mult}(A^2) = H^\infty$. By the previous Corollary $\operatorname{Mult}(A^2) \subset H^\infty$; the other inclusion follows from (2.17).

For the Hardy space the answer is the same; $\operatorname{Mult}(H^2) = H^\infty$, but the argument is not so simple. It is not transparent from the definition (2.14) that every $m \in H^\infty$ is a multiplier. One route to showing m is a multiplier is to develop the boundary value theory for the spaces H^2 and H^∞. We have said that the map of f to f^β takes H^2 to a closed subspace of L^2 of the circle. The same map also takes H^∞ to a closed subspace of L^∞ of the circle. Furthermore, the boundary values work out as expected; if $f \in H^2$ and $m \in H^\infty$ then $(mf)^\beta (e^{i\theta}) = f^\beta(e^{i\theta}) m^\beta(e^{i\theta})$ *a.e.* $d\theta$. One can then detect if m is a multiplier on H^2 by checking that m^β is a multiplier of the space of boundary values of H^2 functions. That holds because multiplication by bounded functions takes $L^2(\mathbb{T})$ to itself, and because the product of two functions with Fourier series of power series type is another such function. Note that this argument uses the fact that, just as with the Bergman space, the Hardy space can be realized as a closed subspace of a Lebesgue space, and its multiplier operators are the restrictions to that subspace of multiplication operators on the larger space. The Dirichlet space cannot be realized that way (Corollary 11.7) and we will see in a moment that $\operatorname{Mult}(\mathcal{D})$ is not all of H^∞. An alternative path to showing that every $m \in H^\infty$ is a multiplier for H^2 involves Carleson measures.

For any Hilbert space of Borel measurable functions on the disk, H, we say that a positive measure μ on the disk is a *Carleson measure* for H, $\mu \in CM(H)$, if there is a $C = C(\mu)$ so that for all $h \in H$

$$\int_{\mathbb{D}} |h|^2 \, d\mu \leq C^2 \|h\|_H^2 . \tag{2.28}$$

That is, the natural inclusion map of H to $L^2(\mathbb{D}, \mu)$ has norm at most C. The norm of the inclusion map, the infimum of all such C, is the Carleson imbedding constant of μ, $\|\mu\|_{CM(H)}$.

If $1 \in H$, as is the case for $H = H^2$, then every Carleson measure is a finite measure. A similar comment holds, and for a similar reason, for the measures we will call Carleson measures in related contexts. We will often use that fact, perhaps sometimes without comment or proof.

Suppose now that we want to show that $m \in H^\infty$ is a multiplier of H^2. Select $f \in H^2$. By Exercise 2.15 we need to show that $mf \in H^2$. By (2.15) we want to show

$$|m(0)f(0)|^2 + \int_{\mathbb{D}} \left|(mf)'(z)\right|^2 (1 - |z|^2) dA < \infty. \tag{2.29}$$

The constant term causes no problem. Using $(mf)' = mf' + m'f$ we can estimate the main term by estimating two pieces. For the first note using (2.15) that if

$m \in H^\infty$, and $f \in H^2$ then $\int_{\mathbb{D}} |m(z)|^2 |f'(z)|^2 (1-|z|^2) dA < \infty$. Hence, to verify (2.29) it remains to show that for some $C(m) < \infty$,

$$\int_{\mathbb{D}} |f(z)|^2 |m'(z)|^2 (1-|z|^2) dA \leq C(m)^2 \|f\|_{H^2}^2 . \tag{2.30}$$

That is, we need to know $d\mu_m = |m'(z)|^2 (1-|z|^2) dA$ is a Carleson measure.

PROPOSITION 2.20. *$m \in \text{Mult}(H^2)$ if and only if $m \in H^\infty$ and $\mu_m \in CM(H^2)$.*

PROOF. The previous discussion shows that such m are multipliers. In the other direction we know from Corollary 2.18 that the multipliers are bounded. Finally, if $m \in \text{Mult}(H^2)$ and $f \in H^2$ then, taking note of (2.15) we have

$$\begin{aligned}
\|f\|_{H^2} &\gtrsim \|mf\|_{H^2} \gtrsim \|(mf)'\|_{L^2(\mathbb{D},(1-|z|^2)dA)} \\
&\gtrsim \|m'f + mf'\|_{L^2(\mathbb{D},(1-|z|^2)dA)} \\
&\gtrsim \left(\|m'f\|_{L^2(\mathbb{D},(1-|z|^2)dA)} - \|mf'\|_{L^2(\mathbb{D},(1-|z|^2)dA)} \right) \\
&\gtrsim \left(\|m'f\|_{L^2(\mathbb{D},((1-|z|^2)dA)} - \|m\|_\infty \|f\|_{H^2} \right).
\end{aligned}$$

Because we know m is bounded this establishes that $\mu_m \in CM(H^2)$, $\square$

Of course this is not satisfying without more insight into the condition $\mu_m \in CM(H^2)$. That condition is a subtle one. However it is known to hold for bounded functions (and some unbounded ones). The proof that $\text{Mult}(H^2) = H^\infty$ is completed with the following non-trivial observation.

THEOREM 2.21. *If $m \in H^\infty$ then $d\mu_m = |m'(z)|^2 (1-|z|^2) dA \in CM(H^2)$.*

We will discuss that theorem in Section 3.7.

COROLLARY 2.22. $\text{Mult}(H^2) = H^\infty$.

On the other hand, it is not true that $\text{Mult}(\mathcal{D}) = H^\infty$. In fact we saw at (2.9) that H^∞ is not even contained in $\mathcal{D}$ which contains $\text{Mult}(\mathcal{D})$. However, as we go forward we will see several results similar to Proposition 2.20; a function is a multiplier if and only if it is bounded and certain derivatives generate Carleson measures. In particular, the argument that led to that proposition applies essentially unchanged for the Dirichlet space.

PROPOSITION 2.23. *$m \in \text{Mult}(\mathcal{D})$ if and only if $m \in H^\infty$ and*

$$d\mu_m = |m'(z)|^2 dA \in CM(\mathcal{D}).$$

That Carleson measure condition on m will be of interest later even if m is not bounded and so we introduce a notation for it. We say that $h \in \text{Hol}(\mathbb{D})$ is in $\mathcal{X}$ if

$$d\mu_h = d\mu_{h,\texttt{Dir}} := |h'(z)|^2 dA \in CM(\mathcal{D}), \tag{2.31}$$

in which case we set $\|h\|_{\mathcal{X}} = \|\mu_h\|_{CM(\mathcal{D})}$.

In Chapter 4 we give the classical characterizations of Carleson measures for H^2 and for $\mathcal{D}$.

2.4.3. The Distance Function. For H a RKHS on X the function δ defined in (2.22) is closely related to the function theory of H. We saw this in Proposition 2.13 and Proposition 2.19, and we will see other examples later. Here we collect some basic facts about δ. Further discussion of δ is in [**50**] and [**280**].

If H is the Hardy space then δ_H has a familiar description;

$$\delta_H(z,w) = \delta_{H^2}(z,w) = \sqrt{1 - \frac{(1-|z|^2)(1-|w|^2)}{|1-\overline{z}w|^2}}.$$

In view of (1.4) and (1.6) we conclude that δ is the pseudohyperbolic distance ρ; that is

$$\delta_{H^2}(z,w) = \left|\frac{z-w}{1-\overline{z}w}\right| = \rho(z,w). \tag{2.32}$$

For a general H, the function $\delta_H(\cdot,\cdot)$ from (2.22) clearly satisfies $0 \leq \delta_H \leq 1$, and, in fact, is again a distance function. It is clearly positive and symmetric. The triangle inequality can be verified by an elementary argument ([**4**, page 128]), or, alternatively, it is a corollary of the second line of the next proposition.

Fix $z, w \in X$. Let P_z and P_w be the self adjoint projections onto the span of k_z and k_w and $P_z^{\perp}$ and $P_w^{\perp}$, the corresponding complementary projections.

PROPOSITION 2.24 (Coburn [**110**]).

$$\begin{aligned}\delta(z,w) &= \sqrt{1-\|P_zP_w\|^2}\\ &= \|P_z - P_w\| = \left\|P_w^{\perp} - P_z^{\perp}\right\|\\ &= \left\|P_z^{\perp}P_w\right\| = \left\|P_zP_w^{\perp}\right\|.\end{aligned}$$

PROOF. The operator P_zP_w is a rank one operator which vanishes on $(\operatorname{span}\{k_w\})^{\perp}$ hence,

$$\left\|P_z^{\perp}P_w\right\| = \left\|P_z^{\perp}P_w\widehat{k_w}\right\| = \left\|\left\langle\widehat{k_w},\widehat{k_z}\right\rangle\widehat{k_z}\right\| = \left|\left\langle\widehat{k_w},\widehat{k_z}\right\rangle\right|$$

which is what we need. Similarly

$$\left\|P_z^{\perp}P_w\right\| = \left\|P_zP_w\widehat{k_w}\right\| = \left\|\widehat{k_w} - \left\langle\widehat{k_w},\widehat{k_z}\right\rangle\widehat{k_z}\right\| = \delta(z,w)$$

as required. The final expression is done the same way after noting that

$$\left\|P_zP_w^{\perp}\right\| = \left\|\left(P_zP_w^{\perp}\right)^*\right\| = \left\|P_w^{\perp}P_z\right\|.$$

For $P_z - P_w = P_w^{\perp} - P_z^{\perp}$ we use several facts about the trace functional on finite rank operators. If L is such an operator then $\operatorname{Trace}(L)$ is the sum of the diagonal elements of the representation of L with respect to any orthonormal basis. In particular if L is self adjoint then it will have real eigenvalues $\{\lambda_i\}_{i=1}^n$ in which case $\|L\| = \sup\{|\lambda_i|\}$ and $\operatorname{Trace}(L) = \sum_i \lambda_i$. Also, if L' is another such operator then $\operatorname{Trace}(LL') = \operatorname{Trace}(L'L)$

We have that P_z and P_w are rank one self adjoint projections, and thus each has trace 1. Hence the difference, $P_z - P_w$, is a rank two self adjoint operator with trace zero. Such an operator must have two eigenvalues, $\pm\lambda$ for some $\lambda \geq 0$. We will be finished if we show $2\delta(z,w)^2 = 2\lambda^2$. We compute

$$\begin{aligned} 2\lambda^2 &= \operatorname{Trace}((P_z - P_w)^2) \\ &= \operatorname{Trace}(P_z + P_w - P_z P_w - P_w P_z) \\ &= 2 - \operatorname{Trace}(P_z P_w) - \operatorname{Trace}(P_w P_z) \\ &= 2 - 2\operatorname{Trace}(P_z P_w). \end{aligned}$$

Here the first line comes from putting $P_z - P_w$ in diagonal form, squaring, and summing the diagonal elements.

We now compute $\operatorname{Trace}(P_z P_w)$. Let V be the span of k_z and k_w. $P_z P_w$ maps V to V and is identically zero on $V^\perp$; hence we can evaluate the trace while regarding $P_z P_w$ as an operator on V. We pick an orthonormal basis for V, $\widehat{k_w}$ and j where j is any unit vector in V orthogonal to k_w. We compute the trace using the matrix representation of the operator in this basis. Note that $P_w \widehat{k_w} = \widehat{k_w}$ and $P_w j = 0$ and hence

$$\begin{aligned} \operatorname{Trace}(P_z P_w) &= \left\langle P_z P_w \widehat{k_w}, \widehat{k_w} \right\rangle + \langle P_z P_w j, j \rangle \\ &= \left\langle P_z \widehat{k_w}, \widehat{k_w} \right\rangle \\ &= \left\langle \left\langle \widehat{k_w}, \widehat{k_z} \right\rangle \widehat{k_z}, \widehat{k_w} \right\rangle \\ &= \left| \left\langle \widehat{k_w}, \widehat{k_z} \right\rangle \right|^2. \end{aligned}$$

which is what we needed. □

Using the notation δ will help emphasize formal similarities between the Hardy space and the Dirichlet space. However $\delta_{\mathcal{D}}$ and δ_{H^2} behave very differently from each other. Near any point each is comparable to a constant multiple of the Euclidean distance. However the constants of comparability depend on the point and on which Hilbert space is considered. At large scales the differences are extreme. The next exercise develops that point.

EXERCISE 2.25.

(1) *For ε, ε' small and positive and $k > 1$; find the approximate value of $\delta_{H^2}(0, 1-\varepsilon)$, of $\delta_{H^2}(1-\varepsilon, 1-\varepsilon')$, and of $\delta_{H^2}(1-\varepsilon, (1-\varepsilon)e^{i\varepsilon'})$.*
(2) *At what rate should $\varepsilon_n \to 0$ to ensure that the sequence of distances of $\delta_{H^2}(1-\varepsilon_n, 1-\varepsilon_{n+1})$ remains bounded away from 0 and 1?*
(3) *Answer the same two questions for $\delta_{\mathcal{D}}$.*

In Chapter 7 we study interpolating sequences. One important property of those sequences is that their points are separated. The function δ gives a natural way to make this precise. Given a RKHS H of holomorphic functions on the disk; we will say that a sequence of points $Z = \{z_n\}_{n=1}^\infty \subset \mathbb{D}$ is separated (relative to H) if

$$\exists \varepsilon > 0 \text{ such that } \forall z, w \in Z,\ z \neq w \text{ we have } \delta_H(z,w) > \varepsilon. \tag{2.33}$$

For the Hardy space this condition states that there is a lower bound on the hyperbolic distance between the points. That formulation, together with some knowledge of hyperbolic geometry, gives an informal understanding of the condition. The metric $\delta_{\mathcal{D}}$ however is less familiar and is very different from δ_{H^2} and an intuitive understanding of the condition (2.33) for the Dirichlet space is less clear. However, again, there is a formulation in terms of the hyperbolic geometry of the disk, without reference to $\mathcal{D}$. That formulation is, perhaps, more intuitive.

Recall that β, defined in (1.7), is the hyperbolic metric on the disk.

PROPOSITION 2.26 ([**220**]). *Given $Z = \{x_n\}_{n=1}^{\infty} \subset \mathbb{D}$, Z satisfies (2.33) for $H = \mathcal{D}$ if and only if it satisfies the following condition:*

$$\forall z, w \in Z, z \neq w, \text{ we have } \beta(0,z) \lesssim \beta(z,w). \tag{2.34}$$

REMARK 2.27. *In this proof, as in many proofs in this area, the points far from the boundary are relatively easy to deal with by straightforward, though perhaps awkward looking, standard approximations. Although those cases are necessary for the proof, they are side issues conceptually, and are frequently left to the reader. The analysis for points near the boundary is the conceptual heart of the proof.*

PROOF. Pick and fix $z, w \in \mathbb{D}$, set $\theta = |\left\langle \hat{k}_z, \hat{k}_w \right\rangle|^2$ and $\rho = \rho(z,w)$. Recalling the definitions we see that we want to show the following. There is a constant $\lambda > 0$ so that

$$\theta < 1 - \lambda \tag{2.35}$$

if and only if (2.34) holds. That condition is that

$$\log\left(\frac{1+|z|}{1-|z|}\right) \lesssim \log\left(\frac{1+\rho}{1-\rho}\right). \tag{2.36}$$

It will be convenient to work instead with the equivalent

$$\log\left(\frac{1}{1-|z|^2}\right) \lesssim \log\left(\frac{1}{1-\rho^2}\right). \tag{2.37}$$

EXERCISE 2.28. *Show (2.36) and (2.37) are equivalent. Outline: If ρ is near one then terms involving $\log(1+\rho)$ and $\log(1+|z|)$ can be absorbed. If ρ is small then so is $|z|$ and the first order Taylor expansion of $\log(1+x)$ can be used.*

We now follow the proof of Lemma 19 of [**220**] to show that (2.35) and (2.37) are equivalent. First suppose (2.35) holds. Using (2.35) and the fact that $2ab \leq a^2 + b^2$ we have

$$\begin{aligned} 2\log\frac{1}{|1-\overline{z}w|} &\leq 2\left|\log\frac{1}{1-\overline{z}w}\right| \leq 2\left|(1-\lambda)^2\log\frac{1}{1-|z|^2}\right|^{1/2}\left|\log\frac{1}{1-|w|^2}\right|^{1/2} \\ &\leq (1-\lambda)^2\log\frac{1}{1-|z|^2} + \log\frac{1}{1-|w|^2}. \end{aligned}$$

Hence, doing a small bit of algebra and invoking (1.6), we obtain

$$\log\left(1-\rho^2\right) \leq \left[(1-\lambda)^2 - 1\right]\log\left(\frac{1}{1-|z|^2}\right).$$

Noting that $(1-\lambda)^2 - 1 < 0$, this gives (2.37).

Suppose now that (2.37) holds and, without loss of generality, that $|w| \leq |z|$. Using (2.37) and (1.6) we find that we can select μ with $0 < \mu < 1$ so that

$$\log\frac{1}{|1-\overline{z}w|^2} \leq \mu\log\frac{1}{1-|z|^2} + \log\frac{1}{1-|w|^2} \tag{2.38}$$

where $\mu = 1 - 1/C'$. Note that if (2.37) holds then it also holds for any larger C'. Hence we can assume $C' > 1$ and thus $0 < \mu < 1$.

Next note that if $0 \leq a \leq b \leq (2-\mu)\, a$ then

$$\mu b + a \leq \mu(2-\mu)\, a + a = \left(1 + 2\mu - \mu^2\right) a < \left(1 + 2\mu - \mu^2\right) a^{1/2} b^{1/2}. \tag{2.39}$$

Thus, if

$$\log \frac{1}{1-|z|^2} \leq (2-\mu) \log \frac{1}{1-|w|^2} \tag{2.40}$$

then we can use (2.38) and (2.40) in inequality (2.39) and obtain, in that case,

$$\log \frac{1}{|1-\bar{z}w|^2} \leq 2\left(\frac{1}{2} + \mu - \frac{\mu^2}{2}\right)\left(\log \frac{1}{1-|z|^2}\right)^{1/2}\left(\log \frac{1}{1-|w|^2}\right)^{1/2}. \tag{2.41}$$

If (2.40) fails then we have

$$\begin{aligned} \log \frac{1}{|1-\bar{z}w|} &\leq \log \frac{1}{1-|z|\,|w|} \leq \log \frac{1}{1-|w|} \\ &\leq \left(\frac{1}{2-\mu}\right)^{1/2}\left(\log \frac{1}{1-|z|^2}\right)^{1/2}\left(\log \frac{1}{1-|w|^2}\right)^{1/2}. \end{aligned}$$

Thus, in either case we have

$$\log \frac{1}{|1-\bar{z}w|} \leq \sigma\left(\log \frac{1}{1-|z|^2}\right)^{1/2}\left(\log \frac{1}{1-|w|^2}\right)^{1/2}$$

for some $\sigma < 1$. That is

$$\left|\operatorname{Re}\left\langle \hat{k}_z, \hat{k}_w \right\rangle\right| \leq \sigma.$$

Furthermore

$$\left|\operatorname{Im}\left\langle \hat{k}_z, \hat{k}_w \right\rangle\right| \leq \frac{\pi}{2\left|\log(1-|z|^2)\right|^{1/2}\left|\log(1-|w|^2)\right|^{1/2}}.$$

If both $|z|$ and $|w|$ are close to 1 then these two estimates combine to give the desired

$$\left|\left\langle \hat{k}_z, \hat{k}_w \right\rangle\right|^2 \leq \sigma' \tag{2.42}$$

with $\sigma' < 1$. It remains to deal with the possibility that one of z, w is not near the boundary; suppose z is not. For each such z we have, for elementary reasons, that, for all allowable choices of w, $|\left\langle \hat{k}_z, \hat{k}_w \right\rangle|^2 \leq \sigma(z) < 1$. Also, there are only finitely many z in this category. Hence we get the desired control using the maximum of $\sigma(z)$ over such z. □

2.5. Other Viewpoints and Related Spaces

We described the Dirichlet space as a Hilbert space of holomorphic functions on the disk with a simple explicit reproducing kernel; and that will be our primary point of view going forward. However there are also other views of the Dirichlet space which bring different insights, technical tools, and questions; and they have also helped shape our selection of topics in this book

A fundamental fact about the Dirichlet space is that functions in it are characterized by their local oscillation. This places the Dirichlet space in the family of Besov spaces, each of which is characterized by the local oscillation of their functions. That type of characterization is also central to our study of Besov spaces of functions on the complex ball in later chapters. The function space BMO which we will encounter several times is also defined by local oscillation, and that leads to some of its similarities with the Dirichlet space.

Another fundamental fact about the Dirichlet space of the disk is that the disk has a transitive group of automorphisms, and each of those automorphism interacts nicely with the measures of local oscillation characterizing Dirichlet space functions. This leads to a homogeneity property which actually characterizes the Dirichlet space [**34**]. It also leads to the extremely useful technical fact that automorphisms can be used to reduce many technical questions about local behavior to questions about behavior near the origin. That technical tool is implicit in much of our later work with function spaces on trees and is explicit and pervasive in our work in later chapters with Besov spaces on the ball.

Other viewpoints on the Dirichlet space can be seen in other function spaces, or families of functions spaces, for which the questions similar to those we consider for the Dirichlet space are also discussed. Some of those spaces will be studied later in this volume; others will not be mentioned again. We discuss some of them very informally here because it is useful to be aware of them as we move forward.

2.5.1. Other Planar Domains. One can define a Bergman space or Dirichlet space or Hardy space associated with a general domain in the plane. The Bergman space of the halfplane and the Hardy space of the halfplane are both frequently studied. However even at this step problems appear. For instance, there are two different spaces with good claims to be the Hardy space of the halfplane and both are studied [**178**, Chapters 5 and 8].

Most of the topics we have discussed have analogs for these more general spaces. When the geometry is simple and the boundary is tame, using a conformal map to transport results to and from function spaces on the disk is valuable, but need not be decisive. When the domain or its boundary is complicated, those complications become basic objects of study.

2.5.2. The Dirichlet, Hardy, Bergman Family. The definitions of the Dirichlet, Hardy, and Bergman space make it clear that $\mathcal{D} \subset H^2 \subset A^2$, and the definitions also suggest a natural scale of spaces containing all three. Specifically, for $-\infty < \alpha < \infty$, consider the scale of Hilbert spaces $\mathcal{D}_\alpha$ of holomorphic functions on the disk, $f(z) = \sum_{n=0}^{\infty} a_n z^n$, normed by

$$\|f\|_{\mathcal{D}_\alpha}^2 := \sum_{n=0}^{\infty} (n+1)^\alpha |a_n|^2 < \infty. \tag{2.43}$$

Thus $\mathcal{D}$, H^2, and A^2 correspond to $\alpha = 1$, 0 and -1 respectively. Following the pattern used in analysis of the Dirichlet space, and recalling the binomial theorem for general powers, we find that the reproducing kernels for the spaces $\mathcal{D}_\alpha$, $\alpha < 1$, are given by

$$k_{\mathcal{D}_\alpha,\zeta}(z) = c_\alpha (1 - \overline{\zeta} z)^{1-\alpha}$$

The parameter values $\alpha = 1$ and $\alpha = 0$ correspond to sharp changes in the properties of functions in the spaces. The change in boundary behavior is one example. For $\alpha > 1$, functions in the function spaces $\mathcal{D}_\alpha$ extend to continuous functions on the closed disk. This is not true for the Dirichlet space which contains, for instance, $f(z) = (\log(1-z))^{1/3}$. The spaces $\mathcal{D}_\alpha$ for the range $1 < \alpha < 0$ are sometimes called generalized Dirichlet spaces. They are each contained in H^2 and hence functions in those spaces have radial boundary values almost everywhere. (In fact, more is true, depending on α, [**334**], [**333**].) The spaces $\mathcal{D}_\alpha$ for $\alpha < 0$ are the generalized Bergman spaces; functions in them need not have limits along any radius. As with the Bergman space, this can be shown using power series with gaps. Gap series are discussed in [**24**] and [**174**, Chapter 1].

The spaces $\mathcal{D}$, H^2, and A^2 have equivalent norms given by (2.10), (2.15), and (2.18). In fact there is an alternative approach to the full family of $\mathcal{D}_\alpha$ spaces which focuses on integral formulas rather than Taylor coefficients. Those formulas are also a starting point for describing the Besov spaces, which we mention next.

2.5.3. Besov Spaces. We have focused on Hilbert spaces of functions. However, many of the topics considered are equally natural for Banach spaces of functions. For instance, many questions studied for the Hardy space H^2 are also considered for the more general Hardy spaces, H^p, $1 < p < \infty$, described in Chapter 3.

The Besov spaces, B_p, $1 < p < \infty$, are a family of spaces which contains the Dirichlet space, $B_2 = \mathcal{D}$. A function, $f \in \operatorname{Hol}(\mathbb{D})$, is in the Besov space $B_p = B_p(\mathbb{D})$ if

$$\|f\|_{B_p}^p := |f(0)|^p + \frac{1}{\pi}\int_{\mathbb{D}} \left|(1-|z|^2)f'(z)\right|^p \frac{dA}{(1-|z|^2)^2} < \infty. \tag{2.44}$$

The general pattern here, as well as in other Besov type spaces, is that the norm is given by a global measure of size applied to a local measure of oscillation. That is clearly visible in this case when we rewrite the norm using the invariant gradient $\tilde{\nabla}$ defined in (1.11) and the invariant area element $d\lambda$ defined in (1.10):

$$\|f\|_{B_p}^p = |f(0)|^p + c\int_{\mathbb{D}} |\tilde{\nabla} f(z)|^p d\lambda(z) < \infty. \tag{2.45}$$

Furthermore, these spaces carry a variety of equivalent norming functionals based on alternative instances of the same paradigm. For instance, if $1 < p < \infty$ and m is a positive integer then $\|\cdot\|_{m,B^p}$ defined by

$$\|f\|_{m,B^p}^p = \left|\sum_{j=0}^{m-1} f^j(0)\right|^p + \frac{1}{\pi}\int_{\mathbb{D}} \left|(1-|z|^2)^m f^{(m)}(z)\right|^p \frac{dA}{(1-|z|^2)^2} \tag{2.46}$$

is an equivalent norm for B_p. This viewpoint is discussed further in Section 10.1. The Besov spaces $B_p(\mathbb{D})$ are presented in detail in [**348**]. The Bloch space B_∞ is a limiting case of the Besov spaces; $f \in B_\infty$ exactly if

$$\|f\|_{\mathtt{Bloch}} := |f(0)| + \sup_{\mathbb{D}} \left|\tilde{\nabla} f(z)\right| = |f(0)| + \left\|\tilde{\nabla} f\right\|_\infty < \infty. \tag{2.47}$$

Many of the topics we discuss here for the Dirichlet space are studied for the full range of spaces, $B_p(\mathbb{D})$, $1 < p < \infty$. In later chapters we will work with the full range of Besov spaces of holomorphic functions on the ball in $\mathbb{C}^n$; $B_p(\mathbb{B}_n)$, $1 < p < \infty$. Those spaces are discussed systematically in [**346**], [**345**].

2.5.4. The Containing Sobolev Space. The theory of the Hardy space H^2 on the circle is closely entwined with the theory of the Lebesgue space L^2. In fact, we have a decomposition $L^2 = H^2 \oplus \overline{H_0^2}$ where the second term is the space of complex conjugates of H^2 functions of mean value 0. Using this, L^2 can be realized as the space of boundary values of a space of harmonic functions in the disk, each such harmonic function being the sum of a holomorphic with H^2 boundary values and the complex conjugate of such a function.

Similarly, the Dirichlet space $\mathcal{D}$ sits inside the Sobolev space $\mathcal{W}^{1/2,2}(\mathbb{T})$, the space of functions on the boundary that can be defined as $\mathcal{W}^{1/2,2}(\mathbb{T}) = \mathcal{D} \oplus \overline{\mathcal{D}_0}$. (The notation is for having 1/2 derivative in $L^2(\mathbb{T})$.) The harmonic extensions of these functions to the disk have square integrable gradients and form a subset of the Hilbert space $\mathcal{W}^{1,2}(\mathbb{D})$, functions with one derivative in $L^2(\mathbb{D})$. Each of the spaces $\mathcal{W}^{1/2,2}(\mathbb{T})$, $\mathcal{W}^{1,2}(\mathbb{D})$, and the space of harmonic functions in $\mathcal{W}^{1,2}(\mathbb{D})$ is sometimes called "the Dirichlet space." Each of these spaces are different and they arise quite naturally in many contexts, some quite unrelated to what we discuss here, some closer to that which we discuss, for example [**234**].

2.5.5. Several Complex Variables. Given a domain $\Omega \subset \mathbb{C}^n$, one can study various spaces of functions contained in $\mathrm{Hol}(\Omega)$. In particular, the Bergman space, defined as $L^2(\Omega, dV) \cap \mathrm{Hol}(\Omega)$, and its kernel function, the Bergman kernel, are basic tools in the study of the relation between the geometry of Ω and the function theory of Ω, [**197**].

2.5.6. The Segal–Bargmann space. The Segal–Bargmann space (also known as the Bargmann space or the Fock space) is the space of entire functions whose modulus is square integrable with respect to the standard Gaussian density; f such that

$$\int_{\mathbb{C}} |f(z)|^2 e^{-|z|^2/2} dA < \infty.$$

In interesting counterpoint to the other spaces of holomorphic functions we mention, the multiplier algebra of this space is trivial, i.e. one dimensional.

There are also variants with a rescaled version of the density, and there are analogous spaces of functions on $\mathbb{C}^n$.

2.5.7. RKHS with a CPP. The Hardy and Dirichlet spaces are RKHS which have the *complete Pick property, CPP*, that is, they have *complete Pick kernels* (*CPK*); and that underpins some of the close analogies between results for the two spaces. The study of spaces with this property is an active research area; a picture of the area as of 2002 is in [**4**].

We will define and discuss the CPP in Chapter 6. For the moment we just mention some other spaces that share this property.

The spaces $\mathcal{D}_0$ and $\mathcal{D}_1$ in the notation of Section 2.5.2 are the Hardy and Dirichlet space respectively. Both of these spaces have the CPP as do the spaces $\mathcal{D}_\alpha$ for $0 < \alpha < 1$. The other $\mathcal{D}_\alpha$ do not. A class of spaces related to this scale is considered in [**39**].

The weighted Dirichlet spaces of the next section also have complete Pick kernels.

For $n = 1, 2, \ldots$ the Drury-Arveson space, DA_n, (also, sometimes, H_n^2; and, in Chapter 17, $B_2^{1/2}(\mathbb{B}_n)$) is the space of holomorphic functions on the unit ball

$\mathbb{B}_n \subset \mathbb{C}$ with the reproducing kernels

$$k_\zeta(z) = \frac{1}{1 - \langle z, \zeta \rangle}, \quad \zeta \in \mathbb{B}_n.$$

The value $n = \infty$ is allowed, with $\mathbb{B}_\infty$ taken to be the unit ball of $\ell^2(\mathbb{Z}_+)$. The reproducing kernel for DA_1 is the same as the reproducing kernel of H^2 and hence the two spaces are the same; however for $n > 1$, DA_n is not the space traditionally called the "Hardy space" in n variables.

The space DA_n has a CPK and, in fact, the spaces DA_n have a central place in the theory of spaces with the CPP; and DA_∞, which contains all the others, has a particularly distinguished position [**4**]. However because DA_∞ is a space of holomorphic functions on an infinite dimensional domain it is the least well understood.

We should not leave the impression that the theory of spaces with a CPK only involves spaces of holomorphic functions. The first order Sobolev space on $[0, 1]$, the space of functions f defined on $[0, 1]$ with

$$\|f\|^2 = \int_0^1 |f(x)|^2 \, dx + \int_0^1 |f'(x)|^2 \, dx < \infty,$$

also has a CPK. Another example with functions that are not holomorphic is the "diameter space" of [**46**]; another is the space on functions on a dyadic tree with kernel function $\Lambda^{d(x \wedge y)}$, $\Lambda > 1$ discussed in [**43**]. Yet another is in [**296**, Theorem 7.3].

2.5.8. Weighted Dirichlet Spaces.

For a positive superharmonic function U on $\mathbb{D}$, $\mathcal{D}(U)$, the weighted Dirichlet space, or Dirichlet type space, with weight U, is the Hilbert space of holomorphic functions on the disk for which

$$\|f\|^2_{\mathcal{D}(U)} = \|f\|^2_{H^2} + \int_{\mathbb{D}} |f'|^2 \, U(z) dA < \infty. \tag{2.48}$$

By convention $\mathcal{D}(1) = H^2$.

Any such U can be written as

$$U(\zeta) = U_{\nu,\mu}(\zeta) = \int_{\mathbb{D}} \log \left| \frac{1 - \bar{z}\zeta}{\zeta - z} \right|^2 \frac{d\nu(z)}{1 - |z|^2} + \int_{\mathbb{T}} \frac{1 - |\zeta|^2}{|\zeta - z|^2} d\mu(z),$$

with ν and μ nonnegative measures, the first supported on the open disk, the second on the unit circle.

The choice $U(z) = (1 - |z|^2)^\beta$ for $0 < \beta < 1$ gives the spaces $\mathcal{D}_{1-\beta}$ of Section 2.5.2, see (2.43).

EXERCISE 2.29. *Compute, or look up, the formula for the Laplacian in polar coordinates and check that this U is superharmonic.*

These weighted Dirichlet spaces play an important role in characterizing two-isometries, see Theorem 11.16. Also, it is a theorem of Shimorin [**308**] that each of these spaces have a CPK, the first step of his proof is in Section 6.4.2. Further information about those spaces is in [**136**].

2.5.9. Spaces on Trees. We will study function spaces on trees, both for their intrinsic interest and as tools and models for studying spaces of holomorphic functions. The tree spaces are considered systematically in Chapters 5 and 16. They are also used in several other sections, and are essential in the proofs of Theorems 13.2, 15.1, and 15.2. Many of these spaces have the CPP [**279**].

2.5.10. Real Variable Analogs. When studying spaces of holomorphic functions there are powerful tools from function theory such as the Cauchy-Riemann equations and power series representations, and those tools can produce results of great subtlety. Inspired by such results, one can look for hypotheses weaker than holomorphy which still allow results of comparable delicacy. The function spaces developed in such programs are sometimes called "real variable versions"; the thought being that tools based on holomorphy are being replaced by "real variable techniques."

For example, instead of the Bergman space, the subspace of $L^2(\mathbb{D})$ consisting of holomorphic functions, we could consider the *harmonic Bergman space* of the harmonic functions in $L^2(\mathbb{D})$ [**59**]. Those functions satisfy the same smoothness and growth conditions as functions in the Bergman space, but without the requirement of holomorphy. The "real variable Besov spaces" introduced and studied in Section 17.4.2 are another example. Those spaces contain functions which are not holomorphic, but in many ways behave similarly to the holomorphic functions in the classical Besov spaces.

The spaces just mentioned are still spaces of smooth functions. However, one can go in a different direction. In the spaces of functions on trees which we consider systematically in Chapter 5 and later in Chapter 16, the functions have no smoothness. What is retained is local geometric data (the tree structure) and local oscillation data (the values of the tree functions at the tree vertices are proxies for normalized derivatives.). However, we will see that those spaces mimic many properties of spaces of holomorphic functions.

One of the triumphs of real variable techniques has been the development of a theory of "real variable Hardy spaces." The Hardy spaces, discussed in the next chapter, are subspaces of the Lebesgue spaces of the circle. Hardy spaces functions have properties not shared by general Lebesgue space functions; for example, a Hardy space function whose boundary function vanishes on a set of positive measure must vanish identically, Corollary 3.3. Classical Hardy space theory was strongly tied to the realization of Hardy space functions as boundary values of holomorphic functions. Over years, the reasons for the more subtle behavior of those functions were isolated and became the basis of the modern real variable Hardy space theory which is a powerful tool in many contexts completely unrelated to holomorphy. The early evolution of the modern theory can be seen in [**145**] and [**119**], both are dated but very readable; a mature version is presented in [**316**]. Some recent work in that area shares resonances with approaches used in this volume, see for instance [**260**].

2.6. Notes and Comments

The article [**54**] and book [**67**] are the foundations of the modern theory of reproducing kernel Hilbert spaces. The full history of those ideas is quite rich and includes earlier work by those authors as well as work by others, including [**344**], [**230**], [**74**], [**231**], and the work of M. M. Djrbashian discussed in the introduction of [**127**]. More information is in [**127**], [**291**], [**4**], [**255**] and the works referenced there.

The Dirichlet space is studied systematically in the book [**136**]. There are also several survey papers about the Dirichlet space, including [**340**], [**287**], [**51**], and [**205**].

The theory of the Hardy space is presented in many books, we list several at the end of the next chapter.

The Bergman space has a smaller role in our story; mainly as an example of things behaving differently. The theory of the Bergman space on the disk is considered in depth in [**127**], [**348**], [**174**], and [**133**].

More information about δ is in [**50**].

CHAPTER 3

Intermezzo: Hardy Spaces

The theory of the Hardy space is a fundamental part of the historical and conceptual landscape for the topics in this book. In this chapter we review that theory. We offer only a few proofs or proof outlines. We focus on topics which we will also consider for the Dirichlet space.

For $1 \leq p \leq \infty$, define the Hardy space $H^p(\mathbb{D})$ to be the Banach space of functions, f, which are holomorphic on the unit disk and for which the norm is given by

$$\|f\|_{H^p(\mathbb{D})} := \sup_{0<r<1} \left\{ \int_0^{2\pi} \left|f(re^{i\theta})\right|^p d\theta \right\}^{1/p}, \quad 1 \leq p < \infty$$
$$\|f\|_{H^\infty(\mathbb{D})} := \sup\left\{|f(z)| : z \in \mathbb{D}\right\}, \quad p = \infty,$$

is finite. (The same definition also defines the Hardy spaces $H^p(\mathbb{D})$, $0 < p < 1$. Those are quasi-Banach spaces but not Banach spaces.) Direct computation shows that a function $f(z) = \sum_{k=0}^\infty a_k z^k$, which is holomorphic in the disk, is in $H^2(\mathbb{D})$ if and only if $\|f\|_{H^2}^2 = \sum_{k=0}^\infty |a_k|^2 < \infty$, thus recapturing our earlier definition of $H^2(\mathbb{D})$. If $p \neq 2$ the spaces are not Hilbert spaces and there there is no simple relationship between membership in the space and the size of the Taylor coefficients $\{a_n\}$.

3.1. Boundary Values and Fourier Series

Functions in the $H^p(\mathbb{D})$ spaces need not extend continuously to the closed disk; for example, $f(z) = \log(1-z)$ is in $H^p(\mathbb{D})$ for $p < \infty$. However, as we said earlier for $H^2(\mathbb{D})$, these functions do have boundary values in a certain sense. For $1 \leq p \leq \infty$, any $f \in H^p(\mathbb{D})$ has radial boundary limits *a.e.* $d\theta$ on the unit circle and those limits define the boundary function f^β,

$$f^\beta(e^{i\theta}) := \lim_{r\to 1^-} f(re^{i\theta}) \ a.e.\ d\theta. \tag{3.1}$$

The same boundary limit function is also obtained using nontangential convergence. That is, for fixed $\alpha > 1$, recalling the definition $\Gamma(\alpha, e^{i\theta})$ in Section 1.6, we have

$$f^\beta(e^{i\theta}) = \lim_{\substack{z\to e^{i\theta} \\ z\in\Gamma(\alpha,e^{i\theta})}} f(z) \ a.e.\ d\theta. \tag{3.2}$$

The map of f to $\beta(f) = f^\beta$ is an isometric map of $H^p(\mathbb{D})$ to a closed subspace of the Lebesgue space $L^p(\mathbb{T})$. We denote that subspace by $H^p(\mathbb{T})$, thus

$$\|f\|_{H^p(\mathbb{D})}^p = \left\|f^\beta\right\|_{H^p(\mathbb{T})}^p = \int_{\mathbb{T}} \left|f^\beta(e^{i\theta})\right|^p \frac{d\theta}{2\pi}. \tag{3.3}$$

It is common to use β to identify the holomorphic function on the disk with the corresponding boundary function on the circle and to use the same notation for both, i.e. $f^\beta = f$. It is also common to continue that path and identify $H^p(\mathbb{D})$ with $H^p(\mathbb{T})$ and write H^p for either. We will do that when it is convenient and unlikely to cause confusion, and perhaps also some other times.

Any $G \in L^1(\mathbb{T})$ can be represented by a Fourier series

$$G(e^{i\theta}) \sim \sum_{k=-\infty}^{\infty} \hat{G}(k) e^{ik\theta}. \tag{3.4}$$

There is then an associated function $\mathcal{P}G$, the *Poisson extension* of G, a harmonic function on the disk, which has boundary function $(\mathcal{P}G)^\beta$ in the sense of (3.1) and with $(\mathcal{P}G)^\beta = G$ *a.e.*. For G as in (3.4), $\mathcal{P}G$ is given by the Poisson integral formula (1.15). Alternatively, $\mathcal{P}G$ can be obtained directly from (3.4). The two formulae are

$$\begin{aligned} \mathcal{P}G(re^{i\theta}) &:= \frac{1}{2\pi} \int_0^{2\pi} G(e^{i\theta}) \frac{1 - r^2}{|re^{i\theta} - e^{i\phi}|^2} d\phi \\ &= \sum_{k=-\infty}^{\infty} \hat{G}(k) r^{|k|} e^{ik\theta}. \end{aligned} \tag{3.5}$$

If $G \in L^1$, then the integral converges absolutely and uniformly on compact subsets of the disk, the series does the same, and the two limits are the same harmonic function. Of course, if G is smooth then $\mathcal{P}G$ is the solution of the Dirichlet problem described in Section 1.5.

If the Fourier series of G is of *power series type*, that is, $\hat{G}(k) = 0$ for all $k < 0$, then $\mathcal{P}G$ given by (3.5) will be a power series which converges in the open disk to a holomorphic function,

$$\mathcal{P}G(z) = \sum_{k=0}^{\infty} \hat{G}(k) z^k.$$

Furthermore, if G is also in L^p, $1 \le p \le \infty$, then $\mathcal{P}G \in H^p$. Thus, in terms of boundary functions, H^p is the subspace of L^p consisting of functions having Fourier series of power series type.

This is straightforward when the functions involved are nice, for instance if G is a trigonometric polynomial, then what we have just said is elementary. However there are complications for general functions. For example, the function

$$f(z) = \frac{1+z}{1-z} \tag{3.6}$$

is not in the H^p spaces just defined but it is in the analogously defined H^p spaces for $0 < p < 1$. When we study its boundary values we find that, except for $\theta = 0$, we have

$$\operatorname{Re} f^\beta(e^{i\theta}) = \operatorname{Re} \lim_{r \to 1} f(re^{i\theta}) = 0.$$

Thus, for this function, the scheme of passing from f to $(\mathcal{P}f)^\beta$ and then back to f fails. This suggests, correctly, that when working with general functions there are important technical details to be filled in and pitfalls to be avoided.

3.2. Zero Sets

Suppose $Z = \{z_n\}_{n=1}^{\infty}$ is a sequence in $\mathbb{D}$ which may include repetitions. We define the *Blaschke product* associated with the sequence Z to be the product

$$B_Z(z) := \prod_{n=1}^{\infty} \frac{|z_n|}{z_n} \frac{z_n - z}{1 - \overline{z}_n z}. \tag{3.7}$$

By convention, the factors on the right are taken to be z if $z_n = 0$.

Each factor is holomorphic on the closed disk, has modulus identically one on the boundary circle, has modulus strictly less than one on the open disk, and has a single zero in the disk. Hence on $\mathbb{D}$ this infinite product either diverges identically to zero or converges to a holomorphic function whose only zeros are the points of Z, with multiplicity taken into account. We can find out which case occurs by testing convergence at some point not in Z. Suppose $z = 0$ is such a point, the general case being easily reduced to this. We find $B_Z(0) = \prod_{n=1}^{\infty} |z_n|$ and thus the necessary and sufficient condition for convergence to a nontrivial product is that $\prod_{n=1}^{\infty} |z_n| > 0$. However, it is customary to present the condition in terms of a sum. B_Z will be a holomorphic function on the disk, bounded by one in modulus, and with zero set exactly Z, if and only if Z satisfies the *Blaschke condition*

$$\sum_{n=1}^{\infty} (1 - |z_n|^2) < \infty. \tag{3.8}$$

Except in trivial cases, $B(z)$ is not holomorphic on the closed disk; however it is of modulus one on the boundary, at least in the extended sense:

$$|B_Z^{\beta}\left(e^{i\theta}\right)| = 1 \quad a.e. \ d\theta. \tag{3.9}$$

Further, it is a consequence of Jensen's formula that the zero set, Z, of a function f in any H^p must satisfy (3.8). In that case, the previous construction gives a function, B_Z, which will have the same zero set as f. Thus $f(z) = B_Z(z)h(z)$, where h is holomorphic in the disk and zero free. It is also true, as suggested by that equality and (3.9), that $h \in H^p$ and its boundary values satisfy $\left|f^{\beta}\right| = \left|h^{\beta}\right|$ *a.e.*. Because h is zero free it has a holomorphic logarithm which satisfies $\operatorname{Re}\log h^{\beta} = \log\left|f^{\beta}\right|$. The analysis of that logarithm is a basic tool in the study of f; in particular, it is one of the foundations of the factorization theorem we now describe.

3.3. The Factorization Theorem

We now introduce the two classes of functions which, in addition to Blaschke products, are used in the factorization theorem.

A function $Q(z) \in H^1$ is called an *outer function* if it is determined by $\left|Q^{\beta}\right|$ through,

$$Q(z) := \gamma \exp\left\{\frac{1}{2\pi}\int_{\mathbb{T}} \frac{e^{i\theta} + z}{e^{i\theta} - z} \log\left|Q^{\beta}(e^{i\theta})\right| d\theta\right\} \tag{3.10}$$

with γ a unimodular constant. In the other direction, given $h(e^{i\theta}) \geq 0$ with $\log h \in L^1$ and $h \in L^p$ we have that

$$R(z) := \exp\left\{\frac{1}{2\pi}\int_{\mathbb{T}} \frac{e^{i\theta} + z}{e^{i\theta} - z} \log h(e^{i\theta}) d\theta\right\} \tag{3.11}$$

is an outer function in H^p which has $\left|R^{\beta}\right| = h$ *a.e.*.

We say that a function, $S(z)$, holomorphic on the disk, is a *singular inner function* if it is zero free and satisfies

$$\sup_{z\in\mathbb{D}} |S(z)| \leq 1 \text{ and } \lim_{r\to 1} \left|S(re^{i\theta})\right| = 1 \ a.e.\ d\theta.$$

It can be shown that the singular inner functions are exactly those functions which can be written as

$$S(z) := \alpha \exp\left\{-\int_{\mathbb{T}} \frac{e^{i\theta}+z}{e^{i\theta}-z} d\mu(\theta)\right\} \tag{3.12}$$

for unique choices of unimodular constant α and nonnegative measure μ which is singular with respect to Lebesgue measure $d\theta$.

PROPOSITION 3.1. *Any analytic function simultaneously in two of the three classes, Blaschke products, singular inner functions, and outer functions, is a constant.*

PROOF. This is a consequence of the definitions and the following facts: a Blaschke product has modulus one on the boundary, (3.9); a zero free Blaschke product is, by definition, identically one; and, by (3.10) an outer function with constant modulus on the boundary is constant. □

It is sometimes convenient to work with the product of the Blaschke product and the singular inner function

$$I_f(z) := B_f(z)S_f(z).$$

Any function that can be written in that form is called an *inner function.* The particular function I_f is called the *inner factor* of f and the representation $f = I_f O_f$ is called the *inner-outer factorization* of f.

There is then the fundamental *factorization theorem* for H^p.

THEOREM 3.2. *Given $f(z)$ in H^p for some p, $1 \leq p \leq \infty$, there is a Blaschke product $B_f(z)$, a singular inner function $S_f(z)$, and an outer function $O_f(z)$ so that*

$$f(z) = B_f(z)S_f(z)O_f(z). \tag{3.13}$$

These functions are unique (up to placement of unimodular constant factors). B_f is the Blaschke product with the same zero set as f; the outer function is given by

$$O_f(z) = \exp\left\{\frac{1}{2\pi}\int_{\mathbb{T}} \frac{e^{i\theta}+z}{e^{i\theta}-z} \log\left|f^\beta(e^{i\theta})\right| d\theta\right\}. \tag{3.14}$$

Conversely, if $u(e^{i\theta})$ is an integrable real-valued function on $\mathbb{T}$ with $\exp u \in L^p(d\theta)$, then

$$O(z) = \exp\left\{\frac{1}{2\pi}\int_{\mathbb{T}} \frac{e^{i\theta}+z}{e^{i\theta}-z} u(e^{i\theta}) d\theta\right\}$$

is an outer function in H^p and for any inner function $I(z)$ the product $O(z)I(z) \in H^p$.

The theorem is sometimes called "Beurling's Factorization Theorem" but the history is richer than that name suggests, see [**178**], [**161**].

COROLLARY 3.3. *If f is in H^p for some p, $1 \leq p \leq \infty$, and $f^\beta = 0$ on a subset of the circle $\mathbb{T}$ of positive measure, then f is the zero function.*

The singular inner function can be defined by rewriting (3.13) as $S_f(z) = f(z)/(B_f(z)O_f(z))$. Alternatively, the representation of $S_f(z)$ in the form (3.12) can be obtained analytically by studying the distributional boundary values of the logarithm of the nonvanishing holomorphic function $f(z)/B_f(z)$, [**161**].

The factor B_f carries the information about the zeros of f. Formula (3.14) shows that O_f is determined by the size of $\left|f^\beta\right|$. The factor S_f is more subtle. Singular inner functions have modulus bounded by one and have no zeros in the disk. However, they tend to zero strongly near the support of μ. The model case is the *atomic* singular inner function obtained when μ is the point mass at $z = 1$;

$$f(z) = \exp\left(-\frac{1+z}{1-z}\right).$$

Straightforward computation shows that $|f(z)|$ tends strongly to 0 as z approaches 1 along the real axis.

One consequence of the factorization which we use later is:

PROPOSITION 3.4. *Given $f \in H^1$ there are $g_1, g_2 \in H^2$ with $\|f\|_{H^1}^{1/2} = \|g_1\|_{H^2} = \|g_2\|_{H^2}$ and $f = g_1 g_2$.*

PROOF. Start with the factorization of f, $f = BSO$. Next, note that O is an exponential and hence has a holomorphic square root. Set $g_1 = BSO^{1/2}$ and $g_2 = O^{1/2}$ and check that everything works. □

3.4. The Shift Operator

The shift operator on H^2, $S = S_{H^2}$, is the operator of multiplication by the coordinate function, z, acting on the Hardy space. It has a combination of properties that are impossible for a linear operator on a finite dimensional space; it is isometric, and yet it is not invertible (the constant function is not in the range). Because S has this and other quintessentially infinite dimensional behavior, and because there are powerful function theoretic tools to study S, this operator has become a central example in operator theory.

A closed subspace V of H^2 is called an *invariant subspace* for S if $f \in V$ implies $Sf \in V$. A central question for any operator is to describe its nontrivial invariant subspaces. An important particular case is describing the cyclic vectors for the operator, those f for which $[f]$, the smallest invariant subspace containing f, is all of H^2.

In 1948, the invariant subspaces of S were described by Beurling in a fundamental paper [**68**].

We will say an inner function is normalized if its first nonvanishing Taylor coefficient is positive.

THEOREM 3.5. *The non-trivial invariant subspaces of the operator S acting on H^2 are exactly the subspaces of the form $IH^2 := \left\{If : f \in H^2\right\}$ for inner functions I. Different normalized inner functions give different invariant subspaces. Given a non-zero $f \in H^2$, $[f] = I_f H^2$. In particular, f is a cyclic vector for S if and only if f is an outer function.*

With this result in hand, many basic facts about the invariant subspaces of S can be obtained easily. For instance, given the theorem, it is trivial that each invariant subspace contains a bounded function. Also, it is not hard to see that the intersection of two nontrivial invariant subspaces contains a third. These and other

other results about the invariant subspaces of S suggest questions for the operator of multiplication by z acting on other function spaces, We consider some of those questions in Chapters 11 and 12.

3.5. Carleson Measures

We defined Carleson measures for the Hardy space in Section 2.4.2 using the inequality (2.28). We saw there, Proposition 2.22, that the problem of describing Carleson measures for H^2 is related to describing the multipliers of H^2. In the next chapter we develop an intrinsic description of those measures; here we just record the description. Recall that for an interval I in $\mathbb{T}$, $T(I)$, the tent over I is defined in Section 1.7.2.

THEOREM 3.6 ([**91**]). *A non-negative measure, μ, on the open disk is a Carleson measure for the Hardy space, $\mu \in CM(H^2)$, if and only if*

$$\sup\left\{\frac{1}{|I|}\mu\left(T(I)\right) : I \text{ an interval in } \mathbb{T}\right\} < \infty.$$

EXERCISE 3.7. *Show that the class of measures described in the theorem would be the same if we replaced the tents $T(I)$ with the Carleson boxes $CB(I)$ defined in Section* 1.7.2.

In Section 2.4.2 we defined Carleson measures for H^p, $p = 2$, by the requirement that there is a $C_2 > 0$ such that, for $p = 2$,

$$\forall h \in H^2, \quad \int_{\mathbb{D}} |h|^2 \, d\mu \leq C_2 \|h\|_{H^2}^2 .$$

We now expand that definition. For each p, $1 \leq p < \infty$, we define Carleson measures for H^p by requiring that there be a $C_p > 0$ so that the previous inequality holds for that p.

EXERCISE 3.8. *For $1 \leq p < \infty$, the class of measures obtained in this way is independent of p.*

3.6. Pick, Interpolation, Corona

Some basic questions in Hardy space theory ask about finding functions which take assigned values at assigned points.

3.6.1. Pick-Nevanlinna.

One classical result is:

THEOREM 3.9 (Pick [**262**], Nevanlinna [**242**]). *Given $\{z_i\}_{i=1}^n$ and $\{w_i\}_{i=1}^n$ in $\mathbb{D}$ a necessary and sufficient condition for the existence of $f \in H^\infty$ with $\|f\|_{H^\infty} \leq 1$, and $f(z_i) = w_i$, $i = 1, \ldots, n$ is that the matrix*

$$\left(\frac{1 - \overline{w}_i w_j}{1 - \overline{z}_i z_j}\right)_{i,j=1}^n$$

be positive definite.

Although this formulation of the result does not involve Hilbert spaces, the result answers a special case of a general question about Hilbert spaces and their multiplier algebras. Suppose H is a RKHS of functions on Z, $\{z_i\} \subset Z$, and $\{w_i\} \subset \mathbb{D}$. How can you know if there an $f \in \operatorname{Mult}(H)$ of norm at most one and taking the values $\{w_i\}$ at the points $\{z_i\}$? If $H = H^2$ then $\operatorname{Mult}(H) = H^\infty$ and the answer is given by the previous theorem. There is an important class of RKHS for which there is a similar answer; we discuss that in Chapter 6.

3.6.2. Interpolation. A related question is describing the sequences on which the values of functions in H^2, or in $\mathrm{Mult}(H^2)$, can be freely specified, subject only to a natural size constraint. Such sequences are called *interpolating sequences.*

Given $Z = \{z_i\}_{i=1}^{\infty} \subset \mathbb{D}$, define the measure μ_Z by

$$d\mu_Z = \sum_{i=1}^{\infty} \|k_{z_i}\|_{H^2}^{-1} \delta_{z_i}.$$

Associated to Z is the restriction map $R_Z = R$ which takes f in H^2 to the function Rf on Z which is defined by $(Rf)(z_i) = f(z_i)$, $i = 1, 2, \ldots$. The sequence Z is called an interpolating sequence for H^2 if R maps H^2 boundedly into and onto $\ell_Z^2 := \ell^2(Z, \mu_Z)$. The sequence is called an interpolating sequence for $\mathrm{Mult}(H^2)$ if R maps $\mathrm{Mult}(H^2)$ into and onto ℓ^∞. However in this case, by Corollary 2.18, "into" is automatic.

If R maps boundedly into ℓ^2, then, by unpacking the definitions, we see that

$$\mu_Z \text{ is a Carleson measure} \tag{3.15}$$

for H^2. Also, starting with Proposition 2.13, it is easy to see that for R to map onto ℓ_Z^2, even if R is not known to be bounded, a necessary condition is that Z be separated, that is, it satisfies (2.33);

$$\inf \{\delta_{H^2}(z_i, z_j) : i \neq i\} > 0. \tag{3.16}$$

EXERCISE 3.10. *Show that* (3.16) *is a necessary condition for interpolation. (Hint: use the closed graph theorem and Proposition* 2.13*).*

In Chapter 7 we show that these two necessary conditions are in fact sufficient for Z to be an interpolating sequence for H^2 or for $\mathrm{Mult}(H^2)$ and, furthermore, analogous results hold for $\mathcal{D}$.

In Chapter 14 we consider certain Banach spaces of holomorphic functions of several variables and their multiplier algebras, and also associated discrete model spaces. A similar pattern holds for those spaces, a separation condition together with a Carleson measure condition characterizes interpolating sequences. However there are other spaces with very different behavior, [**302**].

3.6.3. Corona. Corona problems are questions about multiplier algebras. They are related to interpolation questions both historically and through shared solution techniques.

If H is a RKHS, then $\mathrm{Mult}(H)$ is a commutative Banach algebra and it is natural to ask about its algebraic structure. For instance, given $f \in \mathrm{Mult}(H)$, how can you tell if f is contained in a proper closed ideal? For H^∞, this question is elementary, but the analogous question, how do you know if the set $\{f_i\}_{i=1}^{n}$ is contained in a proper closed ideal, is already difficult for $n = 2$.

An equivalent rephrasing of the question is to ask what conditions ensure that there are $\{g_i\}_{i=1}^{n} \subset H^\infty$ so that $\sum_{i=1}^{n} f_i g_i = 1$. There is a natural necessary condition:

EXERCISE 3.11. *Show that the condition*

$$\inf_{z\in\mathbb{D}}\sum_{i=1}^{n}|f_i(z)|>0 \tag{3.17}$$

is necessary for the existence of $\{g_i\}$.

In general the question of the existence of such $\{g_i\}$, or the more specific question if the analog of (3.17) is sufficient to ensure that existence, is called a *Corona problem* and a resolution is a *Corona theorem.* The Corona theorem for H^∞ was proved by Carleson in 1962 [**89**]. The ideas from Carleson's original proof, such as the "corona construction" were, and continue to be, very influential, but we will not speak of them further. A later proof of the theorem focused on the solvability of certain $\overline{\partial}-$equations with Carleson measure data. That approach has evolved into a flexible general tool and we will see a version of it Chapter 17 where we prove Corona theorems for certain spaces of functions on the ball.

Although we will not develop the ideas here, the same $\overline{\partial}-$equations can be used to give alternative approaches to some of the results we will present about interpolation problems, for example in [**186**], [**343**], and [**38, 40**].

3.7. H^1, BMO, Fefferman's Duality Theorem

For $1<p<\infty$, the dual space of the Lebesgue space $L^p(\mathbb{T})$ is $L^{p'}(\mathbb{T})$, where the conjugate index, p', is defined by $p^{-1}+p'^{-1}=1$. Because of the close relation between the (boundary value) H^p spaces and the Lebesgue spaces, one might expect a similar relationship for the Hardy space. In fact, that is the case and we briefly indicate why.

Because H^2 is a closed subspace of L^2, there is a bounded self-adjoint projection P, mapping L^2 to H^2. This projection can be used to give an isometric identification of the Banach space dual of H^2 with H^2. (A direct appeal to the fact that H^2 is a Hilbert space and thus self-dual would not shed light on the case $p\neq 2$.) The core of the argument goes as follows. Suppose we are given a continuous linear functional, L, on H^2. By the Hahn-Banach theorem, there is a linear functional $\tilde{L}$ on L^2 which agrees with L on H^2 and has the same norm as L. Hence, by the duality results for Lebesgue spaces, there is a $G\in L^2$ with $\|G\|=\|\,\tilde{L}\,\|$ and such that

$$\forall f\in H^2,\ L(f)=\tilde{L}(f)=\langle f,G\rangle_{L^2}.$$

Because f is in H^2, it is fixed by the projection P. Also, for two functions in H^2, the L^2 inner product and the H^2 inner product are the same. Hence, recalling that P is self adjoint, we continue with

$$L(f)=\langle f,G\rangle_{L^2}=\langle Pf,G\rangle_{L^2}=\langle f,P^*G\rangle_{L^2}=\langle f,PG\rangle_{L^2}=\langle f,PG\rangle_{H^2}.$$

Setting $g=PG$, we see that every linear functional L on H^2 is of the form

$$L(f)=L_g(f)=\langle f,g\rangle_{H^2},$$

for some $g\in H^2$. Considering this equation as f varies over H^2, one finds that $\|L\|\leq\|g\|$. Considering the equation for $f=g$, one checks that $\|g\|\leq\|L\|$. Combining these gives $\|g\|=\|L\|$, as required. The converse is immediate and we have an isometric identification of the dual space $\left(H^2\right)^*$ with the space H^2.

A version of this argument holds for H^p for $1 < p < \infty$. We start with the straightforward observation that, for trigonometric polynomials, the behavior of P is the natural one;

$$P\left(\sum_{n=-N}^{N} a_n e^{in\theta}\right) = \sum_{n=0}^{N} a_n e^{in\theta}.$$

This shows that P maps the trigonometric polynomials, a dense subset of L^p, into L^p. Next, we need a deeper fact, this densely defined P extends to be a bounded map of L^p to itself which is the identity on H^p. Furthermore, the adjoint P^* mapping $L^{p'}$ to itself is also given by P. When we now follow the argument of the previous paragraph, two things change. First, the extension of the linear functional L is now a linear functional $\tilde{L}$ acting on $L^{p'}$. Second, the map $P^* = P$ acting on $L^{p'}$ has norm greater than one. Thus, when the details are filled in, this argument gives an identification of the Banach spaces $(H^p)^*$ and $H^{p'}$, but it is not an isometric identification.

This approach fails when $p = 1$. It is true that $(L^1)^* = L^\infty$ but the operator P does not extend to a bounded linear operator on L^1, and the dual of H^1 is not equal to H^∞. However, the dual of H^1 can be realized as a space of holomorphic functions on the disk and is easily seen to be contained in H^1, hence we have full access to the correspondence between the holomorphic functions in the disk and their boundary values. Thus, functions in the dual space, $\left(H^1\right)^*$, can be characterized either in terms of their boundary values or in terms of the corresponding holomorphic functions on the disk. We will offer both characterizations, their equivalence is one of the fundamental results of the theory.

The condition on the boundary value function $f \in (H^1)^*$ is that it be an L^1 function with Fourier series of analytic type, and that it have bounded mean oscillation, which we now define. Given an integrable function f defined on $\mathbb{T}$ and given any interval $I \subset \mathbb{T}$, we define the *mean* of f on I, $m_I(f)$ by

$$m_I(f) := \frac{1}{|I|}\int_I f(e^{i\theta})d\theta.$$

For the same interval, we define the *oscillation* of f about its mean, $\operatorname{Osc}_I(f)$, by

$$\begin{aligned}\operatorname{Osc}_I(f) &:= m_I\left(|f - m_I f|\right)\\ &= \frac{1}{|I|}\int_I \left|f(e^{i\theta}) - m_I(f)\right| d\theta.\end{aligned} \tag{3.18}$$

A function, f, is said to have *bounded mean oscillation* if the seminorm

$$\|f\|_{BMO} := \sup\left\{\operatorname{Osc}_J(f) : J \text{ an interval in } \mathbb{T}\right\} \tag{3.19}$$

is finite. The space of all such functions is denoted BMO (or $BMOA$, for analytic BMO, when one wants to restrict to those functions which are boundary values of analytic functions.) Even though $\|\cdot\|_{BMO}$ is only a seminorm (or a norm on the space of functions modulo constants), it is generally called the "BMO norm."

THEOREM 3.12 ([**146**]). *$(H^1)^* = BMOA$.*

This is a fundamental result in Hardy space theory. Informally the theorem appears to say that any bounded linear functional L on H^1 is given by a $BMOA$

function b through the formula

$$L(f) = L_b(f) = \int_{\mathbb{T}} f(e^{i\theta})\overline{b(e^{i\theta})}d\theta \quad \forall f \in H^1. \tag{3.20}$$

That formula is a common shorthand and it hides a delicate point. Using (3.11) we can see that for any $G \in L^1$, $G > 1$ there is an outer function $f \in H^1$ with $|f| = G$. Hence, for any unbounded $b \in BMOA$, there is an $f \in H^1$ for which the integral in (3.20) is not absolutely convergent. The precise definition of L_b is

$$L_b(f) = \lim_{r \to 1} \int_{\mathbb{T}} f(re^{i\theta})\overline{b(re^{i\theta})}d\theta.$$

In particular, this duality pairing is more subtle than that for the $p > 1$ which involves absolutely convergent integrals. (However, in fairness, the standard proofs that the operator P has a bounded extension to all of L^p also involves integrals that do not converge absolutely.)

An equally basic part of the story is characterizing the behavior in the disk of holomorphic functions whose boundary functions are in $BMOA$. From the same paper:

THEOREM 3.13 ([**146**]). *A function b is in $BMOA$ if and only if $|b'|^2(1 - |z|^2)dA$ is a Carleson measure for the Hardy space H^1.*

Recall that we saw in Exercise 3.8 that the Carleson measures for H^1 are the same as those for H^2. Those measures were described in Theorem 3.6.

It follows immediately from the definitions that $H^\infty \subset BMOA$. Also, one can show directly, or by using the previous theorem, that $\mathcal{D} \subset BMOA$.

EXERCISE 3.14. *Both inclusions are proper; $H^\infty \subsetneqq BMOA$, $\mathcal{D} \subsetneqq BMOA$.*

With this theorem we complete our characterization of $\operatorname{Mult}(H^2)$.

COROLLARY 3.15. $\operatorname{Mult}(H^2) = H^\infty$.

PROOF. By Proposition 2.20 and the previous theorem, $\operatorname{Mult}(H^2) = H^\infty \cap BMOA$. Now take note of the previous exercise. □

Using Theorem 3.13, we can sketch the proof of the easy half of Theorem 3.12. Full details for these results are in [**161**].

For convenience, and because this is only a sketch, we assume that all the holomorphic functions we consider vanish at the origin. We start with $b \in BMOA$ and want to show that

$$L_b(f) = \int_{\mathbb{T}} f(e^{i\theta})\overline{b(e^{i\theta})}d\theta$$

is a bounded functional on H^1. By polarizing (2.15), we can rewrite this as

$$L_b(f) = \int_{\mathbb{D}} f'(z)\overline{b'(z)}(1 - |z|^2)dA,$$

(recall that our functions vanish at the origin). We now invoke Proposition 3.2 and write $f = g_1 g_2$ with $g_1, g_2 \in H^2$ and $\|f\|_{H^1}^{1/2} = \|g_1\|_{H^2} = \|g_2\|_{H^2}$. We continue with

$$\begin{aligned} L_b(f) &= \int_{\mathbb{D}} (g_1 g_2)'(z)\overline{b'}(z)(1 - |z|^2)dA \\ &= \int_{\mathbb{D}} g_1'(z)g_2(z)\overline{b'}(z)(1 - |z|^2)dA + \int_{\mathbb{D}} g_1(z)g_2'(z)\overline{b'}(z)(1 - |z|^2)dA = I + II. \end{aligned}$$

By symmetry we need only consider I. By Cauchy-Schwarz

$$\begin{aligned}
|I| &\leq \left(\int_{\mathbb{D}} |g_1'(z)|^2 (1-|z|^2)dA\right)^{1/2} \left(\int_{\mathbb{D}} |g_2(z)|^2 \left|\overline{b'}(z)\right|^2 (1-|z|^2)dA\right)^{1/2} \\
&\leq \|g_1\|_{H^2} \|g_2\|_{H^2} \left\| \left|\overline{b'}\right|^2 (1-|z|^2)dA \right\|_{CM(H^2)} \\
&\leq \|g_1\|_{H^2} \|g_2\|_{H^2} \|b\|_{BMO} \\
&= \|f\|_{H^1} \|b\|_{BMO} .
\end{aligned}$$

Thus we have that, if $b \in BMOA$, and if we have the previous theorem, then we have a path to proving L_b is a bounded functional on H^1.

3.8. Exponential Integrability, the John-Nirenberg Theorem

Sometimes, if a function is required to satisfy a condition uniformly on all intervals the condition is "self improving"; that is, one can actually conclude that the function satisfies a formally stronger condition on each interval. A classic example of this is the theorem of John and Nirenberg [**185**], which shows that a function in BMO, and thus having local oscillation which is, in the mean, uniformly bounded; actually has local oscillation which is uniformly exponentially integrable.

THEOREM 3.16 (John-Nirenberg [**185**]). *There are $A, B > 0$ so that if $f \in BMO$, $\|f\|_{BMO} \leq 1$ and I is an interval in $\mathbb{T}$, then*

$$\frac{1}{|I|}\int_I \exp\left(A\left|f(t) - m_I(f)\right|\right) dt < B.$$

EXERCISE 3.17. *Show that if f satisfies the conclusion of the previous theorem, then $f \in BMO$.*

The theorem shows, among other things, that the function $\log(1-z)$, which we noted was in BMO, exhibits as large a singularity as is possible for functions in the space.

The elegant classical proof of this theorem works directly with the boundary values of f, see [**161**]. Here we prove a related result using the Carleson measure characterization of BMO.

THEOREM 3.18. *There are $A, B > 0$ so that if $f \in BMOA$, $\|f\|_{BMO} \leq 1$, then*

$$\int_{\mathbb{T}} \exp\left(A\left|f(t) - f(0)\right|\right) dt < B.$$

PROOF. We work with $g(z) = f(z) - f(0)$. The fact that $|g'|^2 (1-|z|^2) \in CM(H^2)$, when applied to the function $1 \in H^2$, shows that $g \in H^2$ and $\|g\|_{H^2} \lesssim 1$ for some absolute constant C.

That fact gives a first step for an inductive estimate of $\|g^n\|_{H^2}$ using (2.15) and Theorem 3.13;

$$\begin{aligned}\left\|g^{n+1}\right\|_{H^2}^2 &\lesssim \int_{\mathbb{D}} \left|\left(g^{n+1}\right)'\right|^2 (1-|z|^2)dA \\ &\approx (n+1)^2 \int_{\mathbb{D}} |g^n(z)|^2 |g'(z)|^2 (1-|z|^2)dA \\ &\lesssim (n+1)^2 \|g^n\|_{H^2}^2 \left\| |g'|^2 (1-|z|^2)dA \right\|_{CM(H^2)}^2 \\ &\lesssim (n+1)^2 \|g^n\|_{H^2}^2 \|g\|_{BMO}^2 .\end{aligned}$$

Hence, by induction,

$$\left\|g^{n+1}\right\|_{H^2}^2 = O\left(C^{n+1}(n+1)!\right)^2 \|g\|_{BMO}^2 = O\left(C^{n+1}(n+1)!\right)^2 .$$

This, together with Stirling's estimate, gives

$$\int_{\mathbb{T}} |g|^{2k} d\theta = O(\left(C^k k!\right)^2) = O(\tilde{C}^{2k}(2k)!).$$

Using Holder's inequality, this estimate for even integers $2k$ can be extended to all nonnegative integers n : $\int_{\mathbb{T}} |g|^n d\theta = O(C^n n!)$.

Using the power series for the exponential, we obtain the estimate

$$\int_{\mathbb{T}} \exp(A|g|)\, d\theta = \sum_{n=1}^{\infty} \frac{A^n}{n!} \int_{\mathbb{T}} |g|^n d\theta = \sum_{n=1}^{\infty} \frac{A^n}{n!} O(C^n n!).$$

If A is small enough the last sum is finite, which is the required conclusion. □

Using the fact that composition with conformal automorphisms of the disk leaves $\|f\|_{BMO}$ essentially unchanged, the previous theorem can be extended to give a result in which the uniform measure on the boundary is replaced by a more localized probability measure (given by the Poisson kernel); thus giving a result with greater resemblance to the invariant version of the John Nirenberg theorem. This is discussed in [**161**, Chapter 6].

We mentioned that some Hardy space proofs apply, essentially unchanged, to a class of RKHS. The proof of Theorem 3.18 is an example.

Suppose $w > 0$, $w \in L^1(\mathbb{D})$ and that J, the set of $h \in \operatorname{Hol}(\mathbb{D})$ for which

$$\|h\|_J^2 = |h(0)|^2 + \int_{\mathbb{D}} |h'(z)|^2 w(z)dA < \infty,$$

is a RKHS. With the model case of $w = 1$, $J = \mathcal{D}$ in mind, we introduce the following definitions. Let $\operatorname{Mult}(J)$ be the multiplier algebra of J and let $CM(J)$ be the set of Carleson measures for J, those μ for which there is a constant such that

$$\int_{\mathbb{D}} |h|^2 d\mu \lesssim \|h\|_J^2 . \tag{3.21}$$

Following (2.31), we let $\mathcal{X}(J)$ be the space of holomorphic functions which generate Carleson measures; that is h for which $|h'|^2 wdA \in CM(J)$.

PROPOSITION 3.19.

(1) $\mathcal{X}(J) \subset J$
(2) $\operatorname{Mult}(J) = H^\infty \cap \mathcal{X}(J)$.

(3) *There is a* $C > 0$ *such that if* $h \in \mathcal{X}(J)$, $\|h\|_J \leq 1$, *then*

$$\exp\left(C\left(h - h(0)\right)\right) \in J.$$

PROOF. The first statement follows from evaluating (3.21) with $\mu = |h'|^2\, wdA$ and $f = 1$. The second holds because the proof of Proposition 2.20 works, essentially unchanged, in this context. Similarly, the third statement holds because the proof of the previous theorem works here. □

In particular, with $w = 1$, these results apply to the spaces $\mathcal{D}$ and $\mathcal{X} = \mathcal{X}(\mathcal{D})$. For the Dirichlet space we had seen the first statement as Proposition 2.23.

EXERCISE 3.20. *Give conditions on* w *that ensure that* J *is a RKHS, that evaluation of functions in* J *at points in the disk is a continuous functional. The proof of the third part of Theorem* 5.13 *may be useful here.*

3.9. Notes and Comments

Most of the material in this chapter is covered in the many books on the Hardy space. Books the authors use frequently include [**178**], [**132**], [**194**], [**161**], [**246**], and [**245**]. The material in the final section is new.

CHAPTER 4

Carleson Measures

Carleson measures were introduced in [**91**] and are now basic tools throughout analysis. In this chapter we show how the viewpoint of the "reproducing kernel thesis" leads to a description of Carleson measures for the Hardy space, we discuss how that approach fails to give the full story for Carleson measures on the Dirichlet space, and we present Stegenga's characterization [**314**] of Dirichlet space Carleson measures.

Some work with Carleson measures only uses their defining property (2.28). For instance that is all that was used in Section 2.4.2 when we established the relation between multipliers and Carleson measures for H^2 and for $\mathcal{D}$. However, it is interesting, useful, and at times crucial, to have intrinsic descriptions of those measures. Those descriptions are the topic of this chapter.

We begin with analysis of Carleson measures for the Hardy space. This topic is covered well and in detail in many books including [**132**], [**161**] and [**245**]. We proceed rather quickly, our goal is to present some techniques of general use and to help elucidate the contrast with the Dirichlet space.

We will then turn to the Dirichlet space. The situation there is more complicated; there are several seemingly different characterizations of Dirichlet space Carleson measures. In this chapter, we prove the classical result of Stegenga which characterizes those measures using capacity. We also take a preliminary look at another approach, but postpone the completion of that discussion until the next chapter.

4.1. The Hardy Space and the Reproducing Kernel Thesis

Recall that a non-negative measure, μ, defined on $\mathbb{D}$ is a Carleson measure (for H^2), $\mu \in CM(H^2)$, if the natural embedding map, $\mathcal{J}$, maps H^2 boundedly into $L^2(\mathbb{D}, \mu)$; that is, there is a constant $C(\mu)$ so that

$$\int_{\mathbb{D}} |\mathcal{J}f(z)|^2 d\mu(z) \leq C(\mu)^2 \|f\|_{H^2}^2 \quad \forall f \in H^2(\mathbb{D}). \tag{4.1}$$

If we view $f \in H^2$ as a holomorphic function on $\mathbb{D}$ then the embedding operator is defined simply by $\mathcal{J}f(z) = f(z)$. If we choose to identify $f \in H^2$ with its boundary value function f^β defined on $\mathbb{T}$ then the formula for the embedding is given by the Poisson integral formula (1.15): writing $P_z(\cdot)$ rather than $P(z, \cdot)$ we have

$$\mathcal{J}(f)(z) = \frac{1}{2\pi} \int_{\mathbb{T}} P_z(\xi) f^\beta(\xi) \, |d\xi| \, .$$

We will generally suppress both "$\mathcal{J}''$ and "β'' below.

Our goal is to give an intrinsic characterization of μ which satisfy (4.1). To see the direction in which we are heading, suppose we use f^β to extend f, at least almost everywhere, on the boundary circle, $\mathbb{T}$. In that case, it is natural to consider at least those measures defined on the closed disk and whose restrictions to the circle are absolutely continuous with respect to arc length measure. (In fact, the discussion below extends to that case.)

Suppose the measure of interest is supported on the circle. We noted in Section 3.1 that

$$\|f\|_{H^2}^2 = \frac{1}{2\pi}\int_{\mathbb{T}} \left|f^\beta\right|^2 d\theta.$$

This is exactly the statement that arclength measure on the circle is a Carleson measure. It then follows immediately that if $w(e^{i\theta})$ is positive and bounded, then $w(e^{i\theta})d\theta$ is also a Carleson measure. If we wanted to show that this boundedness of w was also necessary, we would need to be able to construct functions in H^2 of norm one which had large modulus on any preassigned set on the circle.

EXERCISE 4.1. *Finish the argument above to show that the positive measure $w(e^{i\theta})d\theta$ is a Carleson measure if and only if w is bounded.*

Recall from Section 1.7 that if I is an interval in the circle of length $|I|$ and with center c_I then $T(I)$, the tent over I, is the convex hull of I and the vertex $z(I) = (1 - |I|/2\pi)c_I$. We will write $B(\xi, r)$ for the Euclidean ball with center ξ and radius r. We will prove the following detailed version of Theorem 3.6.

THEOREM 4.2. *Let μ be a non-negative measure in $\mathbb{D}$. Then the following are equivalent:*

(1) *The embedding operator $\mathcal{J} : H^2(\mathbb{D}) \to L^2(\mathbb{D}, \mu)$ is bounded.*
(2) $C(\mu)^2 := \sup_{z\in\mathbb{D}} \left\|\hat{k}_z\right\|_{H^2}^2 = \sup_{z\in\mathbb{D}} \|P_z\|_{L^1(\mu)} < \infty$.
(3) $I(\mu) := \sup\left\{\frac{1}{r}\mu(\mathbb{D}\cap B(\xi, r)) : r > 0, \xi \in \mathbb{T}\right\} < \infty$.
(4) $J(\mu) := \sup\left\{\frac{1}{|I|}\mu\left(T(I)\right) : I \textit{ an interval in } \mathbb{T}\right\} < \infty$.

And if these hold then $\|\mathcal{J}\| \approx C(\mu) \approx I(\mu) \approx \|\mathcal{J}\|$.

If the first condition holds, then there is a uniform bound for $\|\mathcal{J}f\|$ as f ranges over the unit ball of H^2. If this holds, then automatically the second condition, a uniform bound as f ranges over a subset of that unit ball, must also hold.

In some contexts, including this one, the reverse implication also holds; that is (2) implies (1) and hence (2) characterizes Carleson measures. The suggestion that such an equivalence holds in a particular case, that is, the suggestion that testing over kernel functions is the full story, a tempting speculation for operators closely tied to the underlying function theory, is sometimes called the *reproducing kernel thesis*. This theorem is one of the great successes of that viewpoint. For more about the reproducing kernel thesis see [**245**, Section A.5.8].

PROOF OF THEOREM 4.2 (WE CLOSELY FOLLOW [**245**, Section A.5.8]). We just observed that (1) $\Rightarrow$ (2). Next, note that if any of the conditions hold then μ must be a finite measure. This follows by testing the first two conditions on the constant function, by testing (3) for large r, and checking (4) for large intervals. Next, it suffices to consider z near the boundary, r near 0, and $|I|$ near zero respectively in conditions (2), (3), (4); what happens in the other cases is easily controlled using the fact that μ is a finite measure (see Remark 2.27).

We now suppose z is near the boundary. To show that (2) $\Rightarrow$ (3), we select $\xi \in \mathbb{T}$ and, with no loss, suppose $\xi = 1$. Direct computation shows that, for any $z \in \mathbb{D}$, $|\hat{k}_z|^2$ equals the Poisson kernel for z :

$$|\hat{k}_z(w)|^2 = P_z(w) = \frac{1-|z|^2}{|1-\overline{w}z|^2}.$$

Thus the two suprema in (2) are equal. Next, select $z = 1 - r$ with r small. We now want to estimate $\min\{P_z(w) : w \in \mathbb{D} \cap B(1,r)\}$. By geometric analysis this minimum is attained at $w = 1 - r$ and is, essentially, $2/r$. Thus

$$\begin{aligned}\frac{1}{r}\mu(\mathbb{D}\cap B(1,r)) &= \frac{1}{r}\left(\frac{r}{2}\int_{\mathbb{D}\cap B(1,r)} \frac{2}{r}d\mu\right)\\ &\lesssim \int_{\mathbb{D}\cap B(1,r)} P_z(w)d\mu \lesssim \int_{\mathbb{D}} P_z(w)d\mu = \|P_z\|_{L^1(\mu)}.\end{aligned}$$

Hence (2) $\Rightarrow$ (3).

To see that (3) and (4) are equivalent, note that $\mathbb{D} \cap B(1,r)$ both contains and is contained in a tent over an interval I of length $|I| \approx r$ and that a tent over I both contains and is contained in regions of the form $\mathbb{D} \cap B(1,r)$ with $r \approx |I|$.

EXERCISE 4.3. *Draw a sketch to convince yourself of those geometric facts.*

We now prove that (3) $\Rightarrow$ (2). First, suppose $|z| \le 3/4$. Elementary estimates show that for some finite positive A,

$$\sup\{P_z(w) : z, w \in \mathbb{D},\ |z| \le 3/4\} \le A.$$

We noted that (3) implies $\|\mu\| < \infty$. Combining this with the previous estimate, we see

$$\sup\left\{\|P_z(\cdot)\|_{L^1(\mu)} : z \in \mathbb{D},\ |z| \le 3/4\right\} \le A\,\|\mu\| < \infty.$$

Now consider the case $|z| > 3/4$. The method we use in this case is quite general, and straightforward geometric ideas lie behind the notation. A common informal description of what we are about to do is, "Now we break things up into dyadic rings." Readers not familiar with this type of argument should make a sketch.

Pick and fix such a z, without loss of generality z is real and positive. Define

$$B_k := \mathbb{D} \cap B(1, 2^{k+1}(1-|z|^2)) \qquad \forall k \in \mathbb{N}.$$

For w in the ring $R_k = B_{k+1} \setminus B_k$, we have that $|1-w| \ge 2^k(1-|z|^2)$. By the triangle inequality

$$\begin{aligned}|1-w| &\le |1-w\overline{z}| + |wz - w|\\ &\le |1-w\overline{z}| + (1-|z|^2).\end{aligned}$$

Combining these we have

$$|1-w\overline{z}| \ge 2^{k-1}(1-|z|^2).$$

Hence, for $w \in R_k$

$$P_z(w) = \frac{1-|z|^2}{|1-\overline{w}z|^2} \le \frac{1-|z|^2}{4^{k-1}(1-|z|^2)^2} = \frac{1}{4^{k-1}(1-|z|^2)}. \tag{4.2}$$

EXERCISE 4.4. *Show that there is a constant C, independent of k, so that for $w \in R_k$*

$$P_z(w) \gtrsim \frac{1}{4^{k-1}(1-|z|^2)}.$$

That is, the estimate (4.2) doesn't give away much.

Thus,

$$\begin{aligned}
\int_{\mathbb{D}} P_z(w)d\mu(w) &= \int_{B_1} P_z(w)d\mu(w) + \sum_{k=1}^{\infty} \int_{B_{k+1}\setminus B_k} P_z(w)d\mu(w) \\
&\le \int_{B_1} \frac{2^2}{(1-|z|^2)} d\mu(w) + \sum_{k=1}^{\infty} \int_{B_{k+1}\setminus B_k} \frac{1}{4^{k-1}(1-|z|^2)} d\mu(w) \\
&\lesssim \frac{\mu(B_1)}{(1-|z|^2)} + \sum_{k=1}^{\infty} \left(\frac{1}{4}\right)^k \frac{\mu(B_{k+1})}{(1-|z|^2)}.
\end{aligned}$$

From the definition of $I(\mu)$, it follows that $\mu(B_k) \le 2^{k+1}(1-|z|^2)I(\mu)$. Hence we can continue with

$$\begin{aligned}
\int_{\mathbb{D}} P_z(w)d\mu(w) &\lesssim \frac{\mu(B_1)}{(2^2(1-|z|^2))} + \sum_{k=1}^{\infty} (4^{-1})^k (2^{k+2}) \frac{\mu(B_{k+1})}{(2^{k+2}(1-|z|^2))} \\
&\lesssim I(\mu) + I(\mu).
\end{aligned}$$

Combining this with elementary estimates for $|z| \le 3/4$, we have that

$$\int_{\mathbb{D}} P_z(w)d\mu(w) \lesssim I(\mu).$$

Taking the supremum over $z \in \mathbb{D}$ then establishes (1).

We now come to the heart of the proof, showing that (2) $\Rightarrow$ (1). It is here that we see the success of the reproducing kernel thesis. It is enough to show that the formal adjoint $\mathcal{J}^* : L^2(\mathbb{D}, \mu) \to L^2(\mathbb{T})$ given by

$$\mathcal{J}^*(f)(\xi) = \int_{\mathbb{D}} P_z(\xi) f(z) d\mu(z), \tag{4.3}$$

is bounded. With this in mind we compute,

$$\begin{aligned}
\|\mathcal{J}^*(f)\|_{L^2(\mathbb{T})}^2 &= \int_{\mathbb{T}} |\mathcal{J}^*(f)(\xi)|^2 \, |d\xi| \qquad (4.4)\\
&= \int_{\mathbb{T}} \mathcal{J}^*(f)(\xi)\overline{\mathcal{J}^*(f)(\xi)} \, |d\xi| \\
&= \int_{\mathbb{T}} \left(\int_{\mathbb{D}} P_z(\xi) f(z) d\mu(z)\right)\left(\int_{\mathbb{D}} P_{z'}(\xi)\overline{f(z')} d\mu(z')\right) |d\xi| \\
&= \int_{\mathbb{D}}\int_{\mathbb{D}} \left(\int_{\mathbb{T}} P_z(\xi)P_{z'}(\xi)\, |d\xi|\right) f(z)\overline{f(z')} d\mu(z) d\mu(z'). \qquad (4.5)
\end{aligned}$$

The following lemma will be used to prove that $\mathcal{J}^*$ is bounded.

LEMMA 4.5. *(Vinogradov-Senichkin Test) Let $(\mathcal{Z}, \sigma)$ be a measure space and k a non-negative measurable function on $\mathcal{Z} \times \mathcal{Z}$. If*

$$\int_{\mathcal{Z}} k(s,t)k(s,x)d\sigma(s) \lesssim (k(t,x) + k(x,t))$$

for a.e. $(t, x) \in \mathcal{Z} \times \mathcal{Z}$, *then the linear operator* T_k *defined by*

$$T_k(g)(z) = \int_{\mathcal{Z}} k(z', z) g(z') d\sigma(z')$$

is bounded on $L^2(\mathcal{Z}, \sigma)$ *and has norm at most* $2C$.

EXERCISE 4.6. *Prove the Vinogradov-Senichkin Test. (Hint: Estimate*

$$\|T_k(g)\|^2 .)$$

We finish by appealing to the Vinogradov-Senichkin test. To do this, we need to know that the hypotheses of the lemma are satisfied with $L^2(\mathcal{Z}, \sigma) = L^2(\mathbb{D}, \mu)$ and $k(z', z) = P_z(z')$.

We begin by estimating the product of Poisson kernels. Let $z, z', w \in \mathbb{D}$ then we claim that

$$P_z(z')P_w(z') \lesssim (P_z(w)P_w(z') + P_w(z)P_z(z')) .$$

As a first step note that the following inequality holds:

$$\begin{aligned} a^{-1} := |1 - z\overline{w}| &= \left|1 - z\overline{z'} + z\overline{z'} - z\overline{w}\right| \\ &\leq \left|1 - z\overline{z'}\right| + |z| \left|\overline{z'} - \overline{w}\right| \\ &\leq \left|1 - z\overline{z'}\right| + \left|1 - \overline{z'}w\right| \\ &=: b^{-1} + c^{-1}. \end{aligned}$$

Multiplying through by abc we obtain $bc \leq a(b + c)$, which in turn implies that

$$b^2c^2 \leq 2a^2(b^2 + c^2).$$

Inserting the expressions defining a, b, and c and multiplying by $(1 - |z|^2)(1 - |w|^2)$, we find that

$$\begin{aligned} P_z(z')P_w(z') &\leq \frac{2}{|1 - z\overline{w}|^2}(1 - |z|^2)(1 - |w|^2)\left(\frac{1}{\left|1 - z\overline{z'}\right|^2} + \frac{1}{|1 - z'\overline{w}|^2}\right) \\ &= 2\left[\frac{(1 - |w|^2)}{|1 - z\overline{w}|^2}\frac{(1 - |z|^2)}{\left|1 - z\overline{z'}\right|^2} + \frac{(1 - |w|^2)}{|1 - z\overline{w}|^2}\frac{(1 - |z|^2)}{|1 - z'\overline{w}|^2}\right] \\ &= 2(P_w(z)P_z(z') + P_z(w)P_w(z')). \end{aligned}$$

We will now apply the Vinogradov-Senichkin Test. We first show that the kernel $k(z', z) = P_z(z')$ satisfies the hypothesis of the lemma. To this end, we need to estimate

$$\int_{\mathbb{D}} P_z(z')P_w(z')d\mu(z').$$

Using the estimate we just found for the product of two Poisson kernels, we have

$$\begin{aligned} \int_{\mathbb{D}} P_z(z')P_w(z')d\mu(z') &\leq 2\int_{\mathbb{D}} P_z(w)P_w(z') + P_w(z)P_z(z')d\mu(z') \\ &= 2\left[P_z(w)\int_{\mathbb{D}} P_w(z')d\mu(z') + P_w(z)\int_{\mathbb{D}} P_z(z')d\mu(z')\right] \\ &\leq 2C(\mu)^2(P_z(w) + P_w(z)). \end{aligned}$$

The last inequality uses the hypothesis (2). Thus the Vinogradov-Senichkin Test applies and we conclude T, given by

$$Tf(z') = \int_{\mathbb{D}} \left(\int_{\mathbb{T}} P_z(\xi) P_{z'}(\xi) \, |d\xi| \right) f(z) d\mu(z),$$

is bounded. We now note that, by (4.4),

$$\|\mathcal{J}^*(f)\|^2_{L^2(\mathbb{T})} = \langle Tf, f \rangle$$

and hence the boundedness of T implies that of $\mathcal{J}^*$and hence of $\mathcal{J}$. □

EXERCISE 4.7. *In formulating our boundedness criterion we used the functionals $J(\mu)$ and $I(\mu)$ defined using tents and semidisks respectively. Another traditional formulation of the condition uses Carleson boxes introduced in Section* 1.7.2. *That is, given an interval I in the circle, let*

$$CB(I) := \{z \in \mathbb{D} : |z| > 1 - |I| / 2\pi, \ z / |z| \in I\}$$

and

$$K(\mu) := \sup \left\{ \frac{1}{|I|} \mu \left(CB(I) \right) : I \text{ an interval in } \mathbb{T} \right\} < \infty.$$

Show that the theorem also holds with $J(\mu)$ replaced by $K(\mu)$. (Note: This is an elementary exercise in geometry. It does not require reconsidering the heart of the previous proof.)

EXERCISE 4.8. *Describe the Carleson measures for the Bergman space. Note that for the Bergman space the characterization in terms of the measure of boxes $B(I)$, described in Section* 1.7.2, *is equivalent to a characterization in terms of the measure of the top halves $TB(I)$ described there, and hence is also equivalent to a characterization in terms of the measure assigned to a hyperbolic ball of fixed radius.*

4.2. Carleson Measures for the Dirichlet Space

In this section we give Stegenga's characterization of the Carleson measures for the Dirichlet space. We start our analysis by trying the path that succeeded for the Hardy space. As with the Hardy space, the statement that μ is a Carleson measure is equivalent to the boundedness of the embedding map $\mathcal{J}$, but now $\mathcal{J}$ is viewed as acting from $\mathcal{D}$ to $L^2(\mu)$. When we consider the boundedness of $\mathcal{J}$ acting on reproducing kernels we find a condition which is necessary for $\mu \in CM(\mathcal{D})$. However, in contrast to the Hardy space, the condition we obtain from this simple test, sometimes called the *simple condition*, is not sufficient to characterize Carleson measures.

After that we consider using $\mathcal{J}^*$, the adjoint of $\mathcal{J}$, to describe Carleson measures. That approach will lead to a characterization; but it is convenient to postpone completion of that analysis to the chapter after this.

We then proceed in this section to introduce tools from potential theory and prove Stegenga's theorem characterizing $CM(\mathcal{D})$ using capacity.

4.2.1. The Reproducing Kernel Thesis Fails. If $\mu \in CM(\mathcal{D})$, then we must have an analog of the condition

$$\mu(T(I)) \lesssim |I| \tag{4.6}$$

which held for $\mu \in CM(H^2)$. That condition came from the fact that for such μ there must be a uniform bound on $\left\| \mathcal{J}\hat{k}_{H^2,z} \right\|_{L^2(\mu)}$. That is;

$$\sup_{z\in\mathbb{D}} \left\{ \int_{\mathbb{D}} |\hat{k}_{H^2,z}|^2 d\mu \right\} = C^2 < \infty.$$

Similarly, if $\mu \in CM(\mathcal{D})$, then we must have a uniform bound when testing the embedding $\mathcal{J}$ over normalized kernel functions $\hat{k}_{\mathcal{D},z}$;

$$\sup_{z\in\mathbb{D}} \left\{ \int_{\mathbb{D}} |\hat{k}_{\mathcal{D},z}|^2 d\mu \right\} = C^2 < \infty. \tag{4.7}$$

Select an interval I in the boundary, with associated tent $T(I)$ and with $z = z(I)$ the vertex of that tent. Making direct estimates, one finds that, on $T(I)$, $|k_{\mathcal{D},z}|$ is dominated by a multiple of $k_{\mathcal{D},z}(z)$. Hence, if (4.7) holds, then we must have estimates

$$C > \left\{ \int_{\mathbb{D}} |\hat{k}_z|^2 d\mu \right\} > C \left\{ \int_{\mathbb{D}} \chi^2_{T(I)}(\zeta)\, k_z(z) d\mu(\zeta) \right\} \approx k_z(z)\mu(T(z)).$$

Hence our analog of condition (4.6) is that if $\mu \in CM(\mathcal{D})$, then for intervals I

$$\mu(T(I)) \lesssim \frac{1}{1 + \log(2\pi/|I|)}. \tag{4.8}$$

Thus this condition, sometimes called the *simple condition* or the *one box condition*, is a necessary condition for μ to be a Dirichlet space Carleson measure. (The word "box" is a reference to Carleson boxes $CB(I)$ which could have been used instead of the tents, see Exercise 4.7. The "one" is in contrast to the necessary and sufficient condition we develop in Section 4.2.4 which requires working with unions of boxes.)

If the analysis followed the pattern from the Hardy space story then (4.8), which we just showed was necessary, would also be sufficient. However, it is not. The examples that show this are not straightforward, [**314**], [**311**], or Proposition 5.22; and rely on actually knowing the correct characterization of Carleson measures. However, here is a simple computation which already suggests that (4.8) might not be the answer.

Notice that when we split I into n equal size subintervals, $\{I_k\}$, and apply (4.8) to each one, we obtain

$$\begin{aligned} \mu(\textstyle\bigcup_k T(I_k)) &= \sum_k \mu(T(I_k)) \\ &\leq nC\frac{1}{1+\log(2n\pi/|I|)} \approx \frac{n}{1+\log n + \log(2\pi/|I|)}. \end{aligned} \tag{4.9}$$

As n gets larger, the region $\bigcup_k T(I_k)$ gets small, yet the right hand side, our constraint on its μ−measure, gets larger. This would be odd behavior for a sufficient condition, something that does not occur for (4.6).

4.2.2. The Adjoint of the Inclusion Map and a Bilinear Form. Characterizing Carleson measures means giving boundedness criteria for the embedding map $\mathcal{J}$. We know $\mathcal{J}$ is bounded if and only if its adjoint, $\mathcal{J}^*$, is bounded. We used that observation when studying the Hardy space and it will be useful here. We start by developing the formula for $\mathcal{J}^*$.

We have $\mu \in CM(\mathcal{D})$ if and only if $\mathcal{J}^*$ maps $L^2(\mathbb{D}, \mu)$ boundedly into $\mathcal{D}$; that is, $\forall g \in L^2(\mathbb{D}, \mu)$

$$\|\mathcal{J}^*(g)\|_{\mathcal{D}}^2 \lesssim \|g\|_{L^2(\mu)}^2 . \tag{4.10}$$

Recalling (2.7) we see that $\mathcal{J}^*(g)$ is the holomorphic function on the disk given by

$$\begin{aligned}\mathcal{J}^*(g)(z) &= \langle \mathcal{J}^*(g), k_z\rangle_{\mathcal{D}} = \langle g, \mathcal{J}k_z\rangle_{L^2(\mu)} \\ &= \int_{\mathbb{D}} g(w)\overline{k_z(w)}d\mu(w) \\ &= \int_{\mathbb{D}} g(w)k_w(z)\, d\mu(w).\end{aligned}$$

Hence, just as we computed for (4.4),

$$\begin{aligned}\|\mathcal{J}^*(g)\|_{\mathcal{D}}^2 &= \langle \mathcal{J}^*(g), \mathcal{J}^*(g)\rangle_{\mathcal{D}} \\ &= \left\langle \int_{\mathbb{D}} g(w)\overline{k_z(w)}d\mu(w), \int_{\mathbb{D}} g(w')\overline{k_z(w')}d\mu(w')\right\rangle_{\mathcal{D}} \\ &= \int_{\mathbb{D}}\int_{\mathbb{D}} g(w)\overline{g(w')}\, \langle k_w(z), k_{w'}(z)\rangle_{\mathcal{D}}\, d\mu(w)d\mu(w') \\ &= \int_{\mathbb{D}}\int_{\mathbb{D}} g(w)\overline{g(w')}k_w(w')d\mu(w)d\mu(w').\end{aligned}$$

Thus, a necessary and sufficient condition for $\mu \in CM(\mathcal{D})$ is that there is a C so that $\forall g \in L^2(\mu)$

$$\int_{\mathbb{D}}\int_{\mathbb{D}} g(w)\overline{g(w')}k_w(w')d\mu(w)d\mu(w') \lesssim \|g\|_{L^2(\mu)}^2 .$$

We now make several reductions. If this inequality holds for all g, then it certainly holds for real-valued g. If g is real then the quantity on the left, which is positive, is unchanged if $k_w(w')$ is replaced by $\operatorname{Re} k_w(w')$. This gives

$$B(g,g) = \int_{\mathbb{D}}\int_{\mathbb{D}} g(w)g(w')\operatorname{Re} k_w(w')d\mu(w)d\mu(w') \lesssim \|g\|_{L^2(\mu)}^2$$

for all real g and particularly for non-negative g.

Now note that we can reverse these last few observations. Because $B(\cdot,\cdot)$ is given by integration against a real symmetric non-negative kernel, its boundedness for real functions can be established by bounding $B(g,g)$ for non-negative g. Next, on the real L^2 space, one can pass from $\operatorname{Re} k_w(w')$ to $k_w(w')$ without changing the value of $B(g,g)$. That establishes the boundedness of $\mathcal{J}^*$ on real functions, the complex case then follows by linearity. In sum:

PROPOSITION 4.9. *$\mu \in CM(\mathcal{D})$ if and only if there is a constant so that for all $g \geq 0$*

$$\int_{\mathbb{D}}\int_{\mathbb{D}} g(w)g(w')\operatorname{Re} k_w(w')d\mu(w)d\mu(w') \lesssim \|g\|_{L^2(\mu)}^2 . \tag{4.11}$$

4.2.3. The Reproducing Kernel Thesis Extended. The previous discussion was general and a version of the previous proposition holds for Carleson measures on any RKHS. We now specialize that discussion to the Dirichlet space and obtain a geometric condition which is a candidate for characterizing $\mu \in CM(\mathcal{D})$.

We obtained a necessary condition for membership in $CM(H^2)$ by testing the boundedness of $\mathcal{J}$ on reproducing kernels $\{k_z\}$. The kernels played a central role because each $k_z(\cdot)$ is approximately equal to, and hence is a holomorphic proxy for, a multiple of the characteristic functions of the tent $T(z)$. The tents are crucial because they are, essentially, the superlevel sets of the integration kernels in (4.11). The reproducing kernel thesis is the suggestion that the necessary condition obtained this way is also sufficient. That turned out to be correct, although the proof was not straightforward.

We said earlier that applying the reproducing kernel thesis to $\mathcal{J}$ gives a condition (4.8) which was necessary, but not sufficient, for being a Carleson measure. However, we can make further progress by studying the boundedness of the adjoint, $\mathcal{J}^*$, using similar ideas. We will estimate (4.11) for a family of test functions, $\{g_z\}$. Having $\mathcal{J}^*$ be uniformly bounded on the family will be a necessary condition for the boundedness of $\mathcal{J}^*$, and hence, also, a necessary condition for the boundedness of $\mathcal{J}$. Presumably this will be most effective if we select g's which are large where $\operatorname{Re} k_w(w')$ in (4.11) is large. The domain of $\mathcal{J}^*$ is a Lebesgue space, so rather than select g's which approximate characteristic functions, we can start with actual characteristic functions; set $g_z = \chi_{T(z)}$ where $T(z)$ is the tent with vertex z.

In general, conditions for boundedness of an operator obtained by testing on a set of relatively simple functions, often selected to be large where an integral kernel is large, are called *testing conditions*. Based on the analogy with our analysis of $\mathcal{J}$ we regard the suggestion that this new testing condition is sufficient for membership in $CM(\mathcal{D})$ to be an extension of the reproducing kernel thesis.

In the next chapter we use this pattern of analysis to describe the Carleson measures on a discrete model of the Dirichlet space and then use that to develop a criterion for membership in $CM(\mathcal{D})$. Now we just do heuristic computations to identify the form that criterion will take.

We are given μ and want to test the boundedness of $\mathcal{J}^*$ on the set of functions $\{\chi_{T(z)}\}$. A necessary condition for $\mu \in CM(\mathcal{D})$ is that

$$\sup_{z\in\mathbb{D}} \left\|\mathcal{J}^*\chi_{T(z)}\right\|_{\mathcal{D}}^2 \lesssim \left\|\chi_{T(z)}\right\|_{L^2(\mu)}^2. \tag{4.12}$$

The right hand side of the inequality can be evaluated,

$$\left\|\chi_{T(z)}\right\|_{L^2(\mu)}^2 = \mu(T(z)). \tag{4.13}$$

To estimate the Dirichlet norm on the left hand side of (4.12), we need to know the size of $d\mathcal{J}^*(g_z)/d\zeta$. We have

$$\begin{aligned}\left|\frac{d}{d\zeta}\mathcal{J}^*(g_z)(\zeta)\right| &= \left|\frac{d}{d\zeta}\int_{\mathbb{D}} \chi_{T(z)}(w)\, k_w(\zeta)\, d\mu(w)\right| \\ &\leq \int_{T(z)} \left|\frac{d}{d\zeta}k_w(\zeta)\right| d\mu(w).\end{aligned}$$

We then use the crude estimate that

$$\left|\frac{d}{d\zeta}k_w(\zeta)\right| \approx \begin{cases} \frac{1}{1-|\zeta|^2} & \text{if } w \in T(\zeta) \\ 0 & \text{otherwise} \end{cases}$$

which is obtained by differentiating the kernel function and then estimating the expression $|1/(1-\overline{w}\zeta)|$ on $T(\zeta)$.

The Dirichlet norm in (4.12) involves a Hardy space norm and a Dirichlet integral. The Dirichlet integral is the main term, and in this informal analysis we drop the other term. Using the estimate just given for k_w' we estimate the Dirichlet integral with

$$\mathfrak{D}(\mathcal{J}^*(g_z)) \approx \int_{\zeta\in\mathbb{D}} \left(\int_{w\in T(z)\cap T(\zeta)} \frac{1}{1-|\zeta|^2} d\mu(w) \right)^2 dx(\zeta)\, dy(\zeta). \tag{4.14}$$

Suppose we now replace the integration over $\mathbb{D}$ with integration over the smaller region $T(z)$. If $\zeta \in T(z)$, then $T(\zeta) \cap T(z) = T(\zeta)$ and, in that case, the inner integral is easy to evaluate. Recalling that the invariant measure is $d\lambda(\zeta) \approx (1-|\zeta|^2)^{-2}dA$, we continue with

$$\mathfrak{D}(\mathcal{J}^*(g_z)) \gtrsim \int_{T(z)} \frac{\mu(T(\zeta))^2}{(1-|\zeta|^2)^2} dA \approx \int_{T(z)} \mu(T(\zeta))^2 d\lambda(\zeta).$$

We have replaced our earlier estimate with something smaller but we have not lost too much. Informally, the reason is that the test function g_z is supported in the tent $T(z)$ and the operator $\mathcal{J}^*$ produces a new function $\mathcal{J}^*(g_z)$ which has most of its mass in the same tent. That effect will be particularly easy to see when we do the analysis of the Carleson measures for the discrete model of the Dirichlet space.

Accepting this and putting it and (4.13) in (4.12), we have the following candidate to be a characterization of $\mu \in CM(\mathcal{D})$.

CONDITION 4.10. *If there exists a constant so that $\forall z \in \mathbb{D}$*

$$\int_{T(z)} \mu(T(\zeta))^2 d\lambda(\zeta) \lesssim \mu(T(z))$$

then $\mu \in CM(\mathcal{D})$.

In Theorems 5.13 and 5.17 we see that a condition of this sort does characterize Dirichlet space Carleson measures. However, it is convenient to develop that fact in tandem with the analogous result for the discrete model of the Dirichlet space which we consider in the next chapter.

We now develop Stegenga's description of $\mu \in CM(\mathcal{D})$ which, on the surface, is of a very different form than Condition 4.10. We discuss the relation between the two briefly in Sections 9.1.3 and 13.2.2 and in more detail in [**51**] and [**49**].

4.2.4. Capacity, Stegenga's Theorem. If $f(z) := \sum_{n=0}^{\infty} c_n z^n \in H^2$ then $g(z) := \sum_{n=0}^{\infty} (n+1)^{-1/2} c_n z^n \in \mathcal{D}$ and, in fact, $\|g\|_{\mathcal{D}} = \|f\|_{H^2}$. Informally, this looks like a half-order integration is taking the Hardy space to the Dirichlet space. We now describe one of the ways of making that precise.

Recall from Chapter 1 that any $f \in L^2(\mathbb{T})$ is the *a.e.* boundary function of the harmonic function, $\mathcal{P}f$, the Poisson extension of f, described by (1.15):

$$\mathcal{P}f(z) = \frac{1}{2\pi}\int_{\mathbb{T}} f(\xi)\frac{1-|z|^2}{|1-\bar{\xi}z|^2}|d\xi|.$$

Alternatively, working with Fourier series, if $f \sim \sum_{n=-\infty}^{\infty} c_n e^{in\theta}$, then

$$\mathcal{P}f(re^{i\theta}) = \sum_{n=-\infty}^{\infty} c_n r^{|n|} e^{in\theta}.$$

Let θ be the angle variable on the circle, selected in the range $|\theta| \le \pi$ and let $k_{1/2}(\theta) = |\theta|^{-1/2}$. The Fourier series of this function is

$$k_{1/2}(\theta) = \sum_{n=0}^{\infty} \frac{\alpha_n}{(1+|n|)^{1/2}} e^{in\theta} \tag{4.15}$$

for a sequence $\{\alpha_n\}$ of positive numbers which is bounded and bounded away from zero.

EXERCISE 4.11. *Prove this.*

Recall that convolution of functions on the circle has the effect of multiplying their Fourier coefficients. Thus

$$(k_{1/2} * f)(e^{i\theta}) = \sum_{n=-\infty}^{\infty} c_n \frac{\alpha_n}{(1+|n|)^{1/2}} e^{in\theta}$$

and

$$\mathcal{P}(k_{1/2} * f)(re^{i\theta}) = \sum_{n=-\infty}^{\infty} c_n \frac{\alpha_n}{(1+|n|)^{1/2}} r^{|n|} e^{in\theta}.$$

If $f \in H^2$ then its negative Fourier coefficients are zero and thus $\mathcal{P}(k_{1/2} * f)(z)$ is holomorphic. Also, checking the size of its Taylor coefficients, we see that it is in $\mathcal{D}$. The argument can be reversed and we see that

$$f \to \mathcal{P}(k_{1/2} * f) \tag{4.16}$$

is a bounded invertible linear map of H^2 onto $\mathcal{D}$. That is, we have

LEMMA 4.12. *A function $g \in \mathcal{D}$ if and only if $g = \mathcal{P}(f * k_{1/2})$ for $f \in H^2(\mathbb{D})$. Moreover, the $\|g\|_{\mathcal{D}}$ is comparable to $\|f\|_{H^2(\mathbb{D})}$.*

For $n > 0$ this map takes z^n to a controlled positive multiple of $z^n/\sqrt{n}$, so it is a plausible proxy for integration of order one half.

At this point we remove holomorphy from the discussion and move to a real variable formulation. Recalling that every function $h \in L^2$ is a sum $h = f + \overline{g}$ with $f, g \in H^2$, $g(0) = 0$, we see that the same mapping (4.16) is also a bounded invertible linear map of L^2 onto the space of harmonic functions, F, in the disk with finite Dirichlet integral

$$\int_{\mathbb{D}} |\nabla F(z)|^2 \, dA(z) < \infty.$$

Also, if the map is bounded on L^2 then it is certainly bounded when restricted to non-negative functions. Further, if $F \in L^2$ then there are non-negative functions

p_i, $i = 1, 2, 3, 4$ such that $F = p_1 - p_2 + i(p_3 - p_4)$ and $\|F\| \approx \sum_{i=1}^4 \|p_i\|$. Combining these observations we have the following:

LEMMA 4.13. *A non-negative measure μ is a $\mathcal{D}$-Carleson measure if and only if*

$$\int_{\mathbb{D}} \left|\mathcal{P}(f * k_{1/2})(z)\right|^2 d\mu(z) \leq C(\mu)^2 \|f\|^2_{L^2(\mathbb{T})}$$

for all $f \in L^2(\mathbb{T})$ such that $f \geq 0$.

Informally, the previous lemma shows that describing the Carleson measures for $\mathcal{D}$ is related to understanding the size and geometry of sets where functions of the form $\mathcal{P}(f * k_{1/2})$ might be large. Also, considering where the Poisson kernel is large suggests that the only places $\mathcal{P}(f * k_{1/2})$ might be large are in regions of $\mathbb{D}$ near the region in $\mathbb{T}$ where $|f|$ is large. Finally, the discussion which led to (4.9) suggests that when we study that region in $\mathbb{T}$, we consider not just its length, but also its geometry. An effective set function for that purpose is capacity, which we now introduce.

For $E \subset \mathbb{T}$, let $\tau(E)$ denote the corresponding subset of $[-\pi, \pi]$ on the real line $\mathbb{R}$. We define the *capacity* of a set E, $\operatorname{Cap}(E)$, by

$$\operatorname{Cap}(E) := \inf\left\{\|f\|^2_{L^2(\mathbb{T})} : f \geq 0,\ k_{1/2} * f \geq 1 \text{ on } \tau(E)\right\}.$$

Capacity is a set function which takes note of both the size and the geometry of a set. It is a more subtle notion than the Lebesgue measure of a set and is designed to be extremely well suited to relating properties of $f \in L^2(\mathbb{T})$ to those of $\mathcal{P}(f * k_{1/2})$. (If $k_{1/2}$ were replaced by another choice, for instance if $1/2$ in (4.15) were replaced by α, $0 < \alpha < 1$, we would use an similar definition and have an analogous, but different, capacity functional.)

Some properties of this set function are easy to establish; for example, given E_1, E_2 we have

$$\sqrt{\operatorname{Cap}(E_1 \cup E_2)} \leq \sqrt{\operatorname{Cap}(E_1)} + \sqrt{\operatorname{Cap}(E_2)}.$$

However, there are are fundamental differences between measure theory which we use in the Hardy space theory and capacity theory which we need when working with the Dirichlet space. In particular, even for disjoint E_1 and E_2, there is no algebraic relationship between the three terms in the previous inequality; there is no analog of the fact that for Lebesgue measure, $|\cdot|$, $|E_1 \cup E_2| = |E_1| + |E_2|$. The capacity of the union of disjoint sets depends, in complicated ways, on their geometric positions relative to each other.

We will adapt to our context some results about capacity from the book by Adams and Hedberg, [**1**]. That book presents a general development of capacity theory for function spaces including the Dirichlet space and also some of the other spaces mentioned in Section 2.5. Capacity theory more directly related to the Dirichlet space is developed in [**266**] and in [**136**]. Capacity theory on trees, which we will use later, is discussed in [**44**], and [**49**].

For ease of presentation, we will now conflate the circle and line and offer details for functions on the line. (Informally, we can do this without loss because the crucial estimates are only critical for points extremely close to the boundary; and for such points, the pictures, and the geometry of the two situations, are almost identical; see Remark 2.27.) There are technical adjustments to be made when working on the circle and we refer to Stegenga's paper [**314**] for those details.

We begin with some general definitions which are used even in contexts much more general than $\mathbb{R}$ or $\mathbb{T}$. We are given a positive kernel $g(x,y)$ on $\mathbb{R}\times\mathbb{R}$ and a measure ν and now define two potentials that will play a role in what we are studying. Let μ be a non-negative regular sigma finite Borel measure, (and we denote the class of all such measures by $\mathcal{M}^+(\mathbb{R})$) and let f be a ν measurable function. The *potentials* are then

$$\mathcal{G}f(x) := \int_{\mathbb{R}} g(x,y)f(y)d\nu(y), \text{ and}$$
$$\check{\mathcal{G}}\mu(y) := \int_{\mathbb{R}} g(x,y)d\mu(x).$$

The case of primary interest for us is $g(x,y)\approx |x-y|^{-1/2}$ and $d\nu = dx$. Using these two potentials we define the *mutual energy* of μ and f by

$$\begin{aligned}\mathcal{E}_g(f,\mu) &:= \int_{\mathbb{R}} \mathcal{G}f(x)d\mu(x) = \int_{\mathbb{R}} \check{\mathcal{G}}\mu(y)f(y)d\nu(y)\\ &= \int_{\mathbb{R}}\int_{\mathbb{R}} g(x,y)f(y)d\mu(x)d\nu(y).\end{aligned}$$

As a definition, we take the capacity of a set with respect to the kernel g to be

$$\mathrm{Cap}_g(E) := \inf\left\{\|f\|_{L^2(\nu)}^2 : \mathcal{G}f(x)\geq 1 \text{ on } E\right\}$$

It turns out that based just on this definition and then using standard measure theory arguments, one can prove the capacity is a subadditive and monotone function of sets. One can further extend this definition to more general sets (much as in standard measure theory where one first defines Lebesgue measure for "nice" sets and then extends to more general sets.) Another important fact is that there is a unique function f^E such that $f^E\in L^2(\nu)$ and $\mathcal{G}f^E(x)\geq 1$ on E (technically up to a set of capacity zero) and

$$\int_{\mathbb{R}} (f^E)^2 d\nu = \mathrm{Cap}_g(E).$$

This unique function is usually called the *capacitary function*.

We also need the deeper fact that it is possible to compute the capacity of a set via a dual definition. In particular, for compact sets $K\subset\mathbb{R}$, we will have

$$\mathrm{Cap}_g(K)^{1/2} = \sup\left\{\mu(K) : \mu\in\mathcal{M}^+(K), \left\|\check{\mathcal{G}}\mu\right\|_{L^2(\nu)}\leq 1\right\}$$

or, the elementary reformulation

$$\mathrm{Cap}_g(K)^{-1} = \inf\left\{\left\|\check{\mathcal{G}}\mu\right\|_{L^2(\nu)}^2 : \mu\in\mathcal{M}^+(K),\ \mu(K)=1\right\}.$$

One direction is trivial, namely it is easy to show that $\mu(K)\leq \mathrm{Cap}_g(K)^{1/2}$ just using the definitions and Hölder's inequality. The other direction then uses a substantial result, von Neumann's minimax theorem, applied to the functional $\mathcal{E}_g$. Similar to the existence of the capacitary function, there is a capacitary measure which plays an important role. For any $K\subset\mathbb{R}$ compact we have a measure μ^K

such that $f^K = \check{\mathcal{G}}\mu^K$ and

$$\begin{aligned}\mu^K(K) &= \int_{\mathbb{R}} (\check{\mathcal{G}}\mu^K)^2(s)d\nu(s) = \int_{\mathbb{R}} (\check{\mathcal{G}}\mu^K)(s) f^K(s)d\nu(s) \\ &= \int_{\mathbb{R}} g(y,s)d\mu(y) f^K(s)d\nu(s) = \int_{\mathbb{R}} g(y,s)\mathcal{G}f^K(y)d\mu(y) \\ &= \mathrm{Cap}_g(K).\end{aligned} \tag{4.17}$$

The case of interest to us is $g(x,y) = |x-y|^{-1/2}$ and $d\nu = dx$. But, continuing to conflate the circle with the line, we now work on the circle with $\mu \in \mathcal{M}^+(\mathbb{T})$, $\check{\mathcal{G}}\mu = k_{1/2} * \mu$, and $d\nu = dm =$ arc length on the circle. A straightforward computation verifies

$$\begin{aligned}\int_{\mathbb{T}} (\check{\mathcal{G}}\mu)^2 d\nu &= \int_{\mathbb{T}} (k_{1/2} * \mu)(t)(k_{1/2} * \mu)(t)dm(t) \\ &= \int_{\mathbb{T}}\int_{\mathbb{T}} (k_{1/2} * k_{1/2})(r-s)d\mu(r)d\mu(s).\end{aligned}$$

We continue looking at the Fourier coefficients of $k_{1/2} * k_{1/2}$ and see that we can continue with the following approximation

$$\int_{\mathbb{T}} (\check{\mathcal{G}}\mu)^2 d\nu \approx \int_{\mathbb{T}}\int_{\mathbb{T}} \log^+ \frac{1}{|r-s|} d\mu(r)d\mu(s).$$

The right-hand expression is sometimes called the *energy* (or self-energy, or energy integral) of μ, and the definition of capacity is sometimes given using it. The name comes from electrostatics and relates to the potential energy associated with placing a distribution μ of infinite, charged wires perpendicular to the complex plane and intersecting it on the unit circle. Informally, in two dimensional electrostatics $-\log^+ \frac{1}{|r-s|} d\mu(r)d\mu(s)$ is the repulsive force between the infinitesimal charged wires at positions r and s, whose sections have magnitude $d\mu(r)$ and $d\mu(s)$. Hence the double integral represents the energy that would be released if the charges were allowed to fly apart; or, equivalently, the energy required to bring the charges from infinitely far away to the position specified by μ.

As a consequence we have, for K a compact subset of $\mathbb{T}$,

$$\mathrm{Cap}_g(K)^{-1} = \inf\left\{\int_{\mathbb{T}}\int_{\mathbb{T}} \log^+ \frac{1}{|r-s|} d\mu(r)d\mu(s); \mu \in \mathcal{M}^+(K), \mu(K) = 1\right\}. \tag{4.18}$$

Thus, for example, if K is a finite set then the integrand is infinite on a set of positive measure and thus $\mathrm{Cap}_g(K) = 0$.

Care is needed here with names. $\mathrm{Cap}_g(E)$ is certainly a measure of the size and complexity of E. It might be tempting, given the previous formula, to call that quantity the "logarithmic capacity" of E. However, that name is, in fact, generally given to

$$\mathrm{Logcap}(E) = \exp\left(-\frac{1}{\mathrm{Cap}_g(E)}\right).$$

Unfortunately, the literature is not consistent on these usages. The fact that the two quantities agree in the important special case of "negligible" sets; $\mathrm{Logcap}(E) = 0$ if and only if $\mathrm{Cap}_g(E) = 0$; both encourages the confusion and mitigates its consequences.

For a single interval the computation of capacity is not hard.

EXERCISE 4.14. *Show that for I an interval on the circle (or a short interval on the line)*

$$\operatorname{Cap}(I) \approx \frac{1}{1+\log(2\pi/|I|)}.$$

The following lemma, called the strong capacitary inequality, is a basic interface between capacity theory and measure theory. We will just present the particular case of interest. See [**136**].

LEMMA 4.15. *If $f \in L^2(\mathbb{T})$, then*

$$\int_0^\infty \lambda \operatorname{Cap}\left(\left\{\xi \in \mathbb{T} : \left(k_{1/2} * f\right)(\xi) > \lambda\right\}\right) d\lambda \lesssim \|f\|^2_{L^2(\mathbb{T})} .$$

PROOF. Note that it suffices to prove that

$$\sum_{k\in\mathbb{Z}} 2^{2k} \operatorname{Cap}\left(\left\{\xi \in \mathbb{T} : k_{1/2} * f(\xi) > 2^k\right\}\right) \lesssim \|f\|^2_{L^2(\mathbb{T})} .$$

Without loss of generality, we may suppose that f is a non-negative, smooth, compactly supported function. Set

$$E_k = \{\xi : k_{1/2} * f(\xi) \geq 2^k\}.$$

The hypotheses on f imply that E_k is compact and empty for k sufficiently large. Define

$$J = \sum_{k\in\mathbb{Z}} 2^{2k} \operatorname{Cap}\left(\left\{\xi \in \mathbb{T} : k_{1/2} * f(\xi) > 2^k\right\}\right)$$

and let μ_k be the extremal measure for E_k, as described in (4.17). Then we have

$$\begin{aligned} J &= \sum_{k=-\infty}^{\infty} 2^{2k} \int_{\mathbb{T}} d\mu_k \\ &\leq \sum_{k=-\infty}^{\infty} 2^{k} \int_{\mathbb{T}} k_{1/2} * f d\mu_k \\ &= \sum_{k=-\infty}^{\infty} 2^{k} \int f(x)(k_{1/2} * \mu_k)(x) dx \\ &\leq \|f\|_{L^2} \left\| \sum_{k=-\infty}^{\infty} 2^k k_{1/2} * \mu_k \right\|_{L^2} . \end{aligned}$$

We set

$$L = \left\| \sum_{k=-\infty}^{\infty} 2^{2k-1} k_{1/2} * \mu_k \right\|^2_{L^2}$$

and will show that $L \lesssim J$ to complete the proof.

To see that $L \lesssim J$, first set

$$\Lambda(x) = \sum_{k=-\infty}^{\infty} 2^k k_{1/2} * \mu_k(x)$$

and

$$\Lambda_j(x) = \sum_{k=j}^{\infty} 2^k k_{1/2} * \mu_k(x).$$

Clearly we have that $\Lambda_j \in L^2$ (because of the assumptions on f) and $\Lambda_j \to \Lambda$. Now, notice that

$$\Lambda(x)^2 = \lim_{n\to-\infty} \Lambda_n(x)^2 = \lim_{n\to-\infty} \sum_{j=n}^{\infty} \left(\Lambda_j(x)^2 - \Lambda_{j+1}(x)^2\right)$$
$$\leq 2 \sum_{j=-\infty}^{\infty} \Lambda_j(x) \left(\Lambda_j(x) - \Lambda_{j-1}(x)\right)$$
$$= 2 \sum_{j=-\infty}^{\infty} \Lambda_j(x) 2^j k_{1/2} * \mu_j(x),$$

where the passage from the first line to the second used the monotonicity of the Λ_j. Using this, we see that

$$L \lesssim \int_{\mathbb{T}} \sum_{j=-\infty}^{\infty} 2^j \Lambda_j(x)(k_{1/2} * \mu_j(x))dx$$
$$= \sum_{j=-\infty}^{\infty} \sum_{k=j}^{\infty} 2^k 2^j \int_{\mathbb{T}} k_{1/2} * \mu_k(x) k_{1/2} * \mu_j(x)dx$$
$$= \sum_{j=-\infty}^{\infty} \sum_{k=j}^{\infty} 2^k 2^j \int_{\mathbb{T}} (\check{\mathcal{G}}\mu^K)^2 dx$$
$$= \sum_{j=-\infty}^{\infty} \sum_{k=j}^{\infty} 2^k 2^j \operatorname{Cap}(E_k)$$
$$= \sum_{k=-\infty}^{\infty} 2^{2k} \operatorname{Cap}(E_k) = J.$$

□

With these preliminaries out of the way, we now prove the following characterization of the Carleson measures for $\mathcal{D}$ obtained by Stegenga. Recall from Section 1.7.2 that $T(I)$ is the tent over the interval I in the circle.

THEOREM 4.16 (Stegenga, [**314**]). *Suppose μ is a non-negative Borel measure in the disk $\mathbb{D}$. Then the following are equivalent:*

(i) *The embedding*
$$\mathcal{D} \to L^2(\mathbb{D}, \mu)$$
is bounded;

(ii) *For $f \in \mathcal{D}$ we have*
$$\int_{\mathbb{D}} |f(z)|^2 \, d\mu(z) \lesssim \|f\|^2_{\mathcal{D}};$$

(iii) *Uniformly over all families $\{I_j\}$ of disjoint intervals in $\mathbb{T}$ we have*
$$\mu\left(\bigcup_j T(I_j)\right) \lesssim \operatorname{Cap}\left(\bigcup_j I_j\right). \tag{4.19}$$

Moreover, the implied constants in the previous two statements are comparable.

REMARK 4.17. *In fact, it is enough to require (iii) for* finite unions *of intervals. We refer to Stegenga's paper for that technical improvement.*

PROOF OF THEOREM 4.16. The equivalence between (i) and (ii) is an immediate consequence of the definitions. We now focus on showing that (ii) and (iii) are equivalent. First suppose that μ is a $\mathcal{D}$ Carleson measure. Let $E = \bigcup_{j=1}^n I_j$. Now select a test function for the capacity, namely a competitor $f \in L^2(\mathbb{T})$ for the infimum with $f \geq 0$ and $f * k_{1/2} \geq 1$ on $\tau(E)$. A simple computation shows that $\mathcal{P}(\chi_{I_j})(z) \geq 1/4$ for $z \in T(I_j)$. So, we further have that

$$\mathcal{P}(f * k_{1/2})(z) \geq \frac{1}{4} \quad \forall z \in \bigcup_{j=1}^n T(I_j).$$

Thus, by Lemma 4.13, we have that

$$\begin{aligned}\mu\left(\bigcup_{j=1}^n T(I_j)\right) &\lesssim \int_{\mathbb{D}} \left|\mathcal{P}(f * k_{1/2})(z)\right|^2 d\mu(z) \\ &\lesssim \|f\|_{L^2(\mathbb{T})}^2 .\end{aligned}$$

Since f was arbitrary, this implies that for any competitor for the infimum, we have

$$\mu\left(\bigcup_{j=1}^n T(I_j)\right) \lesssim \|f\|_{L^2(\mathbb{T})}^2 .$$

So, in particular, we have that

$$\mu\left(\bigcup_{j=1}^n T(I_j)\right) \lesssim \operatorname{Cap}\left(\bigcup_{j=1}^n I_j\right).$$

Conversely, suppose that for every disjoint collections of intervals $I_j \subset \mathbb{T}$ we know that

$$\mu\left(\bigcup_{j=1}^n T(I_j)\right) \lesssim \operatorname{Cap}\left(\bigcup_{j=1}^n I_j\right).$$

We then need to show that for all $f \in \mathcal{D}$, we have

$$\int_{\mathbb{D}} |f(z)|^2 \, d\mu(z) \lesssim \|f\|_{\mathcal{D}}^2 .$$

Equivalently, via Lemma 4.13, it suffices to show that

$$\int_{\mathbb{D}} \left|\mathcal{P}(f * k_{1/2})(z)\right|^2 d\mu(z) \lesssim \|f\|_{L^2(\mathbb{T})}^2 .$$

Set $u(z) = \mathcal{P}(f * k_{1/2})(z)$. Recall u^*, the nontangential maximal function of u, from Section 1.6. Now, observe that the set

$$S = \{t : u^*(t) > \lambda\}$$

is an open set, and so is a disjoint union of open intervals $\{I_j\}$. Let $T(I_j)$ denote the corresponding tent over I_j. We now claim that we have the inclusion

$$L = \{z \in \mathbb{D} : |u(z)| > \lambda\} \subset \bigcup_j T(I_j) =: R.$$

To see that $L \subset R$ we select a point $z \in L$. By the definition of u^*, we know that $z/|z| \in S$ and hence $z/|z|$ is in one of the I_j. Furthermore, if ω is near $z/|z|$, then the nontangential region with vertex ω will contain z, and thus $u^*(\omega) > \lambda$.

This observation, applied to all possible ω, shows the set where $u^* > \lambda$ includes an interval J on the circle. That interval is centered at $z/|z|$ and has width comparable to the distance from z to the circle. We must have $J \subset I_j$ and hence $T(J) \subset T(I_j)$. We will be finished if we know that $z \in T(J)$. We would like to say that this holds because the midpoint of J is the radial projection of z and the width of J is comparable to the distance from z to the boundary. This does hold, and is easy to see geometrically, but only if we have, in advance, selected the aperture α to be large enough. We assume that has been done. (Absent that assumption, the proof is slightly more technical.)

EXERCISE 4.18. *Make a sketch showing those things that are "easy to see."*

Now we can continue with

$$\begin{aligned}\mu\left(\{z \in \mathbb{D} : |u(z)| > \lambda\}\right) &\leq \mu\left(\bigcup_{j=1}^{n} T(I_j)\right)\\ &\lesssim C(\mu)\,\mathrm{Cap}\left(\bigcup_{j=1}^{n} I_j\right)\\ &= C(\mu)\,\mathrm{Cap}\left(\{t \in \mathbb{R} : u^*(t) > \lambda\}\right).\end{aligned}$$

The second inequality holds by (4.19).

We now apply all these observations with $u = \mathcal{P}(f * k_{1/2})$. Doing so, and recalling the definition of the Hardy-Littlewood maximal function, M, and Theorem 1.4 from Section 1.6, we see that

$$\begin{aligned}\int_{\mathbb{D}} \left|\mathcal{P}(f * k_{1/2})(z)\right|^2 d\mu(z) &= 2\int_0^\infty \lambda\mu\left(\{z : \left|\mathcal{P}(f * k_{1/2})(z)\right| > \lambda\}\right) d\lambda\\ &\lesssim \int_0^\infty \lambda\,\mathrm{Cap}\left(\{\mathcal{P}(f * k_{1/2})^*(\xi) > \lambda\}\right) d\lambda\\ &\lesssim \int_0^\infty \lambda\,\mathrm{Cap}\left(\{M(f * k_{1/2})(\xi) > \lambda\}\right) d\lambda\\ &\lesssim \int_0^\infty \lambda\,\mathrm{Cap}\left(\{Mf * k_{1/2}(\xi) > \lambda\}\right) d\lambda\\ &\lesssim \|Mf\|_{L^2(\mathbb{T})}^2\\ &\lesssim \|f\|_{L^2(\mathbb{T})}^2.\end{aligned}$$

Here, in the second to last inequality, we used Lemma 4.15. Then applying Lemma 4.13 gives that μ is a $\mathcal{D}$ Carleson measure. □

Notice that in the proof we used a geometric argument to establish a relationship between the set in $\mathbb{D}$ where the harmonic function u was large and the set in $\mathbb{T}$ where the maximal function u^* is large. This type of relationship is very general and widely exploited in harmonic analysis.

For any open set $\Omega \subset \mathbb{T}$, set

$$T(\Omega) = \bigcup \{T(I) : I \text{ an interval}, I \subset \Omega\}.$$

Another way to view condition (4.19) is that for any open set $\Omega \subset \mathbb{T}$, we have that

$$\mu(T(\Omega)) \lesssim \mathrm{Cap}(\Omega).$$

It is not true for a general family of disjoint intervals $\{I_j\}$ that $\operatorname{Cap}\left(\bigcup_j I_j\right) = \sum_j \operatorname{Cap}(I_j)$. If it did, then the condition (4.8), which we noted earlier in the chapter is suggested by the reproducing kernel thesis, (and is, in fact, (4.19) for a *single* interval) would imply (4.19). Thus the condition (4.8) would be necessary and sufficient for μ to be a Carleson measure. Although that is not generally true; examples where that fails are given in [**314**], [**311**], and Proposition 5.22; for some simple geometries, the condition (4.8) is necessary and sufficient for μ to be a Carleson measure.

EXERCISE 4.19. *If the support of μ is on the segment $[0,1]$, then μ is a Carleson measure for the Dirichlet space if and only if* (4.8) *holds.*

We introduced Carleson measures for $\mathcal{D}$ in the context of describing multipliers and saw that $f \in \operatorname{Mult}(\mathcal{D})$ if and only if it is bounded and $|f'|^2\, dA$ is a Carleson measure. Stegenga also gives an example that shows (8.16) does not imply (4.19) even for this restricted class of measures.

THEOREM 4.20 (Stegenga, [**314**]). *There exists a function $f \in H^\infty(\mathbb{D})$ such that for all intervals I in the boundary*

$$\int_{T(I)} |f'(z)|^2\, dA(z) \lesssim \frac{1}{1+\log(2\pi/|I|)},$$

*but f is **not** a multiplier of $\mathcal{D}$.*

EXERCISE 4.21. *Look up the example constructed by Stegenga and work through it.*

Stegenga also has a description of boundary values of functions $f \in \mathcal{X}$, that is, f such that $|f'|^2\, dA$ is a Carleson measure. We will return to that in Section 10.3.3 after we have introduced the local Dirichlet integral.

4.3. Notes and Comments

Carleson measures were introduced in [**91**]. The name is now widely used for measures which produce a bounded embedding map or restriction map on a function space. Many examples are in the books [**161**], [**348**], [**316**], [**164**], [**165**], [**294**], and [**232**].

Our presentation of Carleson measures for H^2 follows [**245**].

The attempt to use the reproducing kernel thesis for Dirichlet space Carleson measures is adapted from [**45**].

Our presentation of Stegenga's result follows [**314**]. Analogous results for the scale of Besov spaces was given by Verbitsky in [**336**]. As we said, our discussion of the related capacity theory is sketchy. More detailed presentations are in [**1**], [**266**], and [**136**].

An improvement of Stegenga's result, obtained through an approach closer to one we use the next chapter is given by Kerman and Sawyer in [**191**].

Questions similar to that of characterizing Carleson measures on the Dirichlet space are studied in potential theory under the name "trace inequalities." Some of that work is very close to topics discussed in this chapter and in later chapters, see for instance [**337**].

CHAPTER 5

Analysis on Trees

In this chapter we introduce trees and study functions defined on them. We define $\mathcal{D}(\mathcal{T})$, the Dirichlet space on the tree, $\mathcal{T}$, and describe its Carleson measures, they are measures which satisfy a refined version of Condition 4.10. We also show that the set in the boundary on which a function in $\mathcal{D}(\mathcal{T})$ fails to have radial limits must be a null set for every Carleson measure.

The connection between the spaces $\mathcal{D}(\mathcal{T})$ and $\mathcal{D}$ is quite strong and very useful; partly because of the strong analogies between the geometry of $\mathcal{T}$ induced by $\mathcal{D}(\mathcal{T})$ and aspects of the geometry of the disk which are crucial for $\mathcal{D}$. Using that connection we obtain an alternative characterization of the Carleson measures for $\mathcal{D}$, and we obtain a new proof of Beurling's classical result that functions in $\mathcal{D}$ have radial limits off a set of capacity zero.

In the final sections of this chapter we develop refined results about capacity on trees; results which are used in later chapters to study onto interpolation for the Dirichlet space and in studying Hankel forms on the Dirichlet space. Those uses of tree based results to study holomorphic functions are one of the triumphs of the approach.

Recall from Section 1.7 that both the dyadic Whitney decomposition of the unit disk and the related dyadic decomposition of the unit interval are associated with tree structures. This viewpoint is not only inspiring, but it is useful in proving results on the Dirichlet space, and in later chapters, the Besov spaces. Here we will discuss trees and the analysis on them in some detail, always with an eye to those applications. The spaces of functions on trees are valuable models and useful tools for studying spaces of holomorphic functions. We also find the function spaces on trees intrinsically very interesting.

Trees will recur throughout the book, especially in Chapter 16 where there is an extensive development of tree Besov spaces on the complex ball.

Again, we encourage readers new to these ideas to make sketches along the way.

5.1. Trees, Metrics, Boundaries

Before we embark on our discrete journey, let us briefly recall some metric structures on the unit disk which are relevant to function theory in general, and to the Dirichlet space in particular. The most important one is the hyperbolic metric; the Riemannian metric with length element $ds_\beta^2 = 4\left|dz\right|^2/(1-\left|z\right|^2)^2$. Up to a multiplicative constant, it is the only Riemannian metric on $\mathbb{D}$ which is invariant under biholomorphic transformations of the disk. Some authors prefer the pseudo-hyperbolic metric ρ which is not a Riemannian distance function but is invariant and has the simple formula $\rho(0,z) = |z|$ for $z \in \mathbb{D}$. And, of course, there is the usual Euclidean metric with $ds^2_{\mathtt{Euclid}} = |dz|^2$. In contrast to the previous two, that metric has the virtue of extending naturally to the boundary circle $\mathbb{T}$.

As it often happens, it is easier to begin with the general case and then go to the specific instance of central interest. A *tree* $\mathcal{T} = (V(\mathcal{T}), E(\mathcal{T}))$ is a connected, undirected graph without loops. We are given a countable set of vertices $V(\mathcal{T})$ and a set of edges $E(\mathcal{T})$ such that, for any edge α in $E(\mathcal{T})$, α can be identified with a pair $\{x,y\}$ of vertices from $V(\mathcal{T})$, called the endpoints of the edge. We will generally identify the tree $\mathcal{T}$ with the vertex set $V(\mathcal{T})$ while regarding the set of edges as structural data about $V(\mathcal{T})$.

We assume that any vertex is the endpoint of at most finitely many edges (the graph is locally finite). A finite path in $\mathcal{T}$ is a sequence $\Gamma = \{\alpha_j\}_{j=1}^n$ of edges, such that $\alpha_j = \{x_{j-1}, x_j\}$ for a sequence $\{x_j\}_{j=0}^n$ of vertices. The length of the path Γ is n, the number of the edges in the sequence. When this causes no confusion, we identify a path with its sequence of vertices, writing $\Gamma = \{x_j\}_{j=0}^n$ as a sequence in $V(\mathcal{T})$.

The graph having vertices $V(\mathcal{T})$ and edges $E(\mathcal{T})$ is a tree if

(1) it is connected: for all x, y in $V(\mathcal{T})$ there is a path $\Gamma = \{x_j\}_{j=0}^n$ such that $x_0 = x$ and $x_n = y$;
(2) it is simply connected: the only way a path can start and end at the same point is by backtracking; if Γ is a path with edges $\{\alpha_j\}_{j=1}^n$ and vertices $\{x_j\}_{j=0}^n$ with $x_0 = x_n$ then each edge in the list $\{\alpha_j\}_{j=1}^n$ appears an even number of times.

Given $x \neq y$ in $V(\mathcal{T})$, the *geodesic* $[x,y]$ between x and y is the shortest path starting at x and ending at y. We set by default $[x,x] := \{x\}$, a singleton sequence of vertices (the corresponding sequence of edges is empty!).

The natural, edge-counting distance $d(x,y) \in \mathbb{N}$ between x and y in $V(\mathcal{T})$ is the number of the edges in $[x,y]$. Notice that, as a consequence of the simple connectivity, for any weight $w : E(\mathcal{T}) \to [0, +\infty)$, the weighted distance $d_w(x,y)$ on $V(\mathcal{T})$, defined by minimizing the sum of the weights over paths joining x and y, has the same geodesics as $d = d_1$.

We add structure to the tree by choosing a privileged vertex $o \in V(\mathcal{T})$ as the *root* of $\mathcal{T}$. We define a partial order on $V(\mathcal{T})$, writing $x \leq y$ if $x \in [o,y]$. We say that x^{-1} is the *predecessor*, or *parent*, of x if $x^{-1} < x$ and $d(x^{-1}, x) = 1$. In that case x is said to be a *child* of x^{-1}. In particular, o has no predecessor.

Unless we specifically state otherwise, all the trees we consider will be rooted trees.

Convenient Fiction I: It is convenient to adjoin to each rooted tree a fictitious vertex, the *pre-root*, o^{-1} and fictitious *root edge* $\sigma^* \approx \{o^{-1}, o\}$ connecting the pre-root and root. After including the root edge we have a natural bijection between $E(T)$ and $V(T)$ given by pairing each edge with its vertex farther from o^{-1}. We will sometimes use this bijection to regard functions defined on one of those two sets as also defined on the other.

Further, we will assume that any function defined on the vertex set is defined and equals zero on the pre-root. These conventions, which we use freely and generally without mention below, let us avoid dealing with the root as a special case in various definitions and computations.

Convenient Fiction II: We will often talk as if the trees we consider are dyadic trees; each vertex $x \neq o$ being on three edges, one connecting to its parent x^{-1} and one to each of two children, x_+ and x_-; the root, having no parent, is on only two edges. We do this even when it is clear that the tree actually being considered is not of this type. This allows simpler language and notation, and the results we describe extend mechanically to rooted trees with a universal upper bound for the number of successors at each vertex. We say those trees are of *dyadic type.*

For x, y in $V(\mathcal{T})$, we define their minimum as $x \wedge y = \max([o,x] \cap [o,y]) = \max\left(\left[o^{-1},x\right] \cap \left[o^{-1},y\right]\right)$. Here we regard the geodesics as subsets of $V(\mathcal{T})$ and the maximum is with respect to the partial order on that set. We also use the symbol $d^+(x) = d(o^{-1},\alpha) = 1 + d(o,x)$ to denote the level of x in the rooted tree with the root as level one, the number of edges in the path connecting x to the pre-root. Observe that $d^+(x \wedge y) = (d^+(x) + d^+(y) - 2d(x,y))/2$. These quantities are stable with respect to a change of root. If $\tilde{o}$ is another root, $\tilde{\wedge}$ denotes the corresponding wedge, and $\Delta_x = d(\tilde{o},x) - d(o,x)$, similarly for Δ_y, then:

$$|d(\tilde{o}, x\tilde{\wedge}y) - d(o, x \wedge y)| = |(\Delta_x + \Delta_y)/2| \leq |d(o,\tilde{o})|.$$

EXERCISE 5.1. *Check this.*

The metric d is the discrete version of the hyperbolic metric on the unit disk. The close relation between the two metrics is central to the relationship between the tree Dirichlet space and the classical Dirichlet space. We now introduce a discrete version of the Euclidean metric. To each edge α in $E(\mathcal{T})$ assign the weight $\epsilon(\alpha) = 2^{-d(o,\alpha)-1}$, where $d(o,\alpha)$ is the distance from o to the endpoint of α which is farthest from o. It models the Euclidean distance from α to the boundary. The ϵ-distance between x and y is the minimum over all paths connecting x and y of the sum of the weights $\epsilon(\alpha)$ for all edges in the path. On a tree this number is finite and the unique path for which the minimum is attained is the tree geodesic $[x,y]$. It then holds that

$$\epsilon(x,y) = 2^{-d(o,x\wedge y)} - \frac{1}{2}\left(2^{-d(o,x)} + 2^{-d(o,y)}\right).$$

By our previous remark, the distances associated to different roots $o, \tilde{o}$ are bi-Lipschitz equivalent: $\hat{\epsilon}(x,y) \approx \epsilon(x,y)$, where the bi-Lipschitz constants depend only on $d(o,\tilde{o})$.

Inspection of a half-infinite geodesic starting at the root makes it clear that the metric space $(\mathcal{T},\epsilon) = (V(\mathcal{T}),\epsilon)$ is far from being complete. Each such geodesic $\zeta = \{x_n\}$ is a sequence in $V(\mathcal{T})$ with $x_0 = o$, with all elements distinct, and such that $\{x_{n-1}, x_n\} \in E(\mathcal{T})$ for $n \geq 1$. It is a Cauchy sequence because $\epsilon(x_n, x_{n+l}) \approx 2^{-n}$, but it has no limit in $\mathcal{T}$.

We denote by $\overline{\mathcal{T}}$ the metric completion of $\mathcal{T}$: $\overline{\mathcal{T}}$ is the set of Cauchy sequences in $(\mathcal{T},\epsilon)$, modulo the equivalence relation $\{x_n\} \backsim \{y_n\}$ if $\epsilon(x_n,y_n) \to 0$, with the distance $\epsilon([\{x_n\}],[\{y_n\}]) = \lim_{n\to\infty}\epsilon(x_n,y_n)$. We identify $\mathcal{T}$ as a subset of $\overline{\mathcal{T}}$ by identifying each x in $\mathcal{T}$ with the constant sequence $\{x_n = x\}$ and extend $\leq$ to $\overline{\mathcal{T}}$ in the natural way. We denote by $\partial\mathcal{T} := \overline{\mathcal{T}} \setminus \mathcal{T}$ the *metric boundary* of $\mathcal{T}$. The expression for the metric ϵ on $\mathcal{T}$ is obtained by taking the limit of the metric on $\mathcal{T}$. For ζ, ξ in $\partial\mathcal{T}$, we have in particular that

$$\epsilon(\zeta, \xi) = 2^{-d(o,\zeta\wedge\xi)},$$

and, more generally, for $x, y \in \overline{\mathcal{T}}$, $x \neq y$, $\epsilon(x, y) \approx 2^{-d(o,x\wedge y)}$.

It is useful to have a more concrete model for the boundary elements. Let $\mathrm{Ray}(\mathcal{T})$ be the set of *rays* in $\mathcal{T}$; that is, geodesics of infinite length which have the root as an endpoint.

PROPOSITION 5.2. *For $\zeta = \{x_n\} \in \mathrm{Ray}(\mathcal{T})$, set $A(\zeta) = [\{x_n\}] \in \overline{\mathcal{T}}$. Then, A is a bijection of* $\mathrm{Ray}(\mathcal{T})$ *onto the metric boundary $\partial\mathcal{T}$.*

PROOF. The map is well defined. It is one to one: if $\zeta = \{x_n\}$ and $\xi = \{y_n\}$ are different geodesics with $\zeta \wedge \xi = \mathrm{Max}(\zeta \cap \xi)$, then $\epsilon(x_n, y_n) \gtrsim 2^{-d(o,\zeta\wedge\xi)}$ for $n \geq d(o, \zeta \wedge \xi)$, hence they are not equivalent. We have to show that the map is onto. Cauchy sequences tending to points of $\mathcal{T}$ must be eventually constant. Let $\sigma = \{x_n\}$ be a Cauchy sequence in $(\mathcal{T}, \epsilon)$ which is not eventually constant and, for $n \geq 0$, let $t_n = \mathrm{Max}(\cap_{m\geq n}[o, x_m])$ be the point which is farthest from the root, and common to all geodesics joining the root and the elements in the sequence with index $m \geq n$. The sequence $\{x_n\}$ is Cauchy, hence

$$\begin{aligned}
0 &= \lim_{n\to\infty} \sup_{l>0} \epsilon(x_n, x_{n+l}) \\
&\geq \lim_{n\to\infty} \epsilon(x_n, t_n) \\
&\qquad \text{because there is } l > 0 \text{ such that } x_n \wedge x_{n+l} = t_n \\
&\geq \lim_{n\to\infty} \frac{1}{2} 2^{-d(t_n,o)} \\
&\qquad \text{which is the weight of an edge having } t_n \text{ as the endpoint closest to } o.
\end{aligned}$$

Consider now the sequence $\tau = \{t_n\}$. It is not eventually constant. If it were $t_n = \bar{t}$ for n large enough, then we could find points x_m and x_l with m, l arbitrarily large and such that $x_m \wedge x_l = \bar{t}$. But this implies that either (i) there are points $x_l \neq x_m$ with l, m arbitrarily large and such that $\epsilon(x_l, x_m) \approx 2^{-d(o,\bar{t})}$ (but this contradicts the fact that the sequence is Cauchy); or (ii) $x_l = x_m = \bar{t}$, but this implies that, being Cauchy, the sequence is eventually equal to $\bar{t}$.

We then have the properties: (i) τ increases, $t_n \leq t_{n+1}$; (ii) it is Cauchy, $\epsilon(t_n, t_{n+l}) \approx 2^{-d(t_n,o)}$ if l is large enough (because τ is not eventually constant); (iii) it is equivalent to σ: $\epsilon(x_n, t_n) \lesssim \epsilon(t_n, o) \to 0$ as $n \to \infty$. □

EXERCISE 5.3. *Show that ϵ is an ultrametric on the boundary $\partial\mathcal{T}$:*

$$\epsilon(\zeta, \xi) \leq \max(\epsilon(\zeta, \eta), \epsilon(\eta, \xi)).$$

EXERCISE 5.4. *Show that $(\partial\mathcal{T}, \epsilon)$ is a complete, compact, totally disconnected metric space. Also show that, if each vertex x (except possibly the root) is the endpoint of at least three geodesics, then, in addition, $(\partial\mathcal{T}, \epsilon)$ is perfect: each point in $\partial\mathcal{T}$ is of accumulation point of $\partial\mathcal{T}$.*

EXERCISE 5.5. Root at infinity. *Let $\zeta_0 = \{x_n\}_{n=-\infty}^{+\infty}$ be a distinguished, doubly infinite geodesic in $\mathcal{T}$, which we think of as oriented by the map $n \mapsto x_n$. Define $x \wedge \zeta_o =: x_n$ as the point on ζ_0 which is closest to x with respect to the metric d. If $\alpha = \{p(x), x\}$ is the edge having as endpoint x in the geodesic $[x, x \wedge \zeta_0]$, assign to* *it the weight* $\tilde{\epsilon}(\alpha) = 2^{-d(x,x\wedge\zeta_0)}$. *Use the weight* $\tilde{\epsilon}$ *to define a metric on* $\mathcal{T} = V(\mathcal{T})$. *Show that the completion of the metric space* $(\mathcal{T}, \tilde{\epsilon})$ *is a metric space* $\left(\tilde{T}, \tilde{\epsilon}\right)$ *which is not compact. Show that* $\left(\overline{\mathcal{T}}, \epsilon\right)$ *is in fact the one-point compactification of* $\left(\tilde{T}, \tilde{\epsilon}\right)$, *and that* ζ_0, *thought of as a point in* $\partial\mathcal{T}$, *is the point one has to add to* $\tilde{T}$ *in order to compactify it.* (Remark. $\mathcal{T}$ *is the "upper halfplane" version of the Euclidean structure of the tree. The ideal boundary point "*$+i\infty$*" is the limit of all geodesics paths on which "*y*" increases steadily.*)

5.2. The Disk and the Tree

One of the themes of this volume is the value of passing back and forth between analysis on a tree and analysis on the disk, and later between the tree and the complex ball. For instance we did that in the proof of Theorem 5.13 and will do it again when proving Lemma 14.1. Here we set up a detailed version of the relation between the two viewpoints. There are many possible variations in the details, all serving the same overall themes. Also, the discussion below with the disk extends, with very little change, to the ball.

5.2.1. The Tidy Case. We are still under the aegis of the "convenient fiction" and will discuss the dyadic tree, the notationally minimal model case. However we often have more general possibilities in mind for the tree and the decomposition of the disk.

We can realize the vertices of the dyadic tree $\mathcal{T}$ as the regions of a dyadic decomposition of the unit disk described in Section 1.7.1, or, rather, the indices they depend on. That is, for integer $n \geq 0$ and $1 \leq j \leq 2^n$, let

$$Q_{n,j} = \left\{ z = re^{2\pi i t} \in \mathbb{D} : 2^{-n-1} \leq 1 - r \leq 2^{-n} \ \frac{j-1}{2^n} \leq t \leq \frac{j}{2^n} \right\}.$$

We write $(n, j) = z_{n,j} = \left(1 - 2^{-n-1/2}\right)\exp(2\pi i 2^{-n}(j - 1/2)))$ for the center of $Q_{n,j}$ and $I_{n,j} = \{z/|z| : z \in Q_{n,j}\}$ for its projection on the boundary $\mathbb{T}$ of the disk. Note that $\mathbb{D} = \bigcup_n \bigcup_{j=1}^{2^n} Q_{n,j}$. We then view the vertices of $\mathcal{T}$ as a set of points in the disk; $V(\mathcal{T}) = \{z_{n,j}\} = \{(n, j)\}$ and the root is $o = (0, 1)$ corresponding to the origin. With the exception of the root, each vertex (n, j) belongs to three edges:

$$\{(n-1, [j/2]), (n, j)\}, \{(n, j), (n+1, 2j)\}, \{(n, j), (n+1, 2j+1)\}.$$

In each pair the first entry is the predecessor of the second. The root has no predecessor and hence is only part of two edges.

After conflating tree vertices and points in the disk, some quantities associated to $\mathcal{T}$ are analogs of familiar quantities associated to $\mathbb{D}$. The quantity $d(o, x \wedge y)$ is the tree distance between the root and the point in the tree geodesic between x and y which is closest to the root. It is a model for the hyperbolic distance between the origin and the point closest to the origin on the hyperbolic geodesic connecting x and y, a distance which is on the order of $\log|1 - \overline{x}y|^{-1}$. The quantity $2^{-d(o,x\wedge y)}$, which models $|1 - \overline{x}y|$, is roughly the Euclidean distance between that point and the boundary and hence also, again roughly, the Euclidean distance between x and y.

EXERCISE 5.6. *Check the estimates in the previous paragraph. Make appropriate sketches.*

The dyadic tree is a sufficiently good model of the disk so that some results involving capacity or involving Dirichlet space Carleson measures can be moved from one venue to the other. We will see instances of that, but will not develop the theme systematically; for that we refer to [**49**]. However, we do want to discuss a graph $\mathcal{G}$ which is a useful construct in that development.

The vertices of $\mathcal{G}$ are the boxes of the dyadic decomposition of the disk which we just described. There is an edge connecting two vertices whenever those boxes share a side. We define a distance on $\mathcal{G}$ by starting with the observation that, in the disk, moving from one dyadic box to a neighboring one corresponds to a displacement by approximately one unit in the hyperbolic metric. Thus we consider the corresponding movement in $\mathcal{G}$ to be a step of size one. This produces an analog of the hyperbolic metric. To obtain an analog of the Euclidean metric we start by observing that the Euclidean distance between two adjacent boxes is approximately the distance of either of the boxes to the boundary.

From $\mathcal{G}$ we can now make the dyadic tree $\mathcal{T}$ by keeping the same vertices but removing the edges connecting pairs of vertices which correspond to boxes of the same size. We can set up distance functions on $\mathcal{T}$ using the same weight functions as on $\mathcal{G}$ and they will produce (approximately) the hyperbolic- and Euclidean-like metrics, which we introduced with the dyadic tree. The distances between vertices in $\mathcal{T}$ will in general be greater than distances in $\mathcal{G}$, as there are fewer paths in the dyadic tree. As a consequence, the metric completion of $\mathcal{G}$ with respect to the Euclidean metric will be larger (more Cauchy sequences).

The graph $\mathcal{G}$ models many features of the disk and it apparently models the disk more closely than $\mathcal{T}$ does. The graph functions as an effective intermediate step in the job of transferring results between $\mathcal{T}$ and $\mathbb{D}$; some basic results can be transferred first from $\mathcal{T}$ to $\mathcal{G}$ and then from $\mathcal{G}$ to $\mathbb{D}$. This is done for instance in [**49**].

5.2.2. The More General Case. The construction just described is an elegant way to associate the vertex set of a dyadic tree with a set of points in the disk, but the method is relatively rigid. However, the core idea of breaking the disk into parts that are, very approximately, hyperbolic disks of the same size is rather flexible. Here we describe a way to start with certain types of $A = \{\alpha_i\} \subset \mathbb{D}$ and construct a dyadic type tree $\mathcal{T}$ in such a way that when the vertex set $V(\mathcal{T})$ is identified with A then the tree geometry and disk geometry fit together well.

We regard $\mathbb{D}$ as a metric space with the hyperbolic metric β. (The pseudohyperbolic metric ρ is a monotone function of β so we could use either.) We require that A is neither too dense nor too sparse; there are $C_1, C_2 > 0$ such that

$$\begin{aligned} &\text{the balls } \{B_\beta(\alpha_i, C_1) : \alpha_i \in A\} \text{ are pairwise disjoint,} \\ &\text{the union of the balls } \{B_\beta(\alpha_i, C_2) : \alpha_i \in A\} \text{ covers } \mathbb{D}. \end{aligned} \tag{5.1}$$

Later we will use the fact that if $S = \{z_i\} \subset \mathbb{D}$ is such that there is a lower bound on the hyperbolic distance between points of S then S is contained in such an A. A similar comment will apply for the sets in $\mathbb{B}_n$ which we consider in later chapters.

For convenience we suppose $0 \in A$ and we will identify it with the root $o \in V(\mathcal{T})$. We regard the other elements of A as the remaining elements of $V(\mathcal{T})$. To complete the description of the tree $\mathcal{T}$ we need to define the edge structure.

For each vertex $\alpha \neq o$ we introduce an edge connecting α to its *parent* (predecessor) $\alpha^- \in A$. The point α^- is selected near the radial segment $o\alpha$ in $\mathbb{D}$ and a bit closer to the origin than α is. More precisely, we require that for some $C_3, C_4 > 0$ and all $\alpha \in A$ other than the root, we have

$$C_3 < \beta(o,\alpha) - \beta(o,\alpha^-) < C_4. \tag{5.2}$$

There is often a convenient choice for α^-. Absent that, here is an awkward but explicit scheme that works in general. If α is close to the origin, $\beta(o,\alpha) < 4C_2$, then set $\alpha^- = o$. Otherwise let α_- be an auxiliary point on the radial segment $o\alpha$ in the disk with $\beta(\alpha,\alpha_-) = 2C_2$ and let Δ be the hyperbolic disk centered at α_- with hyperbolic radius C_2. From (5.1) it follows that $\Delta \cap A$ is not empty. We select for α^- the element of $\Delta \cap A$ with the greatest modulus and, if there is a tie, with the greatest argument.

It is not hard to verify the following:

(1) Hyperbolic balls about the $\{\alpha_i\}$ of radius $B > C_2$ give a covering of the disk of bounded depth: There is a $C = C(B, \{C_i\})$ so that

$$\bigcup\nolimits_{\alpha\in A} B_\rho(\alpha, B) = \mathbb{D}, \text{ and}$$
$$\sup_z \operatorname{card}\{\alpha \in A : z \in B_\rho(\alpha, B)\} < C.$$

(2) The tree $\mathcal{T}$ is of dyadic type; it has bounded branching. That is, there is a $C = C(\{C_i\})$ so that

$$\sup_\beta \operatorname{card}\left\{\alpha \in A : \beta = \alpha^-\right\} < C$$

(3) For $\alpha \in A$, the number of edges in the geodesic path in the tree connecting the root to α is comparable to the hyperbolic distance, β, of α from the origin

$$d(o,\alpha) \approx \beta(0,\alpha). \tag{5.3}$$

The constants of comparability can be selected to only depend on the $\{C_i\}$. (This is where (5.2) is used.)

(4) Suppose $o \in Z$, $Z \subset \mathbb{D}$ and Z is separated in the hyperbolic metric β; that is $\exists c > 0, \forall z_1, z_2 \in Z, z_1 \neq z_2$ we have $\beta(z_1, z_2) > c$. Then there is a tree $\mathcal{T} \subset \mathbb{D}$ of the type just described with $Z \subset A = V(\mathcal{T})$. Furthermore A can be chosen with structural constants $\{C_i\}$ which only depend on the separation constant c.

EXERCISE 5.7. *Why do these properties hold?*

Given A, which we can regard as the vertex set of an abstract tree $\mathcal{T}$, we can use holomorphic functions on the disk to construct functions on A. The value $f(\alpha)$ is an informal proxy for the value of a Bergman space function $f(z) \in A^2$ in a small ρ–ball about α, and $(1-|a|^2)^2$ is comparable to the Euclidean area of that ρ–ball. Hence the map taking $f(z) \in A^2$ to the sequence of weighted restrictions $\left\{(1-|a|^2)f(\alpha)\right\}_{\alpha\in A}$ is a bounded map of A^2 into $\ell^2(A)$. The boundedness of this restriction map is a starting point for the relation between spaces of holomorphic functions on the disk and the spaces on functions on trees. In this context such A, vertex sets of an abstract tree realized as point sets inside a complex ball, are sometimes informally referred to as *Bergman trees* with the space $\ell^2(\mathcal{T})$ called the *tree Bergman space.*

EXERCISE 5.8. *Prove that if f is in the Bergman space then*

$$\left\{(1-|a|^2)f(\alpha)\right\}_{\alpha\in A}$$

is in ℓ^2. (There are two basic steps. First $|f(\alpha)|^2$ can be estimated by the mean value of $|f(z)|^2$ over a small ball about α. Second, this can be done so that the union of the small balls is a cover of the disk of bounded depth.)

The dyadic Dirichlet space $\mathcal{D}(\mathcal{T})$ will be defined precisely in a moment. Informally, it is the following. Functions in the Dirichlet space are primitives of functions in the Bergman space: if $G \in \mathcal{D}$ then $G = \int^z g dw$ for some $g \in A^2$. The discrete model for the Dirichlet space, $\mathcal{D}(\mathcal{T})$, is constructed analogously, but using the discrete model for integration. That is, if $f \in \ell^2(\mathcal{T})$, the tree Bergman space, then, with I as a summation operator which serves as a discrete analog of integration, $F = If$ is in $\mathcal{D}(\mathcal{T})$, the tree Dirichlet space. Furthermore, $\mathcal{D}(\mathcal{T})$ consists of all functions obtained this way.

5.3. The Ideal Boundary and Integration

In Proposition 5.2 we saw that $\operatorname{Ray}(\mathcal{T})$ was in bijection with $\partial\mathcal{T}$. There is also a bijection between elements of $\operatorname{Ray}(\mathcal{T})$, half infinite geodesics $\zeta = \{x_n\}_{n=0}^{\infty}$ starting at the root, and sequences of nested closed dyadic arcs on $\mathbb{T}$. This lets us define a map $\Lambda : \partial\mathcal{T} \to \mathbb{T}$, $\zeta = \{x_n\} \mapsto \bigcap_{n\geq 0} I_{x_n}$. The map is surjective, but it fails to be injective precisely at the set of dyadic rationals.

The map Λ can be extended to a map $\Lambda : \overline{\mathcal{T}} \to \overline{\mathbb{D}}$ by assigning to $x \in \mathcal{T}$ the point $\Lambda(x) = z_x$, the center of the dyadic box Q_x.

EXERCISE 5.9. *Show that the map $\Lambda : \left(\overline{\mathcal{T}}, \epsilon\right) \to \left(\overline{\mathbb{D}}, d_{\mathrm{Euc}}\right)$ is Lipschitz. Here d_{Euc} is the Euclidean distance on $\overline{\mathbb{D}}$.*

We would like to use Λ to move functions and measures back and forth between $\overline{\mathcal{T}}$ and $\overline{\mathbb{D}}$; but there are some difficulties. First note that as a map from $\mathcal{T}$ into $\mathbb{D}$ the map has a very small range, so it is not clear how much information can be gained knowing the values of a function at the points $\Lambda(\mathcal{T})$. However, by Theorem 2.13 the functions on $\mathbb{D}$ that we wish to study, namely the functions in $\mathcal{D}$, are essentially constant on each box of the dyadic decomposition, and $\Lambda(\mathcal{T})$ has one point in each dyadic box. Treating the value on each box as if it were constant, and assigning that constant value to a point of $\Lambda(\mathcal{T})$ defines a function on $\Lambda(\mathcal{T})$, and that assignment is the basis for going forward.

(Here the analogy is clearer than the details. By Theorem 2.13 the variation of a function in $\mathcal{D}$ on a dyadic box is controlled by the pseudohyperbolic diameter of the box. All the dyadic boxes have, roughly, the same pseudohyperbolic diameter. To ensure that the variation is small, i.e., the function is "essentially constant", requires using smaller boxes. Although that takes us out of the world of dyadic trees, it stays within the world of dyadic type trees.)

Also, the map from the boundary of $\overline{\mathcal{T}}$ to the circle is not a differentiable map between structures. We deal with that issue, which is a greater concern in the general situation considered in [**49**], by being more explicit and careful than usual when moving measures and integrals back and forth using changes of variables involving Λ.

Because Λ is a continuous map between metric spaces, it can be used to canonically push measures forward from $\partial\mathcal{T}$ to $\mathbb{T}$: $\Lambda_*\mu(E) = \mu\left(\Lambda^{-1}(E)\right)$, whenever E is Borel measurable in $\mathbb{T}$. A special feature of Λ is that we can define a (noncanonical) pull back in $\partial\mathcal{T}$ of a Borel measure on $\mathbb{T}$. This will in turn prove useful when we discuss Carleson measures at the boundary.

PROPOSITION 5.10. *Let ν be a Borel measure on $\mathbb{T}$ and F a closed subset in $\partial\mathcal{T}$. Then, the function $t \mapsto \sharp\left(\Lambda^{-1}(t) \cap F\right)/\sharp\Lambda^{-1}(t)$ is Borel measurable on $\mathbb{T}$.*

Define a Borel measure $\Lambda^*\nu$ on $\partial\mathcal{T}$ by:

$$\Lambda^*\nu(F) := \int_{\mathbb{T}} \frac{\sharp\left(\Lambda^{-1}(t) \cap F\right)}{\sharp\Lambda^{-1}(t)} d\nu(t).$$

Then, $\Lambda^*\nu(F) \approx \nu(\Lambda(F))$ and $\Lambda_*\Lambda^*\nu = \nu$.

If moreover ν is atomless, then $\Lambda^*\nu(F) = \nu(\Lambda(F))$.

PROOF. The proof is easy, because $\sharp\Lambda^{-1}(t) = 1$, except for t in the countable set of dyadic rationals, in which case $\sharp\Lambda^{-1}(t) = 2$. If we also know that ν does not have atoms, then the measure of the set of dyadic rationals is zero. □

COROLLARY 5.11. *The action of Λ_* and Λ^* on integrals of measurable, positive functions is as follows:*

$$\int_{\mathbb{T}} gd(\Lambda_*\mu) = \int_{\partial\mathcal{T}} (g \circ \Lambda)\, d\mu,$$

$$\int_{\partial\mathcal{T}} \varphi d(\Lambda^*\nu) = \int_{\mathbb{T}} \frac{\sum_{\zeta\in\Lambda^{-1}(t)} \varphi(\zeta)}{\sharp\Lambda^{-1}(t)} d\nu(t) = \int_{\mathbb{T}} \varphi\left(\Lambda^{-1}(t)\right) d\nu(t),$$

where the integrand in the final integral is finite ν – a.e. if ν is atomless.

5.4. The Dyadic Dirichlet Space

Let $\mathcal{T}$ be a dyadic type tree with root o. To each vertex $\zeta \in \overline{\mathcal{T}}$ we associate its predecessor set $\Pr(\zeta) = [o, \zeta) \subset E(\mathcal{T})$. Dually, to each $\alpha \in E(\mathcal{T})$ we associate its successor set $S(\alpha) \subseteq \overline{\mathcal{T}}$, $S(\alpha) = \left\{\zeta \in \overline{\mathcal{T}} : \alpha \in \Pr(\zeta)\right\}$ and we set $S(o^*) = \overline{\mathcal{T}}$. Alternatively $\Pr(\zeta) = \{\alpha \in E(\mathcal{T}) : \alpha \leq \zeta\}$; $S(\alpha) = \left\{\zeta \in \overline{\mathcal{T}} : \alpha \leq \zeta\right\}$. Note that when $\mathcal{T}$ is realized as regions inside the disk then $S(\alpha)$ is, roughly, the tent $\mathcal{T}(\Lambda(\alpha))$ described in Section 1.7.2. Recall that $d^+(\alpha) = d(o^{-1}, \alpha) = d(o, \alpha) + 1$, and thus $d^+(\alpha)$ counts the number of edges in the geodesic $\left[o^{-1}, \alpha\right]$ or, equally, the number of vertices in the geodesic $[o, \alpha]$.

Our fundamental example is given by the dyadic tree $\mathcal{T}$ with its vertex set realized as the set of dyadic boxes $\{Q_{j,k}\}$ in the disk, or, when convenient, realized as the set of centers $\{z_{j,k}\}$.

When doing analysis on trees it is often useful, and at times crucial, to emphasize the distinction between a function defined on the set of vertices of a tree and a function defined on the set of edges. We are now abandoning that distinction. We will identify each edge with its further vertex and will conflate functions defined on the edges and functions defined on the vertices. That will let us define the Hardy operator, our discrete model for indefinite integration, and also its inverse, as maps from functions defined on the vertex set to functions defined on the *same* set. This will help emphasize the analogy with indefinite integration and differentiation which have a similar property.

Let φ be a complex-valued function defined on $V(\mathcal{T})$ and suppose $\zeta \in \overline{\mathcal{T}}$. We define the *Hardy operator* I to be the map from functions on $\mathcal{T}$ to functions on $\overline{\mathcal{T}}$given by

$$I\varphi(\zeta) := \sum_{\alpha \in \Pr(\zeta)} \varphi(\alpha). \tag{5.4}$$

In particular, given our conventions, $I\varphi(o) = \varphi(o)$. (The analog of this operator on the tree of nonnegative integers was studied by Hardy.)

The operator I models indefinite integration, and its inverse, Δ, defined as the map from $F = \{F(\alpha)\}$ to

$$\Delta F = \left\{F(\alpha) - F(\alpha^{-1})\right\}, \tag{5.5}$$

is similar to differentiation. Note that our convention that $F(o^{-1}) = 0$ ensures that $\Delta F(o) = F(o)$. By convention we set $\Delta F(o^{-1}) = 0$.

When we view the vertex set $\{\alpha\}$ of $\mathcal{T}$ as a set $\{w_\alpha\}$ inside $\mathbb{D}$ then we can use any $f \in \mathcal{D}$ to construct a function φ on $\mathcal{T}$ by setting $\varphi(\alpha) = (1 - |w_\alpha|^2)|f'(w_\alpha)|$. Thus $\varphi(\alpha)$ is the modulus of the invariant gradient (1.11) of f; the distortion of the holomorphic map $f : (\mathbb{D}, d_{\text{hyp}}) \to (\mathbb{C}, d_{\text{Euc}})$. If the details work well, we can estimate the Dirichlet seminorm as follows;

$$\mathfrak{D}(f) = \frac{1}{\pi}\int_{\mathbb{D}} \left|(1 - |z|^2)f'(z)\right|^2 d\lambda \approx \sum_{\alpha \in \mathcal{T}} \left|(1 - |w_\alpha|^2)f'(w_\alpha)\right|^2 = \|\varphi\|^2_{\ell^2(\mathcal{T})}.$$

The middle approximation is, at its heart, a Riemann sum approximation to the integral, that is eveloped in detail in the proof of Theorem 5.13. It is central, both analytically and conceptually, to the close relation between the Dirichlet space and the dyadic Dirichlet space. It is discussed in more detail, but still informally, in Section 10.2.3 and it is discussed in full detail in the references listed there.

With this background we now define the *dyadic Dirichlet space*, or the *tree Dirichlet space*, $\mathcal{D}(\mathcal{T})$, to be the Hilbert space of functions F on $\mathcal{T}$ for which $\Delta F \in \ell^2(\mathcal{T})$. It is normed by

$$\|F\|^2_{\mathcal{D}(\mathcal{T})} = \|\Delta F\|^2_{\ell^2(\mathcal{T})}.$$

As a space of functions on $V(\mathcal{T})$, $\mathcal{D}(\mathcal{T})$ is a RKHS with inner product

$$\begin{aligned}\langle F, G\rangle_{\mathcal{D}(\mathcal{T})} &= \langle \Delta F, \Delta G\rangle_{\ell^2(\mathcal{T})} \\ &= \sum_{\beta \in \mathcal{T}} \Delta F(\beta)\overline{\Delta G(\beta)}.\end{aligned}$$

The value of the reproducing kernel, $k_\alpha(\beta)$, is the level of $\alpha \wedge \beta$, the number of vertices in the segment $[o, \alpha \wedge \beta]$;

$$k_\alpha(\beta) = d^+(\alpha \wedge \beta) = 1 + d(\alpha \wedge \beta).$$

This can be seen from the following computation:

$$\begin{aligned}\langle F, k_\xi\rangle_{\mathcal{D}(\mathcal{T})} &= \langle \Delta F, \Delta k_\xi\rangle_{\ell^2(\mathcal{T})} = \left\langle \Delta F, \chi_{(\eta \leq \xi)}\right\rangle_{\ell^2(\mathcal{T})} \\ &= \sum_{\eta \leq \xi} \Delta F(\eta) = F(\xi).\end{aligned}$$

Thus for any $f \in \mathcal{D}(\mathcal{T})$, $\alpha \in \mathcal{T}$ we have $\langle f, k_\alpha\rangle = f(\alpha)$ and

$$\|k_\alpha\|^2 = k_\alpha(\alpha) = d^+(\alpha). \tag{5.6}$$

For any $\alpha, \beta \in \mathcal{T}$ we have

$$\delta^2_{\mathcal{D}(\mathcal{T})}(\alpha, \beta) = 1 - \frac{d^+(\alpha \wedge \beta)^2}{d^+(\alpha) d^+(\beta)}. \tag{5.7}$$

The following easy analog of Proposition 2.13 is an oscillation estimate for functions in $\mathcal{D}(\mathcal{T})$.

PROPOSITION 5.12. *If $F \in \mathcal{D}(\mathcal{T})$, $\|F\|_{\mathcal{D}(\mathcal{T})} \leq 1$, $\alpha, \beta \in \mathcal{T}$ then $|F(\beta) - F(\alpha)| \leq d(\alpha, \beta)^{1/2}$.*

PROOF. Let $n = d(\alpha, \beta)$ and let $\alpha = \gamma_0, \gamma_1, \ldots, \gamma_n = \beta$ be the sequence of vertices on the geodesic from α to β. Then $F(\beta) - F(\alpha) = \sum_{i=1}^n \pm \Delta F(\gamma_i)$ (where the signs depend on the location of γ_i in the path). Applying the Cauchy-Schwarz inequality and the definitions completes the proof. □

5.5. Carleson Measures for $\mathcal{D}(\mathcal{T})$ and $\mathcal{D}$

We want to describe Carleson measures for I. That is, we look for positive, Borel measures on $\overline{\mathcal{T}}$ such that the *tree inequality*

$$\int_{\overline{\mathcal{T}}} |I\varphi(\zeta)|^2 \, d\mu(\zeta) \leq ||\mu||^2_{\mathrm{CM}} \sum_{\alpha \in \mathcal{T}} |\varphi(\alpha)|^2, \text{ equivalently} \tag{TI}$$

$$\int_{\overline{\mathcal{T}}} |\Phi(\zeta)|^2 \, d\mu(\zeta) \leq ||\mu||^2_{\mathrm{CM}} \sum_{\alpha \in \mathcal{T}} |\Delta\Phi(\alpha)|^2 \tag{5.8}$$

holds with finite $||\mu||^2_{\mathrm{CM}}$; that is, μ for which

$$||\mu||^2_{\mathrm{CM}} := \sup_{\ell^2(\mathcal{T}) \ni \varphi \neq 0} \frac{\int_{\overline{\mathcal{T}}} |I\varphi(\zeta)|^2 \, d\mu(\zeta)}{\sum_{\alpha \in \mathcal{T}} |\varphi(\alpha)|^2} = ||I||^2_{\mathcal{B}(\ell^2(\mathcal{T}), L^2(\mu))} < \infty.$$

By analogy with earlier usage these measures can also be called the Carleson measures for $\mathcal{D}(\mathcal{T})$. In the language often used in potential theory we are looking for nonnegative measures μ which satisfy a *trace inequality*.

The reader is perhaps worried that, although we want the inequality for all φ in $\ell^2(\mathcal{T})$, the series in $I\varphi(\zeta)$ might not be defined for $\zeta \in \partial\mathcal{T}$. However here, and often below, it suffices for our needs to have the inequality for φ which are real and nonnegative. In that case, accepting $+\infty$ as a value, $I\varphi(\zeta)$ is defined everywhere.

We now establish a quantitative relation between the Hardy operator acting on the tree Bergman space ($= \ell^2(\mathcal{T})$) and the integration operator acting on the Bergman space A^2. This will let us show that a characterization of the Carleson measures for $\mathcal{D}(\mathcal{T})$ leads to a characterization of Carleson measures for $\mathcal{D}$.

We will write I for the integration type operator defined in (5.4) and denote the identity operator acting on a Banach space by Id. For Banach spaces A, B we denote by $\mathcal{B}(A, B)$ the space of bounded linear operators from A to B.

For $f \in \mathrm{Hol}(\mathbb{D})$ we define $V(f)(re^{it})$ to be the radial variation of f up to $re^{i\theta} \in \mathbb{D}$; the Euclidean length of the image under f of the radial segment from 0 to $re^{i\theta}$; $V(f)(re^{it}) = \int_0^r |f'(\rho e^{it})| \, d\rho$. We will say that a non-negative Borel measure ν on $\overline{\mathbb{D}}$ is *Carleson for the Variation* if there is a constant K such that:

$$\int_{\overline{\mathbb{D}}} V(f)^2 d\nu \leq K^2 \|f\|^2_{\mathcal{D}}. \tag{5.9}$$

For $f \in \mathcal{D}$ let X_f denote the set of $e^{i\theta} \in \mathbb{T}$ for which f fails to have a radial limit at $e^{i\theta}$. We will say that a positive Borel measure ν on $\overline{\mathbb{D}}$ is $\mathcal{D}$-Carleson if both

$$\nu(X_f) = 0 \text{ for all } f \in \mathcal{D}, \tag{5.10}$$

and the following embedding holds:

$$\int_{\overline{\mathbb{D}}} |f|^2 \, d\nu \leq CM(\nu)^2 \|f\|_{\mathcal{D}}^2 \,. \tag{5.11}$$

Given $f \in \mathcal{D}$ let Y_f be the set of $e^{i\theta} \in \mathbb{T}$ for which the image under f of the segment $\left[0, e^{i\theta}\right]$ has infinite length. It is easy to check that if $e^{i\theta} \notin Y_f$ then f has a radial limit at $e^{i\theta}$, i.e. $X_f \subset Y_f$. In particular then, if ν on $\overline{\mathbb{D}}$ is *Carleson for the Variation*, each of the sets X_f are null sets for the measure ν when $f \in \mathcal{D}$, and so the left hand side of (5.11) is well-defined as a nonnegative extended real number.

THEOREM 5.13. *Let ν be a positive Borel measure on the closed disk $\overline{\mathbb{D}}$, and let μ be a non-negative Borel measure on the closed tree $\overline{\mathcal{T}}$.*

1. *The following conditions* (a), (b) *and* (c) *on ν are equivalent:*
   - (a) *$\Lambda^*\nu$ is $\mathcal{D}(\mathcal{T})$-Carleson, namely* (TI) *holds with measure $\Lambda^*\nu$;*
   - (b) *ν is Carleson for the variation, namely* (5.9) *hold with measure ν.*
   - (c) *ν is $\mathcal{D}$-Carleson on $\overline{\mathbb{D}}$, namely* (5.10) *and* (5.11) *hold with measure ν.*
   - (d) *Moreover, if any of the above three conditions hold, then*

$$||I||_{\mathcal{B}(\ell^2(\mathcal{T}),L^2(\Lambda^*\nu))} \approx ||V||_{\mathcal{B}(\mathcal{D},L^2(\nu))} \approx \|\mathrm{Id}\|_{\mathcal{B}(\mathcal{D},L^2(\nu))} \,.$$

2. *The following conditions* (a), (b) *and* (c) *on μ are equivalent:*
   - (a) *μ is $\mathcal{D}(\mathcal{T})$-Carleson, namely* (TI) *holds with measure μ;*
   - (b) *$\Lambda_*\mu$ is Carleson for the variation, namely* (5.9) *holds with measure $\Lambda_*\mu$;*
   - (c) *$\Lambda_*\mu$ is $\mathcal{D}$-Carleson on $\overline{\mathbb{D}}$, namely* (5.10) *and* (5.11) *hold with measure $\Lambda_*\mu$.*
   - (d) *Moreover, if any of the above three conditions hold, then*

$$||I||_{\mathcal{B}(\ell^2(\mathcal{T}),L^2(\mu))} \approx ||V||_{\mathcal{B}(\mathcal{D},L^2(\Lambda_*\mu))} \approx \|\mathrm{Id}\|_{\mathcal{B}(\mathcal{D},L^2(\Lambda_*\mu))} \,.$$

The proof of the theorem is simple conceptually. In one direction, to prove that "discrete Carleson" implies "holomorphic Carleson", we write $f(z) - f(0)$ as the sum of differences of values of f at points belonging to adjacent dyadic boxes (rather, we use that idea to deal with the larger $V(f)$). That is, we estimate f by some $I\varphi$, and it turns out that we can estimate $\|\varphi\|_{\ell^2}$ by $\|f\|_{\mathcal{D}}$.

Showing that "holomorphic Carleson" implies "discrete Carleson" is slightly more delicate. We need to produce a holomorphic function with behavior similar to that of a given discrete function. We do this by following the path suggested in the previous chapter; we consider the dual inequality, the one for I_μ^*. Not only does that operator map "unstructured" sequences in $L^2(\mu)$ to holomorphic functions, but when the formulas are written explicitly it is seen that the holomorphy is a consequence of the conjugate holomorphy of the Dirichlet space kernel, $k_z(w)$, as a function of z. We then make estimates using the two facts which were used heuristically in the discussion in Section 4.2.3; (i) the real part of the kernel's derivative is nonnegative; (ii) on $S(z)$, the tent under z, the real part of $\overline{\partial}_z k(z, w)$

is comparable to $(1-|z|^2)^{-1}$. Testing the dual inequality on functions which are nonnegative and are constant on dyadic boxes we obtain what we need, the dual inequality of the trace inequality for I.

PROOF OF THEOREM 5.13. First note that because the Dirichlet space does not have bounded point evaluation at the boundary, the Carleson measures for the Dirichlet space have no atoms on the boundary. Similarly, a Carleson measure for I does not have atoms on the boundary because a point mass at the boundary point ζ would lead to the following false inequality: for any positive $\{\varphi(\alpha)\}$, $\left(\sum_{\alpha\in P(\zeta)}\varphi(\alpha)\right)^2 \lesssim \sum_{\alpha\in P(\zeta)}\varphi(\alpha)^2$. Having established these facts we have access to the full strength of Corollary 5.11.

Now, given a Borel measure ν on $\mathbb{D}$, we extend the definition of $\Lambda^*\nu$ to vertices of $\mathcal{T}$. We now modify the definition of the $Q_{n,jt}$ so that they do not have overlapping boundaries; however we continue the earlier notation, e.g. for $(n,j)\in\mathcal{T}$, let $\tilde{Q}_{n,j} = \left\{z = re^{2\pi it}\in\mathbb{D} : 2^{-n-1} < 1-r \le 2^{-n}\ ,\ \frac{j-1}{2^n} < t \le \frac{j}{2^n}\right\}$, and let $\Lambda^*\nu(\{(n,j)\}) = \nu(\tilde{Q}_{n,j})$. This time, $\Lambda_*\Lambda^*\nu$ concentrates the mass of $\tilde{Q}_{n,j}$ into its center $z_{n,j}$, while $\Lambda^*\Lambda_*\mu = \mu$ for all Borel measures on $\mathcal{T}$.

Taking into account the elementary pointwise inequality for $f\in\mathcal{D}$ with $f(0) = 0$: $|f(z)|^2 \lesssim V(f)(z)^2$, it now suffices to establish the following two estimates:

(1) If $\Lambda^*\nu$ is a Carleson measure for I then ν is what could be called a Carleson measure for the variation operator V. That is for all f in $\mathcal{D}$

$$\int_{\mathbb{D}} V(f)^2\,d\nu \lesssim \|\Lambda^*\nu\|^2_{\mathrm{CM}(\mathcal{T})}\,\|f\|^2_{\mathcal{D}}. \tag{5.12}$$

(2) If $\Lambda_*\mu$ is a Carleson measure for $\mathcal{D}$ then $\mathrm{I}^*_\mu : L^2(\mu)\to\ell^2(\mathcal{T})$, the adjoint of the map $\mathrm{I} : \ell^2(\mathcal{T})\to L^2(\mu)$, is bounded and thus μ is a Carleson measure for I. That is, $\forall\varphi\in L^2(\mu)$

$$\left\|\mathrm{I}^*_\mu\varphi\right\|^2_{\ell^2} \lesssim \|\Lambda_*\mu\|^2_{\mathrm{CM}(\mathcal{D})}\int_{\mathcal{T}}\varphi^2\,d\mu. \tag{5.13}$$

For the first claim, we start by estimating $V(f)$ pointwise. For $z,w\in\mathbb{D}$ let $[z,w]$ be the straight line segment from z to w.

$$\begin{aligned} V(f)(\Lambda\xi) = V(f)(\zeta) &= \int_0^{|\zeta|}\left|f'\left(\rho\frac{\zeta}{|\zeta|}\right)\right|d\rho \\ &\le \sum_{\xi\in\Lambda^{-1}(\zeta)}\sum_{\alpha\in\Pr(\xi)}\int_{Q_\alpha\cap[0,\zeta]}|f'(w)|\,|dw| \\ &\quad\text{where } Q_\alpha \text{ is the dyadic region indexed by } \alpha \\ &\le \sum_{\xi\in\Lambda^{-1}(\zeta)}\sum_{\alpha\in\Pr(\xi)}|Q_\alpha|^{1/2}\left|f'\left(w^{\mathrm{Max}}_\alpha\right)\right| \\ &\quad\text{where } w^{\mathrm{Max}}_\alpha \text{ maximizes } |f'| \text{ in } Q_\alpha \text{ and } |Q_\alpha| = \int_{Q_\alpha} dA. \end{aligned}$$

Integrating,

$$\int_{\overline{\mathbb{D}}} V(f)^2 \, d\nu \le \int_{\overline{\mathbb{D}}} \left(\sum_{\xi \in \Lambda^{-1}(\zeta)} \sum_{\alpha \in \Pr(\xi)} |Q_\alpha|^{1/2} \left| f'\left(w_\alpha^{\mathrm{Max}}\right) \right| \right)^2 d\nu(\zeta)$$

$$\approx \int_{\overline{\mathbb{D}}} \frac{1}{\sharp\left(\Lambda^{-1}(\zeta)\right)} \sum_{\xi \in \Lambda^{-1}(\zeta)} \left(\sum_{\alpha \in \Pr(\xi)} |Q_\alpha|^{1/2} \left| f'\left(w_\alpha^{\mathrm{Max}}\right) \right| \right)^2 d\nu(\zeta)$$

$$= \int_{\overline{\mathcal{T}}} I\varphi(\xi)^2 \, d\Lambda^* \nu(\xi)$$

$$\text{where} \varphi(\alpha) = |Q_\alpha|^{1/2} \left| f'\left(w_\alpha^{\mathrm{Max}}\right) \right|$$

$$\le \|\Lambda^* \nu\|_{\mathrm{CM}(\mathcal{T})}^2 \sum_{\alpha \in \mathcal{T}} \left(|Q_\alpha|^{1/2} \left| f'\left(w_\alpha^{\mathrm{Max}}\right) \right| \right)^2$$

$$= \|\Lambda^* \nu\|_{\mathrm{CM}(\mathcal{T})}^2 \sum_{\alpha \in \mathcal{T}} \left(|Q_\alpha|^{1/2} \left| \frac{\int_{B_\alpha} f'(z) \, dA}{|B_\alpha|} \right| \right)^2 .$$

Here B_α is a disk with center at w_α^{Max}, with Euclidean radius comparable to $|Q_\alpha|^{1/2}$, and which meets no Q_β except possibly those sharing a boundary point with Q_α. We then use the area version of the mean value property for that disk. The geometry of the construction ensures that there is a finite upper bound on the number of Q_β which contain any given $z \in \mathbb{D}$. That bounded overlap allows us to dominate expressions of the type $\sum_\alpha \int_{Q_\alpha}$ by integrals over the entire disk.

In the other direction, elementary hyperbolic geometry estimates show it is possible to obtain a similar covering with pseudohyperbolic balls, all having the same radius ρ_0; and, furthermore there is a ρ_1, $\rho_0 < \rho_1 < 1$ so that the pseudohyperbolic balls with the same centers and radius ρ_1 will be a cover of $\mathbb{D}$ and will have the bounded overlap property just described. Hence, among other things, one can then bound integrals over the disk by sums of integrals over these larger balls.

EXERCISE 5.14. *Draw sketches to elucidate the basic geometric properties of the covering.*

We continue our estimates with

$$\le \|\Lambda^* \nu\|_{\mathrm{CM}(\mathcal{T})}^2 \sum_{\alpha \in \mathcal{T}} |Q_\alpha| \frac{\int_{B_\alpha} |f'(z)|^2 \, dA}{|B_\alpha|} \quad \text{by Jensen's inequality,}$$

$$\approx \|\Lambda^* \nu\|_{\mathrm{CM}(\mathcal{T})}^2 \sum_{\alpha \in \mathcal{T}} \int_{B_\alpha} |f'(z)|^2 \, dA$$

$$\le \|\Lambda^* \nu\|_{\mathrm{CM}(\mathcal{T})}^2 \int_{\mathbb{D}} \sum_{\alpha \in \mathcal{T}} \chi_{Q_\alpha}(z) |f'(z)|^2 \, dA$$

$$\approx \|\Lambda^* \nu\|_{\mathrm{CM}(\mathcal{T})}^2 \int_{\mathbb{D}} |f'(z)|^2 \, dA.$$

Hence we have established (5.12).

To establish (5.13) assume that $\nu = \Lambda_* \mu$ is a Carleson measure for $\mathcal{D}$, and assume that μ is supported on $\partial \mathcal{T}$, the other case being easier. If $k_z(\zeta) = 1 +$

$\log(1-\overline{z}\zeta)^{-1}$ is the reproducing kernel for $\mathcal{D}$ at $z\in\mathbb{D}$, then

$$\begin{aligned}\operatorname{Id}^* g(z) &= \langle \operatorname{Id}^* g, k_z\rangle_{\mathcal{D}} = \langle g, k_z\rangle_{L^2(\nu)}\\ &= \int_{\overline{\mathbb{D}}} g(\zeta)\, k_\zeta(z)\, d\nu(\zeta).\end{aligned}$$

(This kernel k is actually the kernel associated with the norm $\|\cdot\|_{\mathcal{D},\mathtt{alt}}$ of (2.10) rather than the norm (2.1). That leads to slightly cleaner computations and makes no essential difference.)

Now, the operator Id^* is bounded if and only if

$$\begin{aligned}\int_{\mathbb{D}}\left|\frac{d}{dz}\int_{\overline{\mathbb{D}}} g(\zeta)\, k_\zeta(z)\, d\nu(\zeta)\right|^2 dA &= \int_{\mathbb{D}}\left|\int_{\overline{\mathbb{D}}}\frac{g(\zeta)\overline{\zeta}}{1-\overline{\zeta}z}d\nu(\zeta)\right|^2 dA\\ &\lesssim \left\|\operatorname{Id}_{\mathcal{B}(\mathcal{D},L^2(\nu))}\right\|^2\int_{\overline{\mathbb{D}}}|g|^2\, d\nu.\end{aligned}\tag{5.14}$$

There is control for $|\operatorname{Id}^* g(0)|$. For that we have

$$|\operatorname{Id}^* g(0)|^2 = \left|\int_{\overline{\mathbb{D}}} g(\zeta)\, d\nu(\zeta)\right|^2 \le \nu\left(\overline{\mathbb{D}}\right)\|g\|^2_{L^2(\nu)}$$

and

$$\nu\left(\overline{\mathbb{D}}\right) = \frac{\|g\|^2_{L^2(\nu)}}{\|1\|^2_{\mathcal{D}}} \lesssim \|\operatorname{Id}\|_{B(\mathcal{D},L^2(\nu))}.$$

Let $S(z)$ be a tent with vertex z, similar to that described in Section 1.7.2:

$$S(z) := \left\{w\in\overline{\mathbb{D}} : 2(1-|w|)\le 1-|z|\,,\ |\arg(\overline{z}w)|\le 2(1-|z|)\right\}.$$

We use the elementary estimate:

$$\operatorname{Re}\left(\frac{1-|z|^2}{1-\overline{\zeta}z}\right)\ \begin{array}{ll}\ge 0 & \text{on } \mathbb{D}\\ \approx 1 & \text{on } S(z).\end{array}$$

We proceed to estimate the first terms in (5.14) from below, making some assumptions on g,

$$\begin{aligned}\int_{\mathbb{D}}\left|\int_{\overline{\mathbb{D}}}\frac{g(\zeta)\overline{\zeta}}{1-\overline{\zeta}z}d\nu(\zeta)\right|^2 dA &\geq \int_{\mathbb{D}}\operatorname{Re}\left(\int_{\overline{\mathbb{D}}}\frac{g(\zeta)\overline{\zeta}}{1-\overline{\zeta}z}d\nu(\zeta)\right)^2 dA\\ &= \int_{\mathbb{D}}\left(\int_{\overline{\mathbb{D}}}\operatorname{Re}\left(\frac{1-|z|^2}{1-\overline{\zeta}z}\right)g(\zeta)\overline{\zeta}d\nu(\zeta)\right)^2\frac{dA}{(1-|z|^2)^2}\\ &\text{if } h(\zeta)=g(\zeta)\overline{\zeta} \text{ is positive,}\\ &\gtrsim \int_{\mathbb{D}}\left(\int_{S(z)}h(\zeta)\,d\nu(\zeta)\right)^2\frac{dA}{(1-|z|^2)^2}\\ &\approx \int_{\mathbb{D}}\left(\int_{S(z)}h(\zeta)\,d\nu(\zeta)\right)^2\frac{dA}{(1-|z|^2)^2}\\ &= \sum_{\alpha\in E(V)}\int_{Q_\alpha}\left(\int_{S(z)}h(\zeta)\,d\nu(\zeta)\right)^2\frac{dA}{(1-|z|^2)^2}.\end{aligned}$$

We now consider two cases; $\nu(\mathbb{T})=0$ and $\operatorname{supp}(\nu)\subseteq\mathbb{T}$. In the first case we can continue with

$$\begin{aligned}&\gtrsim \sum_{\alpha\in E(V)}\int_{Q_\alpha}\left(\sum_{x\in S(\alpha)}\int_{Q_x}h(\zeta)\,d\nu(\zeta)\right)^2\frac{dA}{(1-|z|^2)^2}\\ &\approx \sum_{\alpha\in E(V)}\left(\sum_{x\in S(\alpha)}\int_{Q_x}h(\zeta)\,d\nu(\zeta)\right)^2\\ &= \sum_{\alpha\in E(V)}\left(\sum_{x\in S(\alpha)}\int_{Q_x}h(\zeta)\,d\nu(\zeta)\right)^2.\end{aligned}$$

Assuming now that $h=h^\sharp(\alpha)$ is constant in Q_x we continue with

$$\begin{aligned}\sum_{\alpha\in E(V)}\left(\sum_{x\in S(\alpha)}\int_{Q_x}h(\zeta)\,d\nu(\zeta)\right)^2 &= \sum_{\alpha\in E(V)}\left(\sum_{x\in S(\alpha)}h^\sharp(\alpha)\Lambda^*\nu(\{\alpha\})\right)^2\\ &= \sum_{\alpha\in E(V)}\left(I^*_{\Lambda^*\nu}h^\sharp(\alpha)\right)^2.\end{aligned}\tag{5.15}$$

In the second case, we continue with

$$\begin{aligned}\sum_{\alpha\in E(V)}\left(\sum_{x\in S(\alpha)}\int_{Q_x}h(\zeta)\,d\nu(\zeta)\right)^2 &\gtrsim \sum_{\alpha\in E(V)}\left(\int_{S(\alpha)}h(\Lambda(\zeta))\,d\Lambda^*\nu(\zeta)\right)^2\int_{Q_\alpha}\frac{dA}{(1-|z|^2)^2}\\ &\text{and if } h^\sharp(\zeta)=h(\Lambda(\zeta)) \text{ we continue with}\\ &\approx \sum_{\alpha\in E(V)}\left(I^*_{\Lambda^*\nu}h^\sharp(\alpha)\right)^2.\end{aligned}\tag{5.16}$$

So, in either case we have

$$\int_{\mathbb{D}} |g|^2 \, d\nu \lesssim \int_{\overline{\mathcal{T}}} h^{\sharp 2} d\Lambda^* \nu.$$

Combining this with (5.14), (5.15), and (5.16), we have then, for any choice of $h^\sharp$, the inequality

$$\sum_{\alpha \in E(V)} \left(\mathrm{I}^*_{\Lambda^* \nu} h^\sharp (\alpha) \right)^2 \lesssim \left\| \mathrm{Id}_{\mathcal{B}(\mathcal{D}, L^2(\nu))} \right\|^2 \int_{\overline{\mathcal{T}}} h^{\sharp 2} d\Lambda^* \nu,$$

which is dual to the trace inequality $\mathrm{I} : \ell^2 \to L^2 (\Lambda^* \nu)$, which then holds with the same constant. □

The partitioning of the disk into subsets that are disjoint (up to sets of measure zero), whose dilates (= controlled expansions) form a cover of bounded depth, and all being roughly spherical in terms of a natural intrinsic distance, is at the core of this proof and, more generally at the center of the analogy between spaces of holomorphic functions and spaces of functions on trees. We are now focusing on the Dirichlet space; but for $f \in \mathcal{D}$ the function we analyze is f' which is in the Bergman space A^2. Hence the natural choice of distance is δ_{A^2} rather than the very different $\delta_{\mathcal{D}}$. The metric δ_{A^2} is equivalent to the pseudohyperbolic metric, $\rho = \delta_{H^2}$, and for convenience we will generally use ρ.

Theorem 5.13 and its proof are rather stable under variations. For instance $\ell^2(\mathcal{T})$ can be replaced by $\ell^p(\mathcal{T}, \omega)$ for $p \neq 2$ and various weights ω.

Exercise 5.15. *State and prove a version of the theorem corresponding to the imbedding* $\mathrm{Id} : B_p \to L^q(\mu)$, *where* $1 < p, q < \infty$ *and* B_p *is the Besov space normed by* (2.44). *Hint: Look for a reference where it is proved that* $B_p^* = B_{p'}$ *under the pairing in* $\mathcal{D}$.

5.5.1. Radial Variation. Recall that for $f \in \mathcal{D}$ we have $X_f \subset Y_f$ where Y_f is the set of $e^{i\theta} \in \mathbb{T}$ for which the image under f of the segment $\left[0, e^{i\theta}\right]$ has infinite length, and X_f is the set of $e^{i\theta} \in \mathbb{T}$ at which f fails to have a radial limit. It is easy to check that if $e^{i\theta} \notin X$ then f has a radial limit at $e^{i\theta}$. The above theorem ensures that Y_f is a null set for every Carleson measure. However that property also characterizes sets of capacity zero, see Theorem 9.6. Thus we obtain the following classical result of Beurling [**69**].

Corollary 5.16. *If* $f \in \mathcal{D}$ *then for each* θ *outside a set of capacity zero* $\lim_{r \to 1} f(re^{i\theta})$ *exists.*

This is a much stronger statement then the existence of the limit *a.e.*, a fact which follows from the discussion in Section 3.1 after noting that $f \in \mathcal{D} \subset H^2$.

5.6. The Tree Characterization of Carleson Measures

We now describe the Carleson measures for $\mathcal{D}(\mathcal{T})$, the measures which satisfy a trace inequality. When that result is combined with Theorem 5.13 we also obtain a characterization of the Carleson measures for $\mathcal{D}$.

In the previous chapter we saw that the simple, one box, condition did not characterize the Carleson measures for the Dirichlet space and we gave a heuristic argument that suggested considering a condition such as Condition 4.10, a condition based on testing the adjoint of the inclusion map. We now prove such a result for the tree Besov space.

THEOREM 5.17. *Let $\mu \geq 0$ be a finite Borel measure on $\overline{\mathcal{T}}$. Then, μ satisfies the trace inequality* (TI) *if and only if it satisfies*

$$\sum_{\beta \geq \alpha} \mu\left(S\left(\beta\right)\right)^2 \leq \left[\mu\right]_{\mathrm{CM}} \mu\left(S\left(\alpha\right)\right) < \infty \tag{TC}$$

with a finite constant $[\mu]_{\mathrm{CM}}$. Moreover, $[\mu]_{\mathrm{CM}} \approx \|\mu\|_{\mathrm{CM}}$.

The trace inequality (TI) says a certain operator is bounded, in which case its adjoint must also be bounded. The condition (TC) is the seemingly weaker statement that the adjoint is bounded on a certain subset of its domain. Hence the heart of the theorem is the converse; that $\|\mu\|_{\mathrm{CM}} \lesssim [\mu]_{\mathrm{CM}}$.

The condition (TC) is sometimes called the *tree condition*. More generally, that name is used generically for conditions in the spirit of Condition 4.10 which are based on testing the adjoint of an operator of interest.

More generally: This theorem characterizes μ for which $\|\mu\|_{\mathrm{CM}} < \infty$; that is, the operator I is bounded from $\ell^2\left(\mathcal{T}, \text{counting measure}\right)$ to $L^2\left(\mu\right)$. There is also a "two weight" version of the theorem which gives a tree condition characterization of pairs (ν, w) for which I is bounded from $\ell^2\left(\mathcal{T}, \nu\right)$ to $L^2\left(w\right)$; [**48**, Theorem 3]. We state and use the more general result later, near (15.59).

PROOF OF THEOREM 5.17. By duality, the trace inequality is equivalent to the dual inequality

$$\sum_{\alpha \in \mathcal{T}} I_\mu^*\left(g\right)^2\left(\alpha\right) \leq \|\mu\|_{\mathrm{CM}} \|g\|_{L^2(\mu)}^2,$$

where $I_\mu^* : L^2\left(\mu\right) \to \ell^2\left(\mathcal{T}\right)$ is the adjoint of $I : \ell^2\left(\mathcal{T}\right) \to L^2\left(\mu\right)$. We will prove this dual inequality.

To find the formula for I^* we compute

$$\begin{aligned}
\sum_{\alpha \in \mathcal{T}} I_\mu^* g\left(\alpha\right) \varphi\left(\alpha\right) &= \langle I_\mu^* g, \varphi \rangle_{\ell^2(\mathcal{T})} = \langle g, I\varphi \rangle_{L^2(\mu)} \\
&= \int_{\overline{\mathcal{T}}} g\left(\zeta\right) \sum_{\alpha \in \mathrm{Pr}(\zeta)} \varphi\left(\alpha\right) d\mu\left(\zeta\right) \\
&= \sum_{\alpha \in \mathcal{T}} \left(\int_{S(\alpha)} g\left(\zeta\right) d\mu\left(\zeta\right)\right) \varphi\left(\alpha\right).
\end{aligned}$$

Therefore,

$$I_\mu^* g\left(\alpha\right) = \int_{\mathcal{T}(\alpha)} g\left(\zeta\right) d\mu\left(\zeta\right).$$

If we test the inequality for I_μ^* on functions of the form $g = \chi_{\mathcal{T}(\alpha)}$, we obtain the necessity statement:

$$\begin{aligned}\|\mu\|_{\mathrm{CM}}\,\mu\left(\mathcal{T}\left(\alpha\right)\right) &= \|\mu\|_{\mathrm{CM}}\left\|\chi_{\mathcal{T}(\alpha)}\right\|^2_{L^2(\mu)}\\ &\geq \sum_\beta I_\mu^*\chi^2_{\mathcal{T}(\alpha)}\left(\beta\right)\\ &= \sum_\beta \mu\left(\mathcal{T}\left(\alpha\right)\cap\mathcal{T}\left(\beta\right)\right)^2\\ &= d\left(\alpha\right)\mu\left(\mathcal{T}\left(\alpha\right)\right)^2 + \sum_{\beta\geq\alpha}\mu\left(\mathcal{T}\left(\beta\right)\right)^2.\end{aligned}$$

For the sufficiency part, we first majorize the left hand side by a maximal function and show a stronger inequality for the maximal function. We define the maximal function $\mathcal{M}_\mu$ by the following. Let $g \geq 0$ be a measurable function on $\overline{\mathcal{T}}$. Recall that $[z, w]$ is the line segment connecting z to w. Set

$$\sigma\left(x\right) = \mu\left(S\left(\left[x^{-1}, x\right]\right)\right)^2.$$

We have

$$\begin{aligned}\sum_{\alpha\in\mathcal{T}} I_\mu^*\left(g\right)^2\left(\alpha\right) &\leq \sum_{x\in V(\mathcal{T})}\max_{\beta\in P(x)}\frac{I_\mu^*\left(g\right)^2\left(\beta\right)}{\mu\left(S\left(\beta\right)\right)^2}\mu\left(S\left(\left[x^{-1}, x\right]\right)\right)^2\\ &=: \sum_{x\in V(\mathcal{T})}\mathcal{M}_\mu g\left(x\right)^2\sigma\left(x\right).\end{aligned}$$

Observe that $\mathcal{M}_\mu$ extends in the obvious way to $\overline{\mathcal{T}}$. Now, the condition on μ implies that for $t > 0$ we have

$$\begin{aligned}&\sigma\left(x : \mathcal{M}_\mu g\left(x\right) > t\right) = \sigma\left(\cup_j S\left(\alpha_j\right)\right)\\ &\qquad\text{where each } \alpha_j = \beta \text{ is minimal with the property that} \frac{I_\mu^*\left(g\right)\left(\beta\right)}{\mu\left(S\left(\beta\right)\right)} > t\\ &\qquad = \sum_j\sum_{\gamma\geq\alpha_j}\mu\left(S\left(\gamma\right)\right)^2\\ &\qquad \leq \sum_j\left[\mu\right]_{\mathrm{CM}}\mu\left(S\left(\alpha_j\right)\right) \text{ by the necessary condition}\\ &\qquad = \left[\mu\right]_{\mathrm{CM}}\mu\left(\zeta : \mathcal{M}_\mu g\left(\zeta\right) > t\right).\end{aligned}$$

Then,

$$\begin{aligned}\sum_{x\in V(\mathcal{T})}\mathcal{M}_\mu g\left(x\right)^2\sigma\left(x\right) &= \int_0^\infty \sigma\left(x : \mathcal{M}_\mu g\left(x\right) > t\right)2tdt\\ &\leq \left[\mu\right]_{\mathrm{CM}}\int_0^\infty \mu\left(\zeta : \mathcal{M}_\mu g\left(\zeta\right) > t\right)2tdt\\ &= \left[\mu\right]_{\mathrm{CM}}\int_{\mathcal{T}}\left(\mathcal{M}_\mu g\right)^2 d\mu.\end{aligned}$$

We obtain the desired L^2 inequality by the standard interpolation argument. The strong L^∞ inequality $\|\mathcal{M}_\mu g\|_{L^\infty(\mu)} \leq \|g\|_{L^\infty(\mu)}$ is obvious, and the weak L^1-inequality, using the maximality of the α_i we have:

$$\begin{aligned}\mu\left(\zeta : \mathcal{M}_\mu g(\zeta) > t\right) &= \sum_j \mu\left(S\left(\alpha_j\right)\right)\\ &\leq \sum_j \frac{1}{t} I^*\left(g\right)\left(\alpha_j\right) = \frac{1}{t}\int_{\cup_j S(\alpha_j)} g d\mu\\ &\leq \frac{1}{t}\int_{\mathcal{T}} g d\mu\end{aligned}$$

holds as well. The desired inequality then follows by interpolation. □

Combining the previous theorem with Theorem 5.13 gives the following characterization of Carleson measures for $\mathcal{D}$. Recall the definition of Λ^* from Proposition 5.10.

COROLLARY 5.18. *The nonnegative measure ν on $\overline{\mathbb{D}}$ is a Carleson measure for $\mathcal{D}$ if and only if the measure $\Lambda^*\nu$ satisfies* (TC).

This can be recast as a description that has a more classical form. Roughly, Condition 4.10 is a good first approximation. However the discrete world is only a grainy approximation to the holomorphic world and hence to get a correct statement Condition 4.10 must be softened a bit. For $z \in \mathbb{D}$ let z^* be the point obtained by moving z a bit closer to the origin: $z^* = .9z$.

COROLLARY 5.19. *The nonnegative measure ν on $\overline{\mathbb{D}}$ is a Carleson measure for $\mathcal{D}$ if and only if*

$$\int_{T(z^*)} \mu(T(\zeta) \cap T(z))^2 d\lambda(\zeta) \lesssim \mu(T(z)).$$

The passage from the first corollary to the second is discussed in detail in [**48**] and [**51**].

Of course these characterizations of Carleson measures for $\mathcal{D}$ and the capacity theoretic characterization in Theorem 4.16 must be equivalent. The direct proof that the two are equivalent reveals an interesting relationship between the Carleson measure theory for $\mathcal{D}(\mathcal{T})$ and capacity theory on $\mathcal{T}$, and then, also, between the Carleson measure theory for $\mathcal{D}$ and the capacity theory on the circle. Theorem 9.6 is an explicit statement. The following exercise gives a glimpse of the connection.

EXERCISE 5.20. *Let $0 < \theta \leq 1$. Show that for a finite Borel measure on $\mathcal{T}$ one has:*

$$\sum_{x,y\in S(\alpha)} \mu(x)\,\mu(y)\,d(x\wedge y)^\theta \approx d(\alpha)^\theta\,\mu(S(\alpha))^2 + \theta\sum_{\beta\geq\alpha} d(\beta)^{\theta-1}\,\mu(S(\beta))^2, \tag{5.17}$$

with constants independent of θ.

Let $\ell^2\left(\mathcal{T}, d^{1-\theta}\right)$ be the weighted ℓ^2 space on $\mathcal{T}$:

$$\|\varphi\|^2_{\ell^2(\mathcal{T},d^{1-\theta})} = \sum_{\alpha\in\mathcal{T}} |\varphi(\alpha)|^2\,d(\alpha)^{1-\theta}.$$

Use the proof of the characterization theorem to show that

$$\theta \sum_{\beta \geq \alpha} d(\beta)^{\theta-1} \mu(S(\beta))^2 \lesssim \mu(S(w))$$

holds uniformly in w for a given positive measure μ if, and only if, the Hardy operator $I : \ell^2(\mathcal{T}, d^{1-\theta}) \to L^2(\mathcal{T}, \mu)$ is bounded, and show that the norm does not crucially depend on θ.

Note that the left hand side of (5.17) is of the general form of an energy integral in capacity theory (for the tree) and the right hand side is related to the criterion for Carleson measures (again, on the tree). That observation is a step in understanding the relationship between the tree description of Carleson measures and the capacity description.

In the previous chapter we saw that the simple condition

$$\mu(S(\alpha)) \leq \|\mu\|_{\mathrm{CM}} \, d(\alpha)^{-1} < \infty \tag{5.18}$$

is necessary for a measure to be a Carleson measure. Hence, by the previous theorem this condition must be implied by the tree condition (TC).

EXERCISE 5.21. *Give a direct proof of that implication. Hint: Pick α in $\mathcal{T}$ with $d(\alpha) \geq 2$ and let $\dot{\alpha}$ be an edge about half-way between the root and α. If μ satisfies* (TC), $\mu|_{S(\alpha)}$ *does as well. Write* (TC) *for* $\mu|_{S(\alpha)}$ *at $\hat{\alpha}$ and deduce* (SC) *for μ at α.*

On the other hand, we noted in the previous chapter that the simple condition was not sufficient for a measure to be a Carleson measure. That follows from the previous theorem and the following construction.

PROPOSITION 5.22. *There is a measure μ on the dyadic, rooted tree, such that (SC) holds, but μ does not satisfy* (TC).

PROOF. We will produce a measure μ having the form

$$d\mu = \sum_{n=0}^{\infty} \sum_{j=1}^{2^n} d(x_{n,j})^{-1} \delta_{x_{n,j}},$$

where the points $x_{n,j} \in V(\mathcal{T})$ will be specified shortly and δ_x is a unit mass at x. We will identify edges and vertices. For $n \geq 0$, pick 2^n points $x_{n,j}$ such that $d(x_{n,j}) = 2^n a_n$, with a_n to be chosen, having the structure of a subtree. That is, $x_{n,2j-1} \wedge x_{n,2j} = x_{n-1,j}$ holds for all n, j. For each $x_{n,j}$ choose then a point $y_{n,j} > x_{n,j}$ such that $d(y_{n,j}) = 2^n b_n$ (with $b_n > a_n$ to be chosen), and that, for $x_{m,k} > x_{n,j}$, one has $y_{n,j} \wedge x_{m,k} = \{x_{n,j}, \hat{x}_{n,j}\}$ is one of the immediate successors of $x_{n,j}$. For simplicity of notation, we suppose that $y_{n,j} \wedge x_{m,k} = x_{n,j}$ (which is not possible in a dyadic tree, but the reader can check that the estimate works the same way in the more rigorous formulation above).

The simple condition (SC) becomes, in the worst scenario:

$$2^{-n}a_{n-1}^{-1} = \left(d\left(x_{n,j}\right)+1\right)^{-1} \gtrsim \sum_{x_{m,k}>x_{n,j}} d\left(y_{m,k}\right)^{-1} = \sum_{l\geq 0} 2^{l}\left(2^{n+l}b_{n+l}\right)^{-1},$$

i.e. $a_{n-1}^{-1} \gtrsim \sum_{l\geq 0} b_{n+l}^{-1}$. The tree condition (TC) is:

$$\begin{aligned}
2^{-n}\sum_{l\geq 0} b_{n+l}^{-1} &\approx \sum_{x_{m,k}>x_{n,j}} d\left(y_{m,k}\right)^{-1} \\
&\gtrsim \sum_{t\geq x_{n,j}} \left(\sum_{x_{m,k}>t} d\left(y_{m,k}\right)^{-1}\right)^{2} \\
&\approx \sum_{l\geq 0} 2^{l}\left(2^{n+l}a_{n+l} - 2^{n+l-1}a_{n+l-1}\right)\left(2^{-n-l}\sum_{j\geq 0} b_{n+l+j}^{-1}\right)^{2} \\
&\text{since changes only occur at the points } x_{m,k}, \\
&= 2^{-n}\sum_{l\geq 0}\left(a_{n+l} - 2^{-1}a_{n+l-1}\right)\left(\sum_{j\geq 0} b_{n+l+j}^{-1}\right)^{2}.
\end{aligned}$$

If we pick $a_n \approx n$ and $b_n \approx n^2$ as $n \to \infty$ then the last expression in the chain of inequalities diverges. This gives examples with (SC) true and (TC) false. □

Larger measures have larger Carleson measure constants: if $0 \leq \mu \leq \nu$ then $\|\mu\|_{\mathrm{CM}} \leq \|\nu\|_{\mathrm{CM}}$. Hence, by the previous theorem, it must also hold that, for some C, $[\mu]_{\mathrm{CM}} \lesssim [\nu]_{\mathrm{CM}}$. We now show that this weak monotonicity of $[\cdot]_{\mathrm{CM}}$ is a direct consequence of the definitions and does not require the previous theorem.

PROPOSITION 5.23. *Suppose the positive Borel measure μ on $\overline{\mathcal{T}}$ satisfies* (TC) *and that $\lambda : \overline{\mathcal{T}} \to [0,1]$ is measurable. Then, $[\lambda\mu]_{\mathrm{CM}} \leq 2\,[\mu]_{\mathrm{CM}}$.*

PROOF. Using the same notation as in the proof of the characterization theorem, but writing $\sigma = \sigma_\mu$ to exhibit the dependence of σ on μ, we have for $\alpha \in \mathcal{T}$:

$$\begin{aligned}
\sigma_{\lambda\mu}(S(\alpha)) &= \sum_{\beta \geq \alpha} \left(\int_{S(\beta)} \lambda d\mu \right)^2 \\
&\leq \sum_{\beta \geq \alpha} (\mathcal{M}_\mu(\lambda)(\beta))^2 \sigma_\mu(\beta) \\
&= \int_0^1 \sigma_\mu(\beta \geq \alpha : \mathcal{M}_\mu \lambda(\beta) > t)\, 2t dt \\
&= \int_0^1 \sigma_\mu(\cup_j S(\alpha_j(t)))\, 2t dt = \int_0^1 \sum_j \sigma_\mu(S(\alpha_j(t)))\, 2t dt
\end{aligned}$$

where each $\alpha_j(t) = \beta$ is minimal in $S(\alpha)$ with $\dfrac{I_\mu^*(\lambda)(\beta)}{\mu(S(\beta))} > t$

$$\leq 2[\mu]_{\text{CM}} \int_0^1 \sum_j \mu(S(\alpha_j))\, t dt$$

$$\leq 2[\mu]_{\text{CM}} \int_0^1 \sum_j I_\mu^* \lambda(\alpha_j)\, dt$$

by minimality of the α_j's

$$\leq 2[\mu]_{\text{CM}}\, I_\mu^* \lambda(\alpha) = 2[\mu]_{\text{CM}} \int_{S(\alpha)} \lambda d\mu,$$

as wished. □

5.7. A Capacity Construction for Interpolation

When studying function spaces, we will sometimes want to construct a function in the space which takes assigned values at specified points. We see this in Chapters 7, 8, and 13 for functions in $\mathcal{D}(\mathcal{T})$ and $\mathcal{D}$; and in later chapters when we study interpolation questions for functions in Besov spaces. One approach is to write the desired function as an informal sum of solutions to simple "atomic" problems, and then show the sum converges. When working in $\mathcal{D}(\mathcal{T})$ it is possible to show convergence by controlling the overlap of the support of the discrete derivatives of the solutions to the simple problems. The techniques for constructing solutions with the required type of localization are based on tree geometry and tree capacity. We will construct the required functions here. We will use the functions in constructions in Chapters 8 and 13.

For $x \in \mathcal{T}$ we write x^{-1} for its immediate predecessor and x_+, x_- for its two immediate successors. We start with a fixed dyadic tree $\mathcal{T}$ and a subset $\{\alpha_j\}_{j=1}^\infty = Z \subset \mathcal{T}$. We suppose that Z satisfies the *tree separation condition*, there is a $c > 0$ such that

$$\forall z, w \in Z, \quad d(z, w) \geq c d(o, z). \tag{5.19}$$

Let E, F be disjoint subsets of $\mathcal{T}$. The capacity $\text{Cap}_\mathcal{T}(E, F)$ of the condenser (E, F) is defined as

$$\text{Cap}_\mathcal{T}(E, F) = \inf\{\|h\|_{\ell^2}^2 : \ Ih|_E = 1, \ Ih|_F = 0\}. \tag{5.20}$$

For Z a sequence in $\mathcal{T}$ set $\gamma(z,Z) = \operatorname{Cap}_{\mathcal{T}}(\{z\}, Z \setminus \{z\})$. We say that Z satisfies the *tree capacity condition* if

$$\exists C,\ \forall z \in Z,\ \gamma(z,Z) \lesssim d^{+}(z)^{-1}. \tag{5.21}$$

DEFINITION 5.24. *We say that $S \subset \mathcal{T}$ is a stopping time or stopping region if every pair of distinct points in S are incomparable in $\mathcal{T}$. Equivalently, for all $x, y \in S$, $x \neq y$, $S(x) \cap S(y) = \phi$.*

For each t the set $S = \{\alpha_j(t)\}$ in the previous proof is a stopping time. As a stopping time it has controlled geometry. Additionally the points of S are selected as places where a functional of interest takes an extremal value. Using these two facts allows reduction of a global question to more straightforward local questions to be answered at each point of S. Exploiting that type of analysis is a *stopping time argument*. The previous proof and parts of the next proof are examples. The technique is widely used in analysis, in particular in analysis on trees.

LEMMA 5.25. *Given a subset Z of $\mathcal{T}$ that satisfies the tree separation condition* (5.19)*, there are functions $H_w = Ih_w$ on $\mathcal{T}$ and a constant C depending only on the constant in* (5.19) *satisfying*

(1) *$H_w(z) = \delta_{w,z}$ for $w, z \in Z$,*
(2) *$\operatorname{supp}(h_w) \cap \operatorname{supp}(h_z) = \phi$ for $z, w \in Z$, $z \neq w$,*
(3) *$\|H_w\|_{\mathcal{D}(\mathcal{T})} \leq 2C\gamma(w,Z)$ for $w \in Z$,*
(4) *If S is a stopping region in $\mathcal{T}$, then $\sum_{\alpha \in S} |h_w(\alpha)| \leq 2\gamma(w,Z)$ for $w \in Z$.*

5.7.1. Extremal Functions. We now prove a string of results on capacity that will culminate in the proof of Lemma 5.25. More precisely, properties (1), (2) and (3) of Lemma 5.25 will follow from Proposition 5.31 below, and property (4) will follow from Proposition 5.33. We use the following notation. If x is an element of the tree $\mathcal{T}$, x^{-1} denotes its immediate predecessor in $\mathcal{T}$. If z is an element of the sequence $Z \subset \mathcal{T}$, $\operatorname{ZPr}(z)$ denotes its predecessor in Z: $\operatorname{ZPr}(z) \in Z$ is the maximum element of $Z \cap [o, z)$, where by convention $o \in Z$.

Let $\Omega \subseteq \mathcal{T}$. A point $x \in \mathcal{T}$ is in the interior of Ω if $x, x^{-1}, x_+, x_- \in \Omega$. A function H is *harmonic* in Ω if, for every point x which is interior in Ω,

$$H(x) = \frac{1}{3}[H(x^{-1}) + H(x_+) + H(x_-)]. \tag{5.22}$$

In that case, if $H = Ih$, then, for the same x we have the martingale property

$$h(x) = h(x_+) + h(x_-). \tag{5.23}$$

PROPOSITION 5.26. *Let $\mathcal{T}$ be a dyadic tree.*

(1) *If E and F are finite, there is an extremal function $H = Ih$ such that $\operatorname{Cap}_{\mathcal{T}}(E,F) = \|h\|^2_{\ell^2}$.*
(2) *The function H is harmonic on $\mathcal{T} \setminus (E \cup F)$.*
(3) *Let $E = \{z\}$, $F = Z \setminus \{z\}$. Then the support of h consists of (at most) three connected components. The upper support consists of the segment $(\operatorname{ZPr}(z), z]$ and of all segments $[\zeta(w), w)$, where $w \in Z$ and $(\zeta(w), w]$ has some intersection with the component of $\mathcal{T} \setminus Z$ lying above z (i.e., the component containing z^{-1}). The lower support consists of all segments $[\zeta(w), w)$, where $w \in Z$ and $(\zeta(w), w]$ has some intersection with one of the (at most) two components of $\mathcal{T} \setminus Z$ lying below z (i.e., the components containing, respectively, z_+ and z_-).*

(4) *The function* h *is positive on* $(\mathrm{ZPr}(z), z]$, *negative on the segments* $[\zeta(w), w)$ *and vanishes everywhere else.*

In classical potential theory capacities can be recovered from the derivative of the Green potential. An analogous result holds here for the capacity of $(z, Z \setminus \{z\})$ with h playing the role of that derivative.

PROPOSITION 5.27. *Let* $\gamma_\pm$, γ_P *be the* ℓ^2*-sum of* h *over the lower and upper components of its support, respectively. Then,*

$$\gamma_\pm = -h(z_\pm), \ \gamma_P = h(z^{-1}).$$

PROOF OF PROPOSITION 5.26 AND PROPOSITION 5.27. We consider first the case of γ_+. Let $\mathcal{T}_z^+$ be the component of $\mathcal{T} \setminus Z$ containing x_+ and let $\overline{\mathcal{T}}_z^+$ be its forward closure, which is obtained by adding to $\mathcal{T}_z^+$ all points $w \in Z$, $w \neq z$, such that $d(w, \mathcal{T}_z^+) = 1$. For each $x \in \mathcal{T}_z^+$, $x_\pm \in \overline{\mathcal{T}}_z^+$. We proceed by induction on the cardinality of $Z \cap \overline{\mathcal{T}}_z^+$.

If $\sharp(Z \cap \overline{\mathcal{T}}_z^+) = 1$, $Z \cap \overline{\mathcal{T}}_z^+ = \{w\}$, then $\mathrm{supp}(h) = (z, w]$ and $h = -d(z, w)^{-1}$ over its support, so that $\gamma_+ = d(z, w) \cdot d(z, w)^{-2} = d(z, w)^{-1} = -h(z_+)$.

Suppose we know that $N \geq 1$ and that the property holds when $\sharp(Z \cap \overline{\mathcal{T}}_z^+) \leq N$ and suppose now that $\sharp(Z \cap \overline{\mathcal{T}}_z^+) = N+1$. Consider $W_z = \{w \wedge w' : w, w' \in Z \cap \overline{\mathcal{T}}_z^+\}$, the subtree of $\overline{\mathcal{T}}_z^+$ generated by $Z \cap \overline{\mathcal{T}}_z^+$. Then, W_z has a minimal element $\zeta \in \mathcal{T}_z^+$ ($\zeta \notin Z$ because $N + 1 \geq 2$). Let $U_+ = S(\zeta_+) \cap Z \cap \overline{\mathcal{T}}_z^+$ and $U_- = S(\zeta_-) \cap Z \cap \overline{\mathcal{T}}_z^+$. Then, the function H goes from 1 to $H(\zeta) > 0$ on $[z, \zeta]$ and from $H(\zeta)$ to 0 as x moves from ζ to $U_+ = S(\zeta_\pm) \cap Z \cap \overline{\mathcal{T}}_z^+$ in $\{\zeta\} \cup X_\pm = \{\zeta\} \cup [S(\zeta_\pm) \cap \overline{\mathcal{T}}_z^+]$. Note that

$$H(\zeta) = 1 - d(z, \zeta)h(z_+) = 1 - d(z, \zeta)h(\zeta).$$

The function $h|_{X_+}$ has minimal ℓ^2 norm with the property that $\sum_{x=\zeta_+}^{w} h(x) = -H(\zeta)$ whenever $w \in U_+$, otherwise we could modify it to obtain a global H with better ℓ^2 norm, contradicting the hypothesis that H was optimal. Also, observe that $\sharp(U_\pm) \leq N$.

Let $H_+ = Ih_+$ be the function such that $H_+(\zeta) = 1$, $H_+|_{U_+} = 0$ and which has minimal ℓ^2 norm with these properties. Similarly, set $H_- = Ih_-$ and $H' = Ih'$, the latter with the conditions $H'(z) = 1$ and $H'(\zeta) = 0$. By the induction hypothesis,

$$\|h_+\|_{\ell^2}^2 = h_+(\zeta_+), \ \|h_-\|_{\ell^2}^2 = h_-(\zeta_-), \ \|h'\|_{\ell^2}^2 = h'(z_+) = h'(\zeta).$$

By uniqueness of the extremal function H and by homogeneity of the minimization problem, we have that, with constants $a_+ = H(\zeta) = a_-$, $a' = 1 - H(\zeta)$,

$$h = a_\pm \cdot h_\pm, \ h = a' \cdot h'.$$

The norm of h is then

$$\begin{aligned}
\|h\|_{\ell^2}^2 &= a_+^2 \|h_+\|_{\ell^2}^2 + a_-^2 \|h_-\|_{\ell^2}^2 + a'^2 \|h'\|_{\ell^2}^2 \\
&= H(\zeta)^2 \cdot [h_+(\zeta_+) + (h_-(\zeta_-)] + (1 - H(\zeta))^2 \cdot h'(\zeta) \\
&= H(\zeta) \cdot [h(\zeta_+) + h(\zeta_-)] + (1 - H(\zeta)) \cdot h(\zeta) \\
&= h(\zeta) = h(z_+),
\end{aligned}$$

as wished. In the fourth equality, we used the harmonicity of $H = Ih$.

Exactly the same argument works for γ_- and a variation thereof gives the desired formula for γ_P. □

REMARK 5.28. *The proof gives a useful formula for computing capacities. Given* $z, \zeta \in \mathcal{T}$, $z < \zeta$ *and given* $U_\pm \subset S(\zeta_\pm)$ *we have*

$$\mathrm{Cap}_{\mathcal{T}}(z, U_+ \cup U_-) = \frac{\mathrm{Cap}_{\mathcal{T}}(\zeta, U_+) + \mathrm{Cap}_{\mathcal{T}}(\zeta, U_-)}{1 + d(z,\zeta)\left\{\mathrm{Cap}_{\mathcal{T}}(\zeta, U_+) + \mathrm{Cap}_{\mathcal{T}}(\zeta, U_-)\right\}}. \tag{5.24}$$

For the definition of capacity, the reader will need to consult (8.12). *To see this note that,*

$$\begin{aligned}\mathrm{Cap}_{\mathcal{T}}(z, U_+ \cup U_-) &= h(z_+) = h(\zeta) = h(\zeta_+) + h(\zeta_-)\\ &= H(\zeta)\left\{h_+(\zeta_+) + h_-(\zeta_-)\right\}\\ &= (1 - d(z,\zeta)h(\zeta_+))\left\{\mathrm{Cap}_{\mathcal{T}}(\zeta, U_+) + \mathrm{Cap}(\zeta, U_-)\right\}\\ &= (1 - d(z,\zeta)\cdot \mathrm{Cap}_{\mathcal{T}}(z, U_+ \cup U_-))\left\{\mathrm{Cap}_{\mathcal{T}}(\zeta, U_+)\right.\\ &\qquad \left.+ \mathrm{Cap}_{\mathcal{T}}(\zeta, U_-)\right\}.\end{aligned}$$

This formula is used in [**42**] *to construct sequences with desired properties.*

5.7.2. Modified Extremal Functions. From now on, we consider a *finite* sequence Z satisfying (5.19). We want to show that the functions $h = h_z$ can be replaced by near extremal functions k_z, with the extra property that $\mathrm{supp}(k_z) \cap \mathrm{supp}(k_w) = \phi$ if $z \neq w$.

We will assume that $Z = \{z_j : j \geq 0\}$ is ordered in such a way that $d^+(z_n) \leq d^+(z_{n+1})$. We will also assume that $z_0 = o = 0$ belongs to Z. We define

$$Z_n := \{z_0, \dots, z_n\},$$

and

$$\mathcal{T}_n := \textstyle\bigcup_{j=1}^n [o, z_j].$$

We also need $\mathcal{T}_\infty := \bigcup_{j=1}^\infty [o, z_j]$, the minimum subset of $\mathcal{T}$ containing Z which is geodesically connected. The *landing point* of $z = z_{n+1}$ is, by definition,

$$\xi(z) := \max\left([o, z] \cap \mathcal{T}_n\right).$$

By construction, $Z \cap (\xi(z), z) = \phi$ and, if $z \neq w$, $[\xi(z), z]$ and $[\zeta(w), w]$ are either disjoint, or they intersect in $\xi(z)$, or in $\xi(w)$. If $i_j \geq i_k$, then $\xi(z_{i_k}) \in [\xi(z_{i_j}), z_{i_j}]$ if and only if $\xi(z_{i_j}) = \xi(z_{i_k})$.

LEMMA 5.29. *Let Z be a sequence satisfying* (5.19). *Then, for some $\eta > 0$ and all z in Z,*

$$d(\xi(z), z) \geq \eta d^+(z). \tag{5.25}$$

PROOF. $d(\xi(z), z) \geq d(w, z)/2 \geq c d^+(z)/2$. The first inequality because $d^+(z) \geq d^+(w)$ by the way we ordered Z; the second by the separation hypothesis. □

As a consequence, by removing a finite number of points from Z we can assume that $d(\xi(z), z) \geq 3$ for all $z \neq o$.

We record some further properties of the functions $H = H_z$.

PROPOSITION 5.30. *The function H is increasing on* $[\mathrm{ZPr}(z), z]$ *and decreasing on all segments of the form* $[\zeta(w), w]$. *Moreover, H is convex on all intervals of the form* $[\zeta(w), w]$ *and on* $[\mathrm{ZPr}(z), z]$.

PROOF. If z lands on a geodesic $[o, w]$ with $w \in Z$ then we have

$$d(\zeta(z), z) \geq \frac{1}{2} d(w, z) \geq \frac{1}{2} d^+(z).$$

The first inequality because $d^+(z) > d^+(w)$, the second by the separation hypothesis. □

For z in Z, let

$$\mathcal{N}(z) = \bigcup_{\substack{w \in Z \\ w \geq \mathrm{ZPr}(z) \\ (w, \mathrm{ZPr}(z)) \cap Z \subseteq \{z\}}} [w, \mathrm{ZPr}(z)).$$

The set $\mathcal{N}(z)$ is the "downward closure" of the connected component $\mathcal{N}_0(z)$ of $\mathcal{T}_\infty \setminus (Z \setminus \{z\})$ containing z. We have

$$\mathcal{N}_0(z) = \mathcal{N}(z) \setminus (Z \setminus \{z\}).$$

All of the interesting action takes place inside $\mathcal{N}(z)$.

The main tool we need is the following. Recall that $\Delta K = k$ if $K = Ik$.

PROPOSITION 5.31. *If* (4.19) *holds, then to each z in Z we can associate a function K_z in $\mathcal{D}$ such that $K_z(w) = \delta_z(w)$ for $w \in Z$, K_z is harmonic on* $\mathrm{supp}(\Delta K_z) \setminus Z$, $\|K_z\|_{\mathcal{D}} \lesssim \gamma(z, Z)$ *and such that* $(\Delta K_z) \cap \mathrm{supp}(\Delta K_w) = \emptyset$ *if $z \neq w$. In fact the K's are pairwise disjoint as well:* $\mathrm{supp}(K_z) \cap \mathrm{supp}(K_w) = \emptyset$ *if $z \neq w$.*

REMARK 5.32. *Because the K's have disjoint support we have an immediate solution, $\sum_{j=1}^{\infty} \xi_j K_{z_j}$, to the problem of interpolating $\xi = \{\xi_j\}_{j=1}^{\infty}$ on the tree $\mathcal{T}$.*

PROOF OF PROPOSITION 5.31. Fix $z \in Z$. By Proposition 5.30 there is a function H such that $\|H\|_{\mathcal{D}}^2 = \gamma(z, Z)$ and $H(z) = 1$, $H|_{Z \setminus \{z\}} = 0$, $H \geq 0$, $\mathrm{supp}(\Delta H) \subseteq \mathcal{N}(z)$ and H is convex on intervals of the form $[w, \xi(w)]$. Let Q_z be the first point x on $[\mathrm{ZPr}(z), z]$ such that $d(x, z) \leq \frac{1}{3} d(\mathrm{ZPr}(z), z)$. For each $w \in \mathcal{N}(z) \cap (Z \setminus \{z\}) \cap S(\xi(z))$, let Q_w be the first point y in $[\xi(w), w)$ such that $d(y, \xi(w)) \geq \frac{1}{3} d(\xi(w), w)$. As in the proof of Proposition 5.30, $\mathcal{N}(z) \cap (Z \setminus \{z\}) \cap S(\xi(z)) = \{z_{i_j} : j \geq 0\}$, where $z = z_{i_0}$, $d^+(z_{i_{j+1}}) \geq d^+(z_{i_j})$ and $\xi(z_{i_{j+1}}) \in \mathcal{T}_{i_j}$. The function K_z is constructed inductively, in such a way that

$$\mathrm{supp}(k_z) \subseteq \mathcal{N}_r(z) := \textstyle\bigcup_j [Q_{z_{i_0}}, Q_{z_{i_j}}], \qquad k_z = \Delta K_z.$$

It is clear that if $z \neq w$ are points in Z, then $\mathcal{N}_r(z) \cap \mathcal{N}_r(w) = \emptyset$.

Step $j = 0$. Construct a new function K_0 as follows. On $[\mathrm{ZPr}(z), Q_z)$, set $K_0 = 0$. For $x \in [Q_z, z]$, set

$$K_0(x) = \frac{H(x) - H(Q_z)}{H(z) - H(Q_z)}.$$

If y is such that $y \wedge z \in [\mathrm{ZPr}(z), z]$, set

$$K_0(y) = \frac{K_0(y \wedge z)}{H(y \wedge z)} H(y).$$

Set $K_0(y) = H(y)$ otherwise. The function K_0 has the following properties (with η from (5.25) and with the parameter $\ell = 0$),

(1) K_ℓ is admissible for z, $K_\ell \geq 0$, K_ℓ is harmonic on $\mathcal{N}_r(z)$, K_ℓ is convex on $[\xi(z_{i_j}), z_{i_j}]$ for $j \geq 1$;

(2) $K_\ell \le H$, pointwise;
(3) if $x \notin [Q_z, z]$, then $|\Delta K_\ell(x)| \le |\Delta H(x)|$;
(4) if $x \in [Q_z, z]$, then $|\Delta K_\ell(x)| \lesssim |\Delta H(x)|$, where $C = 3/\eta$;
(5) $\mathrm{supp}(\Delta K_\ell) \subset [Q_z, z] \bigcup (\bigcup_{j\ge 1} [z_{i_j}, \xi(z_{i_j})])$.

Properties (1), (2), (3) and (5) immediately follow from the construction. We show (4). By convexity,

$$\frac{1 - H(Q_z)}{d^+(z) - d^+(Q_z)} = \frac{H(z) - H(Q_z)}{d^+(z) - d^+(Q_z)} \ge \frac{H(z) - H(o)}{d^+(z) - d^+(o)} = \frac{1}{d^+(z) - 1}$$

hence,

$$H(z) - H(Q_z) \ge \frac{d^+(z) - d^+(Q_z)}{d^+(z) - 1} \ge \frac{\eta}{3}$$

by Lemma 5.29 and by the definition of Q_z. Thus, if $x \in [Q_z, z]$, then $|\Delta K_0(x)| = |\Delta H(x)|/(H(z) - H(Q_z)) \lesssim |\Delta H(x)|$.

Induction. Pick now z_{i_1}. If $\xi(z_{i_1}) \in [\mathrm{ZPr}(z), Q_z)$, let $K_1 = K_0$. If $\xi(z_{i_1}) \in (z, Q_z]$, construct K_1 as we did above, changing first the values of K_0 on $[\xi(z_{i_1}), z_{i_1}]$ in such a way K_1 vanishes on $(Q_{z_{i_1}}, z_{i_1}]$, then adjusting the values elsewhere to make K_1 admissible for z. Observe that $\{x : K_1 \ne K_0\} \cap [\mathrm{ZPr}(z), z] = \emptyset$. The function K_1 has ($\ell = 1$) properties 1, 2 (that is, $K_\ell \le K_{\ell-1}$), 3 (that is, if $x \notin [Q_{z_{i_\ell}}, z_{i_\ell}]$, then $|\Delta K_\ell(x)| \le |\Delta K_{\ell-1}(x)|$), 4 (that is, if $x \in [Q_{z_{i_j}}, z_{i_j}]$, then $|\Delta K_\ell(x)| \lesssim |\Delta K_{\ell-1}(x)| \lesssim |\Delta H(x)|$, and this time $C = 3$ suffices). Property (5) becomes

$$\mathrm{supp}(K_\ell) \subset [Q_z, z] \bigcup (\bigcup_{1\le j\le \ell} [Q_{z_{i_j}}, z_{i_j}]) \bigcup (\bigcup_{j\ge \ell+1} [z_{i_j}, \xi(z_{i_j})]).$$

Moreover, we have that

$$\Delta K_\ell(x) = \Delta K_{\ell-1}(x) \text{ on the set } B_{\ell-1} = \bigcup_{1\le j\le \ell-1} [\xi(z_{i_j}), z_{i_j}].$$

Passing to the limit, we find a function K_z with all the desired properties, since the estimates for $|\Delta K_\ell|$ add up nicely.

□

5.7.3. Stopping Estimates. Let S be a *stopping region* in the tree $\mathcal{T}$ as in Definition 5.24.

PROPOSITION 5.33. *Let $K_z = Ik_z$ be the function in Proposition 5.31 and let S be a stopping region. Then,*

$$\sum_{x\in S} |k_z(x)| \le 2\gamma(z, Z).$$

PROOF. Let $k = k_z$. We know that $k \ge 0$ on $[Q_z, z]$ and that $k \le 0$ elsewhere. Let S be a stopping time in $\mathcal{T}$. Then $S \cap [Q_z, z]$ consists of at most one element x and

$$0 \le k(x) \le k(z) = \gamma_P,$$

since k is convex on $[Q_z, z]$.

Let $x_0 = Q_z, x_1, \dots, x_N = z$ be an ordered enumeration of the points in $[Q_z, z]$. Let $S_j = S \cap (S(x_j) - S(x_{j+1}))$, $j = 0, \dots, N-1$. Without loss of generality, we can assume that $x_{j+1} = x_{j_+}$, so that $S(x_j) - S(x_{j+1}) = S(x_{j\_})$. By the harmonicity of k (5.23) and an easy induction we have that

$$\sum_{x \in S_j} |k(x)| = \sum_{x \in S_j} k(x) \leq -k(x_{j\_}) - - \left[k(x_j) - k(x_{j_+})\right].$$

Summing over j, we have

$$\sum_{j=0}^{N-1} \sum_{x \in S_j} |k(x)| \leq k(z) - k(Q_z) \leq k(z) = \gamma_P.$$

Let $S_\pm = S \cap S(z_\pm)$. Induction and harmonicity show that

$$\sum_{x \in S_\pm} |k(x)| = |k(z_\pm)| = \gamma_\pm.$$

All points in the support of k fall in $S(Q_z) \bigcup \cdots \bigcup S(z)$, hence

$$\sum_{x \in S} |k(x)| \leq 2\gamma_P + \gamma_+ + \gamma_- \leq 2\gamma(z, Z).$$

□

5.8. Capacity and Blowups

In Chapter 13 we will use quantitative estimates of how the capacity of a union of intervals in the boundary of the disk changes when all the individual intervals are expanded; i.e., when we form the *blowup* of the set. It turns out to be effective to study that question in conjunction with the analogous question on the tree, and hence we include that analysis in this chapter, collected in the four lemmas below.

5.8.1. The Disk and the Tree. We begin by recalling and elaborating on some earlier notation, in particular that from Section 1.7.

For an interval I in the circle we let I_m be its midpoint and $z(I) = (1 - |I| / 2\pi)\, I_m$ be the associated index point in the disk. In the other direction let $I(z)$ be the interval such that $z(I(z)) = z$. Let $T(I)$ be the tent over I, the convex hull of I and $z(I)$ and let $T(z) = T(z(I)) = T(I)$. More generally, for any open subset H of the circle $\mathbb{T}$, we define $T(H)$, the *tent region* over H in the disk $\mathbb{D}$, by

$$T(H) := \bigcup_{I \text{ an interval in } H} T(I).$$

We work with the dyadic tree $\mathcal{T}$ of Sections 1.7.1 and 5.2.1. We write $\{x\}$ for the index set for the boxes $\{Q_x\}$ in the dyadic decomposition of $\mathbb{D}$. There is a natural bijection between the boxes and the intervals on the circle $\{I(Q_x)\}$ obtained by radial projection of the boxes. This also induces a bijection with the set of index points associated to those intervals, the set $\{z(I(Q_x))\} \subset \mathbb{D}$. Furthermore $z(I(Q_x)) \in Q_x$. At times we will use the label x to denote the point $z(I(Q_x))$.

Recall Definition 5.24 that $S \subset \mathcal{T}$ is a stopping time if no pair of distinct points in S are comparable in $\mathcal{T}$. Given stopping times $E, F \subset \mathcal{T}$ we say that $F \succ E$ if for every $x \in F$ there is $y \in E$ above x, i.e., with $x > y$. For stopping times $F \succ E$ denote by $\mathcal{G}(E, F)$ the union of all those geodesics connecting a point of $x \in F$ to the point $y \in E$ above it.

The bijections between $\{Q_x\}$, $\{I(Q_x)\}$, and $\{z(I(Q_x))\}$ induce bijections between other sets. We will be particularly interested in three types of sets:

- *stopping times* W in the tree $\mathcal{T}$;
- $\mathcal{T}$*-open subsets* G of the circle $\mathbb{T}$;
- $\mathcal{T}$*-tent regions* Γ of the disk $\mathbb{D}$.

The bijections are given as follows. For W a *stopping time* in $\mathcal{T}$, its associated $\mathcal{T}$*-open set* in $\mathbb{T}$ is the $\mathcal{T}$-shadow $S_{\mathcal{T}}(W) = \cup\{I(x) : x \in W\}$ of W on the circle (this also *defines* the collection of $\mathcal{T}$-open sets). The associated $\mathcal{T}$*-tent region* in $\mathbb{D}$ is $T_{\mathcal{T}}(W) = \cup\{T(I(\kappa)) : \kappa \in W\}$ (this also *defines* the collection of $\mathcal{T}$-tent regions).

At times we will identify a stopping time $W = W_{\mathcal{T}}$ in a tree $\mathcal{T}$ with its associated $\mathcal{T}$-shadow on the circle and its $\mathcal{T}$-tent region in the disk and will use W or $W_{\mathcal{T}}$ to denote any of them. When we do this the exact interpretation will be clear from the context.

Note that for any open subset E of the circle $\mathbb{T}$, there is a unique $\mathcal{T}$-open set $G \subset E$ such that $E \smallsetminus G$ is at most countable. We often informally identify the open sets E and G.

EXERCISE 5.34. *Make enough sketches to ensure you understand the geometric content of the previous definitions and the various bijections and identifications.*

5.8.2. Capacity. Recall from Chapter 4 that for $G \subset \mathbb{T}$ the capacity of G is

$$\operatorname{Cap}_{\mathbb{D}} G = \inf\left\{\|\psi\|_{\mathcal{D}}^2 : \psi(0) = 0, \operatorname{Re}\psi(z) \geq 1 \text{ for } z \in G\right\}.$$

We saw in Theorem 4.16 that $\mu \in CM(\mathcal{D})$ exactly if for any finite collection of disjoint arcs $H = \{I_j\}_{j=1}^N$ in the circle $\mathbb{T}$ we have

$$\mu(T(H)) = \mu\left(\bigcup_{j=1}^N T(I_j)\right) \lesssim \operatorname{Cap}_{\mathbb{D}}\left(\bigcup_{j=1}^N I_j\right). \tag{5.26}$$

We will need to understand how the capacity of a set changes if we expand it in certain ways. For I an open arc and $0 < \rho \leq 1$, let I^ρ be the arc concentric with I having length $|I|^\rho$.

DEFINITION 5.35. *[disk blowup]For G open in $\mathbb{T}$ we call*

$$G_{\mathbb{D}}^\rho = \bigcup_{I \text{ an interval in } G} T(I^\rho)$$

the disk blowup *(of order ρ) of G.*

The important feature of the disk blowup is that it achieves a good geometric separation between $\mathbb{D} \smallsetminus G_{\mathbb{D}}^\rho$ and $G_{\mathbb{D}}^1 = T(G)$. This fact will be crucial in the work in Chapter 13.

LEMMA 5.36. *Given ρ there is a $c = c(\rho)$ and $\varepsilon = \varepsilon(c, \rho)$ so that for all z, w at distance at most ε from the boundary: if G is an open subset of the circle $\mathbb{T}$ and $w \in G_{\mathbb{D}}^1 = T(G)$, $z \notin G_{\mathbb{D}}^\rho$, then $|z - w| \geq c(1 - |w|^2)^\rho$.*

PROOF. For given z let I be the interval with $z(I)$. Because z is near the boundary $T(I) \subseteq D(z(I), |I|^\rho)$. Hence we have the estimates

$$|z - w| \geq |z - z(I)| = |w - z(I)| \geq |I|^\rho - |I| \geq c_1|I|^\rho \geq c_2(1 - |w|^2)^\rho.$$

□

It would be useful to us if we knew there were constants C_ρ, $0<\rho<1$, such that

$$\operatorname{Cap}_{\mathbb{D}}(G_{\mathbb{D}}^{\rho}) \leq C_\rho \operatorname{Cap}_{\mathbb{D}} G. \tag{5.27}$$

and

$$\lim_{\rho\to 1^-} C_\rho = 1. \tag{5.28}$$

In [**72**] Bishop proved (5.27), but not (5.28). Here, rather than give a direct proof of (5.27) we obtain Lemma 5.41, an analog of (5.27) and (5.28) in a tree model. That lemma will play an important role in the proof here, and after we show that tree and disk capacity are comparable, Corollary 13.5, it will yield (5.27) which we will also use. It is an open question whether the disk result (5.28) holds.

Recall from (5.4) and (5.5) that for functions k, K defined on $\mathcal{T}$, we have $Ik(x)=\sum_{y\in[o,x]}k(y)$ and $\Delta K(x)=K(x)-K(x^-)$. Also, by convention, $K(o^-)=0$.

Let $\operatorname{Cap}_{\mathcal{T}}$ be the tree capacity associated with $\mathcal{T}$:

$$\operatorname{Cap}_{\mathcal{T}}(E) = \inf\left\{\|f\|_{\ell^2(\mathcal{T})}^2 : f\geq 0,\ If\geq 1 \text{ on } E\right\}. \tag{5.29}$$

We refer to an f which can be used in evaluating that infimum as *admissible*. More generally, if $E,F\subset\mathcal{T}$ are disjoint stopping times with $F\succ E$, then, recall, the capacity of the pair (E,F), commonly known as a condenser, is given by (5.20). In particular, for any E, $\operatorname{Cap}_{\mathcal{T}}(E)=\operatorname{Cap}_{\mathcal{T}}(\{o\},E)$.

Let $\mathcal{T}_\theta$ be the rotation of the tree $\mathcal{T}$ by the angle θ, and let $\operatorname{Cap}_{\mathcal{T}_\theta}$ be the tree capacity associated with $\mathcal{T}_\theta$ as in (5.29), and extend the definition to open subsets G of the circle $\mathbb{T}$ by,

$$\operatorname{Cap}_{\mathcal{T}_\theta}(G) = \inf\left\{\sum_{\kappa\in\mathcal{T}_\theta} f(\kappa)^2 : f\geq 0.\ If(\beta)\geq 1 \text{ for } \beta\in\mathcal{T}_\theta,\ I(\beta)\subset G\right\}.$$

This is consistent with the definition of tree capacity of a stopping time W in $\mathcal{T}_\theta$; that is, if $G=\cup\{I(\kappa):\kappa\in W\}$ we have

$$\operatorname{Cap}_{\mathcal{T}_\theta}(W) = \operatorname{Cap}_{\mathcal{T}_\theta}(\{o\},W) = \operatorname{Cap}_{\mathcal{T}_\theta}(G).$$

When the angle θ is not important, we will simply write $\mathcal{T}$ with the understanding that all results have analogs with $\mathcal{T}_\theta$ in place of $\mathcal{T}$.

5.8.2.1. *Blowups.*

DEFINITION 5.37 (Stopping Time Blowup). *Given $0\leq\rho\leq 1$ and a stopping time W in a tree $\mathcal{T}$, define the* stopping time blowup $W_{\mathcal{T}}^{\rho}$ *of W in $\mathcal{T}$ as the set of minimal tree elements in $\{R^\rho\kappa:\kappa\in\mathcal{T}_\theta\}$, where $R^\rho\kappa$ denotes the unique element in the tree $\mathcal{T}$ satisfying*

$$\begin{gathered} o\leq R^\rho\kappa\leq\kappa, \\ \rho d^+(\kappa)\leq d^+(R^\rho\kappa)<\rho d^+(\kappa)+1. \end{gathered} \tag{5.30}$$

Clearly $W_{\mathcal{T}}^{\rho}$ is a stopping time in $\mathcal{T}$. Note that $R^1\kappa=\kappa$. The element $R^\rho\kappa$ can be thought of as the "ρ^{th} root of κ" since $|R^\rho\kappa|=2^{-d^+(R^\rho\kappa)}\approx 2^{-\rho d^+(\kappa)}=|\kappa|^\rho$.

If W is a stopping time for $\mathcal{T}$ and $W_{\mathcal{T}}^{\rho}$ is the stopping time blowup of W, then there is a good estimate for the tree capacity of $W_{\mathcal{T}}^{\rho}$ given in Lemma 5.41 below: $\operatorname{Cap}_{\mathcal{T}}(\{o\},W_{\mathcal{T}}^{\rho})\leq\rho^{-2}\operatorname{Cap}_{\mathcal{T}}(\{o\},W)$. Unfortunately, there is not a good condenser estimate of the form $\operatorname{Cap}_{\mathcal{T}}(W_{\mathcal{T}}^{\rho},W)\leq C_\rho\operatorname{Cap}_{\mathcal{T}}(\{o\},W)$; the left side can be infinite when the right side is finite. We now introduce another type of blowup, a tree analog of the disk blowup, for which we do have an effective condenser estimate. We do this using a capacitary extremal function and a comparison principle. Let W be a stopping time in $\mathcal{T}$. By Proposition 5.26, there is a unique extremal function $H = Ih$ such that

$$Ih(x) = H(x) = 1 \text{ for } x \in W, \qquad (5.31)$$
$$\operatorname{Cap}_{\mathcal{T}} W = \|h\|_{\ell^2}^2 .$$

DEFINITION 5.38 (Capacitary Blowup). *Given a stopping time W in $\mathcal{T}$, the corresponding extremal H satisfying* (5.31), *and $0 < \rho < 1$, define the* capacitary blowup $\widehat{W_{\mathcal{T}}^{\rho}}$ *of W by*

$$\widehat{W_{\mathcal{T}}^{\rho}} := \{t \in \mathcal{G}(\{o\}, W) : H(t) \geq \rho \text{ and } H(x) < \rho \text{ for } x < t\}.$$

Clearly $\widehat{W_{\mathcal{T}}^{\rho}}$ is a stopping time in $\mathcal{T}$.

LEMMA 5.39. $\operatorname{Cap}_{\mathcal{T}} \widehat{W_{\mathcal{T}}^{\rho}} \leq \rho^{-2} \operatorname{Cap}_{\mathcal{T}} W$.

PROOF. Let H be the extremal for W in (5.31) and set $h = \Delta H$, $h^{\rho} = \frac{1}{\rho} h$ and $H^{\rho} = \frac{1}{\rho} H$. Then H^{ρ} is a candidate for the infimum in the definition of capacity of $\widehat{W_{\mathcal{T}}^{\rho}}$, and hence by the comparison principle, that is, by the fact that the capacity was defined as the infimum,

$$\operatorname{Cap}_{\mathcal{T}} \widehat{W_{\mathcal{T}}^{\rho}} \leq \|h^{\rho}\|_{\ell^2}^2 = \left(\frac{1}{\rho}\right)^2 \|h\|_{\ell^2}^2 = \rho^{-2} \operatorname{Cap}_{\mathcal{T}} W.$$

□

The next lemma will be used in the proof of the main estimate, (13.18) in Chapter 13 and it requires an upper bound on $\operatorname{Cap}_{\mathbb{D}}(G)$. However, (13.18) is straightforward if $\operatorname{Cap}_{\mathbb{D}}(G)$ bounded away from zero so that restriction is not a problem. In fact, moving forward we will assume, at times implicitly, that $\operatorname{Cap}_{\mathbb{D}}(G)$ is not large.

LEMMA 5.40. $\operatorname{Cap}_{\mathcal{T}}\left(W, \widehat{W_{\mathcal{T}}^{\rho}}\right) \leq \frac{4}{(1-\rho)^2} \operatorname{Cap}_{\mathcal{T}} W$ *provided* $\operatorname{Cap}_{\mathcal{T}} W \leq (1-\rho)^2/4$.

PROOF. Let H be the extremal for W in (5.31). For $t \in \widehat{W_{\mathcal{T}}^{\rho}}$ we have by our assumption,

$$h(t) \leq \|h\|_{\ell^2} \leq \sqrt{\operatorname{Cap}_{\mathcal{T}} W} \leq \frac{1}{2}(1-\rho),$$

and so

$$H(t) = H\left(t^{-}\right) + h(t) \leq \rho + \frac{1}{2}(1-\rho) = \frac{1+\rho}{2}.$$

If we define $\widetilde{H}(t) = \frac{2}{1-\rho}\left\{H(t) - \frac{1+\rho}{2}\right\}$, then $\widetilde{H} \leq 0$ on $\widehat{W_{\mathcal{T}}^{\rho}}$ and $\widetilde{H} = 1$ on W. Thus $\widetilde{H}$ is a candidate for the capacity of the condenser and so by the comparison principle

$$\begin{aligned} \operatorname{Cap}_{\mathcal{T}}\left(W, \widehat{W_{\mathcal{T}}^{\rho}}\right) &\leq \left\|\triangle \widetilde{H}\right\|_{\ell^2\left(\mathcal{G}\left(W_{\mathcal{T}}^{\rho}, W\right)\right)}^2 \leq \left\|\triangle \widetilde{H}\right\|_{\ell^2(\mathcal{T})}^2 \\ &= \left(\frac{2}{1-\rho}\right)^2 \|h\|_{\ell^2(\mathcal{T})}^2 = \frac{4}{(1-\rho)^2} \operatorname{Cap}_{\mathcal{T}} W. \end{aligned}$$

□

We also have good *tree* separation inherited from the stopping time blowup $W_{\mathcal{T}}^{\rho}$. This gives our substitute for (5.27) and (5.28).

LEMMA 5.41. *$W_{\mathcal{T}}^{\rho} \subset \widehat{W_{\mathcal{T}}^{\rho}}$ as open subsets of the circle or, equivalently, as $\mathcal{T}$-tent regions in the disk. Consequently* $\operatorname{Cap}_{\mathcal{T}} W_{\mathcal{T}}^{\rho} \leq \rho^{-2} \operatorname{Cap}_{\mathcal{T}} W$.

PROOF. The restriction of H to a geodesic is a concave function of distance from the root, and so if $o < z < w \in W$, then

$$H(z) \geq \left(1 - \frac{d^{+}(z)}{d^{+}(w)}\right) H(o) + \frac{d^{+}(z)}{d^{+}(w)} H(w) = \frac{d^{+}(z)}{d^{+}(w)} \geq \rho, \ z \in \widehat{W_{\mathcal{T}}^{\rho}},$$

and this proves $W_{\mathcal{T}}^{\rho} \subset \widehat{W_{\mathcal{T}}^{\rho}}$. The inequality now follows from Lemma 5.39. □

5.9. Notes and Comments

Hyperbolic geometry and tree models are parts of the early history of function theory on the disk. There is an interesting discussion in [**162**]. Since then the study of function spaces on trees has flourished, both because of their intrinsic interest and as sources of insight into smooth functions. We will not attempt an overview of this rich history; classic papers include [**94**] and [**304**]. More recent works include [**216**] and [**150**].

Dyadic decompositions and their variants, as well as associated tree models, have many roles in harmonic analysis; for instance [**195**], [**312**], [**261**], [**204**]. Our function theoretic viewpoint here is in a tradition going back to [**117**] and [**282**].

The development of a detailed theory of function spaces on trees as models for the Dirichlet space and its relatives is in a series of papers by the authors. Much of the material in the first part of the chapter is from [**48**], which also considers similar questions for the analogous Banach spaces, and from [**49**]. The material in Section 5.7 is from [**42**]. The material in Section 5.8 is from [**53**], where some of these ideas are taken much further.

The Carleson measure inequality, aka Trace Inequality (TI), is a norm estimate for a map from an ℓ^2 space to an ℓ^2 space. That map is also studied acting between an ℓ^p and an ℓ^q space, $1 < p, q < \infty$. The type of behavior that is observed differs according to if $p < q$, $p = q$, or $p > q$. Analogous questions are also studied for spaces of smooth functions. Some references are [**37, 48**] and [**99**].

When the approach we used to identify the Carleson measures for $\mathcal{D}(\mathcal{T})$ is used in general, it often produces two conditions which are, together, necessary and sufficient for boundedness; one, in the spirit of (5.18) obtained by testing the operator, and one in the spirit of (TC), obtained by testing the adjoint of the operator. In the particular case of $\mathcal{D}(\mathcal{T})$ that pattern, although present, is hidden because the second condition is implied by the first, Exercise 5.21. The presence of a pair of conditions, one from testing the operator and one from testing the adjoint, is quite common and is very similar to the pair of hypotheses in the famous $T(1)$ theorem of David and Journé, as well as in the testing condition characterizations for positive operators in [**295**]. This is discussed in detail in the introductory section of [**237**] and of [**238**].

There are interesting similarities between the work in this chapter and the work by Evans and collaborators [**135**] and by Naimark and Solomyak [**236**] on classical inequalities for Sobolev spaces on trees.

The paper [**88**] has results on the radial variation of functions in a tree version of the Hardy space, as well as references to other work in the area.

CHAPTER 6

The Pick Property

The theorem of Pick and Nevanlinna about interpolation of values by bounded holomorphic functions in the unit disk is more than a hundred years old. For a long time the theorem was seen as a particularly elegant, but relatively isolated, piece of classical function theory. However, more recently it has become clear that the result is just the most accessible manifestation of a fundamental property, the complete Pick property, that is shared by some, but not all, reproducing kernel Hilbert spaces. When that property is present there is a particularly close relationship between the Hilbert space and its multiplier algebra.

In this chapter we define the complete Pick property and discuss criteria for a space to have the property; the Hardy and Dirichlet space do have it, the Bergman space does not. We also develop some of the consequences of having that property, including results about solutions to multiplier extremal problems and results about zero sets.

A brief overview of the role of Pick's theorem is in [**228**]. A systematic presentation of the theory associated with the Pick property is in [**4**].

6.1. Pick Interpolation

Pick's theorem, Theorem 6.4 below, answers a question about existence of a bounded analytic function on the disk which takes specified values at specified points and has modulus bounded by one.

PROBLEM 6.1 (Pick's Interpolation Question). *What is a necessary and sufficient condition, given sets* $Z = \{z_i\}_{i=1}^n \subset \mathbb{D}$ *and* $W = \{w_i\}_{i=1}^n \subset \mathbb{D}$, *for there to be an* $f \in H^\infty$, $\|f\|_{H^\infty} \leq 1$, *which solves the interpolation problem* $f(z_i) = w_i$ $i = 1, \ldots, n$?

Pick's solution of the problem was published in 1916 [**262**]. A few years later Rolf Nevanlinna [**242**] independently found an alternative solution. The problem is now sometimes called Pick's problem and sometimes goes with both names, Pick-Nevanlinna (chronological) or Nevanlinna-Pick (alphabetical).

We now view this question as asking about interpolation by functions in $\operatorname{Mult}(H)$, the multiplier algebra of a RKHS H; in this particular case $H = H^2$ and thus $\operatorname{Mult}(H) = H^\infty$. Given that viewpoint it is natural to study similar questions for other choices of H.

PROBLEM 6.2 (Generalized Pick's Interpolation Question). *Given a RKHS* H *of functions on* X, *what are necessary and sufficient conditions on sets* $Z =$

$\{z_i\}_{i=1}^n \subset X$, $W = \{w_i\}_{i=1}^n \subset \mathbb{D}$ to ensure that there is an f which solves the following interpolation problem:

$$f(z_i) = w_i, \quad i = 1, \ldots, n; \; f \in \mathrm{Mult}(H), \|f\|_{\mathrm{Mult}(H)} \leq 1 \; ? \tag{6.1}$$

The study of this and related questions has been extraordinarily fruitful and influential, and many of the ideas in this volume have intellectual roots in those studies.

There is a necessary condition for there to be such an f and we develop that now. Suppose H is specified and we are given data for the interpolation question, a finite set $Z = \{z_i\} \subset X$ and a finite set $W = \{w_i\} \subset \mathbb{D}$. Let V be the span of the kernel functions $\{k_i\}$. (**Notation:** Here, and at times later, we are writing k_i for k_{H,z_i}.)

THEOREM 6.3. *Let T be the map of $V = \mathrm{span}\{k_j\}$ to itself given by*

$$T\left(\sum_{i=1}^n a_i k_i\right) = \sum_{i=1}^n a_i \overline{w}_i k_i.$$

A necessary condition for there to be an f which satisfies (6.1) is that $\|T\| \leq 1$. Equivalently, a necessary condition is that the matrix

$$\mathrm{Mx}(T) = \left(\left(1 - \overline{w}_j w_i\right) k_j(z_i)\right)_{i,j=1}^n \tag{6.2}$$

be positive semidefinite; $\mathrm{Mx}(T) \geq 0$.

PROOF. Suppose there is such a multiplier m and let M be the operator of multiplication by m acting on H. We have $\|M\| = \|m\|_{\mathrm{Mult}(H)} \leq 1$. Hence the adjoint operator, M^*, satisfies $\|M^*\| \leq 1$. We know from Proposition 2.17 that for any $\zeta \in \mathbb{D}$, $M^* k_\zeta = \overline{m(\zeta)} k_\zeta$. Thus V is an invariant subspace for M^* and the restriction of M^* to V is the operator T of the theorem. Furthermore, the restriction of M^* to V has, *a fortiori,* norm at most one. That gives the first statement.

The fact that the norm of T^* is at most one means that for any scalars $\{a_i\}_{i=1}^n$ we have

$$\left\|\sum_{i=1}^n a_i \overline{w}_i k_i\right\|^2 \leq \left\|\sum_{i=1}^n a_i k_i\right\|^2.$$

When we compute both norms explicitly in terms of inner products, recall that $\langle k_i, k_j\rangle_H = k_i(z_j)$, and rearrange terms; we find that

$$\sum_{i,j=1}^n \left(1 - w_j \overline{w}_i\right) k_i(z_j)\, \overline{a}_j a_i \geq 0.$$

Having this for arbitrary $\{a_i\}_{i=1}^n$ is exactly the condition that $\mathrm{Mx}(T) \geq 0$. □

The matrix $\mathrm{Mx}(T)$ is called the *Pick matrix* of the problem. For the Hardy space it takes the form

$$\mathrm{Mx}(T) = \left(\frac{1 - w_i \overline{w}_j}{1 - z_i \overline{z}_j}\right)_{i,j=1}^n \tag{6.3}$$

In this language Pick's result is:

THEOREM 6.4. *For $H = H^2$, the necessary and sufficient condition for the existence of an f which satisfies (6.1) is that the Pick matrix, given in (6.3), be positive semidefinite.*

PROOF. We will not include a proof here. There are several proofs available, each with its own insights; see [**161**] or [**4**] for proofs and more information. □

EXERCISE 6.5. *The analog of Pick's theorem fails for the Bergman space:* (6.2) *is not sufficient for there to be an f which satisfies* (6.1).

EXAMPLE 6.6. *When $n = 2$ things can be made very explicit. Given $z_1, z_2, w_1, w_2 \in \mathbb{D}$,*

$$\exists m \in \operatorname{Mult}(H^2),\ m(z_i) = w_i, i = 1, 2,\ \|m\|_{H^\infty} \leq 1 \tag{6.4}$$

if and only if the matrix

$$\begin{pmatrix} \frac{1-|w_1|^2}{1-|z_1|^2} & \frac{1-w_1\overline{w}_2}{1-z_1\overline{z}_2} \\ \frac{1-\overline{w}_1 w_2}{1-\overline{z}_1 z_2} & \frac{1-|w_2|^2}{1-|z_2|^2} \end{pmatrix}$$

is positive semidefinite. For this two by two matrix with nonnegative elements on the diagonal, being positive semidefinite is equivalent to the determinant's being nonnegative,

$$\frac{1-|w_1|^2}{1-|z_1|^2}\frac{1-|w_2|^2}{1-|z_2|^2} - \left|\frac{1-w_1\overline{w_2}}{1-z_1\overline{z_2}}\right|^2 \geq 0.$$

Doing a small bit of algebra, noting (1.6) *and taking note of the definition of the pseudohyperbolic metric, ρ, this can be rewritten as*

$$\rho(w_1, w_2) \leq \rho(z_1, z_2). \tag{6.5}$$

In sum, if m is any analytic map of $\mathbb{D}$ into itself then m must be a (not necessarily strict) contraction in the pseudohyperbolic metric. For any $\alpha, \beta \in \mathbb{D}$,

$$\rho\left(m(\alpha), m(\beta)\right) \leq \rho(\alpha, \beta). \tag{6.6}$$

This statement for maps of $\mathbb{D}$ into itself is sometimes called the Schwarz-Pick lemma; we prove a version of it for general RKHS with CPP in Theorem 6.22 *below. The fact that such maps satisfy* (6.6) *is a consequence of Theorem* 6.3 *and does not require Theorem* 6.4. *That theorem ensures that if the data satisfy* (6.5) *then* (6.4) *holds.*

EXERCISE 6.7 (Equality in the Schwarz-Pick lemma). *Show that, given $\alpha \neq \beta$ then there is a multiplier m for which equality holds in* (6.6). *Furthermore this m is unique and is a linear fractional map of the disk onto itself. Hint: use disk automorphisms to reduce to the case $\alpha = m(\alpha) = 0$.*

6.2. The Complete Pick Property

It is now understood that there are classes of RKHS for which the condition (6.2) is sufficient for the interpolation problem to have a solution. Those spaces, a class including the Hardy space and the Dirichlet space, are said to have the *Pick property.* There is a subclass of that class of spaces, again including H^2 and $\mathcal{D}$, for which the matricial extension of (6.2) is a necessary and sufficient for a matricial extension of the interpolation problem to have a solution. Those spaces are said to have the *complete Pick property (CPP).*

REMARK 6.8. *The Pick property is weaker than the complete Pick property and many of our proofs in which we use the CPP actually only require the PP. However we work consistently with the stronger CPP. The main reason is that the current theory for the CPP is more straightforward. For instance there is no known characterization of the PP which is as simple as the condition* (CPK 1) *below which characterizes the CPP. Consequentially, most spaces which are known to have the PP are shown to have it by first showing the space has the stronger CPP. Also, the Hilbert spaces of primary interest in this volume,* H^2, $\mathcal{D}$, *and the Hilbert spaces of holomorphic functions considered in later chapters, all have the CPP. Finally, although most of the time we invoke the CPP when we could have used the PP, there are two important places where that is not so. The CPP is needed in Chapter* 17 *to let us pass from the Baby Corona Theorem to the classical Corona theorem and is central in Chapter* 7 *on interpolation, although it is not known if the Pick property alone suffices there.*

Rather than consider matricial multipliers we proceed ahistorically and present a more recent condition as our definition of the CPP; in fact the equivalence of the definition we use and the original definition is one of the substantial results of the theory.

We will say a space has the complete Pick property or that its kernel function is a *complete Pick kernel* (CPK) exactly if it satisfies an algebraic condition which we now describe.

We consider a RKHS H defined on a set X. We say that a complex-valued function B, defined on $X \times X$ is a positive semidefinite kernel function if for any choice of $\{x_i\}_{i=1}^n \subset X$ the matrix $(B(x_i, x_j))_{i,j=1}^n$ is positive semidefinite. We say that $k_z(w)$, the reproducing kernel for H, is *irreducible* if it satisfies the first two conditions below and an irreducible *complete Pick kernel, CPK*, if it satisfies all three:

(1) For any $x, y \in X$ the functions k_x and k_y are linearly independent.
(2) For any $x, y \in X$, $k_x(y) \neq 0$.
(3) There is a positive semidefinite kernel function B on $X \times X$ with modulus at most 1, and a non vanishing function δ defined on X so that

$$\text{(CPK 1)} \qquad k_x(y) = \frac{\delta(y)\overline{\delta(x)}}{1 - B(y,x)}.$$

Irreducibility excludes examples such as the direct sum of two copies of the Hardy space. Here is the example that shows how property (2) can fail in simple examples with direct sums. Define $2H^2$ to be two disjoint copies of H^2

$$2H^2 := \left\{(f_1, f_2) : f_1, f_2 \in H^2\right\}$$

These are functions on

$$\mathbb{D}_1 \cup \mathbb{D}_2 = \{z_1 : |z_1| < 1\} \cup \{z_2 : |z_2| < 1\}$$

with the evaluations, for $z_i \in \mathbb{D}_i$ given by

$$(f_1, f_2)(z_1) = f_1(z_1), \quad (f_1, f_2)(z_2) = f_2(z_2)$$

the inner product is

$$\langle (f_1, f_2), (g_1, g_2)\rangle_{2H^2} = \langle f_1, f_2\rangle_{H^2} + \langle g_1, g_2\rangle_{H^2}$$

and so the reproducing kernel for, say, $\alpha_1 \in \mathbb{D}_1$ is $K_{\alpha_1} = (k_{\alpha_1}, \mathbf{0})$ where k_{α_1} is the H^2 kernel for the point α_1 and $\mathbf{0}$ is the zero vector in H^2. Hence, for any $\alpha_2 \in \mathbb{D}_2$,

$$K_{\alpha_1}(\alpha_2) = (k_{\alpha_1}, \mathbf{0})(\alpha_2) = \mathbf{0}(\alpha_2) = 0$$

So this example fails property (2).

The kernels we consider will be irreducible and we will regard irreducibility as a standing assumption. If H has an irreducible complete Pick kernel we will say H has the complete Pick property.

It is not difficult to pass from the definition we just gave to the following variation. Suppose H is a RKHS of functions on X having irreducible kernel $k_y(x)$. The kernel has the CPP if and only if there is a $\beta \in X$ and a positive semidefinite kernel function B_β on $X \times X$ with modulus at most 1, so that for all $x, y \in X$

$$k_x(y) = \frac{k_x(\beta)k_\beta(y)}{k_\beta(\beta)} \frac{1}{1 - B_\beta(y, x)}. \tag{CPK 2}$$

If a RKHS has the CPP then we can solve the Pick interpolation problem.

THEOREM 6.9. *If H is a RKHS with a CPK then the necessary condition for the Generalized Pick's Interpolation Question to have a positive solution, given in Theorem 6.3, is also sufficient for the problem to have a solution.*

Although we will use this result going forward, we will not prove it here. It is a special case of [**4**, Theorem 7.28].

6.3. Rescaling

Under what circumstances should we regard two RKHS, H and J, as the same? If H and J have the same dimension they are isometrically isomorphic, but that is independent of their structure as RKHS and so is not the relation we want. A useful notion is that one space is a *rescaling* of the other. Suppose that H is a RKHS of functions on X and J on Y. We say J is a rescaling of H if there is a unitary map U of H to J, and a nonvanishing function c defined on X, and a map $\Phi : X \to Y$ so that for all $x \in X$ we have

$$U(k_{H,x}) = c(x) k_{J,\Phi(x)}.$$

In some contexts it would be said that J is obtained from H by a *gauge transformation.*

EXERCISE 6.10. *This is an equivalence relation on the class of RKHS.*

EXERCISE 6.11. *Suppose γ is a nonvanishing holomorphic function on the disk. Consider the space J of functions $\{\gamma f : f \in H^2\}$ with the inner product*

$$\langle \gamma f, \gamma g \rangle_J = \langle f, g \rangle_{H^2}.$$

Show that J is a RKHS which is a rescaling of H^2.

A common case is $X = Y$ and Φ is the identity map. In that case the two spaces are related by rescaling exactly if there is a nonvanishing function γ defined on X such that $\forall x, y \in X$

$$k_{J,x}(y) = \gamma(y)\overline{\gamma(x)} k_{H,x}(y). \tag{6.7}$$

If, further, k_H is a CPK with representation (CPK 1) then k_J is also a CPK and has a representation of the form (CPK 1) with the same B but with δ replaced by $\tilde{\delta} = \gamma\delta$. In particular J can be constructed so that $\tilde{\delta} = 1$.

EXERCISE 6.12. *Conversely, if H is a RKHS with CPK given by* (CPK 1)*, then replacing δ with $\delta\gamma$ in* (CPK 1) *produces the kernel function $k_{J,x}$ described in* (6.7).

If H and J are related this way then it is straightforward to verify the following facts by working directly with the kernel functions:

(1) $J = \gamma H = \{\gamma f : f \in H\}$,
(2) The inner products for J and H are related as follows: for $h_1, h_2 \in H$, $j_1 = \gamma h_1$, $j_2 = \gamma h_2 \in J$;

$$\langle j_1, j_2 \rangle_J = \langle \gamma h_1, \gamma h_2 \rangle_J = \langle h_1, h_2 \rangle_H .$$

(3) The map $\Delta : H \to J$ given by $\Delta(h) = \gamma h$ is a unitary map of H to J.
(4) The multiplier algebras $\mathrm{Mult}(J)$ and $\mathrm{Mult}\,(H)$ are the same; that is, they are the same sets of functions, both with product given by pointwise multiplication, and they have the same norm.

It is often obvious from descriptions of spaces that one is a rescaling of the other; however, sometimes not.

EXAMPLE 6.13. *Suppose Z is the two dimensional subspace of the Hardy space spanned by kernel functions, k_{z_1} and k_{z_2}, and, similarly, W is spanned by k_{w_1} and k_{w_2}. Show that Z and W are rescalings of each other if and only if the two sets $\{z_1, z_2\}$ and $\{w_1, w_2\}$ are related by a disk automorphism. Show that a similar condition is sufficient for a pair of n dimensional subspaces of this sort to be rescalings of each other.*

EXAMPLE 6.14. *At* (3.3) *we saw that for $f \in H^2$ we had $\|f\|_{H^2}^2 = \int_{\mathbb{T}} \left|f^\beta\right|^2 d\theta$. Suppose we have a positive function on the circle, $1 \le W(e^{i\theta}) \le 2$, and define an equivalent norm on H^2 by $\|f\|_{\mathtt{new}}^2 = \int_{\mathbb{T}} \left|f^\beta\right|^2 W d\theta$. Show that the Hilbert space defined by this new norm is a rescaling of H^2. (Use the factorization theorem to find $\gamma(z)$ so that the new normed space is, isometrically, γH^2). The analogous statement is not generally true for the Bergman space, see* [**50**].

REMARK 6.15. *Some natural quantities associated with an RKHS are unchanged by rescaling. For example the inner product of normalized reproducing kernels stays the same, and hence so does the distance function δ of* (2.22)*. However function values and the norms of reproducing kernels may change. This is awkward because function values and reproducing kernels were both big parts of our story when we introduced the idea of RKHS. However from the viewpoint of equivalence classes of RKHS these same quantities now appear as circumstantial artifacts.*

6.4. Identifying CPKs

Theorem 6.9 is a first indication that knowing H has a CPK is a very valuable piece of information about H. In some cases, for instance if $H = H^2$, it is clear how to find δ and B so that (CPK 1) is satisfied, and thus see that H^2 has a CPK. However it is not clear how to proceed for a general H. Here we describe two situations in which we can go further.

6.4.1. Diagonal Kernels. For many of the spaces of primary interest to us (CPK 1) can be made much more tractable.

Suppose for the remainder of this section that we are considering H, a RKHS of holomorphic functions on the unit disk $\mathbb{D}$. Suppose further, as was the case with $\mathcal{D}$, H^2, and A^2, that the monomials are an orthogonal basis. In such cases the analysis we did in Chapter 2 that led to the formulas for the kernel functions for $\mathcal{D}$, H^2, and A^2 can be repeated. The normalized monomials will be an orthonormal basis (the normalization depending on H). We can then write the kernel as

$$k_z(w) = \sum_{n=0}^{\infty} a_n w^n \overline{z}^n \tag{6.8}$$

where $a_n = \|z^n\|^{-2}$ if $\|z^n\| \neq 0$, $a_n = 0$ otherwise. We will say such a kernel function is *diagonalized* by the monomials.

We want to give conditions on a kernel function of the form (6.8) which ensure that we have a representation of the form (CPK 1). We suppose that $a_0 = \|1\|_H^{-2} > 0$ and that $\sum_{n=0}^{\infty} a_n t^n$ has a positive radius of convergence. In that case the function $\left(\sum_{n=0}^{\infty} a_n t^n\right)^{-1}$ has a power series that is convergent in a neighborhood of the origin. Define $\{c_n\}$ by

$$\left(\sum_{n=0}^{\infty} c_n t^n\right)\left(\sum_{n=0}^{\infty} a_n t^n\right) = 1. \tag{6.9}$$

Having $a_0 > 0$ ensures $c_0 > 0$.

THEOREM 6.16 (Quigen [**265**]). *The space H has a CPK if and only if*

$$c_n \leq 0 \quad \forall n > 0.$$

PROOF. We will show (CPK 1) is satisfied; there is a δ so that

$$B(x, y) = 1 - \frac{\delta(x)\delta(\overline{y})}{k(x, y)}$$

is a positive kernel. By selecting $\delta = a_0^{1/2}$ we reduce to considering the case $a_0 = 1$, in which case $c_0 = 1$ and the signs of the other c_n are unchanged. We then have, with the new c_n,

$$B(w, z) = -\sum_{n \neq 0} c_n w^n \overline{z}^n.$$

With these choices we have the formula (CPK 1). The fact that each c_n is nonpositive ensures that B is nonnegative, as can be seen by considering each $c_n w^n \overline{z}^n$ separately.

Because B is a positive kernel we have, for all x, y, $|B(x, y)|^2 \leq B(x, x)B(y, y)$. Hence the required $|B(x, y)| \leq 1$ is a consequence of that same estimate with $x = y$, and that case is elementary.

The proof in the other direction, if one of the coefficients is positive then no such representation is possible requires more work; it is Theorem 7.33 in [**4**]. □

EXERCISE 6.17. *The Hardy space has a CPK as do the spaces $\mathcal{D}_\alpha$, $0 < \alpha < 1$, defined in* (2.43).

The same is true of the spaces of functions in $\mathbb{C}^n$ with kernel functions $k_z(w) = (1 - \langle w, z\rangle)^{-\alpha}$, $0 < \alpha \leq 1$; and for essentially the same reason. We will consider those spaces in later chapters.

EXERCISE 6.18. *The Bergman space A^2 fails to have a CPK.*

The Dirichlet space also has a CPK. That follows from the previous theorem and the following fact about power series.

LEMMA 6.19 (Kaluza [**188**]). *Suppose the power series $\sum_{n=0}^{\infty} a_n t^n$ and $\sum_{n=0}^{\infty} c_n t^n$ are related as in (6.9). Suppose further that for all n we have $a_n > 0$ and that the sequence of ratios a_{n+1}/a_n, $n = 0, 1, 2, \ldots$ is nondecreasing. Then $c_n \leq 0$ for all $n > 0$.*

PROOF. By hypothesis $a_0 > 0$ so without loss of generality we can suppose $a_0 = 1$. In that case we find $c_1 = -a_1 < 0$. We now proceed by induction. Starting from (6.9), clearing the fractions, multiplying the power series, and then considering the coefficients of t^n and of t^{n+1} we find

$$\begin{aligned} a_0 c_n + a_1 c_{n-1} + \cdots + a_n c_0 &= 0 \\ a_0 c_{n+1} + a_1 c_n + \cdots + a_{n+1} c_0 &= 0. \end{aligned}$$

We now use these two equations to obtain an expression for c_{n+1}. We multiply the first equation by a_{n+1}, the second by a_n, and then subtract the first from the second to obtain

$$a_n a_0 c_{n+1} + (a_1 a_n - a_0 a_{n+1}) c_n + \cdots + (a_n a_n - a_{n-1} a_{n+1}) c_1 + 0 c_0 = 0.$$

From this it follows that

$$\begin{aligned} a_n a_0 c_{n+1} = a_0 a_n \left(\frac{a_{n+1}}{a_n} - \frac{a_1}{a_0} \right) c_n + a_1 a_n \left(\frac{a_{n+1}}{a_n} - \frac{a_2}{a_1} \right) c_{n-1} + \cdots \\ + a_{n-1} a_n \left(\frac{a_{n+1}}{a_n} - \frac{a_n}{a_{n-1}} \right) c_1. \end{aligned}$$

The hypotheses ensure that the a_k are positive and that the expressions in parentheses are non-negative. The induction hypothesis ensures that the c_j on the right hand side are non-positive. Thus $c_{n+1} \leq 0$ and we are done. □

EXERCISE 6.20. *The space $\mathcal{D}$ normed by*

$$\left\| \sum_{n=0}^{\infty} a_n z^n \right\|_{\mathcal{D}}^2 = \sum_{n=0}^{\infty} (n+1) |a_n|^2 = \|a\|_{H^2}^2 + \mathfrak{D}(a)$$

has a CPK. The same space of functions with the alternative, equivalent, norm

$$\left\| \sum_{n=0}^{\infty} a_n z^n \right\|_{\mathcal{D},\mathtt{alt}}^2 = |a_0|^2 + \sum_{n=1}^{\infty} n |a_n|^2 = |a(0)|^2 + \mathfrak{D}(a) \tag{6.10}$$

does not have a CPK.

6.4.2. Shimorin's Lemma. Theorem 6.16 and its straightforward extension to functions of several variables apply to a large variety of frequently studied RKHS, but where that theorem does not apply it is difficult to detect if a kernel function is a CPK. However in [**308**] Shimorin showed that the generalized Dirichlet spaces defined in Section 2.5.8 via a positive superharmonic function U have CPK. The first step in his argument is very general. He shows that the property of having a CPK is stable under certain elementary perturbations of the structure. We present that now.

Suppose H is a RKHS on X with a CPK and with norm and inner product denoted by $\|\cdot\|_{\text{o}}$ and $\langle f, g\rangle_{\text{o}}$, and with kernel functions k_{o}. Suppose $x^* \in X$, and $\alpha > 0$ are given, and are such that for every f in H other than the zero function

$$\|f\|_{\text{n}}^2 := \|f\|_{\text{o}}^2 - \alpha\,|f(x^*)|^2 > 0. \tag{6.11}$$

Thus $\|\cdot\|_{\text{n}}$ is a new inner product built from the old one $\|\cdot\|_{\text{o}}$.

Proposition 6.21 (Shimorin). *If $(H, \|\cdot\|_{\text{o}})$ is a RKHS with a CPK, then $(H, \|\cdot\|_{\text{n}})$ is a RKHS with equivalent norm and with a CPK.*

Proof. The condition (6.11) ensures that the Hermitian form

$$\langle f, g\rangle_{\text{n}} = \langle f, g\rangle_{\text{o}} - af(x^*)\overline{g(x^*)} \tag{6.12}$$

is positive and hence that $\|\cdot\|_{\text{n}}$ is a Hilbert space norm.

By the Cauchy Schwarz inequality $|f(x^*)|^2 \leq \|f\|_{\text{o}}^2 k_{\text{o},x^*}(x^*)$. Evaluating (6.11) for $f = k_{\text{o},x^*}$ and noting that $\|k_{\text{o},x^*}\|_{\text{o}}^2 = k_{\text{o},x^*}(x^*)$ we see that $ak_{\text{o},x^*}(x^*) < 1$. Combining these two facts with (6.11) we find that for any f

$$\|f\|_{\text{o}}^2 \geq \|f\|_{\text{n}}^2 \geq \|f\|_{\text{o}}^2 - a\,\|f\|_{\text{o}}^2\, k_{\text{o},x^*}(x^*) \gtrsim \|f\|_{\text{o}}^2,$$

and thus the two norms are equivalent. In particular, the two spaces have the same set of bounded point evaluations.

We now want to find the new kernel function. Noting Proposition 2.8, the new kernel function $k_{\text{n},y}$ is a scalar multiple of the function f which has (new) norm one and which maximizes $\operatorname{Re} f(y)$. Suppose we have a candidate f for that extremal problem. If we project f onto the two dimensional subspace spanned by k_{o,x^*} and $k_{\text{o},y}$ (orthogonal projection in the $\langle\cdot,\cdot\rangle_{\text{o}}$ structure) then we obtain $\tilde{f}$, a better competitor. That is, we have $f(y) = \tilde{f}(y)$. Also $\left\|\tilde{f}\right\|_{\text{o}} \leq \|f\|_{\text{o}}$ and $f(x^*) = \tilde{f}(x^*)$, hence $\left\|\tilde{f}\right\|_{\text{n}} \leq \|f\|_{\text{n}}$. Thus the search for an extremal function can be confined to candidates of the form $\alpha k_{\text{o},y} + \beta k_{\text{o},x^*}$ for scalars α, β. One then checks by explicit computation that, setting

$$\gamma = \frac{ak_{\text{o},x^*}(x^*)}{1 - ak_{\text{o},x^*}(x^*)},$$

the function

$$k_{\text{n},y}(\cdot) = k_{\text{o},y}(\cdot) + \gamma\frac{k_{\text{o},y}(x^*)}{k_{\text{o},x^*}(x^*)}k_{\text{o},x^*}(\cdot) \tag{6.13}$$

is a reproducing kernel when used with the inner product (6.12).

To show this is a CPK we will use this formula for the kernel and use the fact that k_{o} has a representation (CPK 2). We then do some algebra to conclude that k_{n} has a similar representation.

By evaluating the previous expression at x^* we find, suppressing the subscript y,

$$k_{\mathrm{n}}(x^*) = (1+\gamma)\, k_{\mathrm{o}}(x^*). \tag{6.14}$$

When we rewrite (CPK 2) we obtain

$$k_{\mathrm{o},x}(y) - \frac{k_{\mathrm{o},\alpha}(y)k_{\mathrm{o},x}(\alpha)}{k_{\mathrm{o},a}(\alpha)} = B_{\mathrm{o}}(y,x)k_{\mathrm{o},x}(y).$$

Straightforward calculation using the previous two equations yields

$$k_{\mathrm{n},x}(y) - \frac{k_{\mathrm{n},\alpha}(y)k_{\mathrm{n},x}(\alpha)}{k_{\mathrm{n},a}(\alpha)} = B_{\mathrm{n}}(y,x)k_{\mathrm{n},x}(y) \tag{6.15}$$

with

$$B_{\mathrm{n}} = \frac{B_{\mathrm{o}}}{\gamma+1-\gamma B_{\mathrm{o}}} = \sum_{j=0}^{\infty} \frac{\gamma^j}{(\gamma+1)^{j+1}} B_{\mathrm{o}}^j. \tag{6.16}$$

The equation (6.15) expresses the new kernel k_{n} in the required form (CPK 2). We need to show that B_{n} has the required properties. Each B_{o}^j is a positive definite kernel because the Schur (elementwise) product of positive definite matrices is positive definite. Hence B_{n} which is a convergent linear combination of the B_{o}^j with positive coefficients is itself a positive definite kernel. Using (6.16) the size estimate on B_{n} follows from that for B_{o} after summing the geometric series. □

To go further Shimorin next notes that this construction can be applied repeatedly, allowing modifications of the original norm involving finitely many points. He then obtains very interesting results by studying the limiting behavior of such modifications as the number of point modifications grows. In particular he shows that the weighted Dirichlet spaces described in Section 2.5.8 all have property CPK. Other applications of the lemma are given by Serra in [**303**].

6.5. Consequences of the Pick Property

We now look at some consequences of having a CPK.

6.5.1. The General Schwarz Pick Lemma. We saw in Example 6.6 that, for the Hardy space, the positivity of the two by two Pick matrix was related to the classical Schwarz Pick lemma. The same argument applies more generally.

THEOREM 6.22 (General Schwarz Pick Lemma). *Suppose H is a RKHS on X. Suppose $z_1, z_2 \in X$, $z_1 \neq z_2$; $w_1, w_2 \in \mathbb{D}$ are given. If there is an $m \in \mathrm{Mult}\,(H)$, $\|m\|_{\mathrm{Mult}\,(H)} \leq 1$, with $m(z_i) = w_i$, $i = 1, 2$ then*

$$\rho(w_1, w_2) \leq \delta_H(z_1, z_2). \tag{6.17}$$

If, additionally, H has a CPK then the previous inequality is sufficient for the existence of such a multiplier m.

PROOF. As in Example 6.2 we must find out when the two by two Pick matrix is positive semidefinite. The matrix is

$$M = \begin{pmatrix} (1-\overline{w_1}w_1)\, k_{z_1}(z_1) & (1-\overline{w_1}w_2)\, k_{z_1}(z_2) \\ (1-\overline{w_2}w_1)\, k_{z_2}(z_1) & (1-\overline{w_2}w_2)\, k_{z_2}(z_2) \end{pmatrix}.$$

The w_i are in the disk and hence the diagonal elements are nonnegative. Thus M is positive semidefinite if and only if the determinant is nonnegative. Computing the determinant, writing the inequality, and rearranging gives

$$\frac{(1 \quad \overline{w_1}w_1)(1-\overline{w_2}w_2)}{(1-\overline{w_2}w_1)(1-\overline{w_1}w_2)} \geq \frac{k_{z_2}(z_1)k_{z_1}(z_2)}{k_{z_1}(z_1)k_{z_2}(z_2)}.$$

By (1.6) this is equivalent to

$$1-\rho^2(w_1,w_2) \geq \frac{|\langle k_{z_1},k_{z_2}\rangle|^2}{\|k_{z_1}\|^2\,\|k_{z_2}\|^2}$$

which is in turn equivalent to (6.17). □

Thus, as was the case in Example 6.6, by Theorem 6.3, if there is such a multiplier then (6.17) holds. In the presence of a CPK, Theorem 6.22 applies and if (6.17) holds there must be such a multiplier.

We could have reduced the previous proof to the special case $w_1 = 0$ by using the following Proposition.

PROPOSITION 6.23. *Suppose H is a RKHS on X. Suppose $z_1, z_2 \in X$, $z_1 \neq z_2$; $w_1, w_2 \in \mathbb{D}$ are given. There is an $m \in \operatorname{Mult}(H)$, $\|m\|_{\operatorname{Mult}(H)} \leq 1$, with $m(z_i) = w_i$, $i = 1, 2$ if and only if there is a $\mu \in \operatorname{Mult}(H)$, $\|\mu\|_{\operatorname{Mult}(H)} \leq 1$, $\mu(z_1) = 0$, $\mu(z_2) = \rho(w_i, w_2)$.*

PROOF. Suppose m is given. Define L by

$$L(w) = \alpha\frac{w_1 - w}{1-\overline{w_1}w}$$

where α is the unimodular constant which makes $L(w_2)$ positive. L is a conformal automorphism of the disk and is selected so that $\mu(z) = L(m(z))$ takes the required values. We want to show μ is a multiplier of norm at most 1. If we had $H = H^2$ we would have $\operatorname{Mult}(H) = H^\infty$ and the multiplier norm would be the supremum. In that case, recalling that L is an automorphism we would continue with

$$\|m\|_\infty = \sup|m(z)| \leq 1 \Longrightarrow \|\mu\|_\infty = \sup|L(m(z))| \leq 1$$

and thus μ would be a multiplier of norm at most one.

In general the multiplier norm is more complicated than the supremum norm, but in the general case we can appeal to von Neumann's inequality [**4**]. More specifically, because M_m, multiplication by m, is a contraction on a Hilbert space, and L is holomorphic on the closed disk, we have the operator norm estimate

$$\|L(M_m)\| \leq \|L\|_{H^\infty}\|M_m\| = \|M_m\| = \|m\|_{\operatorname{Mult}(H)} \leq 1.$$

On the other hand, one checks that .

$$L(M_m) = M_{L(m)} = M_\mu.$$

Hence μ is the required multiplier.

For the other half of the proof, repeat the same argument but using L^{-1} to go from μ to m. □

6.5.2. The Gleason Metric. Suppose H is a RKHS on X. The Gleason metric [**163**], $d_{H,G}$, is a metric on $\hat{X}$, the unit ball of the dual space of the multiplier algebra of H. Using the previous proposition we can see the equality of two alternative definitions of $d_{H,G}(\cdot,\cdot)$:

$$\begin{aligned} d_{H,G}(\alpha,\beta) &:= \sup\left\{\rho(\alpha(m),\beta(m)) : m \in \operatorname{Mult}(H), \|m\|_{\operatorname{Mult}(H)} \leq 1\right\} \\ &= \sup\left\{|\beta(m)| \ : m \in \operatorname{Mult}(H), \|m\|_{\operatorname{Mult}(H)} \leq 1,\ \alpha(m) = 0\right\}. \end{aligned}$$

There is a natural inclusion of X into $\hat{X}$ as point evaluations; i.e. for $x \in X$ define $e_x \in \hat{X}$ by $e_x(m) = m(x)$. Using this we can regard $d_{H,G}$ as a metric on X; $d_{H,G}(x,y) = d_{H,G}(e_x, e_y)$. In general $d_{H,G}$ gives smaller distances than the metric δ_H, however, the two are equal if H has a CPK.

COROLLARY 6.24. *For general H, $d_{H,G} \leq \delta_H$. If H has a CPK then the two metrics agree on X.*

PROOF. After unpacking the definitions, the Corollary is a direct consequence of Proposition 2.19. □

6.5.3. Function Spaces on the Disk. In the class of metrics induced on $\mathbb{D}$ by a RKHS with a CPK and consisting of holomorphic functions, the pseudohyperbolic metric δ_{H^2} is the largest.

THEOREM 6.25. *Suppose that H is a RKHS of holomorphic functions on $\mathbb{D}$ which has a CPK, then $\delta_H \leq \delta_{H^2}$. That is, $\forall z, w \in \mathbb{D}$*

$$\delta_H(z,w) \leq \delta_{H^2}(z,w). \tag{6.18}$$

Furthermore, if equality holds for some pair of points then it holds for all points and H is (a rescaling of) H^2.

PROOF. Because both spaces have CPK we can take note of the previous corollary and it suffices to prove the result for the Gleason metrics $d_{H,G}$ and $d_{H^2,G}$. However now the inequality is immediate. By Corollary 2.22 the unit ball of $\operatorname{Mult}(H^2)$ is the unit ball of H^∞. By Corollary 2.18 the unit ball of $\operatorname{Mult}(H)$ is a subset of the unit ball of H^∞. Hence the supremum defining $d_{H,G}$ is taken over a subset of the set used in the supremum defining $d_{H^2,G}$; thus the inequality (5.20).

Suppose now that equality holds for some z, w. Then there is an m in $\operatorname{Mult}(H)$ of norm at most one for which

$$\rho(m(z), m(w)) = d_{H,G}(z,w) = d_{H^2,G}(z,w) = \rho(z,w).$$

By Corollary 2.18 m must be an element of H^∞ of norm at most one. By Exercise 6.7 we conclude that m is a linear fractional map of the disk to itself. Using this multiplier m in Proposition 6.23 one then checks that (6.18) holds for *all* z, w in the disk.

Pick z_0 in $\mathbb{D}$ and consider the Pick problem in $\operatorname{Mult}(H)$ of finding the multiplier of norm one which vanishes at the origin and maximizes $|m(z_0)|$. The classical Schwarz lemma ensures this maximum is at most $|z_0|$. The form of the extremal problem ensures that the maximum is attained. Hence, because the space H has the Pick property, the Pick matrix for the problem must be positive semidefinite but not strictly positive definite. In short, the two by two matrix

$$\begin{pmatrix} k_H(0,0) & k_H(0,z_0) \\ k_H(z_0,0) & (1-|z_0|^2)k_H(z_0,z_0) \end{pmatrix}$$

must have determinant zero. On rearranging that gives

$$k_H(z_0, z_0) = \frac{|k_H(z_0, 0)|^2}{k_H(0, 0)} \frac{1}{1 - |z_0|^2}.$$

Hence, writing $\delta(z_0) = k_H(z_0, 0)/ |k_H(0, 0)|^{1/2}$, discarding the subscript (recalling that z_0 was arbitrary) we have

$$k_H(z, w) = \delta(z)\overline{\delta(w)}k_{H^2}(z, w)$$

for $z = w$. However, given that each side is holomorphic in z and conjugate holomorphic in w that is enough to ensure the equality for all z, w. That equality expresses the claimed rescaling. □

The proof actually establishes several intertwined conclusions. First, even without the assumption of CPK, we have the inequality between the Gleason metrics $d_{H,G} \leq d_{H^2,G}$. Second, if equality holds in that inequality for a single pair of points then it holds for all points and the coordinate function $m(z) = z$ is a multiplier of H of norm one. Third, in the presence of a CPK, the fact that $m(z) = z$ is a multiplier of norm one ensures that H is a rescaling of the Hardy space. This last fact is [**4**, Theorem 5.37] which is the source of the second part of the previous proof; [**4**, Theorem 5.38] considers a variation on these ideas in which the functions in H are not assumed to be holomorphic.

6.5.4. Extremal Problems.

6.5.4.1. *Pairs of Extremal Problems.* Suppose now that H is a RKHS with a CPK. In that case there is a rich relationship between extremal problems for the Hilbert space and those for its multiplier algebra $\text{Mult}(H)$.

Suppose $Z = \{z_i\}_{i=1}^n \subset X$ is a set of distinct points in X and $w \in X \setminus Z$. Let $F_{Z,w}$ denote the function in H which solves the extremal problem

$$\sup \{\text{Re}\, f(w) : f \in H, \|f\|_H = 1, f(z_i) = 0,\ i = 1, \ldots, n\}. \tag{6.19}$$

Hilbert space considerations ensure that the supremum is obtained and that the extremal function is unique. To see this consider first the problem of finding g which vanishes on Z, takes the value 1 at w and has the smallest norm. That problem asks for the g of smallest norm in a closed convex set of a Hilbert space, and by Theorem A.8 that Hilbert space geometry ensures that such g exists and is unique. It is then straightforward to check that $f = g/ \|g\|_H$ solves the original extremal problem.

Let V_Z be the subspace of H consisting of all functions which vanish at all points of Z. Then V_Z is also a RKHS on X and from Proposition 2.8 we see that $k_{Z,w} = F_{Z,w}(w) F_{Z,w}$ is the reproducing kernel in the space V_Z for the point w. That is, $k_{Z,w} \in V_Z$ and $\forall g \in V_Z$, $g(w) = \langle g, k_{Z,w}\rangle$. We write $\hat{k}_{Z,w} = k_{Z,w}/ \|k_{Z,w}\|_H = F_{Z,w}$ for the normalized kernel.

In the classical case of the Hardy space everything can be made explicit.

EXERCISE 6.26. *Show that if $H = H^2$, $Z = \{z_i\}_{i=1}^n \subset \mathbb{D}$, $0 \notin Z$, then $F_{Z,0}$ is a finite Blaschke product with zero set Z. Does the analogous statement hold with V_Z replaced by the most general invariant subspace, IH^2, where I is a fixed inner function?*

There is one aspect of the previous exercise which appears to be particular to H^2 rather than a general RKHS with a CPK. The exercise considered $F_{Z,0}$ but the same argument applies to $F_{Z,w}$ for any $w \notin Z \cup \{0\}$. Thus $F_{Z,0}$ and $F_{Z,w}$ are two Blaschke products with the same zero set. Hence they differ only by a unimodular factor. In particular, for any ζ the value of $|F_{z,w}(\zeta)|$ is independent of w.

We now consider a similar problem, but looking for an extremal function in the multiplier algebra $\mathrm{Mult}(H)$. For the same H, Z, and w, let $M_{Z,w}$ be the multiplier of norm at most one which vanishes at all points of Z and, subject to those constraints, maximizes $\operatorname{Re} M(w)$. Thus

(6.20)

$$\operatorname{Re} M_{Z,w}(w) = \sup\left\{\operatorname{Re} m(w) : m \in \mathrm{Mult}(H), \|m\|_{\mathrm{Mult}(H)} \leq 1, m = 0 \text{ on } Z\right\}.$$

PROPOSITION 6.27. *There is a unique function $M_{Z,w} \in \mathrm{Mult}(H)$ for which the supremum in (6.20) is attained. It is characterized by*

$$M_{Z,w}\hat{k}_w = \hat{k}_{Z,w} = F_{Z,w}. \tag{6.21}$$

PROOF. In the proof we write $z_0 = w$. We know $F_{Z,w}$ is a unit vector and vanishes on Z. Hence, for any scalars $\{\lambda_j\}_{j=0}^n$,

$$|\lambda_0 F_{Z,w}(z_0)| = \left|\left\langle F_{Z,w}, \sum_{i=0}^n \lambda_i k_{z_i}\right\rangle\right| \leq \left\|\sum_{i=0}^n \lambda_i k_{z_i}\right\|_H.$$

Thus, setting $\xi_0 = |F_{Z,w}(z_0)| \,/\, \|k_w\|$ and $\xi_1 = \cdots = \xi_n = 0$ we have

$$0 \leq \left\|\sum_{i=0}^n \lambda_i k_{z_i}\right\|_H^2 - |\lambda_0 F_{Z,w}(z_0)|^2 = \sum_{j,m=0}^n \left(1 - \xi_j \bar{\xi}_m\right) k_{z_j}(z_m) \lambda_j \bar{\lambda}_m. \tag{6.22}$$

Associated with the question of the existence of a multiplier, G, of norm at most one, and with $G(z_j) = \zeta_j$, $j = 0, 1, \ldots, n$ is a Pick matrix M. The previous inequality is exactly the statement that $M \geq 0$. Hence, because we assumed H has the CPP, we can invoke Theorem 6.9 and conclude that there is a multiplier $M_{Z,w}$, with $\|M_{Z,w}\|_{\mathrm{Mult}(H)} \leq 1$, which has

$$M_{Z,w}(w) = \xi_0 = \frac{|F_{Z,w}(z_0)|}{\|k_w\|}, \quad M_{Z,w}(z_i) = 0, \ i = 1, 2, \ldots, n.$$

Consider now the function $h = M_{Z,w}\hat{k}_w$. Because $M_{Z,w}$ has multiplier norm at most one, $\|h\|_H \leq 1$. Also $h(z_i) = 0$, $i = 1, \ldots, n$; hence h is a competitor for the extremal problem which was used to define $F_{Z,w}$. Because of this $|F_{Z,w}(z_0)| = \operatorname{Re} h(w) \leq \operatorname{Re} F_{Z,w}(z_0)$. Thus h is a solution to that extremal problem. However that solution is unique; hence $h(z) = F_{Z,w}(z)$ and we have (6.21). Knowing that F is unique it follows from (6.21) that the solution to the extremal multiplier problem, $M_{Z,w}$, is also unique. □

We know that any multiplier of norm one takes values of modulus at most one. Hence

COROLLARY 6.28. *Given Z, w; for any $\zeta \in X$*

(1) $k_{Z,w}(w) < k_w(w)$,
(2) $\left|\hat{k}_{Z,w}(\zeta)\right| \leq \left|\hat{k}_w(\zeta)\right|$,
(3) $|k_{Z,w}(\zeta)| \leq |k_w(\zeta)|$.

PROOF. By Proposition 2.12 $k_{Z,w}$ is the projection of k_w onto a subspace and hence $k_{Z,w}$ has smaller norm. From Proposition 2.8 we know that for any kernel function j_λ the norm is given by $\|j_\lambda\|_H^2 = j_\lambda(\lambda)$. Combining these facts gives the first statement. The second statement is a consequence of (6.21) and the fact that $M_{Z,w}$, being a multiplier of norm at most one also has supremum at most one, Corollary 2.18. The third inequality is a consequence of the second and the just mentioned inequality between the norms. □

EXERCISE 6.29. *The analog of the corollary fails for the Bergman space.*

Because $M_{Z,w}$ has norm at most one, the operator $M_{Z,w}^* M_{Z,w}$ is a positive operator of norm at most one. Hence any eigenvalue it has must be in the interval $[0,1]$. In fact k_w is an eigenvector of $M_{Z,w}^* M_{Z,w}$ with eigenvalue 1. That is:

COROLLARY 6.30. $M_{Z,w}^* M_{Z,w} k_w = k_w$.

EXERCISE 6.31. *Prove the corollary.*

6.5.4.2. *Two Point Extremal Functions.* The function $F_{Z,w}$ is the renormalized projection of the kernel function k_w onto the subspace $(\bigvee_i \{k_{z_i}\})^\perp$ and hence can be studied using linear algebra; that is done in [**306**], [**294**], and [**4**]. We will not do that here but we will make the details of the previous analysis explicit for $n = 2$.

We are given H a RKHS on X which we suppose has a CPK. We are given x_1, $x_2 \in X$ and we are interested in the multiplier of norm one which vanishes at the first point and is as large as possible at the second. Thus we want to consider M which solves (6.20). It will have

$$\begin{aligned} &M \in \operatorname{Mult}(H), \qquad \|M\|_{\operatorname{Mult}(H)} \le 1, \\ &M(x_1) = 0, \qquad M(x_2) = \delta_H(x_1, x_2). \end{aligned}$$

Following the pattern of the proof of the previous proposition we also consider $F \in H$ which satisfies the associated (6.19). It will be of norm one, vanish at x_1, and maximize $\operatorname{Re} f(x_2)$. By Proposition 2.13 we have a formula for this function. With the obvious notation for the kernel functions

$$F(\cdot) = \frac{k_2(\cdot) - \left\langle k_2, \hat{k}_1 \right\rangle \hat{k}_1(\cdot)}{\|k_2\|_H \, \delta_H(x_1, x_2)}.$$

Combining this with the previous proposition we find

$$M = \frac{1}{\delta(x_1, x_2)} \left(1 - \frac{k_2(x_1)}{k_1(x_1)} \frac{k_1}{k_2} \right).$$

We now introduce some notation and record this result.

DEFINITION 6.32. *Suppose H is an RKHS on X and $z, w \in X$, $z \neq w$. Define $G_{z,w}$ by*

$$G_{z,w}(\zeta) := \delta(z,w)^{-1} \left(1 - \frac{k_w(z) k_z(\zeta)}{k_z(z) k_w(\zeta)} \right). \tag{6.23}$$

PROPOSITION 6.33. *Suppose H is an RKHS on X and $z, w \in X, z \neq w$, and suppose further that H has the CPP. The function $G = G_{z,w}$ satisfies $G \in \operatorname{Mult}(H)$, $\|G\|_{\operatorname{Mult}(H)} = 1$ and $G(z) = 0$. Among all functions $\tilde{G}$ which satisfy those three conditions, G is the unique function which maximizes $\operatorname{Re} \tilde{G}(w)$. That maximum value is $G_{z,w}(w) = \delta(z,w)$.*

Note that the CPK is being used to show there is a G for which the value δ is attained, the uniqueness and the formula follows from that. That existence is a version of the "two point Pick property" which is much easier to satisfy than the CPP, see [**4**, Equation (7.53)] for an example.

In particular, if H is the Hardy space, the Dirichlet space, or, in fact, any space for which $k_0(\cdot) = 1$ then we have

$$G_{z,0}(\zeta) = \frac{1}{\sqrt{1 - \|k_z\|^{-2}}}\left(1 - \frac{k_z(\zeta)}{k_z(z)}\right) = \frac{\hat{k}_z(z) - \hat{k}_z(\zeta)}{\sqrt{k_z(z) - 1}} \tag{6.24}$$

$$G_{z,0}(0) = \delta_H(0, z) = (1 - \|k_z\|^{-2})^{1/2}. \tag{6.25}$$

Equation (6.24) shows a path for reconstructing H from information about Mult (H), see [**171**, Proposition 11.8] for details.

For the Hardy space the analysis of the extremal functions in the Schwarz-Pick lemma leads to Blaschke products of order one; that is, individual Blaschke factors. Thus when (6.23) is specialized to H^2 we must obtain a Blaschke factor. When we in particular specialize (6.24) to H^2 we obtain

$$\begin{aligned} G_{z,0}(\zeta) &= \frac{1}{\sqrt{1 - (1 - |z|^2)}}\left(1 - \frac{(1 - \overline{z}\zeta)^{-1}}{(1 - |z|^2)^{-1}}\right) \\ &= \frac{\overline{z}}{|z|}\frac{z - \zeta}{1 - \overline{z}\zeta} \end{aligned}$$

which is the individual Blaschke factor normalized to be positive at the origin.

6.6. Zero Sets

6.6.1. Zero Sets for $\mathcal{D}$ and Mult$(\mathcal{D})$. If H is a RKHS on $\mathbb{D}$ and $Z \subset \mathbb{D}$, we say that Z is a *zero set* for H if there is a function $f \in H$, not the zero function, with $f = 0$ on Z. We say that Z is an *exact zero set* if f can be chosen to vanish only on Z. We define zero sets and exact zero sets for the multiplier algebra, Mult(H), analogously.

It is trivial that exact zero sets are zero sets, that exact multiplier zero sets are multiplier zero sets, and if $1 \in H$ that every multiplier zero set is a zero set. Other implications however depend on the individual space being considered.

For the Hardy space, H^2, the situation is simple and well understood. A complete characterization of each class of the sets can be read off from Theorem 3.2. In each of the four cases the Blaschke condition (3.8) is necessary and sufficient. Hence, in particular, the four classes are the same.

The analysis for the Dirichlet space is more complicated. Any finite Blaschke product B is in $\mathcal{D}$ and, if fact, by Exercise 2.16, $B \in$ Mult$(\mathcal{D})$. Hence any finite set is in each of the four classes of zero sets. However beyond that the analysis is complicated and intrinsic characterizations of the sets are not known. Nevertheless it is known that the four classes of zero sets are the same. We prove that now.

We start with a general fact for spaces with a CPK.

Proposition 6.34. *Suppose H is a RKHS on a set X and H has a CPK. If Y is a zero set for H then it is a zero set for* Mult(H).

PROOF FOLLOWING [4, Page 138]. Select $w \in X$ and $f(w) = c \neq 0$. Without loss of generality we may suppose $\|f\|_H = 1$, $c > 0$. The function f is a nontrivial competitor for the extremal problem (6.19) and hence $F_{Y,w}$, the solution to the (6.19) must satisfy $F_{Y,w}(w) > 0$. Hence the solution $G_{Y,w}$ to the extremal problem (6.20) will be a multiplier of norm at most one, which vanishes on Y and has $G_{Y,w}(w) > 0$. Thus G is not the zero function and hence Y is a zero set for $\mathrm{Mult}(H)$. □

Next we show how to make a zero set smaller.

LEMMA 6.35. *Suppose $f \in \mathcal{D}$, $\alpha \in \mathbb{D}$ and $f(\alpha) = 0$ and, for convenience that $f'(\alpha) \neq 0$. Set*

$$f_\alpha(z) = f(z) \Big/ \frac{\overline{\alpha}}{|\alpha|} \frac{\alpha - z}{1 - \overline{\alpha} z}$$

(where the bottom factor is taken to be z if $\alpha = 0$). Then $f_\alpha \in \mathrm{Hol}(\mathbb{D})$, $f_\alpha(\alpha) \neq 0$, the zeros of f_α are exactly the remaining zeros of f, in location and multiplicity. Also $f_\alpha(0) = f(0)/|\alpha|$, (or $f'(0)$ if $\alpha = 0$), $|f_\alpha(z)| \geq |f(z)|$ $\forall z \in \mathbb{D}$ and $f_\alpha \in \mathcal{D}$ with $\|f_\alpha\|_{\mathcal{D}} \leq \|f\|_{\mathcal{D}}$. Furthermore, if $f \in \mathrm{Mult}(\mathcal{D})$ then $f_\alpha \in \mathrm{Mult}(\mathcal{D})$ and $\|f_\alpha\|_{\mathrm{Mult}(\mathcal{D})} \leq \|f\|_{\mathrm{Mult}(\mathcal{D})}$.

PROOF. The estimate of $\|f_\alpha\|_{\mathcal{D}}$ is the crucial fact. Recall that $\|f_\alpha\|_{\mathcal{D}}^2 = \|f_\alpha\|_{H^2}^2 + \mathfrak{D}(f_\alpha)$. Now note that f is a Blaschke factor times f_α and thus $|f_\alpha| = |f|$ on the boundary and hence $\|f_\alpha\|_{H^2} = \|f\|_{H^2}$. We now want an inequality for $\mathfrak{D}(f_\alpha)$. Note that $\mathfrak{D}(f_\alpha)$ can be expressed in terms of the Taylor coefficients f_α, and similarly for f. Hence, if $\alpha = 0$ the desired inequality

$$\mathfrak{D}(f_\alpha) \leq \mathfrak{D}(f) \tag{6.26}$$

follows directly from considering those coefficients. Finally, recall that for any $\phi \in \mathrm{Aut}(\mathbb{D})$ and $h \in \mathrm{Hol}(\mathbb{D})$ we have $\mathfrak{D}(h(\phi(z))) = \mathfrak{D}(h(z))$. By choosing the appropriate ϕ we can reduce (6.26) for general α to the case $\alpha = 0$ which we just settled.

To obtain the result about multipliers note that for α as above $f \in \mathrm{Mult}(\mathcal{D})$, and $g \in \mathcal{D}$ we have $f_\alpha g = (fg)_\alpha$ and use the result for $\mathcal{D}$.

The other details are straightforward. □

We return to this topic when looking at the F property in Section 10.3.3.4.

THEOREM 6.36 ([**220**],[**4**, Section 9.2]). *For $\mathcal{D}$ the four classes of sets, zero sets and exact zero sets for the Hilbert space, and zero sets and exact zero sets for the multiplier algebra, are the same.*

PROOF. Taking into account the earlier elementary observations and the previous results, it remains to show, for both $\mathcal{D}$ and its multiplier algebra, that zero sets are exact zero sets. We do that using the previous lemma to modify zero sets one element at a time.

We want to show that if Z is a zero set for $\mathcal{D}$ then it is an exact zero set. Suppose we have a nontrivial $f \in \mathcal{D}$ which vanishes on Z and possibly at other points. We assume that all the zeros of f are simple and not at the origin, the general case requires only minor modifications. Pick any $a_1 \notin Z$ with $f(a_1) = 0$. Using the previous lemma we can find $g_1 \in \mathcal{D}$ with smaller norm and not vanishing at a_1. We can now apply the lemma to g_1 and continue. That is, with $\{a_i\}$ a listing of the zeros of f that are not in Z we construct a sequence of $g_n \in \mathcal{D}$ with decreasing norms. We claim that the limit function g is in $\mathcal{D}$ and has zeros at exactly the points of Z. By the lemma, for any $z \in \mathbb{D}$ the sequence $\{|g_n(z)|\}$ is nondecreasing. By the uniform bounds on $\|g_n\|_{\mathcal{D}}$ the sequence is bounded pointwise. This gives a pointwise limit which must be zero exactly on Z. The existence of a pointwise limit together with the norm bounds ensures that the limit function is in $\mathcal{D}$, Proposition A.18. This g shows that Z is an exact zero set.

Again using the previous lemma, the same pattern of argument gives the result for zero sets of Mult $(\mathcal{D})$. □

This proof actually shows a bit more.

COROLLARY 6.37. *For the Dirichlet space or its multiplier algebra; any subset of an exact zero set is an exact zero set.*

6.6.2. Generalized Blaschke Products. We saw that our extremal functions $G_{z,0}$ were generalizations of Blaschke factors. We now use them to construct generalized Blaschke products. Versions of this construction work for any RKHS with a CPK; however for now we only look at $H = \mathcal{D}_\alpha$, $0 \leq \alpha \leq 1$. That is, H will be H^2, or $\mathcal{D}$ or one of the $\mathcal{D}_\alpha$, $0 < \alpha < 1$ (defined in (2.5.2)). Each of these spaces has 0 as a distinguished point with $k_0 = 1$.

We will work with the functions G of the previous section. Those functions depend on the choice of H but we will suppress that in the notation.

Suppose we are given a sequence $Z = \{z_i\}_{i=1}^\infty \subset \mathbb{D}$ and we construct a generalized Blaschke product; that is, we form the infinite product

$$\mathfrak{B}(\cdot) = \prod_{j=1}^\infty G_{z_j,0}(\cdot). \tag{6.27}$$

Each G has multiplier norm one and hence modulus bounded by one. Given that, general facts about convergence of infinite products of analytic functions ensure that there are exactly two possibilities, either the product diverges to 0, or the product converges, uniformly on compact sets, to a holomorphic function $\mathfrak{B}(\cdot)$ whose zero set is exactly Z^+, the union of the zero sets of the functions $G_{z_j,0}$.

We now want to know which of those two possibilities occur. Suppose for the moment that $0 \notin Z$ and leave the other possibility to the reader. If $0 \notin Z$ we can test which side of the dichotomy we are on by checking to see if the product for $\mathfrak{B}(0)$ converges to a non-zero value. By checking (6.25) we see that the product converges to a nontrivial function if and only if the product $\prod_{i=1}^\infty \delta(0, z_i)$ converges.

PROPOSITION 6.38. *The product* (6.27) *converges to a nontrivial function* $\mathfrak{B}(\cdot)$ *analytic in the disk if and only if*

$$\prod_{i=1}^\infty \delta_H(z_i, 0) > 0, \tag{6.28}$$

or, equivalently

$$\sum_{i=1}^\infty k_{z_i}(z_i)^{-1} < \infty. \tag{6.29}$$

In those cases $\mathfrak{B} \in \operatorname{Mult}(H)$, $\|\mathfrak{B}\|_{\operatorname{Mult}(H)} \leq 1$, *and* $\mathfrak{B}(z) = 0$ *exactly for* $z \in Z^+$.

PROOF. We have discussed why (6.28) is equivalent to the convergence of the product. The standard facts about products and sums ensure that (6.28) is equivalent to (6.29). We also noted that if the product converges then Z^+ is the zero set for $\mathfrak{B}$.

To see that $\mathfrak{B}$ is in the multiplier algebra with norm at most one, note that the individual factors in the product are multipliers of norm at most one, and hence so are the partial products. It remains to note that the limit of a sequence of multipliers of uniformly bounded norms which converges uniformly on compact sets will be a multiplier with the same norm bound. □

EXERCISE 6.39. *Justify the last sentence of the previous proof.*

EXERCISE 6.40. *Check that the same proof works for any Hilbert space of holomorphic functions on the disk, which has the CPP, and in which k_0 is identically one. Formulate and prove a modified result in which the assumption that $k_0 \equiv 1$ is replaced with the existence of $\zeta \in \mathbb{D}$, $0 < A < B < \infty$ such that $A \leq |k_\zeta(z)| \leq B$. (The condition on k_ζ corrects an oversight in* [**294**, Proposition 5.10]*.)*

COROLLARY 6.41. *For H equal to H^2, $\mathcal{D}$, or $\mathcal{D}_\alpha$, $0 < \alpha < 1$; if Z is given then* (6.28)*, or, equivalently,* (6.29) *is sufficient to ensure there is a function $m(z)$, in both H and its multiplier algebra* Mult(H)*, equal to* 0 *on Z, and with no other zeros.*

For $H = H^2$ condition (6.29) *is the* Blaschke condition

$$\sum_{i=1}^{\infty}(1-|z_i|) < \infty. \tag{6.30}$$

For the spaces $\mathcal{D}_\alpha$, $0 < \alpha < 1$, the condition is

$$\sum_{i=1}^{\infty}(1-|z_i|)^{\alpha} < \infty. \tag{6.31}$$

For the Dirichlet space the condition is

$$\sum_{i=1}^{\infty}\left(-\log\left(1-|z_i|\right)\right)^{-1} < \infty. \tag{6.32}$$

PROOF. The statements for the sums are straightforward. As to the zero sets, we have functions which vanish on Z^+ and want functions which vanish on the possibly smaller set Z. If the zero set for each factor $G_{z_j,0}$ is exactly $\{z_j\}$ then the $Z = Z^+$. That is easy to check for $\alpha > 0$. For $\alpha = 0$, the Dirichlet space, we saw in the proofs in the previous section how to go from a function which vanishes on a set to one which vanishes on a subset. □

We have done these computations using 0 as a base point. It is not hard to check that a different choice of base point would lead to different formulas but, at the end, the same conditions on Z.

Condition (6.30) for H^2 or H^∞ is the classical Blaschke condition which is the necessary and sufficient condition for zero sets for functions in those spaces. For the other spaces the conditions (6.31) and (6.32) are sufficient conditions for a set to be a zero set but they are not necessary. We will see in a moment that if all the z_i are real and positive then, in fact, condition (6.30) is necessary and sufficient for there to be an $f \in \mathcal{D}$ whose zero set is exactly Z. This condition is much less restrictive than (6.31) or (6.32). Also, $\mathcal{D} \subset \mathcal{D}_\alpha$, $0 < \alpha < 1$, hence the same f will also show that Z is a zero set for those spaces.

6.6.3. General Zero Sets. It is not known what conditions ensure that $W = \{w_i\} \subset \mathbb{D}$ is a zero set for $\mathcal{D}$. The condition (6.32) is the optimal condition which only depends on the moduli of the $\{w_i\}$. It was shown by Nagel, Rudin, and Shapiro [**235**] that if (6.32) fails there is a choice of $\{e^{i\theta_j}\}$ so that $W' = \{e^{i\theta_j} |w_j|\}$ is not the zero set of any function in $\mathcal{D}$. That was refined by Richter, Ross, and Sundberg [**269**] to show that there was always a choice with $\lim \theta_i = 0$ which would produce a W' with that property.

In the construction by Richter, Ross and Sundberg, the points approach the boundary in a very flat manner; that is, the numbers $\{\theta_j\}$ tend to zero very, very slowly as $|w_i| \to 1$. The opposite case, in which the $w_j' = e^{i\theta_j} |w_j|$ and the $\{\theta_j\}$ tend to zero quickly, is easier to analyze. First consider the extreme case in which all the θ_j are zero. In that case $W' = \{r_i\}$ with the r_i positive and tending to one. Because $\mathcal{D} \subset H^2$ any zero set for $\mathcal{D}$ is also a zero set for H^2 and hence must satisfy the Blaschke condition (3.8). Hence for W' to possibly be a zero set we must have $\sum_{j=1}^{\infty} 1 - r_j < \infty$. The Blaschke product with those zeros is

$$B(z) = \prod_{i=1}^{\infty} \frac{r_j - z}{1 - r_j z}.$$

If this is a finite product then it is certainly in the Dirichlet space; however if it is an infinite product it is not in $\mathcal{D}$ (Corollary 10.38). However, it was noted by Carleson [**92**] that the function $f(z) = (1-z)^2 B(z)$, which also has the zero set W', is in $\mathcal{D}$. To see this, first note that

$$\begin{aligned} |B'(z)| &\leq \sum_{j=1}^{\infty} \frac{1 - r_j}{|1 - r_j z|^2} \\ &\leq \frac{C}{|1 - z|^2}. \end{aligned}$$

The first line uses the fact that the individual factors of B each have modulus at most one. The second line follows from the first by using the Blaschke condition on the numerators and making each denominator smaller. Using that estimate it is easy to check that $f \in \mathcal{D}$ and hence W' is a zero set for $\mathcal{D}$.

It is possible to extend these observations to sets W' which approach a single boundary point nontangentially and, further, to finite unions of such sets. Bogdan [**79**] proved a more subtle result giving a necessary and sufficient condition on a set $W \subset \mathbb{D}$ for it to have the property that every $W' \subset W$ which satisfies the Blaschke condition is a zero set for the Dirichlet space. The condition is that

$$\int_0^{2\pi} -\log \operatorname{dist}\left(W, e^{i\theta}\right) < \infty.$$

We will not prove this here, but will describe a basic preliminary reduction. Suppose that $W' \subset W$ and that there is a Blaschke product, $B_{W'}$, with zero set W'. Taking into account the factorization theorem, Theorem 3.3, and the discussion of the F property in Section 10.3.3.4 we see that, if there is a function in $\mathcal{D}$ which is zero on W', then there is such a function of the form $B_{W'}G$ with G an outer function in $\mathcal{D}$. In the elementary example of points on the axis the choice was $G(z) = (1-z)^2$.

However the general problem of constructing an outer function in $\mathcal{D}$ with specified properties is very hard.

Research continues on describing the general zero sets for the Dirichlet space; [**190**] includes recent research on the question.

We have not discussed boundary zero sets. As discussed for the Hardy space in Chapter 3 and the Dirichlet space in Chapter 9, functions in these spaces do not have boundary values in the classical sense. Their boundary functions, defined as radial limits, are only well defined off an exceptional set. For the Hardy space the exceptional sets are sets of Lebesgue measure zero and those sets are also exactly the sets where the boundary value function can vanish. For $f \in \mathcal{D}$ Theorem 9.1 ensures radial boundary values exist off a set of capacity zero. It is a result of Carleson that a closed set of capacity zero on the boundary can be the exact zero set of a boundary function. We discuss boundary zero sets for Dirichlet space functions further in Chapter 9. A complete description of those sets is not known.

For RKHS without a CPK the situation is more complicated and less is known. For instance, the Bergman space has many similarities to the spaces we have been focused on, but some of the basic results for it are quite different. Here are some of those to give some flavor.

The multiplier algebra of A^2 is H^∞ and hence any multiplier zero set must satisfy the Blaschke condition. However the zero sets for the Bergman space itself need not satisfy that condition. Hence Bergman space zero sets are not the same as the zero sets for its multiplier algebra. Because of this our previous proof that the union of two zero sets is a zero set (based of forming the product of multipliers) does not work. In fact, it was shown by Horowitz, [**180**], that the union of two Bergman space zero sets need not be a Bergman space zero set. On the other hand some results do parallel those for the Hardy space. It is a result of Horowitz from the same paper that every subset of a zero set is a zero set, and it is a result of Hedenmalm [**173**] that every zero set is an exact zero set. These aspects of the theory of the Bergman space are discussed in detail in [**127**] and [**174**].

6.7. The Toeplitz Corona Theorem

The Toeplitz corona theorem, which will be important to us in Chapter 17 is a very useful consequence of the CPP. For that reason we give an informal statement of it here. In Chapter 17 we discuss the theorem more systematically and also apply it. However we do not include a proof.

Suppose B is a Banach space of holomorphic functions and Mult (B) is its multiplier algebra (which is defined the way you would guess). A corona theorem for Mult (B) is a statement about solvability of certain equations in Mult (B). A baby corona theorem for B is the statement that certain similar equations can be solved in B. It is often true that the baby corona theorem is the easier of the two to establish. The Toeplitz corona theorem is the statement that for RKHS with the CPP (not just the Pick property) the baby corona theorem implies the corona theorem.

6.8. Notes and Comments

The book by Agler-McCarthy [**4**] contains the parts of the theory of reproducing kernels with the complete Pick property which we use here. There are many active research topics related to Pick's theorem which we are not discussing, including transfer functions, control theory, the commutant lifting theorem, and matricial multipliers. Beyond what is mentioned in [**4**], the short [**228**] offers a quick overview of some of those topics; the books [**152**], [**148**], and [**155**] give a sample of some of those further directions.

The authors' understanding of the history of Pick's theorem in the context of the Dirichlet space, comes from [**4**], [**228**], and [**302**]. The history of the area is made more opaque by the fact that the influential papers [**3**], [**220**], and [**72**] were never published.

The discussion of generalized Blaschke products is based on [**294**] which in turn evolved from the work by Shapiro and Shields in [**306**] and later analysis in [**220**], [**302**], and [**4**].

Additional work on zero sets is in [**100, 222, 223**] and [**224**].

CHAPTER 7

Interpolation

For holomorphic functions there is an inherent tension between specifying the values a function takes at a sequence of points and controlling the size of the function. The interpolating sequences for a function space are those sequences that are sufficiently scattered so that this tension can be overcome. Characterizing those sequences is an important part of understanding the fine structure of the function space. In this chapter we present the classical results about interpolating sequences for the Hardy space and its multiplier algebra. We also present newer results about similar questions for the Dirichlet space and its multiplier algebra. The Pick property plays a role in the analysis.

7.1. The Basic Questions

Given a space of functions A on a set X and a sequence $Z \subset X$, we ask which functions on Z are the restrictions to Z of functions in A? In particular, we want to characterize the *interpolating sequences,* those Z for which the set of restrictions is, in some natural sense, as large as possible.

If, for instance, $X = \mathbb{D}$, $A = H^\infty$, and Z is a finite set in $\mathbb{D}$, then any function on Z is the restriction of a function in A. If we ask further that the interpolating function have norm at most one then this is the Pick interpolation problem which we discussed in Chapter 6. Now we look at a different variation. Suppose Z is an infinite sequence. The restriction to Z of any function in A will be bounded and we ask for which Z it is true that every bounded function on Z can be obtained this way.

In this chapter and the next, we consider that problem for H^∞ as well as related questions for the Hardy space, the Dirichlet space, and the tree model of the Dirichlet space. The results for H^2 and $\text{Mult}(H^2) = H^\infty$ are classical [**91**], [**307**]; those for the $\mathcal{D}$, $\text{Mult}(\mathcal{D})$, and $\mathcal{D}(\mathcal{T})$, the tree model of the Dirichlet space, are more recent, [**220**], [**72**], [**77**], [**48**, Theorem 26], [**41**, Theorem 4], [**75**] and [**42**]. A great deal of information about interpolation is in the book of Seip [**302**]. Because both H^2 and $\mathcal{D}$ are RKHS with a CPK, some of the results for the two spaces are strikingly similar, and the proofs we offer for those two spaces are the same until the final steps. The results for $\mathcal{D}$ and $\mathcal{D}(\mathcal{T})$ are similar to each other, but the proofs are not.

Our main tools in this chapter are the theory of RKHS and of complete Pick kernels. In the next chapter and in Chapters 15 and 16, we will see alternative, more constructive, techniques that can be used on the problems here as well as analogous problems in Banach spaces. In those situations we make crucial use of analysis on function spaces on trees in proving results about spaces of holomorphic functions.

Similar problems are also considered for spaces without the CPP, for instance the Bergman space and the Fock space. The results for those spaces have a very different form from the ones here. A good introduction to that larger picture is in [**302**].

7.1.1. Multiplier Interpolation. Suppose we have a RKHS, H, of functions on a set X and a sequence $\{z_n\}_{n=1}^{\infty} = Z \subset X$. Let R be the restriction map taking functions on X to functions on Z. We know that every $f \in \mathrm{Mult}(H)$ is a bounded function, and hence Rf is a bounded function. The multiplier interpolation question asks for which Z is it true that every bounded function on Z can be obtained this way? In other words, given H and $\mathrm{Mult}(H)$, for which Z is it true that R, which automatically maps into $\ell^\infty(Z)$, maps onto all of $\ell^\infty(Z)$? If it does, then Z is called, perhaps somewhat oddly, a *multiplier interpolating sequence*.

There are natural necessary conditions for Z to be a multiplier interpolating sequence. For instance, if the set of restrictions is all of $\ell^\infty(Z)$ then for any $z_i \in Z$, there is f_i in $\mathrm{Mult}(H)$ with $f_i(z_j) = \delta_{ij}$. With such functions in hand we see that Z must be a zero set for $\mathrm{Mult}(H)$. For instance, if $H = H^2$ then $\mathrm{Mult}(H) = H^\infty$ and the zero sets for H^∞ must satisfy the Blaschke condition (3.8). This shows that $\left\{1 - n^{-1}\right\}_{n=1}^{\infty}$ is not a multiplier interpolating sequence but doesn't rule out $\left\{1 - n^{-2}\right\}_{n=1}^{\infty}$.

Furthermore, if a pair of points in Z is very close together then any f which takes very different values at the two points would have very large norm. This is a potential obstacle to having interpolation with good norm control. To make this idea quantitative note that if the restriction map is surjective then Theorem A.11 ensures that there is a constant C so that all of the f_i can be chosen to have norm at most C. We can then apply Proposition (2.19) to the functions $\left\{C^{-1} f_i\right\}$ and conclude that the sequence must be separated. That is

$$\exists \varepsilon > 0, \forall i, j,\ i \neq j;\ \delta_H(z_i, z_j) > \varepsilon \tag{7.1}$$

is a necessary condition on Z for it to be a multiplier interpolation sequence. This is enough to show that, for H^∞, $Z = \left\{1 - n^{-2}\right\}$ is not a multiplier interpolating sequence; but it allows that $\{1 - 2^{-n}\}$ might be one (which, in fact, it is).

EXERCISE 7.1. *The use of the notation δ_H emphasizes the analogy between the results for the Hardy space and those for the Dirichlet space. However, the two metrics, δ_{H^2} and $\delta_{\mathcal{D}}$, differ in detail. Show that, for $\{z_i\} = Z \subset (0,1)$, having the numbers $\{1 - z_i\}$ tend to zero at least at an exponential rate is a necessary and sufficient condition for the sequence to be separated with respect to δ_{H^2}. The condition is not sufficient for separation with respect to $\delta_{\mathcal{D}}$. Give a condition that is sufficient for that.*

7.1.2. Hilbert Space Interpolation. When we study the action of R on H, new issues arise. It is less clear what the natural target space should be. By Proposition 2.8, if $f \in H$, then the sequence of numbers $\left\{\|k_i\|_H^{-1} f(z_i)\right\}$ is bounded. Further, for $H = H^2$ or $\mathcal{D}$, by Proposition 2.10 this sequence tends to zero. With these facts as background, we ask if this sequence is always square summable and/or if all square summable sequences can be obtained this way.

Given $Z = \{z_i\} \subset X$, we will write $\{k_i\}$ rather than $\{k_{z_i}\}$ for the associated kernel functions. Define the measure μ_Z by

$$d\mu_Z = \sum_{i=1}^{\infty} k_i(z_i)^{-1} \delta_{z_i} = \sum_{i=1}^{\infty} \|k_i\|_H^{-2} \delta_{z_i}. \tag{7.2}$$

There are then three questions one can ask:

(into) Is $R(H) \subset \ell^2(Z, \mu_Z)$?

(into and onto) Is $R(H) = \ell^2(Z, \mu_Z)$?

(onto) Is $R(H) \supset \ell^2(Z, \mu_Z)$?

Both the second and third question could claim to be the precise formulation of the question, "Is the interpolation always possible?" For multiplier interpolation, the distinction is moot because the restriction operator automatically maps into ℓ^∞, and hence the analogs of the second and third questions automatically have the same answer. Those two questions also have the same answer for the Hardy space $H = H^2$; but in that case it is a nontrivial fact, specific to the Hardy space, that if R satisfies (onto), then it also satisfies (into). In contrast for $H = \mathcal{D}$, there are Z for which the second and third questions have different answers; the map R satisfies (onto) but not (into).

For Hilbert spaces, a sequence is called an *interpolating sequence*, IS, if the question (into and onto) has an affirmative answer. Such sequences will be discussed in this chapter. Sequences for which there is a positive answer to (onto) even if the answer to (into) is negative or is not known, are called *onto interpolating sequences*, OIS. We will discuss those sequences in the next chapter.

Some facts about these sequences are straightforward and hold in great generality. For instance, any finite sequence will be both an IS and an OIS; or, if $H \subset \mathrm{Hol}(\mathbb{D})$, then neither class will contain any sequence with an interior accumulation point. However full characterization of the IS or OIS depend strongly of the specific H being considered. In this chapter and the next we develop ideas that have been used successfully in studying interpolation for H^2, $\mathcal{D}$, and related spaces. We study related problems using very different methods in Chapters 15 and 16. The book by Seip [**302**] is an excellent overview of questions about interpolating sequences.

There are two conditions which are easily seen to be necessary for the equality $R(H) = \ell^2(Z, \mu_Z)$. First, if the two spaces are equal, then, by Theorem A.1 the map R will be bounded from H into $\ell^2(Z, \mu_Z)$;

$$\int_{\mathbb{D}} |f|^2 \, d\mu_Z = \sum_{i=1}^{\infty} \|k_i\|_H^{-2} |f(z_i)|^2 \lesssim \|f\|_H^2 . \tag{7.3}$$

Thus, by definition, μ_Z is a Carleson measure for H.

Second, we noted a moment ago that if Z is a multiplier interpolating sequence, then the sequence must be separated. That same argument shows that the separation condition (7.1) must hold for any IS. Again, if the restriction map is a bounded map onto the ℓ^2 space then, by Theorem A.11, there must be functions $\{g_i\} \subset H$ with uniformly bounded norm and $g_i(z_j) = \delta_{ij}$. Hence, by an appeal to Proposition 2.13 and a bit of analysis, there must be a lower bound on $\delta_H(z_i, z_j)$. That is (7.1) must hold for the sequence.

7.2. Interrelated Conditions

There are a number of implications between statements related to interpolating sequences, and we collect some of them here. These results are based on general functional analysis and do not involve specific properties of H beyond its being a RKHS. In particular, here we are not requiring that H have the CPP.

Recall that a set of vectors $\{v_n\}$ in a Hilbert space J is said to be a *Riesz basis* for J if $\forall j \in J$, $j = \sum_n a_n v_n$ and $\|j\|_{\ell^2}^2 \approx \sum_n |a_n|^2$; it is a slightly weaker condition than being an orthonormal basis. A set of vectors is said to be a *Riesz basic sequence* if it is a Riesz basis for its closed linear span. That is, $\{w_n\} \subset J$ is a Riesz basic sequence if for any sequence $\{a_n\}$ we have

$$\left\| \sum_n a_n w_n \right\|_J^2 \approx \sum_n |a_n|^2 .$$

Suppose H and $\{z_i\} = Z \subset X$ are given.

THEOREM 7.2. *Consider the following*

(1) *Z is a multiplier interpolating sequence.*
(2) *The reproducing kernels $\{k_i\}$ are an unconditional basic sequence in H. That is, whenever $|a_i| \leq |b_i|$ $\forall i$, then*

$$\left\| \sum_i a_i k_i \right\|_H \lesssim \left\| \sum_i b_i k_i \right\|_H .$$

(3) *The normalized reproducing kernels $\left\{ \hat{k}_i \right\}$ are a Riesz basic sequence in H.*

$$\left\| \sum_i a_i \hat{k}_i \right\|_H^2 \approx \sum_i |a_i|^2 .$$

(4) *Z is an interpolating sequence; that is,* (into and onto) *has an affirmative answer.*
(5) *Z is separated and μ_Z is a Carleson measure for H; that is,* (7.3) *and* (7.1) *hold.*

We have

$$(1) \implies (2) \iff (3) \iff (4) \implies (5).$$

PROOF. **(1) $\implies$ (2):** For $i = 1, 2, \ldots$, define $\gamma(z_i)$ by requiring $\overline{\gamma(z_i)} b_i = a_i$. The hypotheses in (2) ensure $\|a(z_i)\|_{\ell^\infty} \leq 1$. Hence by condition (1), there is an $m \in \operatorname{Mult}(H)$ of norm at most C with $m(z_i) = \gamma(z_i)$, $i = 1, 2, \ldots$. Hence M^*, the adjoint of the operator of multiplication by m, has norm at most C. Also, we know from Proposition 2.17 that for any z, $M^*(k_z) = \overline{m(z)} k_z$. Hence

$$\begin{aligned} \left\| \sum_i a_i k_i \right\|_H &= \left\| \sum_i \overline{\gamma(z_i)} b_i k_i \right\|_H = \left\| M^* \left(\sum_i b_i k_i \right) \right\|_H \\ &\leq \|M^*\| \left\| \sum_i b_i k_i \right\|_H \lesssim \left\| \sum_i b_i k_i \right\|_H . \end{aligned}$$

(2) $\Longrightarrow$ (3): We begin with a preliminary observation. Suppose $\{h_n\} \subset H$ with $\{\|h_n\|_H\} \in \ell^2$. For $e^{i\theta} \in \mathbb{T}$ set $\phi(\theta) = \left\|\sum_n e^{in\theta} h_n\right\|_H^2$. It is a straightforward exercise that ϕ is a continuous function of θ. Its mean value is given by

$$\frac{1}{2\pi}\int_0^{2\pi} \phi(\theta)d\theta = \sum_{n,m} \langle h_n, h_m\rangle_H \frac{1}{2\pi}\int_0^{2\pi} e^{i(n-m)\theta}d\theta = \sum_n \|h_n\|_H^2 .$$

Hence, by the intermediate value theorem there is an α for which $\phi(\alpha) = \left\|\sum e^{in\alpha} h_n\right\|_H^2 = \sum \|h_n\|_H^2$.

Suppose now $\{h_n\} = \{a_n \hat{k}_n\}$ and select that α. Recalling that the $\hat{k}_i$ are unit vectors and using (2) twice we find

$$\begin{aligned}\sum_n |a_n|^2 = \left\|\sum_n e^{in\alpha} a_n \hat{k}_n\right\|_H^2 &\lesssim \left\|\sum_n a_n \hat{k}_n\right\|_H^2 \\ &\lesssim \left\|\sum_n e^{in\alpha} a_n \hat{k}_n\right\|_H^2 = \sum_n |a_n|^2 .\end{aligned}$$

The estimates for the middle term are the required estimates for (3).

(3) $\Longrightarrow$ (2): Straightforward.

(3) $\Longrightarrow$ (4): First we show the map T of $f \in H$ to $Tf = \left\{\|k_i\|_H^{-1} f(z_i)\right\}$ is a map into $\ell^2(\mathbb{Z}^+)$.

$$\begin{aligned}\left\|\left\{f(z_i)\|k_i\|_H^{-1}\right\}\right\|_{\ell^2} &= \sup\left\{\left|\sum_i f(z_i)\|k_i\|_H^{-1} a_i\right| : \sum_i |a_i|^2 = 1\right\} \\ &= \sup\left\{\left|\left\langle f, \sum_i a_i\hat{k}_i\right\rangle_H\right| : \sum_i |a_i|^2 = 1\right\} \\ &\leq \sup\left\{\|f\|_H \left\|\sum_i a_i\hat{k}_i\right\|_H : \sum_i |a_i|^2 = 1\right\} \\ &\lesssim \sup\left\{\|f\|_H \left(\sum_i |a_i|^2\right)^{1/2} : \sum_i |a_i|^2 = 1\right\} \\ &= \|f\|_H .\end{aligned}$$

We now want to show that the same map is surjective. By a consequence of Corollary A.15 it suffices to show that the adjoint, T^*, is bounded below. One checks directly that

$$T^*(\{a_i\}) = \sum_i a_i\hat{k}_i. \tag{7.4}$$

Now we use (3):

$$\|T^*(\{a_i\})\|_H^2 = \left\|\sum_i a_i\hat{k}_i\right\|_H^2 \gtrsim \sum_i |a_i|^2 = \|\{a_i\}\|_{\ell^2}^2 .$$

(4) $\Longrightarrow$ (3)**:** Suppose $b = \{b_i\} \in \ell^2(\mathbb{Z}^+)$. Because we have an interpolating sequence, there is $B \in H$, $B(z_i) = b_i \|k_i\|_H$ for $i = 1, 2, \ldots$, $\|B\|_H \lesssim \|b\|_{\ell^2}$. Set $C = \sum_i c_i \hat{k}_i$. We then have

$$\left|\sum_i b_i \bar{c}_i\right| = |\langle B, C\rangle_H| \le \|B\|_H \|C\|_H \lesssim \|b\|_{\ell^2} \|C\|_H ,$$

and taking the supremum over $b \in \ell^2(\mathbb{Z}^+)$ gives the left hand inequality.

In the other direction, we know the restriction map is bounded and thus the map of T just defined is bounded from H to $\ell^2(\mathbb{Z}^+)$. Hence the adjoint, T^*, is bounded; $\|T^*(\{c_i\})\|_H \lesssim \sum_i |c_i|^2$, From (7.4), $T^*(\{c_i\}) = C$, and hence this is the required right hand inequality.

(4) $\Longrightarrow$ (5)**:** We showed this in the discussion in the previous section.

□

7.2.1. CPP and Multiplier Interpolation. We now show that, for both the Hardy space and the Dirichlet space, the interpolating sequences are the same as multiplier interpolating sequences.

For the Hardy space and for the Dirichlet space, the fact that the two types of interpolating sequences are the same is mediated by the CPP and does not require that the sequences be characterized. However, an analogous result holds in some cases where there is no analog of the theory of Pick kernels. In particular, we will see in Chapter 15 that a similar result holds for certain Banach spaces of holomorphic functions in n variables. Yet for those spaces, the interpolating sequences and the multiplier interpolating sequences are characterized independently and then the characterizations are observed to be identical. In those cases, it is not known how to show the two sets of sequences are the same without first characterizing each of them.

THEOREM 7.3. *If, in the situation of Theorem* 7.2*, additionally, H has a CPK, then the interpolating sequences and the multiplier interpolating sequences are the same. In the terminology of Theorem* 7.2*,* (1) $\Longleftrightarrow$ (4)*.*

PROOF. From Theorem 7.2, we know that (1) $\Longrightarrow$ (4). We need to show that (4) $\Longrightarrow$ (1). We want to ensure that, given $\{b_i\} \in \ell^\infty$, we can find a multiplier m with $m(z_i) = b_i$, $i = 1, 2, \ldots$ with control of $\|m\|_{\text{Mult}(H)}$. Alternatively, it is enough to show that there is a bounded multiplication operator M such that $M^* k_i = \bar{b}_i k_i$. Because we have a kernel with the CPP this will happen if and only if such an M^* would be bounded on the closed span of the $\{k_i\}$. That is the case exactly if there is C such that, given $h = \sum_i a_i k_i$, we have a norm estimate for the purported value of $M^* h$ which is $\sum_i \bar{b}_i a_i k_i$. That is, we want

$$\left\|\sum_i \bar{b}_i a_i k_i\right\|_H \lesssim \left\|\sum_i a_i k_i\right\|_H$$

to hold if $\{b_i\} \in \ell^\infty$. This last statement is a consequence of the unconditional basic sequence statement, condition (2) of Theorem 7.2. Hence, the proof will follow once we know that (4) $\Longrightarrow$ (2), and that implication is included in the previous theorem. □

7.2.2. The Gram and Dual Gram Matrices. An important role in the previous analysis was played by the map from sequences $\{a_i\}$ to Hilbert space vectors $\sum_i a_i k_i$. The Gram matrix and the dual Gram matrix are tools for studying such maps.

Suppose H is a RKHS on X, Z is a sequence in X and $\{k_{z_i}\}_{z_i \in Z} = \{k_i\}_{z_i \in Z}$ is the associated set of kernel functions. Let $\hat{K} = \left\{\hat{k}_i\right\}_{i=1}^{\infty}$ be the sequence of normalized kernels. The *Gramian* or *Gram matrix* of $\hat{K}$ or, equivalently, the Gram matrix of the sequence Z, Gram(Z), is the matrix

$$\text{Gram}(Z) = \text{Gram}(\hat{K}) = G_{\hat{K}} = \left(\left\langle \hat{k}_i, \hat{k}_j \right\rangle_H\right). \tag{7.5}$$

LEMMA 7.4. *The Gram matrix* (7.5) *is bounded on ℓ^2 if and only if μ_Z is a Carleson measure for H.*

PROOF. To see this, let $T : H \to \ell^2$ be the normalized restriction map $Tf = \left\{\|k_j\|_H^{-1} f(z_j)\right\}_{j=1}^{\infty}$. Then μ_Z is a Carleson measure for H if and only if T is bounded. But $T^* \{\xi_j\}_{j=1}^{\infty} = \sum_{j=1}^{\infty} \xi_j \hat{k}_j$, and so the matrix representation of TT^* relative to the standard basis $\{\mathbf{e}_j\}_{j=1}^{\infty}$ of ℓ^2 is the Gramian:

$$\left(\langle TT^* \mathbf{e}_i, \mathbf{e}_j \rangle_{\ell^2}\right)_{i,j=1}^{\infty} = \left(\left\langle T\hat{k}_i, \mathbf{e}_j \right\rangle_{\ell^2}\right)_{i,j=1}^{\infty} = \left(\left\langle \hat{k}_i, \hat{k}_j \right\rangle_H\right)_{i,j=1}^{\infty}.$$

Finally, recall that T is bounded if and only if TT^* is bounded. $\square$

The sequence $\{f_i\}_{i=1}^{\infty} \subset H$ of minimal norm solutions to the set of equations $\left\langle f_i, \hat{k}_j \right\rangle_H = \delta_{ij}$ $i, j = 1, 2, \ldots$ is called the *minimal dual system*. Such a system need not exist; but if it does, then the matrix

$$G_{\hat{K}}^{\mathtt{dual}} = (f_{ij}) = \left(\langle f_i, f_j \rangle_H\right) \tag{7.6}$$

is called the *dual Gramian* of the system. (Note that the vectors $\{f_i\}$ are not normalized.)

EXERCISE 7.5. *Given a finite dimensional Hilbert space H and a basis $\{h_j\}_{j=1}^n$, find an expression for the minimal dual system $\{f_j\}_{j=1}^n$.*

EXERCISE 7.6. *Suppose S is a bounded invertible linear map on a Hilbert space H and $\{e_i\}$ is an orthonormal basis of H. Let $K = \{Se_i\}$ and let $\hat{K}$ be the corresponding set of normalized vectors. Compute $G_{\hat{K}}$ and $G_{\hat{K}}^{\mathtt{dual}}$.*

There may not be a minimal dual system. For instance there is none if one of the k_i is in the closed linear span of the others. Issues of this type are more complicated for functions of several variables; see, for instance the discussion in Section 15.4.2 of adding a point to an interpolating sequence.

EXERCISE 7.7. *Let $H^2(\mathbb{D}^2)$ be the Hardy space of the bidisk, the RKHS of holomorphic functions on $\mathbb{D}^2$ with power series representations $\sum_{n,m \in \mathbb{Z}} a_{nm} z_1^n z_2^m$ satisfying $\sum_{n,m \in \mathbb{Z}} |a_{nm}|^2 < \infty$. Let $Z = \{(1 - 1/n, 0) : n = 1, 2, \ldots\}$ and consider the kernels $\{k_z\}_{z \in Z}$. Show that this set does not admit a minimal dual system.*

However this issue is not a problem in the cases of primary interest to us.

EXERCISE 7.8. *Show that if* $H = H^2$ *or* $\mathcal{D}$ *and if* Z *is a zero set, then the set of kernel functions* $\{k_z\}_{z\in Z}$ *does admit a minimal dual system.*

Assuming there is a minimal dual system the entries of the dual Gram matrix can be expressed using quantities we have already seen. For $A \subset Z$, let H_A be the subspace of H consisting of functions which vanish on A. For $z_i \in Z \setminus A$, let k_i^A be the reproducing kernel for z_i in the space H_A. Let F_i^A be the function which maximizes $\operatorname{Re} F(z_i)$ among all F in H_A of norm one. The space H_A is a RKHS; hence Proposition 2.8 applies and the reproducing kernels, k_i^A, can be described using the extremal function F_i^A:

$$k_i^A = F_i^A(z_i) F_i^A \tag{7.7}$$

$$F_i^A(z_i)^2 = k_i^A(z_i) = \left\|k_i^A\right\|_H^2 . \tag{7.8}$$

For $n = 1, 2, \ldots$, set $Z(n) = Z \setminus \{z_n\}$.

LEMMA 7.9. *If there is a minimal dual system* $\{f_n\}$ *then it is given by, for* $n = 1, 2, \ldots$

$$f_n = \frac{\|k_n\|_H}{F_n^{Z(n)}(z_n)} F_n^{Z(n)} = \frac{\|k_n\|_H}{\left\|k_n^{Z(n)}\right\|_H^2} k_n^{Z(n)}. \tag{7.9}$$

The entries of the associated dual Gram matrix are

$$\langle f_n, f_m\rangle_H = \frac{\|k_n\|_H \|k_m\|_H}{\left\|k_n^{Z(n)}\right\|_H^2 \left\|k_m^{Z(m)}\right\|_H^2} \left\langle k_n^{Z(n)}, k_m^{Z(m)} \right\rangle_H . \tag{7.10}$$

If, furthermore, H *has the CPP (or even the PP) this can be rewritten as*

$$\langle f_n, f_m\rangle_H = \frac{1}{F_n^{Z(n)}(z_n) F_m^{Z(m)}(z_m)} \left\langle M_n^{Z(n)} k_n, M_m^{Z(m)} k_m \right\rangle_H . \tag{7.11}$$

PROOF. The first statement is a consequence of the following computation which uses the reproducing property of the $\{k_n\}$.

$$\left\langle f_n, \hat{k}_m \right\rangle_H = \frac{\|k_n\|_H}{\left\|k_n^{Z(n)}\right\|_H^2} \left\langle k_n^{Z(n)}, \hat{k}_m \right\rangle_H = \frac{\|k_n\|_H}{\left\|k_n^{Z(n)}\right\|_H^2} \frac{k_n^{Z(n)}(z_m)}{\|k_m\|_H} = \delta_{nm}.$$

The second statement follows from the first.

For the third statement we recall from Proposition 6.27, that if H has the Pick property then $F_i^A = M_i^A \hat{k}_i$, and hence also $k_i^A = M_i^A(z_i) M_i^A k_i$. With these facts the third statement follows from the second. □

We now have the following theorem.

THEOREM 7.10. *The following are equivalent:*

(1) *Condition (3) of Theorem 7.2 holds; the normalized reproducing kernels* $\left\{\hat{k}_i\right\}$ *are a Riesz basic sequence;*

$$\left\|\sum_i a_i \hat{k}_i\right\|_H^2 \approx \sum_i |a_i|^2 . \tag{7.12}$$

(2) *The Gram matrix is bounded and bounded below,*

$$\sum_{i,j} \left\langle \hat{k}_i, \hat{k}_j \right\rangle_H a_i \overline{a}_j \approx \sum_i |a_i|^2 . \tag{7.13}$$

(3) *The Gram matrix is bounded, there is a minimal dual system* $\{f_i\}$, *and the dual Gram matrix bounded. That is,*

$$\sum_{i,j} \left\langle \hat{k}_i, \hat{k}_j \right\rangle_H a_i \overline{a}_j \lesssim \sum_i |a_i|^2 \tag{7.14}$$

$$\sum_{i,j} \langle f_i, f_j \rangle_H a_i \overline{a}'_j \lesssim \sum_i |a_i|^2 . \tag{7.15}$$

PROOF. Pick and fix the sequence of scalars $\{a_i\}$.

The left sides of (7.12) and (7.13) are equal by definition, and hence those two statements are equivalent. Inequality (7.14) is the same as one of the two estimates in (7.13). Hence it remains to show that (7.15) is equivalent to the other estimate in (7.12) or, equivalently the corresponding estimate in (7.13).

If (7.15) holds; that is, the dual Gram matrix is bounded, then, for some constant, we have

$$\left\| \sum_i a_i f_i \right\|_H^2 \lesssim \sum_i |a_i|^2 .$$

Using that estimate, we find

$$\begin{aligned}
\sum_i |a_i|^2 = \left\langle \sum_i a_i \hat{k}_i, \sum a_j f_j \right\rangle_H &\leq \left\| \sum_i a_i \hat{k}_i \right\|_H \left\| \sum_j a_j f_j \right\|_H \\
&\lesssim \left\| \sum_i a_i \hat{k}_i \right\|_H \left(\sum_i |a_i|^2 \right)^{1/2},
\end{aligned}$$

which is the required estimate in (7.13).

In the other direction, if the Gram matrix is bounded below, then we have the estimates

$$\begin{aligned}
\left\| \sum_j a_j f_j \right\|_H &= \sup \left\{ \left| \left\langle \sum_j a_j f_j, \sum_j b_j \hat{k}_j \right\rangle_H \right| : \left\| \sum_j b_j \hat{k}_j \right\|_H \leq 1 \right\} \\
&\leq \sup \left\{ \sum_j a_j \overline{b}_j : \sum_j |b_j|^2 \lesssim 1 \right\} \\
&\lesssim \left(\sum_j |a_j|^2 \right)^{1/2},
\end{aligned}$$

which is (7.15), the desired boundedness for the dual Gramian. The first inequality follows from the fact that the dual Gram matrix is bounded below. In showing this, we took into account that the $\{f_i\}$ were a minimal system. Knowing that ensures that each f_j is in the closed span of $\{k_j\}$, and hence, when computing the norm of $F = \sum_i a_i f_i$, it was sufficient to pair F with vectors in that span. $\square$

7.3. Interpolation in H^2 and $\mathcal{D}$

We now specialize to the Hardy space and the Dirichlet space. We will show in both cases that if a sequence Z satisfies a Carleson measure condition and is separated, that is, if condition (5) of Theorem 7.2 is satisfied, then the sequence is an interpolating sequence for the Hilbert space and for the multiplier algebra.

We have already developed many of the tools we need. Theorem 7.3 ensures that if Z is an interpolating sequence, then we also have a multiplier interpolating sequence. To show Z is an interpolating sequence, in the language of Theorem 7.2, we want to show that condition (5) implies condition (4). By virtue of that theorem, it suffices to show that condition (5) implies condition (3). Theorem 7.10 ensures that condition (3) is equivalent to knowing both the Gram matrix and the dual Gram matrix are bounded. Furthermore, noting Lemma 7.4, condition (5) implies the boundedness of the Gram matrix. Hence, in sum, our task is reduced to proving the boundedness of the dual Gram matrix. We will do that with separate proofs for the Hardy space and the Dirichlet space.

The proofs we present are not the original ones; they are more recent proofs based on the RKHS theory. Although these arguments are elegant, they do not display the many beautiful analytical techniques developed in this area, including those of Carleson [**91**], Wolff [**161**, Chapter 8], Shapiro-Shields [**307**], Jones [**186**], Marshall-Sundberg [**220**], and Böe [**77**], [**75**]. We will see some of those ideas, implicitly or explicitly, in the next chapter when we consider "onto interpolation", in Chapter 15 where we give a constructive proof of an interpolation theorem for Besov spaces, and in Chapter 17 when we use $\overline{\partial}$ techniques to prove corona theorems.

7.3.1. The Hardy Space.

We have $H = H^2$ and a fixed $Z = \{z_n\}_{n=1}^{\infty} \subset \mathbb{D}$.

THEOREM 7.11. *If Z is separated and μ_Z is a Carleson measure for the Hardy space, that is* (7.3) *and* (7.1) *hold, then Z is an interpolating sequence for H^2 and for* Mult $(H^2) = H^\infty$.

As we just said, it remains to show that the dual Gram matrix is bounded. At this point, we use the specifics of the Hardy space and explicitly compute the dual Gram matrix.

PROOF. For $n = 1, 2, \ldots$, let b_n be the Blaschke factor associated with z_n,

$$b_n(z) = \frac{|z_n|}{z_n}\frac{z_n - z}{1 - \overline{z}_n z}.$$

Let B be the associated Blaschke product, $B = \prod_{k=1}^{\infty} b_k$. For $n = 1, 2, \ldots$ let $B_n = B/b_n = \prod_{k \neq n} b_k$. At the end of this proof we obtain a positive lower bound for $|B_n(z_n)|$ and having such a bound ensures that these infinite products converge.

In the Hardy space we can write down the exact solution to the extremal problems which define $M_n^{Z(n)}$; $M_n^{Z(n)} = \beta_n B_n$ with $\beta_n = |B_n(z_n)| / B_n(z_n)$ insuring $M_n^{Z(n)}(z_n) > 0$.

EXERCISE 7.12. *Prove that this is the solution.*

Now, using (6.21), we see that $F_n^{Z(n)}(z_n) = \beta_n B_n(z_n) \|k_n\|_{H^2}$. Thus, by (7.11), the matrix entries of interest are

$$\begin{aligned}
\langle f_n, f_m\rangle_{H^2} &= \frac{1}{\beta_n\overline{\beta}_m B_n(z_n)\|k_n\|_{H^2} B_m(z_m)\|k_m\|_{H^2}} \langle B_n k_n, B_m k_m\rangle_{H^2} \\
&= \alpha(n,m)\left\langle k_n \textstyle\prod_{j\neq n} b_j, k_m \textstyle\prod_{j\neq m} b_j\right\rangle_{H^2} \\
&= \alpha(n,m)\langle k_n b_m, k_m b_n\rangle_{H^2}.
\end{aligned}$$

Here $\alpha(n,m)$ is just a shorthand for the number preceding the inner product. The passage from the second line to the third uses the fact that the inner product can be computed as an L^2 inner product on the boundary values and, on the boundary, for any j, $b_j\overline{b}_j = 1$. Hence Blaschke factors which appear on both sides in the inner product can be canceled.

We now use partial fractions and evaluate

$$\begin{aligned}
\langle k_n b_m, k_m b_n\rangle_{H^2} &= \left\langle k_n b_m, \frac{1}{1-\overline{z}_m z}\frac{|z_n|}{z_n}\frac{z_n - z}{1-\overline{z}_n z}\right\rangle_{H^2} \\
&= \frac{|z_n|}{\overline{z}_n}\left\langle k_n b_m, \overline{\left(\frac{z_m\overline{z}_n - 1}{z_m - z_n}\right)} k_m + \overline{\left(\frac{1-|z_n|^2}{z_m - z_n}\right)} k_n\right\rangle_{H^2}.
\end{aligned}$$

Using the fact that the $k's$ are reproducing kernels and recalling that $b_m(z_m) = 0$, we continue with

$$\begin{aligned}
\langle k_n b_m, k_m b_n\rangle_{H^2} &= \frac{|z_n|}{\overline{z}_n}\frac{|z_m|}{z_m}\frac{1-|z_n|^2}{z_m - z_n}\frac{z_m - z_n}{1-\overline{z}_m z_n}\frac{1}{1-|z_n|^2} \\
&= \frac{|z_n|}{\overline{z}_n}\frac{|z_m|}{z_m}\frac{1}{1-\overline{z}_m z_n} \\
&= \frac{|z_n|}{\overline{z}_n}\frac{|z_m|}{z_m}\overline{\langle k_n, k_m\rangle_{H^2}}.
\end{aligned}$$

Thus the matrix elements are

$$\begin{aligned}
\langle f_n, f_m\rangle_{H^2} &= \frac{1}{\beta_n B_n(z_n)\|k_n\|_{H^2}\overline{\beta}_m B_m(z_m)\|k_m\|_{H^2}}\frac{|z_n|}{\overline{z}_n}\frac{|z_m|}{z_m}\overline{\langle k_n, k_m\rangle_{H^2}} \\
&= \left[\frac{1}{B_n(z_n)}\frac{|z_n|}{\beta_n\overline{z}_n}\right]\left[\frac{1}{B_m(z_m)}\frac{|z_m|}{\overline{\beta}_m z_m}\right]\overline{\left\langle \hat{k}_n, \hat{k}_m\right\rangle_{H^2}}.
\end{aligned} \tag{7.16}$$

Let $R(n)$ be the row matrix whose n^{th} entry is the expression in the first square brackets in the previous line, and let $C(m)$ be the column matrix $C = R^T$. Using that notation, we want to show that the matrix $\left(R(n)\overline{\left\langle \hat{k}_n, \hat{k}_m\right\rangle_{H^2}} C(m)\right)$ is bounded. The middle entries are the conjugates of the entries of the Gram matrix which we know is bounded. Hence those entries correspond to a bounded matrix. Multiplication by R and C will produce a new bounded matrix if R and C have bounded entries. Recall that the factors $|z_n|/\beta_n\overline{z}_n$ are unimodular. Hence we will be done if we show that

$$\inf_n |B_n(z_n)| = \inf_n \left|\textstyle\prod_{j\neq n} b_j(z_n)\right| > 0.$$

Recalling that $|b_j(z_n)|^2 = 1-\delta^2(z_j, z_n)$, we see that the separation condition ensures that the individual factors in this product are uniformly bounded away from zero. For such products, we can obtain a lower estimate which is uniform in j if we can find a uniform upper bound on $\sum_{n\neq j}(1-|b_j(z_n)|)$. It is enough to bound the larger quantity $\sum_n(1-|b_j(z_n)|^2)$. By (1.6), the sum equals $\sum_n \left|\left\langle \hat{k}_n, \hat{k}_j \right\rangle_{H^2}\right|^2$. Now note that the Gram matrix is bounded and hence its rows and its columns are square summable with an upper bound independent of which row or column is considered. This gives us our required uniform bound, and we are done. Alternatively, one can appeal to the Carleson measure condition to see that this sum is controlled. □

7.3.2. The Dirichlet Space. In this section, we consider the Dirichlet space and quantities such as $F_n^{Z(n)}$ are to be interpreted in that space.

THEOREM 7.13. *If Z is separated and μ_Z is a Carleson measure for the Dirichlet space, that is* (7.3) *and* (2.33) *hold, then Z is an interpolating sequence for $\mathcal{D}$ and for* Mult $(\mathcal{D})$.

The first proofs of this were due to Marshall-Sundberg [**220**] and Bishop [**72**] in 1994. Both papers were widely circulated but remain unpublished. The first published proof is by Böe in 2002 [**77**]. The proof we give in Chapters 15 and 16 characterizing interpolating sequences for Besov spaces of functions of several variables is similar in spirit to that proof in [**77**].

The proof in [**77**], which gives results for the family of Besov spaces, does not simplify very much when restricted to the Dirichlet space, the one Hilbert space in the family. On the other hand, the simpler Hilbert space based argument we will present here, which is from [**75**], has yet to give insights into a proof for the Besov spaces which are not Hilbert spaces.

PROOF OF THEOREM. The discussion in Section 7.2.2 applies and hence we have formula (7.10) for the entries of the dual Gram matrix $\Gamma = (\langle f_n, f_m\rangle_{\mathcal{D}})$:

$$\langle f_n, f_m\rangle_{\mathcal{D}} = \frac{\|k_n\|_{\mathcal{D}}\,\|k_m\|_{\mathcal{D}}}{\left\|k_n^{Z(n)}\right\|_{\mathcal{D}}^2 \left\|k_m^{Z(m)}\right\|_{\mathcal{D}}^2} \left\langle k_n^{Z(n)}, k_m^{Z(m)}\right\rangle_{\mathcal{D}}. \tag{7.17}$$

In contrast to the Hardy space, we do not know how to write explicit formulas for these terms. Instead, we proceed by estimating the $\langle f_n, f_m\rangle_{\mathcal{D}}$ beginning with the inner product on the right hand side of the previous equation.

For $n, m = 1, 2, \ldots$, let $Z(n,m) = Z\setminus\{z_n, z_m\}$; and thus

$$Z(n,m) = Z(m,n) \subset \begin{matrix} Z(n) \\ Z(m) \end{matrix} \subset Z.$$

Corollary 2.12 tells us how to go from $k_n^{Z(n,m)}$, which is the kernel function the space $H_{Z(n,m)}$ for evaluating at z_m, and obtain the kernel function for the same point z_m but with respect to the space $H_{Z(m)}$, which is the subspace of $H_{Z(n,m)}$ of functions which also vanish at x_n. We obtain

$$k_m^{Z(m)} = k_m^{Z(n,m)} - \frac{k_m^{Z(n,m)}(z_n)}{k_n^{Z(n,m)}(z_n)} k_n^{Z(n,m)}.$$

Hence

$$\begin{aligned}
\left\langle k_n^{Z(n)}, k_m^{Z(m)} \right\rangle_{\mathcal{D}} &= \left\langle k_n^{Z(n)}, k_m^{Z(n,m)} - \frac{k_m^{Z(n,m)}(z_n)}{k_n^{Z(n,m)}(z_n)} k_n^{Z(n,m)} \right\rangle_{\mathcal{D}} \\
&= k_n^{Z(n)}(z_m) - \frac{k_m^{Z(n,m)}(z_n)}{k_n^{Z(n,m)}(z_n)} k_n^{Z(n)}(z_n) \\
&= 0 - \frac{k_n^{Z(n)}(z_n)}{k_n^{Z(n,m)}(z_n)} \overline{k_m^{Z(n,m)}(z_n)}. \qquad (7.18)
\end{aligned}$$

To see this, note first that $k_n^{Z(n)} \in H_{Z(n)} \subset H_{Z(n,m)}$, and hence the first line can be evaluated using the fact that $k_n^{Z(n,m)}$ and $k_m^{Z(n,m)}$ are kernel functions in that larger space. Moving to the third line uses the fact that $k_n^{Z(n)} \in H_{Z(n)}$, which ensures $k_n^{Z(n)}(z_m) = 0$.

Now note that by the first statement in Corollary 6.28, the fraction in the last line is between 0 and 1. Thus we have $\left|\left\langle k_n^{Z(n)}, k_m^{Z(m)} \right\rangle_{\mathcal{D}}\right| \leq \left|k_m^{Z(n,m)}(z_n)\right|$ and, combining this with (7.17), we obtain

$$|\langle f_n, f_m \rangle_{\mathcal{D}}| \leq \frac{\|k_n\|_{\mathcal{D}} \|k_m\|_{\mathcal{D}}}{\left\|k_n^{Z(n)}\right\|_{\mathcal{D}}^2 \left\|k_m^{Z(m)}\right\|_{\mathcal{D}}^2} \left|k_m^{Z(n,m)}(z_n)\right|. \qquad (7.19)$$

Kernel functions with respect to a subspace are obtained by projecting a kernel function for the entire space, hence for any n, $\left\|k_n^{Z(n)}\right\|_{\mathcal{D}} \leq \|k_n\|_{\mathcal{D}}$. We want an estimate in the other direction. That will be a quantitative formulation of the informal idea that, for each n, z_n is well separated from $Z^{(n)}$. To obtain those estimates, we use (6.21), which is a consequence of the CPP, to move from questions about subspace kernels, which are solutions to Hilbert space extremal problems, to questions about solutions to multiplier extremal problems. We estimate the extremal multiplier by writing a generalized Blaschke product, which is a competitor for the extremal problem, and evaluating it. The convergence of that Blaschke product is ensured by the boundedness of the Gram matrix, which is a consequence of the Carleson measure hypothesis.

In detail,

$$\begin{aligned}
\left\|k_n^{Z(n)}\right\|_{\mathcal{D}}^2 &= k_n^{Z(n)}(z_n) \\
&= M_n^{Z(n)}(z_n) M_n^{Z(n)}(z_n) k_n(z_n) = M_n^{Z(n)}(z_n)^2 \|k_n\|_{\mathcal{D}}^2 \\
&\geq B_{Z(n),z_n}^2(z_n) \|k_n\|_{\mathcal{D}}^2 \\
&= \|k_n\|_{\mathcal{D}}^2 \prod_{j \neq n} \left(1 - \left|\left\langle \hat{k}_j, \hat{k}_n \right\rangle_{\mathcal{D}}\right|^2\right) \\
&\gtrsim \|k_n\|_{\mathcal{D}}^2 .
\end{aligned}$$

The passage to the second line uses (6.21). The third line contains the generalized Blaschke product of (6.12), the inequality holds in that line because $M_n^{Z(n)}$ is the solution to an extremal problem for which $B_{Z(n),z_n}$ is also a competitor. The next line is the evaluation of $B_{Z(n),z_n}^2(z_n)$. To show the final inequality, it suffices to bound the product away from zero, independently of n, That bound follows from knowing that $\sum_j |\left\langle \hat{k}_j, \hat{k}_n \right\rangle|^2$ has a bound independent of n. As we noted in the proof of the previous theorem, that estimate for the sum is a consequence of the boundedness of the Gram matrix.

Using that estimate for n and again for m, we continue (7.19) with

$$|\langle f_n, f_m \rangle_{\mathcal{D}}| \lesssim \frac{\left| k_m^{Z(n,m)}(z_n) \right|}{\|k_n\|_{\mathcal{D}} \|k_m\|_{\mathcal{D}}}. \tag{7.20}$$

By the second part of Corollary 6.28, we know $\left| k_m^{Z(n,m)}(z_n) \right| \le |k_m(z_n)|$. Combining this with (7.20) gives

$$|\langle f_n, f_m \rangle_{\mathcal{D}}| \lesssim \frac{|k_m(z_n)|}{\|k_n\|_{\mathcal{D}} \|k_m\|_{\mathcal{D}}} = \left| \left\langle \hat{k}_m, \hat{k}_n \right\rangle_{\mathcal{D}} \right|.$$

Recall that if a matrix $A = (\alpha_{ij})$ satisfies $|\alpha_{ij}| \le \beta_{ij}$, and if the matrix (β_{ij}) defines a bounded operator, then so does the matrix A. Hence the proof will be finished when we prove the following lemma which is elementary for the Dirichlet space and whose analog for the Hardy space is false. □

LEMMA 7.14. *If Z satisfies* (7.3), *then the matrix*

$$\left(\left| \left\langle \hat{k}_i, \hat{k}_j \right\rangle_{\mathcal{D}} \right| \right)$$

defines a bounded operator.

PROOF OF LEMMA. Recalling the formula for the kernel function, we must show that if $\{a_n\} \in \ell^2$, then we can bound

$$\left| \sum_{n,m} \frac{a_n}{\|k_n\|_{\mathcal{D}}} \frac{a_m}{\|k_m\|_{\mathcal{D}}} \left| \frac{1}{z_n \overline{z}_m} \log \left(\frac{1}{1 - z_n \overline{z}_m} \right) \right| \right|.$$

However, one can easily check that there are positive constants A, B so that for $z, w \in \mathbb{D}$,

$$\left| \frac{1}{z\overline{w}} \log \left(\frac{1}{1 - z\overline{w}} \right) \right| \le A + B \operatorname{Re} \frac{1}{z\overline{w}} \log \left(\frac{1}{1 - z\overline{w}} \right). \tag{7.21}$$

We then have

$$\begin{aligned}
&\left| \sum_{n,m} \frac{a_n}{\|k_n\|_{\mathcal{D}}} \frac{a_m}{\|k_m\|_{\mathcal{D}}} \left| \frac{1}{z_n \overline{z}_m} \log \left(\frac{1}{1 - z_n \overline{z}_m} \right) \right| \right| \\
&\quad \le A \sum_{n,m} \frac{|a_n|}{\|k_n\|_{\mathcal{D}}} \frac{|a_m|}{\|k_m\|_{\mathcal{D}}} \\
&\qquad + B \sum_{n,m} |a_n|\, |\overline{a}_m| \operatorname{Re} \frac{1}{\|k\mathcal{D}_n\|_{\mathcal{D}} \|k_m\|} \frac{1}{z_n \overline{z}_m} \log \left(\frac{1}{1 - z_n \overline{z}_m} \right) \\
&\quad = A \left(\sum_n \frac{|a_n|}{\|k_n\|_{\mathcal{D}}} \right)^2 + B \operatorname{Re} \sum_{n,m} |a_n|\, |\overline{a}_m| \left\langle \hat{k}_n, \hat{k}_m \right\rangle_{\mathcal{D}}.
\end{aligned}$$

The first term in the previous line can be estimated by

$$\left(\sum_n |a_n|^2 \right)^{1/2} \left(\sum_n \|k_n\|_{\mathcal{D}}^{-2} \right)^{1/2};$$

the first factor of which is under control because $\{a_n\} \in \ell^2$ and the second factor is seen to be finite because it is the integral of the function 1 against the Carleson measure μ_Z. The second term is under control because, by the Carleson measure hypothesis, the matrix $\left(\left\langle \hat{k}_n, \hat{k}_m \right\rangle_{\mathcal{D}}\right)$ is bounded on ℓ^2. □

7.3.3. Other Spaces. In [**75**] Böe pointed out that the proof, the one we just gave for Dirichlet space interpolating sequences, applies directly to some other spaces.

Suppose we are given H, a RKHS on X with a CPK and $Z = \{z_i\} \subset X$. We denote the associated Gram matrix $\operatorname{Gram}(Z)$ and we denote by $|\operatorname{Gram}|(Z)$ the matrix obtained by replacing each entry of $\operatorname{Gram}(Z)$ with its modulus. Thus

$$\operatorname{Gram}(Z) = \left(\left\langle \hat{k}_i, \hat{k}_j \right\rangle_H\right) \text{ and } |\operatorname{Gram}|(Z) = \left(\left|\left\langle \hat{k}_i, \hat{k}_j \right\rangle_H\right|\right).$$

In the previous proof we used the facts that the Dirichlet space has a CPK and then we used (7.21) to conclude that

$$|\operatorname{Gram}|(Z) \text{ is bounded if } \operatorname{Gram}(Z) \text{ is bounded.} \tag{7.22}$$

In general, if (7.22) holds for H, or even just for a particular Z, then the previous path can be followed to show Z is an interpolating sequence. For instance, (7.22) is automatic if we know $\left|\left\langle \hat{k}_z, \hat{k}_w \right\rangle_H\right| \lesssim \operatorname{Re}\left\langle \hat{k}_z, \hat{k}_w \right\rangle_H$. This condition holds for the spaces D_α, $0 < \alpha < 1$, mentioned in Section 2.5.2, which have kernels $(1 - \bar{z}w)^{-\alpha}$ as well as for spaces of holomorphic functions on the ball with kernels $(1 - \langle w, z \rangle)^{-\alpha}$. Also, (7.22) is automatic if the entries of the Gram matrix are always real and positive; for instance in the "diameter space" of [**46**] and the space on functions on a dyadic tree with kernel function $\Lambda^{d(x \wedge y)}$, $\Lambda > 1$ discussed in [**43**]. On the other hand there is no such estimate for the Hardy space, nor for its higher dimensional generalizations, the Drury-Arveson spaces defined in Section 6.2.

In particular, this path does not lead to a proof of the following conjecture of Seip, that the pattern we have seen holds for *all* H which have a CPK and, in fact for all those with Pick kernels.

CONJECTURE 7.15. [**302**, page 33] *If H is a RKHS which has the Pick property, then Z is an interpolating sequence for H if and only if it is an interpolating sequence for* $\operatorname{Mult}(H)$ *if and only if both Z is separated and μ_Z is a Carleson measure.*

Late Remark: See the notes at the end of this chapter.

7.4. Notes and Comments

There are many paths through the proof of the classical Hardy space interpolation theorems, for instance those in [**178**], [**161**], [**246**], [**4**], [**302**], and [**294**]. Our presentation is a synthesis of those and others.

The equivalence of the second and third condition in Theorem 7.2 is the Köthe-Toeplitz theorem; the equivalence the third and fourth is part of Bari's theorem, [**246**, Chapter 6].

The relation between Carleson's interpolation theorem for H^∞ and Pick's theorem is discussed in [**193**] and in [**220**]. The Mathematics Reviews review of [**193**, MR1771759] also has information about formula (7.16) relating the Gram matrix and dual Gram matrix for Hardy space interpolation.

The first study of interpolating sequences for Mult($\mathcal{D}$) is in Axler's paper [**57**].

Theorem 7.13 describes interpolating sequences for $\mathcal{D}$. A similar result for $\mathcal{D}(\mathcal{T})$ is in [**48**]. However that result is proved constructively, not with Hilbert space techniques.

Interpolation questions similar to those in this chapter are also studied extensively for Banach spaces of functions. In particular, fundamental results for one variable Besov spaces were given by Böe in [**77**] and for the Bloch space by Böe and Nicolau in [**78**]. We consider interpolation for Besov spaces in several variables in Chapter 15.

After this manuscript was completed Aleman, Hartz, McCarthy, and Richter posted a preprint [**15**] containing a proof of Conjecture 7.15 for the case of H with the CPP. Beyond the ideas we have seen here, their proof makes fundamental use of the resolution of the Paving Conjecture which used the recent work by Marcus, Spielman and Srivastava of [**218**] combined with earlier results of J. Anderson [**25**].

Part 2

The Dirichlet Space: Selected Topics

In this part we look at several questions about the Dirichlet space and its functions. We study the onto interpolating sequences (another topic which we do in parallel with analysis on the tree model), the refined behavior of the boundary values of functions, the uses of norming functionals based on oscillation, the invariant subspace theory for the shift operator on the Dirichlet space, and the boundedness criterion for Hankel forms on the Dirichlet space.

Work in the first part emphasized the strong analogies between the Hardy space theory and the Dirichlet space theory. In this part the techniques and results diverge more from the Hardy space case; we use more function theory and less Hilbert space theory.

These topics are less thoroughly developed and many are in areas of active research, hence we encounter fundamental open questions. These include: what is the geometric description of the onto interpolating sequences? What are the possible zero sets for functions in the Dirichlet space (both in the disk and on the boundary)? What are the cyclic vectors for the Dirichlet shift? In fact these specific questions just mark the edges of larger unexplored regions.

CHAPTER 8

Onto Interpolation

An onto interpolating sequence is one for which the associated restriction map is surjective but perhaps not bounded. A crucial difference between the Hardy and Dirichlet spaces is the vastly increased separation that interpolating sequences of the Dirichlet space enjoy. One striking manifestation of this phenomenon is that the Dirichlet space has onto interpolating sequences that are not interpolating sequences.

We begin this chapter looking at the relation between onto interpolation and the boundedness of the dual Gram matrix. That provides a framework for showing that in H^2 every onto interpolating sequence is an interpolating sequence, and for discussing Bishop's analytic characterization of onto interpolating sequences for $\mathcal{D}$.

We then give a geometric characterization of onto interpolating sequences for the dyadic Dirichlet space $\mathcal{D}(\mathcal{T})$. Finally there is a discussion of possible geometric descriptions of onto interpolating sequences for $\mathcal{D}$.

Onto interpolating sequences are less well studied than interpolating sequences and not as well understood. One reason is that in H^2, where interpolating sequences were first studied, every OIS is an IS, deflecting attention from describing OIS. The results we describe here for OIS are less crisp than those in the previous chapter and for several topics we include discussions but not complete proofs; referring instead to the papers [**72**] and [**42**] which, although unpublished, are available online.

8.1. The Dual Gram Matrix and OIS

We will continue to use notation and terminology of Section 7.2.2. In particular, we have a RKHS H of functions on a set X, a sequence $\{z_i\} = Z \subset X$. We write k_i for k_{z_i} and set $\hat{k}_i = k_i / \|k_i\|_H$. The Gram matrix associated with that sequence is $G = \left(\left\langle \hat{k}_i, \hat{k}_j \right\rangle_H\right)$. Further, if there is a minimal dual system, we denote it by $\{f_i\}$, in which case the matrix $G^{\mathrm{dual}} = \left(\langle f_i, f_j \rangle_H\right)$ is the dual Gram matrix of the system. Recall that if we are working with H^2 or $\mathcal{D}$, and Z is a zero sequence, then the existence of a minimal dual system is automatic.

Recall that the associated measure is

$$d\mu_Z = \sum_{j=1}^{\infty} \left\|k_{z_j}\right\|_H^{-2} \delta_{z_j}$$

and an associated restriction map from H to $\ell^2(Z, \mu_Z)$.

As with a IS, an OIS must be separated. If Z is an OIS for H then Theorem A.11 ensures that any interpolation problem can be solved with natural norm bounds; that is

$$\forall \tau \in \ell^2(Z, \mu_Z),\ \exists g \in H,\ \|g\|_H \lesssim \|\tau\|_{\ell^2(Z,\mu_Z)},\ Rg = \tau. \tag{8.1}$$

Using this we see that, given $z, w \in Z$ there is a function $f \in H$ with $f(z) = 0$, $f(w) = \|k_w\|_H$, and $\|f\|_H$ under control. Using this f with Proposition 2.8 we find that Z is separated. That is

$$\exists \varepsilon > 0,\ \forall z, w \in Z,\ z \neq w;\ \delta_H(z, w) > \varepsilon. \tag{8.2}$$

In Theorem 7.10 we saw relations between G, $G^{\mathtt{dual}}$, and interpolation problems. We continue that theme with the following:

THEOREM 8.1. *Given H, X, Z as above; suppose there is a dual Gram matrix $G^{\mathtt{dual}}$. Then $G^{\mathtt{dual}}$ is bounded if and only if Z is an onto interpolation sequence.*

PROOF. Suppose $G^{\mathtt{dual}}$ is bounded and we want to solve the interpolation problem associated to the sequence $\alpha = \{\alpha_i\} \in \ell^2(Z, \mu_Z)$. Consider the function

$$h_\alpha = \sum_i \frac{\alpha_i}{\|k_i\|_H} f_i. \tag{8.3}$$

We compute $h_\alpha(z_i) = \langle h_\alpha, k_i\rangle_H = \alpha_i$ and hence h_α is a formal solution to the interpolation problem. To see that it is an actual solution we check that $h_\alpha \in H$ by checking that its norm is finite. We have

$$\|h_\alpha\|_H^2 = \langle h_\alpha, h_\alpha\rangle_H = \sum_{i,j} \frac{\alpha_i}{\|k_i\|_H} \frac{\overline{\alpha}_j}{\|k_i\|_H} \langle f_i, f_j\rangle_H .$$

Now note that $\left\{\alpha_i \|k_i\|_H^{-1}\right\}$ is a sequence in unweighted ℓ^2 and by hypothesis the matrix $\left(\langle f_i, f_j\rangle_H\right)$ is bounded on that space. Thus $h_\alpha \in H$ and we have solved the interpolation problem.

In the other direction, suppose we wish to show that the matrix is bounded. It suffices to show it is bounded on finitely supported sequences. We select such a sequence $\{\beta_i\}$ and set $\alpha_i = \beta_i \|k_i\|_H$. The sum in (8.3) is finite and hence we can use it to define h_α. We have $h_\alpha(z_i) = \alpha_i$ for all i, and h_α is orthogonal to functions which vanish at all points of Z. Hence h_α is the minimal norm solution to this interpolation problem. On the other hand from (8.1) we know that the interpolation problem has a solution j_α with $\|j_\alpha\|_H \lesssim \|\alpha\|_{\ell^2(Z,\mu_Z)} \approx \|\beta\|_{\ell^2}$. But h_a is the minimal norm solution so

$$\sum_{i,j} \beta_i \overline{\beta}_j \langle f_i, f_j\rangle_H = \|h_\alpha\|_H^2 \leq \|j_\alpha\|_H^2 \lesssim \|\beta\|_{\ell^2}^2$$

as required. □

8.1.1. OIS for H^2.

COROLLARY 8.2. *The onto interpolating sequences for H^2 are the same as the interpolating sequences.*

PROOF. An IS is automatically an OIS. For the other direction, suppose Z is an OIS. By Theorem 7.11 we need to show that Z is separated and satisfies a Carleson measure condition. We saw at (8.2) that Z is separated so to finish we need to establish the Carleson measure condition.

It is immediate from the definitions, and was verified in the previous chapter, that $\mu_Z \in CM(H^2)$ if and only if G is bounded. Thus we will be finished if we show G is bounded. By the previous theorem we know G^{dual} is bounded. We now use the relation between G^{dual} and G that we developed in the proof of Theorem 7.11.

In that proof we computed the entries of G^{dual} and in doing that we needed to know that certain infinite Blaschke products converged. We obtained the required convergence using the assumption that G was bounded. Here we take a different path. Using (8.1) it is easy to see that Z must be a zero sequence for H^2. By the factorization theorem such zero sets must satisfy the Blaschke condition and that is enough to ensure that the products used in the proof of Theorem 7.11 are convergent. Hence we can follow the computations in that proof and get to (7.16) and the discussion that follows. There we saw that the passage from the Gram matrix to the dual Gram matrix involved multiplying by a row matrix and its transpose, and the elements of those matrices were seen, after some thought, to be bounded. We now want to go in the other direction, from the boundedness of the dual Gram matrix to the boundedness of the Gram matrix. The form of argument is the same but now the entries of the relevant row matrix are the reciprocals of those we saw earlier and thus their moduli are given by Blaschke products and hence are bounded. Hence the boundedness of G follows from that of G^{dual} and we are done. □

8.1.2. Bishop's Theorem, OIS for $\mathcal{D}$. We have seen that the boundedness of G^{dual} is equivalent to having an OIS. Certainly the boundedness of G^{dual} implies a uniform bound on the diagonal elements of that matrix:

$$\Phi := \sup \langle f_i, f_i \rangle_H = \sup \|f_i\|_H^2 < \infty. \tag{8.4}$$

In [**72**] Bishop investigated the relation between OIS and the formally weaker condition WIS which we now define. If we apply the condition (8.1) to the standard unit vectors in $\ell^2(Z, \mu_Z)$ we see that

$$\forall z, w \in Z,\ \exists h_z \in H,\ h_z(w) = \begin{cases} k_z(z)^{1/2} & w = z \\ 0 & w \neq z \end{cases},\ \|h_z\|_H \leq C. \tag{8.5}$$

This condition is superficially weaker than being an OIS. It was studied for various RKHS by Schuster and Seip in [**300**]. We follow them in calling sequences which satisfy the condition *weak interpolating sequences* for H.

We now suppose that $H = \mathcal{D}$ and we continue the other notation. In particular recall that $\mu_Z = \sum_i \|k_{z_i}\|^{-2} \delta_{z_i}$.

THEOREM 8.3 ([**72**]). *Z is an OIS for $\mathcal{D}$ if and only if any, and hence all, of the following equivalent conditions hold.*

1. *Z is a WIS,*
2. $\Gamma := \inf_i M_i^Z(z_i) > 0$,
3. $\sup_i \|f_i\|_{\mathcal{D}} < \infty$,
4. *Z is a WIS, and for some C the $\{h_{z_n}\}$ can be chosen with $\|h_{z_n}\|_{\mathcal{D}} \leq C \|k_n\|_{\mathcal{D}}$.*

Conditions (1)-(4) are a global separation condition, stronger than (8.2) and more subtle. For instance, the condition (8.2) is the local condition stating that no k_i is close to being in the span of **any particular** one of the other k_j, On the other hand condition (2), or the equivalent condition (3), is the requirement that no k_i is close to being in the span of the union of **all** of the others, k_j.

EXERCISE 8.4. *Prove the equivalence of Conditions* (1)-(4) *in the theorem. This is a "soft" result which follows from the definitions, the previous formulas, and basic facts about kernel functions. The proof does not require further analysis using function theory or the CPP.*

EXERCISE 8.5. *Formulate the natural analogous conditions for the Hardy space and show they are equivalent.*

PROOF DISCUSSION. The equivalence of the conditions is Exercise 8.4.

If Z is an OIS then it is automatically a WIS and hence all four conditions hold. The heart of Bishop's proof is the reverse implication, showing that Condition (4) implies Z is an OIS. If we start with the functions $\{h_{z_i}\}$ from (4) and want to interpolate the sequence of values $\{\alpha_i\}$ then the sum $h = \sum_i \alpha_i h_{z_i}$ is a formal solution to the interpolation problem. However that series may not converge. Bishop then constructs a sequence of functions $\{j_n\}$ with $j_n(z_n) = 1$ and $|j'(z)|$ small away from z_n, and sets $r_i = h_{z_i} j_i$. The details are arranged so that $r = \sum_i \alpha_i r_i$, which is another formal solution to the interpolation problem, is a convergent series in $\mathcal{D}$, and hence a solution to the interpolation problem. The construction of the required j_i is done using substantial results from function theory. □

(Also, see the notes at the end of this Chapter.)

8.1.3. OIS for H^2 revisited. If $H = H^2$, the Hardy space, then an analogous result holds; given Z, the uniform boundedness of the diagonal entries of the dual Gram matrix, $\Phi < \infty$, ensures it is an OIS. One way to prove that is by using the condition $\Phi < \infty$ to derive the conditions necessary to invoke Theorem 7.11, and thus obtain the stronger fact that Z is an IS, roughly as in the proof of Corollary 8.2. The following proof due to Shapiro and Shields [**306**] gives an alternative path to the OIS result, not passing through consideration of the Gram matrix or Carleson measures.

In particular we can work with Condition (3). Recall from Exercise 6.26 that for the Hardy space the extremal multipliers M_n^Z are given by Blaschke products; in the notation of Section 7.3.1. $M_n^Z = B_n$. For $n = 1, 2, \ldots$ set $\gamma_n = |B_n(z_n)|$. Hence in this context Condition (2) becomes

$$\Gamma = \inf_n |B_n(z_n)| = \inf_n \gamma_n > 0. \tag{8.6}$$

THEOREM 8.6. *Z is an OIS for H^2 if and only if $\Gamma > 0$.*

PROOF. If Z is an OIS then we have an estimate for the norms of the entries of the minimal dual system and hence the analogs of the six conditions hold, in particular $\Gamma > 0$.

In the other direction we saw in the proof of Theorem 8.1 that to show a sequence is an OIS it suffices to exhibit functions $\{j_n\} \in H^2$ which satisfy two conditions:

$$j_n(z_m) = \delta_{nm} \|k_n\|_{H^2} \text{ and}$$
$$\text{the matrix } J := (\langle j_n, j_m \rangle_{H^2}) \text{ is bounded on } \ell^2.$$

Recall the functions B and B_n that were used in the proof of Theorem 7.11. For $n = 1, 2, \ldots$ define

$$g_n(z) := \frac{1}{\gamma_n} B_n(z)\hat{k}_n = \frac{(1-|z_n|^2)^{1/2}}{\gamma_n}\left(B(z)\frac{1}{z-z_k}\right)$$
$$h_n(z) := \frac{(g_n(z))^2}{\|k_n\|_{H^2}} = \frac{(1-|z_n|^2)^{3/2}}{\gamma_n^2}\left(B(z)\frac{1}{z-z_n}\right)^2.$$

By simply evaluating the functions we see that $g_n(z_m) = h_n(z_m) = \delta_{nm}\|k_m\|_{H^2}$ and so either set of functions might work as a choice for the $j's$. For each n we set $j_n = h_n$; we will discuss the $g_n's$ after the proof. We compute

$$\begin{aligned}\langle h_n, h_m\rangle_{H^2} &= \frac{(1-|z_n|^2)^{3/2}}{\gamma_n^2}\frac{(1-|z_m|^2)^{3/2}}{\gamma_m^2}\left\langle\left(\frac{B(z)}{z-z_n}\right)^2, \left(\frac{B(z)}{z-z_m}\right)^2\right\rangle_{H^2} \\ &= \frac{(1-|z_n|^2)^{3/2}(1-|z_m|^2)^{3/2}}{\gamma_n^2\gamma_m^2}\left\langle\frac{1}{(z-z_n)^2}, \frac{1}{(z-z_m)^2}\right\rangle_{H^2} \\ &= \frac{(1-|z_n|^2)^{3/2}(1-|z_m|^2)^{3/2}}{\gamma_n^2\gamma_m^2}\frac{1+z_n\overline{z_m}}{(1+z_n\overline{z_m})^3},\end{aligned}$$

the first equality because Blaschke factors on both sides of an $L^2(d\theta)$ inner product can be canceled, the second by, for instance, the Cauchy integral formula. We continue with

$$\begin{aligned}|\langle h_n, h_m\rangle_{H^2}| &\lesssim \frac{(1-|z_n|^2)(1-|z_m|^2)}{|1-z_n\overline{z_m}|^2}\frac{(1-|z_n|^2)^{1/2}(1-|z_m|^2)^{1/2}}{|1-z_n\overline{z_m}|} \\ &\lesssim \frac{(1-|z_n|^2)(1-|z_m|^2)}{|1-z_n\overline{z_m}|^2} \\ &= (1-\delta_{H^2}(z_n, z_m)).\end{aligned}$$

The first estimate is obtained using the hypothesis that $\Gamma > 0$ and the trivial estimate on $1 + z_n\overline{z_m}$. The next estimate holds because the second fraction is at most one. The final line follows from (1.6) and the definition of δ_{H^2}.

Our hypothesis ensures that the quantities $\prod_{n\neq m}\delta_{H^2}(z_n, z_m)$ are uniformly bound below, hence by standard facts about sums and products we also have a uniform bound on $\sum_{n\neq m}(1-\delta_{H^2}(z_n, z_m))$. Those are the sums of the entries in the rows of J, and J is a symmetric matrix with positive entries. Hence by the elementary case of Schur's lemma, Theorem B.2 with the choice $h = 1$, J is bounded and we are done. □

Combining this result with Corollary 8.2 we have the following.

COROLLARY 8.7. *The condition $\Gamma > 0$ implies μ_Z is a Carleson measure.*

It is interesting, perhaps surprising, that we used the functions $\{h_n\}$ in the proof of the Theorem. The simpler functions $\{g_n\}$ are scalar multiples of the minimal dual system and hence are a tempting choice for the $j's$. However that choice leads to the question of boundedness of a matrix that is essentially the Gram matrix itself. It is possible, but not at all straightforward, to go from $\Gamma > 0$ to the boundedness of the matrix that shows up. This smart choice of $\{h_n\}$ avoids those difficulties.

In some sense Bishop's proof of Theorem 8.3 for the Dirichlet space has a similar flavor. The natural first candidates for the interpolating functions do not have sufficient almost orthogonality to see easily that the formal solutions to the interpolation problems are convergent sums. Bishop modifies those functions, creating the analog of a wiser choice for the $j's$, and obtains modified functions for which the convergence can be demonstrated directly.

We have seen that for H^2 the IS the OIS are the same. The situation is more complicated for weighted H^2 spaces, [**229**].

8.1.4. More Generally. We see from (8.6) that the condition $\Gamma > 0$ can be formulated in terms of the geometry of Z and hence the previous corollary gives alternative insights into the geometric properties of Carleson measures for H^2. Analogous results for $\mathcal{D}$ are not known, but with the possibility of such results in mind we collect some straightforward observations about the generalized Blaschke products studied in Section 6.6.2. As in that section, we restrict attention to the spaces $H = \mathcal{D}_\alpha$, $0 \le \alpha \le 1$, $H = H^2$, or $H = \mathcal{D}$.

Given a separated sequence $Z = \{z_i\} \subset \mathbb{D}$ we consider the product $\mathfrak{B}(\cdot) = \prod_{z_j \in Z} G_{z_j,0}(\cdot)$ and the partial products $\mathfrak{B}_s(\cdot) = \prod_{z_j \in Z, z_j \neq z_s} G_{z_j,0}(\cdot)$. Here $\dot{G}$, and hence also the $\mathfrak{B}$, are specific to the space H being considered. We are interested in the positivity $\theta := \inf_s |\mathfrak{B}_s(z_s)|$.

A few facts are immediate. First, if $H = H^2$ then the generalized Blaschke products are the classical Blaschke products, and thus θ is equal to Γ of (8.6). A similar relationship between the $\mathfrak{B}_s(z_s)$ and the δ_H distances holds in general and using it we can rewrite θ to emphasize the analogy with Γ;

$$\theta = \inf_s |\mathfrak{B}_s(z_s)| = \inf_{n \ge 1} \prod_{m:\ m \neq n} \delta_H(z_m, z_n).$$

Also, if $H = \mathcal{D}$ then $\theta \le \Gamma$, where Γ is from (2). That holds because those $M_i^Z(z_i)$ are solutions to the maximum problem for which $\mathfrak{B}_s(z_s)$ is a competitor. An analogs result holds for $H = \mathcal{D}_\alpha$.

Suppose $A = (a_{ij})_{i,j=1}^\infty$ is an infinite matrix. We write $\|A\|_{2,\infty}$ for the mixed matrix norm of A defined by

$$\|A\|_{2,\infty}^2 = \max\left\{\sup_{i \ge 1} \sum_{j=1}^\infty |a_{ij}|^2,\ \sup_{j \ge 1} \sum_{i=1}^\infty |a_{ij}|^2\right\}.$$

We have the following (which is a corrected version of Proposition 5.12 of [**294**].)

PROPOSITION 8.8. *Suppose H and Z are as above and let*

$$G = \left(\left\langle \widehat{k}_{z_m}, \widehat{k}_{z_n}\right\rangle_H\right)_{m,n=1}^\infty$$

be the associated Gram matrix of normalized kernels. Then $\|G\|_{2,\infty} < \infty$ if and only if $\theta > 0$. If G is a bounded operator on $\ell^2(\mathbb{Z}_+)$, equivalently, if μ_Z is a Carleson measure, then $\|G\|_{2,\infty} < \infty$.

PROOF. We have

$$\prod_{n:\ n \neq m} d(z_m, z_n)^2 = \prod_{n:\ n \neq m} \left(1 - \frac{|k_{z_m}(z_n)|^2}{\|k_{z_m}\|_H^2 \|k_{z_n}\|_H^2}\right) \ge c > 0$$

if and only if

$$\sum_{n:\ n\neq m}\frac{\left|k_{z_m}(z_n)\right|^2}{\|k_{z_m}\|_H^2\,\|k_{z_n}\|_H^2}\lesssim 1<\infty.$$

But the above sum can be rewritten as

$$\sum_{n:\ n\neq m}\left|\left\langle\frac{k_{z_m}}{\|k_{z_m}\|_H},\frac{k_{z_n}}{\|k_{z_n}\|_H}\right\rangle_H\right|^2=\sum_{n:\ n\neq m}\left|\left\langle\widehat{k}_{z_m},\widehat{k}_{z_n}\right\rangle_H\right|^2=\left\|\left\{\left\langle\widehat{k}_{z_m},\widehat{k}_{z_n}\right\rangle\right\}_{n=1}^{\infty}\right\|_{\ell^2}^2,$$

and the proof of the first statement is now evident. For the second statement note that G being a bounded matrix automatically implies a uniform bound for the ℓ^2 norm of its row vectors and its column vectors, i.e. that $\|G\|_{2,\infty}<\infty$. □

REMARK 8.9. *Note that if $\mu_Z=\sum_{j=1}^{\infty}\left\|k_{z_j}\right\|_H^{-2}\delta_{z_j}$ is a Carleson measure for H, then the Gram matrix G is bounded on ℓ^2, hence 'sup-square' bounded. The converse statement holds for the Gram matrix of the Hilbert space $H^2(\mathbb{D})$, but remains an open question in general.*

8.2. The Tree Dirichlet Space, $\mathcal{D}(\mathcal{T})$

In this section we will consider IS and OIS for $\mathcal{D}(\mathcal{T})$. The known results for IS are complete and are quite similar to those for $\mathcal{D}$, although the proofs are not. We will discuss some of what is known about OIS for $\mathcal{D}(\mathcal{T})$, the full story there is not known. After that we return to considering $\mathcal{D}$, and the framework we developed for $\mathcal{D}(\mathcal{T})$ will be useful.

8.2.1. IS for $\mathcal{D}(\mathcal{T})$. We recall the dyadic Dirichlet space, $\mathcal{D}(\mathcal{T})$, from Section 5.4. We will use the notation from that section, some of which we recall now.

We start with a rooted tree $\mathcal{T}$ and we suppose that the branching number of $\mathcal{T}$ is bounded above. Recalling "A Convenient Fiction" from Chapter 5, we will talk about it as a dyadic tree. The tree carries a partial order, $\leq$, induced by declaring the root to be the minimal element. It has a distance function, $d(\alpha,\beta)$, which is the number of edges on the unique edge path without repetitions connecting α and β. Also, recall the notation, for any vertex α, $d^+(\alpha)=1+d(o,\alpha)$.

For functions defined on $\mathcal{T}$ we recall the discrete differentiation operator Δ given by $(\triangle f)(\beta)=f(\beta)-f(\beta^{-1})$; and the discrete integration operator I, given by $(If)(\beta)=\sum_{o\leq\gamma\leq\beta}f(\gamma)$. By direct computation both ΔI and $I\Delta$ are the identity operator.

The space $\mathcal{D}(\mathcal{T})$ is the RKHS on $\mathcal{T}$ consisting of all complex-valued functions F on $\mathcal{T}$ for which the norm given by

$$\|F\|_{\mathcal{D}(\mathcal{T})}^2:=\sum_{\beta\in\mathcal{T}}\left|\triangle F(\beta)\right|^2 \tag{8.7}$$

is finite. It has the inner product

$$\langle F,G\rangle_{\mathcal{D}(\mathcal{T})}:=\sum_{\beta\in\mathcal{T}}\triangle F(\beta)\,\overline{\triangle G(\beta)},$$

and kernel functions $k_\alpha(\beta)=d^+(\alpha\wedge\beta)$ which has $\Delta k_\alpha(\beta)=\chi_{(\beta\leq\alpha)}$ and $\|k_\alpha\|_{\mathcal{D}(\mathcal{T})}^2=k_\alpha(\alpha)=d^+(\alpha)$.

For a sequence $Z\subset\mathcal{T}$ let R be the map which restricts functions on $\mathcal{T}$ to functions on Z and let μ_Z be the measure given by (7.2). A measure μ defined on $\mathcal{T}$ is called a Carleson measure for $\mathcal{D}(\mathcal{T})$ if the natural restriction map taking $\mathcal{D}(\mathcal{T})$ to $\ell^2(\mathcal{T},\mu)$ is bounded;

$$\sum_{\beta\in\mathcal{T}} |f(\beta)|^2 \mu(\beta) \lesssim \sum_{\beta\in\mathcal{T}} |\triangle f(\beta)|^2 . \tag{8.8}$$

An intrinsic characterization of these measures is given in Theorem 5.17. As before, knowing μ_Z is a Carleson measure is equivalent to knowing R is bounded. Also as before, knowing that R is surjective ensures that Z is a weak interpolating sequence. In this case the statement is:

(8.9)

$$\forall\alpha,\beta\in Z, \exists f_\alpha \in \mathcal{D}(\mathcal{T}), \text{with } f_\alpha(\beta) = \begin{cases} k_\alpha(\alpha)^{1/2} & \beta=\alpha \\ 0 & \beta\neq\alpha \end{cases} \text{ and } \|f_\alpha\|_{\mathcal{D}(\mathcal{T})} \leq C.$$

Again, this ensures a separation condition,

$$\exists c>0,\ \forall\alpha,\beta\in Z, \alpha\neq\beta;\ \delta_{\mathcal{D}(\mathcal{T})}(\alpha,\beta) > c. \tag{8.10}$$

When studying interpolating sequences for $\mathcal{D}$ we had two different looking versions of the separation condition; (2.33) and (2.34). The condition above is the analog of the first of those. The analog of the second is obtained by using the oscillation result, Proposition 5.12, for the functions $\{f_z\}$ from (8.9) and noting that for those functions

$$|f_z(z) - f_z(w)|^2 = k_z(z) \approx \beta(0,z) \approx d(o,z).$$

From this we obtain an alternative description of the separation condition,

$$\exists C>0,\ \forall\alpha,\beta\in Z, \alpha\neq\beta;\ d(0,\alpha) < Cd(\alpha,\beta). \tag{8.11}$$

A sequence Z is called an IS for $\mathcal{D}(\mathcal{T})$ exactly if R is a bounded map of $\mathcal{D}(\mathcal{T})$ onto $\ell^2(\mathcal{T},\mu_Z)$. Just as with H^2 and $\mathcal{D}$, such sequences are characterized by a separation condition together with a Carleson measure condition.

THEOREM 8.10. *A sequence $Z\subset\mathcal{T}$ is an interpolating sequence for $\mathcal{D}(\mathcal{T})$ if and only if and Z is separated and the associated measure μ_Z is a Carleson measure for $\mathcal{D}(\mathcal{T})$; that is, if and only if Z satisfies* (8.11), *or, equivalently,* (8.10) *and μ_Z defined in* (7.2) *satisfies* (8.8).

PROOF. This is proved in [**48**, Theorem 26]. Also, it is a special case of Theorem 16.7 whose proof is outlined in Section 16.1.1. □

8.2.2. OIS for $\mathcal{D}(\mathcal{T})$. For the tree space, $\mathcal{D}(\mathcal{T})$, the weak interpolation condition (8.9), which we saw was necessary for a sequence to be an OIS, is also sufficient. Most of the work for proving that was done in Lemma 5.25 in Chapter 5 and in this section we provide the final details. However, before doing that we restate condition (8.9) in the language of capacity on trees.

Recall that for E,F disjoint subsets of $\mathcal{T}$ the capacity $\operatorname{Cap}(E,F)$ of the condenser (E,F) is defined as

$$\operatorname{Cap}_{\mathcal{T}}(E,F) := \operatorname{Cap}(E,F) = \inf\{\|h\|_{\ell^2}^2 :\ Ih|_E = 1,\ Ih|_F = 0\}. \tag{8.12}$$

For Z a sequence in $\mathcal{T}$ and $z\in Z$ set $\gamma(z,Z) := \operatorname{Cap}_{\mathcal{T}}(z, Z\setminus\{z\})$. We say that Z satisfies the *tree capacity condition* if (8.13) holds;

$$\forall z\in Z,\ \gamma(z,Z) \lesssim d^+(z)^{-1}. \tag{8.13}$$

Notice that this condition is equivalent to being a weak interpolating sequence. It is straightforward to pass back and forth from functions which satisfy (8.9) to functions which are allowable competitors for computing $\gamma(z, Z)$.

THEOREM 8.11. *A sequence Z is an onto interpolating sequence for $\mathcal{D}(\mathcal{T})$ if and only if it satisfies the weak interpolation condition* (8.9)*; equivalently if and only if it satisfies the capacity condition* (8.13).

One attractive feature of this result is that the quantities $\gamma(z, Z)$ can be evaluated with a recursive algebraic algorithm using geometric data about Z. This is crucial for developing some of the details in [**42**], see also [**49**].

The proof of the theorem uses the construction, for each z, of a function which is approximately extremal for the problem of computing $\gamma(z, Z)$ and such that the supports of the (discrete) derivatives of those functions are disjoint. It is then straightforward to combine these functions to solve interpolation problems.

PROOF. We have already discussed that any OIS must satisfy the tree capacity condition. Conversely, if (8.13) holds uniformly then so does (8.9) and hence also properties (1), (2) and (3) of Lemma 5.25. Write $Z = \{\alpha_j\}_{j=1}^{\infty}$. Given $\xi = \{\xi_j\}_{j=1}^{\infty} \in \ell^2(\mu)$, set $f = \sum_{j=1}^{\infty} \xi_j H_{\alpha_j}$. By property (1) $f(\alpha_j) = \xi_j$ for all $j \geq 1$. By properties (2) and (3), $f \in \mathcal{D}(\mathcal{T})$ and $\|f\|_{\mathcal{D}(\mathcal{T})} \lesssim \|\xi\|_{\ell^2(\mu)}$. □

8.3. The Dirichlet Space, $\mathcal{D}$

The description in Theorem 8.3 involves analytic capacity and as such is more analytic than geometric. In this section we describe work focusing on the geometric structure of OIS. We report some progress but the results are not decisive. Full details are in [**42**].

Of course our analysis of OIS for $\mathcal{D}(\mathcal{T})$ involved detailed work with the tree $\mathcal{T}$. In this section we adapt the tree notation and language to study OIS for $\mathcal{D}$. That lets use and highlight some strong analogies between those two situations. For instance, the tree capacity condition (8.13) is an analog of condition (5) in Theorem 8.3 (although we avoided capacitary language in that case). A useful, although ahistorical, way to view the analogy is that this analysis for $\mathcal{D}$ is the result of adapting the arguments from $\mathcal{D}(\mathcal{T})$ to the holomorphic category. In that context sharp geometric localization is not possible, and substitutes and approximations must be found.

Now we consider an OIS Z for the Dirichlet space. As we noted in Section 5.2.2 we can regard Z as sitting inside a tree $\mathcal{T}$ which sits in $\mathbb{D}$ and we will use some of the language of the dyadic Dirichlet space $\mathcal{D}(\mathcal{T})$.

Elision: Having placed Z in $\mathcal{T}$ we will use some of the language of the dyadic Dirichlet space $\mathcal{D}(\mathcal{T})$ to study Z. However that entails certain distinctions which we find it convenient to ignore. For instance, when studying Z we consider the measure μ_Z on $\mathbb{D}$ defined by (7.2). For the same Z, thought of as the vertex set of $\mathcal{T}$ we might consider $\widetilde{\mu_Z}$, again defined by (7.2) but now using the kernel function for $\mathcal{D}(\mathcal{T})$. It is a measure defined on $\mathcal{T}$. The formulas are

$$d\mu_Z := \sum_{z_i \in Z} k_{z_i}^{\mathcal{D}}(z_i)^{-1}\delta_{z_i};\ k_{z_i}^{\mathcal{D}}(z_i) \approx \beta(0, z_i)$$

$$d\widetilde{\mu_Z} := \sum_{z_i \in Z} k_{z_i}^{\mathcal{D}(\mathcal{T})}(z_i)^{-1}\delta_{z_i};\ k_{z_i}^{\mathcal{D}(\mathcal{T})}(z_i) \approx d^+(z_i).$$

However both measures are supported on Z and, taking note of 5.3, we see that we can move from one to the other by multiplying by a function which is bounded and bounded away from zero, with constants only depending on the containing tree $\mathcal{T}$. In short, for the purposes of this chapter we can work with the two interchangeably or talk as if they were the same. We will do that and we will not mention it further. Also, we will make similar elisions of, for instance, $d^+(\cdot)$ and $1+\beta(0,\cdot)$; and we will continue these practices in later chapters when studying the Besov spaces.

8.3.1. Geometric Conditions. From Theorem 8.3 we know that Z being a WIS for $\mathcal{D}$ is necessary and sufficient for Z to be an OIS for $\mathcal{D}$. However it is not straightforward to apply the WIS criterion to specific Z. In particular it is not clear how to use that result to answer the question raised by Bishop of whether there is an OIS with $\|\mu_Z\| = \infty$. We obtain a positive answer to that question later with Corollary 8.25 using an alternative approach to describing OIS for $\mathcal{D}$.

We will now give geometric conditions which ensure a sequence is an OIS and which can be verified easily in some cases. However, although the conditions ensure a sequence is OIS they are not complete characterizations of those sequences.

However, although this is a condition in terms of purely geometric data it is still not easy to work with. For further conditions we return to our analysis of OIS for the tree space $\mathcal{D}(\mathcal{T})$.

We saw in Exercise 5.8 that the map which takes $f \in \mathrm{Hol}\,(\mathbb{D})$ to the sequence $\left\{(1-|z|^2)f(z)\right\}_{z\in\mathcal{T}}$ is a bounded map of the Bergman space to $\ell^2(\mathcal{T})$. Using that it is not hard to see that the restriction map taking $F \in \mathrm{Hol}\,(\mathbb{D})$ to the sequence $\{F(z)\}_{z\in\mathcal{T}}$ maps the Dirichlet space $\mathcal{D}$ boundedly into the tree Dirichlet space $\mathcal{D}(\mathcal{T})$, see e.g. [**42**, Section 7.2].

EXERCISE 8.12. *Why is that so? (You need to bound a sum of terms of the type* $|F(z) - F(w)|^2$ *with* z *close to* w.*)*

This implies that if Z is an OIS for the Dirichlet space, and if Z sits inside a tree $\mathcal{T}$ in the way we described, then Z will be an OIS for the tree space $\mathcal{D}(\mathcal{T})$. Hence, by the previous theorem, Z must satisfy the tree capacity condition (8.13) or, equivalently the weak interpolation condition (8.9). This gives nontrivial information about OIS for the Dirichlet space.

However this condition is not decisive, the tree geometry is only an imperfect reflection of the holomorphic geometry. For example, given $\alpha \in Z$ then, noting (5.3), the quantity $d^+(\alpha)$ is essentially independent of the choice of $\mathcal{T}$; however given $\alpha, \beta \in Z$ the value of $d(\alpha,\beta)$ depends in an essential way on $\mathcal{T}$. In fact the tree capacity condition (8.13) is a condition on the *pair* $(Z,\mathcal{T})$.

We suppose Z is a separated sequence, not just in the hyperbolic metric but in the stronger sense of (8.11) or, equivalently (8.10). We know that condition is necessary for having Z be an OIS and we now keep it as a standing assumption as we look for sufficient conditions for Z to be an OIS.

Recall that for any $\alpha \in \mathcal{T}$ the *successor set* of α is $S(\alpha) := \{\beta \in \mathcal{T} : \beta \geq \alpha\}$. This set is, roughly, the intersection of the tree with the tent $T(\alpha)$ in $\mathbb{D}$ having vertex α. The Carleson measures for $\mathcal{D}(\mathcal{T})$, which are defined by (8.8), are characterized in Theorem 5.17; namely, a measure μ is a $\mathcal{D}(\mathcal{T})$ Carleson measure if and only if

$$\sum_{\beta\in\mathcal{T}} \mu(S(\beta))^2 \lesssim \mu(S(\alpha)) < \infty. \tag{8.14}$$

Using (5.6) we can rewrite (7.2) as

$$d\mu_Z = \sum_{\alpha\in Z} d^+(\alpha)^{-1}\delta_\alpha$$

and hence the condition that μ_Z be a Carleson measure is

$$\sum_{\beta\in\mathcal{T}}\left(\sum_{\gamma\in S(\beta)} d^+(\gamma)^{-1}\right)^2 \lesssim \sum_{\gamma\in S(\alpha)} d^+(\gamma)^{-1}. \tag{8.15}$$

EXERCISE 8.13. *If $Z = \{z_k\}$ is contained in a single tree geodesic and $d^+(z_k) \approx k^{-2}$ then* (8.15) *holds.*

8.3.1.1. *The Simple Condition.* The condition (8.15) is the type of condition we are looking for; it is formulated using geometric data about the metric measure space $(\mathcal{T}, d, \mu_Z)$.

Assuming Z is separated, having μ_Z satisfy (8.15) is necessary and sufficient for Z to be an IS. If we are looking for conditions that might characterize OIS that are not IS we should consider conditions weaker. A natural candidate for a weaker condition is the *simple condition*:

$$\mu(S(\alpha)) \lesssim d^+(\alpha)^{-1}. \tag{8.16}$$

For μ_Z the condition can be written as

$$\sum_{\beta\in S(\alpha)} d^+(\beta)^{-1} \lesssim d^+(\alpha)^{-1}. \tag{8.17}$$

Condition (8.16) is a straightforward consequence of (8.15).

EXERCISE 8.14. *Show that.*

In particular, evaluating the condition with $\alpha = o$ we see that μ must be a finite measure.

If Z is a separated sequence and μ_Z satisfies the simple condition we will also say that Z satisfies the *simple condition*. We met a similar condition earlier, condition (4.8) which arose in Section 4.2.1 where we tried to characterize Carleson measures for the Dirichlet space by testing boundedness of the embedding operator on kernel functions. That condition was also called the simple condition and was additionally called the one box condition. As noted there, it is true but not obvious that that condition is strictly weaker than being a Dirichlet space Carleson measure. The construction in Proposition 5.22 shows that (8.16) is not equivalent to (8.14). Other examples are in [**72**], [**42**] and the last section of this chapter.

THEOREM 8.15. *Suppose Z is a separated sequence which satisfies the simple condition. Then Z is an OIS for the Dirichlet space $\mathcal{D}$.*

PROOF. This theorem is proved by Bishop in [**72**, Theorem 1.3] with a separate proof from Theorem 8.3. Also, it was noted by Böe in [**77**, Corollary 5.1] that his constructive proof describing IS used only the simple condition, not the Carleson measure condition, in constructing interpolating functions, and hence showed that sequences which satisfy the simple condition are OIS. □

Similar results hold for OIS for some spaces of functions on the ball, and for similar reasons [**47**, Remark 5.27].

COROLLARY 8.16. *The OIS for the Dirichlet space are not the same as the IS.*

8.3.1.2. *The Weak Simple Condition.* Thinking about the structure of the tree integration operator, I, suggests that the simple condition is not a natural candidate to be a sufficient condition for an OIS.

Suppose a sequence Z is given. Consider what we need to do if, given $\alpha \in Z$, we want to find a function f_α of the type described in (8.9). If $f_\alpha = I\Delta f_\alpha$ is to have the required large positive value at α there must be positive values of Δf_α on the tree path $[o, \alpha]$. Further, for each $\beta \in Z \cap S(\alpha)$ such that the open interval (α, β) is disjoint from $Z \cap S(\alpha)$, there must be some negative values of Δf_α on that path to ensure that $f_\alpha(\beta) = 0$. Once those values are in place Δf_α can be set to zero at the other points in $S(\alpha)$ and the structure of the tree integration will ensure that f_α will also vanish at any other points of $Z \cap S(\alpha)$. Furthermore, setting those values equal zero will decrease the norm of f_α. In short, the points of $Z \cap S(\alpha)$ other than the (α, β) just described do not contribute to the obstruction to constructing the required f_α.

This suggests considering a weaker variation on ideas which we used in defining the simple condition, a condition which, for each α, only takes note of the points of $Z \cap S(\alpha)$ which are visible from α with an unobstructed view. In other words, a condition at α that only involves $\beta \in Z \cap S(\alpha)$ for which the interval $[\alpha, \beta]$ contains no additional points of Z. Here is such a condition. We say that a sequence Z, or its associated measure μ_Z satisfies the *weak simple condition*, WSC, if $\forall \alpha \in \mathcal{T}$

$$\sum_{\substack{\beta \in S(\alpha) \\ [\alpha,\beta] \cap Z \cap S(\alpha) = \{\alpha,\beta\}}} d^+(\beta)^{-1} \lesssim d^+(\alpha)^{-1}. \tag{8.18}$$

Note that, in contrast to (8.16), the condition (8.18) does not itself ensure that $\|\mu\| < \infty$.

8.4. The Weak Simple Condition and OIS

8.4.1. The Building Blocks and the Space They Generate. We describe two results about OIS for $\mathcal{D}$ and prove one of them. Both results say that if Z is a separated sequence that satisfies the WSC and also an additional condition then Z is an OIS. For the first result the additional condition is that $\|\mu_Z\| < \infty$; for the second result the additional condition is a geometric condition on the sequence, that it be "tree-like."

Those results show the usefulness of the WSC and the WSC shows up again in Theorem 8.19 as part of a necessary and sufficient condition for describing OIS for a space closely related to $\mathcal{D}$. However the WSC is a flawed condition for studying OIS for $\mathcal{D}$ because it is not preserved on passing to subsequences. A subsequence of an OIS is automatically an OIS and, once the including tree is fixed, then a subsequence of a separated sequence is separated. Hence we are looking for a condition on sequences in trees which is preserved on passage to subsequences. The WSC is not such a condition.

The proofs of the theorems we referred to are constructive. The required interpolating functions are built using linear combinations of basic building blocks, similar to building blocks introduced by Böe in [**77**, Corollary 5.1]. We begin with a discussion of the building blocks.

Suppose Z is given and fixed. In [**77**, Corollary 5.1] Böe showed how to construct, for each $w \in Z$, a holomorphic function φ_w with values very close to one on the tent $T(w)$ and very close to zero on an enlargement of that tent. In this chapter we will use those functions and variations on them as building blocks in explicitly constructing solutions to certain interpolation problems. In Chapter 15 we will use extensions of these ideas to functions of several variables to construct solutions to certain interpolation problems on the ball.

A basic role is played by the operators Γ_s, defined for $s > -1$ by

$$\Gamma_s g(z) := \int_{\mathbb{D}} \frac{g(\zeta)(1-|\zeta|^2)^s}{(1-\overline{\zeta}z)^{1+s}} dA(\zeta).$$

For such s the operator P_s given by

$$P_s g(z) := c_s \int_{\mathbb{D}} \frac{g(\zeta)(1-|\zeta|^2)^s}{(1-\overline{\zeta}z)^{2+s}} dA(\zeta),$$

is a skew (i.e. non-selfadjoint) bounded projection of $L^2(\mathbb{D}, dA)$ to A^2 [**346**, Theorem 2.11]. Hence it is nearly true that the operator Γ_s satisfies $\partial_z \Gamma_s g = P_s g$. Thus Γ_s produces a holomorphic function with derivative equal to the projection of a given function. The operators Γ_s and P_s are used in moving from rough functions g with useful properties to holomorphic functions with closely related properties. In other approaches to interpolation problems a somewhat similar role is taken by formulas for solving $\overline{\partial}$ equations, for instance in [**40**].

To obtain our building block, φ_w, $w \in \mathbb{D}$, we apply Γ_s to a function which is supported near w, and which is essentially constant on that support. The details of the construction and properties can be found in [**77**] and [**42**]. Here we just collect the relevant properties of φ_w.

Suppose $1 - c/5 < \beta < 1$ where c is as in (8.10). Given $\beta < \rho < \alpha < 1$, we will use the cutoff function $c_{\rho,\alpha}$ defined by

$$c_{\rho,\alpha}(\gamma) := \begin{cases} 0 & \text{for} \quad \gamma < \rho \\ \frac{\gamma-\rho}{\alpha-\rho} & \text{for} \quad \rho \le \gamma \le \alpha \\ 1 & \text{for} \quad \alpha < \gamma. \end{cases} \tag{8.19}$$

The next lemma fully exploits the extra separation enjoyed by onto interpolating sequences in the Dirichlet space. A multivariable version is given later as Lemma 15.16.

LEMMA 8.17 ([**77**, Lemma 4.1]). *Suppose $s > -1$, c is as in (8.10), and $1 - c/5 < \beta < 1$. There are β_1, ρ and α satisfying $\beta < \beta_1 < \rho < \alpha < 1$ such that for every $w \in \mathbb{D}$, we can find a function g_w so that*

$$\varphi_w(z) := \Gamma_s g_w(z) = \int_{\mathbb{D}} \frac{g_w(\zeta)(1-|\zeta|^2)^s}{(1-\overline{\zeta}z)^{1+s}} dA(\zeta)$$

satisfies

$$\begin{cases} \varphi_w(w) & = & 1 \\ \varphi_w(z) & = & c_{\rho,\alpha}(\gamma_w(z)) + O\left(\left(\log \frac{1}{1-|w|^2}\right)^{-1}\right), & z \in V_w^{\beta} \\ |\varphi_w(z)| & \lesssim & \left(\log \frac{1}{1-|w|^2}\right)^{-1} \left(1-|w|^2\right)^{(\rho-\beta_1)(1+s)}, & z \notin V_w^{\beta_1}, \end{cases} \tag{8.20}$$

where $\gamma_w(z)$ *is defined by*

$$\left|z - \frac{w}{|w|}\right| = \left(1 - |w|^2\right)^{\gamma_w(z)},$$

$\frac{w}{|w|}$ *is radial projection of* w *to* ∂D, *and* $c_{\rho,\alpha}$ *is as in* (8.19). *Furthermore we have the estimate*

$$\int_{\mathbb{D}} |g_w(\zeta)|^2 \, dA(\zeta) \lesssim \left(\log \frac{1}{1 - |w|^2}\right)^{-1}. \tag{8.21}$$

We will also need the following lemma from [**220**], (see also [**47**].) For $w \in \mathbb{D}$ and c, β as above define

$$V_w := V_w^\beta = \left\{ z \in \mathbb{D} : \left|z - \frac{w}{|w|}\right| \le (1 - |w|^2)^\beta \right\}. \tag{8.22}$$

LEMMA 8.18. *Suppose the separation condition in* (8.11) *holds. Then for every* β *satisfying* $1 - c/5 < \beta < 1$ *there is* $\eta > \beta\eta > 1$ *such that if* $V_{z_i}^\beta \cap V_{z_j}^\beta \neq \phi$ *and* $|z_j| \ge |z_i|$, *then* $z_i \notin V_{z_j}^\beta$ *and*

$$(1 - |z_j|) \le (1 - |z_i|)^\eta. \tag{8.23}$$

There are choices of parameters in the previous lemmas; we assume that allowable choices have been made once and for all. We will also need further constraints on the parameters:

$$\beta < \alpha < 2\frac{\beta\eta}{\eta + 1}, \quad s > \frac{\alpha - \beta}{\rho - \beta}. \tag{8.24}$$

We have the following useful consequence of Lemma 8.18. If $\sigma > 0$ and μ satisfies (8.18), then

$$\sum_{z_j \ge z_k} (1 - |z_j|)^\sigma \le C_\sigma (1 - |z_k|)^\sigma. \tag{8.25}$$

Indeed, if $\mathcal{G}_1(z_k) = \{\alpha_m^1\}$ consists of the minimal elements in $[S(z_k) \setminus \{z_k\}] \cap Z$, $\mathcal{G}_2(z_k) = \cup_m \mathcal{G}_1(\alpha_m^1)$, etc., we have using (8.23) and (8.18) ,

$$\begin{aligned}
\sum_{z_j \ge z_k} (1 - |z_j|)^\sigma &= (1 - |z_k|)^\sigma + \sum_{\ell=1}^\infty \sum_{\beta \in \mathcal{G}_{\ell-1}(z_k)} \sum_{\alpha \in \mathcal{G}_1(\beta)} (1 - |\alpha|)^\sigma \\
&\le (1 - |z_k|)^\sigma + C_\delta \sum_{\ell=1}^\infty \sum_{\beta \in \mathcal{G}_{\ell-1}(z_k)} \sum_{\alpha \in \mathcal{G}_1(\beta)} (1 - |\alpha|)^{\sigma-\delta} \left(\log \frac{1}{1 - |\alpha|}\right)^{-1} \\
&\le (1 - |z_k|)^\sigma + C_\delta \sum_{\ell=1}^\infty \sum_{\beta \in \mathcal{G}_{\ell-1}(z_k)} (1 - |\beta|)^{(\sigma-\delta)\eta} C \left(\log \frac{1}{1 - |\beta|}\right)^{-1} \\
&\le (1 - |z_k|)^\sigma + C_\delta \sum_{z_j > z_k} (1 - |z_j|)^{(\sigma-\delta)\eta}.
\end{aligned}$$

Now we can choose $\delta > 0$ so small that $(\sigma - \delta)\eta - \sigma = \theta > 0$, and R such that $C_\delta (1 - R)^\theta = \frac{1}{2}$, so that for $|z_k| \ge R$ we have

$$C_\delta \sum_{z_j > z_k} (1 - |z_j|)^{(\sigma-\delta)\eta} \le \left\{ C_\delta \sup_{j \ge 1} (1 - |z_j|)^\theta \right\} \sum_{z_j > z_k} (1 - |z_j|)^\sigma \le \frac{1}{2} \sum_{z_j > z_k} (1 - |z_j|)^\sigma.$$

Thus $\sum_{z_j \geq z_k} (1-|z_j|)^\sigma \leq 2(1-|z_k|)^\sigma$, proving (8.25) for $|z_k| \geq R$. Now the number of points z_k in the ball $B(0,R)$ depends only on R and the separation constant C in (8.11), and it is now easy to obtain (8.25) in general.

8.4.2. Separated Weak Simple with $\|\mu_Z\| < \infty$ implies OIS. We define the *Böe space*, $\mathcal{D}_Z$, to be the closed linear span in $\mathcal{D}$ of the functions $\{\varphi_w\}_{w \in Z}$. It follows from results in [**42**] that for an appropriate cofinite subset $\{\zeta_j\}_{j=1}^\infty$ of Z we have

$$\mathcal{D}_Z := \left\{ \varphi = \sum_{j=1}^\infty a_j \varphi_{\zeta_j} : \left\| \{a_j\}_{j=1}^\infty \right\|_{\ell^2(\mu)} < \infty \right\}, \tag{8.26}$$
$$\|\varphi\|_{\mathcal{D}_Z} \approx \left\| \{a_j\}_{j=1}^\infty \right\|_{\ell^2(\mu)}.$$

We will say Z is an onto interpolating sequence for $\mathcal{D}_Z$ if it is an onto interpolating sequence for the Dirichlet space $\mathcal{D}$ and if, further, the interpolating functions can all be selected from $\mathcal{D}_Z$. In the case that μ_Z is finite, the next theorem completely characterizes onto interpolation for $\mathcal{D}_Z$ in terms of separation and the weak simple condition. The proof is in [**42**].

THEOREM 8.19. *Let $Z \subset \mathbb{D}$ and suppose $\|\mu_Z\| < \infty$. Then Z is an onto interpolating sequence for the Böe space $\mathcal{D}_Z$ if and only if both the separation condition* (8.10) *and the weak simple condition* (8.18) *hold.*

COROLLARY 8.20. *Let $Z \subset \mathbb{D}$ and suppose $\|\mu_Z\| < \infty$. If Z satisfies the separation condition* (8.10) *and the simple condition* (8.18) *then Z is an onto interpolating sequence for the Dirichlet space $\mathcal{D}$.*

Example B in Section 8.4.4 is a sequence which is covered by these results but not by Theorem 8.15.

For future use we include one small part of the proof of Theorem 8.19 as given in [**42**]. This arises in the proof that the weak simple condition (8.18) is necessary for onto interpolation for the Böe space $\mathcal{D}_Z$, which requires additional tools, including the fact that the Böe functions $\{\varphi_{z_j}\}_{j=1}^\infty$ corresponding to a separated sequence $Z = \{z_j\}_{j=1}^\infty$ in the disk $\mathbb{D}$ form a Riesz basis for the Böe space $\mathcal{D}_Z$, at least in the presence of a mild summability condition on Z. It is interesting to note that for a separated sequence Z in $\mathbb{D}$, the set of Dirichlet reproducing kernels $\{k_{z_j}\}_{j=1}^\infty$ form a Riesz basis if and only if μ_Z is $\mathcal{D}$-Carleson, [**77**], a condition much stronger than the mild summability used for the Böe functions. This points to an essential advantage of the set of Böe functions $\{\varphi_{z_j}\}_{j=1}^\infty$ over the set of corresponding normalized reproducing kernels $\left\{k_{z_j}(z_j)^{-1} k_{z_j}\right\}_{j=1}^\infty$. The feature of Böe functions responsible for this advantage is the fact that the supports of the functions g_{z_i} are pairwise disjoint.

PROPOSITION 8.21. *Let $Z = \{z_j\}_{j=1}^\infty \subset \mathbb{D}$ satisfy the separation condition* (8.11) *and the mild summability condition $\sum_{j=1}^\infty (1-|z_j|^2)^\sigma < \infty$ for all $\sigma > 0$. Then there is a finite subset S of Z such that $\{\varphi_{z_j}\}_{z_j \in Z \setminus S}$ is a Riesz basis for the closed linear span $\mathcal{D}_Z$ of $\{\varphi_{z_j}\}_{j=1}^\infty$ in the Dirichlet space $\mathcal{D}$.*

Proof: A sequence of Böe functions $\{\varphi_{z_j}\}_{j=1}^{\infty}$ is a Riesz basis if

$$\left\|\{a_j\}_{j=1}^{\infty}\right\|_{\ell^2(\mu)}^2 \lesssim \left\|\sum_{j=1}^{\infty} a_j \varphi_{z_j}\right\|_{\mathcal{D}}^2 \lesssim \left\|\{a_j\}_{j=1}^{\infty}\right\|_{\ell^2(\mu)}^2 \tag{8.27}$$

holds for all sequences $\{a_j\}_{j=1}^{\infty}$ and some implied constant. Here $d\mu = \sum_{j=1}^{\infty} \left\|\varphi_{z_j}\right\|_{\mathcal{D}}^{-2} \delta_{z_j}$ and $\mu(z_j) = \left\|\varphi_{z_j}\right\|_{\mathcal{D}}^{-2} \approx d(z_j)^{-1}$. The inequality on the right follows from (15.49) and the disjoint supports of the g_{z_i} - see the argument use to prove (15.38) above - so we concentrate on proving the leftmost inequality in (8.27) for an appropriate set of Böe functions. We begin with

$$\begin{aligned}
\left\|\sum_{j=1}^{\infty} a_j \varphi_{z_j}\right\|_{\mathcal{D}}^2 &= \int_{\mathbb{D}} \left|\sum_{j=1}^{\infty} a_j \varphi'_{z_j}(z)\right|^2 dA(z) \\
&= \sum_{j,k=1}^{\infty} a_j \overline{a_k} \int_{\mathbb{D}} \varphi'_{z_j}(z) \overline{\varphi'_{z_k}(z)} dA(z) \\
&= \sum_{j=1}^{\infty} |a_j|^2 \mu(z_j) + \sum_{j \neq k} a_j \overline{a_k} \int_{\mathbb{D}} \varphi'_{z_j}(z) \overline{\varphi'_{z_k}(z)} dA(z).
\end{aligned}$$

We now claim that by discarding finitely many points of Z, we have

$$\left|\sum_{j \neq k} a_j \overline{a_k} \int_{\mathbb{D}} \varphi'_{z_j}(z) \overline{\varphi'_{z_k}(z)} dA(z)\right| < \frac{1}{2} \sum_{j=1}^{\infty} |a_j|^2 \mu(z_j). \tag{8.28}$$

Indeed, we will estimate (8.28) using the following derivative estimates for Böe functions in the unit disk.

LEMMA 8.22. *Let $\varphi_w(z)$ be as in Lemma 8.17 and continuing the notation from (8.22) we have*

$$\left\{\begin{array}{llll}
|\varphi'_w(z)| \lesssim & \left(1-|w|^2\right)^{-\alpha}, & & z \in V_w^{\alpha} \\
|\varphi'_w(z)| \lesssim & \left|z - w|w|^{-1}\right|^{-1} & \leq \left(1-|w|^2\right)^{-\alpha}, & z \in V_w^{\rho} \setminus V_w^{\alpha} \\
|\varphi'_w(z)| \lesssim & \frac{(1-|w|^2)^{\rho(1+s)}}{\left|z-w|w|^{-1}\right|^{2+s}} & \leq \left(1-|w|^2\right)^{-\rho}, & z \notin V_w^{\rho}.
\end{array}\right. \tag{8.29}$$

PROOF. This follows readily from the formula

$$\varphi_w(z) = \Gamma_s g_w(z) = \int_{\mathbb{D}} \frac{g_w(\zeta)\left(1-|\zeta|^2\right)^s}{\left(1-\overline{\zeta} z\right)^{1+s}} dA(\zeta),$$

together with the estimate in [**48**, Equation (5.45)],

$$|g_w(\zeta)| \lesssim \left(\log \frac{1}{1-|w|^2}\right)^{-1} \left|\zeta - w|w|^{-1}\right|^{-1}, \quad \zeta \in \mathbb{D},$$

and the fact that the support of g_w lives in the annular sector $\mathcal{S}$ centered at $w|w|^{-1}$ given as the intersection of the annulus

$$\mathcal{A} = \mathcal{A}_w = \left\{\zeta \in \mathbb{D} : \left(1-|w|^2\right)^{\alpha} \leq \left|\zeta - w|w|^{-1}\right| \leq \left(1-|w|^2\right)^{\rho}\right\}$$

and the $45°$ angle cone $\mathcal{C}(w)$ with vertex at $w\left|w\right|^{-1}$. Note that the cone $\mathcal{C}(w)$ corresponds to the geodesic in the Bergman tree $\mathcal{T}$ joining the root to the "boundary point" $w\left|w\right|^{-1}$. □

The estimate we will prove is, for $j \neq k$,

$$\begin{aligned}
\text{(8.30)} \qquad \left|\langle \varphi_{z_j}, \varphi_{z_k}\rangle_{\mathcal{D}}\right| &= \left|\int_{\mathbb{D}} \varphi'_{z_j}(z)\,\overline{\varphi'_{z_k}(z)}dA(z)\right| \\
&\lesssim \left(1-|z_j|^2\right)^{\sigma}\left(1-|z_k|^2\right)^{\sigma}\mu(z_j)\,\mu(z_k),
\end{aligned}$$

for some $\sigma > 0$. We may assume that $1-|z_j|^2 \leq 1-|z_k|^2$ and write

$$\begin{aligned}
\left|\int_{\mathbb{D}} \varphi'_{z_j}(z)\,\overline{\varphi'_{z_k}(z)}dA(z)\right| &\leq \int_{\mathbb{D}} \left|\varphi'_{z_j}(z)\right|\left|\varphi'_{z_k}(z)\right| dA(z) \\
&= \left\{\int_{V_{z_j}} + \int_{\mathbb{D}\setminus V_{z_j}} +\right\} \left|\varphi'_{z_j}(z)\right|\left|\varphi'_{z_k}(z)\right| dA(z) \\
&= I + II.
\end{aligned}$$

To estimate II we use Lemma 8.22 to obtain

$$\begin{aligned}
II &\lesssim \int_{\mathbb{D}\setminus V_{z_j}} \frac{\left(1-|z_j|^2\right)^{\rho(1+s)}}{\left|1-\overline{z}z_j\right|^{2+s}}\left(1-|z_k|^2\right)^{-\alpha} dA(z) \\
&\approx \left(1-|z_k|^2\right)^{-\alpha}\left(1-|z_j|^2\right)^{\rho(1+s)} \int_{\mathbb{D}\setminus V_{z_j}} \frac{dA(z)}{\left|1-\overline{z}z_j\right|^{2+s}} \\
&\lesssim \left(1-|z_k|^2\right)^{-\alpha}\left(1-|z_j|^2\right)^{\rho(1+s)}\left(1-|z_j|^2\right)^{-\beta s} \\
&\lesssim \left(1-|z_j|^2\right)^{\rho(1+s)-\beta s-\alpha}.
\end{aligned}$$

Using (8.24) we see that the exponent $\rho(1+s)-\beta s-\alpha$ is positive, and using $1-|z_j|^2 \leq 1-|z_k|^2$ we easily obtain (8.30).

To estimate I we consider two cases. In the case that $V_{z_j} \cap V_{z_k} \neq \phi$, we have from Lemma 8.22 and the estimate $\left|V_{z_j}\right| \lesssim \left(1-|z_j|^2\right)^{2\beta}$ that

$$\begin{aligned}
I &\lesssim \sup_{\mathbb{D}}\left|\varphi'_{z_j}\right| \sup_{\mathbb{D}}\left|\varphi'_{z_k}\right|\left|V_{z_j}\right| \\
&\lesssim \left(1-|z_j|^2\right)^{-\alpha}\left(1-|z_k|^2\right)^{-\alpha}\left(1-|z_j|^2\right)^{2\beta} \\
&\lesssim \left(1-|z_j|^2\right)^{2\beta-\alpha}\left(1-|z_k|^2\right)^{-\alpha},
\end{aligned}$$

and now Lemma 8.18 yields that for some constant

$$I \lesssim \left(1-|z_j|^2\right)^{2\beta-\alpha}\left(1-|z_j|^2\right)^{-\frac{\alpha}{\eta}} = C\left(1-|z_j|^2\right)^{2\beta-\alpha-\frac{\alpha}{\eta}}.$$

Now using (8.24) we see that the exponent $2\beta-\alpha-\frac{\alpha}{\eta}$ is positive and we again obtain (8.30). On the other hand, if $V_{z_j}\cap V_{z_k}=\phi$, then we use

$$\begin{aligned}
I &\lesssim \sup_{\mathbb{D}}\left|\varphi'_{z_j}\right| \sup_{V_{z_j}}\left|\varphi'_{z_k}\right|\left|V_{z_j}\right| \\
&\lesssim \left(1-|z_j|^2\right)^{-\alpha}\frac{\left(1-|z_k|^2\right)^{\rho(1+s)}}{\left|1-\overline{z_j}\cdot z_k\right|^{2+s}}\left(1-|z_j|^2\right)^{2\beta} \\
&\lesssim \left(1-|z_j|^2\right)^{2\beta-\alpha}\left(1-|z_k|^2\right)^{\rho(1+s)-\beta(2+s)} \\
&\lesssim \left(1-|z_j|^2\right)^{\varepsilon}\left(1-|z_k|^2\right)^{\rho(1+s)-\beta(2+s)+2\beta-\alpha-\varepsilon},
\end{aligned}$$

upon using $\left(1-|z_j|^2\right)^{2\beta-\alpha-\varepsilon}\leq\left(1-|z_k|^2\right)^{2\beta-\alpha-\varepsilon}$ in the last line. Now choosing

$$s>\frac{\alpha-\rho+\varepsilon}{\rho-\beta},$$

the exponent $\rho(1+s)-\beta(2+s)+2\beta-\alpha-\varepsilon$ is positive, and once more we obtain (8.30).

Now we can estimate the left side of (8.28) by (8.30) and (8.25) to obtain

$$\begin{aligned}
\left|\sum_{j\neq k}a_j\overline{a_k}\int_{\mathbb{D}}\varphi'_{z_j}(z)\overline{\varphi'_{z_k}(z)}dA(z)\right| &\lesssim \sum_{j\neq k}|a_ja_k|\left(1-|z_j|^2\right)^{\sigma}\left(1-|z_k|^2\right)^{\sigma}\mu(z_j)\,\mu(z_k) \\
&\lesssim \left\{\sum_k\left(1-|z_k|^2\right)^{\sigma}\mu(z_k)\right\}\sum_{j=1}^{\infty}|a_j|^2\mu(z_j) \\
&< \frac{1}{2}\sum_{j=1}^{\infty}|a_j|^2\mu(z_j)
\end{aligned}$$

if $\sum_k\left(1-|z_k|^2\right)^{\sigma}\mu(z_k)$ is sufficiently small, which can be achieved by discarding a sufficiently large finite subset F from Z. This shows that $\left\{\varphi_{z_j}\right\}_{z_j\in Z\setminus F}$ is a Riesz basis. However, if $w\in F$ is *not* in the closed linear span of the Riesz basis $\left\{\varphi_{z_j}\right\}_{z_j\in Z\setminus F}$, then it is immediate that $\left\{\varphi_{z_j}\right\}_{z_j\in Z\setminus F}\cup\{\varphi_w\}$ is also a Riesz basis. We can continue adding Böe functions φ_w with $w\in G\subset F$ so that $\left\{\varphi_{z_j}\right\}_{z_j\in Z\setminus F}\cup\{\varphi_w\}_{w\in G}$ is a Riesz basis, and such that *all* of the remaining Böe functions φ_w with $w\in F\setminus G$ lie in the closed linear span of the Riesz basis $\left\{\varphi_{z_j}\right\}_{z_j\in Z\setminus F}\cup\{\varphi_w\}_{w\in G}$. This completes the proof of Proposition 8.21 with $S=F\setminus G$.

8.4.3. Separated Weak Simple with Treelike Implies OIS.

8.4.3.1. *Tree-Like Sequences.* In this section, for convenience of discussion, we will consider sequences $Z=\{z_k\}$ which are subsequences of the standard dyadic tree $\mathcal{T}_2$ described in Section 5.2.1. The extension to the more general case is straightforward. Also, in this section, for $z_k\in Z$ we will use the notation Az_k for the predecessor of z_k in the tree structure induced on Z by $\mathcal{T}_2$.

The building blocks for constructing interpolating functions on the tree take the value one on a part of the boundary and zero on the complementary part. The analogous holomorphic functions on the disk also have additional regions where they take intermediate values. When trying to construct solutions to interpolation problems, these regions of intermediate behavior can pile up and prevent extending to the disk the construction which works on the tree. In the previous theorem the fact that $\|\mu_Z\| < \infty$ helped limit the effect of these overlaps. We now introduce an alternative way to control the overlaps, introducing a geometric condition on Z which ensures that those regions are not an issue; or, rather, are a small enough issue so they can be controlled. Although this is an ad hoc condition which is not very general, it allows straightforward construction of interesting examples. In particular it can be used to construct an OIS for which μ_Z has infinite mass, answering a question of Bishop.

Suppose that Z satisfies the separation condition (8.10), the weak simple condition (8.18), and the following tree-like condition: there is $\beta \in (1 - c/5, 1)$ where c is the constant in (8.10) such that, recalling that for ζ, $T(\zeta)$ is the tent in the disk with vertex ζ on the boundary,

$$(8.31)\quad z_j \in T(z_k) \text{ whenever } |z_j| \geq |z_k|, \left| z_j - \frac{z_k}{|z_k|} \right| \leq (1-|z_k|^2)^\beta \text{ and } z_j, z_k \in Z.$$

DEFINITION 8.23. *Suppose that Z satisfies* (8.10) *with constant $c > 0$ and $\beta \in (1 - c/5, 1)$. Then we define the β-Böe tree structure on Z, or simply the Böe tree structure on Z when β is understood, by declaring that $z_j \in Z$ is a child of $z_i \in Z$ provided $z_j \in V_{z_i}^\beta$ and $|z_i| \leq |z_j|$ and there is no other $z_k \in Z$ with $z_j \in V_{z_k}^\beta$ and $z_k \in V_{z_i}^\beta$.*

Thus for a separated sequence Z there is the β-Böe tree structure on Z, as well as the Bergman tree structure on Z that is inherited from the usual Bergman tree structure on $\mathcal{T}_2$. The Bergman structure is finer than the Böe structure in the sense that the set of Bergman children is always contained in the set of Böe children. We now see that the tree-like condition on Z simply asserts that the Böe and Bergman tree structures coincide on Z. Now if Z is such a subtree of $\mathcal{T}_2$, i.e. if $z \in Z$ is a Böe child of $w \in Z$, then z actually lies in the Bergman successor set $S(w)$ of w, then the value of $c_{\rho,\alpha}(\gamma_w(z))$ in Lemma 8.17 is 1, which is exploited in (8.35) below. The advantage when assuming (8.31) is that we may use the elementary construction in (8.33) below of a holomorphic function $M\xi$ approximating the *integrated* sequence $(I\xi)_j = \sum_{z_i \leq z_j} \xi_i$ on Z. This permits us to interpolate the *difference* sequence $\triangle\xi$ using the operator $\triangle M$, whose kernel is well localized. Of course in the absence of (8.31), the values $c_{\rho,\alpha}(\gamma_w(z))$ may lie in $[0, 1)$ and then $M\xi$ will *not* in general be a good approximation to $I\xi$ on Z.

Under these conditions Z is an OIS for the Dirichlet space $\mathcal{D}$ even if the measure μ is *infinite*.

THEOREM 8.24. *Suppose $Z = \{z_j\}_{j=1}^\infty \subset \mathbb{D}$ is a subtree of $\mathcal{T}_2$ that satisfies the separation condition* (8.10) *and is tree-like; that is, if c is the constant in* (8.10), *that there is $\beta \in (1 - c/5, 1)$ satisfying* (8.31). *Then Z is onto interpolating for the Böe space $\mathcal{D}_Z$ if and only if the weak simple condition* (8.18) *holds.*

This result, together with Example A in Section 8.4.4 gives an affirmative answer to a question raised by Bishop in [**72**].

COROLLARY 8.25. *There are OIS Z for which $\|\mu_Z\| = \infty$.*

PROOF OF THE THEOREM. To see the necessity of (8.18) when Z is onto interpolating for the Böe space $\mathcal{D}_Z$, we note that a subtree of a dyadic tree has branching number at most 2, and it follows easily from the separation condition that

$$\sum_{j=1}^{\infty}(1-|z_j|^2)^{\sigma}<\infty$$

for all $\sigma>0$. Thus Proposition 8.21 can be applied together with the argument used above to prove necessity of (8.18) in the case $\|\mu_Z\|<\infty$.

To establish sufficiency we first fix $\{\xi_j\}_{j=1}^{\infty}$ with

$$\left\|\left\{\frac{\xi_j}{\|k_{z_j}\|_{\mathcal{D}}}\right\}_{j=1}^{\infty}\right\|_{\ell^2}=1.$$

Recall that $\|k_{z_j}\|_{\mathcal{D}}\approx\left(\log\frac{1}{1-|z_j|^2}\right)^{1/2}$ and that we have $Z\subset\mathcal{T}_2$. We note that (8.25) holds here - in fact the proof is simpler using the separation condition (8.10) and the assumption that Z is a subtree of $\mathcal{T}_2$ (and hence has branching number at most 2). In particular we obtain that $\sum_{j=1}^{\infty}(1-|z_j|)^{\sigma}<\infty$. Thus given any $\sigma>0$, we can discard all points from Z that lie in some ball $B(0,R)$, $R<1$, and reorder the remaining points so that

$$\left(\log\frac{1}{1-R^2}\right)^{-1},\ 1-R^2,\ \sum_{j=1}^{\infty}(1-|z_j|)^{\sigma}<\varepsilon. \tag{8.32}$$

(This reduction by discarding finitely many points is straightforward in one dimension but more subtle in higher dimensions, see the proof of Lemma 15.19 below.) We now suppose in addition that the sequence $Z=\{z_j\}_{j=1}^{J}$ is finite, and obtain an appropriate estimate independent of $J\geq 1$. Given a sequence of complex numbers $\xi=\{\xi_j\}_{j=1}^{J}$ we define a holomorphic function $M\xi$ on the ball by

$$M\xi(z):=\sum_{j=1}^{J}\xi_j\varphi_{z_j}(z),\qquad z\in\mathbb{D}, \tag{8.33}$$

where $\varphi_w(z)$ is as in Lemma 8.17. View μ as the measure assigning the mass $\left(\log\frac{1}{1-|z_j|^2}\right)^{-1}$ to the point $j\in\{0,1,2,\dots,J\}$. We have

$$\left\|\{\xi_j\}_{j=1}^{J}\right\|_{\ell^2(\mu)}\approx\left\|\left\{\frac{\xi_j}{\|k_{z_j}\|_{\mathcal{D}}}\right\}_{j=1}^{J}\right\|_{\ell^2},$$

for any complex sequence $\{\xi_j\}_{j=1}^{J}$.

Recall that Az_j denotes the predecessor of z_j in the forest structure on Z defined in (15.41) above (we identify z_k here with k there). Recall also that $\mathcal{C}(w)$ is denoting the cone with vertex at $\frac{w}{|w|}$ opening 45 degrees. We will use another useful consequence of Lemma 8.18:

$$1-|Az_j|^2\leq\left(1-|z_\ell|^2\right)^{\eta},\qquad\text{for } z_j\in V_{z_\ell}\setminus\mathcal{C}(z_\ell). \tag{8.34}$$

Indeed, if $z_j\in V_{z_\ell}\setminus\mathcal{C}(z_\ell)$, then $Az_j\neq z_\ell$ and $|Az_j|\geq|z_\ell|$ by the construction in (15.41). Then $V_{z_\ell}\cap V_{Az_j}$ contains z_j and is thus nonempty, and Lemma 8.18 now shows that $1-|Az_j|^2\leq\left(1-|z_\ell|^2\right)^{\eta}$.

Now define a linear map T from $\ell^2(\mu)$ to $\ell^2(\mu)$ by

$$\begin{aligned} T\xi &:= \triangle(M\xi)\mid_Z = \{M\xi(z_k) - M\xi(Az_k)\}_{j=1}^J \\ &= \left\{\sum_{j=1}^J \xi_j\left[\varphi_{z_j}(z_k) - \varphi_{z_j}(Az_k)\right]\right\}_{j=1}^J . \end{aligned}$$

Let $\mathcal{R}$ denote the set of all roots of maximal trees in the forest. In the event that $z_k \in \mathcal{R}$, then Az_k is not defined and our convention is to define $\varphi_{z_j}(Az_k) = 0$. We claim that T is a bounded invertible map on $\ell^2(\mu)$ with norms independent of $J \geq 1$. To see this it is enough to prove that $\mathbb{I} - T$ has small norm on $\ell^2(\mu)$ where $\mathbb{I}$ denotes the identity operator. We have

$$\begin{aligned} (\mathbb{I} - T)\xi &= \left\{\xi_k - \sum_{j=1}^J \xi_j\left[\varphi_{z_j}(z_k) - \varphi_{z_j}(Az_k)\right]\right\}_{k=1}^J \\ &= \{\xi_k\varphi_{z_k}(Az_k)\}_{k=1}^J - \left\{\sum_{j:j\neq k} \xi_j\left[\varphi_{z_j}(z_k) - \varphi_{z_j}(Az_k)\right]\right\}_{k=1}^J \end{aligned}$$

since $\varphi_{z_k}(z_k) = 1$.

We will now estimate the kernel $K(k,j)$ of the operator $\mathbb{I} - T$ and then use Schur's lemma to obtain the required norm estimate. We have on the diagonal,

$$|K(k,k)| = \begin{cases} |\varphi_{z_k}(Az_k)| \leq (1 - |z_k|^2)^{(\rho-\beta_1)(1+s)} & \text{if } z_k \notin \mathcal{R} \\ 0 & \text{if } z_k \in \mathcal{R}, \end{cases}$$

by the third estimate in (8.20).

Suppose now that $z_k \notin \mathcal{R}$ and $j \neq k$. Lemma 8.17 shows that $\left|\varphi'_{z_j}(\zeta_k)\right| \leq (1 - |z_j|^2)^{-\alpha}$ and the definition of V_{z_j} shows that $|z_k - Az_k| \leq (1 - |Az_k|^2)^{\beta}$. Thus if $1 - |Az_k|^2 \leq (1 - |z_j|^2)^{\eta}$, then

$$\begin{aligned} |K(k,j)| = \left|\varphi_{z_j}(z_k) - \varphi_{z_j}(Az_k)\right| &\leq \left|\varphi'_{z_j}(\zeta_k)\right| |z_k - Az_k| \\ &\lesssim (1 - |z_j|^2)^{-\alpha}(1 - |Az_k|^2)^{\beta} \\ &\lesssim (1 - |z_j|^2)^{\eta(\beta-\delta)-\alpha}(1 - |Az_k|^2)^{\delta}, \end{aligned}$$

where the exponent $\eta(\beta - \delta) - \alpha$ is positive if we choose δ small enough, since $\alpha < 1 < \beta\eta$ by Lemma 8.18.

Suppose instead that $1 - |Az_k|^2 > (1 - |z_j|^2)^{\eta}$. Then $Az_k \notin V_{z_j}$ by Lemma 8.18. If $z_k \notin \mathcal{C}(z_j)$, then $z_k \notin V_{z_j}$ by (8.34), and this time we use the third estimate in (8.20) to obtain

$$\begin{aligned} |K(k,j)| = \left|\varphi_{z_j}(z_k) - \varphi_{z_j}(Az_k)\right| &\leq \left|\varphi_{z_j}(z_k)\right| + \left|\varphi_{z_j}(Az_k)\right| \\ &\lesssim (1 - |z_j|^2)^{(\rho-\beta_1)(1+s)} \\ &\lesssim (1 - |z_j|^2)^{(\rho-\beta_1)(1+s)-\delta}(1 - |Az_k|^2)^{\delta/\rho}. \end{aligned}$$

On the other hand, if $z_k \in \mathcal{C}(z_j)$, then $|z_k| \geq |z_j|$ and our hypothesis (8.31) implies that $z_k \in S(z_j)$. Then we have

$$|K(k,j)| = \left|\varphi_{z_j}(z_k) - \varphi_{z_j}(z_j)\right| \leq \left(\log \frac{1}{1-|z_j|^2}\right)^{-1} \tag{8.35}$$

by the first two estimates in (8.20) since $c_{\rho,\alpha}\left(\gamma_{z_j}(z_k)\right) = 1$ in Lemma 8.17 if $z_k \in S(z_j)$.

Finally, we consider the case when $z_k \in \mathcal{R}$ and $j \neq k$. The third estimate in (8.20) shows that

$$|K(k,j)| = \left|\varphi_{z_j}(z_k)\right| \lesssim (1-|z_j|^2)^{(\rho-\beta_1)(1+s)},$$

where the exponent $(\rho - \beta_1)(1+s)$ can be made as large as we wish by taking s sufficiently large. Combining all cases we have in particular the following estimate for some $\sigma_1, \sigma_2 > 0$:

$$|K(k,j)| \lesssim \begin{cases} (1-|z_j|^2)^{\sigma_1}(1-|Az_k|^2)^{\sigma_2}, & \text{if } z_k \notin \mathcal{R} \text{ and } z_k \notin \mathcal{C}(z_j) \\ \left(\log \frac{1}{1-|z_j|^2}\right)^{-1} & \text{if } z_k \notin \mathcal{R} \text{ and } z_k \in \mathcal{C}(z_j) \\ (1-|z_j|^2)^3, & \text{if } z_k \in \mathcal{R}. \end{cases}$$

Now we obtain that $\mathbb{I} - T$ is bounded on $\ell^2(\mu)$ and has small norm by Schur's test . It is here that we use the assumption that μ satisfies the weak simple condition (8.18). With $\xi \in \ell^2(\mu)$ and $\eta \in \ell^2(\mu)$, we have

$$\begin{aligned} \left|\langle(\mathbb{I}-T)\xi,\eta\rangle_{\ell^2(\mu)}\right| &= \left|\sum_k \left(\sum_j K(k,j)\xi_j\right)\overline{\eta_k}\mu(k)\right| \\ &\lesssim \sum_j \sum_{k\notin\mathcal{R},z_k\notin\mathcal{C}(z_j)} (1-|z_j|^2)^{\sigma_1}(1-|Az_k|^2)^{\sigma_2}|\xi_j||\eta_k|\mu(k) \\ &\quad + \sum_j \sum_{k\notin\mathcal{R},z_k\in\mathcal{C}(z_j)} \left(\log\frac{1}{1-|z_j|^2}\right)^{-1} |\xi_j||\eta_k|\mu(k) \\ &\quad + \sum_j \sum_{k\in\mathcal{R}} (1-|z_j|^2)^3|\xi_j||\eta_k|\mu(k), \end{aligned}$$

and since $\mu(j) \leq (1-|z_j|^2)^{\varepsilon}$, we have with $\sigma_1' = \sigma_1 - \varepsilon$,

$$\begin{aligned} \left|\langle(\mathbb{I}-T)\xi,\eta\rangle_{\mu}\right| &\lesssim \sum_j \sum_{k\notin\mathcal{R},z_k\notin\mathcal{C}(z_j)} (1-|z_j|^2)^{\sigma_1'}(1-|Az_k|^2)^{\sigma_2}|\xi_j|\mu(j)|\eta_k|\mu(k) \\ &\quad + \sum_j \sum_{k\notin\mathcal{R},z_k\in\mathcal{C}(z_j)} |\xi_j|\mu(j)|\eta_k|\mu(k) \\ &\quad + \sum_j \sum_{k\in\mathcal{R}} (1-|z_j|^2)^2|\xi_j|\mu(j)|\eta_k|\mu(k). \end{aligned}$$

By Schur's test , Theorem B.1, it suffices to show

$$\mu(Ak)+\sum_{j=1}^{J}(1-|z_j|^2)^{\sigma_1'}\mu(j)<C\varepsilon<1, \tag{8.36}$$
$$\sum_{k:z_k\in\mathcal{C}(z_j)}\mu(k)+\sum_{k\notin\mathcal{R}}(1-|Az_k|^2)^{\sigma_2}\mu(k)<C\varepsilon<1,$$
$$\sum_{k\in\mathcal{R}}(1-|z_j|^2)^3\mu(k)<C\varepsilon<1.$$

Now (8.32) yields

$$\sum_{j=1}^{J}(1-|z_j|^2)^{\sigma_1'}\mu(j)\lesssim\sum_{j}(1-|z_j|^2)^{\sigma_1''}<C\varepsilon,$$

and combined with the weak simple condition (8.18), we have

$$\begin{aligned}\sum_{k=1}^{J}(1-|Az_k|^2)^{\sigma_2}\mu(k)&=\sum_{\ell}\left(1-|z_\ell|^2\right)^{\sigma_2}\left(\sum_{z_k\in\mathcal{C}(z_\ell)}\mu(k)\right)\\&\lesssim\sum_{\ell}\left(1-|z_\ell|^2\right)^{\sigma_2}\mu(\ell)\\&\lesssim\sum_{\ell}\left(1-|z_\ell|^2\right)^{\sigma_2'}<C\varepsilon.\end{aligned}$$

Finally we write the annulus $B(0,1)\setminus B(0,R)$ as a pairwise disjoint union $\bigcup_{i=1}^{N}B_i$ of Carleson boxes of "size" R where $N\approx\left(1-R^2\right)^{-1}$. Then

$$\sum_{z_k\in B_i:k\in\mathcal{R}}\mu(k)\lesssim\left(1+\log\frac{1}{1-R^2}\right)^{-1}\lesssim 1$$

by the weak simple condition (8.18), and thus the left side of the final estimate in (8.36) satisfies

$$\begin{aligned}\sum_{k\in\mathcal{R}}(1-|z_j|^2)^2\mu(k)&\leq\left(1-R^2\right)^2\sum_{i=1}^{N}\sum_{z_k\in B_i:k\in\mathcal{R}}\mu(k)\\&\lesssim\left(1-R^2\right)^2N\\&\lesssim\left(1-R^2\right)<C\varepsilon,\end{aligned}$$

by (8.32) as required.

Thus T^{-1} exists uniformly in J. Now we take $\xi\in\ell^2(\mu)$ and set $\eta=\triangle\xi$. Here we use the convention that $\xi(A\alpha)=0$ if α is a root of a tree in the forest Z. By the weak simple condition we have the estimate

$$\begin{aligned} \|\eta\|_{\ell^2(\mu)}^2 &= \sum_j |\eta_j|^2 \mu(j) = \sum_j |\xi_j - \xi_{Aj}|^2 \mu(j) \\ &\lesssim \sum_j |\xi_j|^2 \mu(j) + \sum_\ell |\xi_\ell|^2 \left(\sum_{z_j \in \mathcal{C}(z_\ell)} \mu(j) \right) \\ &\lesssim \sum_j |\xi_j|^2 \mu(j) + \sum_\ell |\xi_\ell|^2 \mu(\ell) \\ &\lesssim \|\xi\|_{\ell^2(\mu)}^2 . \end{aligned} \tag{8.37}$$

Then let $h = M\left(T^{-1}\eta\right)$ so that

$$\triangle h \mid_Z = \triangle \left(MT^{-1}\eta\right) \mid_Z = TT^{-1}\eta = \eta = \triangle \xi.$$

Thus the holomorphic function h satisfies

$$h \mid_Z = \xi.$$

Finally, from (8.21) and then (8.37) we have the Besov space estimate [**42**],

$$\begin{aligned} \|h\|_{\mathcal{D}}^2 &\lesssim \sum_{j=1}^J \left|\left(T^{-1}\eta\right)_j\right|^2 \int_{\mathbb{D}} \left|\left(1 - |\zeta|^2\right) g_w(\zeta)\right|^2 d\lambda(\zeta) \\ &\lesssim \sum_{j=1}^J \left|\left(T^{-1}\eta\right)_j\right|^2 \left(\log \frac{1}{1 - |w|^2}\right)^{-1} \\ &\lesssim \left\|T^{-1}\eta\right\|_{\ell^2(\mu)}^2 \lesssim \|\eta\|_{\ell^2(\mu)}^2 \lesssim \|\xi\|_{\ell^2(\mu)}^2 . \end{aligned}$$

Since all of this is uniform in J we may let $J \to \infty$ and use a normal families argument to complete the proof of Theorem 8.24. Indeed, if $h_J \in \mathcal{D}$ satisfies

$$\begin{aligned} \|h_J\|_{\mathcal{D}} &\leq C, \quad 1 \leq j \leq J, \\ h_J(z_j) &= \xi_j, \quad 1 \leq j \leq J, \end{aligned}$$

then $|h_J(z)| \lesssim \left(1 + \log \frac{1}{1-|z|^2}\right)^{1/2} \|h_J\|_{\mathcal{D}}$ shows that $\{h_J\}_{J=1}^\infty$ is a normal family on the disk. If the subsequence $\{h_{J_k}\}_{k=1}^\infty$ converges uniformly on compact subsets of $\mathbb{D}$, then the limit $h = \lim_{k\to\infty} h_{J_k}$ satisfies

$$\begin{aligned} \|h\|_{\mathcal{D}} &= \left(\int_{\mathbb{D}} \left|\left(1 - |z|^2\right) h'(z)\right|^2 d\lambda(z)\right)^{1/2} \\ &\leq \liminf_{k\to\infty} \left(\int_{\mathbb{D}} \left|\left(1 - |z|^2\right) h'_{J_k}(z)\right|^2 d\lambda(z)\right)^{1/2} \\ &\lesssim 1, \\ h_J(z_j) &= \xi_j, \quad 1 \leq j < \infty. \end{aligned}$$

□

8.4.4. Examples of Sequences in the Disk. We present two examples of sequences. The first is a tree-like OIS having infinite mass. The second improves on the known theorems for sequences having finite mass, by showing that there is a finite mass sequence that satisfies the weak simple condition, but not the simple condition.

8.4.4.1. *Example A: An OIS with $\|\mu_Z\| = \infty$.* Here we construct a subtree with infinite measure to which Theorem 8.24 applies. This yields an example of an onto interpolating sequence of infinite mass, thus resolving a question of Bishop in [**72**].

Let $a, b > 1$ satisfy $\left[a^{k+1}b\right] \geq \left[a^k b\right] + 1$ for all $k \geq 0$ (in particular this will hold if $(a-1)\,b \geq 2$), and define a Cantor-like sequence Z,

$$Z := Z_{a,b} = \bigcup_{k=0}^{\infty} \left\{z_j^k\right\}_{j=1}^{2^k} \subset \mathcal{T}_2,$$

as follows. Pick a point $z_1^0 = 0$ of $\mathcal{T}_2$ satisfying $d^+\left(z_1^0\right) = [b]$. Then choose 2^1 points $\left\{z_1^1, z_2^1\right\} \subset \mathcal{T}_2$ that are successors to distinct children of z_1^0 and having $d^+\left(z_j^1\right) = [ab]$, $1 \leq j \leq 2^1$, and $\beta\left(z_i^1, z_j^1\right) \gtrsim [ab]$ for $i \neq j$. Then choose 2^2 points $\left\{z_1^2, z_2^2, z_3^2, z_4^2\right\} \subset \mathcal{T}_2$ that are successors to distinct children of the points in $\left\{z_1^1, z_2^1\right\}$ and having $d^+\left(z_j^2\right) = \left[a^2 b\right]$, $1 \leq j \leq 2^2$, and $\beta\left(z_i^2, z_j^2\right) \gtrsim \left[a^2 b\right]$ for $i \neq j$. Having constructed 2^k points $\left\{z_j^k\right\}_{j=1}^{2^k} \subset \mathcal{T}_2$ in this way, we then choose 2^{k+1} points $\left\{z_j^{k+1}\right\}_{j=1}^{2^{k+1}} \subset \mathcal{T}_2$ that are successors to distinct children of the points in $\left\{z_j^k\right\}_{j=1}^{2^k}$ and having $d^+\left(z_j^{k+1}\right) = \left[a^{k+1}b\right]$, $1 \leq j \leq 2^k$, and $\beta\left(z_i^{k+1}, z_j^{k+1}\right) \gtrsim \left[a^{k+1}b\right]$ for $i \neq j$. Note that the condition $\left[a^{k+1}b\right] \geq \left[a^k b\right] + 1$ allows for the existence of such points. Then $Z = \bigcup_{k=0}^{\infty} \left\{z_j^k\right\}_{j=1}^{2^k}$ satisfies the separation condition (8.10) with constant roughly $a-1$ and condition (8.31) with β close to 1, and the associated measure μ_Z satisfies the weak simple condition (8.18) with constant 2. Thus Theorem 8.24 applies to show that Z is onto interpolating for $\mathcal{D}$. Yet the total mass of the measure μ_Z satisfies

$$\|\mu_Z\| = \sum_{k=0}^{\infty} 2^k \left[a^k b\right]^{-1} \approx \frac{1}{b} \sum_{k=0}^{\infty} \left(\frac{2}{a}\right)^k = \infty$$

if $a \leq 2$.

8.4.4.2. *Example B: Weak Simple does not imply Simple .* We now use the previous example to construct a separated sequence W in the disk with *finite* measure $\mu = \mu_W$ satisfying the weak simple condition but not the simple condition. This yields an example of a sequence which fails the simple condition, but to which Theorem 8.20 applies.

With notation as in the previous section, we choose $a = 2$ for convenience, let $b, N \in \mathbb{N}$ be large integers, and replace the sequence $Z_{2,b}$ above with the truncated sequence $Z_{2,b,N} = \bigcup_{k=0}^{N} \left\{z_j^k\right\}_{j=1}^{2^k}$. Then $Z_{2,b,N}$ satisfies the separation condition (8.10) with constant roughly 1, the associated measure $\mu_{Z_{2,b,N}}$ satisfies the weak simple condition (8.18) with constant 2, and the total mass of $\mu_{Z_{2,b,N}}$ is about N/b:

$$\left\|\mu_{Z_{2,b,N}}\right\| = \sum_{k=0}^{N} 2^k \left[2^k b\right]^{-1} \approx \frac{1}{b} \sum_{k=0}^{N} \left(\frac{2}{2}\right)^k \approx \frac{N}{b}.$$

On the other hand the constant $C\left(\mu_{Z_{2,b,N}}\right)$ in the simple condition (8.16) for $\mu_{Z_{2,b,N}}$ satisfies

$$C\left(\mu_{Z_{2,b,N}}\right) \gtrsim N, \tag{8.38}$$

since

$$\frac{N}{b} \approx \left\|\mu_{Z^*_{2,b}}\right\| = \sum_{\alpha \geq z_0} \mu_{Z^*_{2,b}}(\alpha) \lesssim \frac{1}{d^+(z_0)} = \frac{C}{b}.$$

It is now an easy exercise to choose sequences of parameters $\{b(n)\}_{n=1}^{\infty}$ and $\{N(n)\}_{n=1}^{\infty}$, and initial points $\left\{z_1^0(n)\right\}_{n=1}^{\infty}$ so that the corresponding sequences $Z_{2,b(n),N(n)} = \bigcup_{k=0}^{N(n)} \left\{z_j^k(n)\right\}_{j=1}^{2^k}$ satisfy

$$\left\|\mu_{Z_{2,b(n),N(n)}}\right\| \approx \frac{N(n)}{b(n)} \leq 2^{-n} \tag{8.39}$$

and

$$\lim_{n\to\infty} N(n) = \infty, \tag{8.40}$$

along with the nested property

$$z_1^0(n+1) \geq z_1^{b(n)}(n), \qquad n \geq 1. \tag{8.41}$$

Then the union $W = \bigcup_{n=1}^{\infty} Z_{2,b(n),N(n)}$ satisfies the separation condition and the associated measure μ_W is finite by (8.39), satisfies the weak simple condition by (8.41), yet fails the simple condition by (8.38) and (8.40).

8.4.5. An example of a sequence satisfying the tree-capacitary condition but not the weak simple condition. We give an example of a sequence $Z = \{z_n\}_{n\geq 0}$ which satisfies the tree-capacitary condition (8.13), hence the natural separation condition (5.19), but not the weak simple condition.

Fix a point $z_0 \in \mathcal{T}$ with $d(z_0)$ large. Let N and $b = \frac{1}{\beta}$ be positive integers. Fix a geodesic segment $[z_0 = w_0, w_1, w_2, \ldots, w_N]$ in $\mathcal{T}$ and choose for each $1 \leq n \leq N$ a point z_n satisfying

$$\begin{aligned} d(w_n, z_n) &= b, \\ w_{n+1} &\notin [w_n, z_n], \qquad 1 \leq n < N. \end{aligned}$$

Thus z_n and w_{n+1} lie on different branches below w_n and $z_n \wedge w_{n+1} = w_n$. Set $Z_{N-j+1} = \{z_n\}_{n=j}^{N}$ for $1 \leq j \leq N$. We have $\mathrm{Cap}(w_n, \{z_n\}) = \beta$ for each n. Then the formula in (5.24) above shows that, for $j = 1, \ldots, N$

$$\begin{aligned} \mathrm{Cap}(w_{j-1}, Z_N) &= \frac{1}{1 + \frac{1}{\mathrm{Cap}(w_j, \{z_j\}) + \mathrm{Cap}(w_j, Z_{N-j})}} \\ &= \frac{1}{1 + \frac{1}{\beta + \mathrm{Cap}(w_j, Z_{N-j})}} \\ &= \varphi_\beta\left(\mathrm{Cap}(w_j, Z_{N-j})\right), \end{aligned}$$

where the function $\varphi_\beta(x) = \frac{1}{1+\frac{1}{\beta+x}}$ is strictly increasing. So, if we take $\gamma_0 = 0$ and note that $\gamma_{N+1} = \varphi_\beta(\gamma_N)$ we see that $\gamma_N < \gamma_{N+1}$ for all $N \geq 0$. If $\gamma_\infty =$

$\lim_{N\to\infty}\gamma_N$ denotes the corresponding infinite continued fraction,

$$\gamma_\infty = \cfrac{1}{1+\cfrac{1}{\beta+\cfrac{1}{1+\cfrac{1}{\beta+\cfrac{1}{\ddots}}}}},$$

then $\mathrm{Cap}_{z_0}(Z_N) = \gamma_N < \gamma_\infty$. Since $\gamma_\infty = \varphi_\beta(\gamma_\infty) = \frac{1}{1+\frac{1}{\beta+\gamma_\infty}}$ we compute that

$$\gamma_\infty = \frac{\sqrt{\beta^2+4\beta}-\beta}{2} < \sqrt{\beta}$$

since $0<\beta<1$. Altogether then,

$$\mathrm{Cap}(z_0, Z_N) < \sqrt{\beta}, \qquad N \geq 1.$$

Now we fix $N = b = d(z_0)^2$ so that

$$\mathrm{Cap}(z_0, Z_N) < \sqrt{\frac{1}{d(z_0)^2}} = \frac{1}{d(z_0)}.$$

From this we obtain that

$$Z \equiv \{z_0\} \bigcup Z_N = \{z_n\}_{n=0}^N$$

satisfies the tree capacity condition (8.13) with constant C (for $n \geq 1$ the capacity $\mathrm{Cap}\left(z_n, \{z_m\}_{m\neq n}\right)$ is easily seen to be bounded by $Cd(z_0)^{-2}$ since the distance from z_n to w_n is $d(z_0)^2$, and the geodesics from z_n to another point of Z must pass through w_n). Observe that (8.13) implies the natural separation condition:

$$\forall z, w \in Z, \quad d(z,w) \geq cd(o,z)$$

On the other hand, the weak simple condition constant for Z is quite large since $\{z_n\}_{n=1}^N$ are the children of z_0 and

$$\sum_{n=1}^N \mu(z_n) = \sum_{n=1}^N \frac{1}{d(z_n)} = \sum_{n=1}^{d(z_0)^2} \frac{1}{d(z_0)+n+d(z_0)^2}$$
$$\approx \log \frac{d(z_0)+2d(z_0)^2}{d(z_0)+1+d(z_0)^2} \approx \log 2,$$

which is much larger than $\mu(z_0) = \frac{1}{d(z_0)}$ if $d(z_0)$ is large.

EXERCISE 8.26. *Let Z be a sequence in the unit disk and suppose that each dyadic box $Q_{n,j}$ (see Section* 1.7.1*) contains at most a point of Z. Show that Z, identified with a subsequence, still called Z, of the corresponding dyadic tree (see Section* 5.2.1*). Show that, if Z is onto interpolating for $\mathcal{D}$, then Z is onto interpolating in $\mathcal{D}(\mathcal{T})$. In particular, condition* (8.13) *is necessary for Z to be onto interpolating for $\mathcal{D}$. Hint: The argument used to deal with the radial variation in Theorem* 5.13 *is helpful.*

8.4.6. Relations Between the Conditions. Other related examples are in [**42**]. In particular the recursive scheme for computing tree capacities is used there to construct an example of an OIS for $\mathcal{D}(\mathcal{T}_2)$ that fails not only the weak simple condition, but also fails to be contained in any separated sequence satisfying the weak simple condition.

8.5. Notes and Comments

In showing that the OIS for the Hardy space are automatically IS, we adapted the presentation in [**178**].

The work about onto interpolating sequences for $\mathcal{D}$ is from [**72**] and [**42**]. The second paper also contains the results for $\mathcal{D}(\mathcal{T})$.

After this book was completed, a manuscript by N. Chalmoukis was posted, [**104**] which includes a proof of Theorem 8.3.

CHAPTER 9

Boundary Values

In Chapter 5 we saw that functions in $\mathcal{D}$ have radial boundary values off a set of capacity zero. In this chapter we begin by discussing that result and other types of convergence of functions in $\mathcal{D}$ to their boundary value function. We then study those boundary functions. We consider their size, their zero sets, and the associated sets of uniqueness (determining sets). Proofs are provided for only some of the topics.

Facts that hold except on a set of measure zero are said to hold *almost everywhere, a.e.*; properties which hold off a set of capacity zero are said to hold *quasieverywhere.*

9.1. Finding the Boundary Function

9.1.1. Radial Limits. We introduced $\mathcal{D}$ as a space of functions on the disk. However we know $\mathcal{D} \subset H^2$ and hence, recalling results from the Hardy space theory, any $f \in \mathcal{D}$ has radial limits *a.e.*. In fact, more is true. The story starts with the classic results of Beurling [**69**] who showed that every $f \in \mathcal{D}$ has radial limits quasieverywhere. He also considered points at which the boundary function, which we denoted by f^β is sufficiently smooth so that it could be recaptured by "differentiating its integral"; that is, points $e^{i\theta}$ for which

$$\lim\operatorname{avg} f^\beta(e^{i\theta}) = \lim_{h\to 0}\frac{1}{2h}\int_{-h}^{h} f^\beta(e^{i(\theta+t)})dt$$

exists. The classical Lebesgue differentiation theorem ensures that this happens *a.e.*.

THEOREM 9.1 ([**69**]). *Given $f \in \mathcal{D}$, there is an $S \subset \mathbb{T}$ with* $\operatorname{Cap}(S) = 0$ *such that, for all $e^{i\theta} \notin S$, the radial limit defining the boundary function f^β and the limit* $\lim\operatorname{avg} f^\beta$ *both exist and the two are equal.*

This relation between boundary smoothness and radial smoothness is an instance of a common pattern. If $f \in \operatorname{Hol}(\mathbb{D})$ has, in some sense, boundary values f^β on the circle, then the smoothness of $f(z)$ as z approaches $e^{i\theta} \in \mathbb{T}$ is intimately tied to the smoothness of f^β at $e^{i\theta}$. If f were, in fact, holomorphic in the ball $B(e^{i\theta}, \varepsilon)$, then the Cauchy-Riemann equations would give an explicit relation between radial smoothness, measured by $\frac{\partial}{\partial r}f$, and the angular smoothness measured by $\frac{\partial}{\partial\theta}f$. The art is in finding ways in which vestiges of that fact survive as $\varepsilon \to 0$. An interesting introduction to this theme is in [**126**].

In fact, Beurling went further still and considered the set of boundary points at which the radial variation function was finite. We return to that topic in a moment.

Beurling's theorem is sharp in the following sense.

THEOREM 9.2 (Carleson [**93**]). *Given a closed set $E \subset \mathbb{T}$ with* $\operatorname{Cap}(E) = 0$, *there is an $f \in \mathcal{D}$ such that $\lim_{r \to 1} f(re^{i\theta})$ does not exist for any $e^{i\theta} \in E$.*

9.1.2. Radial Variation. Beurling obtained Theorem 9.1 as a consequence of a result on radial variation [**69**]. We discussed that radial variation result in Section 5.5.1 and we return to it now, but with different notation. We say that f has *finite radial variation* in the direction $e^{i\theta}$ if the curve $\{f(re^{i\theta}) : 0 < r < 1\}$ has finite length: $L_f(e^{i\theta}) := \int_0^1 \left|f'(re^{i\theta})\right| dr < \infty$. It is easy to see that if $L_f(e^{i\theta}) < \infty$ then f has a radial boundary value at the point $e^{i\theta}$. Also, it is not hard to see that if $f \in \mathcal{D}$, then $L_f(e^{i\theta})$ is finite *a.e.* $d\theta$.

EXERCISE 9.3. *Verify the last two statements.*

Beurling proved the following theorem.

THEOREM 9.4. *Given $f \in \mathcal{D}$, $L_f(e^{i\theta}) < \infty$ for all $e^{i\theta}$ outside an exceptional set $S \subset \mathbb{T}$ having* $\operatorname{Cap}(S) = 0$.

A similar statement for $f \in H^2$ fails completely. Existence of $f \in H^2$ with $L_f = \infty$ *a.e.* was established by Rudin [**290**]. In fact there are examples where L_f is identically ∞, [**84**]. For a discussion of discrete analogs of that Hardy space behavior see [**88**].

One of our themes is that function spaces on trees can be useful both as models and as tools for studying holomorphic functions. The study of boundary values of functions on the tree and on the disk is a good example. In Section 5.5. we used functions in the tree model of the Dirichlet space to obtain Theorem 5.13, which includes the following.

THEOREM 9.5. *Given $f \in \mathcal{D}$, $L_f(e^{i\theta}) < \infty$ for all $e^{i\theta} \notin S$ with the exceptional set S having $\mu(S) = 0$ for all $\mu \in CM(\mathcal{D})$.*

We will give an informal recapitulation of the proof of this theorem in a moment and will note a core similarity with Beurling's proof of Theorem 9.4. We will also discuss why the sets of capacity zero are the same as the null sets for every $\mathcal{D}$ Carleson measure, the fact that ensures the equivalence of the two previous results.

9.1.3. Trees. We start with a tree $\mathcal{T}$ placed in the disk in the manner described in Section 5.2.2 and we suppose that the parameters C_1 and C_2 are rather small. However, having said that, it may be easier to picture the following discussion by thinking of the model in Section 5.2.1.

The "integration" operator I acts on functions $\mathcal{T}$ by summation along geodesics

$$If(\beta) = \sum_{o \le \gamma \le \beta} f(\gamma).$$

The space $\ell^2(\mathcal{T})$ is the dyadic model of the Bergman space, and the space $\{If : f \in \ell^2(\mathcal{T})\}$ ("primitives of functions in the Bergman space") is the dyadic Dirichlet space, $\mathcal{D}(\mathcal{T})$. Let Γ be the collection of semi-infinite geodesics which start at the root o. These are in one to one correspondence with the points of $\partial\mathcal{T}$, the boundary of $\mathcal{T}$. We will use a parameter θ to label the points of $\partial\mathcal{T}$ and also to index the associated semi-infinite geodesics $\gamma_\theta = (o, \gamma_1^\theta, \gamma_2^\theta, \ldots, \theta) \in \Gamma$.

We then have three results:

First, given $F \in \mathcal{D}(\mathcal{T})$, there is an exceptional set $X \subset \partial\mathcal{T}$ which is so small that $\mu(X) = 0$ for every $\mu \in CM(\mathcal{D}(\mathcal{T}))$ and such that F has finite variation along γ_θ for every $\theta \in \partial\mathcal{T}\setminus X$. To see that, note that $\Delta F \in \ell^2(\mathcal{T})$ and hence $|\Delta F| \in \ell^2(\mathcal{T})$. Thus $F^+ = I\,|DF| \in \mathcal{D}(\mathcal{T})$. The variation of F along any $\gamma \in \Gamma$ is majorized by (actually, is equal to) F^+, so the choice $X = \{\theta : F^+(\theta) = \infty\}$ has the desired properties.

Second, one can put a tree, $\mathcal{T}$, inside the disk in such a way that for any $G \in \mathcal{D}$, we can majorize the radial variation of G by the variation along geodesics using $If \in \mathcal{D}(\mathcal{T})$. That function is defined, roughly, by setting at $\alpha \in \mathcal{T}$, $f(\alpha) = |G'(z)|\,(1-|z|^2)$ for some z close to α.

Combining the previous two results, one can conclude that $G \in \mathcal{D}$ has finite radial variation in the direction $e^{i\theta}$, except for a set of $e^{i\theta}$ which is a null set for every $\mathcal{D}(\mathcal{T})$-Carleson measure. The third fact we need, which connects the previous statement to Beurling's theorem, is a relation between Carleson measures and capacities.

THEOREM 9.6. *Suppose R is a closed subset of $\partial\mathcal{T}$ and S is the corresponding set in $\mathbb{T}$ (in the sense of the map Λ of Section 5.3). Then*

(1)
$$\mathrm{Cap}_{\mathcal{T}}(R) \approx \mathrm{Cap}_{\mathbb{D}}(S),$$

(2)
$$\mathrm{Cap}_{\mathbb{D}}(S) \approx \sup \frac{\mu(S)}{\|\mu\|_{CM(\mathcal{D})}} \tag{9.1}$$

where the supremum is over all $\mu \in CM(\mathcal{D})$ which are supported on S.

In particular, S is a null set for every Carleson measure, or, equivalently, S does not support a nontrivial Carleson measure if and only if $\mathrm{Cap}(S) = 0$.

PROOF. This result follows by combining the following facts. First, there is the result of Benjamini and Peres [**65**] comparing capacity on the tree and capacity on the disk. The informal conclusion is that the two are equal. Second is the relation between $CM(\mathcal{D}(\mathcal{T}))$ and $CM(\mathcal{D})$ described in Theorem 5.13. That result says that, informally, $CM(\mathcal{D}(\mathcal{T}))$ and $CM(\mathcal{D})$ are the same. Finally, we need the fact from [**44**, Section 5] that a version of (9.1) holds for measures on $\mathcal{T}$; that is capacity on $\partial\mathcal{T}$ can be computed using Carleson measures, and, in particular, capacity zero corresponds to being a null set for every Carleson measure.

These results are discussed further in [**53**], [**49**], and Chapter 13. □

COROLLARY 9.7. $\mathrm{Cap}_{\mathbb{D}}(S) = 0$ *if and only if* $\mu(S) = 0$ *for all* $\mu \in CM(\mathcal{D})$.

Even the outline we just gave of the proof of Theorem 9.4 is longer than Beurling's original proof. However, the proof we discussed, based on underlying hyperbolic geometry and on modulus of continuity estimates, can be extended to contexts where the functions studied are not holomorphic or even harmonic.

It is interesting to note that, at their hearts, Beurling's proof and the proof of the tree version are not that different. The tree proof uses the fact that one can make the "discrete derivative" positive and obtain a majorant which is still inside the dyadic Dirichlet space. Beurling's proof also makes the derivative positive, that is it dominates $f = \int f'$ by $M_f = \int |f'|$. Although M_f is not in the space of holomorphic functions, it can be studied using the function theory associated to functions in $\mathcal{D}$.

9.1.4. Broader Approach Regions. The boundary function of f, f^β, is defined using radial limits. There is a rich story of what happens when radial approach to the boundary is replaced by approach through regions with tighter contact with the boundary. Those results only depend on local behavior and it is more convenient to discuss them in the upper halfplane picture. Instead of considering convergence along radii of function values to boundary values on the circle, we will talk about convergence on vertical lines of function values to boundary values on the real axis. That is, we set $S_0 = \{0 + iy : 0 < y\}$ and for f defined on the upper halfplane, we consider, for each $t \in \mathbb{R}$, the boundary limit, $f^\beta(t)$, defined by

$$f^\beta(t) := \lim_{y\to 0,\ z\in S_0} f(z+t) = \lim_{y\to 0,\ z\in S_0+t} f(z). \tag{9.2}$$

The upper halfplane analog of the previous theorem ensures that the limit exists for all t outside a set of capacity zero.

What happens when we replace S_0 with a larger region? The crucial factor is the size of the approach region close to the boundary, and one way to investigate that is to consider approach regions with varying degrees of tangency to the boundary. As a first step we could consider nontangential limits such as those obtained by replacing S_0 by $S_1 = \{x + iy : |x| < y\}$ (this region is tangent of order 0) and, again, asking for the size of the exceptional set on which the limit fails to exist. In this case, again, the exceptional set has capacity zero. We can go further and consider the regions with positive order algebraic contact with the axis:

$$\text{algebraic contact, } \beta > 1 \quad S_{a,\beta} = \{x + iy : |x|^\beta < y\}.$$

The regions $S_{\alpha,2}$ are the type of regions used in orocyclic convergence in Proposition 10.17. Going further still, we could consider:

$$\begin{array}{ll} \text{subexponential contact, } 0 < \gamma < 1 & S_{s,\gamma} = \{x + iy : \exp\left(-1/|x|^\gamma\right) < y\}, \\ \text{exponential contact} & S_e = \{x + iy : \exp\left(-1/|x|\right) < y\}. \end{array}$$

For each choice of S and each $t \in \mathbb{R}$, one considers the analog of (9.2). For each S, the question is the size of the set on which the limit might fail to exist.

For regions which make algebraic contact the result is the same as for nontangential regions, there is quasieverywhere convergence. For the subexponential contact regions, there is again convergence for t outside a small exceptional set S. However, in this case, we cannot require $\mathrm{Cap}(S) = 0$; rather we have the weaker statement $\mathrm{Cap}_\gamma(S) = 0$ where Cap_γ is a functional whose definition involves the parameter γ. Early results of this type are due to Kenny [**192**]; more recent results are due to Twomey, [**334**] and [**333**] are samples. At the end of the scale of results is the result of Nagel, Rudin and Shapiro [**235**] showing that, in the case of exponential contact, there is convergence for t outside a set of measure zero. The result of [**235**], specialized to the Dirichlet space, is proved in full in [**136**, Chapter 3].

It is straightforward to extend the idea of radial convergence in a tree to a version of convergence using these larger regions. Let Γ be the set of half infinite geodesics in $\mathcal{T}$ parameterized by their terminal point $\theta \in \partial\mathcal{T}$. That is, if $\gamma \in \Gamma$ has endpoint θ, we realize the geodesic as a sequence of vertices $\gamma = \gamma_\theta = (o, \gamma_1^\theta, \gamma_2^\theta, \ldots, \theta)$. For a function f defined on $\mathcal{T}$, we studied its limiting behavior by majorizing the variation of f along γ using $M_0 = I(|\Delta f|)$. We then observed that $M_0 \in \mathcal{D}(\mathcal{T})$ and went on from there.

Each geodesic is the analog of a set $S_0 + t$ in (9.2) and convergence along $S_0(\gamma_\theta)$ is an analog of radial convergence. We form analogs of the nontangential convergence regions in the halfplane by replacing $S_0(\gamma_\theta)$ by larger neighborhoods of γ_θ, $S_{\texttt{bigger}}(\gamma_\theta)$. The boundary limit of interest will then be

$$f^{\texttt{bigger}}(\theta) := \lim_{d(o,\lambda)\to\infty,\ \lambda\in S_{\texttt{bigger}}(\gamma_\theta)} f(\gamma).$$

An analog of a nontangential approach region is obtained by selecting a positive integer j and setting

$$S_{+j}(\gamma_\theta) = \gamma_\theta^{+j} = \left\{\sigma \in \mathcal{T} : \exists\lambda \in \gamma_\theta^{+j},\ d(\sigma,\lambda) \le j\right\}.$$

In the metric geometry of the tree this is a tubular neighborhood; however the region in the disk, or the halfplane, marked by these vertices is a nontangential wedge.

EXERCISE 9.8. *Make enough sketches to understand the previous comments. Should convergence along $S_0(\gamma)$ be regarded as an analog of radial convergence or of nontangential convergence?*

Suppose we want to study the $S_j(\gamma)$ limits of an $F \in \mathcal{D}(\mathcal{T})$. Fix θ and we want to study limits of the form $\lim_n F(\alpha_n^\theta)$ where $\{\alpha_n^\theta\}$ satisfies

$$d(o,\alpha_n^\theta) = n \text{ and } d(\alpha_n^\theta,\gamma_\theta) \le j. \tag{9.3}$$

We again build function, F^+ which is a majorant for the variation of F. More specifically, we want to majorize, for each fixed θ, the variation along all sequences $\{F(\alpha_n^\theta)\}$ where $\{\alpha_n^\theta\}$ satisfies (9.3). For $\alpha \in \gamma_\theta$ set

$$\phi(\alpha) = \max\{|\Delta F(\beta)|\ ; \beta \in \mathcal{T}, d(\alpha,\beta) < 3j\}, \tag{9.4}$$

and set $F^+ = I\phi$. Drawing a sketch we see that, with θ fixed, and for any $\{\alpha_n^\theta\}$ which satisfies (9.3) we have, for all $\beta \in \gamma_\theta$,

$$\text{Variation}(F,\{\alpha_n^\theta\})(\beta) = \sum_{m\le d(o,\beta)} \left|F(\alpha_m^\theta) - F(\alpha_{m-1}^\theta)\right| \le 3jI\phi(\beta).$$

We can finish the argument as before if we know that $\sum_{\alpha\in T}|\phi(a)|^2 < \infty$. We know $\{\Delta F(\beta)\} \in \ell^2(\mathcal{T})$, and we know that each $\phi(\alpha)$ is one of the values of $|\Delta F(\beta)|$ with β satisfying $d(\alpha,\beta) < 3j$. Hence it will suffice to bound how frequently each $|\Delta F(\beta)|$ is used. That is, we will be finished if we can bound $\text{card}(\beta) = |\{\alpha : d(\alpha,\beta) < 3j\}|$. However this holds with a constant which only depends on the upper bound of the branching number of the tree and on j.

EXERCISE 9.9. *Check the details of the previous sketch of a proof.*

COROLLARY 9.10. *Theorem* 9.1 *continues to hold if radial convergence is replaced by nontangential convergence.*

One can consider approach regions $S_{\texttt{bigger}}$ which are larger still. If we follow the previous argument we are led to a situation where the upper bound on the number of times a particular β can be selected depends on $d(o,\beta)$. This would lead to a function $\{\phi(a)\}$ which was not in $\ell^2(\mathcal{T})$ but, instead, was in a weighted ℓ^2 space. One would then use the theory of function spaces on weighted trees from [**48**] to follow the previous arguments and obtain exceptional sets which were null sets for a modified type of capacity. There are some discussions of this approach in [**44**, Section 6].

The book by DiBiasi [**125**] has a detailed discussion of types of boundary convergence, both for harmonic functions (and hence also holomorphic ones) and for functions defined on trees.

9.2. How Big is the Boundary Function?

9.2.1. Beurling's Theorem. At (2.1.2), we defined the space $\boldsymbol{D}$ as

$$\boldsymbol{D} = \{f \in \operatorname{Hol}(\mathbb{D}) : f(0) = 0,\ \mathfrak{D}(f) < \infty\}$$

with norm $\mathfrak{D}(f)^{1/2}$. The same techniques we used to find the reproducing kernel for $\mathcal{D}$ can be used for $\boldsymbol{D}$, and we find that the reproducing kernel in the space $\boldsymbol{D}$ for the point ω in the disk is function $\kappa_\omega(z) = -\log(1-\overline{\omega}z)$.

In Chapter 2, we saw that the Dirichlet space was contained in BMO. Combined with the John-Nirenberg theorem (Theorem 3.16), this is enough to conclude that for any $f \in \boldsymbol{D}$, there is a positive ε so that, using the same notation for the boundary function, $\exp(\varepsilon\,|f|)$ is integrable on $\mathbb{T}$. However, much more is true.

THEOREM 9.11 ([**70**]). *Suppose $0 \le \alpha < 1$. There is a $C(\alpha) < \infty$ so that for all $f \in \boldsymbol{D}$ having $\|f\|_{\boldsymbol{D}} \le 1$,*

$$\int_{\mathbb{T}} \exp(\alpha\,|f(e^{i\theta})|^2)d\theta \le C(\alpha).$$

PROOF. We are given $f \in \boldsymbol{D}$, $\|f\|_{\boldsymbol{D}} \le 1$. We have,

$$\begin{aligned}(9.5)\quad \int_{\mathbb{T}} \exp(\alpha\,|f|^2)d\theta &= \sum_n \frac{\alpha^n}{n!}\int_{\mathbb{T}} |f|^{2n}\,d\theta = \sum_n \frac{\alpha^n}{n!}\,\|f^n\|_{H^2}^2 \\ &\approx \sum_n \frac{\alpha^n}{n!}\int_{\mathbb{D}} |(f^n)'|^2\,(1-|z|^2)dA \\ &= \sum_n \frac{\alpha^n n^2}{n!}\int_{\mathbb{D}} \left\{(1-|z|^2)\,|f^{n-1}(z)|^2\right\}|f'(z)|^2\,dA \\ &\le \sum_n \frac{\alpha^n n^2}{n!}\left(\sup_{z\in\mathbb{D}}\left\{(1-|z|^2)\,|f^{n-1}(z)|^2\right\}\right)\int_{\mathbb{D}} |f'(z)|^2\,dA.\end{aligned}$$

Passage from the first line to the second used (2.15) for an alternative description of H^2 norms. Because $\|f\|_{\boldsymbol{D}} \le 1$, we can estimate the integral in the last line by 1. Also, we know from (2.8) that at any point z with $|z| = r$ we have $|f(z)|^2 \le (-\log(1-r^2))$. Hence we can continue with

$$\begin{aligned}(9.6)\quad \int_{\mathbb{T}} \exp(\alpha\,|f|^2)d\theta &\lesssim \sum_n \frac{\alpha^n n^2}{n!}\sup_{0\le r\le 1}\,(1-r^2)\,(-\log(1-r^2))^{n-1}. \\ &= \sum_n \frac{\alpha^n n^2}{n!}e^{-n+1}\,(n-1)^{n-1} \\ &\approx \sum_n \frac{1}{\sqrt{2\pi n}}\alpha^n =: C(\alpha).\end{aligned}$$

We obtained the second line by explicitly solving the extremal problem in the first line. We then used Stirling's approximation, $n! \approx n^n e^{-n}\sqrt{2\pi n}$, and the approximation $(1 + A/n)^n \approx e^A$. □

A central aspect of this result is the uniformity; that is, for a fixed α, the integral has a bound independent of f. Using this result we can see that, for any $f \in \boldsymbol{D}$, $\exp(|f|^2)$ is integrable (equivalently, the previous theorem holds for all α but with constant $C(\alpha, f)$ rather than $C(\alpha)$). To see this, note that by working with the power series we can split any given f as $f = p + s$ with p a polynomial and $\mathfrak{D}(s)$ very small. Then, using the fact that p is bounded, we can reduce the question of the integrability of $\exp(|f|^2)$ to the question of the integrability of $\exp(|s|^2)$; a question covered by the previous theorem if $\mathfrak{D}(s) < 1$.

EXERCISE 9.12. *Fill in the details of the argument in the previous paragraph.*

If Theorem 9.11 holds for a particular α, then it automatically holds for smaller α. The theorem is false for $\alpha > 1$. That can be seen using the normalized kernel functions for the space $\boldsymbol{D}$.

EXERCISE 9.13. *Verify that the κ_ω defined above are the kernel functions for $\boldsymbol{D}$. Use those functions to show that Theorem 9.11 fails for $\alpha > 1$.*

We saw in Proposition 3.19 that if $f \in \mathcal{X}$ then, for some $\alpha > 0$, we have $\exp(\alpha f) \in \boldsymbol{D}$. The previous result can then be applied to a normalized version of $\exp(\alpha f)$ to conclude that for some $\beta > 0$, f must satisfy a form of double exponential integrability;

$$\int \exp(\beta\,|\exp \alpha f|^2) < \infty$$

That result and related ones are studied in the thesis of Lin [**209**].

9.2.2. The Sharp Result of Chang and Marshall.

Chang and Marshall proved a version of Beurling's theorem for the optimal case $\alpha = 1$ [**105**].

THEOREM 9.14 (Chang and Marshall, [**105**]). *There is a constant so that if $f \in \boldsymbol{D}$, $\|f\|_{\boldsymbol{D}} \leq 1$, then*

$$\int_{\mathbb{T}} \exp(|f(e^{i\theta})|^2)\,d\theta \lesssim 1. \tag{9.7}$$

We will not prove that theorem here, we will comment on the proof in [**105**] as well as more recent related work.

9.2.2.1. *The Proof.* We will just describe the beginning of the path that Chang and Marshall use in proving Theorem 9.14.

A crucial fact in the proof of Theorem 9.11 was that at any point z we have $|f(z)|^2 \leq -\log(1 - |z|^2)$. If, for a given f, we had the stronger estimate that for some small $\varepsilon > 0$ and all z

$$|f(z)|^2 < -(1-\varepsilon)\log(1 - |z|^2). \tag{9.8}$$

then that same proof also shows that (9.7) holds with a constant C that depended on ε, but was independent of f. Hence, in order to complete the proof of Theorem 9.14, it suffices to obtain the estimate in (9.7) under the hypothesis that (9.8) fails.

If (9.8) fails then there is a point $a \in \mathbb{D}$ where equality holds.

$$|f(a)|^2 = (1-\varepsilon)\left(-\log(1 - |a|^2)\right).$$

In this case, we can use the following Hilbert space observation: The function f is close to the maximum allowable value at a, a value attained uniquely by the normalized reproducing kernel $\hat{\kappa}_a$. Hence, we expect that f is close to $\hat{\kappa}_a$. More precisely we define a remainder function r by $f(z) = \arg(f(a))\sqrt{1-\varepsilon}\hat{\kappa}_a(z) + r(z)$. We compute that $r(a) = 0$, and thus $r \perp \hat{\kappa}_a$. Hence, using the Pythagorean Theorem and the hypothesis on the size of f, we see that $\|r\|^2 \leq \varepsilon$. Hence $f = \hat{\kappa}_a + \tilde{r}$ with $\|\tilde{r}\| \leq \sqrt{2\varepsilon}$. The heart of the proof in [**105**] is estimating the integral for such an f, with ε small, and obtaining a final estimate which is independent of f and ε.

9.2.2.2. *Search for the Extremal.* Given Theorem 9.14, it is natural to wonder what is the smallest possible value of C, if the extreme value is attained, and by what function? Experience with other function theoretic extremal problems suggest such an extremal function might be quite simple and have substantial symmetry. In this particular case, it is speculated that $f(z) = z$ is the extremal. Substantial partial results have been attained, but basic questions remain open. [**33**], [**108**], [**225**], [**32**], [**31**].

9.2.2.3. *Alternate Proofs and Extensions.* The ideas used by Chang and Marshall are inside the framework we are using in this volume. The same is true about the later, shorter, proof by Marshall [**219**]. There are also related developments which open into areas we are not considering. That work shows the range of analytical ideas that can be used with Dirichlet space functions and also the ability of ideas used for the Dirichlet space to be extended into more general contexts. We will just say a few words and offer some references.

Given a nonnegative function F defined on a domain $\Omega \subset \mathbb{C}$, one can ask if F has a harmonic majorant; is there a harmonic function G defined on Ω such that $F \leq |G|$ in Ω? At its heart, this is a question about the size of F and, more specifically, the relation between the regions where F is large and the shape of Ω. Many integrability questions can be reduced to questions about existence of harmonic majorants and there is a very rich analytical toolkit for studying such problems. In [**143**], Essen proves a very general result about existence of such majorants which has as a special case that if Ω is a domain in $\mathbb{C}$ which contains the origin and has area at most π, then the function $F(z) = \exp(|z|^2)$ has a harmonic majorant on Ω. This fact implies Theorem 9.14.

Analysis of nonconstant holomorphic functions in the plane makes frequent use of the fact that such functions are conformal off the small set where the derivative is zero. On the other hand, except for a few relatively elementary functions, there are no conformal maps between higher dimensional spaces. However, there are very rich theories of maps such as quasiregular maps which share, at least approximately, some of the basic geometric properties of conformal maps. Generalizations of, and analogs for, Theorems 9.11 and 9.14 have been proved in several of these contexts, [**263**], [**251**], [**11**].

9.2.2.4. *Related Results.* Theorem 9.14 gives sharp information about integrals of the form $\int_{\mathbb{T}} \exp(\alpha\, |f|^2)$ for $f \in \boldsymbol{D}$. For the same f, Saito [**292**] and Burbea [**87**] have sharp estimates for integrals of the type $\int_{\mathbb{D}} |\exp(\alpha f)|^2$. The proofs of their results are more straightforward and are entirely based on RKHS theory.

In some ways the Dirichlet space is similar to the function space $BMOA$, which we described in Section 3.7, and to the Bloch space defined by (2.47). There are very sharp results concerning exponential integrability of functions in both of those two spaces. Although those results are not known to imply or be implied by the previous theorems, the resemblance between the results is intriguing. This is discussed in

[**105**, Section 5.1] and in the Introduction in Hedenmalm's paper [**172**]. Theorem 3.18 is also in the same ill-defined family of related results.

9.3. Where Can the Boundary Function Vanish?

9.3.1. Zero Sets on the Boundary. In Chapter 6 we considered zero sets and exact zero sets in $\mathbb{D}$, for functions in $\mathcal{D}$, and for functions in $\mathrm{Mult}(\mathcal{D})$. We have seen that those functions have boundary values quasieverywhere on the circle, and hence we can also consider the zero sets in the circle. Less is known about these sets and we restrict our attention to zero sets for $\mathcal{D}$, putting aside consideration of exact zero sets or zero sets for functions in $\mathrm{Mult}(\mathcal{D})$.

For $f \in \mathcal{D}$, we will use the same notation, f, to denote its *boundary function* which is defined quasieverywhere by taking radial limits. A *boundary zero set* for $\mathcal{D}$ is a set $E \subset \mathbb{T}$ with the property that there is an $f \in \mathcal{D}$, not the zero function, which has boundary values zero quasieverywhere on E. If E is not a zero set and if $f, g \in \mathcal{D}$ agree on E, we must have $f = g$. Those sets, which uniquely determine functions in $\mathcal{D}$, are called (boundary) *sets of uniqueness*. In short, $E \subset \mathbb{T}$ is a zero set if and only if it is not a set of uniqueness. We will give a characterization of sets of uniqueness in $\mathbb{T}$, and hence also of zero sets in a moment. However, the characterization is in terms of a capacity functional which is not well understood. Before going to that we describe some of the more geometric facts that are known about zero sets.

9.3.1.1. *Geometric Description.* The Dirichlet space is contained in the Hardy space and so we know from Corollary 3.3 that a zero set cannot have positive measure. However the story is richer than that.

It was shown by Carleson [**93**] that if $S \subset \mathbb{T}$ has $\mathrm{Cap}(S) = 0$, then S is a zero set and if, further, S is closed then it is an exact zero set. In fact one can give a proof of Theorem 9.2 so that the function produced, f, has $\mathrm{Re}\, f \geq 0$ and $f(re^{i\theta}) = -\infty$ for $e^{i\theta}$ in the exceptional set S, insuring that $g = \exp(-f) \in \mathcal{D}$ and has boundary values 0 on S.

In Section 6.6.3 we introduced *Carleson sets.* A closed set $E \subset \mathbb{T}$ of measure zero is *Carleson set* if

$$\int_{\mathbb{T}} \frac{d\theta}{\mathrm{dist}(e^{i\theta}, E)} < \infty. \tag{9.9}$$

Here, $\mathrm{dist}(z, A)$ denotes the distance from a point z to the set A. Equivalently, writing the complement of E as a union of disjoint open intervals, $\mathbb{T} \setminus E = \cup I_n$ we find that E is a Carleson set exactly if

$$\sum_n |I_n| \log(2\pi/|I_n|) < \infty.$$

EXERCISE 9.15. *Show those two conditions are equivalent.*

It was shown, again by Carleson [**93**], that any Carleson set is an exact zero set for the Dirichlet space. These two classes, the closed sets of capacity zero and the Carleson sets are independent; neither class contains the other.

More information about boundary zero sets is in [**136**, Sections 4.3 and 4.4].

9.3.1.2. *Analytic Description.* We now give description, due independently to Malliavin [**217**] and Hamilton [**170**], of the boundary zero sets. (Actually those authors discussed *uniqueness sets.*)

EXERCISE 9.16. *Prove the last statement.*

Their work uses a type of capacity functional which is different from the ones we have looked at earlier; this one is defined using a complex valued kernel and complex measures. The theory of this capacity is much less well developed, but it arises naturally in the question we are about to consider. For a complex measure μ supported on $S \subset \mathbb{T}$, we set

$$\tilde{\mathcal{E}}(S,\mu) := \int_S \int_S \log \frac{1}{(1-\overline{w}z)} d\mu(z)\overline{d\mu(w)}$$

and define the associated functionals $\tilde{\mathcal{E}}(S)$ and $\widetilde{\mathrm{Cap}}(S)$ by

$$\tilde{\mathcal{E}}(S) := \inf \left\{ \begin{array}{c} \tilde{\mathcal{E}}(S,\mu);\ \mu \text{ a complex measure} \\ \text{supported on } S,\ \int d\mu = 1 \end{array} \right\}$$
$$\widetilde{\mathrm{Cap}}(S) = 1/\tilde{\mathcal{E}}(S).$$

We will follow the presentation of Hamilton.

THEOREM 9.17 ([**217**], [**170**]). *Suppose $E \subset \mathbb{T}$ has $\mathrm{Cap}(E) > 0$, and hence also $\widetilde{\mathrm{Cap}}(E) > 0$; then E is a zero set for $\mathcal{D}$ if and only if $\widetilde{\mathrm{Cap}}(E) < \infty$. Equivalently there is a constant c such that*

$$\left| \int_E d\mu \right|^2 \leq c \int_E \int_E \log \frac{1}{1-z\overline{w}} d\mu(z)\overline{d\mu(w)} \tag{9.10}$$

for every complex measure μ supported by E.

A key tool will be the following lemma for $E \subset \mathbb{T}$.

LEMMA 9.18 ([**170**]). *Set $E_0 = E \cup \{0\}$. A measurable function $\tilde{f}$ on E_0 is the restriction to E_0 of a function $f \in \mathcal{D}$ if and only if there is a $c > 0$ such that, for any complex scalar a_0 and complex measure μ supported on E; setting $d\mu_0 = a_0\delta_0 + d\mu$,*

$$\left| \int \tilde{f}\, d\overline{\mu}_o \right|^2 \leq c^2 \int_{\mathbb{D}} \left(1 + \log \frac{1}{1-z\overline{w}}\right) d\mu_0(z)\overline{d\mu_0(w)}. \tag{9.11}$$

In that case, f can be chosen with $\|f\|_{\mathcal{D}} = c$.

Here, δ_0 is the unit mass at the origin.

Giving the proof of the lemma would require a digression into capacity theory. Instead, we will discuss the lemma informally and refer to Hamilton's paper for the proof.

Although the lemma considers μ on the boundary, we gain some insight into its content by considering measures of the form

$$d\mu := a_0\delta_0 + \sum_{i=1}^{n} a_i\delta_{z_i}$$

with the $\{a_n\}$ complex scalars and $E = \{z_n\} \subset \mathbb{D}$. In this modified case, the function $\tilde{f}$ would be an assignment of values at the points of $E \cup \{0\}$. Writing $z_0 = 0$, the left hand side of (9.11) would be

$$\left| \sum_{i=0}^{n} \tilde{f}\,(z_i)\overline{a_i} \right|^2,$$

and the right hand side would be

$$c^2 \sum_{i-0}^{n} \sum_{j-0}^{n} \left\{ 1 + \log \frac{1}{1 - z_i \overline{z_j}} \right\} a_i \overline{a_j}.$$

Now note that the quantities in braces are the reproducing kernels for the Dirichlet space normed by $\|\cdot\|^2_{\mathcal{D},\mathtt{alt}}$, see Exercise 2.1. Hence, setting $K = \sum_i a_i k_{z_i}$, we find that the right hand side of (9.11) is $c^2 \|K\|^2_{\mathcal{D},\mathtt{alt}}$, and, further, writing

$$L(K) = \sum_{i=1}^{n} \overline{\tilde{f}\,(z_i)} a_i, \tag{9.12}$$

we see that the lemma states that the linear functional L, defined on the subspace of $\mathcal{D}$ spanned by the $K's$, extends to a linear functional on all of $\mathcal{D}$ if and only if it is bounded on that subspace. The extension, being a bounded linear functional on the Hilbert space $\mathcal{D}$, must be given by taking an inner product with some f in $\mathcal{D}$ and f can be chosen so that the norm of f is the norm of the functional on the space of $K's$.

Assuming Lemma 9.18, it is straightforward to prove Theorem 9.17.

PROOF OF THEOREM 9.17. That $\operatorname{Cap}(E) > 0$ ensures $\widetilde{\operatorname{Cap}}(E) > 0$ can be seen by comparing (4.18) and the definition of $\widetilde{\operatorname{Cap}}$.

Suppose now that E is a zero set for $\mathcal{D}$. In that case the set

$$\mathcal{D}_E := \left\{ f \in \mathcal{D} : f(0) = 1 \text{ and } f(e^{i\theta}) = 0 \quad \forall e^{i\theta} \in E \right\}$$

is not empty. In fact $\mathcal{D}_E$ is a closed set. This is a consequence of the weak capacitary inequality, see the discussion in the proof of Theorem 12.21. Set

$$m = \inf \left\{ \|f\|_{\mathcal{D}} : f \in \mathcal{D}_E \right\}.$$

Because $f(0) = 1$ we know $m \geq 1$. We now show $m > 1$. Otherwise there is a sequence $\{f_k\} \subset \mathcal{D}_E$ with $\|f_k\|_{\mathcal{D}} \to 1$. Recalling that $f_k(0) = 1$, we see that we must have $\int_{\mathbb{D}} |f_k'(z)|^2 \, dA(z) \to 0$. This implies that $f_k \to 1$ pointwise, and hence by Proposition A.18, $f_k \to 1$ in norm. However, $\{f_k\} \subset \mathcal{D}_E$ which is closed; hence $1 \in \mathcal{D}_E$. This contradiction eliminates the possibility that $m = 1$.

Because E is a zero set for $\mathcal{D}$, there exists a function $f \in \mathcal{D}$, not the zero function, such that $f(e^{i\theta}) = 0$ for all $e^{i\theta} \in E$. By Lemma 9.18, we have that

$$|a_0|^2 \leq m^2 \left(\left| a_0 + \int_E d\mu \right|^2 + \int_E \int_E \log \frac{1}{1 - z\overline{w}} d\mu(z) \overline{d\mu(w)} \right) \tag{9.13}$$

for all $a_0 \in \mathbb{C}$ and all admissible measures μ. Indeed, because $f \in \mathcal{D}_E$, we have that

$$\int_E f d\overline{\mu}_0 = \overline{a}_0 + \int_E f d\mu = \overline{a}_0.$$

This gives the left hand side of (9.13). For the right hand side, observe that

$$d\mu_0(z) \overline{d\mu_0(w)} = |a_0|^2 \delta_{z=0} \delta_{w=0} + a_0 \delta_{z=0} \overline{d\mu(w)} + \overline{a_0} \delta_{w=0} d\mu(z) + d\mu(z) \overline{d\mu(w)}.$$

Hence integration of the function $\left(1+\log\frac{1}{1-z\overline{w}}\right)$ against the measure $d\mu_0(z)\overline{d\mu_0(w)}$ produces

$$\begin{aligned}
&\int_{\mathbb{D}}\left(1+\log\frac{1}{1-z\overline{w}}\right)d\mu_0(z)\overline{d\mu_0(w)}\\
&\quad=|a_0|^2+a_0\int_E\overline{d\mu(w)}+\overline{a_0}\int_E d\mu(z)\\
&\qquad\quad+\int_E d\mu(z)\int_E\overline{d\mu(w)}+\int_E\int_E\log\frac{1}{1-z\overline{w}}d\mu(z)\overline{d\mu(w)}\\
&\quad=\left|a_0+\int_E d\mu\right|^2+\int_E\int_E\log\frac{1}{1-z\overline{w}}d\mu(z)\overline{d\mu(w)}.
\end{aligned}$$

Algebra coupled with (9.13) provides

$$0\le\left(1-\frac{1}{m^2}\right)|a_0|^2+2\mathrm{Re}\left(\overline{a_0}\int_E d\mu\right)+\left|\int_E d\mu\right|^2+\int_E\int_E\log\frac{1}{1-z\overline{w}}d\mu(z)\overline{d\mu(w)}\tag{9.14}$$

for all $a_0\in\mathbb{C}$. Letting $a_0=-\frac{m^2}{m^2-1}\int_E d\mu$ in the above gives

$$\begin{aligned}
0\le&\left(1-\frac{1}{m^2}\right)\left(\frac{m^2}{m^2-1}\right)^2\left|\int_E d\mu\right|^2-2\frac{m^2}{m^2-1}\left|\int_E d\mu\right|^2\\
&+\left|\int_E d\mu\right|^2+\int_E\int_E\log\frac{1}{1-z\overline{w}}d\mu(z)\overline{d\mu(w)}.
\end{aligned}$$

Now algebraic rearrangement proves

$$(m^2-1)\int_E\int_E\log\frac{1}{1-z\overline{w}}d\mu(z)\overline{d\mu(w)}\ge\left|\int_E d\mu\right|^2,$$

which is the necessary condition we need.

For the converse, now suppose that there exists a constant $c>0$ such that

$$\left|\int_E d\mu\right|^2\le c\int_E\int_E\log\frac{1}{1-z\overline{w}}d\mu(z)\overline{d\mu(w)}\tag{9.15}$$

for every complex measure μ supported by E. If we could show that this implied

$$0\le\frac{c}{c+1}|a_0|^2+2\mathrm{Re}\left(\overline{a_0}\int_E d\mu\right)+\left|\int_E d\mu\right|^2+\int_E\int_E\log\frac{1}{1-z\overline{w}}d\mu(z)\overline{d\mu(w)}\tag{9.16}$$

for all $a_0\in\mathbb{C}$, then we would have that

$$|a_0|^2\le c\left(\left|a_0+\int_E d\mu\right|^2+\int_E\int_E\log\frac{1}{1-z\overline{w}}d\mu(z)\overline{d\mu(w)}\right)\tag{9.17}$$

held for all $a_0\in\mathbb{C}$. But, again by an application of Lemma 9.18, we would have that there exists $0\ne f\in\mathcal{D}$ such that f has boundary values equal to 0 on E. Thus $\mathcal{D}_E$ is a zero set.

However, note that under our hypothesis we have that

$$\left|\int_E d\mu\right|^2+\int_E\int_E\log\frac{1}{1-z\overline{w}}d\mu(z)\overline{d\mu(w)}\ge\frac{c+1}{c}\left|\int_E d\mu\right|^2,$$

and so (9.16) follows easily since

$$\frac{c}{c+1}|a_0|^2 + 2\mathrm{Re}\left(\overline{a_0}\int_E d\mu\right) + \frac{c+1}{c}\left|\int_E d\mu\right|^2$$
$$= \left|\frac{\sqrt{c}}{\sqrt{c+1}}a_0 + \frac{\sqrt{c+1}}{\sqrt{c}}\int_E d\mu\right|^2 \geq 0.$$

□

We have chosen to follow Hamilton's presentation and have used the functional $\widetilde{\mathrm{Cap}}$ which is adapted to the RKHS with norm $\|f\|^2_{\mathcal{D},\mathtt{alt}}$. However that RKHS is the same set of functions as $\mathcal{D}$ and hence the conclusion also applies to functions in $\mathcal{D}$. Alternatively we could have adapted Hamilton's definition of $\widetilde{\mathrm{Cap}}$ to the space $\mathcal{D}$ and followed exactly the same proof strategy. Of course the functionals $\widetilde{\mathrm{Cap}}$ and its modified version can be shown directly to have the same null sets.

9.4. Notes and Comments

The theorem of Malliavin-Hamilton presented in this chapter is a necessary and sufficient condition for sets on the boundary to be a uniqueness set or zero set. However it would be very interesting to have a more transparently geometric description of those sets. Recent work on this and related questions is described in [**136**].

CHAPTER 10

Alternative Norms and Applications

The Dirichlet space, and more generally Besov spaces, carry a variety of equivalent norming functionals. After a brief discussion of the general pattern, we develop two specific instances for the Dirichlet space. The first is based, roughly, on integrals over the disk of divided differences. The second is based on integrals of divided differences on the boundary circle. In both cases we also give several applications of those norming functionals. The unifying theme of the applications is that many questions become (more) straightforward if one finds a convenient choice of norm. However there is an overhead cost, the necessity of establishing that various norms are equivalent.

We first defined $\mathcal{D}$, H^2, and A^2 using power series coefficients and some of the results in previous chapters flow from those definitions and computations with reproducing kernels. However other developments used realization of the norms through integral formulas. As an example, the fact that functions in H^∞ are multipliers of H^2 is not an obvious consequence of the power series definition, but it is an easy consequence of the realization of H^2 as a subspace of L^2.

In this chapter we present additional integral formulas for norms of the Dirichlet space. These formulas are more complicated than the ones we have been using; but they carry information about the geometric behavior of functions and are well suited for some interesting questions.

We begin with some general comments.

10.1. Measuring Oscillation in $\mathcal{D}$

The Dirichlet space is a Besov space. Norms for Besov spaces are frequently built by summing or integrating quantities measuring smoothness, variability, or oscillation. The possible choices; sums or integrals, integration on the domain or on the boundary, measuring smoothness with derivatives or divided differences or through oscillation etcetera, lead to a large variety of norming functionals. Work is required to show that various choices give equivalent norms, but finding a convenient choice of norm can be of great help in a specific situations.

The study of norming functionals on Besov spaces has a rich history; an interesting early discussion is in the book by Peetre [**256**]. Recent considerations of these issues for spaces of holomorphic functions include [**127**], [**347**], [**346**], [**345**], and the series of books [**332**].

After defining $\mathcal{D}$ using power series we noted that an equivalent norm could be based on the Dirichlet integral:

$$\mathfrak{D}(f) = \int_{\mathbb{D}} |f'(z)|^2 \, dA. \tag{10.1}$$

The quantity $|f'(z)|$ measures the oscillation of f near the point z. The naturalness of the formula for combining those measures is emphasized when we rewrite (10.1) using the invariant gradient, $\tilde{\nabla}$, from (1.11) and the invariant measure $d\lambda$ (1.10);

$$\mathfrak{D}(f) = \int_{\mathbb{D}} |\tilde{\nabla} f|^2 d\lambda.$$

We saw similar measures for the other Besov spaces in (2.45). In Chapter 14 we introduce Besov spaces of holomorphic functions of several variables and will use similar expressions, but involving higher derivatives.

Variance, a quantity from probability theory, is another measure of oscillation. Given μ a probability measure on X and f a function on X, we define the expectation of f, $E(f) = E(f, \mu)$ and the variance of f, $\operatorname{Var}(f) = \operatorname{Var}(f, \mu)$ by

$$\begin{aligned} E(f) &:= \int_X f d\mu \\ \operatorname{Var}(f) &:= E(|f - E(f)|^2) \\ &= E(|f|^2) - |E(f)|^2. \end{aligned} \tag{10.2}$$

Thus $\operatorname{Var}(f)$ is the average oscillation of f about its average, and the mass distribution of μ specifies the regions being emphasized in computing the averages. For instance, if we were studying a function f on the boundary of the disk, and for a given $z \in \mathbb{D}$, wanted to quantify the oscillation of the boundary values of f near the interval $I(z)$ we could choose for our measure the normalized arc length on $I(z)$.

Another plausible choice would be to use μ_z given by

$$d\mu_z = \frac{1 - |z|^2}{|1 - e^{-i\theta} z|^2} \frac{d\theta}{2\pi},$$

As we discussed in Section 1.5, μ_z is used in the Poisson integral formula to evaluate the harmonic extension of the boundary function f to the point z; $E(f, \mu_z) = PI(f)(z)$. In particular, if f is the boundary value of a function that is holomorphic, and thus harmonic, we have

$$E(f, \mu_z) = PI(f)(z) = f(z).$$

Hence, for such an f

$$\operatorname{Var}(f, \mu_z) = PI(|f|^2)(z) - |PI(f)(z)|^2 = PI(|f|^2)(z) - |f(z)|^2 \tag{10.3}$$

We can use this variation measure to build a functional equivalent to $\mathcal{D}(f)^2$:

$$|f(0)|^2 + \int_{\mathbb{D}} \operatorname{Var}(f, \mu_z) d\lambda. \tag{10.4}$$

(The integration is with respect to the invariant measure $d\lambda$. Part of the craft of working with Besov space norms is mastering the bookkeeping involved in such normalizations.) That fact is a special case of a general result of Böe [**76**], however this particular case has a longer history. We will see this functional again, later in this chapter, when we consider the Garsia norm of the space BMO.

Another approach to quantifying local oscillation of a function on $\mathbb{T}$, or on $\mathbb{D}$, would be to use the size of the difference quotient. For f defined on $\mathbb{T}$ we would study the size of

$$\frac{f(e^{it})-f(e^{is})}{e^{it}-e^{is}}. \tag{10.5}$$

That is a very productive approach and we study it extensively in the second part of this chapter. For holomorphic f defined on $\mathbb{D}$ it is convenient to use a variation on (10.5). For $z,w\in\mathbb{D}$ we consider

$$\frac{f(z)-f(w)}{1-\overline{w}z}. \tag{10.6}$$

As to the choice of $1-\overline{w}z$ in the denominator, note that for points near the boundary, the main case for determining the size of the quantity, $|1-\overline{w}z|\approx|w-z|$ and so this expression is roughly the same size as the difference quotient. The actual formula is dictated by the invariance considerations which we will see in a moment.

In the next section we will show that for appropriate σ, τ the functional

$$I(\sigma,\tau)\,(f):=\int_{\mathbb{D}}\int_{\mathbb{D}}\left|\frac{f(z)-f(w)}{1-\overline{w}z}\right|^2\frac{(1-|z|^2)^{\sigma}(1-|w|^2)^{\tau}}{|1-\overline{w}z|^{2+\sigma+\tau}}dA(z)dA(w).$$

is equivalent to the Dirichlet integral. For $\sigma=\tau>-1$ this functional was studied by Arazy, Fisher, and Peetre **[36]** who showed that, setting $\sigma=\tau=\beta$, there is a $c_\beta\neq 0$ such that, for $f\in\operatorname{Hol}(\mathbb{D})$

$$I(\beta,\beta)\,(f)=c_\beta\mathfrak{D}\,(f)\,.$$

Hence, the functional $I(\beta,\beta)$ can be used in defining the norm on $\mathcal{D}$. In the next section we show that $I(\sigma,\tau)$, $\sigma\neq\tau$, can also be used that way.

We can rewrite $I(\beta,\beta)$ in a way that clarifies its invariance properties. Recall that the measure $d\lambda$ is $\operatorname{Aut}(\mathbb{D})$ invariant and that the invariant pseudohyperbolic metric ρ satisfies the identity

$$1-\rho^2(x,y)=\frac{(1-|x|^2)(1-|y|^2)}{|1-\overline{x}y|^2}.$$

Hence we can rewrite $I(\beta,\beta)$ as

$$I(\beta,\beta)=\int_{\mathbb{D}}\int_{\mathbb{D}}|f(z)-f(w)|^2\left(1-\rho^2(z,w)\right)^{2+\beta}d\lambda(z)d\lambda(w).$$

Thus we have a measure of the "typical" distance between the value of f at z and at w, and this measure has the fundamental invariance that for any $\phi\in\operatorname{Aut}(\mathbb{D})$, $I(\beta,\beta)(f)=I(\beta,\beta)(f\circ\phi)$.

We can refine this formula and insert a numerical factor K_β^2 chosen so that $d\mu_\beta(z)=K_\beta(1-|z|^2)^\beta dA$ is a probability measure. We could then take the limit

$$\lim_{\beta\to-1}K_\beta(1-|z|^2)^\beta dA=\frac{d\theta}{2\pi}.$$

To see this note that the measures on the left are rotationally invariant probability measures with most of their mass near the circle. Hence, recalling that $z\overline{z}=1$ for z on the boundary, we have

$$J(f):=\lim_{\beta\to-1}K_\beta^2 I(\beta,\beta)=\int_0^{2\pi}\int_0^{2\pi}\left|\frac{f(e^{it})-f(e^{is})}{e^{it}-e^{is}}\right|^2\frac{ds}{2\pi}\frac{dt}{2\pi}. \tag{10.7}$$

It is a famous result of Douglas [**128**] that $J(f) = \mathfrak{D}(f)$. That representation of the Dirichlet integral is the conceptual starting point for the analysis of the Dirichlet norm by Richter and Sundberg [**274**] which we present in the second part of this chapter.

10.2. Interior Differences

THEOREM 10.1. *Fix $\sigma, \tau > -1$. For any $f \in \mathrm{Hol}(\mathbb{D})$*

$$\mathfrak{D}(f) \approx \int_{\mathbb{D}} \int_{\mathbb{D}} |f(z) - f(w)|^2 \frac{(1-|z|^2)^\sigma (1-|w|^2)^\tau}{|1-\overline{w}z|^{4+\sigma+\tau}} dA(z) dA(w).$$

The theorem was proved in [**284**] and includes earlier results from [**36**], the proof we give here follows that in [**73**].

PROOF. First note that for $z, w \in \mathbb{D}$

$$(1-|z|)(1-|w|) \leq |1-\overline{z}w| . \tag{10.8}$$

Hence, if $\sigma \geq \tau$ then

$$\begin{aligned} \frac{(1-|z|^2)^\sigma (1-|w|^2)^\sigma}{|1-\overline{w}z|^{4+2\sigma}} &\leq \frac{(1-|z|^2)^\sigma (1-|w|^2)^\tau}{|1-\overline{w}z|^{4+\sigma+\tau}} \\ &\leq \frac{(1-|z|^2)^\tau (1-|w|^2)^\tau}{|1-\overline{w}z|^{4+2\tau}}. \end{aligned}$$

Hence it suffices to prove the theorem in the case $\sigma = \tau$, a case treated in [**36**] by a more intricate argument. (In fact a version of the result holds in the case $\sigma = \tau = 0$ with the disk replaced by a much more general domain; see [**35**].)

Recall that $\varphi_w(z) = (z-w)/(1-\overline{w}z)$ is the conformal automorphism of $\mathbb{D}$ which interchanges w and 0. It satisfies

$$1 - |\varphi_w(z)|^2 = \frac{(1-|z|^2)(1-|w|^2)}{|1-\overline{w}z|^2}, \quad |\varphi_w'(z)| = \frac{1-|\varphi_w(z)|^2}{1-|z|^2}.$$

The first equality is a restatement of (1.6); the second follows from the first after explicitly computing $\varphi_w'(z)$.

Making the change of variables $\zeta = \varphi_w(z)$ we find

$$I(\sigma,\sigma) = \int_{\mathbb{D}} (1-|w|^2)^{-2} \int_{\mathbb{D}} |(f\circ\varphi_w)(\zeta) - (f\circ\varphi_w)(0)|^2 (1-|\zeta|^2)^\sigma dA(\zeta) dA(w).$$

We now use the fact that for $h \in \mathrm{Hol}(\mathbb{D})$

$$\int_{\mathbb{D}} |h(z) - h(0)|^2 (1-|z|^2)^\sigma dA(z) \lesssim \int_{\mathbb{D}} |h'(z)|^2 (1-|z|^2)^{\sigma+2} dA(z).$$

Using polar coordinates one sees that it suffices to verify this for monomials. The estimate for monomials can be obtained by two integrations by parts. Using it we continue with

$$I \lesssim \int_{\mathbb{D}} (1-|w|^2)^{-2} \int_{\mathbb{D}} |(f\circ\varphi_w)'(\zeta)|^2 (1-|\zeta|^2)^{\sigma+2} dA(\zeta) dA(w).$$

We now reverse the change of variables, $z = \varphi_w(\zeta)$. (Recall that $\varphi_w(\varphi_w(z)) = z$.) That leads to

$$\begin{aligned} I &\lesssim \int_{\mathbb{D}} (1-|w|^2)^\sigma \int_{\mathbb{D}} |f'(z)|^2 \frac{(1-|z|^2)^{\sigma+2}}{|1-\overline{w}z|^{4+2\sigma}} dA(z)dA(w). \\ &= \int_{\mathbb{D}} |f'(z)|^2 (1-|z|^2)^{\sigma+2} \left(\int_{\mathbb{D}} \frac{(1-|w|^2)^\sigma}{|1-\overline{w}z|^{4+2\sigma}} dA(w) \right) dA(z) \\ &\lesssim \int_{\mathbb{D}} |f'(z)|^2 \, dA(z). \end{aligned}$$

here in the last inequality we used Lemma B.5.

To obtain an estimate in the other direction we start with the reproducing formula in Corollary B.7

$$f(z) = c_\sigma \int_{\mathbb{D}} f(w) \frac{(1-|w|^2)^{2+\sigma}}{(1-\overline{w}z)^{4+\sigma}} dA(w).$$

From this we obtain

$$\begin{aligned} f'(z) &= d_\sigma \int_{\mathbb{D}} \overline{w} f(w) \frac{(1-|w|^2)^{2+\sigma}}{(1-\overline{w}z)^{5+\sigma}} dA(w), \\ f'(0) &= d_\sigma \int_{\mathbb{D}} \overline{w} f(w) (1-|w|^2)^{2+\sigma} dA(w) \\ &= d_\sigma \int_{\mathbb{D}} \overline{w} \left(f(w) - f(0) \right) (1-|w|^2)^{2+\sigma} dA(w). \end{aligned}$$

Thus

$$|f'(0)|^2 \lesssim \int_{\mathbb{D}} |f(w) - f(0)|^2 (1-|w|^2)^{2+\sigma} dA(w).$$

Replacing f by $f \circ \varphi_z$ we obtain

$$(1-|z|^2)^2 |f'(z)|^2 \lesssim \int_{\mathbb{D}} |(f\circ\varphi_z)(w) - f(z)|^2 (1-|w|^2)^{2+\sigma} dA(w).$$

Using this, and making the change of variables $w = \varphi_z(\zeta)$, we obtain

$$\begin{aligned} &\int_{\mathbb{D}} |f'(z)|^2 \, dA(z) \\ &\lesssim \int_{\mathbb{D}} \frac{1}{(1-|z|^2)^2} \int_{\mathbb{D}} |(f\circ\varphi_z)(w) - f(z)|^2 (1-|w|^2)^{2+\sigma} dA(w) dA(z). \\ &\lesssim \int_{\mathbb{D}} \frac{1}{(1-|z|^2)^2} \int_{\mathbb{D}} |f(\zeta) - f(z)|^2 |\varphi_z'(\zeta)|^2 \left(1 - |\varphi_z(\zeta)|^2\right)^{2+\sigma} dA(\zeta) dA(z) \\ &= \int_{\mathbb{D}} \int_{\mathbb{D}} |f(\zeta) - f(z)|^2 \frac{(1-|z|^2)^\sigma (1-|\zeta|^2)^\sigma}{|1-\overline{z}\zeta|^{4+2\sigma}} \frac{(1-|z|^2)^2(1-|\zeta|^2)^2}{|1-\overline{z}\zeta|^4} dA(\zeta) dA(z) \\ &\lesssim \int_{\mathbb{D}} \int_{\mathbb{D}} |f(\zeta) - f(z)|^2 \frac{(1-|z|^2)^\sigma (1-|\zeta|^2)^\sigma}{|1-\overline{z}\zeta|^{4+2\sigma}} dA(\zeta) dA(z), \end{aligned}$$

where we used (10.8) to go to the last line. This completes the proof. □

Notice that in the proof we used automorphisms to go to and from an analytically more convenient formulation in which a distinguished point was relocated at the origin.

10.2.1. Spreading of Carleson Measures. Recall that a Carleson measure for $\mathcal{D}$ is a positive measure μ on $\mathbb{D}$ such that for all f in $\mathcal{D}$

$$\int_{\mathbb{D}} |f|^2 \, d\mu \lesssim \|f\|_{\mathcal{D}}^2 .$$

Because the integrand has some intrinsic smoothness it is perhaps not surprising that if μ is a Carleson measure then an appropriately smeared version of μ will also be a Carleson measure. The following proposition is a particular instance of that, another is in [**77**, Lemma 2.4]. We will see multivariate versions of the same idea later, for example in Lemma 15.14.

PROPOSITION 10.2. *Given* $\beta > -1$, $b > \max\{1, (\beta+3)/2\}$, *set*

$$L_\beta^2 := L^2(\mathbb{D}, (1-|z|^2)^\beta dA(z)).$$

Consider the linear operator

$$Tf(w) := \int_{\mathbb{D}} f(z) \frac{(1-|z|^2)^{b-2}}{|1-\overline{z}w|^b} dA(z).$$

If $v \in L_\beta^2$ *is a function such that the measure* $d\mu_v = |v(z)|^2 (1-|z|^2)^\beta dA(z)$ *is a Carleson measure for* $\mathcal{D}$, *then the measure* $d\mu_{Tv} = |Tv(z)|^2 (1-|z|^2)^\beta dA(z)$ *is also a Carleson measure for* $\mathcal{D}$.

PROOF. The measure $d\mu_v$ is a Carleson measure for $\mathcal{D}$ exactly if the multiplication operator M_v defined by $M_v f = vf$ is a bounded map from $\mathcal{D}$ into L_β^2. Also, we know from Theorem B.6 that T is a bounded map of L_β^2 to itself. Hence, to show M_{Tv} is bounded it suffices to show that the difference $M_{Tv} - TM_v$ is bounded. We compute

$$\begin{aligned} |(M_{Tz} - TM_v)\, g(w)|^2 &= \left| \int_{\mathbb{D}} v(z) \left(g(w) - g(z)\right) \frac{(1-|z|^2)^{b-2}}{|1-\overline{z}w|^b} dA(z) \right|^2 \\ &\lesssim \|v\|_{L_\beta^2}^2 \int_{\mathbb{D}} |(g(w) - g(z))|^2 \, \frac{(1-|z|^2)^{2b-4-\beta}}{|1-\overline{z}w|^{2b}} dA(z). \end{aligned}$$

Hence,

$$\begin{aligned} &\|(M_{Tz} - TM_v)\, g\|_{L_\beta^2}^2 \\ &\lesssim \|v\|_{L_\beta^2}^2 \int_{\mathbb{D}} \int_{\mathbb{D}} |(g(w) - g(z))|^2 \, \frac{(1-|z|^2)^{2b-4-\beta}}{|1-\overline{z}w|^{2b}} (1-|w|^2)^\beta dA(z) dA(w) \\ &\lesssim \|g\|_{\mathcal{D}}^2 . \end{aligned}$$

For the final inequality we used the previous theorem. □

10.2.2. Hankel Forms.

10.2.2.1. *The Hardy Space.* Several different operators on the Hardy space are called Hankel operators and the associated theories are roughly equivalent. Here we take a quick look at three of them.

We view H^2 as a space of boundary values, a closed subspace of $L^2(\mathbb{T})$. Let P be the orthogonal projection from L^2 to H^2 and $P^\perp = I - P$ the complementary projection. Let $\overline{H^2}$ be the space of complex conjugates of functions in H^2 and let

$\overline{P}$ be its projection. Suppose $b \in L^2$. We begin with densely defined operations and then consider conditions on b which ensure the operators are bounded.

We define the bilinear Hankel form $H_b : H^2 \times H^2 \to \mathbb{C}$ by

$$H_b(f, g) := \langle fg, b\rangle_{H^2}. \tag{10.9}$$

Define the conjugate linear Hankel operator $\mathfrak{h}_b : H^2 \to H^2$ by

$$\mathfrak{h}_b(f) := P(b\overline{f}). \tag{10.10}$$

Define the linear Hankel operator $h_b : H^2 \to (H^2)^\perp$ by

$$h_b(f) := P^\perp(\overline{b}f), \tag{10.11}$$

For example, set $b = k_\zeta$, the Hardy space reproducing kernel for ζ and let $j_\zeta(e^{i\theta}) = \zeta e^{-i\theta}/(1 - e^{-i\theta}) = \overline{k_\zeta - k_\zeta(0)}$ (which reproduces the values of conjugate holomorphic functions of mean zero). Then we have

$$\begin{aligned} H_b(f, g) &= f(\zeta)g(\zeta) \\ \mathfrak{h}_b(f) &= \overline{f(\zeta)}k_\zeta \\ h_b(f) &= f(\zeta)j_\zeta. \end{aligned}$$

In each of the three cases it is immediate that $b \in H^\infty$ ensures the operator is bounded. In fact, if b is bounded then the estimates to show the operator is bounded hold even if each function in the estimate is replaced by its modulus. However even weaker conditions on b also suffice to ensure boundedness and understanding those conditions is one of the goals. Before considering those details note that the same condition for boundedness holds in all three cases. For H_b to be bounded we would need to know that

$$\sup_{\|f\|_{H^2} = \|g\|_{H^2} = 1} |\langle fg, b\rangle_{H^2}| < \infty. \tag{10.12}$$

For the second, $\mathfrak{h}_b$, to be bounded we would need to know that

$$\sup_{\|f\|_{H^2} = \|g\|_{H^2} = 1} \left|\langle g, P(b\overline{f})\rangle_{H^2}\right| < \infty.$$

However, after noting that

$$\langle gf, b\rangle_{H^2} = \langle g, b\overline{f}\rangle_{H^2} \tag{10.13}$$

and that $P^*(g) = P(g) = g$ we see that those two conditions are the same. The other case is only slightly less mechanical.

EXERCISE 10.3. *The boundedness condition for h_b is the same as for the other two.*

EXERCISE 10.4. *The operators are bounded if $b \in L^\infty$.*

We can improve that last condition with the following observation: adding a conjugate holomorphic function of mean zero to b does not change any of the operators. Hence an improved sufficient condition for boundedness is that b is a holomorphic function and there is a conjugate holomorphic function d such that $h = b + d \in L^\infty$, or, equivalently, that there is $h \in L^\infty$ with $b = P(h)$. It is a classical result of Nehari [**239**] that this is the full story. We will come to that through an indirect argument which, along the way will exhibit a relationship between Hankel operators and Carleson measures.

A sufficient condition for the boundedness of, say, H_b is that b generates a Carleson measure for H^2 in the sense described in Chapter 2. That is, H_b will be bounded if

$$d\mu_b = d\mu_{b,\mathtt{Hardy}} = |b'(z)|^2 (1-|z|^2)dA \in CM(H^2). \tag{10.14}$$

To see this, pick f, g, unit vectors in H^2 having norm one. We want to establish an upper bound for

$$\langle fg, b\rangle_{H^2} = \int_{\mathbb{D}} (fg)' \overline{b}' (1-|z|^2)dA.$$

(Here we are using the pairing associated with (2.15).) Differentiating fg produces two similar terms, by symmetry it suffices to estimate one of them. We do that by

$$\begin{aligned}
&\left|\int_{\mathbb{D}} fg'\overline{b}'(1-|z|^2)dA\right|^2 \\
&\quad\leq \left(\int_{\mathbb{D}} |g'|^2 (1-|z|^2)dA\right)\left(\int_{\mathbb{D}} |f|^2 |b'|^2 (1-|z|^2)dA\right) \\
&\quad\lesssim \|g\|_{H^2}^2 \|f\|_{H^2}^2 .
\end{aligned}$$

The first inequality is by Cauchy-Schwarz, the second comes from the definition of the norm and the Carleson measure condition on μ_b.

We noted in Theorem 3.13 that this Carleson measure condition on the holomorphic function b was equivalent to having $b \in BMOA$ which, by Theorem 3.12 is $\left(H^1\right)^*$, the dual of H^1. To finish our argument and see that this condition on b is, in fact, necessary, we want to show that if the Hankel form is bounded then $b \in \left(H^1\right)^*$. For this we appeal to Proposition 3.4 which states that

$$\left\{fg : f, g \in H^2, \|f\|_{H^2} \leq 1, \|g\|_{H^2} \leq 1\right\} = \left\{h : h \in H^1, \|h\|_{H^1} \leq 1\right\}. \tag{10.15}$$

With this, knowing that (10.12) holds is equivalent to knowing that

$$\sup_{h\in H^1, \|h\|_{H^1}=1} \left|\int_{\mathbb{T}} h\bar{b}d\theta\right| < \infty,$$

which states that $b \in (H^1)^*$, which is what we wanted. Functional analysis can then be used to show this is equivalent to Nehari's result that H_b is bounded if and only if there is a conjugate holomorphic g such that $b + g$ is bounded.

10.2.2.2. *The Bergman Space, Big and Small Hankel Operators.* The operators we described were defined using products of functions. There is no natural notion of products for elements of an abstract Hilbert space H; but there is one if H is a RKHS. In that case analogs of the previous operators are again objects of study. We now look briefly at the case $H = A^2$, the Bergman space.

We generalize (10.9) and (10.10) to the Bergman space by making the natural changes in interpretation of the notation. That is, let $\langle \cdot, \cdot \rangle_{L^2(dA)}$ denote the inner product of $L^2(\mathbb{D}, dA)$ and let P be the Bergman projection, the self adjoint projection of L^2 onto A^2. Also, because we again have (10.13), the boundedness criteria for the two will be the same.

There are several, fundamentally different, ways to generalize the definition (10.11) to the Bergman space. The reason the operator in (10.11) fit well with the other two Hardy space operators is that the two spaces $\overline{H^2}$ and $\left(H^2\right)^\perp$, the space of complex conjugates of functions in H^2 and the orthogonal complement of

H^2, are almost the same space; the second has codimension one in the first. The analogous spaces in the Bergman context are $\overline{A^2}$, the space of complex conjugates of functions in A^2, and $\left(A^2\right)^{\perp}$, the orthogonal complement of A^2. These two spaces are not close to being the same, $\overline{zA^2}$ has infinite codimension in $\left(A^2\right)^{\perp}$. The map of the Bergman space into L^2 constructed using formula (10.11), with the projection $P_{A^{2\perp}} = I - P$ is studied, and so is the one built using the projection $P_{\overline{A^2}}$; sometimes each is given the name *Bergman space Hankel operator*. Because the space $\left(A^2\right)^{\perp}$ is a much bigger space than $\overline{A^2}$ the operator built using $P_{A^{2\perp}}$ is sometimes called the *big Hankel operator*, the other being called the *small or little Hankel operator*.

The boundedness criteria for the Bergman space bilinear form analogous to (10.9), for the operator analogous to (10.10), and for the Bergman space little Hankel operator are the same as each other and are similar to what we saw for H^2. It is automatic that having $b \in H^\infty$ is sufficient for boundedness and it is straightforward that having $b = P(h)$ with P the Bergman projection and $b \in L^\infty$ is also sufficient. The argument involving Carleson measures can also be repeated showing that for b holomorphic on the disk and $|b'|^2 (1 - |z|^2)^2 \in CM(A^2)$ then the operators will be bounded. Also as before, the condition obtained this way is also necessary for boundedness, but that requires additional analysis. Furthermore there are analogs of Theorems 3.12 and 3.13. The space of holomorphic functions b for which $|b'|^2 (1 - |z|^2)^2 \in CM(A^2)$ can be identified as the *Bloch space*, mentioned in (2.17), the space of holomorphic functions on the disk for which $\sup_{z\in\mathbb{D}}(1 - |z|^2)|b'(z)| < \infty$ is bounded. That same space can also be identified as the dual space of the space $A^1 := \mathrm{Hol}(\mathbb{D}) \cap L^1(\mathbb{D}, dA)$.

In sum, for holomorphic b the small Hankel operator on the Bergman space is bounded if and only if b is in the Bloch space equivalently, if and only if it generates a Carleson measure, if and only if it is the Bergman projection of a bounded function, if and only if it generates a bounded linear functional on A^1. All this is in [**348**].

A new issue shows up for Bergman space Hankel operators. For the Hardy space we used Proposition 3.4, an immediate consequence of the factorization theorem, to connect Hardy space Hankel operators with the dual of H^1. The facts about factorization of functions in the Bergman space A^1 are more complicated. There is a factorization result due to Horowitz [**180**] and there is a *weak factorization theorem* which represents a function not as a product but as a sum of products.

THEOREM 10.5 ([**118**]). *There is an absolute constant so that given* $f \in A^1$, $\|f\|_{A^1} = 1$, $\exists \{g_i\}, \{h_i\} \subset A^2$ *with*

$$f = \sum_{i=1}^{\infty} g_i h_i \text{ and } \sum_{i=1}^{\infty} \|g_i\|_{A^2} \|h_i\|_{A^2} \lesssim 1.$$

The analog of that theorem holds for the Hardy and Bergman spaces of the ball in $\mathbb{C}^n$, [**118**].

The next exercise shows that the weak factorization result meshes well with the description of Hankel forms. That relationship is discussed a bit more in Section 13.3.1.1.

EXERCISE 10.6. *Show that, for* $b \in \mathrm{Hol}(\mathbb{D})$ *the bilinear form on* A^2 *given by the analog of* (10.9) *is bounded if and only if* $b \in \left(A^1\right)^*$.

10.2.2.3. *The Dirichlet Space.* To define a Hankel form for the Dirichlet space in the style of (10.9) we would want a form $B(f,g)$ that is a linear function of the product fg. Formally, for a given b, we set

$$B_b(f,g) = \langle fg, b\rangle_{\mathcal{D}}, \tag{10.16}$$

and we suppose for now that b is holomorphic. Setting $f = 1$ we see that if this form is bounded then b must be in $\mathcal{D}$. In particular $b \in H^\infty$ is not sufficient to ensure that the form is bounded. However we again have that if b generates a Carleson measure, in this case $|b'|^2\, dA \in CM(\mathcal{D})$, then $(f,g) \to \langle fg, b\rangle_{\mathcal{D}}$ is a bounded map $\mathcal{D}\times\mathcal{D} \to \mathbb{C}$. The proof is the same as for the analogous fact in the Hardy space. It is also true in this case, as for the Hardy space and Bergman space, that the condition is necessary. However that is not straightforward; we prove it in Chapter 13.

Alternatively we can look for an analog of the operator (10.10) or its associated bilinear form. We would want a conjugate linear map $f \to P(b\overline{f})$ of the Dirichlet space to itself. We do not yet have a definition for the projection P, but if we did then the boundedness of that operator would be equivalent to the boundedness of the bilinear form

$$\tilde{B}_b(g,f) = \left\langle g, P(b\overline{f})\right\rangle_{\mathcal{D}}. \tag{10.17}$$

Furthermore, if our projection were self adjoint we would have

$$\tilde{B}_b(g,f) = \left\langle g, b\overline{f}\right\rangle_{\mathcal{D}}.$$

which is formally the same as (10.13). Now there are questions. What is the space on which the projection P acts? What is P? How do we interpret the previous inner product which involves functions which are not holomorphic?

The Hardy and Bergman space each have a natural enveloping Lebesgue space and that gives a natural choice for the projections. That is not true for the Dirichlet space and hence the identification of the appropriate projection is not straightforward.

EXERCISE 10.7. *Show that the Dirichlet space inner product is not the restriction to the Dirichlet space of a Lebesgue space inner product. Hint: Use Corollary* 11.7 *in Chapter* 11.

One standard choice for a Hilbert space containing $\mathcal{D}$ is the space $W^{1,2} = W^{1,2}(\mathbb{D})$. The notation refers to the fact that the functions have one derivative in $L^2(\mathbb{D})$. It is the Hilbert space formed by the completion of the pre-Hilbert space of smooth functions on the closed disk $\overline{\mathbb{D}}$ with the inner product given by

$$\langle F, G\rangle_{W^{1,2}} = F(0)\overline{G(0)} + \int_{\mathbb{D}} \partial F\overline{\partial G} + \overline{\partial}F\overline{\overline{\partial}G}\, dA.$$

There is a natural inclusion of $\mathcal{D}$ into $W^{1,2}$ as a closed subspace. We will not need much information about $W^{1,2}$ but do note that $\overline{\mathcal{D}}_0$, the space of complex conjugates of functions in $\mathcal{D}$ which have mean zero is another closed subspace and is orthogonal to $\mathcal{D}$. The sum, $\overline{\mathcal{D}}_0 \oplus \mathcal{D}$ consists of all harmonic functions h on the disk with finite Dirichlet integral, $\int_{\mathbb{D}} |\nabla h|^2\, dA < \infty$, a space sometimes called the Dirichlet space and sometimes called the *harmonic Dirichlet space.* The orthogonal projection $\mathcal{P}$ of $W^{1,2}$ onto $\mathcal{D}$ is the projection we will use, it is realized by the formula

$$\mathcal{P}F(\zeta) := \langle F, k_\zeta\rangle_{W^{1,2}},$$

where k_ζ is the Dirichlet space reproducing kernel for evaluating at a point $\zeta \in \mathbb{D}$.

We now proceed informally to obtain a bilinear form in the spirit of (10.17) which uses the projection $\mathcal{P}$. We know $k_\zeta(z)$ is holomorphic in z and so $\overline{\partial} k_\zeta = 0$. Hence,

$$\mathcal{P}F(\zeta) = \left(\int_{\mathbb{D}} F dA\right) + \int_{\mathbb{D}} \partial F \overline{\partial k_\zeta} dA.$$

Being informal we drop the constant term in the previous formula. We start with $f \in \mathcal{D}$ and a symbol function $b \in \operatorname{Hol}(\mathbb{D})$ and form the operator

$$\mathcal{P}\left(b\overline{f}\right)(\zeta) = \int_{\mathbb{D}} b' \overline{f} \overline{\partial k_\zeta} dA.$$

The bilinear form we want to consider is

$$\begin{aligned}
\tilde{B}_b(g,f) &= \left\langle g, \mathcal{P}\left(b\overline{f}\right)\right\rangle_{\mathcal{D}} = \left\langle g', \mathcal{P}\left(b\overline{f}\right)'\right\rangle_{A^2} \\
&= \int_{\mathbb{D}} g' \overline{\left(\partial_t \mathcal{P}\left(b\overline{f}\right)\right)} dA(t) \\
&= \int_{\mathbb{D}} g'(t) \overline{\left(\partial_t \int_{\mathbb{D}} b'(w)\overline{f}(w)\overline{\partial_w k_t(w)} dA(w)\right)} dA(t) \\
&= \int_{\mathbb{D}} g'(t) \overline{\left(\int_{\mathbb{D}} b'(w)\overline{f}(w)\overline{\overline{\partial}_t \partial_w k_t(w)} dA(w)\right)} dA(t).
\end{aligned}$$

Now we simplify this expression. We will use the kernel function given in Exercise 2.1; other choices give more awkward formulas but, at the end, an equivalent result. Using that kernel function a straightforward computation shows that

$$\overline{\partial}_t \partial_w k_t(w) = \frac{c}{\left(1 - \overline{t}w\right)^2},$$

(a scalar multiple of) the Bergman space reproducing kernel for evaluating at t. We have

$$\int_{\mathbb{D}} g'(t) \frac{1}{\left(1 - \overline{t}w\right)^2} dA(t) = g'(w).$$

Ignoring c we have

$$\begin{aligned}
\tilde{B}_b(g,f) &:= \int_{\mathbb{D}} g'(w) \overline{b'(w)\overline{f}(w)} dA(w) \\
&= \left\langle g', b'\overline{f}\right\rangle_{A^2} \\
&= \left\langle fg', b'\right\rangle_{A^2}. \qquad (10.18)
\end{aligned}$$

After this informal preamble we define the asymmetric Hankel form $\tilde{B}_b(f,g)$ by that last line,

$$\tilde{B}_b(g,f) = \left\langle fg', b'\right\rangle_{A^2},$$

and we will study the properties of $\tilde{B}_b$. Note that

$$B_b(f,g) = \tilde{B}_b(f,g) + \tilde{B}_b(g,f). \qquad (10.19)$$

and hence some types of information can be passed easily from $\tilde{B}_b$ to B_b. For instance, any condition on b which ensures $\tilde{B}_b$ is bounded will also ensure that B_b is bounded.

In [**111**] a bilinear form B on a RKHS H of functions on the disk is called a Hankel form of order 0 if the bilinear form ΓB defined by $\Gamma B(f,g) = B(zf,g) -$

$B(f, zg)$ is the zero form. The forms of higher order are then defined inductively, a bilinear form B is a *Hankel form of order* $n+1$ if ΓB is a Hankel form of order n.

EXERCISE 10.8. *$\tilde{B}_b$ is a Hankel form of order one; B_b is a Hankel form of order zero.*

10.2.2.4. *The Asymmetric Dirichlet Hankel Form.* One reason the function space $BMOA$ plays an important role in the theory of the Hardy space is the close relation between $BMOA$ functions and Carleson measures for H^2 which was discussed in Chapter 2. Given holomorphic b we have a Carleson measure condition for its membership in $BMOA$;

$$\|b\|_{BMO} \approx \left\| |b'|^2 (1-|z|^2) dA \right\|_{CM(H^2)}.$$

There is a similar relationship between the Bloch space mentioned is Section 10.2.2.2 and Bergman space Carleson measures,

$$\|b\|_{Bloch} \approx \left\| |b'|^2 (1-|z|^2)^2 dA \right\|_{CM(A^2)}.$$

At (2.31) we introduced a Dirichlet space analog, saying, now by definition, that a holomorphic b is in $\mathcal{X}$ exactly if an associated measure is in $CM(\mathcal{D})$;

$$\|b\|_{\mathcal{X}} \approx \left\| |b'|^2 \, dA \right\|_{CM(\mathcal{D})}.$$

Given these analogies it is natural to suspect that $b \in \mathcal{X}$ is the condition which characterizes the boundedness of $\tilde{B}_b$. We show that here, following the argument from Rochberg-Wu [**284**]. Related results were obtained earlier by Coifman and Murai [**116**].

THEOREM 10.9. *Suppose b is holomorphic in the disk. The bilinear form $\tilde{B}_b$ is bounded on $\mathcal{D} \times \mathcal{D}$ if and only if $b \in \mathcal{X}$. Equivalently, with P the Bergman projection, the mapping*

$$f \to \mathfrak{P}_b f(w) := P(b'\overline{f}) = \overline{\int_{\mathbb{D}} \frac{b'(z)\overline{f(z)}}{(1-\overline{z}w)^2} dA(z)}$$

is a bounded map from $\mathcal{D}$ into $L^2(\mathbb{D}, dA)$ if and only if $b \in \mathcal{X}$.

PROOF. We now need to show that $b \in \mathcal{X}$ is necessary for the boundedness of $\mathfrak{P}_b$. Supposing $\mathfrak{P}_b$ is bounded, we need to show that $\forall h \in \mathcal{D}$, $\|b'h\|_{L^2} \lesssim \|h\|_{\mathcal{D}}$. By an elementary reduction we suppose $h(0) = 0$, after which we write

$$b'(w)\overline{h(w)} - \overline{\mathfrak{P}_b h(w)} = \int_{\mathbb{D}} \frac{b'(z)(\overline{h(w)} - \overline{h(z)})}{(1-\overline{z}w)^2} dA(z).$$

We want to estimate the first term on the left. We are assuming the second expression on the left is bounded, hence we can obtain $\|b'\overline{h}\|_{L^2(\mathbb{D})} \lesssim \|h\|_{\mathcal{D}}$ by estimating the right hand side. By Cauchy-Schwarz

$$\left| \int_{\mathbb{D}} \frac{b'(z)(\overline{h(w)} - \overline{h(z)})}{(1-\overline{z}w)^2} dA(z) \right|^2 \leq \int_{\mathbb{D}} |b'(z)|^2 \, dA(z) \int_{\mathbb{D}} \frac{\left|\overline{h(w)} - \overline{h(z)}\right|^2}{|1-\overline{z}w|^4} dA(z).$$

Hence,

$$\begin{aligned}\left\|b'h-\overline{\mathfrak{P}_b h}\right\|^2 &\leq \|b\|_{\mathcal{D}}^2 \int_{\mathbb{D}}\int_{\mathbb{D}} \frac{\left|\overline{h(w)}-\overline{h(z)}\right|^2}{\left|1-\overline{z}w\right|^4} dA(z)dA(w)\\ &\lesssim \|b\|_{\mathcal{D}}^2 \|h\|_{\mathcal{D}}^2 .\end{aligned}$$

To go to the last line we used an equivalent norm for $\mathcal{D}$ from Theorem 10.1. Thus, recalling that $\mathcal{X} \subset \mathcal{D}$, and our assumption that $\mathfrak{P}_b$ is bounded we have

$$\left\|b'h\right\|^2 \lesssim \|b\|_{\mathcal{X}}^2 \|h\|_{\mathcal{D}}^2 ,$$

as required. □

EXERCISE 10.10.

(1) *Check that if $b \in \mathcal{X}$ then $\tilde{B}_b$ is bounded.*

(2) *Show the boundedness of $\tilde{B}_b$ is equivalent to the boundedness of $\mathfrak{P}_b$.*

A vectorial version of $\tilde{B}_b$ is studied in [**16**].

10.2.3. Decomposition Theorems. Informally, a *decomposition theorem* for a space of holomorphic functions on the disk, $\mathcal{S}$, is a set of building block functions, $\{b_i\}$, indexed by a "hyperbolic lattice" $\{z_i\} \subset \mathbb{D}$ (defined below), and a solid sequence space $\mathbf{S}$ (that is, if $\{a_i\} \in \mathbf{S}$ and $\forall i\ |b_i| \leq |a_i|$ then $\|\{b_i\}\|_{\mathbf{S}} \leq \|\{a_i\}\|_{\mathbf{S}}$), so that two conclusions hold. First, if $f \in \mathcal{S}$ then there is a sequence of scalars $\{\lambda_i\} \in \mathbf{S}$, with each λ_i determined mainly by behavior of f near z_i, so that $f = \sum_i \lambda_i b_i$. Second, if $g = \sum_i \mu_i b_i$ with $\{\mu_i\} \in \mathbf{S}$ then $g \in \mathcal{S}$. All of this with natural norm estimates.

The passage from the study of functions in $\mathcal{S}$ to analysis of sequences in $\mathbf{S}$ is not the same as passing from a function space to a tree model, but it is based on the same set of analogies. In some sense the two approaches are complementary. In Chapter 5 we reduced questions such as describing Dirichlet space Carleson measures to questions about functions on trees. In the other direction, the formula $g = \sum_i \mu_i b_i$ together with careful choices for $\{\mu_i\}$ (which can be viewed as a function on a tree) allows construction of functions with desired properties. We saw that in Chapter 8 where we constructed functions for onto interpolation and will use that approach again in Chapter 13 where we construct functions for interpolation; functions which are large in prescribed regions and small away from those regions.

In this section we will state a decomposition theorem for $\mathcal{D}$ and outline the proof of a similar theorem for $\mathcal{X}$. That proof includes an application of the earlier result about spreading of Carleson measures. We will be a bit informal. More detail is in [**117**], [**282**], [**348**, Theorem 9.6.2], and [**284**, Theorem 3'].

One pattern for proving decomposition theorems is to start with a representation of f by an integral formula, for instance the Bergman reproducing formula, (2.20) if $f \in A^2$ then

$$f(w) = \frac{1}{\pi}\int_{\mathbb{D}} f(z)\frac{1}{(1-\overline{z}w)^2} dA. \tag{10.20}$$

We now form a Riemann sum for the integral based on an (approximate) hyperbolic lattice. Given small positive d, we will say that a set $Z = \{z_i\} \subset \mathbb{D}$ is a $d-$lattice if it is separated and it leaves no large gaps. More precisely,

$$\forall z_i, z_j \in Z,\ z_i \neq z_j,\ \delta_{H^2}(z_i, z_j) > d$$
$$\forall \zeta \in \mathbb{D}\ \exists z_i \in Z,\ \delta_{H^2}(\zeta, z_i) < 10d. \tag{10.21}$$

Associated with Z we can find a decomposition of the disk of the type described in Section 5.2.2. We have $\mathbb{D} = \bigcup D_i$ with D_i smooth boundaries and disjoint interiors, and $B_\delta(z_i, d/20) \subset D_i \subset B_\delta(z_i, 20d)$. This ensures that $|D_i|$, the Lebesgue measure of D_i, satisfies $|D_i| \approx (1 - |z_i|^2)^2$. The approximation operator A defined by

$$\begin{aligned} Af(w) &:= \sum_i f(z_i)\frac{1}{(1-\overline{z}_i w)^2}\,|D_i| \\ &= \sum_i \frac{|D_i|\, f(z_i)}{(1-|z_i|^2)}\frac{(1-|z_i|^2)}{(1-\overline{z}_i w)^2}; \end{aligned} \tag{10.22}$$

Af is a Riemann sum approximation for the integral representation of f in (10.20). The numbers $\{\Lambda_i\} = \left\{(1-|z_i|^2)^{-1}\,|D_i|\, f(z_i)\right\}$ are the first approximation to $\{\lambda_i\}$ and $\|\{\Lambda_i\}\|_{\ell^2}$ is a Riemann sum approximation to $\|f\|$. Also note that the functions $(1-|z_i|^2)(1-\overline{z}_i w)^{-2}$ are unit vectors in the Bergman space.

To obtain good estimates for the accuracy of this approximation, crucial use is made of the fact that the integral operator T defined by

$$Tf(w) = \frac{1}{\pi}\int_{\mathbb{D}} f(z)\frac{1}{|1-\overline{z}w|^2}\,dA$$

is bounded; Theorem B.6. Using T one proves

LEMMA 10.11. *With d defined by* (10.21) *and the approximation operator A given by* (10.22) *we have the pointwise estimate*

$$|f(w) - Af(w)| \lesssim dTf(w)$$

In particular we can make the approximation very good by choosing d very small. Once it is known that the approximation is accurate then iterating the approximation scheme gives the desired decomposition.

The boundedness of T also provides the crucial estimates for showing that the function g has the desired properties.

These ideas can be combined to prove a decomposition theorem for functions in the Bergman space. We state the result for the Dirichlet space but differentiation of (10.23) will recapture the Bergman space formulation.

THEOREM 10.12. *There is $d_0 > 0$ such that given a $d-$lattice $Z = \{z_i\} \subset \mathbb{D}$ with $d < d_0$*

(1) *Given $f \in \mathcal{D}$ we can write*

$$f(z) = \sum_{j=1}^{\infty} a_j \frac{(1-|z_j|^2)}{(1-\overline{z}_j z)} \tag{10.23}$$

with $\sum_{j=1}^{\infty} |a_j|^2 \lesssim \|f\|_{\mathcal{D}}^2$.

(2) *If $\sum_{j=1}^{\infty} |a_j|^2 < \infty$ then $f(z)$ given by* (10.23) *is in $\mathcal{D}$ and $\|f\|_{\mathcal{D}} \lesssim \|a\|_{\ell^2}$.*

A general result which includes this as a special case is proved in [**348**, Theorem 9.6.2].

We now outline the proof of a decomposition theorem for $\mathcal{X}$. That theorem is an evolved version of the previous theorem and is analogous to the decomposition theorem for $BMOA$ in [**283**].

THEOREM 10.13. *There is a $d_0 > 0$ so that given a d lattice $Z = \{z_i\} \subset \mathbb{D}$ with $d < d_0$:*

(1) *Given $b \in \mathcal{X}$ we can write*

$$b(z) := \sum_{j=1}^{\infty} a_j \frac{(1 - |z_j|^2)}{(1 - \overline{z}_j z)} \tag{10.24}$$

with

$$\left\| \sum_{j=1}^{\infty} |a_j|^2 \delta_{z_j} \right\|_{CM(\mathcal{D})} \lesssim \|b\|_{\mathcal{X}}^2$$

(2) *If*

$$\left\| \sum_{j=1}^{\infty} |a_j|^2 \delta_{z_j} \right\|_{CM(\mathcal{D})} < \infty \tag{10.25}$$

then $b(z)$ given by (10.24) is in $\mathcal{X}$ and $\|b\|_{\mathcal{X}} \lesssim \left\| \sum_{j=1}^{\infty} |a_j|^2 \delta_{z_j} \right\|_{CM(\mathcal{D})}$.

PROOF. We start with the second statement. By the Theorem 13.1, to show $b \in \mathcal{X}$ it suffices to show that the antilinear map $f \to \mathcal{P}(b\overline{f})$ takes $\mathcal{D}$ to $\mathcal{D}$. We represent b as a sum using the previous theorem. We compute

$$\begin{aligned}
\frac{\partial}{\partial w}\mathcal{P}(b\overline{f})(w) &= \int_{\mathbb{D}} \frac{b'(z)\overline{f(z)}}{(1-\overline{z}w)^2} dA(z) \\
&= \sum_j a_j \overline{z}_j (1-|z|^2) \int \frac{1}{(1-\overline{z}_j z)^2} \overline{\frac{f(z)}{(1-z\overline{w})^2}} dA(z) \\
&= \sum_j a_j \overline{z}_j (1-|z|^2) \frac{\overline{f(z_j)}}{(1-\overline{z}_j w)^2}.
\end{aligned}$$

Here we passed to the third line after noting that the integral in the second line was an A^2 inner product involving the A^2 reproducing kernel. We want this to be in A^2 with appropriate estimates. We now use the Bergman space version of Theorem 10.12, the result obtained from that theorem by termwise differentiation. We see that our function of interest is in the Bergman space if $\sum_j |a_j|^2 |f(z_j)|^2 < \infty$. That estimate is exactly what is ensured by (10.25).

For the first part, suppose we are given $g \in \mathcal{D}$. The assumption $b \in \mathcal{X}$ ensures that $b'g \in A^2$, that is $\int_{\mathbb{D}} |b'gdA|^2 < \infty$. We estimate that integral using a Riemann sum associated to the partition $\mathbb{D} = \bigcup D_i$. Using that estimate we see that

$$\left\{ b'(z_j)g(z_j) \left(1 - |z_i^2|\right) \right\} \in \ell^2.$$

Having this for arbitrary $g \in \mathcal{D}$ is exactly the statement that

$$d\mu := \sum_i |b'(z_i)|^2 \left(1 - |z_i^2|\right)^{-1} |D_i| \, \delta_{z_i}.$$

is a $\mathcal{D}$-Carleson measure. We have

$$\begin{aligned}(Ab')(z) &= \sum_i |b'(z_i)| \left(1-|z_i^2|\right)^{-1} |D_i| \frac{(1-|z_i|^2)}{(1-\overline{z}_i z)} \\ &= \sum_i |b'(z_i)|\, |D_i| \frac{1}{(1-\overline{z}_i z)}.\end{aligned}$$

To finish we will show that A is invertible on $\mathcal{X}$; having that and then writing $b' = A(A^{-1}b')$ will be the required representation of b and complete the proof. We will establish the invertibility of A by showing that $I - A$ is a contraction on $\mathcal{X}$.

From Lemma 10.11 we have the pointwise estimate

$$|b'(z) - A(b')(z)| \lesssim dTb'(z).$$

Hence the norm of the $\mathcal{D}$-Carleson measure $|b' - A(b')|^2 \, dA$, which is the multiplier norm of the operator of multiplication by $b' - Ab'$ acting from $\mathcal{D}$ to A^2, is dominated by Cd times the Carleson measure norm of $|Tb'|^2 \, dA$. We know from the earlier result on spreading of Carleson measures, Proposition 10.2, that this last Carleson measure norm is dominated by $C \|b\|_{\mathcal{X}}$. Hence taking d sufficiently small we ensure that $I - A$ is a contraction on $\mathcal{X}$. □

10.3. The Local Dirichlet Integral

Functions in the Dirichlet space are in the Hardy space and hence can be reconstructed from their boundary values. Thus the full story of a Dirichlet space function is carried in its boundary values. In this section we develop tools which let us extract information from that boundary value function. This is a different focus than regarding $\mathcal{D}$ as a RKHS and it will lead to a complementary collection of results and insights.

Recall the factorization theorem, Theorem 3.2, that says that any $f \in H^2$ can be written, essentially uniquely, as $f = B_f S_f O_f$ where

- B_f is the Blaschke factor of f. It has zeros at $\{z_n\}_{n=1}^{\infty}$, the zero sequence of f, and has the representation

$$B_f(z) := \prod_{n=1}^{\infty} \frac{|z_n|}{z_n} \frac{z_n - z}{1 - \overline{z}_n z}, \tag{10.26}$$

  with the convention that $|z_n|/z_n = 1$ if $z_n = 0$.
- S_f is the singular inner function

$$S_f(z) := \exp\left\{ -\int_{\mathbb{T}} \frac{e^{i\theta}+z}{e^{i\theta}-z} d\mu_f(\theta) \right\} \tag{10.27}$$

  where μ_f is a non-negative measure on the circle which is singular with respect to Lebesgue measure, and
- setting $u_f(e^{i\theta}) = \log\left|f(e^{i\theta})\right|$, O_f is the outer function

$$O_f(z) := \exp\left\{ \frac{1}{2\pi} \int_{\mathbb{T}} \frac{e^{i\theta}+z}{e^{i\theta}-z} u_f(e^{i\theta}) d\theta \right\}. \tag{10.28}$$

Conversely, if f has such a representation and $O_f \in L^2(\mathbb{T})$ then $f \in H^2$ and $\|f\|_{H^2} = \|O_f\|_{L^2}$.

Using this result it is sometimes clear how to relate the behavior of $f \in H^2$ to the behavior of its factors. However that relation is often less transparent for $f \in \mathcal{D}$. Here is an example.

Using the factorization theorem the following proposition is easy.

PROPOSITION 10.14. *If $f \in H^2$, $f = B_f S_f O_f$ as above, then $f/S_f = B_f O_f \in H^2$ and*

$$\left\| \frac{f}{S_f} \right\|_{H^2} = \|f\|_{H^2}.$$

It is also true in the Dirichlet space that dividing out the singular inner factor keeps you in the space:

THEOREM 10.15. *If $f \in \mathcal{D}$, $f = B_f S_f O_f$ as above, then $f/S_f = B_f O_f \in \mathcal{D}$ and*

$$\left\| \frac{f}{S_f} \right\|_{\mathcal{D}} \leq \|f\|_{\mathcal{D}}.$$

But this second theorem is not a direct consequence of the factorization theorem and is less straightforward to prove. It is, however, an easy consequence of formula (10.30) below. We now develop formula (10.30) and also a localized version of that formula, and give some applications of those ideas. Those ideas will also play a major role in Chapter 12 when we study invariant subspaces of the Dirichlet shift.

Recall that the Dirichlet integral, $\mathfrak{D}(f) = \frac{1}{\pi}\int_{\mathbb{D}} |f'|^2 \, dA$ is a seminorm on $\mathcal{D}$. A conceptual starting point for studying the boundary values of $f \in \mathcal{D}$ is Douglas's formula which evaluates $\mathfrak{D}(f)$ using the boundary value function of f:

$$\mathfrak{D}(f) = \frac{1}{\pi}\int_{\mathbb{D}} |f'(z)|^2 \, dA = \int_{\mathbb{T}}\int_{\mathbb{T}} \left| \frac{f(e^{it}) - f(e^{is})}{e^{it} - e^{is}} \right|^2 \frac{dt}{2\pi}\frac{ds}{2\pi}. \tag{10.29}$$

We saw this double integral earlier, in (10.7). It is striking that this is an equality, not just an equivalence, and that fact was crucial in Douglas's work. However, for our uses, equivalence in (10.29) would suffice, both here and for many other formulas in this section.

In [**90**] Carleson gave a refined statement relating the boundary values of f to $\mathfrak{D}(f)$ through data from the factorization theorem. Suppose $f = B_f S_f O_f$ as above and, again, set $u = u_f = \log|f|$ on the boundary. Recall that $P_\zeta(e^{i\theta})$ is the Poisson kernel for evaluating a harmonic function at $\zeta \in \mathbb{D}$; $P_\zeta(e^{i\theta}) = \left(1 - |\zeta|^2\right)/|\zeta - e^{i\theta}|^2$. Carleson's formula is

$$\begin{aligned}\mathfrak{D}(f) = &\int_{\mathbb{T}} \sum_j P_{z_j}(e^{is}) \left|f(e^{is})\right|^2 \frac{ds}{2\pi} \\ &+ \int_{\mathbb{T}}\int_{\mathbb{T}} \frac{2}{|e^{it} - e^{is}|^2} d\mu_f(t) \left|f(e^{is})\right|^2 \frac{ds}{2\pi} \\ &+ \int_{\mathbb{T}}\int_{\mathbb{T}} \frac{(e^{2u(e^{it})} - e^{2u(e^{is})})(u(e^{it}) - u(e^{is}))}{|e^{it} - e^{is}|^2} \frac{dt}{2\pi}\frac{ds}{2\pi}.\end{aligned} \tag{10.30}$$

We now develop a more refined analysis of the boundary function and its relation to $\mathfrak{D}(f)$; an analyses which will have both (10.29) and (10.30) as simple corollaries. The form of Douglas's formula (10.29) suggests our approach. We will analyze $\mathfrak{D}(f)$ by regarding that double integral as a repeated integral and then studying the inner integral carefully. Specifically, we define the *local Dirichlet integral* of f at $\zeta \in \mathbb{T}$, $\mathfrak{D}_\zeta(f)$, by

$$\mathfrak{D}_\zeta(f) := \int_{\mathbb{T}} \left| \frac{f(e^{it}) - f(\zeta)}{e^{it} - \zeta} \right|^2 \frac{dt}{2\pi},$$

with $\mathfrak{D}_\zeta(f) = \infty$ as a possible value. This quantity was used by Stegenga in his study of Carleson measures for the Dirichlet space and we discuss his result in Section 10.3.3.2. For $\zeta \in \mathbb{T}$ he found $\mathfrak{D}_\zeta$ to be a useful measure of the boundary oscillation of f, with an emphasis on oscillation near ζ. In [**273**] and [**272**] Richter and Sundberg go much further, regarding $\mathfrak{D}_\zeta$ as a function on its own and developing it into a powerful tool. We will follow their presentation.

It is convenient to introduce two auxiliary functions.

$$\Phi(x,y) := \begin{cases} e^x - e^y - e^y(x-y) & x, y \in \mathbb{R} \\ e^x & x \in \mathbb{R},\ y = -\infty, \end{cases} \tag{10.31}$$

$$\phi(t) := \begin{cases} e^{2t} - 1 - 2t, & t \in \mathbb{R} \\ +\infty & t = \pm\infty. \end{cases} \tag{10.32}$$

It can be checked using calculus that ϕ is nonnegative and convex with minimum $\phi(0) = 0$. Then, noting that $\Phi(x,y) = e^y \phi\left((x-y)/2\right)$, one sees that Φ is nonnegative.

One basic result in [**273**] is a localized version of (10.30) which evaluates the local Dirichlet integral.

THEOREM 10.16. *Suppose $f \in H^2$, $f = B_f S_f O_f$ as above and set $u = u_f = \log|O_f| = \log|f|$ then*

$$\begin{aligned} \mathfrak{D}_\zeta(f) = &\sum_j P_{z_n}(\zeta)\,|O_f(\zeta)|^2 \\ &+ \left(\int_{\mathbb{T}} \frac{2}{|e^{it}-\zeta|^2} d\mu_f(t)\right) |O_f(\zeta)|^2 \\ &+ \int_{\mathbb{T}} \frac{\Phi(2u(e^{it}), 2u(\zeta))}{|e^{it}-\zeta|^2} \frac{dt}{2\pi}. \end{aligned} \tag{10.33}$$

10.3.1. Preliminary Analysis. We begin by developing some basic facts about $\mathfrak{D}_\zeta$.

Suppose $f \in H^2$. This ensures that f has *a.e.* boundary values on the circle and we will identify f with that boundary value function. Hence for that f and any $\varepsilon > 0$

$$\int_{|e^{it}-\zeta|>\varepsilon} \left| \frac{f(e^{it}) - f(\zeta)}{e^{it} - \zeta} \right|^2 \frac{dt}{2\pi} < \infty.$$

Thus the finiteness of $\mathfrak{D}_\zeta(f)$ is determined by the smoothness of f near ζ.

PROPOSITION 10.17. *For $f \in H^2$, $\zeta \in \mathbb{T}$, there is at most one complex number α such that*

$$\mathfrak{D}_\zeta(f) = \int_{\mathbb{T}} \left| \frac{f(e^{it}) - \alpha}{e^{it} - \zeta} \right|^2 \frac{dt}{2\pi} < \infty.$$

If there is such an α then

(1) $\lim_{r\to 1} f(r\zeta)=\alpha$, *thus, by definition of the boundary function, $f(\zeta)=\alpha$,*
(2) *For $k>0$, letting $\mathcal{O}_{\kappa,\zeta}$ be the oricycle*

$$\mathcal{O}_{\kappa,\zeta}:=\left\{z\in\mathbb{D}:|z-\zeta|^2<k(1-|z|^2)\right\}.$$

We have orocyclic convergence $\lim_{z\to\zeta,z\in\mathcal{O}_{\kappa,\zeta}} f(z)=f(\zeta)$, and
(3) *The Fourier series of f converges at ζ to the value $f(\zeta)$.*

PROOF. Certainly there is at most one such α. If there is such an α then the finiteness of the integral ensures that $\left(f(e^{it})-\alpha\right)/\left(e^{it}-\zeta\right)$ is the boundary value of an H^2 function which we will call $g(z)$. Thus

$$f=\alpha+(e^{it}-\zeta)g.$$

Hence

$$|f(z)-\alpha|^2=|(z-\zeta)\,g(z)|^2=\left(\frac{|z-\zeta|^2}{1-|z|^2}\right)(1-|z|^2)\,|g(z)|^2.$$

The assumption that $z\in\mathcal{O}_{\kappa,\zeta}$ ensures that the first factor on the right is bounded. By Proposition 2.10 the second factor tends to 0 as $|z|\to 1$. Thus the left hand side tends to zero as z approaches ζ from within $\mathcal{O}_{\kappa,\zeta}$. This is the second statement and the second implies the first, thus $f(\zeta)=\alpha$. $\square$

EXERCISE 10.18. *The third statement will not be used, it was included as background information. Prove it. First express the Taylor coefficients of f in terms of those of g. Use those to find compact expression for the partial sum of the series for f, then take a limit.* [**287**].

We now want to relate boundary integrals of $\mathfrak{D}_\zeta(f)$ to integrals of $|f'|^2$ on the disk. For a finite nonnegative Borel measure μ on $\mathbb{T}$ we let $\phi_\mu(z)$ be the harmonic function on the disk which has μ as its "boundary values." That is,

$$\phi_\mu(z)=\int_{\mathbb{T}}P_z(\zeta)d\mu(\zeta)=\int_{\mathbb{T}}\frac{1}{2\pi}\frac{1-|z|^2}{\left|1-\bar{\zeta}z\right|}d\mu(\zeta).$$

PROPOSITION 10.19. *For $f\in H^2$*

$$\int_{\mathbb{T}}\mathfrak{D}_\zeta(f)d\mu(\zeta)=\int_{\mathbb{D}}|f'(z)|^2\,\phi_\mu(z)\frac{dA}{\pi}. \tag{10.34}$$

PROOF. We will verify (10.34) for δ_ζ, the point mass at ζ. The general result then follows by integrating that special case with respect to $\mu(\zeta)$ and then using Fubini's theorem. By the obvious rotation invariance it suffices to consider $\zeta=1$. We now want to show that, for general $f\in H^2$,

$$\int_{\mathbb{T}}\left|\frac{f(e^{it})-f(1)}{e^{it}-1}\right|^2\frac{dt}{2\pi}=\int_{\mathbb{D}}|f'(z)|^2\frac{1-|z|^2}{|1-z|^2}\frac{dA}{\pi}. \tag{10.35}$$

More precisely, we want to show that if either side is finite then so is the other and the two are equal.

First we consider polynomial f. Multiplying out the quadratic expressions we see that it is enough to consider monomials and verify that for any nonnegative integers k,j we have

$$\int_{\mathbb{T}}\frac{e^{ikt}-1}{e^{it}-1}\overline{\frac{e^{ijt}-1}{e^{it}-1}}\frac{dt}{2\pi}=\int_{\mathbb{D}}\left(z^k\right)'\overline{\left(z^j\right)'}\frac{1-|z|^2}{|1-z|^2}\frac{dA}{\pi}.$$

We evaluate the two sides separately, first the left:

$$LHS = \frac{1}{2\pi}\int_{\mathbb{T}}\left(\sum_{r=0}^{k-1} e^{irt}\right)\overline{\left(\sum_{s=0}^{j-1} e^{isl}\right)}dt = \min(k,j).$$

For the other side we go to polar coordinates:

$$RHS = 2kj\int_0^1 r^{k+j-1}\left(\frac{1}{2\pi}\int_0^{2\pi} e^{i(k-j)t}\frac{1-|z|^2}{|1-z|^2}dt\right)dr.$$

We evaluate the inner integral by recalling that $\frac{1-|z|^2}{|1-z|^2} = \sum_n r^{|n|}e^{int}$ and we find

$$RHS = 2kj\int_0^1 r^{k+j-1}r^{|k-j|}dr = \frac{2kj}{k+j+|k-j|} = \min(k,j).$$

For general f, first suppose that the left side of (10.35) is finite. Thus $g(z) = (f(z)-f(1))/(z-1) \in H^2$. Select a sequence of polynomials $\{g_n\}$ which converge to g in H^2. The sequence of polynomials $\{f_n\}$ given by $f_n(z) = f(1)+(z-1)g_n(z)$ converges pointwise to f. Also, $\mathfrak{D}_1(f_n - f) = \|g_n - g\|_{H^2}^2 \to 0$. Hence $\mathfrak{D}_1(f_n) \to \mathfrak{D}_1(f)$.

Now consider the seminorm N defined by

$$N(h)^2 = \frac{1}{\pi}\int_{\mathbb{D}}\left|\frac{h'(z)}{1-z}\right|^2(1-|z|^2)dA.$$

The polynomial case of the proposition gives us that $N(f_n) = \mathfrak{D}_1(f_n)$ for all n. We need to show that $N(f_n) \to N(f)$. The polynomial case also ensures that $\{f_n\}$ is an N-Cauchy sequence. We noted that $f_n \to f$ pointwise. Combining those two gives $N(f_n) \to N(f)$; Proposition A.18.

Suppose now the right hand side of (10.35) is finite. It then follows from (2.15) that the function $h = \int_0^\zeta (f'/(1-z))dz$ is in H^2. Select $\{h_n\} \subset H^2$, a sequence of polynomials converging in norm to h and set

$$r_n(z) = \frac{\int_0^z (1-\zeta)h_n'(\zeta)d\zeta - \int_0^1 (1-\zeta)h_n'(\zeta)d\zeta}{z-1}.$$

By the polynomial version of (10.35)

$$\begin{aligned}\|r_n - r_m\|_{H^2}^2 &= \int_{\mathbb{T}}\left|r_n(e^{it}) - r_m(e^{it})\right|^2\frac{dt}{2\pi}\\ &= \int_{\mathbb{D}}|(h_n-h_m)'|^2(1-|z|^2)\frac{dA}{\pi} \lesssim \|h_n - h_m\|_{H^2}^2.\end{aligned}$$

Thus $\{r_n\}$ is a Cauchy sequence in H^2 and has some limit, r. Hence, in particular, there is pointwise convergence at the origin,

$$\int_0^1 (1-\zeta)h_n'(\zeta)d\zeta = r_n(0) \to r(0).$$

Now consider the equation

$$(z-1)r_n(z) = \int_0^z (1-\zeta)h_n'(\zeta)d\zeta - r(0) - [r_n(0) - r(0)].$$

As $n \to \infty$ the numbers in the square brackets tend to 0, and h_n converges uniformly to h on compact subsets. Thus, in the limit, the equation becomes

$$\begin{aligned}(z-1)\,r(z) &= \int_0^z (1-\zeta)\,h'(\zeta)d\zeta - r(0)\\ &= \int_0^z f'(\zeta)d\zeta - r(0)\\ &= f(z) - (f(0) + r(0)).\end{aligned}$$

Thus there is a constant α such that $(f-\alpha)\,/\,(z-1) \in H^2$. Hence, by the previous proposition the left side of (10.35) is finite. Once we know that then, by the previous part of this proof, it equals the right hand side □

10.3.2. Proof of the Theorem. We will prove Theorem 10.16 in three steps. The first step is to show that it suffices to prove the result separately for inner functions and for outer functions. We then prove the theorem in those two cases. We will use the conventions that $0 \cdot (\pm\infty) = 0$.

10.3.2.1. *Separating The Inner and Outer Factors.*

PROPOSITION 10.20. *Suppose $\zeta \in \mathbb{T}$, $f \in H^2$, and ϕ is an inner function. Then*

(1) *If $\mathcal{D}_\zeta(f) < \infty$ then*

$$\mathcal{D}_\zeta(\phi f) = \mathcal{D}_\zeta(\phi)\,|f(\zeta)|^2 + \mathcal{D}_\zeta(f). \tag{10.36}$$

(2) *If $\mathcal{D}_\zeta(f) = \infty$ then $\mathcal{D}_\zeta(\phi f) = \infty$.*

In particular, if $f(\zeta) = 0$ then $\mathcal{D}_\zeta(\phi f) = \mathcal{D}_\zeta(f)$ even if $\mathcal{D}_\zeta(\phi) = \infty$.

LEMMA 10.21. *Suppose $\lambda \in \mathbb{D}$, $f \in H^2$, then, writing $z \in \mathbb{T}$ as $z = e^{i\theta}$,*

$$\begin{aligned}\left\|\frac{f(z)-f(\lambda)}{z-\lambda}\right\|_{H^2}^2 &= \int_{\mathbb{T}} \frac{\left|f(e^{i\theta})\right|^2}{\left|1-\overline{\lambda}e^{i\theta}\right|^2}\frac{d\theta}{2\pi} - \frac{|f(\lambda)|^2}{1-|\lambda|^2}\\ &= \int_{\mathbb{T}} \frac{\left|f(e^{i\theta})\right|^2 - |f(\lambda)|^2}{\left|1-\overline{\lambda}e^{i\theta}\right|^2}\frac{d\theta}{2\pi}.\end{aligned}$$

PROOF OF THE LEMMA. The singularity of $(f(z)-f(\lambda))\,/\,(z-\lambda)$ at λ is removable. We compute the H^2 norm by integration on $\mathbb{T}$. There we have $|f(z)-f(\lambda)|^2 = \left|f(e^{i\theta})\right|^2 - 2\operatorname{Re} f(e^{i\theta})\overline{f(\lambda)} + |f(\lambda)|^2$ and $|z-\lambda|^2 = \left|1-\overline{\lambda}e^{i\theta}\right|^2$. Using these formulas and recalling that integration against the Poisson kernel $P_\lambda(\theta) = (1-|\lambda|^2)/|1-\overline{\lambda}e^{i\theta}|^2$ evaluates harmonic functions at λ, we obtain the right hand side of the first line. Another evaluation gives the second line. □

Recall from (1.20) that for $\zeta \in \mathbb{T}$ and $\alpha > 0$ the nontangential convergence region $\Gamma_\alpha(\zeta)$ is

$$\Gamma_\alpha(\zeta) = \{\lambda \in \mathbb{D} : |\lambda - \zeta| < \alpha\,(1-|\lambda|)\};$$

and we say $f(\lambda)$ converges to L nontangentially if $f(\lambda) \to L$ as $\lambda \to \zeta$ with $\lambda \in \Gamma_\alpha(\zeta)$.

LEMMA 10.22. *Suppose $\zeta \in \mathbb{T}$, $f \in H^2$. For $\lambda \in \mathbb{D}$ set*

$$g_\lambda(z) = \frac{f(z)-f(\lambda)}{z-\lambda}.$$

(1) *If* $\mathcal{D}_\zeta(f) < \infty$ *then for each* $\alpha > 0$ *and* $\lambda \in \Gamma_\alpha(\zeta)$ *we have that* $\|g_\lambda\|_{H^2}^2 < (1+\alpha)^2 \mathcal{D}_\zeta(f)$ *and* $\|g_\lambda\|_{H^2}^2 \to \mathcal{D}_\zeta(f)$ *as* $\lambda \to \zeta$ *nontangentially.*
(2) *If there is a sequence* $\{\lambda_n\}$ *in* $\mathbb{D}$ *which converges unrestrictedly to* ζ *and for which* $\{\|g_{\lambda_n}\|_{H^2}\}$ *is bounded then* $\mathcal{D}_\zeta(f) < \infty$.

In particular $\|g_\lambda\|_{H^2}^2 \to \mathcal{D}_\zeta(f)$ *as* $\lambda \to \zeta$ *nontangentially even if* $\mathcal{D}_\zeta(f) = \infty$.

PROOF. Because $\mathcal{D}_\zeta(f) < \infty$ we know that $g(z) = (f(z) - f(\zeta))/(z-\zeta)$ is in H^2 and $\|g\|_{H^2}^2 = \mathcal{D}_\zeta(f)$. By doing the algebra we find

$$g_\lambda(z) = g(z) + (\lambda - \zeta)\frac{g(z) - g(\lambda)}{z - \lambda}. \tag{10.37}$$

Now note that starting with the previous lemma we have

$$\left\|\frac{g(z) - g(\lambda)}{z-\lambda}\right\|_{H^2}^2 = \int_{\mathbb{T}} \frac{\left|g(e^{i\theta})\right|^2}{\left|1 - \overline{\lambda}e^{i\theta}\right|^2}\frac{d\theta}{2\pi} - \frac{|g(\lambda)|^2}{1-|\lambda|^2} \leq \frac{1}{(1-|\lambda|)^2}\|g\|_{H^2}^2. \tag{10.38}$$

The inequality comes from dropping the negative term and estimating the denominator of the integrand. Starting with (10.37) and using the previous estimate we obtain, for $\lambda \in \Gamma_\alpha(\zeta)$,

$$\|g_\lambda\|_{H^2} \leq \left(1 + \frac{|\lambda - \zeta|}{1 - |\lambda|}\right)\|g\|_{H^2}^2 \leq (1+\alpha)\sqrt{\mathcal{D}_\zeta(f)},$$

which is the first claim. Furthermore, using (10.38) we have.

$$\|g_\lambda - g\|_{H^2}^2 = |\lambda - \zeta|^2 \left\|\frac{g(z) - g(\lambda)}{z - \lambda}\right\|_{H^2}^2 \leq \int_{\mathbb{T}} \frac{|\lambda - \zeta|^2}{\left|1 - \overline{\lambda}e^{i\theta}\right|^2}\left|g(e^{i\theta})\right|^2 \frac{d\theta}{2\pi}.$$

Thus, as $\lambda \to \zeta$, $\lambda \in \Gamma_\alpha(\zeta)$, the fraction in the integral is uniformly bounded and tends to zero for almost all θ. Hence, by dominated convergence, the left side tends to zero, finishing the first part of the proof.

For the second part, suppose we have a sequence $\{\lambda_n\}$ as described. The sequence $\{g_{\lambda_n}\}$ is bounded and hence has a subsequence which converges weakly to a function $g \in H^2$. Weak convergence implies pointwise convergence and by evaluating the pointwise limit at, say, $z = 0$ we find that there is an α such that $f(\lambda_n) \to \alpha$. Thus g must equal the pointwise limit; $g = (f - \alpha)/(z - \zeta)$. We know $g \in H^2$ and hence, by Proposition 10.17, $\mathcal{D}_\zeta(f) < \infty$. □

PROOF OF PROPOSITION 10.20. Fix $\lambda \in \mathbb{D}$. Recalling that $|\phi| = 1$ on $\mathbb{T}$ we compute

$$\begin{aligned}
\left\|\frac{(\phi f)(z) - (\phi f)(\lambda)}{z - \lambda}\right\|_{H^2}^2 &= \int_{\mathbb{T}} \frac{\left|(\phi f)(e^{i\theta})\right|^2 - |(\phi f)(\lambda)|^2}{\left|1 - \overline{\lambda}e^{i\theta}\right|^2}\frac{d\theta}{2\pi} \\
&= \int_{\mathbb{T}} \frac{\left|\phi(e^{i\theta})\right|^2 - |\phi(\lambda)|^2}{\left|1 - \overline{\lambda}e^{i\theta}\right|^2}\frac{d\theta}{2\pi}|f(\lambda)|^2 \\
&\quad + \int_{\mathbb{T}} \frac{\left|f(e^{i\theta})\right|^2 - |f(\lambda)|^2}{\left|1 - \overline{\lambda}e^{i\theta}\right|^2}\frac{d\theta}{2\pi}.
\end{aligned}$$

We now apply Lemma 10.21 to both integrals and find

$$\left\|\frac{\phi f - (\phi f)(\lambda)}{z - \lambda}\right\|_{H^2}^2 = \left\|\frac{\phi - \phi(\lambda)}{z - \lambda}\right\|_{H^2}^2 |f(\lambda)|^2 + \left\|\frac{f - f(\lambda)}{z - \lambda}\right\|_{H^2}^2.$$

The final comment in Lemma 10.22 shows that we can now let λ approach ζ nontangentially and obtain

$$\mathcal{D}_\zeta(\phi f) = \mathcal{D}_\zeta(\phi)\,|f(\zeta)|^2 + \mathcal{D}_\zeta(f)$$

as required. □

10.3.2.2. *Angular Derivatives and the Proof for Inner Functions.* With the previous proposition in hand, proving (10.33) is reduced to proving the formula separately for f inner and f outer.

PROOF. Suppose now that f is inner. In this case we use the classical theory of angular derivatives. If f is a holomorphic map of $\mathbb{D}$ into itself, $\zeta \in \mathbb{T}$, and there is an $\omega \in \mathbb{T}$ so that the nontangential limit

$$\lim_{\lambda\to\zeta \text{ nontangentially}} \frac{f(\lambda)-\omega}{\lambda-\zeta} = \alpha$$

exists, then f is said to have an *angular derivative* at ζ; $f'(\zeta) = \alpha$. It was shown classically, as part of the Julia-Carathéodory theorem, that this will happen if and only if we have the unrestricted (i.e. not necessarily nontangential) limit

$$\liminf_{\lambda\to\zeta} \frac{1-|f(\lambda)|}{1-|\lambda|} = d < \infty, \tag{10.39}$$

in which case $d = |\alpha| = |f'(\zeta)|$. See [**124**] for this formulation. □

A study of the angular derivative and related issues using the theory of RKHS is given in [**293**].

PROPOSITION 10.23. *If f is a holomorphic map of $\mathbb{D}$ into itself and if for some $\zeta \in \mathbb{T}$, f has a finite angular derivative $f'(\zeta)$ at ζ, then $\mathcal{D}_\zeta(f) \le |f'(\zeta)|$. Furthermore, if f is inner then $\mathcal{D}_\zeta(f) < \infty$ if and only if f has a finite angular derivative and, in that case, $\mathcal{D}_\zeta(f) = |f'(\zeta)|$.*

PROOF. Using first Lemma 10.21, then the fact that $|f| \le 1$ on $\mathbb{D}$, and then the formula for the Poisson kernel P_λ; we have

$$\begin{aligned}
\left\| \frac{f(z)-f(\lambda)}{z-\lambda} \right\|_{H^2}^2 &= \int_{\mathbb{T}} \frac{|f(e^{i\theta})|^2 - |f(\lambda)|^2}{|1-\overline{\lambda}e^{i\theta}|^2} \frac{d\theta}{2\pi}. \\
&\le \int_{\mathbb{T}} \frac{1-|f(\lambda)|^2}{|1-\overline{\lambda}e^{i\theta}|^2} \frac{d\theta}{2\pi} \\
&= \frac{1-|f(\lambda)|^2}{1-|\lambda|^2} \\
&= \frac{1-|f(\lambda)|}{1-|\lambda|} \frac{1+|f(\lambda)|}{1+|\lambda|}.
\end{aligned}$$

If f has a finite angular derivative at ζ then, taking note of (10.39), there must be a sequence $\{\lambda_n\} \to \zeta$ along which $|f(\lambda_n)| \to 1$. The final comment in Lemma 10.22 now gives the first statement. If, furthermore, f is inner then equality holds in the previous estimate and that gives the second statement. □

We now need information about the angular derivatives of inner functions. They were studied classically by M. Riesz [**278**] and Frostman [**157**]. A more modern treatment is given by Ahern and Clark in [**9**]. We obtain the two terms in Theorem 10.16 which correspond to the inner factor of f by combining the previous proposition with the following fact about angular derivatives which can be found in [**9**].

PROPOSITION 10.24. *Suppose that ϕ is the inner function $\phi = B_f S_f$ where the two factors are as in* (10.26) *and* (10.27). *For ζ in $\mathbb{T}$ we have*

$$|\phi'(\zeta)| = \sum_n P_{z_n}(\zeta) + \int_{\mathbb{T}} \frac{2}{|e^{it} - \zeta|^2} d\sigma(t).$$

10.3.2.3. *The Proof for Outer Functions.*

PROOF. It remains to verify (10.33) for outer functions. Suppose $f \in H^2$, $f \neq 0$ and f is outer. Thus $\log |f(\lambda)| = u(\lambda)$ is the harmonic function with nontangential boundary values $u(e^{i\theta}) = \log \left| f(e^{i\theta}) \right|$. For $|\lambda| \leq 1$ set

$$I(\lambda) = \int_{\mathbb{T}} \frac{e^{2u(e^{i\theta})} - e^{2u(\lambda)} - 2e^{2u(\lambda)}(u(e^{i\theta}) - u(\lambda))}{\left|1 - \overline{\lambda} e^{i\theta}\right|^2} \frac{d\theta}{2\pi}. \tag{10.40}$$

We need to show that for all $\zeta \in \mathbb{T}$, $I(\zeta) = \mathcal{D}_\zeta(f)$. We have

$$\begin{aligned} I(\lambda) &= \int_{\mathbb{T}} \frac{e^{2u(e^{i\theta})} - e^{2u(\lambda)}}{\left|1 - \overline{\lambda} e^{i\theta}\right|^2} \frac{d\theta}{2\pi} - 2e^{2u(\lambda)} \int_{\mathbb{T}} \frac{(u(e^{i\theta}) - u(\lambda))}{\left|1 - \overline{\lambda} e^{i\theta}\right|^2} \frac{d\theta}{2\pi} \\ &= \int_{\mathbb{T}} \frac{e^{2u(e^{i\theta})} - e^{2u(\lambda)}}{\left|1 - \overline{\lambda} e^{i\theta}\right|^2} \frac{d\theta}{2\pi} + 0 = \int_{\mathbb{T}} \frac{\left|f(e^{i\theta})\right|^2 - |f(\lambda)|^2}{\left|1 - \overline{\lambda} e^{i\theta}\right|^2} \frac{d\theta}{2\pi} \qquad (10.41) \\ &= \left\| \frac{f - f(\lambda)}{z - \lambda} \right\|_{H^2}^2. \end{aligned}$$

Here the first line is an algebraic rewriting, the passage to the second uses the fact that $\left|1 - \overline{\lambda} e^{i\theta}\right|^{-2}$ is a multiple of the Poisson kernel for evaluating at λ. The passage to the third line is by Lemma 10.21.

Our observation after (10.32) that $\Phi \geq 0$ ensures that $I(\lambda)$ is nonnegative, although possibly infinite. Hence, by Fatou's lemma and Lemma 10.22, if $u(\lambda) \to u(\zeta)$ as $\lambda \to \zeta$ nontangentially, then

$$I(\zeta) \leq \liminf I(\lambda) = \mathcal{D}_\zeta(f).$$

Also, if $\mathcal{D}_\zeta(f)$ is finite then $f(\lambda) \to f(\zeta)$ and hence $u(\lambda) \to u(\zeta)$. Thus, if $u(\lambda)$ does not approach $u(\zeta)$ then $\mathcal{D}_\zeta(f) = \infty$ and we still have our inequality.

We now need to establish the reverse inequality. That is trivial if $I(\zeta) = \infty$. To finish we show that if for some ζ, $I(\zeta) < \infty$ then $u(\lambda) \to u(\zeta)$ as $\lambda \to \zeta$ nontangentially and, furthermore, $\mathcal{D}_\zeta(f) \leq I(\zeta)$. If $u(\zeta) = -\infty$ then (10.40) can be simplified and we have

$$I(\zeta) = \int_{\mathbb{T}} \frac{e^{2u(e^{i\theta})}}{\left|1 - \overline{\zeta} e^{i\theta}\right|^2} \frac{d\theta}{2\pi} = \left\| \frac{f}{z - \zeta} \right\|_{H^2}^2 = \mathcal{D}_\zeta(f).$$

On the other hand if $u(\zeta)$ is finite we will see that $u(\lambda) \to u(\zeta)$ as $\lambda \to \zeta$ in $\Gamma_\alpha(\zeta)$. Recalling our auxiliary function ϕ from (10.32) we can write

$$I(\zeta) = e^{2u(\zeta)} \int_{\mathbb{T}} \frac{\phi\left(u(e^{i\theta}) - u(\zeta)\right)}{\left|1 - \overline{\zeta} e^{i\theta}\right|^2} \frac{d\theta}{2\pi}.$$

Also

$$\begin{aligned}
\phi\left(u(e^{i\theta}) - u(\zeta)\right) &= \phi\left(\int_{\mathbb{T}} P_\lambda\left(e^{i\theta}\right)\left(u(e^{i\theta}) - u(\zeta)\right)\frac{d\theta}{2\pi}\right) \\
&\leq \int_{\mathbb{T}} P_\lambda\left(e^{i\theta}\right)\phi\left[\left(u(e^{i\theta}) - u(\zeta)\right)\right]\frac{d\theta}{2\pi} \\
&= \left(1 - |\lambda|^2\right)\int_{\mathbb{T}} \frac{\phi\left[\left(u(e^{i\theta}) - u(\zeta)\right)\right]}{\left|1 - \overline{\zeta} e^{i\theta}\right|^2} \frac{\left|1 - \overline{\zeta} e^{i\theta}\right|^2}{\left|1 - \overline{\lambda} e^{i\theta}\right|^2}\frac{d\theta}{2\pi} \\
&\leq (1+\alpha)^2\left(1 - |\lambda|^2\right)\int_{\mathbb{T}} \frac{\phi\left[\left(u(e^{i\theta}) - u(\zeta)\right)\right]}{\left|1 - \overline{\zeta} e^{i\theta}\right|^2}\frac{d\theta}{2\pi}.
\end{aligned}$$

Here the passage from the first line to the second uses Jensen's inequality and the passage to the final line uses the definition of $\Gamma_\alpha(\zeta)$ and Lemma 10.22. Taking the limit as $\lambda \to \zeta$ now gives what we want.

Having established that $u(\lambda) \to u(\zeta)$, we now study $I(\zeta) - I(\lambda)$. Taking (10.41) into account we have the following expressions for $I(\zeta) - I(\lambda)$,

$$\begin{aligned}
I(\zeta) - I(\lambda) = &\int_{\mathbb{T}} \frac{e^{2u(e^{i\theta})} - e^{2u(\zeta)} - 2e^{2u(\zeta)}\left(u(e^{i\theta}) - u(\zeta)\right)}{\left|1 - \overline{\zeta} e^{i\theta}\right|^2}\frac{d\theta}{2\pi} \\
&- \int_{\mathbb{T}} \frac{e^{2u(e^{i\theta})} - e^{2u(\lambda)}}{\left|1 - \overline{\lambda} e^{i\theta}\right|^2}\frac{d\theta}{2\pi} \\
= &\int_{\mathbb{T}} \frac{e^{2u(\zeta)}\phi\left(u(e^{i\theta}) - u(\zeta)\right)}{\left|1 - \overline{\zeta} e^{i\theta}\right|^2}\left(1 - \frac{\left|1 - \overline{\zeta} e^{i\theta}\right|^2}{\left|1 - \overline{\lambda} e^{i\theta}\right|^2}\right)\frac{d\theta}{2\pi} \\
&+ \int_{\mathbb{T}} \frac{e^{2u(\zeta)}\phi\left(u(e^{i\theta}) - u(\zeta)\right)}{\left|1 - \overline{\lambda} e^{i\theta}\right|^2} - \frac{e^{2u(e^{i\theta})} - e^{2u(\lambda)}}{\left|1 - \overline{\lambda} e^{i\theta}\right|^2}\frac{d\theta}{2\pi} \\
= &\int_{\mathbb{T}} e^{2u(\zeta)}\phi\left(u(e^{i\theta}) - u(\zeta)\right)\left(1 - \frac{\left|1 - \overline{\zeta} e^{i\theta}\right|^2}{\left|1 - \overline{\lambda} e^{i\theta}\right|^2}\right)\frac{d\theta}{2\pi} \\
&+ \int_{\mathbb{T}} \frac{e^{2u(\zeta)}\phi\left(u(\lambda) - u(\zeta)\right)}{\left|1 - \overline{\lambda} e^{i\theta}\right|^2}\frac{d\theta}{2\pi}.
\end{aligned}$$

The first line used (10.41). The passage to the second line is algebraic. The passage to the third line uses the definition of ϕ, algebraic cancellation, and the fact that, because u is harmonic, $\int u(e^{i\theta})/\left|1 - \overline{\lambda} e^{i\theta}\right|^2 d\theta = \int u(\lambda)/\left|1 - \overline{\lambda} e^{i\theta}\right|^2 d\theta$.

By the dominated convergence theorem the first integral tends to zero as $\lambda \to \zeta$ nontangentially. The second integral has positive integrand. This gives an upper bound for $I(\lambda)$ for λ near ζ and hence, by Lemma 10.22, we know that $I(\lambda) \to \mathcal{D}_\zeta(f)$ as $\lambda \to \zeta$ nontangentially. Thus $\mathcal{D}_\zeta(f) \geq I(\zeta)$ and that inequality completes the proof of Theorem 10.16. □

10.3.3. Consequences.

10.3.3.1. *Approximations and Truncations.* The previous results and techniques can be used to establish majorization and approximation tools. We do some of that now and return to those themes in Chapter 12 where we consider the invariant subspaces of the Dirichlet shift.

Approximations: When studying $f \in \mathrm{Hol}(\mathbb{D})$ it is often convenient to be able to approximate f by functions that are better behaved, for instance by f_r where, for $0 \le r < 1$, $f_r(z) = f(rz)$. For example, if $f \in \mathcal{D}$ then $f_r \in \mathcal{D} \cap \mathrm{Hol}\left(\overline{\mathbb{D}}\right)$ and we have norm convergence $f_r \to f$ as $r \to 1$. We now develop tools to do approximation in the function spaces $\mathfrak{D}_\zeta$.

PROPOSITION 10.25. *Fix $\zeta \in \mathbb{T}$, $f \in H^2$, $r < 1$; then*

$$\left(1 - r^2\right)^2 \mathfrak{D}_\zeta(f_r) \le \|f\|_{H^2}^2 .$$

PROOF. Suppose $f(z) = \sum_{n=0}^\infty f_n z^n$. For ζ fixed, we want to compute the H^2 norm of the following function of z

$$\begin{aligned} \frac{f(rz) - f(r\zeta)}{z - \zeta} &= \sum_{n=0}^\infty f_n r^n \frac{z^n - \zeta^n}{z - \zeta} \\ &= \sum_{n=0}^\infty f_n r^n \sum_{k=0}^{n-1} z^k \zeta^{n-k-1} \\ &= \sum_{k=0}^\infty \left(\sum_{n=k+1}^\infty f_n r^n \zeta^{n-k-1} \right) z^k . \end{aligned}$$

We estimate the coefficient of z^k by

$$\begin{aligned} \left| \sum_{n=k+1}^\infty f_n r^n \zeta^{n-k-1} \right|^2 &\le \left(\sum_{n=k+1}^\infty |f_n|^2 \right) \sum_{n=k+1}^\infty r^{2n} \\ &\le \|f\|_{H^2}^2 \frac{r^{2k+2}}{1 - r^2} \end{aligned}$$

and thus

$$\begin{aligned} \mathfrak{D}_\zeta(f_r) &= \left\| \frac{f(rz) - f(r\zeta)}{z - \zeta} \right\|^2 \\ &\le \|f\|_{H^2}^2 \sum_{k=0}^\infty \frac{r^{2k+2}}{1 - r^2} \\ &= \|f\|_{H^2}^2 \frac{r^2}{\left(1 - r^2\right)^2}, \end{aligned}$$

which gives what we want. □

PROPOSITION 10.26. *Suppose $\mathfrak{D}_\zeta(f) < \infty$, $r < 1$, then $\mathfrak{D}_\zeta(f_r) \le 4\mathfrak{D}_\zeta(f)$.*

PROOF. That $\mathfrak{D}_\zeta(f) < \infty$ ensures there is $g \in H^2$ with $f(z) = f(\zeta) + (z - \zeta)\, g(z)$ and $\|g\|^2 = \mathfrak{D}_\zeta(f)$. Thus, for $0 < r < 1$

$$\mathfrak{D}_\zeta(f_r) = \left\| \frac{f(rz) - f(r\zeta)}{z - \zeta} \right\|_{\mathcal{D}}^2 = \left\| r g_r - \zeta(1 - r) \frac{g_r - g_r(\zeta)}{z - \zeta} \right\|_{\mathcal{D}}^2 .$$

Using the fact that $\|\alpha+\beta\|^2 \leq 2\|\alpha\|^2 + 2\|\beta\|^2$ and the previous proposition it is straightforward to complete the proof. □

PROPOSITION 10.27. *Suppose $\mathfrak{D}_\zeta(f) < \infty$ and $g \in H^\infty$. We have $\mathfrak{D}_\zeta(fg) < \infty$ if and only if $f(\zeta) = 0$ or $\mathfrak{D}_\zeta(g) < \infty$. Furthermore*

$$\mathfrak{D}_\zeta(fg) \leq 2\left(\|g\|_\infty^2 \mathfrak{D}_\zeta(f) + |f(\zeta)|^2 \mathfrak{D}_\zeta(g)\right), \text{ and}$$
$$|f(\zeta)|^2 \mathfrak{D}_\zeta(g) \leq 2\left(\|g\|_\infty^2 \mathfrak{D}_\zeta(f) + \mathfrak{D}_\zeta(fg)\right).$$

PROOF. Set

$$A(z) = \frac{fg(z) - fg(\zeta)}{z-\zeta} = g\frac{f - f(\zeta)}{z-\zeta} + f(\zeta)\frac{g-g(\zeta)}{z-\zeta}. \tag{10.42}$$

The hypotheses ensure that the first summand on the right is in H^2. Thus A is in H^2 if and only if the second term on the right is in H^2. That is a reformulation of the first statement to be proved.

To estimate $\mathfrak{D}_\zeta(fg) = \|A\|_{H^2}^2$ we use the right side of (10.42) and the natural estimates. If we then rewrite our equation $A = B + C$, as $C = A - B$ and use the same argument we obtain the desired estimate for $|f(\zeta)|^2 \mathfrak{D}_\zeta(g)$. □

PROPOSITION 10.28. *Suppose $f \in \mathcal{D}$, $g \in H^\infty$, and $gf \in \mathcal{D}$ Then there is a sequence $\{r_n\}$ converging to 1 for which $g_{r_n}f$ converges in norm to gf*

PROOF. Certainly $g_r f$ converges pointwise to gf. Hence, noting Proposition A.18 it suffices to show $\|g_r f\|_{\mathcal{D}}$ is uniformly bounded. That norm involves an H^2 norm and a boundary integral of local Dirichlet integrals. The H^2 term is easy to deal with. We now estimate the local Dirichlet integrals, $\mathfrak{D}_\zeta$. We will find $A, B > 0$ which may depend on f and g but not on ζ or r, so that

$$\mathfrak{D}_\zeta(g_r f) \leq A\mathfrak{D}_\zeta(f) + B\mathfrak{D}_\zeta(gf).$$

Integrating over the circle will then give the desired conclusion.

Using Proposition 10.27, then Proposition 10.26 and then Proposition 10.27 again we have

$$\begin{aligned}
\mathfrak{D}_\zeta(g_r f) &\leq 2\left(\|g_r\|_\infty^2 \mathfrak{D}_\zeta(f) + |f(\zeta)|^2 \mathfrak{D}_\zeta(g_r)\right)\\
&\leq 2\left(\|g_r\|_\infty^2 \mathfrak{D}_\zeta(f) + 4|f(\zeta)|^2 \mathfrak{D}_\zeta(g)\right)\\
&\leq 2\left(\|g_r\|_\infty^2 \mathfrak{D}_\zeta(f) + 8(\|g\|_\infty^2 \mathfrak{D}_\zeta(f) + \mathfrak{D}_\zeta(fg)\right)\\
&= A\mathfrak{D}_\zeta(f) + B\mathfrak{D}_\zeta(gf),
\end{aligned}$$

as required. □

Auxiliary Functions: We collect some computational properties of our auxiliary function Φ. Recall that Φ was defined in (10.31) as

$$\Phi(x,y) = \begin{cases} e^x - e^y - e^y(x-y) & x, y \in \mathbb{R} \\ e^x & x \in \mathbb{R},\ y = -\infty. \end{cases}$$

For real numbers a, b, set $a \vee b := \max\{a, b\}$ and $a \wedge b := \min\{a, b\}$.

PROPOSITION 10.29. *For* $x_1, x_2 \in \mathbb{R}$, $y_1, y_2 \in \mathbb{R} \cup \{\infty\}$,

$$\Phi(x_1 \vee x_2, y_1 \vee y_2) \leq \Phi(x_1, y_1) \vee \Phi(x_2, y_2), \tag{10.43}$$

$$\Phi(x_1 \wedge x_2, y_1 \wedge y_2) \leq \Phi(x_1, y_1) \vee \Phi(x_2, y_2). \tag{10.44}$$

PROOF. If $x_1 \vee x_2 = x_1$ and $y_1 \vee y_2 = y_1$ the conclusions are automatic; similarly if $x_1 \vee x_2 = x_2$ and $y_1 \vee y_2 = y_2$. We now consider the case of $x_1 \vee x_2 = x_1$ and $y_1 \vee y_2 = y_2$, the other case then follows by symmetry. Also, because of the continuity of Φ, the result for $y_2 = \infty$ will follow from the result for finite y_2.

We now use the monotonicity properties of Φ. If $x_1 \leq y_2$ then, noting that $\partial_x \Phi(t, y_2) \leq 0$ for $x_2 \leq t \leq x_1 \leq y_2$, we conclude that $\Phi(x_1, y_2) \leq \Phi(x_2, y_2)$. If, on the other hand, $x_1 \geq y_2$ then, noting that $\partial_y \Phi(x_1, t) \leq 0$ for $y_1 \leq t \leq y_2 \leq x_1$, we see that $\Phi(x_1, y_2) \leq \Phi(x_1, y_1)$. This completes the proof of (10.43).

The proof of (10.44) is similar. If $x_2 \leq y_1$ we use the fact that $\partial_y \Phi(x_2, t) \geq 0$ for $x_2 \leq y_1 \leq t \leq y_2$ to see that $\Phi(x_2, y_1) \leq \Phi(x_2, y_2)$ In the final case, $x_2 \geq y_1$, we note $\partial_x \Phi(t, y_1) \geq 0$ for $y_1 \leq x_2 \leq t \leq x_1$ to see that $\Phi(x_2, y_1) \leq \Phi(x_1, y_1)$. □

For $x, y \in \mathbb{R}$ set

$$g(x) := \begin{cases} 2x & \text{if} \quad x < 0 \\ x & \text{if} \quad x \geq 0 \end{cases} \tag{10.45}$$

$$G(x, y) := 4\Phi(x, y) - \Phi(g(x), g(y)). \tag{10.46}$$

PROPOSITION 10.30. $\forall x, y \in \mathbb{R}$, $G(x, y) \geq 0$.

PROOF. If $x, y \geq 0$ then $G(x, y) = 3\Phi(x, y) \geq 0$.
If $x, y < 0$ then we compute

$$\operatorname{sgn}(\partial_y G) = \operatorname{sgn}(4e^y(1 - e^y)(y - x)) = \operatorname{sgn}(y - x).$$

This, together with the fact that $G(x, x) = 0$, lets us conclude that $G \geq 0$ in this quadrant, which completes the proof in this case.

Now suppose $x < 0 \leq y$. In this region $\partial_y G = e^y(3y - 2x) \geq 0$ and thus G is a non-decreasing function of y. Also, from the previous case we know $G(x, 0) \geq 0$. Hence G is positive in this region.

Finally we consider $y < 0 \leq x$. We know from the first case that in this region $G(x, 0) \geq 0$. Also, in this region

$$\partial_y G(x, y) = 4e^y(y - x) - 2e^{2y}(2y - x) \leq 4e^y(y - x) \leq 0.$$

and hence $G(x, y)$ decreases with y. These two together ensure $G \geq 0$ in this region. □

Truncations: For f, g outer functions in H^1 we define $f \vee g$ to be the outer function whose modulus on the boundary is $|(f \vee g)(\zeta)| = |g(\zeta)| \vee |f(\zeta)|$. We define the outer function $f \wedge g$ by requiring that $|(f \wedge g)(\zeta)| = |g(\zeta)| \wedge |f(\zeta)|$ on the boundary. These constructions interact nicely with the functionals $\mathfrak{D}_\zeta$.

PROPOSITION 10.31. *Suppose f and g are outer functions in $\mathcal{D}$. For every $\zeta \in \mathbb{T}$ we have*

$$\mathfrak{D}_\zeta(f \vee g) \leq \mathfrak{D}_\zeta(f) + \mathfrak{D}_\zeta(g),$$
$$\mathfrak{D}_\zeta(f \wedge g) \leq \mathfrak{D}_\zeta(f) + \mathfrak{D}_\zeta(g).$$

Consequently

$$\|f\vee g\|_{\mathcal{D}}^2 \le \|f\|_{\mathcal{D}}^2 + \|g\|_{\mathcal{D}}^2, \quad (10.47)$$
$$\|f\wedge g\|_{\mathcal{D}}^2 \le \|f\|_{\mathcal{D}}^2 + \|g\|_{\mathcal{D}}^2.$$

PROOF. The cases of $\mathfrak{D}_\zeta(f)=\infty$ or $\mathfrak{D}_\zeta(g)=\infty$ are immediate. Otherwise, by (10.33)

$$\mathfrak{D}_\zeta(f\vee g) = \int_{\mathbb{T}} \frac{\Phi(2\left\{\log\left|f\left(e^{it}\right)\right|\vee \log\left|g(e^{it})\right|\right\}, 2\left\{\log\left|f\left(\zeta\right)\right|\vee\log\left|g(\zeta)\right|\right\})}{\left|e^{it}-\zeta\right|^2}\frac{dt}{2\pi}.$$

We now use the first conclusion of Proposition 10.29 to dominate that by

$$\int_{\mathbb{T}} \frac{\left\{\Phi(2(\log\left|f\left(e^{it}\right)\right|,\log\left|f\left(\zeta\right)\right|)\right\}\vee\left\{\Phi(2\left(\log\left|g(e^{it})\right|\right),\log\left|g(\zeta)\right|)\right)\right\}}{\left|e^{it}-\zeta\right|^2}\frac{dt}{2\pi}.$$

This estimate, together with the observation that if $\alpha,\beta\ge 0$ then $\alpha\vee\beta\le\alpha+\beta$ yields the first part of the conclusion. Integrating that first estimate gives the part of (10.47) which involves $\mathfrak{D}(\cdot)$. The part of (10.47) involving H^2 norms is a consequence of the inequality $|f|^2\vee|g|^2\le|f|^2+|g|^2$.

The work for the other estimates is similar. □

For an outer function $F\in H^1$ and $n>0$ we define the truncation, F_n to be the outer function $F_n := F\wedge n$. We extend the notation to all $f\in H^1$ by letting the truncation act on the outer factor. That is, if f has the inner-outer factorization $f=I_fO_f$ then we set $f_n = I_f(O_f)_n$.

PROPOSITION 10.32. *Suppose $f\in\mathcal{D}$; then*

1. *Fix $n>0$, For $f=I_fO_f$ we have $|f_n(z)|\le|f(z)|$ for all $z\in\mathbb{D}$ and $\|f_n\|_{\mathcal{D}}\le\|f\|_{\mathcal{D}}$.*
2. *There is a sequence $\{g_n\}\subset\mathcal{D}\cap H^\infty$ with $|g_n(z)|\le|f(z)|$ for all n, all $z\in\mathbb{D}$, $\|g_n\|_{\mathcal{D}}\le\|f\|_{\mathcal{D}}$ for all n, and $g_n\to f$ in norm.*

PROOF. The pointwise estimate in the first statement holds because the two functions have the same inner factor and there is a pointwise inequality between their outer factors. That inequality is a consequence of the pointwise inequality of the boundary values of the harmonic functions $\log|O_f|$ and $\log|\left(O_f\right)_n|$.

The norm estimate is a straightforward consequence of the following estimate.

$$\begin{aligned}\mathfrak{D}_\zeta(f_n) &= \mathfrak{D}_\zeta\left(I_f\left(O_f\wedge n\right)\right)\\ &= \mathcal{D}_\zeta(I_f)\left|O_f\wedge n(\zeta)\right|^2 + \mathcal{D}_\zeta(O_f\wedge n)\\ &\le \mathcal{D}_\zeta(I_f)\left|O_f(\zeta)\right|^2 + \mathcal{D}_\zeta(O_f)+\mathcal{D}_\zeta(n)\\ &= \mathcal{D}_\zeta(f).\end{aligned}$$

Here the passage to the second line uses (10.36), the passage to the following line uses the definition of $O_f\wedge n$ and the previous two propositions. The final line comes from noting that $\mathcal{D}_\zeta(n)=0$ and using (10.36) again.

Now that we know the sequence $\{f_n\}_{n=1}^\infty$ is bounded we can find a weakly convergent subsequence $\{h_n\}$. The pointwise limit of the f_n is f and hence the weak limit of the h_n must be f. Mazur's lemma, Theorem A.17, states that in such a case there is a sequence $\{g_n\}$ of convex combinations of the $h_n's$ which converge in norm to the same limit. Because the g_n are convex combinations they inherit the required additional properties from the h_n. □

Note that if $f \in H^1$ and $K > 0$ then $(1/f)_K \in H^\infty$. To see this, first note that it suffices to verify it for f which are outer. The outer function $(1/f)_K$ is defined by having modulus $G = |f|^{-1} \wedge K$ on the boundary. For G to be the modulus of the boundary value of an H^∞ function it suffices to have G bounded above, which it is, and to have $\log^- G \in L^1(d\theta)$ (3.3). That last fact is a consequence of knowing $|f| \in L^1(d\theta)$.

Proposition 10.33. *For $\zeta \in \mathbb{T}, K > 0$, $f \in H^2$, f outer*

$$\mathfrak{D}_\zeta \left(\left(f^{-1} \right)_K f^2 \right) \le 4\mathfrak{D}_\zeta \left(f \right).$$

Proof. There is nothing to do if the right side is infinite.

Now we consider the case $K = 1$. We always have $\left| \left(f^{-1} \right)_1 f^2 \right| \le |f|$. Hence, tracing through the definitions we find that if $f(\zeta) = 0$ we have $\mathfrak{D}_\zeta \left(\left(f^{-1} \right)_1 f^2 \right) \le \mathfrak{D}_\zeta \left(f \right) \le 4\mathfrak{D}_\zeta \left(f \right)$.

Suppose now $f(\zeta) \neq 0$. For $\xi \in \mathbb{T}$, set $w(\xi) = \log |f(\xi)|$. Again working through the definitions we see that

$$\log \left| \left(f^{-1} \right)_1 f^2 \right| (\xi) = g \left(w(\xi) \right)$$

where g was defined in (10.45).

Taking note of formula (10.33) for outer functions, we see that it suffices to show that, $\forall \xi \in \mathbb{T}$,

$$\Phi(2g\left(w(\xi)\right), 2g\left(w(\zeta)\right)) \le 4\Phi(2w(\xi), 2w(\zeta)),$$

or, equivalently, using the notation from (10.46), that $G(2w(\xi), 2w(\zeta)) \ge 0$. That, however is Proposition 10.30.

We finish by showing that, by homogeneity, the case $K = 1$ implies the general case. We compute

$$\begin{aligned} \mathfrak{D}_\zeta \left(\left(f^{-1} \right)_K f^2 \right) &= \mathfrak{D}_\zeta \left(K^{-1} \left((Kf)^{-1} \right)_1 (Kf)^2 \right) = K^{-2} \mathfrak{D}_\zeta \left(\left((Kf)^{-1} \right)_1 (Kf)^2 \right) \\ &\le 4K^{-2} \mathfrak{D}_\zeta \left(Kf \right) = 4\mathfrak{D}_\zeta \left(f \right). \end{aligned}$$

The inequality used the case $K = 1$.

□

10.3.3.2. *Boundary Values and Carleson Measures.* Recall that Mult $(\mathcal{D})$ is the space of multipliers of $\mathcal{D}$, $CM(\mathcal{D})$ is the space of Carleson measures on $\mathcal{D}$, and

$$\mathcal{X} = \left\{ f \in \operatorname{Hol}(\mathbb{D}) : |f'|^2 \, dA \in CM(\mathcal{D}) \right\}$$

with the norm

$$\|f\|_{\mathcal{X}}^2 = |f(0)|^2 + \left\| |f'|^2 \, dA \right\|_{CM(\mathcal{D})}.$$

We encountered $\mathcal{X}$ when we showed that $\operatorname{Mult}(\mathcal{D}) = H^\infty \cap \mathcal{X}$ and again when studying asymmetric Hankel forms. It will reappear in Chapter 13 when we study the symmetric Hankel form.

By testing the Carleson condition against the constant function we see that $\mathcal{X} \subset \mathcal{D} \subset H^2$. Hence functions in $\mathcal{X}$ have boundary values *a.e.* $d\theta$ and can be recovered from those values. This leads to the question of how membership in $\mathcal{X}$, which in defined in terms of behavior *inside* $\mathbb{D}$, can be characterized in terms of the boundary values? The analogous question for multipliers was answered by Stegenga [**314**, Theorem 2.7] using $\mathfrak{D}_\zeta$ and a result similar to this holds for $\mathcal{X}$. Set

$$\widetilde{\mathcal{X}} = \left\{ f \in \operatorname{Hol}(\mathbb{D}) : \mathfrak{D}_\zeta(f) \, |d\zeta| \in CM(\mathcal{D}) \right\}.$$

PROPOSITION 10.34.

$$\mathcal{X} = \widetilde{\mathcal{X}}; \quad \operatorname{Mult}(\mathcal{D}) = H^\infty \cap \mathcal{X} = H^\infty \cap \widetilde{\mathcal{X}}.$$

PROOF. We know that the first two spaces in the second equation are equal. Hence we will be done if we establish $\mathcal{X} = \widetilde{\mathcal{X}}$. The statement $f \in \widetilde{\mathcal{X}}$ is the statement that there as a constant, depending on f, such that, for any $g \in \mathcal{D}$,

$$\int |g|^2 \mathfrak{D}_\zeta(f) \, |d\zeta| \lesssim \|g\|^2 .$$

We now use (10.34) to rewrite the left hand side as an integral over the disk:

$$\int |g|^2 \mathfrak{D}_\zeta(f) \, |d\zeta| = \int_{\mathbb{D}} PI(|g|^2) \, |f'|^2 \, dA, \tag{10.48}$$

where $PI(h)$ is the Poisson integral of a boundary function h. On the other hand the statement $f \in \mathcal{X}$ is the statement that there is a constant depending on f such that, for any $g \in \mathcal{D}$,

$$\int_{\mathbb{D}} |g|^2 \, |f'|^2 \, dA \lesssim \|g\|^2 . \tag{10.49}$$

We will be done if we can get good comparison of the left hand side of (10.49) with the right hand side of (10.48). To do that we compare $\left|g^2(z)\right|$ with $PI(|g|^2)$. First we represent $\left|g^2(z)\right|$ using the Poisson integral formula, bring in absolute values and invoke Jensen's inequality, that is,

$$\left|g^2(z)\right| = |PI(g(z))|^2 \le |PI(|g|)(z)|^2 \le PI(|g|^2)(z).$$

This gives

$$PI(|g|^2)(z) - |PI(|g|)(z)|^2 \ge 0.$$

On the other hand

$$\sup_{z \in \mathbb{D}} \left\{ PI(|g|^2)(z) - |PI(|g|)(z)|^2 \right\} < \infty. \tag{10.50}$$

This fact is more subtle. The supremum in the previous line is the square of the Garsia seminorm, a seminorm which is finite on the space BMO; and, in fact, could be used to define BMO [**193**]. Combining that with the fact that $g \in \mathcal{D} \subset BMO$ allows us to complete the proof. □

An aside: The result about BMO is not necessary for the proof; a direct proof that (10.50) holds for $g \in \mathcal{D}$ is not difficult. We introduced the Garsia seminorm because it is part of the story we were discussing in Section 10.1 of using variation measures to define spaces. Continuing that story, and using notation from (10.3), we set $\operatorname{StdDev}(f)(z) = \sqrt{\operatorname{Var}(f)(z)}$. (In probability theory the standard deviation is the square root of the variance.) With the definition, the fact that BMO can be defined using the Garsia seminorm is the statement that f is in BMO exactly if

$$\operatorname{StdDev}(f)(z) \in L^\infty(\mathbb{D}, d\lambda).$$

This is in the same spirit as the theorem of Böe mentioned in Section 10.1 that f is in $\mathcal{D}$ exactly if

$$\operatorname{StdDev}(f)(z) \in L^2(\mathbb{D}, d\lambda).$$

In fact, for $1 < p < \infty$ having $\operatorname{StdDev}(f)(z) \in L^p(\mathbb{D}, d\lambda)$, characterizes f in the Besov space B^p [**76**] which were defined in Section 2.5.3.

It would be interesting to have a characterization of functions in $\widetilde{\mathcal{X}}$ in terms of local oscillation, in the spirit of the characterization of BMO by (10.50).

10.3.3.3. *The Formulas of Douglas and Carleson.*

COROLLARY 10.35. *If $f \in \mathcal{D}$ then Douglas's formula* (10.29) *holds.*

PROOF. This only requires Proposition 10.19. The formula (10.29) is the special case of that proposition in which μ is arc length measure on the circle. □

The following is an interesting further consequence. However we should note that it is much weaker than the refined boundary convergence results that we discuss briefly in Section 9.1.4.

COROLLARY 10.36. *At almost every boundary point $\zeta, \mathcal{D}_\zeta(f) < \infty$ and f has an orocyclic limit.*

PROOF. Given Douglas's formula, the integrand must be finite *a.e.*. We saw in Proposition 10.17 that there is an orocyclic limit at any ζ at which $\mathcal{D}_\zeta(f) < \infty$. □

COROLLARY 10.37. *If $f \in \mathcal{D}$ then Carleson's formula,* (10.30) *holds.*

PROOF. By Douglas's formula we know $\mathfrak{D}(f) = \int_{\mathbb{T}} \mathfrak{D}_\zeta(f) \, |d\zeta| \, /2\pi$. When we use Theorem 10.16 we obtain three terms. The first two are terms in (10.30). The third term is, recalling that $u = \log |f|$,

$$I = \int_{\mathbb{T}} \int_{\mathbb{T}} \frac{e^{2u(e^{it})} - e^{2u(e^{is})} - 2e^{2u(e^{it})}(u(e^{it}) - u(e^{is}))}{|e^{it} - e^{is}|^2} \frac{dt}{2\pi} \frac{ds}{2\pi}$$

We need to manipulate this expression slightly to get the third term of (10.30). We write $I = I/2 + I/2$, thus

$$I = \int_{\mathbb{T}} \int_{\mathbb{T}} = \frac{1}{2} \int_{\mathbb{T}} \int_{\mathbb{T}} + \frac{1}{2} \int_{\mathbb{T}} \int_{\mathbb{T}} .$$

In the second summand we interchange the order of integration and then relabel the integration variables; $s \to t, t \to s$. We then continue with

$$\begin{aligned} I = & \frac{1}{2} \int_{\mathbb{T}} \int_{\mathbb{T}} \frac{e^{2u(e^{it})} - e^{2u(e^{is})} - 2e^{2u(e^{it})}(u(e^{it}) - u(e^{is}))}{|e^{it} - e^{is}|^2} \frac{dt}{2\pi} \frac{ds}{2\pi} \\ & + \frac{1}{2} \int_{\mathbb{T}} \int_{\mathbb{T}} \frac{e^{2u(e^{is})} - e^{2u(e^{it})} - 2e^{2u(e^{is})}(u(e^{is}) - u(e^{it}))}{|e^{it} - e^{is}|^2} \frac{dt}{2\pi} \frac{ds}{2\pi} \\ = & \int_{\mathbb{T}} \int_{\mathbb{T}} \frac{(e^{2u(e^{it})} - e^{2u(e^{is})})(u(e^{it}) - u(e^{is}))}{|e^{it} - e^{is}|^2} \frac{dt}{2\pi} \frac{ds}{2\pi} . \end{aligned}$$

The last step was algebraic combination of the two integrals. The final line is the required expression for (10.30). □

Every $f \in \mathcal{D}$ has a factorization $f = B_f S_f O_f$ where the three factors are given by (10.26), (10.27), and (10.28). The H^2 norm of f only depends on O_f. The situation with the $\mathcal{D}$ norm is more complicated but the following pair of consequences of (10.30) gives some insight into the roles of the factors.

COROLLARY 10.38. *The only inner functions in $\mathcal{D}$ are finite Blaschke products.*

PROOF. If $f \in \mathcal{D}$ is inner then u is identically zero and O_f is identically one. In that case (10.30) becomes

$$\mathfrak{D}(f) = \int_{\mathbb{T}} \sum_j P_{z_j}(e^{is}) \frac{ds}{2\pi} + \int_{\mathbb{T}} \left(\int_{\mathbb{T}} \frac{2}{|e^{it} - e^{is}|^2} \frac{ds}{2\pi} \right) d\mu_f(t).$$

Each P_{z_i} is a probability measure and hence the first term counts the number of zeros of the Blaschke product; hence that must be finite. In the second term, the inner integral is identically infinite. Hence that term is finite only if μ_f is the zero measure. □

The following observation is due to Ross [**287**].

COROLLARY 10.39. *With the notation of the theorem, and for any ζ in the circle*

$$\mathfrak{D}_\zeta(f) = [\mathfrak{D}_\zeta(O_f)] + [\mathfrak{D}_\zeta(B_f O_f) - \mathfrak{D}_\zeta(O_f)] + [\mathfrak{D}_\zeta(S_f O_f) - \mathfrak{D}_\zeta(O_f)].$$

The corollary is an elementary algebraic consequence of Theorem 10.16. It is noteworthy because it expresses $\mathfrak{D}_\zeta(f)$ as a sum of three positive terms, each having a conceptual description. The first term is the contribution of the outer factor. Once that outer factor is fixed, the further increase in $\mathfrak{D}_\zeta(\cdot)$, due to inclusion of a Blaschke factor and a singular inner factor, are given by the second and third terms.

10.3.3.4. *The F Property and Zero Sets.* Any $f \in \mathcal{D}$, in fact any $f \in H^1$, has a factorization $f = B_f S_f O_f$. Recalling that an inner function is any product of a Blaschke factor and a singular function, we see that f also has an inner-outer factorization, $f = I_f O_f$, where the inner factor, I_f is given by $I_f = B_f S_f$.

Some properties of a function behave in predictable ways when an inner factor is removed. For instance the number of zeros in the disk decreases. Also, the moduli of the boundary values are unchanged, and hence the H^2 norm is unchanged. We say that a space $\mathcal{Y} \subset H^1$ has the F *property* if it is always possible to remove an inner factor and stay in the space. That is, if $If \in \mathcal{Y}$ for some inner function I then $f \in \mathcal{Y}$. In particular, if $\mathcal{Y}$ has the F property then for any $g \in \mathcal{Y}$, the outer factor of g is in $\mathcal{Y}$. For instance, H^2 has the F *property.*

Recall the spaces $\mathcal{X}$ and $\widetilde{\mathcal{X}}$ discussed in Section 10.3.3.2. Also consider, for any nonnegative measure μ on $\mathbb{T}$, the Dirichlet type space $\mathcal{D}(\mu)$ consisting of all $f \in \text{Hol}(\mathbb{D})$ for which the norm given by

$$\|f\|^2_{\mathcal{D}(\mu)} = \|f\|^2_{H^2} + \int_{\mathbb{T}} \mathfrak{D}_\zeta(f) d\mu(\zeta)$$

is finite. Thus, if μ is normalized Lebesgue measure then $\mathcal{D}(\mu) = \mathcal{D}$, if μ is the zero measure then $\mathcal{D}(\mu) = H^2$.

Recall we earlier set

$$\phi_\mu(z) = \int_{\mathbb{T}} P_z(\zeta) d\mu(\zeta).$$

Then, by Proposition 10.19, we also have that

$$\|f\|^2_{\mathcal{D}(\mu)} = \|f\|^2_{H^2} + \int_{\mathbb{D}} |f'(z)|^2 \phi_\mu(z) \frac{dA}{\pi}.$$

COROLLARY 10.40. *Suppose μ is a nonnegative measure on $\mathbb{T}$.*

(1) *The spaces $\mathcal{D}$ and $\mathcal{D}(\mu)$ have the F property.*

(2) *The spaces* Mult $(\mathcal{D})$ *and* Mult $(\mathcal{D}(\mu))$ *have the F property.*
(3) *The space* $\mathcal{X} = \widetilde{\mathcal{X}}$ *has the F property.*

PROOF. Suppose $f \in \mathcal{D}$, I is an inner function, and ζ is on the circle. Theorem 10.16 states that $\mathfrak{D}_\zeta(f)$ is a sum of three terms. The term involving the outer factor of f will not change when f is replaced by If. The other two terms will both become larger. Thus

$$\mathfrak{D}_\zeta(f) \leq \mathfrak{D}_\zeta(If). \tag{10.51}$$

Integrating this estimate and noting that the $\|f\|_{H^2} = \|If\|_{H^2}$ gives the first statement. The second statement follows because, in general, if X has the F property then so does its multiplier algebra. For the third property, having $g \in \widetilde{\mathcal{X}}$ is equivalent to having $\mathfrak{D}_\zeta(g)d\theta$ be a Carleson measure. The fact (10.51) ensures that this measure becomes smaller when an inner factor is removed. Hence the modified measure is also a Carleson measure and thus the modified function is in $\widetilde{\mathcal{X}}$. Finally $\mathcal{X} = \widetilde{\mathcal{X}}$ by Proposition 10.34. □

EXERCISE 10.41. *Check that if X has the F property then so does its multiplier algebra.*

Knowing a space X has the F property is a basic tool for studying zero sets for functions in X. For instance here is the proof of Theorem 6.36 based on the F property instead of the explicit computations involving division by Blaschke factors.

COROLLARY 10.42.
(1) *Any subset of a zero set for* Mult $(\mathcal{D})$ *is an exact zero set for* Mult $(\mathcal{D})$.
(2) *Theorem* 6.36 *holds: $\mathcal{D}$ and* Mult $(\mathcal{D})$ *have the same zero sets.*
(3) *The union of any two exact zero sets for $\mathcal{D}$ is an exact zero set for $\mathcal{D}$.*

PROOF. Suppose S is zero set for Mult $(\mathcal{D})$ and $R \subset S$ is the subset of interest. Select $m \in$ Mult $(\mathcal{D})$, not identically zero and having zero set $T \supset S$. We know Mult $(\mathcal{D}) \subset H^\infty$ and hence T satisfies the Blaschke condition (3.8). The Blaschke product B_T associated with T factors as the product of Blaschke products with zero sets R and $T \setminus R$; $B_T = B_R B_{T\setminus R}$. We know that $m = B_T O$ for some outer function O. Because Mult $(\mathcal{D})$ has the F property we conclude that

$$\frac{m}{B_{T\setminus R}} = \frac{B_T O}{B_{T\setminus R}} = B_R O \in \text{Mult}\,(\mathcal{D}).$$

This gives an element of Mult $(\mathcal{D})$ with zero set exactly R, establishing the first claim.

Given the first claim, the earlier proof of Theorem 6.36 goes through without need for the lemma concerning division by Blaschke factors.

For the third claim, first note that the product of two multipliers is a multiplier, hence the multiplier analog of the third statement is true. Combining this observation with the second statement yields the third statement. □

A similar result holds for the Hardy spaces; there it is a straightforward consequence of the factorization theorem and the basic theory of Blaschke products. The analog of the first statement also holds for the Bergman space, a result of Hedenmalm [**173**]. On the other hand the zero sets for Bergman space functions need not satisfy the Blaschke condition, hence the second statement fails. That the third statement also fails was first shown by Horowitz [**180**].

EXERCISE 10.43. *Suppose H is a RKHS of holomorphic functions on the disk, that H has a CPP, and that H has the F property. To what extent can the previous proof be adjusted to give an analogous result for H?*

10.4. Notes and Comments

Most of the material in the first half of the chapter is based on [**284**] and [**73**].

The decomposition theorem for Bergman space functions, Theorem 10.12 is part of a family of results for holomorphic functions [**117**], [**282**], There are also similar results for functions which are not assumed to be holomorphic [**156**]. More recent views of these ideas are in the paper by Feichtinger and Voigtlaender [**147**], the series of books by Triebel [**330**–**332**], and the references in those sources.

Most of the material in the second part of the chapter is based on a pair of papers by Richter and Sundberg [**274**] and [**273**].

A more general version of Proposition 10.34, one which applies to the spaces described in Section 2.5.8, is given in [**13**, Theorem IV.1.9].

CHAPTER 11

Shift Operators and Invariant Subspaces

In this chapter and the next we consider the shift operator, the operator of multiplication by the coordinate function, on spaces of holomorphic functions on the disk. In this chapter we recall basic facts about operators and summarize some of the basic results about the shift operator on the Hardy space. We also introduce Richter's results on two–isometries which will be fundamental in our discussion of the Dirichlet shift. The following chapter is devoted to the shift operator on the Dirichlet space.

The theory of shift operators on function spaces in general, and on the spaces we are considering in particular, is a rich and rapidly evolving topic. We will only tell a few parts of that story and our presentation will not be self contained.

The shift operator on a space of holomorphic functions on the disk is the operator of multiplication by the coordinate function. Our focus will be on S, the shift operator on the Dirichlet space $\mathcal{D}$. However for background and for comparisons we begin with a discussion of the Hardy shift, S_{H^2}, the operator of multiplication by z on the Hardy space, H^2; and the Bergman shift, S_{A^2}. In each of the three cases the operator is unitarily equivalent to an operator on a one sided sequence space (the space of Taylor coefficients) which acts by inserting 0 in the initial spot and shifting each other entry one place to the right, whence the name. We will pass freely between the two viewpoints, operators on functions and the operators on sequences, generally using the same name and notation for both. For the Hardy space the associated sequence space is $\ell^2(\mathbb{Z}_{\geq 0})$. For the other two a weighted ℓ^2 space is needed.

EXERCISE 11.1. *Fill in the details above. What are the relevant weighted spaces? What are the unitary maps? Also, S_{A^2} is unitarily equivalent to a "weighted shift" on the unweighted space $\ell^2(\mathbb{Z}_{\geq 0})$. Fill in the details of that equivalence and of the analogous one for $S = S_{\mathcal{D}}$.*

The Jordan normal form of an operator (or matrix) on a finite dimensional vector space is a realization of the operator as a sum of relatively easily described parts called "Jordan blocks." The situation for operators on Hilbert space is more complicated.

EXERCISE 11.2. *Write the matrix of the shift operator on the Hardy space in terms of the monomial basis. What properties does this operator share with a finite dimensional Jordan block? What properties does it fail to share.*

We will be particularly interested in the invariant subspaces of these operators. Recall that for a bounded linear operator, T, on a Hilbert space, H, an invariant subspace, V, is a closed subspace of H which is mapped into itself by T, i.e. $\forall v \in V$, $Tv \in V$. The name is sometimes restricted, sometimes without mention, to nontrivial subspaces; $V \neq H$, $V \neq \{0\}$. A fundamental question for any operator T is if it has any nontrivial invariant subspaces. (The question if *every* bounded operator on Hilbert space has an invariant subspace is the *Invariant Subspace Problem* which has been open for more than fifty years.) For instance, for each of our three shifts the subspace of functions which vanish at the origin is a nontrivial invariant subspace.

More generally, with T given, for any f in H one wants a description of $[f]$, the invariant subspace generated by f, that is, the smallest invariant subspace containing f. In the extreme case this invariant subspace is the whole space, $[f] = H$. In that case f is called *cyclic* (for T), and describing the cyclic vectors is one goal of the analysis of T. If there is such a cyclic vector then the operator T is said to be cyclic.

The invariant subspace $[f]$ is the closure in the norm topology of the set $\text{Poly}(T)(f) = \{p(T)f : p \text{ a polynomial}\}$. In particular, if H is a space of holomorphic functions on the disk and T is the shift operator on that space, then $\text{Poly}(T)(f) = \{pf : p \text{ a polynomial}\}$. If, also, $1 \in H$ then $\text{Poly}(T)(1) = \{p : p \text{ a polynomial}\}$ in which case the question if 1 is cyclic is the question of the density of the polynomials. Thus, the constant function is a cyclic vector for the Bergman shift, the Hardy shift, and the Dirichlet shift and each of those operators are cyclic.

On the other hand let $\ell_2^2 = \ell_2^2(\mathbb{Z}_{\geq 0}) = \ell^2(\mathbb{Z}_{\geq 0}, \mathbb{C}^2)$ be the space of sequences of vectors in $\mathbb{C}^2$, $b = (b_0, b_1, b_2, \ldots)$ with $\|b\|^2 = \sum_n \|b_n\|^2 < \infty$. This b can be regarded as the sequence of $\mathbb{C}^2$-valued Taylor coefficients of functions in the Hardy space of $\mathbb{C}^2$-valued functions, $H^2\left(\mathbb{D}, \mathbb{C}^2\right)$. There is an associated shift operator $S_{H^2}^{(2)}$ which can be realized by shifting the entries in the sequences one step to the right and inserting an initial 0, or by multiplication by the coordinate function on the space of vector valued holomorphic functions $H^2\left(\mathbb{D}, \mathbb{C}^2\right)$. This operator is not cyclic (Why not?).

11.1. Classes of Operators

11.1.1. Unitary and Normal Operators. A basic model of an separable infinite dimensional Hilbert space is the one-sided sequence space $\ell^2 = \ell^2(\mathbb{Z}_{\geq 0})$; the space of sequences $A = (a_0, a_1, a_2, \ldots)$ for which $\|A\|_{\ell^2}^2 = \sum_{n=0}^{\infty} |a_n|^2 < \infty$. A basic linear operator on this space is the unilateral shift, S_{H^2}, which shifts each entry one place to the right; $S_{H^2}A = (0, a_0, a_1, a_2, \ldots)$.

EXERCISE 11.3. *Verify that this operator is unitarily equivalent to multiplication by z on H^2.*

This simple operator exhibits quintessentially infinite dimensional behavior. The operator S_{H^2} is isometric, for all v $\|S_{H^2}v\|_{\ell^2} = \|v\|_{\ell^2}$. An isometric linear operator acting on a finite dimensional space must be invertible; however, in contrast, S_{H^2} is not invertible. Specifically, it fails to be surjective; the vector $(1, 0, 0, 0, \ldots)$ is not in its range.

If T is any isometric operator on a Hilbert space then it must hold that $T^*T = I$. (Here T^* denotes the adjoint of T.) If, further, H is finite dimensional, then the left inverse, T^*, must be the inverse; hence $T^* = T^{-1}$ and we can continue with

$T^*T = TT^* = I$. Operators which satisfy this last equation, even on an infinite dimensional space, are called *unitary*. The operator S_{H^2}, having no inverse, is, *a fortiori*, not unitary.

Recall that an operator is called *normal* if it commutes with its adjoint. Every unitary operator is normal, but not conversely. The operator S_{H^2} is not normal; we can check that by computing both $S_{H^2}^*S_{H^2}$ and $S_{H^2}S_{H^2}^*$. Computation shows that $S_{H^2}^*$ acts by discarding the first entry of a sequence and shifting each of the remaining entries one place to the left; $S_{H^2}^*A = (a_1, a_2, a_3, \ldots)$. Thus $S_{H^2}^*S_{H^2} = I$ (reflecting the fact that S_{H^2} is an isometry) and, with just a moment's thought, $S_{H^2}S_{H^2}^* = I - P_0$ where P_0 is the projection onto the span of the vector $(1, 0, 0, \ldots)$. Thus $S_{H^2}S_{H^2}^*$ has a nontrivial kernel and, hence, does not equal $S_{H^2}^*S_{H^2}$; thus S_{H^2} is not normal.

Although we will not pursue this direction, the operator $S_{H^2}^*$ has been and continues to be actively studied [**109**], as have the analogously defined operators on other spaces. The operator $S_{H^2}^*$ is sometimes called the $*-$*shift* (because it is the adjoint) and sometimes called the *backwards shift* (describing its action on ℓ^2). However on $\mathcal{D}$, and in general, the backwards shift is not the adjoint of the shift. Both of those operators are studied, the former perhaps more than the latter.

EXERCISE 11.4. *Verify that the adjoint of the shift on ℓ^2 is the backwards shift. When the natural unitary map is used to identify ℓ^2 and H^2 the shift operator corresponds to M_z, multiplication by the coordinate function z. With that identification what is the formula for the $*-$shift?*

EXERCISE 11.5. *Compute the adjoint of the shift S on the Dirichlet space, both in terms of its action on the sequence of coefficients and as an operator on holomorphic functions.*

11.1.2. Subnormal Operators. Although S_{H^2} is not normal; it is an important member of a closely related class of operators, *subnormal operators*. An operator T on a Hilbert space K is said to be subnormal if there is a larger Hilbert space H which contains K as a closed subspace, and a normal operator R on H, such that for all k in K, $Rk = Tk$. In particular, K is an invariant subspace for R. We see that $T = S_{H^2}$ is subnormal by setting $K = \ell^2(\mathbb{Z}_{\geq 0})$, $H = \ell^2(\mathbb{Z})$, and letting R be is the bilateral shift, the operator on $\ell^2(\mathbb{Z})$ which acts by moving each component of the sequence (in this case, a doubly infinite sequence) one place to the right. The Bergman shift, S_{A^2}, is also subnormal, as is seen by the choices $K = L^2(\mathbb{D}, dA)$, $H = A^2$ and with R the operator of multiplication by z acting on $L^2(\mathbb{D})$.

The Dirichlet shift, S, is not subnormal. One way to see this is to use the unitarily equivalent version of S, which is realized as an operator on a sequence space, and then use Berger's criterion for a weighted shift to be subnormal. A proof of the criterion is in [**120**, Chapter 3], and that book is also a good introduction to the theory of subnormal operators. Here we just state that result.

Given a sequence of positive numbers, $\Delta = \{\delta_n\}_{n=0}^{\infty}$ we say that the operator S_Δ is the weighted shift with weight sequence Δ if there is a Hilbert space H with orthonormal basis $\{e_n\}_{n=0}^{\infty}$ such that the operator S_Δ is defined by $S_\Delta e_n = \delta_n e_{n+1}$. There is no loss in taking H to be $\ell^2(\mathbb{Z}_{\geq 0})$ with S_Δ given by

$$S_\Delta(b_0, b_1, b_2, \ldots) = (0, \delta_0 b_0, \delta_1 b_1, \delta_2 b_2, \ldots).$$

We will use the notation

$$\gamma_n = \begin{cases} 0 & \text{if } n = 0 \\ \delta_0^2\delta_1^2 \cdots \delta_{n-1}^2 & \text{if } n > 0. \end{cases}$$

Thus S, S_{H^2}, and S_{A^2} are (unitarily equivalent to) the weighted shifts with, for $n = 0, 1, 2, \ldots$

$$\delta_n = \left(\frac{n+1}{n+2}\right)^{-1/2}, \ \gamma_n = (n+1)^1;$$

$$\delta_n = \left(\frac{n+1}{n+2}\right)^{0} = 1, \gamma_n = (n+1)^0 = 1$$

$$\delta_n = \left(\frac{n+1}{n+2}\right)^{1/2}, \ \gamma_n = (n+1)^{-1}$$

respectively.

THEOREM 11.6 (Berger [**120**, Chapter 3]). *The weighted shift S_A is subnormal if and only if the numbers $\{\gamma_n^2\}$ are a moment sequence. That is, if and only if there is a nonnegative measure, μ, supported on $\left[0, \|S_A\|^2\right]$ such that, for $n = 0, 1, \ldots$*

$$\gamma_n = \int_0^{\|S_A\|^2} x^n d\mu(x) = \int_0^{\infty} x^n d\mu(x).$$

COROLLARY 11.7. *The Dirichlet shift, S, is not subnormal.*

EXERCISE 11.8. *Check that the theorem gives the correct answer for S_{H^2} and S_{A^2}. Prove the corollary. (Suppose there is a μ and estimate γ_{j+k} for j fixed and k very large.)*

11.1.3. Isometries. The simple properties of S_{H^2} which we noted are almost enough to characterize S_{H^2} (up to unitary equivalence). We first look at examples to see how close we are to a characterization.

We know that S_{H^2} is an isometry and is not invertible; but there are many other operators with the same property. For instance, pick any auxiliary Hilbert space H and consider the operator $T = S_{H^2} \bigoplus I_H$ acting on $\ell^2(\mathbb{Z}_{\geq 0}) \bigoplus H$ (I_H denoting the identity operator on H). This is a noninvertible isometry but is not equivalent to S_{H^2}. That can be seen, for instance, by noting that for any $v \in \ell^2(\mathbb{Z}_{\geq 0})$, $\left\|S_{H^2}^{*n} v\right\| \to 0$. However for any nonzero $h \in H$ we have $\|T^{*n}(0 \oplus h)\| = \|0 \oplus h\| \nrightarrow 0$. Furthermore, letting $\dim(H) = k$ vary we obtain an infinite family of examples, and the kth example has T^* acting isometrically on a subspace of dimension exactly k; hence no two of the examples are unitarily equivalent.

To eliminate such examples we focus on isometries which are not invertible, and, further, are not invertible on any nontrivial subspace. This excludes the previous examples which are invertible on, for example, $\{0\} \bigoplus H$. However, we have still not characterized S_{H^2}. Consider the operator $S_{H^2}^{(2)}$ which we defined a moment ago, the shift operator on $\ell^2(\mathbb{Z}_{\geq 0}, \mathbb{C}^2)$, a space of sequences of elements in $\mathbb{C}^2$. This operator is an isometry which is not invertible on any nontrivial subspace. However $S_{H^2}^{(2)}$ is not unitarily equivalent to S_{H^2}. One way to see this is to compute the codimension of the ranges of the operators:

$$\operatorname{codim}_{\ell^2} S_{H^2}\ell^2 = \dim(\ell^2 \ominus S_{H^2}\ell^2) = 1$$
$$\neq 2 = \dim(\ell^2_2 \ominus S^{(2)}_{H^2}\ell^2_2) = \operatorname{codim}_{\ell^2_2} S^{(2)}_{H^2}\ell^2_2.$$

The operator $S^{(2)}_{H^2}$, which is called the shift of multiplicity two, is easily seen to be unitarily equivalent to a direct sum of two copies of S_{H^2}, the shift of multiplicity one.

That last variation brings us to the end of the story. Defining $S^{(n)}_{H^2}$, $n = 1, 2, \ldots, \infty$ analogously, we have:

THEOREM 11.9 (Halmos [**168**]). *If T is an isometry on a Hilbert space H which is not unitary on any nontrivial subspace, then there is an integer n (possibly infinite) such that T is unitarily equivalent to $S^{(n)}_{H^2}$ acting on ℓ^2_n.*

That T is not unitary on any nontrivial subspace is implied by the condition

$$\bigcap_{n>0} T^n H = \{0\}\,. \tag{11.1}$$

Operators that satisfy (11.1) are called *analytic;* a reference to the fact that multiplication by the coordinate function on spaces of analytic functions on the disk have this property. Halmos' theorem states that any analytic isometry is equivalent to a (finite or countable) direct sum of copies of S_{H^2}. In particular, if T is an analytic isometry and $\dim\ker(T^*) = 1$ then T is unitarily equivalent to S_{H^2}.

11.2. The Hardy Shift

If we regard S_{H^2}, the Hardy space shift, as acting on ℓ^2 then it is easy to find invariant subspaces. The subspace $V_1 = \left\{A \in \ell^2 : a_0 = 0\right\}$ is a non-trivial invariant subspace, as are the subspaces V_n, $n = 2, 3, \ldots$ consisting of vectors which begin with strings on n zeros. It is also possible to find some cyclic vectors. Clearly the vector $(1, 0, 0, \ldots)$ is cyclic and the vector $(0, 1, 0, 0, \ldots)$ is not. Going a bit further:

EXERCISE 11.10. *Show that the vector $v_{a,b} = (a, b, 0, 0, \ldots)$ is cyclic if $|a| > |b|$ and is not cyclic if $|a| < |b|$. What happens when $|a| = |b|$?*

The full story about the operator S_{H^2} is much richer, and that richness begins to unfold beautifully when we change viewpoints from the shift acting on $\ell^2(\mathbb{Z}_{\geq 0})$ to multiplication by z acting on H^2. That is, when we associate $A = (a_0, a_1, \ldots) \in \ell^2(\mathbb{Z}_{\geq 0})$ with $\mathbf{A}(z) = \sum_{k\geq 0} a_k z^k \in H^2$ then the shift, S_{H^2} (using intentionally ambiguous notation), is given by $S_{H^2}(\mathbf{A})(z) = z\mathbf{A}(z)$. In this picture S_{H^2} acts by multiplication and hence the operator is sometimes denoted by M or M_z.

11.2.1. Simple Things. Viewing the shift as a multiplication operator on a space of functions, suggests analogies between the invariant subspace theory of the shift and ideal theory for algebras of analytic functions on the disk. For instance, knowing if a function is in an invariant subspace is similar to knowing if a function is in an ideal. Also, the question whether a function f is cyclic is similar to the question if the ideal generated by that function is the entire algebra. Given this, it is natural to speculate that if f has a zero inside the disk then f is not cyclic. Conversely, perhaps if the values $|f(z)|$ stay away from zero (whatever that may mean) then f will be cyclic. Here are some elementary results in line with these thoughts.

In the polynomial algebra the constant function, 1, is not in any proper ideal. In H^2 the function 1 is not in any nontrivial invariant subspace. Also, each $w \in \mathbb{C}$ corresponds to a maximal ideal in the polynomial algebra; the ideal of polynomials which vanishes at w or, equivalently, the ideal with generator $f(z) = z - w$. For the shift on H^2 each $w \in \mathbb{D}$ corresponds to an invariant subspace of codimension one, $\{f \in H^2 : f(w) = 0\}$ and this subspace is generated by $z - w$. Verify this with the following exercise.

EXERCISE 11.11.
(1) *Show that if* $w \in \mathbb{D}$ *then* $\{f \in H^2 : f(w) = 0\}$ *is a closed subspace and equals* $\{f \in H^2 : f(z) = (z - w)h(z),\ h \in H^2\}$.
(2) *Show* f *is not cyclic by giving an example of a nonzero* $h \perp [f]$.
(3) *Show that if* $h, k \in [f]^{\perp}$ *then there is a linear combination of* h *and* k *that is in* $[f]$; *hence* $\dim [f]^{\perp} \leq 1$. *Together with the previous statement this shows* $\operatorname{codim} [f] = \dim [f]^{\perp} = 1$.

This analysis can be continued. Given $\{w_i\}_{i=1}^{n} \subset \mathbb{D}$, a set of not necessarily distinct points, the polynomial $p_W(z) = \prod_{i=1}^{n}(z - w_i)$ generates an ideal of codimension n in the polynomial algebra and that ideal can also be described as consisting of all polynomials which vanish (to the appropriate order) at all the w_i. The same polynomial generates a shift invariant subspace of H^2 which is also of codimension n, and it too can be characterized in terms of location and orders of zeros. In fact this process gives all the shift invariant subspaces of H^2 having finite codimension.

Our analogy begins to fray when we consider the function $f(z) = z - w$ with $|w| > 1$. That function generates a nontrivial ideal in the polynomial algebra, yet it is cyclic for the shift on the Hardy space.

EXERCISE 11.12. *Show that if* $|w| > 1$ *then* $f(z) = z - w$ *is cyclic. Show that, in fact, given* $g \in H^2$ *then there is a* $k \in H^2$ *such that* $g(z) = (z - w)k(z)$.

The analysis to this point has been sufficiently simple so that what we have said is not actually particular to the Hardy space. The same functions that generate the invariant subspaces of finite codimension for the shift on the Hardy space also generate all the finite codimension invariant subspaces for the Bergman shift or the Dirichlet shift. Also, in each of the three cases, if $|w| > 1$ then $f(z) = z - w$ is cyclic. Differences between the three cases appear when behavior near the boundary of the disk is an issue. Membership in these spaces place subtle constraints on behavior near the boundary; constraints which vary from space to space. That behavior near the boundary requires more careful analysis even in the most straightforward examples.

EXERCISE 11.13. *Suppose we want to describe the invariant subspace of* S_{H^2} *generated by* $f(z) = z - 1$.

(1) *There is no analog of Exercise* 11.12 *to show that* f *is cyclic.*
(2) *Show that there is no nonzero* h *in* $[f]^{\perp}$ *and thus* f *is cyclic.*
(3) *Is the function* $f(z) = z - 1$ *a cyclic vector for the shift operator on the Dirichlet space? For the shift operator on the Bergman space?*

11.2.2. Beurling's Theorem. A major step in merging the function theoretic and operator theoretic viewpoints for the Hardy space was Beurling's transformative 1948 paper [**68**] in which he gave a complete description of the invariant subspaces of the shift operator on the Hardy space.

Recall Theorem 3.2 from Chapter 2 which says that any f in H^2 has an inner-outer factorization $f = I_f O_f$. If I_f is normalized (its first nonvanishing Taylor coefficient is positive), then the factorization is unique. Using this notation we have the following.

THEOREM 11.14. *The non-trivial invariant subspaces of the operator S_{H^2} acting on H^2 are exactly the subspaces of the form $IH^2 = \{If : f \in H^2\}$ for inner functions I. If we require I to be normalized then the representation is unique.*

Given a non-zero $f \in H^2$, $f = I_f O_f$, then the smallest invariant subspace containing f, is $[f] = I_f H^2$. In particular, f is a cyclic vector for S_{H^2} if and only if f is an outer function.

With this result in hand, many facts about the invariant subspaces of S_{H^2} can be obtained easily. For instance, every such subspace has the *codimension one property*; that is, if V is any invariant subspace then $\dim(V \ominus zV) = 1$. Also, every invariant subspace is of the form mH^2 where m is a multiplier of H^2. Because all multipliers are bounded, this ensures that each shift invariant subspace contains a bounded function. In contrast there are subspaces of the Bergman space which are invariant for the Bergman shift and which contain no nontrivial bounded function. Finally, for inner functions I and J we have $IH^2 \cap JH^2 \supset IJH^2$ and hence the intersection of any two invariant subspaces contains a third.

Each of these properties suggest questions about the invariant subspaces of other shift operators. We will show later that the analog of each of these three properties hold for S, the shift operator on the Dirichlet space. None of them hold for S_{A^2}, the shift operator on the Bergman space.

11.2.3. The Projection. Let P denote the orthogonal projection of L^2 onto H^2. Using Fourier series it is easy to check that P can be realized by

$$PF(z) = \langle F, k_z \rangle_{L^2} = \frac{1}{2\pi} \int_{\mathbb{T}} F(\zeta) \frac{1}{1 - \overline{\zeta} z} d\theta.$$

Beurling's theorem ensures that any invariant subspace, V, of S_{H^2} is of the form $\Theta H^2 = \{\Theta f : f \in H^2\}$ for some inner function $\Theta = \Theta_V$. We now derive the formula for the projection P_V of H^2 onto V.

Let M_Θ be the map of H^2 to itself given by multiplication by Θ, $M_\Theta f = \Theta f$. The adjoint, M_Θ^*, is given by $PM_{\overline{\Theta}}$ where $M_{\overline{\Theta}}$ is the map of H^2 to $L^2(d\theta/2\pi)$ given by multiplication by $\overline{\Theta}$. That is verified by the computation:

$$\begin{aligned}\left\langle (M_\Theta)^* f, g \right\rangle_{H^2} &= \langle f, M_\Theta g \rangle_{H^2} = \langle f, M_\Theta g \rangle_{L^2(d\theta/2\pi)} \\ &= \langle M_{\overline{\Theta}} f, g \rangle_{L^2(d\theta/2\pi)} = \langle PM_{\overline{\Theta}} f, g \rangle_{H^2}.\end{aligned}$$

We now claim that P_V, the orthogonal projection of H^2 onto V, is given by

$$P_V = M_\Theta M_\Theta^* = M_\Theta P M_{\overline{\Theta}}. \tag{11.2}$$

To verify this we must show that for any $f \in H^2$, we have $P_V f \in V$ and $f - P_V f \perp V$. The first statement is clear. For the second, select $f \in H^2$ and $\Theta g \in V$. We have

$$\begin{aligned}\langle f - P_V f, \Theta g\rangle_{H^2} &= \langle f, \Theta g\rangle_{H^2} - \langle P_V f, \Theta g\rangle_{H^2}\\ &= \langle f, \Theta g\rangle_{H^2} - \langle M_\Theta M_\Theta^* f, \Theta g\rangle_{H^2}\\ &= \langle f, \Theta g\rangle_{H^2} - \langle \Theta M_\Theta^* f, \Theta g\rangle_{H^2}\\ &= \langle f, \Theta g\rangle_{H^2} - \langle M_\Theta^* f, g\rangle_{H^2}\\ &= \langle f, \Theta g\rangle_{H^2} - \langle f, M_\Theta g\rangle_{H^2}\\ &= \langle f, \Theta g\rangle_{H^2} - \langle f, \Theta g\rangle_{H^2} = 0,\end{aligned}$$

as required.

Although the previous proofs are simple and rather formal, they are far from being arguments about generic RKHS. The role of an inner function Θ is a consequence of Beurling's theorem. The identification $M_\Theta^* = PM_{\overline{\Theta}}$ uses the realization of H^2 as a subspace of a Lebesgue space. The passage from the third line to the fourth in the previous computation used the fact that M_Θ is an isometric multiplier, a fact that uses the structure of inner functions and the fact that the Hardy space sits in a Lebesgue space.

We know from Proposition (2.12) that $k_{V,\zeta}$ the reproducing kernel for the space V which evaluates a function in V at ζ, is given by $P_V k_\zeta$. By the previous computation, that equals $M_\Theta PM_{\overline{\Theta}} k_\zeta$. Recalling that $k_\zeta(z) = (1 - \overline{\zeta}z)^{-1}$ we see that

$$k_{V,\zeta}(z) = P_V k_\zeta(z) = \frac{\overline{\Theta(\zeta)}\Theta(z)}{1 - \overline{\zeta}z},$$

$$P_v f(\zeta) = \left\langle f(z), \frac{\overline{\Theta(\zeta)}\Theta(z)}{1 - \overline{\zeta}z}\right\rangle_{H^2}.$$

Finally, the projection onto $V^\perp$, the orthogonal complement of V in H^2, is given by $P_{V^\perp} = P_V^\perp = I - P_V$. Hence for any f in H^2 (or in L^2)

$$P_V^\perp f(\zeta) = \left\langle f(z), \frac{1 - \overline{\Theta(\zeta)}\Theta(z)}{1 - \overline{\zeta}z}\right\rangle_{H^2}.$$

Given an inner function Θ the space $V^\perp$ just described is called the *model space* associated with Θ. Model spaces are studied for many reasons [**159**].

11.2.4. The Lattice. The set of invariant subspaces of an operator, for example S_{H^2}, has a natural lattice structure. One lattice operation is forming the intersection of two invariant subspaces; the second is forming the join, the closure of the vector sum of the two subspaces. Each gives a new invariant subspace, the first is the largest invariant subspace contained in both, the second is the smallest invariant subspace containing both. This lattice is denoted $\mathrm{Lat}(S_{H^2}, H^2)$ or just $\mathrm{Lat}(S_{H^2})$.

Using Beurling's theorem together with basic facts about inner functions it is straightforward to describe the lattice structure of $\mathrm{Lat}(S_{H^2}, H^2)$. We just collect the basic observations.

The set of inner functions is a lattice with the operations of least common multiple (LCM) and greatest common divisor (GCD). To compute the LCM of a pair of inner functions I and J we work with the Blaschke factors and the singular inner factors separately, the same holds for the GCD. The Blaschke factor of the

LCM is the Blaschke product with zero set equal to the union of the zero sets of I and J. The GCD is formed similarly but using the intersection of the zero sets. Singular inner functions are determined by non-negative singular measures. The lattice structure for them is induced by forming the largest measure smaller than the given two, and the smallest measure larger than both. Both processes produce new measures which are, again, non-negative singular measures.

Beurling's theorem establishes a lattice preserving correspondence between $\mathrm{Lat}(S_{H^2}, H^2)$ and the set of inner functions. In short, the two lattices are the same.

11.3. The Dirichlet Shift

Notice, for instance in the previous two subsections, how valuable inner functions and Beurling's theorem are in studying invariant subspaces of the Hardy shift. The situation is more complicated for the Dirichlet space.

11.3.1. Two-Isometries. Halmos' result, Theorem 11.9, has been very influential. Basing a study of S_{H^2} on the fact that it is a cyclic analytic isometry has been a powerful viewpoint in studying the operator itself, and has been at the heart of many generalizations. Helson's book [**177**] is an early example. More generally, this theme pervades Nikolskii's book [**245**]. The Dirichlet shift is not an isometry but it does have a closely related property. It is a cyclic analytic two-isometry. That fact and the viewpoints that come with it have become a basic tool in the study of the Dirichlet shift, beginning with the results of Richter [**267**].

A linear operator T acting on a Hilbert space H is an isometry exactly if

$$T^*T - I = 0.$$

We will say T is a *two-isometry* if it satisfies the following equivalent conditions:

$$T^{*2}T^2 - 2T^*T + I = 0, \tag{11.3}$$

or, equivalently, for all h in H,

$$\left\|T^2h\right\|^2 - \|Th\|^2 = \|Th\|^2 - \|h\|^2. \tag{11.4}$$

Any isometry is a two-isometry. Also, if T is a two-isometry on H then the restriction of T to any $T-$ invariant subspace of H is a two-isometry. Two-isometries were introduced by Agler and they are studied in [**2**] and [**5–7**] but for reasons quite different from the ideas we explore here. The Dirichlet shift, S, is not an isometry (consider $S(1)$). However it is a cyclic analytic two-isometry. It is analytic because (11.1) holds, and cyclic because the polynomials are dense in $\mathcal{D}$ and hence the constant function is a cyclic vector. To see that it is a two-isometry we verify (11.4). The monomials are a complete orthogonal set and $S_{\mathcal{D}}$ preserves that orthogonality. Hence we can verify (11.4) by checking it for monomials. For $h = z^n$ the equation (11.4) reduces to

$$(n+3) - (n+2) = (n+1) - n.$$

EXERCISE 11.15. *Check that the Hardy space shift, S, is a two-isometry. Find all weighted shifts which are two-isometries. Can you describe those two isometries as M_z, multiplication by z, on spaces of holomorphic functions on the disk with a norm given in terms of the Taylor coefficients?*

11.3.2. The Shift on $\mathcal{D}(\mu)$. For the moment we denote by M_z the operator of multiplication by z acting on any Hilbert space of functions analytic on the disk. Recall the Dirichlet type spaces $\mathcal{D}(\mu)$ defined in Chapter 10; given a nonnegative measure μ on the circle, $\mathcal{D}(\mu)$ is defined as the space of $g \in \mathrm{Hol}(\mathbb{D})$ for which

$$\|g\|_{\mathcal{D}(\mu)}^2 = \|g\|_{H^2}^2 + \int_{\mathbb{T}} \mathcal{D}_\zeta(g) d\mu\,(\zeta) < \infty.$$

Also we saw in Proposition 10.19 the integral on the right equals an integral over the disk:

$$\int_{\mathbb{T}} \mathcal{D}_\zeta(g) d\mu = \int_{\mathbb{D}} |g'|^2\, PI(d\mu) dA.$$

Here $PI(\mu)$ denotes the Poisson integral of the boundary measure $\mu(\zeta)$.

THEOREM 11.16 ([**267**, Theorems 7.1, 9.1]).

(1) *Suppose T is an operator on H with $\dim(H \ominus TH) = 1$. Then T is a cyclic analytic two-isometry if and only if there is a positive finite Borel measure μ on $\mathbb{T}$ such that T is unitarily equivalent to M_z acting on $\mathcal{D}(\mu)$.*

(2) *If W is a proper invariant subspace of M_z acting on $\mathcal{D}(\mu)$ which satisfies*

$$\dim(W \ominus M_z W) = 1, \tag{11.5}$$

then, selecting $f \in W \ominus M_z W$ with $\|f\|_{\mathcal{D}} = 1$, and setting $d\mu_f = |f|^2\, d\mu$, we have $W = f\mathcal{D}(\mu_f)$ and, for all $g \in \mathcal{D}(\mu_f)$, $\|fg\|_{\mathcal{D}(\mu)} = \|g\|_{\mathcal{D}(\mu_f)}$. That is, $\forall g \in \mathcal{D}(\mu_f)$ we have $fg \in W$ and

$$\begin{aligned} \|fg\|_{H^2}^2 + \int_{\mathbb{T}} \mathcal{D}_\zeta(gf) d\mu &= \|g\|_{H^2}^2 + \int_{\mathbb{T}} \mathcal{D}_\zeta(g)\, |f|^2\, d\mu \\ &= \|g\|_{H^2}^2 + \int_{\mathbb{D}} |g'|^2\, PI(|f|^2\, d\mu) dA. \end{aligned}$$

Furthermore, every element of W is obtained this way.

We will not prove this theorem but we will use it in our analysis of the invariant subspaces of $\mathcal{D}$. In the next chapter we will use the first part of the theorem to show that any W in $\mathrm{Lat}(M_z, \mathcal{D})$ satisfies (11.5) and then use the second part of the theorem, in the special case $\mathcal{D}(\mu) = \mathcal{D}$, to develop information about W.

This theorem brings into view a fundamental tool in the study of shift operators in general and the Dirichlet shift in particular. Given a RKHS of holomorphic functions on the disk, and given V, an invariant subspace for the operator of multiplication by z, we define the *extremal function* of V to be a unit vector in $f \in V \ominus M_z V$. In some ways these functions generalize inner functions. In particular the previous theorem and corollary show that those f are contractive multipliers which are generators of the associated invariant subspaces.

EXERCISE 11.17. *If $H = H^2$ and V is a shift invariant subspace then the extremal function for V is a contractive multiplier (in fact, an inner function) which generates V.*

The full story on extremal functions is rich and rapidly evolving. In the next chapter we will see aspects of that theory in the Dirichlet space. Similarly defined functions also play a role in Bergman space theory, see [**174**], or [**127**]; and, more generally, for a large class of RKHS, for example [**309**].

11.4. The Bergman Shift

We will be focusing on the invariant subspace theory of the Dirichlet shift and have discussed the Hardy shift to provide historical context and to highlight analogies. Results for the shift on the Bergman space provide interesting contrasts with those results. Here are two examples.

One basic difference between A^2 and both H^2 and $\mathcal{D}$ is that the Bergman space contains functions whose zero sets do not satisfy the Blaschke condition. This has immediate consequences for invariant subspaces of S_{A^2}. Suppose $f \in A^2$, has a zero set, Z, which does not satisfy the Blaschke condition. Every function in $V_f = [f]$ must vanish on Z, but no nontrivial function in a Hardy space can vanish on Z. Hence $[f]$ is disjoint from the Hardy spaces. In particular $[f]$ contains no bounded functions and thus no multipliers. This contrasts with the shift invariant subspaces of H^2 and $\mathcal{D}$ which are generated by multipliers.

We noted in the previous section that the invariant subspaces of the Hardy space had the codimension-one property; namely, if $V \in \operatorname{Lat}(S_{H^2}, H^2)$ then $\dim (V \ominus S_{H^2}V) = 1$. We will see that a similar statement holds for the shift operator on the Dirichlet space. For the Bergman space the situation is completely different. A striking result of [**66**] is that $\dim (V \ominus S_{A^2}V)$ can take any positive integer value or can be infinite. One immediate consequence of this is that not every $V \in \operatorname{Lat}(S_{A^2}, A^2)$ is singlely generated.

11.5. Notes and Comments

Nikolskii's books [**246**], [**245**] are basic references for the classical theory of the shift operator. An early, and very influential, step toward an operator theoretic viewpoint was given by Halmos in [**168**]. Analogs, extensions, and generalizations of classical results about the shift pervade modern operator theory. The works referenced in the bibliography give some entree to that larger picture.

The Bergman space and its shift operator are actively studied; the books [**174**] and [**127**] are a good introduction.

The algebra of bounded functions in the Dirichlet space is another interesting context in which to consider the topics of this chapter and the next; see, for instance, [**270**].

CHAPTER 12

Invariant Subspaces of the Dirichlet Shift

In this chapter we present some of what is known about S, the shift operator on $\mathcal{D}$, and its invariant subspaces. We will see that each invariant subspace is generated by a single function and we develop relationships between $f \in \mathcal{D}$ and the invariant subspace, $[f]$, which it generates.

A fundamental question is characterizing the f which are cyclic, that is, $[f] = \mathcal{D}$. The question is open; we will discuss some of what is known and the conjectured answer.

In this chapter we write S for the shift operator on $\mathcal{D}$, the operator of multiplication by z acting on functions in $\mathcal{D}$. We see in Theorem 11.16 that results about the shift operator acting on the Dirichlet space and results for the shift operators which act on the weighted Dirichlet spaces $\mathcal{D}(\mu)$ are closely intertwined. Nevertheless, we will restrict our attention to the classical case of S and $\mathcal{D}$. More general results can be seen in [**268**], [**267**]. [**274**], [**273**] and the papers going forward from those.

The book [**136**] is an excellent source of further information about the shift invariant subspaces of $\mathcal{D}$ and $\mathcal{D}(\mu)$, and in particular about cyclic vectors for those operators.

12.1. Overview

A few results about invariant subspaces of the $\mathcal{D}-$shift can be obtained using results about the shift on the Hardy space. For instance, if $f \in \mathcal{D}$ with the inner-outer factorization $f = Ih$ then $[f]$, the shift invariant subspace of $\mathcal{D}$ generated by f, satisfies

$$[f] = [Ih] \subset \mathcal{D} \cap IH^2. \tag{12.1}$$

To see this note that $[f]$ is certainly contained in $\mathcal{D}$. For the other inclusion note that

$$\begin{aligned}[f] = [Ih] &= \text{closure in } \mathcal{D} \text{ of } \{pIh : p \text{ a polynomial}\} \\ &\subset \text{closure in } H^2 \text{ of } \{pIh : p \text{ a polynomial}\} = IH^2.\end{aligned}$$

In particular, if f has an inner factor then $1 \notin [f]$ and hence f is not cyclic.

COROLLARY 12.1. *If $f \in \mathcal{D}$ is cyclic for the shift then it is outer.*

However the Hardy space techniques are of limited use and even when results for the $\mathcal{D}-$shift are very similar to those for the Hardy shift as the methods of proof are often quite different.

A basic tool in our analysis is Richter's result, Theorem 11.16. To access that we need to know that the invariant subspaces of the $\mathcal{D}-$shift have the codimension one property. We prove that in the next section. After that, with Theorem 11.16 in hand, we know that any V in $\mathrm{Lat}(S,\mathcal{D})$, the lattice of invariant subspace of the $\mathcal{D}$–shift, is singly generated. We then study the relation between the generator and the subspace it generates, with our earlier analysis of the local Dirichlet integral as a major tool. In Section 12.4 we study generators which are outer functions. In Section 12.6 we consider the generators which have both inner and outer factors.

If f is cyclic for S it must be an outer function. Although being outer is a complete characterization of cyclic vectors for the shift on H^2 the situation for S is more complicated. In the final section we discuss cyclic vectors for the Dirichlet shift.

12.2. The Codimension One Property

LEMMA 12.2. *Suppose $V \in \mathrm{Lat}(S,\mathcal{D})$ and $\phi \in H^\infty$, then $\phi V \cap \mathcal{D} \subset V$.*

PROOF. Recall that for $0 \le r < 1$ we write $\phi_r(z) = \phi(rz)$. We know from Exercise 2.16 that ϕ_r is the limit in the multiplier norm of a sequence of polynomials in S. Thus for any f in V, $\phi_r f \in V$. Now note that the set $\{\phi_r f\}$ is bounded in norm. This follows from Propositions 10.26 and 10.27, and here we use the fact that $\phi \in H^\infty$. Hence, as in Proposition A.18, the limit ϕf is in V. □

A subspace W is said to be a *multiplier invariant subspace* if it is invariant under multiplication by all multipliers: $\forall w \in W$, $\phi \in \mathrm{Mult}\,(\mathcal{D})$; $\phi w \in W$. Although this is formally a stronger requirement than W being invariant with respect to multiplication by z, the two notions agree for the Dirichlet space.

COROLLARY 12.3. *$V \in \mathrm{Lat}(S,\mathcal{D})$ if and only if $V \in \mathrm{Lat}(M_\phi,\mathcal{D})$ for all $\phi \in \mathrm{Mult}\,(\mathcal{D})$.*

The nontrivial part of the proof of the corollary is covered by the previous lemma.

The next corollary supports the informal idea that membership in any particular shift invariant subspace is determined by knowing where a function is small.

COROLLARY 12.4. *Suppose $V \in \mathrm{Lat}(S,\mathcal{D})$. If $f \in V$, $g \in \mathcal{D}$, and $|g(z)| \le |f(z)|$ then $g \in V$.*

PROOF. Apply the previous lemma to $\phi = g/f$. □

THEOREM 12.5. *If $\{0\} \ne V \in \mathrm{Lat}(S,\mathcal{D})$ then $\dim\,(V \ominus zV) = 1$.*

PROOF. It is straightforward that $\dim\,(V \ominus zV) \ge 1$. We need to establish the other inequality.

In order to access Lemma 12.2 and its corollary we will pass to an equivalent problem where we can find a bounded auxiliary function h. Let k be the greatest nonnegative integer such that $V \subset z^k\mathcal{D}$. The fact that V is nontrivial and that M_z is analytic (11.1) ensures that k is finite. Define W by $V = z^k W$. Note that $W \subset \mathrm{Lat}(S,\mathcal{D})$ and it is straightforward to see that

$$\dim\,(V \ominus zV) = \dim\left(z^k W \ominus z^{k+1} W\right) \le \dim\,(W \ominus zW).$$

We now need to show the right hand quantity is at most one. To do that is suffices to show that if we have any $g \in W$ with $g(0) = 0$ then $g \in zW$. We know $g = zf$ for some $f \in \mathcal{D}$ hence we are reduced to showing that $f \in W$.

The construction of W ensures that it contains a function $h \in W$ with $h(0) \neq 0$. Hence we can suppose that $h(0) = 1$ and, furthermore, using Proposition 10.32, that h is bounded.

By Proposition 10.32 we can find $\{g_n\} \subset \mathcal{D}$, each bounded by f in modulus and converging to f in norm. Consider the decomposition of each g_n given by

$$g_n = g_n h + (zf)\left[\frac{g_n(1-h)}{zf}\right].$$

The functions g_n and h are both in the algebra $H^\infty \cap \mathcal{D}$ and hence their product is in $\mathcal{D}$. Thus, by subtraction, the second summand on the right is also in $\mathcal{D}$. The g_n are each bounded and h is in W, hence by Lemma 12.2 their product is in W. Also, zf is in W and the quantity in the square brackets is bounded, hence, using that lemma again, the second product is in W. Thus g_n, the sum of the two, is in W. To finish note that the g_n converge to f and W is closed, thus $f \in W$. □

12.3. Invariant Subspaces are Generated by Multipliers

THEOREM 12.6. *Suppose* $\{0\} \neq V \in \operatorname{Lat}(S, \mathcal{D})$. *Then:*

(1) *(V has the codimension one property)* $\dim(V \ominus SV) = 1$,
(2) *(ϕ is a multiplier) For* $\phi \in V \ominus SV$, $\phi \in \operatorname{Mult}(\mathcal{D})$

$$\|\phi\|_{\operatorname{Mult}(\mathcal{D})} \lesssim \max\{\|\psi\|_{\mathcal{D}},\ \|\psi\|_{H^\infty}\} < \infty.$$

(3) *(ϕ is a contractive multiplier) If ϕ is normalized by* $\|\phi\|_{\mathcal{D}} = 1$ *then* $\|\phi\|_{\operatorname{Mult}(\mathcal{D})} = 1$.
(4) *(V is generated by ϕ)* $V = [\phi]$; *in fact* $V = \phi\mathcal{D}(|\phi|^2\, d\theta)$ *and* $\forall g \in \mathcal{D}$, $\|g\phi\|_{\mathcal{D}} = \|g\|_{\mathcal{D}(|\phi|^2 d\theta)}$, *in particular*

$$\mathfrak{D}(g\phi) = \int_{\mathbb{D}} \left|(g\phi)'\right|^2 dA = \int_{\mathbb{T}} \mathfrak{D}(g)\, |\phi|^2\, d\theta.$$

(5) *(*$\operatorname{Lat}(S, \mathcal{D})$ *is cellularly indecomposable.) If* $\{0\} \neq V, W \in \operatorname{Lat}(S, \mathcal{D})$ *then* $V \cap W \neq \{0\}$.

PROOF. Statement (1) is the previous theorem. Once we have that then we have access to Theorem 11.16. We now use that theorem but with different notation. In place of f, μ, and μ_f we now write ϕ, $d\theta/2\pi$, and $d\mu_\phi = |\phi|^2\, d\theta/2\pi$.

We now establish (2). Recall the notation from Chapter 10. For functions f, g, I in H^2 with f, g outer, I inner, and for $N \geq 0$ we have: $f \vee g$ is the outer function which has boundary values $|f \vee g| = |f| \vee |g|$, $f \wedge g$ is the outer function which has boundary values $|f \wedge g| = |f| \wedge |g|$, $f_N = f \wedge N$, and $(If)_N = I \cdot f_N$.

We separate the next step as a separate lemma. □

LEMMA 12.7 ([**13**, Lemma 4.8]). *Suppose* $\psi \in \mathcal{D}$, $\|\psi\|_{\mathcal{D}} = 1$ *and for* $n = 1, 2, 3, \ldots$, $\psi \perp z^n\psi$; *then* $\|\psi\|_\infty \leq 1$.

PROOF OF LEMMA. By first considering the case of polynomial ϕ we see that, if $\phi\psi \in [\psi]$ then $\langle \phi\psi, \psi\rangle = \phi(0)$, and that

$$|\phi(0)|^2 \leq \|\phi\psi\|_{\mathcal{D}}^2.$$

Pick and fix $t > 0$, For $\phi = 1 \wedge (t/\psi)$ we have $\phi\psi = \psi \wedge t$, and thus, by Proposition 12.8, $\phi\psi \in [\psi]$. We can then use the previous estimate to obtain

$$\begin{aligned}
|(1 \wedge (t/\psi))(0)|^2 &\leq \|(1 \wedge (t/\psi))\psi\|_{\mathcal{D}}^2 = \|\psi \wedge t\|_{\mathcal{D}}^2 \\
&= \|\psi \wedge t\|_{H^2}^2 + \int_{\mathbb{T}} \mathcal{D}_\zeta(\psi \wedge t)\,|d\zeta| \\
&\leq \|\psi \wedge t\|_{H^2}^2 + \int_{\mathbb{T}} \mathcal{D}_\zeta(\psi)\,|d\zeta| \\
&= \|\psi \wedge t\|_{H^2}^2 + 1 - \|\psi\|_{H^2}^2 \\
&= 1 + \int_{\mathbb{T}} \left(|\psi \wedge t|^2 - |\psi|^2\right)|d\zeta|.
\end{aligned}$$

In the passage from the second line to the third we used Proposition 10.31 and the fact that $\mathcal{D}_\zeta$ of a constant function is zero.

Now note that on the set $\{\zeta : |\psi| \leq t\}$ we have $|\psi \wedge t|^2 = |\psi|^2$. Hence

$$\begin{aligned}
|(1 \wedge (t/|\psi|))(0)|^2 - 1 &\leq \int\limits_{|\psi|>t} \left(\frac{t^2}{|\psi|^2} - 1\right)|\psi|^2\,|d\zeta| \\
&= t^2 \int\limits_{|\psi|>t} \left(1 - \frac{|\psi|^2}{t^2}\right)|d\zeta| \\
&\leq t^2 \int\limits_{|\psi|>t} \log\frac{t^2}{|\psi|^2}\,|d\zeta| \\
&= t^2 \int\limits_{\mathbb{T}} \log\left|\left(1 \wedge (t/|\psi|)^2\right)\right||d\zeta| \\
&= t^2 \log(1 \wedge (t/|\psi|))^2(0).
\end{aligned}$$

Here we passed from the second line to the third using the fact that, for all x, $1 - x \leq \log(1/x)$. After rewriting that line we passed to the final line using the mean value property of harmonic functions. Using $x - 1 \geq \log x$ on the left hand side, we obtain

$$\log\left(1 \wedge (t/\psi)^2\right)(0) \leq t^2 \log\left(1 \wedge (t/\psi)^2\right)(0).$$

The quantity $\log\left(1 \wedge (t/\psi)^2\right)(0)$ is the mean value of a nonpositive boundary function and hence nonpositive. If the previous inequality is to hold for some $t > 1$ then that nonpositive quantity must be zero; and hence its boundary function must vanish identically, which happens only if $\{\zeta : |\psi(\zeta)| > t\}$ is a null set. □

PROOF OF THEOREM, CONTINUED. To establish (2) first note that by homogeneity it suffices to verify the estimate with $\|\phi\|_{\mathcal{D}} = 1$. In that case, by the previous lemma we also have $\|\phi\|_{H^\infty} \leq 1$. Thus we need to show that $\phi \in \operatorname{Mult}(\mathcal{D})$ and $\|\phi\|_{\operatorname{Mult}(\mathcal{D})} \leq 1$; that is, to show that for any $g \in \mathcal{D}$, $\|\phi g\|_{\mathcal{D}} \leq \|g\|_{\mathcal{D}}$.

Because ϕ is bounded, $g \in \mathcal{D} \subset \mathcal{D}(|\phi|^2\, d\theta/2\pi)$. By Theorem 11.16,

$$\|\phi g\|_{\mathcal{D}}^2 = \|g\|_{\mathcal{D}(|\phi|^2 d\theta)}^2 = \|g\|_{H^2}^2 + \int_{\mathbb{T}} \mathcal{D}_\zeta(g)\, |\phi|^2\, d\theta/2\pi$$

$$\leq \|g\|_{H^2}^2 + \int_{\mathbb{T}} \mathcal{D}_\zeta(g) d\theta/2\pi = \|g\|_{\mathcal{D}},$$

as required.

At this point, to establish (3) we only need to show $\|\phi\|_{\mathrm{Mult}(\mathcal{D})} \geq 1$. That follows by testing the operator of multiplication by ϕ on a constant function,

Statement (4) is contained in Theorem 11.16.

Our remaining task is to show $\mathrm{Lat}\,(S, \mathcal{D})$ is cellularly indecomposable.; that is, given $V, \tilde{V} \in \mathrm{Lat}\,(S, \mathcal{D})$, $V, \tilde{V} \neq \{0\}\,, \mathcal{D}$ there is a $W \in \mathrm{Lat}\,(S, \mathcal{D})$, $W \neq \{0\}$,with $W \subset V \cap \tilde{V}$. To do that it suffices to find a nonzero $g \in V \cap \tilde{V}$. Select a nonzero $F \in V$ and a nonzero $f \in [F] \ominus z\,[F] \subset V$. By the previous parts of this proof we see that f is a multiplier. Similarly, for any other nontrivial $\tilde{V} \in \mathrm{Lat}\,(S, \mathcal{D})$ we can find $\tilde{f} \in \tilde{V}$ which is a multiplier. Hence by the second corollary above $f\tilde{f} \in [f] \subset V$ and similarly $f\tilde{f} \in \tilde{V}$. Thus $V \cap \tilde{V}$ is nonempty. □

12.4. Outer Generators: Joins, Meets, and Powers

Now that we know every invariant subspace is singly generated we focus on the relation between the invariant subspace and its generator. In this section we consider outer generators.

The analogy between invariant subspaces and ideals in algebras of functions suggests that a critical factor in describing the invariant subspace is knowing where the generator is small and how small. The invariant subspaces generated by finite Blaschke products, described in Exercise 12.17 are elementary examples. A less subtle, but still straightforward instance is given by the following proposition which is a slight variation of Lemma 12.4.

PROPOSITION 12.8. *If $f, g \in \mathcal{D}$,*

$$\text{for all } z \in \mathbb{D},\ |g(z)| \lesssim |f(z)| \tag{12.2}$$

then $[g] \subset [f]$. If f is outer then it suffices to have

$$\text{for all } \zeta \in \mathbb{T},\ |g(\zeta)| \lesssim |f(\zeta)|\,. \tag{12.3}$$

PROOF. If f is outer then (12.3) implies (12.2). Thus we can assume (12.2) holds. In that case we know $k = g/f \in H^\infty$. Recall that for $0 < r < 1$, $k_r(z) = k(rz)$. From Exercise 2.16 we know k_r is a multiplier. Hence, taking note of the first of the two corollaries above we know $k_r f \in [f]$. We now use Proposition 10.28 with our function k as the g in that proposition. We conclude that $k_r f$ converges weakly to kf. The subspace $[f]$ is closed and hence weakly closed; thus $kf \in [f]$ as required. □

EXERCISE 12.9. *$V \in \mathrm{Lat}(S, \mathcal{D})$ contains an outer function if and only if any generator of V is an outer function. In particular if $V = \mathcal{D}$ then any generator is outer.*

For $G := \{g_1, g_2, \ldots\} \subset \mathcal{D}$, we denote by $[g_1, g_2, \ldots]$ the smallest invariant subspace containing G.

THEOREM 12.10. *If $f, g \in \mathcal{D}$ are outer then*

$$[f, g] = [f \vee g].$$

PROOF. We know that on the boundary circle $|f|$, $|g| \leq |f \vee g|$. Hence by Proposition 12.8, f, $g \in [f \vee g]$; and thus $[f, g] \subset [f \vee g]$.

To prove the opposite inclusion first note that, by Theorem 12.6, that there is a $\phi \in \mathcal{D}$ such that $[f \vee g] = [\phi] = \phi\mathcal{D}(|\phi|^2\, d\theta)$. We just saw $f \in [f \vee g]$, hence, noting the previous exercise, we see that ϕ is outer.

We need to show that $f \vee g \in \phi\mathcal{D}(|\phi|^2\, d\theta)$, or, equivalently that $\phi_1 = (f \vee g)/\phi \in \mathcal{D}(|\phi|^2\, d\theta)$. Working through the definitions we find that $\phi_1 = f_1 \vee g_1$ where $f_1 = f/\phi$, $g_1 = g/\phi$. We know $f \in \phi\mathcal{D}(|\phi|^2\, d\theta)$ and hence $f_1 \in \mathcal{D}(|\phi|^2\, d\theta)$, and similarly $g_1 \in \mathcal{D}(|\phi|^2\, d\theta)$. Hence, by Proposition 10.31, $\phi_1 \in \mathcal{D}(|\phi|^2\, d\theta)$ and thus $\phi\phi_1 \in \phi\mathcal{D}(|\phi|^2\, d\theta)$. However $\phi\phi_1 = \phi\left((f \vee g)/\phi\right) = f \vee g$, and hence we are done. □

PROPOSITION 12.11. *Suppose $f \in \mathcal{D}$ can be written $f = \phi/\psi$ with ϕ, $\psi \in H^\infty \cap \mathcal{D}$, $\|\phi\|_\infty, \|\psi\|_\infty \leq 1$ and $\phi, \psi, 1/\psi \in \mathcal{D}$. Then $[f] = [\phi]$, and, also, ψ and $1/\psi$ are cyclic.*

Given any f it is possible to find ϕ and ψ so that this proposition applies. We will use that in the proof of Theorem 12.14.

We begin with a lemma which we will use to control some approximations.

LEMMA 12.12. *Suppose $\phi \in H^\infty$, $\omega \in H^2$, $N > 0$, $\zeta \in \mathbb{T}$. Then*

$$\mathcal{D}_\zeta(\omega_N\phi) \leq 4\mathcal{D}_\zeta(\omega\phi) + 6\,\|\phi\|_\infty^2\, \mathcal{D}_\zeta(\omega).$$

PROOF OF THE LEMMA. If the right hand side is infinite there is nothing further to do. If it is finite we note that

$$\frac{\omega_N\phi - (\omega_N\phi)(\zeta)}{z-\zeta} = \omega_N\frac{\phi - \phi(\zeta)}{z-\zeta} + \phi(\zeta)\frac{\omega_N - \omega_N(\zeta)}{z-\zeta}$$

and

$$\omega\frac{\phi - \phi(\zeta)}{z-\zeta} = \frac{\omega\phi - (\omega\phi)(\zeta)}{z-\zeta} - \phi(\zeta)\frac{\omega - \omega(\zeta)}{z-\zeta}.$$

Hence

$$\begin{aligned}
\mathcal{D}_\zeta(\omega_N\phi) &\leq 2\left(\left\|\omega_N\frac{\phi-\phi(\zeta)}{z-\zeta}\right\|_{H^2}^2 + \left\|\phi(\zeta)\frac{\omega_N - \omega_N(\zeta)}{z-\zeta}\right\|_{H^2}^2\right)\\
&\leq 2\left(\left\|\omega\frac{\phi-\phi(\zeta)}{z-\zeta}\right\|_{H^2}^2 + \|\phi\|_\infty^2\left\|\frac{\omega_N - \omega_N(\zeta)}{z-\zeta}\right\|_{H^2}^2\right)\\
&\leq 2\left(\left\|\frac{\omega\phi-(\omega\phi)(\zeta)}{z-\zeta} - \phi(\zeta)\frac{\omega-\omega(\zeta)}{z-\zeta}\right\|_{H^2}^2 + \|\phi\|_\infty^2\,\mathcal{D}_\zeta(\omega)\right)\\
&\leq 2\left(2\left\{\mathcal{D}_\zeta(\omega\phi) + \|\phi\|_\infty^2\,\mathcal{D}_\zeta(\omega)\right\} + \|\phi\|_\infty^2\,\mathcal{D}_\zeta(\omega)\right)\\
&= 4\mathcal{D}_\zeta(\omega\phi) + 6\,\|\phi\|_\infty^2\,\mathcal{D}_\zeta(\omega).
\end{aligned}$$

In passage from the first line to the second we used the fact that $|\omega_N| \leq |\omega|$. In passing to the third line we used algebra and Proposition 10.31, which shows that $\mathcal{D}_\zeta(\omega_N) \leq \mathcal{D}_\zeta(\omega)$. □

PROOF OF THE PROPOSITION. First note that $\phi = f\omega$ and hence $|\phi| \leq |f|$. Thus, by Proposition 12.8 $[\phi] \subset [f]$.

We now invoke the previous lemma with $\omega = 1/\psi$. We obtain

$$\mathcal{D}_\zeta((\psi^{-1})_N\phi) \leq 4\mathcal{D}_\zeta(f) + 6\,\|\psi\|_\infty^2\,\mathcal{D}_\zeta(1/\psi).$$

Integrating with respect to $d\theta/2\pi$ we obtain, for some constant,

$$\left\|(\psi^{-1})_N\phi\right\|_{\mathcal{D}}^2 \lesssim \left(\|f\|_{\mathcal{D}}^2 + \|1/\psi\|_{\mathcal{D}}^2\right),$$

that is, the sequence $R = \left\{(\psi^{-1})_N\phi\right\}$ is a bounded subset of $\mathcal{D}$. Further, taking note of Proposition 12.8 again, we see that $R \subset [\phi]$. Finally, it is easy to see that, as N approaches infinity, then pointwise, $(\psi^{-1})_N\phi \to \psi^{-1}\phi = f$.

We have established that f is the pointwise limit of a norm bounded sequence in $[\phi]$ and hence, by Proposition A.18, $f \in [\phi]$ which ensures the other inclusion.

We now show that ψ and $1/\psi$ are cyclic. We know $|1/\psi| \geq 1$ and hence by Proposition 12.8 $1/\psi$ is cyclic. Each $(\psi^{-1})_N\psi \in [\psi]$ and an argument similar to that used earlier in the proof shows $(\psi^{-1})_N\psi \to 1$ weakly which, by Proposition A.18, ensures that $1 \in [\psi]$. □

If we set $\phi = 1$ we obtain the following:

COROLLARY 12.13. *Suppose $f, 1/f \in \mathcal{D}$. If, also, f is bounded then f is cyclic, that is $[f] = \mathcal{D}$.*

We will see later, in Corollary 12.29, that the conclusion holds even without the hypothesis that f is bounded.

THEOREM 12.14. *Suppose f is an outer function in $\mathcal{D}$ and $\alpha > 0$. If $f^\alpha \in \mathcal{D}$; then $[f] = [f^\alpha]$.*

PROOF. First write $f = \phi/\psi$ where $\phi = f \wedge 1$ and $\psi = 1/(f \vee 1)$. It is immediate that $\|\phi\|_\infty, \|\psi\|_\infty \leq 1$. Also, by Proposition 10.31, $\phi, 1/\psi \in \mathcal{D}$. Furthermore, noting $\|\psi\|_\infty \leq 1$ and hence also $|\psi'| = |\psi|^2\left|(1/\psi)'\right| \leq \left|(1/\psi)'\right|$, we see that ψ is in $\mathcal{D}$. Thus, by Proposition 12.11, $[f] = [\phi]$. Similarly $[f^\alpha] = [\phi^\alpha]$. Hence we may, and do, suppose f is bounded.

Consider the case $\alpha = 2$. With f bounded, Proposition 12.8 ensures that $\left[f^2\right] \subset [f]$. To prove the other inclusion, $[f] \subset \left[f^2\right]$, it suffices, by Proposition A.18, to show that the set of functions

$$R = \left\{\left(f^{-1}\right)_N f^2 : N \in \mathbb{R}\right\}$$

is contained in $\left[f^2\right]$, is norm bounded, and that, as $N \to \infty$ the functions converge pointwise to f. To do this we follow the pattern of the proof of Proposition 12.11. It follows from Lemma 12.12 and Proposition 12.8 that $\left(f^{-1}\right)_N f^2 \in \left[f^2\right]$ and that $\left(f^{-1}\right)_N f^2 \to 1$ pointwise; the required norm bound is a consequence of Proposition 10.33.

Suppose now that $\alpha > 1$ and pick an integer n so that $2^n \geq \alpha > 1$. Taking note of the fact that f is bounded we see $f^{2^n} \in \mathcal{D}$. We have

$$[f] \subset [f^{2^n}] \subset [f^\alpha] \subset [f].$$

The first inclusion is a consequence of the n-fold application of the $\alpha = 2$ case theorem which we just proved. Because $2^n \geq \alpha > 1$ and f is bounded, the second and third inclusions are consequences of Theorem 12.8. This gives the desired equality for $\alpha > 1$. If $\alpha < 1$ then the required equality follows from applying what has been done so far to the functions g and g^β where $g = f^a$ and $\beta = 1/\alpha > 1$. □

THEOREM 12.15. *Suppose $f, g \in \mathcal{D}$ are outer then*

$$[f \wedge g] = [f] \cap [g] .$$

If, furthermore, $fg \in \mathcal{D}$ then we can continue with

$$[f \wedge g] = [f] \cap [g] = [fg] .$$

PROOF. We begin by showing that if $fg \in \mathcal{D}$ then $[f] \cap [g] \subset [fg]$. Invoking Theorem 12.8 several times we find that $p = (f \wedge 1)(g \wedge 1) \in [f] \cap [g]$. Theorem 12.6 ensures that there is an $h \in \mathcal{D} \cap H^\infty$ which is a generator of the invariant subspace $[f] \cap [g]$. Noting that p is outer, and recalling Exercise 12.9, we see that h must be outer. Without loss we suppose $\|h\|_\infty = 1$.

We now show $h^2 \in [fg]$, which, by Theorem 12.14 will ensure $[f] \cap [g] = [h] = [h^2] \subset [fg]$.

By assumption there are polynomials p_n such that $p_n f$ converges to f in norm. Hence $\{p_n f\}$ is a bounded set, and hence so is the set of truncated functions $\{(p_n f)_1\}$. Also $(p_n f)_1 (z) \to h(z)$ for each z in $\mathbb{D}$. To see that consider $r_n = (p_n f)_1 / (p_n f)$. Each $\|r_n\|_\infty \le 1$ and hence it suffices to show $r_n(0) \to 1$. Showing that is an exercise in the use of truncations.

We now use Proposition A.18. We have established uniform boundedness and pointwise convergence of the $(p_n f)_1$ to h. Similarly there are q_n and then $(q_n g)_1$; uniformly bounded functions converging pointwise to h. Hence the sequence of functions $\{(q_n g)_1 (p_n f)_1\}$ converges pointwise to h^2. Recalling that $\mathcal{D} \cap H^\infty$ is a Banach algebra we see that $\{(q_n g)_1 (p_n f)_1\}$ is a bounded set in $\mathcal{D}$. Hence, by the Proposition, h^2 is in $[fg]$.

To finish the proof we note that the following chain of inclusions holds for any outer f, g.

$$\begin{aligned}
[f \wedge g] &\subset [f] \cap [g] \\
&= [f \wedge 1] \cap [g \wedge 1] \\
&\subset [(f \wedge 1)(g \wedge 1)] \\
&\subset [(f \wedge 1) \wedge (g \wedge 1)] \\
&\subset [f \wedge g] .
\end{aligned}$$

The first inclusion and also the last two, follow from Proposition 12.8. The equality at the second line is a consequence of Proposition 12.11. To pass from the second line to the third note that $(f \wedge 1), (g \wedge 1) \in \mathcal{D} \cap H^\infty$ which is an algebra. Hence $(f \wedge 1)(g \wedge 1) \in \mathcal{D} \cap H^\infty \subset \mathcal{D}$ and hence the desired inclusion follows from the case of the theorem which is already proved. □

EXERCISE 12.16. *A step in the proof was described as an "exercise in truncations." Do the exercise. (Hint:* [**86**, Lemma 6], [**272**, Lemma 2.8].*)*

12.5. Inner Generators

We have looked at invariant subspaces with a generator, f, which is an outer function. The case of f an inner function is elementary because, as we noted in Corollary 10.38, the only inner functions in $\mathcal{D}$ are the finite Blaschke products; and in that case everything is straightforward. Verify that with the following exercise.

EXERCISE 12.17. *Show that if B is a finite Blaschke product then $[B] = B\mathcal{D}$. More generally, for $f \in \mathcal{D}$, $[Bf] = B[f]$.*

12.6. Mixed Generators

Suppose now that a generator, f, has both an inner and outer factor, with the factorization being $f = Ih$. We noted at (12.1) that

$$[f] = [Ih] \subset \mathcal{D} \cap IH^2.$$

It is also true that

$$[f] = [Ih] \subset \mathcal{D} \cap I[h].$$

To see this note that if we select $g \in [Ih]$ then certainly $g \in \mathcal{D}$. Also, if $g \in [Ih]$ then there is a sequence of polynomials $\{p_n\}$ such that $p_n Ih$ converges to g in Dirichlet norm. That implies $p_n Ih$ converges to g in H^2. From that it follows that $g = IG$ for some G in H^2. We now have that $p_n Ih$ converges to IG in Dirichlet norm. We saw in Corollary 10.40 that the Dirichlet space has the F property, namely removing inner factors decreases Dirichlet norms. Hence $p_n h$ converges to G in Dirichlet norm and thus $G \in [h]$. We have $g = IG \in [h]$, as required

In fact the intersections in the two previous displays are the same.

PROPOSITION 12.18. *Suppose $h \in \mathcal{D}$ is outer and I is inner. (We do not assume $Ih \in \mathcal{D}$.) Then*

$$\begin{aligned}\mathcal{D} \cap I[h] &= [h] \cap IH^2 \\ &= [h] \cap \left(\mathcal{D} \cap IH^2\right).\end{aligned}$$

Furthermore the right hand side is trivial, i.e. equals $\{0\}$, if and only if either $[h]$ is trivial or $\left(\mathcal{D} \cap IH^2\right)$ is trivial.

PROOF. First note that $[h] \subset \mathcal{D}$ and hence the two right hand sides are equal.

To show the left hand side is contained in the right, first note that $I[h] \subset IH^2$. Also, if g is in the left hand side then $g = Ik$ for some k in $[h]$. By Proposition 12.8 this ensures that $g \in [k] \subset [h]$. These two facts establish the desired inclusion.

If g is in the right hand side then $g \in [h] \subset \mathcal{D}$. We now want to show, further, that if $g = Ik$ with $k \in H^2$ then, in fact, $g \in I[h]$. We know h is outer, hence, noting Exercise 12.9, j, the generator of $[h]$ produced by Theorem 12.6 is outer. Furthermore, by that theorem $[h] = j\mathcal{D}(|j|^2\, d\theta)$ and so we must have $g = jr$ for some $r \in \mathcal{D}(|j|^2\, d\theta)$. We know that the inner function I divides $g = jr$, and we know that j is outer. Hence I divides r and, by the F property for local Dirichlet spaces, Corollary 10.40, $r/I \in \mathcal{D}(|j|^2\, d\theta)$. Thus $g = Ijr/I \in Ij\mathcal{D}(|j|^2\, d\theta) = I[h]$; as required.

The final comment follows from the fact that the right hand side, as written, is the intersection of two invariant subspaces. By item (5) in Theorem 12.6, the invariant subspace lattice of the $\mathcal{D}$−shift is cellularly indecomposable. That property states that if the intersection is trivial then at least one of the terms is trivial. □

With this result in hand, Richter and Sundberg give a pair of results on the role of inner factors in generators of invariant subspaces [**273**, Theorem 5.3]. The first focuses on the case in which the generator is known, the second on the case in which the invariant subspace is given. The proofs are similar in spirit to the analyses we have been doing, and we refer to their paper for the proofs.

DEFINITION 12.19. *Given M, a closed subspace of $\mathcal{D}$, we say that an inner function J divides M if J divides m for all $m \in M$. We will say that J is the greatest common inner divisor of M if J divides M and; for any inner function I which divides M, I divides J.*

THEOREM 12.20.

(1) *Suppose I is an inner function, h is an outer function, and $Ih \in \mathcal{D}$. Then*
$$[Ih] = \mathcal{D} \cap I[h] = [h] \cap IH^2.$$

(2) *Suppose $V \in \mathrm{Lat}(S, \mathcal{D})$ and let I be the greatest common inner divisor of V. There is an outer function h in $\mathcal{D}$ with Ih in $\mathcal{D}$ such that*
$$V = [Ih] = \mathcal{D} \cap I[h] = [h] \cap IH^2.$$

12.7. Cyclic Vectors

12.7.1. Which Vectors are Cyclic? The cyclic vectors for the shift on the Hardy space are exactly the outer functions in the Hardy space. The cyclic vectors for the shift on the Dirichlet space are outer functions in the Dirichlet space; however that is not the full story. In this section we describe some of what is known.

Roughly a cyclic vector cannot be too small too often. For instance, if $f \in \mathcal{D}$ is zero at some $\zeta \in \mathbb{D}$ then $k_\zeta \perp [f]$ and f is not cyclic. On the other hand the vector 1 is cyclic and hence, taking note of Proposition 12.8, if $|f|$ is bounded away from zero then f is cyclic. What remains is the delicate case, f is zero free but is small near, or on, part of the boundary. An extreme version of being small near the boundary is if f has a singular inner factor. In that case f tends strongly to zero near the support of the associated singular measure, and in that case we know from Corollary 12.1 that such f cannot be cyclic.

To discuss the boundary values of f we recall $f^\beta(e^{i\theta}) = \lim_{r \nearrow 1} f(re^{i\theta})$ and we define
$$Z(f) := \{e^{i\theta} \in \mathbb{T} : f^\beta(e^{i\theta}) = 0\}.$$
(Such sets are only well defined up to sets of capacity zero and we are being casual about mentioning that, here and elsewhere.) Because $\mathcal{D} \subset H^2$, we know from Corollary 3.3 that for any nontrivial $f \in \mathcal{D}$ the Lebesgue measure of $Z(f)$ is zero. If, further, f is cyclic then much more is true.

THEOREM 12.21. *If f is cyclic then* $\mathrm{Cap}\,(Z(f)) = 0$.

PROOF. Let $E = Z(f)$ and set $\mathcal{D}_E = \left\{g \in \mathcal{D} : g^\beta = 0 \text{ quasieverywhere on } E\right\}$. This space contains f and is invariant under multiplication by z. We claim it is closed To see this, suppose $\{g_n\} \subset \mathcal{D}_E$ and $g_n \to g$. Thus for $t > 0$ and any n;
$$\begin{aligned}\mathrm{Cap}\left\{E \cap \left\{\left|g^\beta\right| > t\right\}\right\} &= \mathrm{Cap}\left\{E \cap \left\{\left|g^\beta - g_n^\beta\right| > t\right\}\right\} \\ &\leq \mathrm{Cap}\left\{\left|g^\beta - g_n^\beta\right| > t\right\} \\ &\lesssim \|g - g_n\|_{\mathcal{D}}^2 / t^2.\end{aligned}$$
The last line is the so-called weak capacitary inequality. That result is weaker than Lemma 4.15 but is still substantial, see [**136**, Theorem 3.2.5].

Passing to the limit, first in n then in t, we find $\mathrm{Cap}\left\{E \cap \left\{\left|g^\beta\right| > 0\right\}\right\} = 0$ and thus $g \in \mathcal{D}_E$.

We now have that $\mathcal{D}_E$ is a closed invariant subspace which contains the cyclic vector f. Hence $1 \in \mathcal{D}_E$ and thus $E \subset Z(1) = \varnothing$ which is the desired conclusion. □

It was conjectured by Brown and Shields in 1984, [**86**], that this is the full story.

CONJECTURE 12.22 (Brown-Shields Conjecture). *$f \in \mathcal{D}$ is cyclic if and only if it is outer and* $\operatorname{Cap} Z(f) = 0$.

To put this in context we have

THEOREM 12.23.

(1) *Given E, a closed subset of the circle with* $\operatorname{Cap}(E) = 0$, *there is an outer function $f \in \mathcal{D}$ which extends continuously to the closed disk, has $Z(f) = E$, and is cyclic.*
(2) *There is a closed subset E of the circle with* $\operatorname{Cap}(E) > 0$ *and an outer function $f \in \mathcal{D}$ which extends continuously to the closed disk, has $Z(f) = E$, and is not cyclic.*

PROOF DISCUSSION. The first statement is a result of Brown and Cohn [**85**] extending an earlier result of Carleson. In the second statement the fact that f is not cyclic follows from the previous theorem as soon as we have E and f with the required properties. For E we can select a Cantor set of measure zero, positive capacity, and which is a Carleson set in the sense described in Section 9.3.1. Given such an E the construction of f is given by Carleson [**93**]. □

12.7.2. Using the Previous Results. There has been substantial work related to the Brown-Shields conjecture. There is some discussion in [**287**] and [**136**, Chapter 8] tells the story as of 2014. In this section and the next we just give a small taste.

Some results about cyclic vectors follow directly from the work in the previous sections and additional results can be obtained using those techniques. Here are some samples.

COROLLARY 12.24. *If, for some $\alpha > 0$, f, $f^\alpha \in \mathcal{D}$ then f is cyclic if and only if f^α is cyclic.*

PROOF. If either f or f^α is cyclic then it is outer, in which case both are outer. Given that, the conclusion follows from Theorem 12.14. □

For any f we have defined $Z(f)$. It is convenient to also define the possibly larger set

$$\underline{Z}(f) := \{e^{i\theta} \in \mathbb{T} : \liminf_{r \nearrow 1} |f(re^{i\theta})| = 0\}. \tag{12.4}$$

We saw in Proposition 12.8 that if f, g are outer functions with $|g| \leq |f|$ on the disk then $[g] \subset [f]$.

COROLLARY 12.25.

(1) *Suppose $f, g \in \mathcal{D}$ are outer and $|g| \leq |f|$ on the disk; if g is cyclic then f is cyclic.*
(2) *If $f \in \mathcal{D}$ and for some $c > 0$ we have $|f| > c$ on the disk, then f is cyclic.*
(3) *If $f \in \mathcal{D}$ is outer and $\underline{Z}(f) = \varnothing$, then f is cyclic.*

PROOF. The first statement follows from Corollary 12.4 and the definitions. The second follows from the first with the choice $g = c$. For the third, note that, f being outer, it has no zeros in the disk. The assumption on $\underline{Z}(f)$ ensures that $|f|$

is bounded away from zero on a neighborhood of the boundary of the disk. Those two facts together are enough to ensure $|f|$ is bounded away from zero on the disk; hence the second statement can be applied. □

It may be $\operatorname{Cap} \underline{Z}(f) = 0$ is enough to ensure f is cyclic. The third part of the previous corollary deals with the case that $\underline{Z}(f)$ is empty. We consider the case of $\underline{Z}(f)$ a singleton later, in Theorem 12.31.

In Theorem 12.15 we saw that if $f, g, fg \in \mathcal{D}$ and f, g are outer, then $[f] \cap [g] = [fg]$.

COROLLARY 12.26. *For $f, g \in \mathcal{D} \cap H^\infty$, f and g are both cyclic if and only if fg is cyclic.*

PROOF. Note that fg is outer if and only if both f and g are. Also, by Corollary 12.1, cyclic functions are outer. Thus the only case that needs further analysis is if all three are outer. If f, g are in $\mathcal{D} \cap H^\infty$, then, again recalling that $\mathcal{D} \cap H^\infty$ is an algebra, $fg \in \mathcal{D} \cap H^\infty$. Hence, if f and g are cyclic we can invoke Proposition 12.15 and conclude fg is cyclic. In the other direction, suppose fg cyclic. We know f is bounded, hence $|fg(z)| \leq \|f\|_\infty |g(z)|$. Thus, by Proposition 12.8, $[fg] \subset [g]$ and hence g is cyclic. Similarly f is also cyclic. □

Another result in line with the informal thought that if f is not too small then it is cyclic is the result of Richter and Sundberg [**273**, Corollary 2.7] who showed that if $f, 1/f \in \mathcal{D}$ then f is cyclic. A later improvement of that result, due to Aleman [**13**] and, independently, Seco [**301**], quantifies smallness by requiring $\log f \in \mathcal{D}$.

THEOREM 12.27 (Aleman [**13**], Seco [**301**]). *If f and $\log f \in \mathcal{D}$ then f is cyclic.*

PROOF. We first claim that f must be outer. This will follow if we show that $\log |f|$ is the Poisson integral of its boundary values. However $\log f = g \in \mathcal{D} \subset H^2$. Hence $\log |f| = \operatorname{Re} g$ and the required integral representation of such a function is a consequence of Hardy space theory. Hence, by Theorem 12.14 we know that for all positive σ, $[f] = [f^\sigma]$. Certainly $\lim_{\sigma \to 0} f^\sigma = 1$ pointwise. If we also had convergence in Dirichlet norm we would know $1 \in [f]$ and thus that f was cyclic. We now show that is the case.

$$\|f^\sigma - 1\|^2 = \|f^\sigma - 1\|_{H^2}^2 + \int_{\mathbb{D}} \left|(f^\sigma - 1)'\right|^2 dA$$

The first term on the right goes to zero by the dominated convergence theorem. We need to show the second term goes to zero as $\sigma \to 0$. We have

$$\begin{aligned}
\int_{\mathbb{D}} \left|(f^\sigma - 1)'\right|^2 dA &= \sigma^2 \int_{\mathbb{D}} \left|f^{\sigma-1} f'\right|^2 dA - \sigma^2 \int_{\mathbb{D}} \left|f^{-1} f'\right|^2 |f|^{2\sigma} dA \\
&= \sigma^2 \left(\int_{|f|\le 1} \left|f^{-1} f'\right|^2 |f|^{2\sigma} dA + \int_{|f|>1} \left|f^{-1} f'\right|^2 |f|^{2\sigma} dA \right) \\
&\le \sigma^2 \left(\int_{|f|\le 1} \left|f^{-1} f'\right|^2 dA + \int_{|f|>1} |f'|^2 |f|^{2\sigma-2} dA \right) \\
&\le \sigma^2 \left(\int_{\mathbb{D}} \left|f^{-1} f'\right|^2 dA + \int_{\mathbb{D}} |f'|^2 dA \right) \\
&\le \sigma^2 (\|\log f\|_{\mathcal{D}}^2 + \|f\|_{\mathcal{D}}^2).
\end{aligned}$$

□

The hypotheses in this theorem are implied by those in the Richter-Sundberg result.

LEMMA 12.28. *If f and $1/f \in \mathcal{D}$ then $\log f \in \mathcal{D}$.*

PROOF. The required integrability of $(\log f)'$ follows from the observation $\left|(\log f)'\right|^2 = |f'| \left| (f^{-1})' \right|$. To obtain the required H^2 estimate for $\log f$ note that from Hardy space theory (the boundedness on L^2 of the conjugate function operation) it suffices to bound the L^2 norm of $\operatorname{Re} \log f$. That follows directly from $|\operatorname{Re} \log f| \le |f| + |1/f|$. □

COROLLARY 12.29 ([**273**]). *If f, $1/f \in \mathcal{D}$ then f is cyclic.*

12.7.3. Constructing Outer Functions. For many questions about functions in $\mathcal{D}$ it is very useful to be able to find outer functions in $\mathcal{D}$ with specific properties. For instance, we noted in Section 6.6.3 that knowing if $Z \subset \mathbb{D}$ is a zero set for $\mathcal{D}$ comes down to knowing if, with B_Z the Blaschke product for Z, we can find an outer function $O_Z \in \mathcal{D}$ such that $B_Z O_Z \in \mathcal{D}$. Flexible tools for constructing outer functions are also of great value in studying cyclic vectors. In this section we give a sample of that.

An outer function G is determined by the specification of $|G|$ on the boundary, and knowing that $G \in H^2$ only requires $|G| \in L^2$. Knowing $G \in \mathcal{D}$ is more delicate, but Carleson's formula for the Dirichlet integral, Corollary 10.37, and its localized version, Theorem 10.16, make explicit the role of $|G|$ in determining if $G \in \mathcal{D}$, and are thus natural tools for constructing outer functions with desired properties.

Here is an argument due to El-Fallah, Kellay, and Ransford [**139**] for constructing outer functions in $\mathcal{D}$.

For E a closed subset of $\mathbb{T}$, and for $\zeta \in \mathbb{T}$, let $d(\zeta, E)$ denote the linear distance from ζ to E.

LEMMA 12.30 (Fusion Lemma). *Suppose $E \subset \mathbb{T}$ is a closed set of measure zero, that $\mathbb{T} \setminus E = U_1 \cup U_2$ for disjoint open sets U_i, $i = 1, 2$, and furthermore for $i = 1, 2$ we have outer functions $f_i \in \mathcal{D}$, and constants C_i such that for all $\zeta \in \mathbb{T}$,*

$|f_i(\zeta)| \leq C_i d(\zeta, E)$. *(In particular the* $|f_i|$ *are bounded.) Then there are constants* $A = A(C_1, C_2)$, $B = B(C_1, C_2)$ *such that if* f *is the outer function which is defined by* $|f| = |f_i|$ *on* U_i. *Then* $f \in \mathcal{D}$ *and*

$$\mathfrak{D}(f) \leq \mathfrak{D}(f_1) + \mathfrak{D}(f_2) + A + B \log |f_1(0) f_2(0)|^{-1}.$$

PROOF. Recall formula (10.30) that says that if f is an outer function in H^2 then,

$$\mathfrak{D}(f) = \int_{\mathbb{T}} \int_{\mathbb{T}} \frac{(|f(\lambda)|^2 - |f(\zeta)|^2)(\log |f(\lambda)| - \log |f(\zeta)|)}{|\lambda - \zeta|^2} \frac{dt}{2\pi} \frac{ds}{2\pi}.$$

We need to estimate the integral. We break the region of integration into four parts; $U_i \times U_j$ for $i, j = 1, 2$. For $i = 1, 2$ the integral over $U_i \times U_i$ is bounded by $\mathfrak{D}(f_i)$.

The two remaining cases are symmetrical; we consider points $(\lambda, \zeta) \in U_1 \times U_2$. In that case $f(\lambda) = f_1(\lambda)$ and $f(\zeta) = f_2(\zeta)$. Because the two points are in different components of $\mathbb{T} \backslash E$ there must be a semicircular arc containing λ, ζ, and some point $\varepsilon \in E$ between the two. On this semicircle the chordal distance is uniformly comparable to the arc length distance. Hence, using the triangle inequality for arc length, there is an absolute $\alpha > 0$ so that

$$d(\lambda, \zeta) \geq \alpha \left(d(\lambda, \varepsilon) + d(\varepsilon, \zeta) \right) \geq \alpha \left(d(\lambda, E) + d(\zeta, E) \right),$$

with the second inequality because $d(\cdot, E)$ is an infimum. Thus

$$\left| \frac{(|f_1(\lambda)|^2 - |f_2(\zeta)|^2)}{|\lambda - \zeta|^2} \right| \lesssim \frac{d(\lambda, E)^2 + d(\zeta, E)^2}{|d(\lambda, E) + d(\zeta, E)|^2}.$$

Hence the contribution of $U_1 \times U_2$ to the integral dominated by a multiple of

$$\int_{U_1} \int_{U_2} \left| u(e^{it}) \right| + \left| u(e^{is}) \right| \frac{dt}{2\pi} \frac{ds}{2\pi}.$$

Recalling the submean value property of the subharmonic function $|u|$ and doing minor bookkeeping completes the proof. □

Here is a sample of using this construction to study cyclic vectors. The previous result applies to any E which is a closed set of measure zero. Here we only use the minimal version with E a singleton.

If f is an outer function in $\mathcal{D}$ then we know from Corollary 12.25 that if $\underline{Z}(f)$, defined in (12.4), is empty, then f is cyclic. We now obtain the same conclusion under the assumption that $\underline{Z}(f)$ is a singleton.

THEOREM 12.31 ([**175**][**272**]). *If* f *is outer and* $\underline{Z}(f)$ *is a single point then* f *is cyclic.*

PROOF. The Fusion Lemma was formulated for bounded functions so we first replace f by $g = f \wedge 1$ which is bounded and, by Proposition 12.15 has $[g] = [f]$.

Pick $\varepsilon_n \to 0$ and for each n let g_n be the function constructed using the lemma with the data: $|f_1(\zeta)| = \left|(\zeta - e^{i\varepsilon_n})(\zeta - e^{-i\varepsilon_n}) g(\zeta)\right|$ on $U_{1,n} = \left\{ e^{i\alpha} : |\alpha| < \varepsilon_n \right\}$ and $|f_2(\zeta)| = \left|(\zeta - e^{i\varepsilon_n})(\zeta - e^{-i\varepsilon_n})\right|$ on $U_{2,n} = \mathbb{T} \smallsetminus \overline{U_{1,n}}$.

The lemma ensures that the resulting functions g_n are in the Dirichlet space and that their norms are uniformly bounded. The construction ensures $|g_n| \leq C_n |g|$ on the boundary, and hence, by Proposition 12.8 $g_n \in [g]$. By the dominated convergence theorem $g_n \to (z-1)^2$. Having established $(z-1)^2$ is the pointwise limit of a norm bounded sequence of functions in $[g]$ we conclude, by Proposition A.18, that $(z-1)^2 \in [g]$.

To finish the proof we need to show $(z-1)^2$ is cyclic. That can be done by extending the proof in Exercise 1 that $(z-1)$ is cyclic, or by using the fact that $(z-1)$ is cyclic, together with Theorem 12.14. □

12.8. Notes and Comments

This is a rich and active area and we have just looked at some pieces. Much more is known, using the techniques in this chapter as well as ones we did not mention. There is discussion of developments in the final section of [**139**]. The book [**136**] has a rich presentation of these topics as of 2014. A sample of more recent work is [**141**].

An interesting, and rather different, approach to some of the issues discussed here is given in [**144**].

CHAPTER 13

Bilinear Forms on $\mathcal{D}$

A Hankel form is a bilinear functional whose value on a pair of functions is a linear function of their product. The characterization of the bounded Hankel forms on H^2 by Nehari in 1957 [**239**], namely that they are bounded exactly if the symbol function generates a Carleson measure for H^2, is a basic part of the rich interrelationship between H^2, H^1, $CM(H^2)$ and $(H^1)^* = BMOA$. In this chapter we give an answer to an old conjecture of the second author and show that the bounded Hankel forms on the Dirichlet space are those whose symbol functions generate Carleson measures for $\mathcal{D}$; the same as the boundedness criterion we obtained in Section 10.2.2.3 for asymmetric Hankel forms. We also then discuss how both of those results are related to weak factorization statements.

In the final section of this chapter we briefly discuss related linear operators on the Dirichlet space and other spaces, as well as some further classes of operators that are studied on the Dirichlet space.

13.1. Preliminaries

In Section 10.2.2.3 we introduced the asymmetric bilinear Hankel forms on the Dirichlet space, $\tilde{B}_b$, and the symmetric form, B_b. Given a holomorphic symbol function b and $f, g \in \mathcal{D}$, the asymmetric Hankel form $\tilde{B}_b$ on $\mathcal{D}$ was defined by

$$\tilde{B}_b(g, f) := \left\langle g', b'\overline{f} \right\rangle_{L^2(\mathbb{D})} = \langle fg', b' \rangle_{A^2} ;$$

and the symmetric Hankel form B_b on $\mathcal{D}$ by

$$B_b(f, g) := \langle fg, b \rangle_{\mathcal{D}} = \langle f'g + fg', b' \rangle_{A^2} . \tag{13.1}$$

Clearly

$$B_b(f, g) = \tilde{B}_b(f, g) + \tilde{B}_b(g, f), \tag{13.2}$$

and hence some types of information can be passed easily from $\tilde{B}_b$ to B_b. For instance, any condition on b which ensures $\tilde{B}_b$ is bounded will also ensure that B_b is bounded.

Also recall the space $\mathcal{X}$ which plays a role in the Dirichlet space theory similar to that of $BMOA$ in the Hardy space theory. A holomorphic function b is in $\mathcal{X}$ if the measure

$$d\mu_b(z) = |b'(z)|^2 \, dA(z)$$

is a Carleson measure for the Dirichlet space. We norm this space by

$$\|b\|_{\mathcal{X}} := \|\mu_b\|_{CM(\mathcal{D})} .$$

As with the "norm" of $BMOA$, this functional is actually only a seminorm. Let $\mathcal{X}_0$ be the closure in $\mathcal{X}$ of the space of polynomials.

We have the following boundedness and compactness criteria for $\tilde{B}_b$.

THEOREM 13.1. *Suppose b is holomorphic in the disk. The asymmetric bilinear form $\tilde{B}_b$ is bounded on $\mathcal{D} \times \mathcal{D}$ if and only if $b \in \mathcal{X}$. Moreover, the asymmetric form is compact if and only if $b \in \mathcal{X}_0$.*

The boundedness criterion was proved in Section 10.2.2.4. The compactness result then follows easily, by the same argument as we use below for the symmetric form.

Given the previous comments, this theorem is enough to show that $b \in \mathcal{X}$ is a sufficient condition for boundedness of B_b. Our main result in this chapter is the converse statement, $b \in \mathcal{X}$ is necessary in order for B_b to be bounded. That is not immediate from what we have said. Equation (13.2) does not foreclose the possibility that for some b the left hand side is a bounded form while the right hand side of a pair of unbounded forms which have substantial cancellation.

The proof we give is from [**53**]. A shorter proof is given by Cascante and Ortega in [**97**]. That proof is based on a different set of background ideas than the ones we are emphasizing. Richter and Sundberg develop related ideas in [**271**].

THEOREM 13.2.

(1) *$B_b : \mathcal{D} \times \mathcal{D} \to \mathbb{C}$ is bounded if and only if $b \in \mathcal{X}$. Moreover,*

$$\|B_b\|_{\mathcal{D}\times\mathcal{D}} \approx \|b\|_{\mathcal{X}} .$$

(2) *B_b is compact if and only if $b \in \mathcal{X}_0$.*

It is straightforward to see that if the first part of the theorem holds then so does the second. Suppose B_b is compact. For any holomorphic function $k(z)$ on $\mathbb{D}$ and r, $0 < r < 1$, set $S_r k(z) = k(rz)$. A computation with monomials verifies that

$$B_{S_r b}(f, g) = B_b(S_r f, S_r g).$$

As $r \to 1$, S_r converges strongly to I. Using this and the fact that B_b is compact we obtain $\lim \|B_{S_r b} - B_b\|_{\mathcal{D}\times\mathcal{D}\to\mathbb{C}} = 0$. Hence, by the first part of the theorem $\lim \|S_r b - b\|_{\mathcal{X}} = 0$. The Taylor coefficients of $S_r b$ decay geometrically and elementary estimates show $\|z^n\|_{\mathcal{X}} = O(n)$, hence $S_r b \in \mathcal{X}_0$ and thus $b \in \mathcal{X}_0$.

In the other direction note that if b is a polynomial then B_b is finite rank and hence compact. By definition the polynomials are dense in $\mathcal{X}_0$. If we select $\{b_n\}$, a sequence of polynomials which converge in norm to $b \in \mathcal{X}_0$, then, by the first part of the theorem B_b is the norm limit of the B_{b_n} and hence B_b is compact.

13.2. The Boundedness Criterion

Now we turn to the boundedness statement. That $\|B_b\|_{\mathcal{D}\times\mathcal{D}\to\mathbb{C}} \lesssim \|b\|_{\mathcal{X}}$ follows from (13.2) and the analogous result for $\tilde{B}_b$; or, alternatively, the simple argument used for $\tilde{B}_b$ adapts easily.

To complete the proof of Theorem 13.2 we must show that, given b, if B_b is bounded then $d\mu_b = |b'(z)|^2 \, dA(z)$ is a $\mathcal{D}$-Carleson measure. We will do this by showing that μ_b satisfies Stegenga's capacitary characterization of Carleson measures, Theorem 4.16. Analysis of that condition will let us focus attention on a certain set V in $\mathbb{D}$ and the relative sizes of $\int_V |b'|^2 \, dA(z)$ and the capacity of the set

$\overline{V} \cap \partial\overline{\mathbb{D}}$. To compare these quantities we construct V_{exp}, an expanded version of the set V which satisfies two conflicting conditions. First, V_{exp} is not much larger than V, either when measured by $\int_{V_{\text{exp}}} |b'|^2 \, dA(z)$ or by the capacity of the $\overline{V_{\text{exp}}} \cap \partial\overline{\mathbb{D}}$. Second, $\mathbb{D} \setminus V_{\text{exp}}$ is well separated from V, in a way that allows the interaction of quantities supported on the two sets to be controlled. Once this is done we construct a function $\Phi_V \in \mathcal{D}$ which is approximately one on V and which has Φ_V' approximately supported on $\mathbb{D} \setminus V_{\text{exp}}$. Using Φ_V we build functions f and g with the property that

$$|B_b(f,g)| = \int_V |b'(z)|^2 \, dA(z) + \text{ error.}$$

The technical estimates on Φ_V allow us to show that the error term is small. The boundedness of B_b then gives the control of $\int_V |b'|^2$ we need to show that μ_b is a Carleson measure.

13.2.1. Capacity and Blowups. To create V_{exp} we will start with a boundary set associated with V, expand that set, and then take the union of the tents over the expanded set. When we do this we will want estimates for the capacities of the various sets we encounter. We develop those estimates making crucial use of the theory of capacity for the tree Dirichlet space; results developed in Section 5.8.

13.2.2. Approximate Extremals and Capacity Estimates. Recall that Proposition 5.26 involves a function $\{h(\kappa)\}_{\kappa \in \mathcal{T}}$ defined on the tree $\mathcal{T}$ and an extremal function $H = Ih$. Recall that I is the "tree integration operator" defined in (5.4). We now define and study Φ, a holomorphic approximation to H.

We will use a parameter $s > -1$, and will place additional constraints on s later; all of which are satisfied if $s > -1$. We regard the trees such as $\mathcal{T}$ to be sitting inside $\mathbb{D}$ and will use the same name, for instance κ, for a vertex of the tree and the corresponding point of the disk.

Define $\varphi_\kappa(z) := \left(\frac{1-|\kappa|^2}{1-\overline{\kappa}z}\right)^{1+s}$ and

$$\Phi(z) := \sum_{\kappa \in \mathcal{T}} h(\kappa)\varphi_\kappa(z) = \sum_{\kappa \in \mathcal{T}} h(\kappa)\left(\frac{1-|\kappa|^2}{1-\overline{\kappa}z}\right)^{1+s}. \tag{13.3}$$

Note that for $\tau \in \mathcal{T}$

$$\sum_{\kappa \in \mathcal{T}} h(\kappa) I\delta_\kappa(\tau) = I\left(\sum_{\kappa \in \mathcal{T}} h(\kappa)\delta_\kappa\right)(\tau) = Ih(\tau) = H(\tau), \tag{13.4}$$

and so

$$\Phi(z) - H(z) = \sum_{\kappa \in \mathcal{T}} h(\kappa)(\varphi_\kappa - I\delta_\kappa)(z). \tag{13.5}$$

We are going to show that the difference in the previous equation is small. That is a quantitative version of the informal idea that functions on the tree are approximations to functions in the Dirichlet space. Alternatively (or, perhaps, equivalently) the functions φ_κ are another version of the building block functions used in decomposition theorems such as Theorem 10.12 and 10.13.

In (13.4) we are viewing h as a sum of point masses. We will obtain our holomorphic approximation by applying an integral operator to h, but it is more convenient to formulate the analysis after the point masses are defocused. Define g by

$$g(\zeta) := \sum_{\kappa \in \mathcal{T}} h(\kappa) \frac{1}{|B_\kappa|} \frac{\left(1-\overline{\zeta}\kappa\right)^{1+s}}{\left(1-|\zeta|^2\right)^s} \chi_{B_\kappa}(\zeta), \tag{13.6}$$

and B_κ is the Euclidean ball centered at κ with radius $\gamma(1-|\kappa|)$ where γ is a sufficiently small positive constant, its size will be implicit in the proof.

Still supposing that $s > -1$, define Γ_s by

$$\Gamma_s f(z) := \int_{\mathbb{D}} f(\zeta) \frac{\left(1-|\zeta|^2\right)^s}{\left(1-\overline{\zeta}z\right)^{1+s}} dA(\zeta). \tag{13.7}$$

We have used this operator before, in Section 8.4, and will use it here for the same purpose. For the function g in (13.6), $\Gamma_s f$ is a holomorphic function whose derivative is (almost) a holomorphic projection of f. It is part of the chain passing information from h and g, rough functions with desired properties, to a holomorphic function with analogous properties. We will see an analogous operator, for functions of several variables, in Chapter 15 where it is used for a similar purpose.

We defined Φ in (13.3) and we also constructed g so that $\Phi = \Gamma_s g$; we will use both presentations.

The function Φ satisfies the following estimates.

PROPOSITION 13.3. *Set* $F = \widehat{E_{\mathcal{T}}^{\rho}}$ *and write* $E = \{w_k\}$. *Suppose* $z \in \mathbb{D}$ *and* $s > -1$. *Then there are constants* C *and* c *so that*

$$\left\{ \begin{array}{lcll} |\Phi(z) - \Phi(w_k)| & \lesssim & \mathrm{Cap}_{\mathcal{T}}(E,F), & z \in T(w_k) \\ \mathrm{Re}\,\Phi(w_k) & \geq & c > 0, & k \geq 1 \\ |\Phi(w_k)| & \leq & C, & k \geq 1 \\ |\Phi(z)| & \lesssim & \mathrm{Cap}_{\mathcal{T}}(E,F), & z \notin F. \end{array} \right. \tag{13.8}$$

COROLLARY 13.4. *If* $s > -1$ *then* $\Phi = \Gamma_s g$. *If, furthermore* $s > 1/2$ *then we can select* g *so that*

$$\|\Phi\|_{\mathcal{D}}^2 \leq \int_{\mathbb{D}} |g(\zeta)|^2 \, dA \lesssim \mathrm{Cap}_{\mathcal{T}}(E,F). \tag{13.9}$$

In fact, although we will not need the fact, the inequality on the right can be attained even if we only require $s > -1/2$.

PROOF. From (13.5) we have

$$\begin{aligned} |\Phi(z) - H(z)| &\leq \sum_{\kappa \in [o,z]} |h(\kappa)(\varphi_\kappa(z) - 1)| + \sum_{\kappa \notin [o,z]} |h(\kappa)\varphi_\kappa(z)| \\ &= I(z) + II(z). \end{aligned}$$

We also have that h is nonnegative and supported in $V_G^{\gamma} \setminus V_G^{\alpha}$. We first show that

$$II(z) \leq \sum_{\kappa \notin [o,z]} h(\kappa) \left| \frac{1-|\kappa|^2}{1-\overline{\kappa}z} \right|^{1+s} \lesssim \mathrm{Cap}(E,F).$$

Select A, $1 < A < 1.17$, and let

$$\Omega_j := \left\{ \kappa \in \mathcal{T} : A^{-j-1} < \left| \frac{1-|\kappa|^2}{1-\overline{\kappa}z} \right| \leq A^{-j} \right\}.$$

We now show that for every j the set Ω_j is a union of two stopping times for $\mathcal{T}$. Let Ω_j^1 be the subset of Ω_j of points whose distance from the root is odd and set $\Omega_j^2 = \Omega_j \setminus \Omega_j^1$. We will show both are stopping times; i.e. if for $r = 1, 2$, $\kappa \in \Omega_j^r$, $\lambda \in \mathcal{T}$, and $\kappa \in [o, \lambda)$, then $\lambda \notin \Omega_j^r$.

Set $\delta\kappa = \lambda - \kappa$. We have

$$\begin{aligned} \left|\frac{1-\overline{\lambda}z}{1-|\lambda|^2}\right| &= \frac{1-|\kappa|^2}{1-|\lambda|^2}\left|\frac{1-\left(\overline{\kappa+\delta\kappa}\right)z}{1-|\kappa|^2}\right| \\ &= \frac{1-|\kappa|^2}{1-|\lambda|^2}\left|\frac{1-\overline{\kappa}z}{1-|\kappa|^2} - \frac{\overline{\delta\kappa}z}{1-|\kappa|^2}\right| \\ &\geq \frac{1-|\kappa|^2}{1-|\lambda|^2}\left\{\left|\frac{1-\overline{\kappa}z}{1-|\kappa|^2}\right| - \frac{\left|\overline{\delta\kappa}z\right|}{1-|\kappa|^2}\right\}. \end{aligned} \tag{13.10}$$

By the construction of the tree $(1-|\kappa|^2) \approx 2^s(1-|\lambda|^2)$ for some positive integer s, and if κ and λ are in the same Ω_j^r then $s \geq 2$. Also, by the construction of $\mathcal{T}$, we have

$$\frac{\left|\overline{\delta\kappa}z\right|}{1-|\kappa|^2} \leq \frac{\sqrt{2}\,(1-|\kappa|)\,|z|}{1-|\kappa|^2} \lesssim \frac{\sqrt{2}}{2},$$

and hence we continue with

$$\left|\frac{1-\overline{\lambda}z}{1-|\lambda|^2}\right| \geq 4\left(A^j - \frac{\sqrt{2}}{2}\right).$$

We are done if for each j, $A^{j+1} \leq 4\left(A^j - \sqrt{2}/2\right)$. That holds because $A < 1.17 < 4(1-\sqrt{2}/2)$.

Now, by the stopping time property in Proposition 5.33, we have

$$\sum_{\kappa\in\Omega_j} h(\kappa) \lesssim \operatorname{Cap}_{\mathcal{T}}(E, F),\ j \geq 0.$$

Altogether we then have

$$II(z) \leq \sum_{j=0}^{\infty}\sum_{\kappa\in\Omega_j} h(\kappa)\,A^{-j(1+s)} \lesssim \operatorname{Cap}_{\mathcal{T}}(E, F).$$

If $z \in \mathbb{D} \setminus F$ then $I(z) = 0$ and $H(z) = 0$ and we have

$$|\Phi(z)| = |\Phi(z) - H(z)| \leq II(z) \lesssim_s \operatorname{Cap}_{\mathcal{T}}(E, F),$$

which is the fourth line in (13.8).

If $z \in T(w_j)$, then for $\kappa \notin [o, w_j]$ we have

$$|\varphi_\kappa(w_j)| \lesssim |\varphi_\kappa(z)|,$$

and for $\kappa \in [o, z]$ we have

$$|\varphi_\kappa(z) - \varphi_\kappa(w_j)| = \left|\left(\frac{1-|\kappa|^2}{1-\overline{\kappa}z}\right)^{1+s} - \left(\frac{1-|\kappa|^2}{1-\overline{\kappa}w_j}\right)^{1+s}\right| \lesssim \frac{|z-w_j|}{1-|\kappa|^2}.$$

Thus for $z \in T\left(w_j^\alpha\right)$,

$$\begin{aligned}
\left|\Phi(z) - \Phi(w_j)\right| &\leq \sum_{\kappa \in \left[o, w_j^\alpha\right]} h(\kappa) \left|\varphi_\kappa(z) - \varphi_\kappa(w_j)\right| + C \sum_{\kappa \notin [o,z]} h(\kappa) \left|\varphi_\kappa(z)\right| \\
&\lesssim \sum_{\kappa \in \left[o, w_j^\alpha\right]} h(\kappa) \frac{\left|z - w_j\right|}{1 - |\kappa|^2} + II(z) \\
&\lesssim \operatorname{Cap}_{\mathcal{T}}(E, F).
\end{aligned}$$

Going to the last line we used that $h(\kappa) \lesssim \operatorname{Cap}_{\mathcal{T}}(E, F)$ and that $\sum_{\kappa \in [o, w_j]} \frac{|z - w_j|}{1 - |\kappa|^2} \approx \frac{|z - w_j|}{1 - |w_j|^2} = O(1)$. This proves the first line in (13.8).

Moreover, we note that for $s = 0$ and $\kappa \in [o, w_j]$,

$$\operatorname{Re} \varphi_\kappa(w_j) = \operatorname{Re} \frac{1 - |\kappa|^2}{1 - \overline{\kappa} w_j} = \operatorname{Re} \frac{1 - |\kappa|^2}{\left|1 - \overline{\kappa} w_j\right|^2} \left(1 - \kappa \overline{w_j}\right) \gtrsim 1 > 0.$$

A similar result holds for $s > -1$ provided the Bergman tree $\mathcal{T}$ is constructed sufficiently thin depending on s. It then follows from $\sum_{\kappa \in [o, w_j]} h(\kappa) = 1$ that

$$\begin{aligned}
\operatorname{Re} \Phi(w_j) &= \sum_{\kappa \in [o, w_j]} h(\kappa) \operatorname{Re} \varphi_\kappa(w_j) + \sum_{\kappa \notin [o, w_j]} h(\kappa) \operatorname{Re} \varphi_\kappa(w_j) \\
&\geq c \sum_{\kappa \in [o, w_j]} h(\kappa) - C \operatorname{Cap}_{\mathcal{T}}(E, F) \gtrsim 1 > 0.
\end{aligned}$$

We trivially have

$$\left|\Phi(w_j)\right| \leq I(z) + II(z) \lesssim \sum_{\kappa \in [o, w_j]} h(\kappa) + \operatorname{Cap}_{\mathcal{T}}(E, F) \lesssim 1,$$

and this completes the proof of (13.8).

Now we prove (15.31). From property (1) of Proposition 5.26 we obtain

$$\begin{aligned}
\int_{\mathbb{D}} |g(\zeta)|^2 \, dA &= \int_{\mathbb{D}} \left| \sum_{\kappa \in \mathcal{T}} h(\kappa) \frac{1}{|B_\kappa|} \frac{\left(1 - \overline{\zeta} \kappa\right)^{1+s}}{(1 - |\zeta|^2)^s} \chi_{B_\kappa}(\zeta) \right|^2 dA \\
&= \sum_{\kappa \in \mathcal{T}} |h(\kappa)|^2 \frac{1}{|B_\kappa|^2} \int_{B_\kappa} \frac{\left|1 - \overline{\zeta} \kappa\right|^{2+2s}}{(1 - |\zeta|^2)^{2s}} dA \\
&\approx \sum_{\kappa \in \mathcal{T}} |h(\kappa)|^2 \approx \operatorname{Cap}_{\mathcal{T}}(E, F).
\end{aligned}$$

Finally (13.9) follows from (15.31) and [**77**, Lemma 2.4]. □

COROLLARY 13.5. *Let G be a finite union of arcs in the circle $\mathbb{T}$. Then*

$$\operatorname{Cap}_{\mathbb{D}}(G) \approx \operatorname{Cap}_{\mathcal{T}}(G), \tag{13.11}$$

where $\operatorname{Cap}_{\mathbb{D}}$ denotes the capacity on the circle $\mathbb{T}$ used by Stegenga which we discussed in Chapter 4.

PROOF. To prove the inequality $\lesssim$ in (13.11) we use Proposition 13.3 to obtain a test function for estimating the Stegenga capacity of G. We take $F = \{o\}$ and $E = G$ in Proposition 13.3. Let c and C be the constants in Proposition 13.3, and suppose that $\operatorname{Cap}(E,F) \leq c/(3C)$. Set $\Psi(z) = \frac{3}{c}(\Phi(z) - \Phi(0))$. Then $\Psi(0) = 0$ and

$$\begin{aligned}\operatorname{Re}\Psi(z) &= \frac{3}{c}\{\operatorname{Re}\Phi(z) - \operatorname{Re}\Phi(0)\} \\ &\geq \frac{3}{c}\{c - 2C\operatorname{Cap}_{\mathcal{T}}(E,F)\} \geq 1,\ z \in G.\end{aligned}$$

By definitions (5.20) and (13.9) we have that for $G \subset \mathbb{T}$

$$\begin{aligned}\operatorname{Cap}_{\mathbb{D}}(G) &\leq \|\Psi\|_{\mathcal{D}}^2 = \left(\frac{3}{c}\right)^2 \|\Phi\|_{\mathcal{D}}^2 \\ &\leq \left(\frac{3}{c}\right)^2 C\ \operatorname{Cap}_{\mathcal{T}}(E,F) \lesssim \operatorname{Cap}_{\mathcal{T}} E \\ &\approx \operatorname{Cap}_{\mathcal{T}} G.\end{aligned}$$

To obtain the opposite inequality we use $\psi \in \mathcal{D}$, an extremal function for computing $\operatorname{Cap}_{\mathbb{D}} G$. For $R > 0$, $z \in \mathbb{D}$ let $B(z,R)$ be the hyperbolic disk of radius R centered at z. Pick R large enough so that for all $\kappa \in \mathcal{T} \setminus \{o\}$ we have $B(\kappa,R) \supset \operatorname{convexhull}(B_\kappa \cup B_{\kappa^{-1}})$. Our candidate for estimating $\operatorname{Cap}_{\mathcal{T}}$ is given by setting $h(o) = 0$ and

$$h(\kappa) = (1 - |\kappa|^2)\sup\{|\psi'(z)| : z \in B(\kappa,R)\};\ \kappa \in \mathcal{T} \setminus \{o\}.$$

We have the pointwise estimate

$$\begin{aligned}\operatorname{Re}\psi(\beta) \leq |\psi(\beta)| &\leq \sum_{\kappa \in [o,\beta]} \left|\psi(\kappa) - \psi\left(\kappa^{-1}\right)\right| \\ &\leq \sum_{\kappa \in [o,\beta]} \left|\kappa - \kappa^{-1}\right| \sup\left\{|\psi'(z)| : z \in \text{segment}\left(\kappa,\kappa^{-1}\right)\right\} \\ &\lesssim \sum_{\kappa \in [o,\beta]} h(\kappa) \approx Ih(\beta).\end{aligned}$$

We have the norm estimate, with $z(\kappa)$ denoting the appropriate point in $B(\kappa,R)$,

$$\begin{aligned}\|h\|_{\ell^2(\mathcal{T})}^2 &= \sum_{\kappa \in \mathcal{T}} \left(1 - |\kappa^2|\right)^2 |\psi'(z(\kappa))|^2 \\ &\lesssim \sum_{\kappa \in \mathcal{T}} \frac{\left(1 - |\kappa^2|\right)^2}{|B(\kappa,R)|} \int_{B(\kappa,R)} |\psi'(z)|^2 \, dA \\ &\lesssim \sum_{\kappa \in \mathcal{T}} \int_{B(\kappa,R)} |\psi'(z)|^2 \, dA \\ &\lesssim \int_{\mathbb{D}} |\psi'(z)|^2 \, dA \lesssim \|\psi\|_{\mathcal{D}}^2.\end{aligned}$$

Here the first inequality uses he submean value property for the subharmonic function $|\psi'(z)|^2$, the second uses straightforward estimates for $|B(\kappa,R)|$, and the next estimate holds because the $B(\kappa,R)$ are approximately disjoint; namely $\sum \chi_{B(\kappa,R)}(z) \lesssim 1$. Recalling definition (5.29) we find

$$\operatorname{Cap}_{\mathcal{T}} G \lesssim \left\|\frac{1}{c}\psi\right\|_{\mathcal{D}}^2 = \frac{C}{c^2}\operatorname{Cap}_{\mathbb{D}} G.$$

□

Recall that $T(E) = \bigcup_{I \subset E} T(I)$. Now define M by

$$M := \sup_{E \text{ open } \subset \mathbb{T}} \frac{\int_{\mathbb{T}} \mu_b (T_\theta (E))\, d\theta}{\int_{\mathbb{T}} \operatorname{Cap}_\theta (E)\, d\theta}. \tag{13.12}$$

COROLLARY 13.6. *We have* $\|\mu_b\|^2_{CM(\mathcal{D})} \approx M$.

PROOF. Using Corollary 13.5, we have

$$\begin{aligned} M &\lesssim \sup_{E \text{ open } \subset \mathbb{T}} \frac{\int_{\mathbb{T}} \mu_b (T(E))\, d\theta}{\int_{\mathbb{T}} \operatorname{Cap}_{\mathbb{D}} (E)\, d\theta} \\ &\approx \sup_{E \text{ open } \subset \mathbb{T}} \frac{\mu_b (T(E))}{\operatorname{Cap}_{\mathbb{D}} (E)} \approx \|\mu_b\|^2_{CM(\mathcal{D})}, \end{aligned}$$

where the final comparison is Stegenga's theorem. Conversely, one can verify using an argument in the style of the one in (13.15) below that for $0 < \rho < 1$,

$$\begin{aligned} \mu_b (E) &\lesssim \int_{\mathbb{T}} \mu_b (T_\theta (E^\rho_{\mathbb{D}}))\, d\theta \\ &\lesssim M \int_{\mathbb{T}} \operatorname{Cap}_\theta (E^\rho_{\mathbb{D}})\, d\theta \\ &\approx M \operatorname{Cap}_{\mathbb{D}} (E^\rho_{\mathbb{D}}) \\ &\lesssim M \operatorname{Cap}_{\mathbb{D}} (E). \end{aligned}$$

Here the third line uses (13.11) with $E^\rho_{\mathbb{D}}$ and $\mathcal{T}(\theta)$ in place of G and $\mathcal{T}$, and the final inequality follows from (5.28), Lemma 5.41 and (13.11). Thus from Stegenga's theorem we obtain

$$\|\mu_b\|^2_{CM(\mathcal{D})} \approx \sup_{E \text{ open } \subset \mathbb{T}} \frac{\mu_b (E)}{\operatorname{Cap}_{\mathbb{D}} (E)} \lesssim M.$$

□

Given $0 < \delta < 1$, let G be an open set in $\mathbb{T}$ such that

$$\frac{\int_{\mathbb{T}} \mu_b (T_\theta (G))\, d\theta}{\int_{\mathbb{T}} \operatorname{Cap}_\theta (G)\, d\theta} \geq \delta M. \tag{13.13}$$

We need to know that $\mu_b(V^\beta_G \setminus V_G)$ is small compared to $\mu_b (V_G)$. This crucial step of the proof is where we use the asymptotic capacity estimate contained in Lemma 5.41.

PROPOSITION 13.7. *Given $\varepsilon > 0$ we can choose $\delta = \delta(\varepsilon) < 1$ in* (13.13) *and $\beta = \beta(\varepsilon) < 1$ so that for any G satisfying* (13.13), *we have, with $V^\beta_G = G^\beta_{\mathbb{D}}$ and $V_G = G^1_{\mathbb{D}} = T(G)$,*

$$\mu_b(V^\beta_G \setminus V_G) \leq \varepsilon \mu_b (V_G). \tag{13.14}$$

PROOF. Let $G^\rho(\theta) = G^\rho_{\mathcal{T}_\theta}$. Lemma 5.41 shows that $\operatorname{Cap}_\theta (G^\rho (\theta)) \leq \rho^{-2} \operatorname{Cap}_\theta (G)$, for $0 \leq \theta < 2\pi$, $0 < \rho < 1$, and if we integrate on $\mathbb{T}$ we obtain

$$\int_{\mathbb{T}} \operatorname{Cap}_\theta (G^\rho (\theta))\, d\theta \leq \rho^{-2} \int_{\mathbb{T}} \operatorname{Cap}_\theta (G)\, d\theta.$$

From (13.12) and (13.13) we thus have

$$\begin{aligned}\int_{\mathbb{T}} \mu_b\left(T_\theta\left(G^\rho\left(\theta\right)\right)\right) d\theta &\leq M \int_{\mathbb{T}} \operatorname{Cap}_\theta\left(G^\rho\left(\theta\right)\right) d\theta \\ &\leq M\rho^{-2} \int_{\mathbb{T}} \operatorname{Cap}_\theta\left(G\right) d\theta \\ &\leq \frac{1}{\delta\rho^2} \int_{\mathbb{T}} \mu_b\left(T_\theta\left(G\right)\right) d\theta.\end{aligned}$$

It follows that

$$\begin{aligned}\int_{\mathbb{T}} \mu_b\left(T_\theta\left(G^\rho\left(\theta\right)\right) \setminus T_\theta\left(G\right)\right) d\theta &= \int_{\mathbb{T}} \mu_b\left(T_\theta\left(G^\rho\left(\theta\right)\right)\right) d\theta - \int_{\mathbb{T}} \mu_b\left(T_\theta\left(G\right)\right) d\theta \\ &\leq \left(\frac{1}{\delta\rho^2} - 1\right) \int_{\mathbb{T}} \mu_b\left(T_\theta\left(G\right)\right) d\theta.\end{aligned}$$

Now with $\eta = (\rho+1)/2$,

$$\begin{aligned}\int_{\mathbb{T}} \mu_b\left(T_\theta\left(G^\rho\left(\theta\right)\right) \setminus T_\theta\left(G\right)\right) d\theta &= \int_{\mathbb{T}} \int_{T_\theta(G^\rho(\theta)) \setminus T_\theta(G)} d\mu_b\left(z\right) d\theta \\ &\geq \int_{\mathbb{T}} \int_{T(G^\rho(\theta)) \setminus T(G)} d\mu_b\left(z\right) d\theta\end{aligned} \tag{13.15}$$

$$= \int_{\mathbb{D}} \left\{ \frac{1}{2\pi} \int_{\{\theta: z \in T(G^\rho(\theta)) \setminus T(G)\}} d\theta \right\} d\mu_b\left(z\right) \geq \frac{1}{2} \int_{T\left(G^\eta_{\mathbb{D}}\right) \setminus T(G)} d\mu_b\left(z\right),$$

since every $z \in T\left(G^\eta_{\mathbb{D}}\right)$ lies in $T\left(G^\rho\left(\theta\right)\right)$ for at least half of the θ's in $[0, 2\pi)$. Here we may assume that the components of $G^\rho_{\mathbb{D}}$ have small length since otherwise we trivially have $\int_{\mathbb{T}} \operatorname{Cap}_{\mathcal{T}(\theta)}\left(G\right) d\theta \gtrsim 1 > 0$. We continue with

$$M \leq \frac{1}{c} \int_{\mathbb{D}} d\mu_b(z) = \frac{1}{c} \left\|b\right\|^2_{\mathcal{D}} \leq \frac{C}{c} \left\|B_b\right\|^2_{\mathcal{D} \times \mathcal{D} \to \mathbb{C}}. \tag{13.16}$$

Combining the above inequalities, using $\rho = 2\eta - 1$, $1/2 \leq \rho < 1$, and choosing $\delta = \eta$, we obtain

$$\begin{aligned}\mu_b\left(T\left(G^\eta_{\mathbb{D}}\right) \setminus T\left(G\right)\right) &\leq 2\left(\frac{1}{\delta\rho^2} - 1\right) \int_{\mathbb{T}} \mu_b\left(T_\theta\left(G\right)\right) d\theta \\ &= 2\left(\frac{1}{\eta\left(2\eta-1\right)^2} - 1\right) \int_{\mathbb{T}} \mu_b\left(T_\theta\left(G\right)\right) d\theta \\ &\lesssim \left(1-\eta\right) \int_{\mathbb{T}} \mu_b\left(T_\theta\left(G\right)\right) d\theta,\end{aligned}$$

for $3/4 \leq \eta < 1$. Recalling that $V^\eta_G = T\left(G^\eta_{\mathbb{D}}\right)$ and that for all θ we have $T_\theta\left(G\right) \subset T\left(G\right) = V_G$ this becomes

$$\mu_b\left(V^\eta_G \setminus V_G\right) \lesssim \left(1-\eta\right) \int_{\mathbb{T}} \mu_b\left(T_\theta\left(G\right)\right) d\theta \lesssim \left(1-\eta\right) \mu_b\left(V_G\right), \quad \frac{3}{4} \leq \eta < 1.$$

Hence given $\varepsilon > 0$ it is possible to select δ and β so that (13.14) holds. □

13.2.3. Schur Estimates and a Bilinear Operator on Trees. The following lemma is proved using a bilinear version of Schur's well-known theorem. That version of Schur's result and the lemma are both proved in Appendix B; see Section B.6.2.

LEMMA 13.8. *Suppose $\mathcal{A}$ and $\mathcal{B}$ are subsets of $\mathcal{T}$, $h \in \ell^2(\mathcal{A})$ and $k \in \ell^2(\mathcal{B})$. Suppose further that for some α, $1/2 < \alpha < 1$, $\mathcal{A}$ and $\mathcal{B}$ satisfy the separation condition: $\forall \kappa \in \mathcal{A}$, $\gamma \in \mathcal{B}$ we have*

$$|\kappa - \gamma| \geq (1 - |\gamma|^2)^\alpha. \tag{13.17}$$

Then the bilinear map of (h, k) to functions on the disk given by

$$T(h,k)(z) := \left(\sum_{\kappa \in \mathcal{A}} h(\kappa) \frac{(1-|\kappa|^2)^{1+s}}{|1-\overline{\kappa}z|^{2+s}} \right) \left(\sum_{\gamma \in \mathcal{B}} k(\gamma) \frac{(1-|\gamma|^2)^{1+s}}{|1-\overline{\gamma}z|^{1+s}} \right)$$

is bounded from $\ell^2(\mathcal{A}) \times \ell^2(\mathcal{B})$ to $L^2(\mathbb{D})$.

EXERCISE 13.9. *The condition* (13.17) *is a separation condition, Make sketches of those regions, including a sketch for the case which is not allowed, $\alpha = 1$.*

REMARK 13.10. *For $h \in \ell^2(\mathcal{A})$ and $b \in \ell^2(\mathcal{B})$ set*

$$H(z) := \sum_{\kappa \in \mathcal{A}} h(\kappa) \frac{(1-|\kappa|^2)^{1+s}}{(1-\overline{\kappa}z)^{2+s}}, \quad B(z) := \sum_{\gamma \in \mathcal{B}} b(\gamma) \frac{(1-|\gamma|^2)^{1+s}}{(1-\overline{\gamma}z)^{1+s}}.$$

By [**346**, Theorem 2.30] *H is in the Bergman space A^2 and B is in the Dirichlet space D. However those two facts are not, themselves, enough to ensure HB is in A^2 because, as we saw in Section* 2.1.2, *$\mathcal{D}$ contains unbounded functions and, as we saw in Section* 2.4.2, *multipliers of A^2 are bounded. The lemma shows that if $\mathcal{A}$ and $\mathcal{B}$ are separated then HB is in A^2.*

13.2.4. The Main Bilinear Estimate. To complete the proof we will show that μ_b is a $\mathcal{D}$-Carleson measure by verifying Stegenga's condition (5.26); that is, we will show that for any finite collection of disjoint arcs $\{I_j\}_{j=1}^N$ in the circle $\mathbb{T}$ we have

$$\mu_b \left(\bigcup_{j=1}^N T(I_j) \right) \lesssim \operatorname{Cap}_{\mathbb{D}} \left(\bigcup_{j=1}^N I_j \right).$$

In fact we will see that it suffices to verify this for the sets $G = \cup_{j=1}^N I_j$ described in (13.13) that are almost extremal for (13.12). We will prove the inequality

$$\mu_b(V_G) \lesssim \|B_b\|^2_{\mathcal{D}\times\mathcal{D}\to\mathbb{C}} \operatorname{Cap}_{\mathbb{D}}(G). \tag{13.18}$$

Once we have this, Corollary 13.5 yields

$$M = \frac{\int_{\mathbb{T}} \mu_b(T_\theta(G))\, d\theta}{\int_{\mathbb{T}} \operatorname{Cap}_\theta(G)\, d\theta} \leq \frac{\mu_b(V_G)}{\int_{\mathbb{T}} \operatorname{Cap}_\theta(G)\, d\theta} \lesssim \|B_b\|^2_{\mathcal{D}\times\mathcal{D}\to\mathbb{C}}.$$

By Corollary 13.6, $\|\mu_b\|^2_{\mathcal{D}-Carleson} \approx M$ which then completes the proof of Theorem 13.2.

We now turn to (13.18). Let $1/2 < \beta < \beta_1 < \gamma < \alpha < 1$ with additional constraints to be added later. Suppose G satisfies (13.13) with $\varepsilon > 0$ to be chosen below. We define in succession the following regions in the disk,

$$\begin{aligned} V_G &= T_{\mathcal{T}}(G), \\ V_G^\alpha &= G_{\mathbb{D}}^\alpha, \\ V_G^\gamma &= \widehat{(V_G^\alpha)_{\mathcal{T}}^{\gamma/\alpha}}, \\ V_G^\beta &= (V_G^\gamma)_{\mathbb{D}}^{\beta/\gamma}. \end{aligned}$$

Thus V_G is the $\mathcal{T}$-tent associated with G, which is the same as V_G^1, the 1-blowup of G; V_G^α is a disk blowup of G, V_G^γ is a $\mathcal{T}$-capacitary blowup of V_G^α, and V_G^β is a disk blowup of V_G^γ. In particular $V_G \subset V_G^\alpha \subset V_G^\gamma \subset V_G^\beta$. Using the natural bijections described in Section 5.8.1, we write

$$V_G = \{w_k\}_k \text{ and } V_G^\alpha = \{w_k^\alpha\}_k \text{ and } V_G^\gamma = \{w_k^\gamma\}_k \text{ and } V_G^\beta = \{w_k^\beta\}_k, \tag{13.19}$$

with $w_k, w_k^\alpha, w_k^\gamma, w_k^\beta \in \mathcal{T}$. Following earlier notation we write $E = V_G^\alpha$ and $F = V_G^\gamma$.

The hypothesis that B_b is bounded lets us estimate $B_b(f,g)$ for any $f, g \in \mathcal{D}$. We now use that fact for well chosen f and g. Let Φ be as in (13.9), we then have the estimates in Proposition 13.3 and Corollary 13.4 that can be applied to Φ. Set $g = \Phi^2$; in particular note that g is approximately equal to χ_{V_G}. The function f defined by

$$f(z) := \Gamma_s\left(\frac{1}{(1+s)\overline{\zeta}}\chi_{V_G} b'(\zeta)\right)(z) = \int_{V_G} \frac{b'(\zeta)\left(1-|\zeta|^2\right)^s}{\left(1-\overline{\zeta}z\right)^{1+s}} \frac{dA(\zeta)}{(1+s)\overline{\zeta}} \tag{13.20}$$

will be approximately $b'\chi_{V_G}$.

We now analyze $B_b(f,g)$. From (13.20) we obtain

$$\begin{aligned} f'(z) &= \int_{V_G} \frac{b'(\zeta)\left(1-|\zeta|^2\right)^s}{\left(1-\overline{\zeta}z\right)^{2+s}} dA(\zeta) \\ &= b'(z) - \int_{\mathbb{D}\setminus V_G} \frac{b'(\zeta)\left(1-|\zeta|^2\right)^s}{\left(1-\overline{\zeta}z\right)^{2+s}} dA(\zeta) \\ &= b'(z) + \Lambda b'(z), \end{aligned}$$

where the last term is defined by

$$\Lambda b'(z) = -\int_{\mathbb{D}\setminus V_G} \frac{b'(\zeta)\left(1-|\zeta|\right)^s}{\left(1-\overline{\zeta}z\right)^{2+s}} dA(\zeta). \tag{13.21}$$

We have

(13.22)

$$\begin{aligned} B_b(f,g) &= \left(f\Phi^2\overline{b}\right)(0) + \int_{\mathbb{D}} \left(f'(z)\Phi(z) + 2f(z)\Phi'(z)\right)\Phi(z)\overline{b'(z)}dA(z) \\ &= \left(f\Phi^2\overline{b}\right)(0) + \int_{\mathbb{D}} |b'(z)|^2\Phi(z)^2 dA(z) \\ &\quad + 2\int_{\mathbb{D}} \Phi(z)\Phi'(z)f(z)\overline{b'(z)}dA(z) + \int_{\mathbb{D}} \Lambda b'(z)\overline{b'(z)}\Phi(z)^2 dA(z) \\ &= (1)+(2)+(3)+(4). \end{aligned}$$

Now we write

$$\begin{aligned} (2) &= \int_{\mathbb{D}} |b'(z)|^2\Phi(z)^2 dA(z) \\ &= \left\{\int_{V_G} + \int_{V_G^\beta \smallsetminus V_G} + \int_{\mathbb{D}\smallsetminus V_G^\beta}\right\} |b'(z)|^2\Phi(z)^2 dA(z) \\ &= (2_A)+(2_B)+(2_C). \end{aligned} \tag{13.23}$$

The main term is (2_A). By (13.8) it satisfies

$$\begin{aligned} (2_A) &= \mu_b(V_G) + \int_{V_G} |b'(z)|^2(\Phi(z)^2-1)dA(z) \\ &= \mu_b(V_G) + O(\|B_b\|^2_{\mathcal{D}\times\mathcal{D}\to\mathbb{C}}\operatorname{Cap}_{\mathcal{T}}(E,F)). \end{aligned} \tag{13.24}$$

Rearranging this and using (13.22) and (13.23) we find

$$\begin{aligned} \mu_b(V_G) \lesssim \|B_b\|^2\operatorname{Cap}_{\mathcal{T}}(E,F) + |B_b(f,g)| \\ + |(1)| + |(2_B)| + |(2_C)| + |(3)| + |(4)| \end{aligned} \tag{13.25}$$

Using the boundedness of B_b and Corollary 13.4 we have

$$\begin{aligned} |B_b(f,g)| &= \left|B_b(f,\Phi^2)\right| = |B_b(f\Phi,\Phi)| \\ &\leq \|B_b\|_{\mathcal{D}\times\mathcal{D}\to\mathbb{C}}\|f\Phi\|_{\mathcal{D}}\|\Phi\|_{\mathcal{D}} \\ &\lesssim \|B_b\|_{\mathcal{D}\times\mathcal{D}\to\mathbb{C}}\|f\Phi\|_{\mathcal{D}}\sqrt{\operatorname{Cap}_{\mathcal{T}}(E,F)}. \end{aligned} \tag{13.26}$$

For (1) we use the elementary estimates on b, f and Φ to find:

$$|(1)| \lesssim \|b\|^2_{\mathcal{D}}\operatorname{Cap}_{\mathcal{T}}(E,F) \lesssim \|B_b\|^2_{\mathcal{D}\times\mathcal{D}\to\mathbb{C}}\operatorname{Cap}_{\mathcal{T}}(E,F).$$

For (2_B) we use (13.14) to obtain

$$|(2_B)| \lesssim \mu_b\left(V_G^\beta \smallsetminus V_G\right) \lesssim \varepsilon\mu_b(V_G). \tag{13.27}$$

Using (13.8) once more, we see that (2_C) satisfies

$$\begin{aligned} |(2_C)| &\lesssim \int_{\mathbb{D}\smallsetminus V_G^\beta} |b'(z)|^2(\operatorname{Cap}_{\mathcal{T}}(E,F))^2 dA \\ &\lesssim \|B_b\|^2_{\mathcal{D}\times\mathcal{D}\to\mathbb{C}}\operatorname{Cap}_{\mathcal{T}}(E,F). \end{aligned} \tag{13.28}$$

Putting these estimates into (13.25) we obtain

$$\begin{aligned} \mu_b(V_G) \lesssim (\|B_b\|^2_{\mathcal{D}\times\mathcal{D}\to\mathbb{C}}\operatorname{Cap}_{\mathcal{T}}(E,F) \\ + \|B_b\|_{\mathcal{D}\times\mathcal{D}\to\mathbb{C}}\|f\Phi\|_{\mathcal{D}}\sqrt{\operatorname{Cap}_{\mathcal{T}}(E,F)} + |(3)| + |(4)|). \end{aligned} \tag{13.29}$$

We now estimate (3) using the Arithmetic Mean-Geometric Mean Inequality. For any $\varepsilon > 0$ we have

$$\begin{aligned}|(3)| &\leq 2\int_{\mathbb{D}} |\Phi(z)\, b'(z)|\, |\Phi'(z)\, f(z)|\, dA(z) \\ &\leq \varepsilon \int_{\mathbb{D}} |\Phi(z)\, b'(z)|^2\, dA(z) + \frac{C}{\varepsilon}\int_{\mathbb{D}} |\Phi'(z)\, f(z)|^2\, dA(z) \\ &= (3_A) + (3_B).\end{aligned}$$

Using the decomposition and the argument surrounding term (2) we obtain

$$\begin{aligned}|(3_A)| &\leq \varepsilon \left\{ \int_{V_G} + \int_{V_G^\beta \setminus V_G} + \int_{\mathbb{D}\setminus V_G^\beta} \right\} |\Phi(z)\, b'(z)|^2\, dA(z) \\ &\lesssim \varepsilon \left(\mu_b(V_G) + C\, \|B_b\|^2_{\mathcal{D}\times\mathcal{D}\to\mathbb{C}} \operatorname{Cap}_{\mathcal{T}}(E, F) \right).\end{aligned} \tag{13.30}$$

To estimate term (3_B) we use

$$\begin{aligned}|f(z)| &\leq \left| \Gamma_s \left(\frac{1}{(1+s)\,\overline{\zeta}} \chi_{V_G}\, b'(\zeta) \right)(z) \right| \\ &\leq \int_{V_G} \frac{(1-|\zeta|^2)^s}{\left|1-\overline{\zeta} z\right|^{1+s}}\, |b'(\zeta)|\, dA(\zeta) \\ &\approx \sum_{\gamma\in\mathcal{T}\cap V_G} \frac{(1-|\gamma|^2)^{1+s}}{\left|1-\overline{\gamma} z\right|^{1+s}} \int_{B_\gamma} |b'(\zeta)|\, (1-|\zeta|^2)\, d\lambda(\zeta) \\ &:= \sum_{\gamma\in\mathcal{T}\cap V_G} \frac{(1-|\gamma|^2)^{1+s}}{\left|1-\overline{\gamma} z\right|^{1+s}}\, k(\gamma),\end{aligned}$$

where

$$\sum_{\gamma\in\mathcal{T}\cap V_G} k(\gamma)^2 \approx \sum_{\gamma\in\mathcal{T}\cap V_G} \int_{B_\gamma} |b'(\zeta)|^2\, (1-|\zeta|^2)^2 d\lambda(\zeta) = \int_{V_G} |b'(\zeta)|^2\, dA(\zeta).$$

We now use the separation of $\mathbb{D}\setminus V_G^\alpha$ and V_G. The facts that $\mathcal{A} = \operatorname{supp}(h) \subset \mathbb{D}\setminus V_G^\alpha$ and $\mathcal{B} = \mathcal{T}\cap V_G \subset V_G$, together with Lemma 5.36, ensure that (13.17) is satisfied and hence we can use Lemma 13.8 and the representation of Φ in (13.3) to continue with

$$(3_B) = \int_{\mathbb{D}} |\Phi'(z)\, f(z)|^2\, dA(z) \lesssim \left(\sum_{\kappa\in\mathcal{A}} h(\kappa)^2 \right) \left(\sum_{\gamma\in\mathcal{B}} k(\gamma)^2 \right).$$

We also have from Corollary 13.4 that

$$\left(\sum_{\kappa\in\mathcal{A}} h(\kappa)^2 \right) \left(\sum_{\gamma\in\mathcal{B}} k(\gamma)^2 \right) \lesssim \operatorname{Cap}_{\mathcal{T}}(E, F)\, \|B_b\|^2_{\mathcal{D}\times\mathcal{D}\to\mathbb{C}}.$$

Altogether we then have

$$(3_B) \lesssim \operatorname{Cap}_{\mathcal{T}}(E, F)\, \|B_b\|^2, \tag{13.31}$$

and thus also

$$|(3)| \leq \varepsilon \mu_b(V_G) + C\, \|B_b\|^2_{\mathcal{D}\times\mathcal{D}\to\mathbb{C}} \operatorname{Cap}_{\mathcal{T}}(E, F). \tag{13.32}$$

We begin our estimate of term (4) by

$$|(4)| = \left| \int_{\mathbb{D}} \Lambda b'(z) \overline{b'(z)} \Phi(z)^2 \, dA(z) \right| \tag{13.33}$$
$$\leq \sqrt{\int_{\mathbb{D}} |b'(z)\,\Phi(z)|^2 \, dA(z)} \sqrt{\int_{\mathbb{D}} |\Lambda b'(z)\,\Phi(z)|^2 \, dA(z)}.$$

Setting $(4_A) = \|\Phi \Lambda b'\|^2_{L^2(\mathbb{D})}$ the previous estimate is

$$|(4)| \leq \sqrt{(3_A)/\varepsilon} \sqrt{(4_A)}.$$

We now estimate (4_A) with a lemma:

LEMMA 13.11.

$$(4_A) = \int_{\mathbb{D}} |\Phi(z)\,\Lambda b'(z)|^2 \, dA(z) \tag{13.34}$$
$$\lesssim \mu_b \left(V_G^\beta \setminus V_G \right) + \|B_b\|^2_{\mathcal{D}\times\mathcal{D}\to\mathbb{C}} \operatorname{Cap}_{\mathcal{T}}(E, F).$$

PROOF. From (13.21) we obtain

$$|(4_A)| \leq \int_{\mathbb{D}} |\Phi(z)|^2 \left| \left\{ \int_{V_G^\beta \setminus V_G} + \int_{\mathbb{D} \setminus V_G^\beta} \right\} \frac{b'(\zeta)(1 - |\zeta|)^s}{(1 - \overline{\zeta} z)^{2+s}} dA(\zeta) \right|^2 dA(z)$$
$$\lesssim \int_{\mathbb{D}} |\Phi(z)|^2 \left(\int_{V_G^\beta \setminus V_G} \frac{|b'(\zeta)|(1 - |\zeta|)^s}{|1 - \overline{\zeta} z|^{2+s}} dA(\zeta) \right)^2 dA(z)$$
$$+ \int_{\mathbb{D}} |\Phi(z)|^2 \left| \int_{\mathbb{D} \setminus V_G^\beta} \frac{b'(\zeta)(1 - |\zeta|)^s}{(1 - \overline{\zeta} z)^{2+s}} dA(\zeta) \right|^2 dA(z)$$
$$= (4_{AA}) + (4_{AB}).$$

Theorem B.6 shows that

$$(4_{AA}) \leq \int_{\mathbb{D}} \left(\int_{V_G^\beta \setminus V_G} \frac{|b'(\zeta)|(1 - |\zeta|)^s}{|1 - \overline{\zeta} z|^{2+s}} dA(\zeta) \right)^2 dA(z)$$
$$\lesssim \int_{V_G^\beta \setminus V_G} |b'(\zeta)|^2 \, dA(\zeta) \approx \mu_b \left(V_G^\beta \setminus V_G \right).$$

We write the second integral as

$$(4_{AB}) = \left\{ \int_{V_G^\gamma} + \int_{\mathbb{D} \setminus V_G^\gamma} \right\} |\Phi(z)|^2 \left| \int_{\mathbb{D} \setminus V_G^\beta} \frac{b'(\zeta)(1 - |\zeta|)^s}{(1 - \overline{\zeta} z)^{2+s}} dA(\zeta) \right|^2 dA(z)$$
$$= (4_{ABA}) + (4_{ABB}),$$

where by Theorem B.6 again,

$$(4_{ABB}) \lesssim \operatorname{Cap}_{\mathcal{T}}(E, F)^2 \int_{\mathbb{D}} |b'(\zeta)|^2 \, dA(\zeta)$$
$$\lesssim \|B_b\|^2 \operatorname{Cap}_{\mathcal{T}}(E, F)^2$$
$$\lesssim \|B_b\|^2 \operatorname{Cap}_{\mathcal{T}}(E, F).$$

Here we are assuming that $\operatorname{Cap}_{\mathcal{T}}(E,F)$ is less than 1, otherwise we have the result we want through easy estimates. Finally, with $\beta<\beta_1<\gamma<\alpha<1$, Theorem B.6 also shows that the term (4_{ABA}) satisfies the following estimate. Recall that $V_G^\gamma=\cup J_k^\gamma$ and $w_j^\gamma=z\left(J_k^\gamma\right)$ We set $A_\ell=\left\{k: J_k^\gamma\subset J_\ell^{\beta_1}\right\}$ and define $\ell(k)$ by the condition $k\in A_{\ell(k)}$. From Lemma 5.36 we have that, with $\rho=\beta_1/\gamma$, side length$(J_k^\gamma)\leq$ side length$(J_\ell^{\beta_1})^{1/\rho}$. Hence

$$
\begin{aligned}
(4_{ABA}) &\lesssim \int_{V_G^\gamma}\left(\int_{\mathbb{D}\setminus V_G^\beta}\frac{\left|b'(\zeta)\right|(1-|\zeta|)^s}{\left|1-\overline{\zeta}z\right|^{2+s}}dA(\zeta)\right)^2 dA(z)\\
&\approx \sum_k\int_{J_k^\gamma}\left|J_k^\gamma\right|\left(\int_{\mathbb{D}\setminus V_G^\beta}\frac{\left|b'(\zeta)\right|(1-|\zeta|)^s}{\left|1-\overline{\zeta}w_k^\gamma\right|^{2+s}}dA(\zeta)\right)^2 dA(z)\\
&\approx \sum_k\frac{\left|J_k^\gamma\right|}{\left|J_{\ell(k)}^{\beta_1}\right|}\left|J_{\ell(k)}^{\beta_1}\right|\left(\int_{\mathbb{D}\setminus V_G^\beta}\frac{\left|b'(\zeta)\right|(1-|\zeta|)^s}{\left|1-\overline{\zeta}w_k^\gamma\right|^{2+s}}dA(\zeta)\right)^2\\
&\approx \sum_\ell\frac{\sum_{k\in A_\ell}\left|J_k^\gamma\right|}{\left|J_\ell^{\beta_1}\right|}\int_{J_\ell^{\beta_1}}\left(\int_{\mathbb{D}\setminus V_G^\beta}\frac{\left|b'(\zeta)\right|(1-|\zeta|)^s}{\left|1-\overline{\zeta}z\right|^{2+s}}dA(\zeta)\right)^2 dA(z)\\
&\lesssim \left|V_G^{\beta_1}\right|^{\varepsilon(\gamma-\beta_1)}\int_{V_G^{\beta_1}}\left(\int_{\mathbb{D}\setminus V_G^\beta}\frac{\left|b'(\zeta)\right|(1-|\zeta|)^s}{\left|1-\overline{\zeta}z\right|^{2+s}}dA(\zeta)\right)^2 dA(z)\\
&\lesssim \left|V_G^{\beta_1}\right|^{\varepsilon(\gamma-\beta_1)}\|b\|_{\mathcal{D}}^2\lesssim\|B_b\|_{\mathcal{D}\times\mathcal{D}\to\mathbb{C}}^2\operatorname{Cap}_{\mathcal{T}}(E,F).
\end{aligned}
$$

We continue from (13.33). We know that $|(4)|\leq\sqrt{(3_A)/\varepsilon}\sqrt{(4_A)}$. We estimate (3_A) using (13.30) and estimate (4_A) using (13.34). After that we continue by using (13.14);

$$
\begin{aligned}
|(4)| &\lesssim \sqrt{\mu_b(V_G)+\|B_b\|_{\mathcal{D}\times\mathcal{D}\to\mathbb{C}}^2\operatorname{Cap}_{\mathcal{T}}(E,F)}\\
&\qquad\times\sqrt{\mu_b\left(V_G^\beta\setminus V_G\right)+\|B_b\|_{\mathcal{D}\times\mathcal{D}\to\mathbb{C}}^2\operatorname{Cap}_{\mathcal{T}}(E,F)}\\
&\lesssim \sqrt{\mu_b(V_G)+\|B_b\|_{\mathcal{D}\times\mathcal{D}\to\mathbb{C}}^2\operatorname{Cap}_{\mathcal{T}}(E,F)}\\
&\qquad\times\sqrt{\varepsilon\mu_b(V_G)+\|B_b\|_{\mathcal{D}\times\mathcal{D}\to\mathbb{C}}^2\operatorname{Cap}_{\mathcal{T}}(E,F)}\\
&\leq\sqrt{\varepsilon}\mu_b(V_G)+\sqrt{\mu_b(V_G)}\sqrt{\|B_b\|_{\mathcal{D}\times\mathcal{D}\to\mathbb{C}}^2\operatorname{Cap}_{\mathcal{T}}(E,F)}\\
&\qquad+\|B_b\|_{\mathcal{D}\times\mathcal{D}\to\mathbb{C}}^2\operatorname{Cap}_{\mathcal{T}}(E,F).
\end{aligned}
\tag{13.35}
$$

Now, recalling that $f'=b'+\Lambda b'$,

$$
\begin{aligned}
\|\Phi f\|_{\mathcal{D}}^2 &\lesssim \int\left|\Phi'(z)f(z)\right|^2 dA(z)+\int\left|\Phi(z)\left(b'(z)+\Lambda b'(z)\right)\right|^2 dA(z)\\
&\lesssim (3_B)+\frac{1}{\varepsilon}(3_A)+(4_A).\\
&\lesssim \mu_b(V_G)+\|B_b\|_{\mathcal{D}\times\mathcal{D}\to\mathbb{C}}^2\operatorname{Cap}_{\mathcal{T}}(E,F),
\end{aligned}
\tag{13.36}
$$

by (13.34) and the estimates (13.30) and (13.31) for (3_A) and (3_B).

Using Proposition 13.7 and the estimates (13.32), (13.35) and (13.36) in (13.29) we obtain

$$\begin{aligned}\mu_b(V_G) &\leq \sqrt{\varepsilon}\mu_b(V_G) + C\|B_b\|^2_{\mathcal{D}\times\mathcal{D}\to\mathbb{C}} \operatorname{Cap}_{\mathcal{T}}(E,F) \\ &\quad + C\sqrt{\|B_b\|^2_{\mathcal{D}\times\mathcal{D}\to\mathbb{C}} \operatorname{Cap}_{\mathcal{T}}(E,F)}\sqrt{\mu_b(V_G)} \\ &\leq \sqrt{\varepsilon}\mu_b(V_G) + C\|B_b\|^2_{\mathcal{D}\times\mathcal{D}\to\mathbb{C}} \operatorname{Cap}_{\mathcal{T}}(E,F).\end{aligned}$$

We absorb the first term into the left side. Now using Lemma 5.40, Lemma 5.41 again, and Corollary 13.5 we obtain

$$\operatorname{Cap}_{\mathcal{T}}(E,F) \lesssim \operatorname{Cap}_{\mathbb{D}} G.$$

Finally we have

$$\mu_b(V_G) \lesssim \|B_b\|^2_{\mathcal{D}\times\mathcal{D}\to\mathbb{C}} \operatorname{Cap}_{\mathcal{T}}(E,F) \lesssim \|B_b\|^2_{\mathcal{D}\times\mathcal{D}\to\mathbb{C}} \operatorname{Cap}_{\mathbb{D}} G,$$

which is (13.18). □

13.3. Related Operators

In Chapter 10 we briefly discussed Hankel forms on the Hardy, Dirichlet, and Bergman space, and we looked in detail at the asymmetric Hankel forms on the Dirichlet space. In this chapter we have been considering the symmetric Hankel form on the Dirichlet space. We now take a moment for an informal look at some other variations on those ideas. We follow that with a brief discussion of other families of operators that are also frequently studied on function spaces with reproducing kernels.

These are large topics and our goal here is limited to letting the reader know that the topics are out there. We offer some superficial comments and a few references to help in further exploration. The main criterion for selecting these particular references is that they were at the top of the authors' memory stacks.

13.3.1. Beyond Hankel Forms.

We have considered Hankel forms on $\mathcal{D}$, H^2, and A^2. Hankel forms are also defined and studied on many other spaces of holomorphic functions of one or several complex variables. Some references are [**118**], [**184**], [**199**], [**80**], [**149**], [**200**]. There are also several interesting classes of linear operators that are closely related, either behaviorally or historically, to Hankel forms. We will take a very brief look at a few.

Additionally there are several interesting classes of linear operators that are closely related, either behaviorally or historically, to Hankel forms and operators.

13.3.1.1. *Weak Factorization.* In the discussion in Section 10.2.2 we defined the Hankel bilinear form $B_b^{H^2}$ on the Hardy space. Given a holomorphic symbol function b we set, for $f, g \in H^2$

$$B_b^{H^2}(f,g) - \langle fg, b\rangle_{H^2}.$$

We then outlined a proof of the following result.

THEOREM 13.12 (Nehari). *$B_b^{H^2} : H^2 \times H^2 \to \mathbb{C}$ is bounded if and only if $b \in (H^1)^*$. Moreover,*

$$\left\|B_b^{H^2}\right\|_{H^2\times H^2\to\mathbb{C}} \approx \|b\|_{(H^1)^*}.$$

At the heart of Nehari's proof is the following observation. The boundedness of $B_b^{H^2}$ is equivalent to having a bound for $\langle fg, b\rangle_{H^2}$ as f and g range over the unit ball of H^2 and, by Proposition 10.15 this is equivalent to having a bound for $\langle h, b\rangle$ as h ranges over the unit ball of H^1. Thus the condition of interest on b is that it be in the dual space $\left(H^1\right)^*$. In current formulations this result is combined with the later identification $(H^1)^* = BMO$ (Theorem 3.12).

In Section 10.2.2.2 we discussed a similar question for the Bergman space; what condition on b ensures that the form

$$B_b^{A^2}(f, g) = \langle fg, b\rangle_{A^2}. \tag{13.37}$$

is bounded? For this case, if we try to follow the Hardy space argument there is an obstacle. Although the product of unit vectors in A^2 is in the unit ball of A^1 it is not clear if that gives the full ball; however, see [**179**]. We went around that obstacle using a *weak factorization theorem* for A^1, Theorem 10.5; which states that every function in A^1 is a sum of products of functions in A^2 with good norm control. That is a sufficient tool to show that, for a holomorphic symbol b, $B_b^{A^2}$ is bounded if and only if b is in the dual space $\left(A^1\right)^* = BLOCH$.

(The previous paragraph is not a historical description. The identification of the b for which $B_b^{A^2}$ is bounded is due to Anderson, Clunie and Pommerenke [**24**] and was obtained by different methods.)

In fact these arguments can be reversed. A boundedness criterion for a class of Hankel operators can be used to obtain a weak factorization result for the predual of the associated space of symbols. With this general fact in mind we define the weak product of two Banach spaces of functions. Suppose $\mathcal{A}$ and $\mathcal{B}$ are Banach spaces of functions defined on a domain X, for example they could both be RKHS on X. Define the *weak product*, or *weakly factored space* $\mathcal{A} \odot \mathcal{B}$ to be the completion of finite sums $f = \sum a_i b_i$; $\{a_i\} \subset \mathcal{A}$, $\{b_i\} \subset \mathcal{B}$ using the norm

$$\|f\|_{\mathcal{A}\odot\mathcal{B}} := \inf\left\{\sum \|a_i\|_{\mathcal{A}} \|b_i\|_{\mathcal{B}} : f = \sum a_i b_i\right\}.$$

With this definition the boundedness result for Hardy space Hankel forms amounts to

$$\left(H^2 \odot H^2\right)^* = BMO \tag{13.38}$$

and the corresponding statement for the Bergman space is

$$\left(A^2 \odot A^2\right)^* = BLOCH.$$

(See Exercise 10.6.) Similarly the first statement in Theorem 15.1 is equivalent to a weak factorization of the predual of $\mathcal{X}$;

$$(\mathcal{D} \odot \mathcal{D})^* = \mathcal{X}. \tag{13.39}$$

More precisely the following statements hold.

COROLLARY 13.13. *For $b \in \mathcal{X}$ set $\Lambda_b h = B_b^{\mathcal{D}}(h, 1)$, then $\Lambda_b \in (\mathcal{D} \odot \mathcal{D})^*$. Conversely, if $\Lambda \in (\mathcal{D} \odot \mathcal{D})^*$ there is a unique $b \in \mathcal{X}$ so that for all polynomials h we have $\Lambda h = B_b^{\mathcal{D}}(h, 1) = \Lambda_b h$. In both cases $\|\Lambda_b\|_{(\mathcal{D}\odot\mathcal{D})^*} \approx \|b\|_{\mathcal{X}}$.*

PROOF. If $b \in \mathcal{X}$ and $f \in \mathcal{D} \odot \mathcal{D}$, say $f = \sum_{i=1}^{\infty} g_i h_i$ with $\sum_{i=1}^{\infty} \|g_i\|_{\mathcal{D}} \|h_i\|_{\mathcal{D}} \leq \|f\|_{\mathcal{D}\odot\mathcal{D}} + \varepsilon$, then

$$|\Lambda_b f| = \left|\sum_{i=1}^{\infty} \langle g_i h_i, b\rangle_{\mathcal{D}}\right| = \left|\sum_{i=1}^{\infty} B_b^{\mathcal{D}}(g_i, h_i)\right|$$
$$\leq \left\|B_b^{\mathcal{D}}\right\| \sum_{i=1}^{\infty} \|g_i\|_{\mathcal{D}} \|h_i\|_{\mathcal{D}} \leq \left\|B_b^{\mathcal{D}}\right\| \left(\|f\|_{\mathcal{D}\odot\mathcal{D}} + \varepsilon\right).$$

It follows that $\Lambda_b f = \langle f, b\rangle_{\mathcal{D}}$ defines a continuous linear functional on $\mathcal{D} \odot \mathcal{D}$ with $\|\Lambda_b\| \leq \left\|B_b^{\mathcal{D}}\right\|$.

Conversely, if $\Lambda \in (\mathcal{D} \odot \mathcal{D})^*$with norm $\|\Lambda\|$, then for all $f \in \mathcal{D}$

$$|\Lambda f| = |\Lambda (f \cdot 1)| \leq \|\Lambda\| \|f\|_{\mathcal{D}} \|1\|_{\mathcal{D}} = \|\Lambda\| \|f\|_{\mathcal{D}}.$$

Hence there is a unique $b \in \mathcal{D}$ such that $\Lambda f = \Lambda_b f$ for $f \in \mathcal{D}$. Finally, if $f = gh$ with $g, h \in \mathcal{D}$ we have

$$\left|B_b^{\mathcal{D}}(g, h)\right| = |\langle gh, b\rangle_{\mathcal{D}}| = |\Lambda_b f| = |\Lambda f|$$
$$\leq \|\Lambda\| \|f\|_{\mathcal{D}\odot\mathcal{D}} \leq \|\Lambda\| \|g\|_{\mathcal{D}} \|h\|_{\mathcal{D}},$$

which shows that $B_b^{\mathcal{D}}$ extends to a continuous bilinear form on $\mathcal{D} \odot \mathcal{D}$ with $\left\|B_b^{\mathcal{D}}\right\| \leq \|\Lambda\|$. By Theorem 15.1 we conclude $b \in \mathcal{X}$ and collecting the estimates that $\|\Lambda\| = \|\Lambda_b\|_{(\mathcal{D}\odot\mathcal{D})^*} \approx \left\|B_b^{\mathcal{D}}\right\| \approx \|b\|_{\mathcal{X}}$. □

Based on the analogy between 13.38 and 13.39 we can regard $\mathcal{D} \odot \mathcal{D}$ as a type of H^1 space and $\mathcal{X}$ as a type of BMO space. However in contrast to the situation with H^1 (or A^1), an intrinsic description of $\mathcal{D}\odot\mathcal{D}$, independent of the factorization, is not known.

Similar analysis can be done for asymmetric Hankel forms on the Dirichlet space, but now using Theorem 13.1. Define the space $\partial^{-1}(\partial\mathcal{D} \odot \mathcal{D})$ to be the completion of the space of functions f which have $f' = \sum_{i=1}^{N} g_i' h_i$ (and thus $f = \partial^{-1} \sum_i (\partial g_i) h_i)$) using the norm

$$\|f\|_{\partial^{-1}(\partial\mathcal{D}\odot\mathcal{D})} := \inf\left\{\sum_i \|g_i\|_{\mathcal{D}} \|h_i\|_{\mathcal{D}} : f' = \sum_{i=1}^{N} g_i' h_i\right\}.$$

COROLLARY 13.14.

$$\left(\partial^{-1}(\partial\mathcal{D} \odot \mathcal{D})\right)^* = \mathcal{X}.$$

PROOF. The argument has the same form as the previous proof. Here are some details.

By Theorem 13.1, $b \in \mathcal{X}$, if and only if, with P the Bergman projection, the operator $f \to P(b'\overline{f})$ is a bounded map from $\mathcal{D}$ to A^2. This holds if and only if there is a constant with

$$\left|\left\langle P\left(b'\overline{f}\right), G\right\rangle_{A^2}\right| \lesssim \|f\|_{\mathcal{D}} \|G\|_{A^2}.$$

Now note that on the left side we can drop the projection; on the right side we can write $G = g'$ with $g \in \mathcal{D}$ and $\|G\|_{A^2} = \|g\|_{\mathcal{D}}$. Thus we have

$$\|f\|_{\mathcal{D}} \|g\|_{\mathcal{D}} \gtrsim \left|\left\langle b'\overline{f}, g'\right\rangle_{A^2}\right| = |\langle b', fg'\rangle_{A^2}| = |\langle b, \mathcal{I}(fg')\rangle_{\mathcal{D}}|.$$

Here $\mathcal{I}(fg')$ is defined by $(\mathcal{I}(fg'))' = fg'$. Thus we have $b \in \mathcal{X}$ if and only if b is in the dual space of $\partial^{-1}(\partial\mathcal{D} \odot \mathcal{D})$, which is what we wanted. □

A refined version of the results in Section 4.2.4 shows that half order differentiation maps $\mathcal{D}$ to H^2. In the obvious notation we have;

COROLLARY 13.15.

$$\partial(\mathcal{D}\odot\mathcal{D}) = \partial\mathcal{D}\odot\mathcal{D}, \tag{13.40}$$

$$\partial(\mathcal{D}\odot\mathcal{D}) \neq \partial^{1/2}\mathcal{D}\odot\partial^{1/2}\mathcal{D}.$$

PROOF. The first statement comes from combining the two previous corollaries. To see the second, note that $\partial^{1/2}\mathcal{D}\odot\partial^{1/2}\mathcal{D} = H^2\odot H^2 = H^1$ and that $f(z) = (\log(1-z))^{1/3}$ satisfies $f' \in \partial\mathcal{D} \subset \partial(\mathcal{D}\odot\mathcal{D})$, $f' \notin H^1$. □

13.3.1.2. *Commutators.* There is a very close relation between the Hardy space Hankel operator and the commutator $[M_{\bar{b}}, P^\perp]$, where $M_{\bar{b}}$ is the operator on L^2 of multiplication by the function $\bar{b}$ and $P^\perp$ is the projection of L^2 onto $(H^2)^\perp$. The commutator is certainly bounded if b is bounded. Less obvious is that there is a subtle cancellation which allows the operator to be bounded even for some unbounded b. In particular it is bounded if $b \in BMO$. More specifically, consider the Hardy space Hankel operator described in (10.11):

$$h_b(f) = P^\perp(\bar{b}f).$$

To make this preliminary version of the story slightly cleaner, we restrict that operator to the subspace H_0^2, functions in H^2 with mean zero and we suppose b is holomorphic. For $f \in H_0^2$ we have $P^\perp f = 0$. Hence we can rewrite h_b as the commutator of $M_{\bar{b}}$ and $P^\perp$

$$[M_{\bar{b}}, P^\perp] f = P^\perp M_{\bar{b}} f - M_{\bar{b}} P^\perp f = P^\perp(\bar{b}f) - \bar{b}\cdot 0 = h_b(f).$$

Thus, from the Hankel operator theory, this (restricted version of the) commutator is bounded if and only $b \in BMOA$. With a bit more background (such as knowing that P is bounded from BMO to $BMOA$) this line of argument leads to

THEOREM 13.16. $[M_b, P] : L^2(\mathbb{T}) \to L^2(\mathbb{T})$ *is bounded if and only if* $b \in BMO$.

That result opens into a large and rich story going back to [**118**] of the relation between Hankel type operators, commutators between integral operators and multiplication operators, and statements about weak factorization. The fundamental theme is that certain bilinear operators similar to, but more subtle than, the pointwise product have mapping properties better than those suggested by the size of the arguments. The pointwise product maps $L^2(\mathbb{R}^n) \times L^2(\mathbb{R}^n)$ to $L^1(\mathbb{R}^n)$. The version of Theorem 13.16 for functions defined on $\mathbb{R}^n$ is that certain bilinear operators map that space into the Hardy space $H^1(\mathbb{R}^n)$, the predual of $BMO(\mathbb{R}^n)$. That technical improvement has many consequences. For instance, special cases of the result yield improved integrability estimates for solutions of partial differential equations. Developments of the story in that direction can be seen in [**112**] and [**210**].

13.3.1.3. *Operators on Potential Space.* Dirichlet space functions are those whose derivative of order 1/2 is in the Hardy space, and topics similar to those we have looked at on the Dirichlet space are also studied for spaces of functions having a fractional order derivative in Lebesgue spaces. Some of that work is inspired by the work on spaces of holomorphic functions, some is not. Examples include [**183**], [**116**], [**337**], [**271**], [**227**]. That last paper was one of the inspirations for the work in [**53**] and its influence is also seen in [**97**] and [**271**].

13.3.1.4. *Paraproducts and Volterra Operators.* Paraproducts are a large, informally defined, class of bilinear expressions. They are characterized by mapping properties that resemble those of products but are also influenced by the oscillation, or lack of oscillation, of the arguments. With a broad interpretation of current terminology, the Hardy space Hankel operator $h_b(f)$, viewed as a function of the pair (b, f), is a paraproduct as is the commutator operator $[M_b, P]\, f$ of the previous theorem.

Paraproducts were introduced by Calderón in a function theoretic context, see the discussion in [**315**, page 1136]. Transplanting Calderón's definition from the halfplane to the disk, for $b, f \in \mathrm{Hol}(\mathbb{D})$ we define the *paraproduct* Π by

$$\Pi_b(f) := \int_0^z b'(\zeta) f(\zeta) d\zeta. \tag{13.41}$$

The name paraproduct reflects the relation of Π to a product: $bf = \Pi_b(f) + \Pi_f(b)$. This particular operator and very broad families of generalizations of it are standard technical tools in many areas of real analysis [**232**].

The operator (13.41) and similar operators are also studied for spaces of holomorphic functions but, for historical reasons, they have a different name. In some contexts indefinite integration is called the *Volterra operator* and operators such as (13.41) are often called *Volterra operators* or *integration operators*, for instance [**12**, **252**].

The operator $\Pi_b(\cdot)$ acting on H^2 was studied by Pommerenke in [**264**], and the first part of the following result is from that paper.

THEOREM 13.17. *The map $f \mapsto \Pi_b(f)$ is*

(1) *a bounded map of H^2 to itself if and only if $b \in BMOA$,*
(2) *a bounded map of $\mathcal{D}$ to itself if and only if $b \in \mathcal{X}$.*

EXERCISE 13.18. *Prove the theorem. It straightforward using the Carleson measure characterizations of $BMOA$ and $\mathcal{X}$ (but less so if other characterizations of $BMOA$ are used).*

The fact that bilinear forms B_b and $\tilde{B}_b$, both defined on $\mathcal{D}$, have similar boundedness criteria is summarized by statement (13.40); $\partial\left(\mathcal{D} \odot \mathcal{D}\right) = \partial\mathcal{D} \odot \mathcal{D}$. The relationship between the Hankel operator h_b and the paraproduct Π_b acting on H^2 is similar. They have similar boundedness criteria, and that is summarized by the statement $\partial\left(H^2 \odot H^2\right) = \partial H^2 \odot H^2$.

13.3.2. Other Operators. We have studied the shift operator acting on $\mathcal{D}$ and Hankel operators. There are many other classes of operators which act on $\mathcal{D}$ and/or the containing Sobolev space $\mathcal{W}^{1/2,2}(\mathbb{T})$ (recall the definition in Section 2.5.4) which are also studied systematically. We mention some of those here very briefly.

13.3.2.1. *Multiplication Operators.* The only multiplication operator on $\mathcal{D}$ which we considered systematically was the shift operator. However for every $\phi \in \mathrm{Mult}(\mathcal{D})$ there is a multiplication operator M_ϕ and those operators are studied systematically; for example [**201**], [**13**], [**15**], and [**212**].

13.3.2.2. *The Backward Shift, the $*$-Shift.* The backward shift is the map of $f(z) \in \mathrm{Hol}(\mathbb{D})$ to $g(z) = (f(z) - f(0))/z$. The Taylor coefficients of g are those of f shifted one place to the left. On the Hardy space this gives S^*, the adjoint of the shift operator, sometimes called the $*$-shift. The story is not as simple for the

Dirichlet space. The backward shift and the adjoint of the Dirichlet shift are close to each other, but they are not the same.

EXERCISE 13.19. *Give a precise version of the previous statement.*

A great deal is known about the backwards shift acting on the Dirichlet space and related spaces; see [**17**] as well as [**287**, Section 8] and the references there. The basic reference for the backwards shift acting on the Hardy space is the book [**109**].

13.3.2.3. *Toeplitz Operators.* A Toeplitz operator on H^2 is a multiplication operator taking H^2 into $L^2(\mathbb{T})$ followed by orthogonal projection of $L^2(\mathbb{T})$ back to H^2. Alternatively, Toeplitz operators on H^2 are sums of holomorphic multiplication operators acting on H^2 and the adjoints of such operators. Just as with Hankel operators, equivalent characterizations of Toeplitz operators on H^2 suggest different generalizations of the idea to $\mathcal{D}$, and several versions of Toeplitz operators are studied for the Dirichlet space. Here are a few references; [**285**], [**342**], [**131**], [**18**], [**107**], [**202**], [**215**].

The multiplication operators on the Hardy and Bergman space are subnormal and this gives tools for studying their adjoints. This fails on $\mathcal{D}$, and partly for this reason the various Toeplitz operators on the Dirichlet space are less well understood than the analogs on the Hardy space or Bergman space.

13.3.2.4. *Composition Operators.* The composition of two holomorphic functions is holomorphic and this fact can be used to define operators called *composition operators*, [**305**], [**124**].

Composition operators on the Dirichlet space: Composition operators on the Dirichlet and on Dirichlet type spaces are studied in various places including [**187**], [**102**], [**103**], and [**203**].

Composition operators on the Sobolev space: Going beyond holomorphic functions leads to further possibilities [**281**]. Here we describe very briefly a particularly interesting instance of this in the context of the spaces $\mathcal{D}$ and $\mathcal{W}^{1/2,2}(\mathbb{T})$.

Suppose we have Φ, an orientation preserving homeomorphism of $\mathbb{T}$. Composition with this homeomorphism, C_Φ, mapping a given $F(e^{i\theta})$ to $F(\Phi(e^{i\theta}))$ certainly takes $\mathcal{D}$, regarded as functions on the circle, to some space of functions defined on the circle. It is then natural to ask if the map is an isomorphism of a space $\mathcal{D}$ to itself. It is not hard to show that this happens if and only if Φ is the boundary function of a conformal automorphism of $\mathbb{D}$, the boundary value of a Mobius map. The question of conditions on Φ which ensure that C_Φ is a bounded map of $\mathcal{D}$ into the containing Sobolev space $\mathcal{W}^{1/2,2}(\mathbb{T})$ has a more interesting and more subtle answer. It is a result of Nag and Sullivan [**234**] that this happens if and only if Φ is the boundary value of a quasiconformal homeomorphism of the disk.

Starting with this result one can construct related operators such as the map of $\mathcal{D}$ to itself given by C_Φ followed by the orthogonal projection of $\mathcal{W}^{1/2,2}(\mathbb{T})$ to $\mathcal{D}$. In addition to their intrinsic interest these mappings can play a role wherever quasiconformal homeomorphisms arise, for instance [**181**] and [**297**].

Also, there is an intriguing resemblance between the operator C_Φ and the operators defined on the Hardy space which were used by Coifman and Meyer in one of their approaches to the boundedness of the Cauchy integral on curves [**115**].

Finally, perhaps surprisingly, there is very interesting work using mappings of trees to study discrete analogs of mappings such as Φ, see, for instance, [**240**], [**241**].

13.4. Notes and Comments

The theory of Hankel operators on the Hardy space is well established and much of it is presented in [**257**]. The theory of the small Hankel operator on the Bergman space goes back to [**24**], see also [**118**, Section 5], [**58**], and [**348**]. An analogous theory can be developed in many function spaces, for instance [**184**], [**80**]. The theory of the big Hankel operator is not as well developed and that material is more scattered; the classic paper on those operators include [**58**] and [**36**], more recent developments include [**158**]. The characterization of Hankel forms on the dyadic Dirichlet space is in [**41**].

The space $\mathcal{D} \odot \mathcal{D}$ and related weakly factored spaces are discussed further in [**52**], [**213**], [**214**], [**271**], [**276**], and [**341**]. The last two papers include discussion of a relation between Hankel operators and invariant subspaces.

There is also related work for spaces of holomorphic functions, and for real variable potential spaces, which features more potential theoretic ideas; [**227**], [**97**], [**95**], [**96**].

Part 3

Besov Spaces on the Ball

The Dirichlet space is a Besov space and hence results about the Dirichlet space suggest natural questions for Besov spaces of one or several variables. However much less in known about those questions. In this part of the book we develop some function theoretic operator theory of Besov spaces of holomorphic functions on the ball.

The presentation is organized around two basic questions; first, describing the interpolating sequences (for the space, for its multiplier algebra, and for the associated tree models) and, second, proving corona type theorems for the spaces and for their multiplier algebras.

Our approach to the interpolation questions is constructive. We build explicit interpolating functions from pieces that are, roughly, holomorphic projections of local characteristic functions. The analysis of the construction requires a refined understanding of local oscillation of functions in the Besov spaces. To obtain that we introduce a variation on the invariant gradient, one designed so that its iterates interact well with automorphisms of the ball. We also introduce and use new tree models of the Besov spaces. These elaborate models are sufficiently subtle to accurately model the behavior of higher derivatives of holomorphic functions in the context of the nonisotropic geometry of the ball. In all cases we again find the interpolating sequences are characterized by a separation condition together with a Carleson measure condition.

The proof of the corona type theorems for the Besov space $B_p^\sigma(\mathbb{B}_n)$ follows the traditional pattern of using the Koszul complex to reduce the problem to solving a system of $\bar{\partial}$ equations with norm estimates, and then dealing with those equations. We solve the equations by doing careful analysis of the kernel functions for the solution operators, and then studying those operators on the Besov type spaces which arise in the analysis. An interesting contrast with other related work is that the final operator boundedness statements only use estimates on the modulus of the kernel functions. For the $B_p^\sigma(\mathbb{B}_n)$ that are, in fact, Hilbert spaces with complete Pick reproducing kernels, the theory of Pick kernels then insures that we also obtain classical multiplier algebra corona theorems for their multiplier algebras

Little is known about these topics beyond what is in these chapters. For instance it is not known why for each of these spaces the interpolating sequences and the multiplier interpolating sequences are the same? We saw in Part I that they were automatically the same in the presence of the complete Pick property. However at this level of generality we separately characterize the two types of sequences and then observe that the two characterizations are identical. We also do not know which $B_p^\sigma(\mathbb{B}_n)$ have a multiplier algebra corona theorem? The results mentioned in the previous paragraph do not even cover all the Hilbert spaces in the family.

The function spaces we are considering and closely related spaces (Hardy Sobolev spaces) are also studied using viewpoints different from those we are presenting here, with focus on potential theory, capacity theory, and boundary value functions. References include [**10**, **64**, **98**, **113**] and [**114**].

CHAPTER 14

Besov Spaces on Balls and Trees

We now study Besov spaces, $B_p(\mathbb{B}_n)$, of holomorphic functions on $\mathbb{B}_n \subset \mathbb{C}^n$, the unit ball in complex $n-$space. This will require that we expand our focus in several ways.

– We have been considering Hilbert spaces of holomorphic functions. The Besov spaces are Banach spaces, hence we leave behind the tools from the theory of RKHS. However the Besov spaces do have reproducing kernels and those evaluation functionals are an important tool.

– For holomorphic functions on $\mathbb{B}_n$ the radial derivative behaves very differently from the derivative in tangential directions. This distinction is fundamental to analysis of holomorphic functions on the ball. Many of the technical details below are designed to work effectively with that distinction or to work around it.

– One way to deal with this asymmetry is to use elements of the automorphism group, $\mathrm{Aut}(\mathbb{B}_n)$, to reduce questions about local behavior to questions about behavior near the origin where the picture is symmetric. We do that often.

– In previous chapters we worked with a space of functions on a tree which was a useful model for the Dirichlet space. We will do the same here for the Besov spaces, but, in n dimensions, the straightforward extension of the ideas used for the Dirichlet space does not give us as good a model, and more sophisticated constructions are used.

– In addition to working with scalar functions we will also work with vector-valued functions. This requires more notation but not much additional work.

We begin this chapter by recalling classical material about analysis and geometry on the complex ball, summarizing needed material from [**288**], [**346**], and [**345**]. We then introduce the Besov spaces B_p and the first of several families of Besov type spaces on trees. We develop basic functional analytical results for B_p, for $CM(B_p)$, and $\mathrm{Mult}(B_p)$. One complication that arises is that the spaces B_p are defined using derivatives and differentiation doesn't interact cleanly with automorphisms . To overcome this we introduce an " almost invariant holomorphic derivative" for dealing with the complex geometry associated to local analysis in $\mathbb{B}_n$.

Much of this material in this chapter and the next two has appeared in [**47**]. However there are new results here making that picture more complete; in particular the restriction on the range of p in [**47**, Theorem 5.2] has been removed.

Before going on we should comment on something we are not doing. The unit ball, $\mathbb{B}_n \subset \mathbb{C}^n$,

$$\mathbb{B}_n := \left\{ z = (z_1, \ldots, z_n) : z \in \mathbb{C}^n, \ \sum_{i=1}^{n} |z_i|^2 < 1 \right\},$$

is a natural analog of the unit disk $\mathbb{D} = \mathbb{B}_1 \subset \mathbb{C}$. Hence questions about function theory and operator theory in subspaces of $\mathrm{Hol}(\mathbb{B}_n)$ are natural extensions of our earlier themes. However $\mathbb{B}_n$ is not the only domain in $\mathbb{C}^n$ with a good claim to being the natural generalization on the unit disk. The unit polydisk, $\mathbb{D}^n \subset \mathbb{C}^n$,

$$\mathbb{D}^n := \left\{ z = (z_1, \ldots, z_n) : z \in \mathbb{C}^n, \ \max_{1 \le i \le n} |z_i| < 1 \right\}$$

is another candidate and the topics we will be discussing are also studied for subspaces of $\mathrm{Hol}(\mathbb{D}^n)$. However we will say nothing about those spaces. The function theory and related operator theory on those spaces have a substantially different character from the theories on the ball; a distinction which was already apparent in the two classic books by Rudin, [**289**] and [**288**].

14.1. The Ball

14.1.1. Automorphisms and their Actions. We recall some basic definitions and properties from [**288**] and [**346**]. For $a \in \mathbb{B}_n$ let P_a denote the orthogonal projection onto the one-dimensional complex subspace $\mathbb{C}a$ generated by a, and let $Q_a = I - P_a$ denote the orthogonal projection onto $(\mathbb{C}a)^\perp$, the orthogonal complement of $\mathbb{C}a$. Thus

$$P_a z = \frac{z \cdot \overline{a}}{|a|^2} a, \quad Q_a z = (I - P_a) z = z - \frac{z \cdot \overline{a}}{|a|^2} a. \tag{14.1}$$

Define a biholomorphic involutive automorphism of the ball $\mathbb{B}_n$ by ([**288**, page 25])

$$\begin{aligned} \varphi_a(z) &:= \frac{a - P_a z - (1 - |a|^2)^{1/2} Q_a z}{1 - z \cdot \overline{a}}, \\ &= \frac{a - \frac{z \cdot \overline{a}}{|a|^2} a - (1 - |a|^2)^{1/2} \left(z - \frac{z \cdot \overline{a}}{|a|^2} a \right)}{1 - z \cdot \overline{a}}, \end{aligned} \tag{14.2}$$

for $z \in \mathbb{B}_n$. We have $\varphi_a(0) = a$, $\varphi_a(a) = 0$ and $\varphi_a \circ \varphi_a = I$. The full group of biholomorphic automorphisms of $\mathbb{B}_n$, $\mathrm{Aut}(\mathbb{B}_n)$, consists of all maps $U\varphi_a$ where U is a unitary transformation and $a \in \mathbb{B}_n$.

We also have the following identities [**288**, Theorem 2.2.2]:

$$\begin{aligned} \varphi_a'(0) &= -(1 - |a|^2) P_a - (1 - |a|^2)^{1/2} Q_a, \\ \varphi_a'(a) &= -(1 - |a|^2)^{-1} P_a - (1 - |a|^2)^{-1/2} Q_a, \\ 1 - \overline{\varphi_a(w)} \cdot \varphi_a(z) &= \frac{(1 - \overline{a} \cdot a)(1 - \overline{w} \cdot z)}{(1 - \overline{w} \cdot a)(1 - \overline{a} \cdot z)}, \\ 1 - |\varphi_a(z)|^2 &= \frac{(1 - |a|^2)(1 - |z|^2)}{|1 - \overline{a} \cdot z|^2}, \end{aligned} \tag{14.3}$$

and [**288**, Theorem 2.2.6]

$$J\varphi_a(z) := |\det \varphi_a'(z)|^2 = \left(\frac{1-|a|^2}{|1-\overline{a}\cdot z|^2}\right)^{n+1},$$

where $J\varphi_a(z)$ denotes the real Jacobian of φ_a at z.

14.1.2. Measures and Distances. An $\mathrm{Aut}(\mathbb{B}_n)$ invariant measure on $\mathbb{B}_n$ is given in [**288**, Theorem 2.2.6] and has the formula:

$$d\lambda_n(z) := (1-|z|^2)^{-n-1} dV(z).$$

The invariance of $d\lambda_n$ follows from the above Jacobian formula and the last identity in (14.3).

We will also make frequent use of the measures, for $\alpha > -1$,

$$d\nu_\alpha(z) := (1-|z|^2)^\alpha dV(z) = (1-|z|^2)^{n+1+\alpha} d\lambda_n(z). \tag{14.4}$$

For these measures, and in fact for any measure μ on $\mathbb{B}_n$ and $E \subset \mathbb{B}_n$, we sometimes write $|E|_\mu$ for the $\mu-$measure of E.

The pseudohyperbolic metric, $\rho(z,w)$, on $\mathbb{B}_n$ is given by $\rho(z,w) = |\varphi_z(w)|$, [**346**, Corollary 1.22]. It satisfies the triangle inequality; for $z, \zeta, w \in \mathbb{B}_n$,

$$\rho(w,z) \leq \rho(\xi, z) + \rho(w, \xi), \tag{14.5}$$

and, in fact, also satisfies the stronger version (1.5), [**134**]. The Bergman (hyperbolic) metric $\beta(z,w)$ is given by [**346**, Proposition 1.21],

$$\beta(z,w) := \frac{1}{2}\log\frac{1+|\varphi_z(w)|}{1-|\varphi_z(w)|} = \frac{1}{2}\log\frac{1+\rho(z,w)}{1-\rho(z,w)}, \qquad z, w \in \mathbb{B}_n. \tag{14.6}$$

Both ρ and β are invariant under composition by elements of $\mathrm{Aut}(\mathbb{B}_n)$

14.1.3. Anisotropic Geometry. We see from (14.6) that Bergman metric balls centered at the origin are Euclidean balls. Given $r > 0$, let t be such that the Bergman ball centered at the origin with Bergman radius r, $B_\beta(0,r)$, is the same as $B(0,t)$, the Euclidean ball centered at the origin with Euclidean radius t. Using the previous formulas

$$1 - t^2 = \frac{4}{e^{2r} + 2 + e^{-2r}},$$

and thus we have the approximations

$$t \approx \frac{r}{\sqrt{2}} \text{ for } 0 < r < 1,$$
$$1 - t^2 \approx 4e^{-2r} \text{ for } 1 < r < \infty.$$

Because the Bergman metric is $\mathrm{Aut}(\mathbb{B}_n)$ invariant, the Bergman metric balls $B_\beta(a,r)$ of radius r at the point $a \in \mathbb{B}_n$ satisfy

$$B_\beta(a,r) = \varphi_a(B_\beta(0,r)).$$

In one complex dimension, a β ball in the disk is a Euclidian ball and, although the formulas relating the centers and radii are slightly awkward, there are convenient estimates to go back and forth. This is not true in dimension $n > 1$. There the β-balls are ellipsoids; the exact formulas are given in terms of involutions and projections by

$$\varphi_a\left(B\left(0,t\right)\right)=\left\{z\in\mathbb{B}_n:\frac{\left|P_az-c_a\right|^2}{t^2\rho_a^2}+\frac{\left|Q_az\right|^2}{t^2\rho_a}<1\right\}, \tag{14.7}$$

where

$$c_a=\frac{\left(1-t^2\right)a}{1-t^2\left|a\right|^2},\ \rho_a=\frac{1-\left|a\right|^2}{1-t^2\left|a\right|^2}.$$

Of course we have

$$\begin{aligned}B_\beta\left(a,r\right) &= \left\{\zeta\in\mathbb{B}_n:\beta\left(a,\zeta\right)<r\right\}\\ &= \left\{\zeta\in\mathbb{B}_n:\frac{1}{2}\ln\frac{1+\left|\varphi_a\left(\zeta\right)\right|}{1-\left|\varphi_a\left(\zeta\right)\right|}<r\right\}\\ &= \left\{\zeta\in\mathbb{B}_n:\left|\varphi_a\left(\zeta\right)\right|<\frac{e^{2r}-1}{e^{2r}+1}\right\}\\ &= \left\{\varphi_a\left(z\right)\in\mathbb{B}_n:\left|z\right|=\left|\varphi_a\left(\varphi_a\left(z\right)\right)\right|<\frac{e^{2r}-1}{e^{2r}+1}\right\}\\ &= \varphi_a\left(B\left(0,\frac{e^{2r}-1}{e^{2r}+1}\right)\right),\end{aligned}$$

and so with

$$t=\frac{e^{2r}-1}{e^{2r}+1},\ \text{equivalently } r=\frac{1}{2}\ln\left(\frac{1+t}{1-t}\right),$$

we conclude that

$$B_\beta\left(a,r\right)=\varphi_a\left(B\left(0,t\right)\right)$$

is the ellipsoid $\frac{\left|P_az-c_a\right|^2}{t^2\rho_a^2}+\frac{\left|Q_az\right|^2}{t^2\rho_a}<1$. In particular if the radius r is fixed (and hence $t=\frac{e^{2r}-1}{2}$ is also fixed), then as a moves to the boundary the ratio of the denominators in (14.7) tends to 0 and thus the eccentricity of the ellipse tends to the maximum possible value of 1.

For $0<a<1$, $A=(a,0,0,...0)$, and r a bit less than 1, the Bergman ball $B=B_\beta\left(A,r\right)$ is approximately a product of a disk D and a cube C; $B\sim D\times C$. The disk D lies in the $(z_1,0,...,0)$ plane and is centered at A. The cube C lies in the $(0,z_2,...,z_n)$ hyperplane where it is centered at the origin. The radius of D is $r(1-\left|a\right|^2)$, comparable to Euclidean distance from A to the boundary of $\mathbb{B}_n$ in the radial direction; the side length of C is the much larger quantity $r(1-\left|a\right|^2)^{1/2}$ which is comparable to the Euclidean distance from A to the boundary in any direction orthogonal to the radius through A. The Euclidean volume of B, its Lebesgue measure, satisfies

$$\begin{aligned}\left|B_\beta\left(a,r\right)\right| &\approx \operatorname{Vol}_{2n}B\approx\operatorname{Area}D\times\operatorname{Vol}_{2n-2}C\\ &\approx\left[r(1-\left|a\right|^2)\right]^2\left[r(1-\left|a\right|^2)^{1/2}\right]^{2n-2}=r^{2n}\left(1-\left|a\right|^2\right)^{n+1}.\end{aligned}$$

This change from round balls in one dimension to ellipsoidal ones in higher dimensions is not just a difference in details in certain formulas; the change reflects the appearance of fundamental new geometric issues in the study of functions of several complex variables. Most notably, holomorphic functions are generically smoother in the tangential direction than in the radial direction. This is an interaction of a geometric fact, a complex disk in the ball centered at a point near the boundary and lying in a radial complex line is much smaller than a complex disk in the ball, with the same center, with a tangential orientation; and the analytical fact that analytic functions are smoother at points deeper into the interior of their domain.

We will analyze functions in $\mathrm{Hol}(\mathbb{B}_n)$ at interior points of the ball $\mathbb{B}_n$. However the functions we consider have radial boundary values *a.e.* on the boundary sphere $\partial\mathbb{B}_n$. That sphere carries its own anisotropic metric σ, and, although we will not do analysis on $\partial\mathbb{B}_n$, we will encounter the formula for σ. For z, w set $\sigma(w,z) = |1 - w\overline{z}|^{1/2}$. In fact that function gives a metric on the closed ball $\overline{\mathbb{B}_{n.}}$. The triangle inequality, for $z, \zeta, w \in \overline{\mathbb{B}_{n.}}$,

$$\sigma(w,z) \leq \sigma(\xi,z) + \sigma(w,\xi), \tag{14.8}$$

is proved in [**288**, Proposition 5.1.2].

14.1.4. Differential Operators. For $k = 1, \dots, n$ we set

$$\partial_k := \frac{\partial}{\partial z_k} = \frac{1}{2}\left(\frac{\partial}{\partial x_k} - i\frac{\partial}{\partial y_k}\right), \qquad \overline{\partial}_k := \frac{\partial}{\partial \overline{z}_k} = \frac{1}{2}\left(\frac{\partial}{\partial x_k} + i\frac{\partial}{\partial y_k}\right).$$

The $\mathrm{Aut}(\mathbb{B}_n)$ invariant gradient, $\widetilde{\nabla}$, and invariant Laplacian, $\widetilde{\Delta}$ are given in terms of the Euclidean ∇ and Δ as follows. For $a \in \mathbb{B}_n$ and f holomorphic on $\mathbb{B}_n$ we set

$$\begin{aligned}(\widetilde{\nabla} f)(a) &:= \nabla(f \circ \varphi_a)(0) \\ (\widetilde{\Delta} f)(a) &:= \Delta(f \circ \varphi_a)(0)\end{aligned} \tag{14.9}$$

Even if $f(\cdot)$ is holomorphic it is not generally true that $(\widetilde{\nabla} f)(\cdot)$ is holomorphic. This leads to computational awkwardness when working with higher derivatives. We avoid those difficulties by introducing, in Section 14.5, a useful alternative to $\widetilde{\nabla}$.

Radial differentiation, R, is the operator given by

$$Rf := \sum_{k=1}^{n} z_k \frac{\partial}{\partial z_k}. \tag{14.10}$$

14.1.5. Boxes and Cubes. In order to facilitate the imposition of a tree structure, we give a more refined construction of Carleson boxes than the dyadic decomposition of Chapters 1 and 5 or than that given in [**346**, Theorem 2.23]. Let β be the Bergman metric on the unit ball $\mathbb{B}_n$ in $\mathbb{C}^n$. The metric balls of radius 1 will essentially play the role of an upper half of a Carleson tent, or top half of a Carleson box. Note that for each $r > 0$ the set

$$\mathcal{S}_r := \partial B_\beta(0,r) = \{z \in \mathbb{B}_n : \beta(0,z) = r\}$$

is a Euclidean sphere (with different radius) centered at the origin. In fact, by [**346**, Equation (1.4)] we have $\beta(0,z) = \tanh^{-1}|z|$, and so

$$\begin{aligned}1 - |z|^2 &= 1 - \tanh^2 \beta(0,z) \\ &= \frac{4}{e^{2\beta(0,z)} + 2 + e^{-2\beta(0,z)}} \\ &\approx 4e^{-2\beta(0,z)}\end{aligned} \tag{14.11}$$

for $\beta(0,z)$ large.

For each $r > 0$ we now use an abstract construction to decompose the metric space (S_r, β) into disjoint pieces of roughly the same size and shape. The construction is similar to those discussed in Sections 1.7, 5.2, and 10.2.3.

LEMMA 14.1. *Let (X, d) be a separable metric space and $\lambda > 0$. There is a denumerable set of points $E = \{x_j\}_{j=1}^J$, and $J = \infty$ is allowed, and a corresponding set of Borel subsets Q_j of X satisfying*

$$
\begin{aligned}
&X = \textstyle\bigcup_{j=1}^J Q_j, \\
&Q_i \cap Q_j = \phi, \qquad i \neq j, \\
&B(x_j, \lambda) \subset Q_j \subset B(x_j, 2\lambda), \qquad j \geq 1.
\end{aligned}
\tag{14.12}
$$

We refer to the sets Q_j as unit *cubes* centered at x_j. (These are not quite cubes, but in the analogous function theory on $\mathbb{R}^n$ it is common to use $\mathbb{R}^n \times \mathbb{R}_+$ as a phase space and there to use actual cubes for the localized pieces.)

PROOF. Let $E = \{x_j\}_{j=1}^J$ be a maximal λ-separated subset of X; that is

$$
\begin{aligned}
d(x_i, x_j) &\geq \lambda, \qquad i \neq j, \\
d(x, E) &< \lambda, \qquad x \in X.
\end{aligned}
$$

Let $A = \bigcup_{j=1}^J B(x_j, \lambda)$ and define inductively

$$
Q_j = B(x_j, 2\lambda) \setminus \{A \cup (\cup_{i<j} Q_i)\}, \qquad j \geq 1.
$$

It is routine to verify that these cubes Q_j satisfy (14.12). □

14.1.6. The Bergman Tree. Now fix $\theta, \lambda > 0$, which we will refer to as *structural constants* for the Bergman tree. (We will make various choices of θ and λ below.) For $N \in \mathbb{N}$, apply Lemma 14.1 with parameter λ to the metric space $(\mathcal{S}_{N\theta}, \beta)$ to obtain points $\{z_j^N\}_{j=1}^J$ and unit cubes $\{Q_j^N\}_{j=1}^J$ in $\mathcal{S}_{N\theta}$ satisfying (14.12). For $z \in \mathbb{B}_n$, let $P_r z$ denote the radial projection of z onto the sphere $\mathcal{S}_r$ (not to be confused with the orthogonal projection P_a defined above). We now define subsets K_j^N of $\mathbb{B}_n$ by $K_1^0 = \{z \in \mathbb{B}_n : \beta(0, z) < \theta\}$ and

$$
K_j^N := \left\{z \in \mathbb{B}_n : N\theta \leq \beta(0, z) < (N+1)\theta,\ P_{N\theta} z \in Q_j^N\right\}, \qquad N \geq 1, j \geq 1.
$$

We define corresponding points $c_j^N \in K_j^N$ by

$$
c_j^N := P_{(N+1/2)\theta}\left(z_j^N\right),
$$

and set $c_1^0 = 0$. Although the K_j^N are not literally cubes, we will refer to them as cubes. We say the cube K_j^N is centered at c_j^N.

We regard the collection of unit cubes $\{K_j^N\}_{N \geq 0, j \geq 1}$ as the vertex set of a tree, $\mathcal{T}_n$; and declare K_i^{N+1} to be child of K_j^N, written $K_i^{N+1} \in \mathcal{C}(K_j^N)$, if the projection $P_{N\theta}(z_i^{N+1})$ of z_i^{N+1} onto the sphere $\mathcal{S}_{N\theta}$ lies in Q_j^N. In the case $N = 0$, we declare every cube K_j^1 to be a child of the root cube K_1^0. We will sometimes abbreviate this construction by writing $\mathcal{T}_n = \{K_j^N\}$.

We will typically write α, β, γ etc. as indices for generic vertices; thus K_α is some K_j^N. We will write c_α for the center of K_a and, at times, write α for c_α.

We also define the *dimension* $n(\mathcal{T})$ of an arbitrary tree $\mathcal{T}$.

DEFINITION 14.2. *The* upper dimension $\overline{n}(\mathcal{T})$ *of a tree* $\mathcal{T}$ *is given by*

$$\overline{n}(\mathcal{T}) := \limsup_{\ell\to\infty} \log_2 (N_\ell)^{1/\ell},$$

where

$$N_\ell := \sup_{\alpha\in\mathcal{T}} \operatorname{card}\{\beta \in \mathcal{T} : \beta > \alpha \text{ and } d(\beta) = d(\alpha) + \ell\},$$

along with a similar definition for the lower dimension $\underline{n}(\mathcal{T})$ *using* $\liminf_{\ell\to\infty}$ *in place of* $\limsup_{\ell\to\infty}$ *and* inf card *in place of* sup card. *If the upper and lower dimensions coincide, we denote their common value, called the* dimension *of* $\mathcal{T}$, *by* $n(\mathcal{T})$.

Note that if $\mathcal{T}$ is a homogeneous tree with branching number N, then $N = (N_\ell)^{1/\ell}$ for all $\ell \geq 1$. The choice of base 2 for the logarithm then yields the relationship $N = 2^n$, consistent with the common usage that the linear tree has dimension 0 and the dyadic tree has dimension 1.

LEMMA 14.3. *The tree* $\mathcal{T}_n$, *constructed above with parameters* λ *and* θ, *satisfies the following properties.*

(1) *The ball* $\mathbb{B}_n$ *is a pairwise disjoint union of the cubes* K_α, $\alpha \in \mathcal{T}_n$, *and there are positive constants* C_1 *and* C_2 *depending on* λ *and* θ *such that*

$$B_\beta(c_\alpha, C_1) \subset K_\alpha \subset B_\beta(c_\alpha, C_2), \qquad \alpha \in \mathcal{T}_n, .$$

(2) $\bigcup_{\beta\geq\alpha} K_\beta$ *is "comparable" to the Carleson tent* V_{c_α} *associated to the point* c_α, *where*

$$V_z = \{w \in \mathbb{B}_n : |1 - \overline{w}\cdot Pz| \leq 1 - |z|\},$$

and Pz *denotes radial projection of* z *onto the unit sphere* $\partial\mathbb{B}_n$.

(3) *The invariant volume of* K_α *is bounded between positive constants depending on* λ *and* θ, *but independent of* $\alpha \in \mathcal{T}_n$. *In particular, for* $0 < \lambda, \theta \leq 1$ *we have* $|K_\alpha|_{\lambda_n} \approx \lambda^{2n-2}\theta^2$ *and*

$$|K_\alpha| \approx \lambda^{2n-2}\theta^2(1 - |c_\alpha|^2)^{n+1} \approx \lambda^{2n-2}\theta^2 e^{-2d(\alpha)\theta(n+1)}.$$

(4) *The dimension* $n(\mathcal{T}_n)$ *of the tree* $\mathcal{T}_n$ *is* $2\theta n/\ln 2$.

(5) *For any* $S > 0$ *there is a constant* $C = C(S, \lambda, \theta, n)$ *such that the balls* $B_\beta(c_\alpha, S)$ *satisfy the finite overlap condition*

$$\sum_{\alpha\in\mathcal{T}_n} \chi_{B_\beta(c_\alpha,S)}(z) \leq C, \qquad z \in \mathbb{B}_n.$$

PROOF. Property (1) follows easily from the construction of K_j^N, and property (2) is then a consequence of the formula for the metric β.

Property (3) follows since using (14.11), K_α is comparable to a rectangle, two of whose side lengths (those in the complex radial directions) are $\theta(1 - |c_\alpha|^2) = \theta e^{-2\theta d(\alpha)}$, while the remaining $n - 2$ side lengths (those in the complex tangential directions) are $\lambda(1 - |c_\alpha|^2)^{1/2} \approx \lambda e^{-\theta d(\alpha)}$. Thus

$$\begin{aligned}|K_\alpha| &\approx \left[\theta(1 - |c_\alpha|^2)\right]^2 \left[\lambda(1 - |c_\alpha|^2)^{1/2}\right]^{2n-2} \\ &\approx \lambda^{2n-2}\theta^2(1 - |c_\alpha|^2)^{n+1} \approx \lambda^{2n-2}\theta^2 e^{-2d(\alpha)\theta(n+1)},\end{aligned}$$

and then

$$|K_\alpha|_{\lambda_n} = \int_{K_\alpha} \frac{dV}{(1-|z|^2)^{n+1}} \approx \frac{1}{(1-|c_\alpha|^2)^{n+1}} |K_\alpha| \approx \lambda^{2n-2}\theta^2.$$

The final two properties follow from volume comparisons. Indeed, given $\alpha \in \mathcal{T}_n$ and $\ell \geq 1$, let $\{\beta_j\}_{j=1}^{N_\ell}$ be an enumeration of the descendants ℓ generations beyond α:

$$\mathcal{C}^{(\ell)}(\alpha) = \{\beta \in \mathcal{T}_n : \beta > \alpha \text{ and } d(\beta) = d(\alpha) + \ell\} = \{\beta_j\}_{j=1}^{N_\ell}.$$

From property (3) and (14.11) we have that $\lambda^{2n-2}\theta^2 \approx |K_\gamma|_{\lambda_n} \approx e^{2d(\gamma)\theta(n+1)} |K_\gamma|$, and since $\left|\bigcup_{j=1}^{N_\ell} K_{\beta_j}\right| \approx e^{-2\ell\theta} |K_\alpha|$, we obtain

$$\begin{aligned}
\lambda^{2n-2}\theta^2 e^{-2\ell\theta} e^{-2d(\alpha)\theta(n+1)} \approx e^{-2\ell\theta} |K_\alpha| &\approx \left|\bigcup_{j=1}^{N_\ell} K_{\beta_j}\right| = \sum_{j=1}^{N_\ell} |K_{\beta_j}| \\
&\approx \sum_{j=1}^{N_\ell} \lambda^{2n-2}\theta^2 e^{-2d(\beta_j)\theta(n+1)} \\
&= N_\ell \lambda^{2n-2}\theta^2 e^{-2(d(\alpha)+\ell)\theta(n+1)}.
\end{aligned}$$

Thus there are positive constants c and C depending only on n and θ such that

$$c \leq \frac{N_\ell}{e^{2n\ell\theta}} \leq C, \qquad \alpha \in \mathcal{T}_n, \ell \geq 1,$$

and it follows that $n(\mathcal{T}) = \log_2 e^{2n\theta}$, completing the proof of property (4).

The finite overlap property (5) is obtained as follows. Let $z \in K_\alpha$. If $z \in \cap_{j=1}^N B_\beta(c_{\alpha_j}, R)$, then since β is a metric,

$$\bigcup_{j=1}^N B_\beta(c_{\alpha_j}, C_1) \subset B_\beta(c_\alpha, R + 2C_1).$$

Since the balls $B_\beta(c_{\alpha_j}, C_1)$ are pairwise disjoint by property (1), we thus have

$$N \approx \sum_{j=1}^N |B_\beta(c_{\alpha_j}, C_1)|_{\lambda_n} \leq |B_\beta(c_\alpha, R + 2C_1)|_{\lambda_n} = |B_\beta(0, R + 2C_1)|_{\lambda_n},$$

and we note that the final term is independent of $\alpha \in \mathcal{T}_n$, hence also of $z \in \mathbb{B}_n$. □

14.1.6.1. *Comments on the Construction.* In our use of sets $\{K_\alpha\}$, for instance to write an integral over the ball as a sum of integrals over the $\{K_\alpha\}$, the distinction between these sets their interiors, and their closures will not be relevant. At times we are casual about that distinction, both with the $\{K_\alpha\}$ and with related families of sets.

The previous construction simplifies greatly when $n = 1$, the spheres S_r are circles, and the cubes Q_j^N can be taken to be circular arcs of equal length. The construction is then a variant of the decomposition in Section 1.7. In this case the V's we defined are the Carleson tents over intervals of the boundary circle, as introduced in Section 1.7.

The choice $\theta = \ln 2/2$ yields $\dim(\mathcal{T}_n) = n$ and the convenient equivalence $1 - |z|^2 \approx 2^{-d(\alpha)}$ for $z \in K_\alpha$. For $n > 1$ a geometric analysis would show that the V's are, roughly, the geometric tents over the anisotropic disks in $\partial\mathbb{B}_n$ defined using the metric σ in Section 14.1.5.

14.2. The Besov Spaces

We will study analytic Besov spaces, sometimes called Besov-Sobolev spaces, on the unit ball $\mathbb{B}_n$ in $\mathbb{C}^n$. For $1 < p < \infty$ and $m > n/p$ the Besov space $B_p = B_p(\mathbb{B}_n)$ consists of those functions f, holomorphic on $\mathbb{B}_n$, for which

$$\int_{\mathbb{B}_n} \left|(1-|z|^2)^m f^{(m)}(z)\right|^p d\lambda_n(z) < \infty. \tag{14.13}$$

Here dV is Lebesgue measure on $\mathbb{C}^n$, $d\lambda_n(z) = (1-|z|^2)^{-n-1}dV(z)$ is the invariant measure on the ball, and $f^{(m)}$ is the mth order complex derivative of f. We will see that, as our definition suggests, the space defined is independent of m as long as m is sufficiently large.

If $n = 1$ then the choice $m = 1$ is allowable for all $p > 1$ and we recapture the definition of single variable Besov spaces given in Section 2.5.3. In particular, if $n = 1$ and $p = 2$ we have the classical Dirichlet space and the choices of $m > 1$ give alternative norming functionals for that space.

We saw in Chapter 10 that, even for the Dirichlet space, there are other, different looking, norming functionals and they were at times very useful. For the spaces we are introducing now there are many measures of oscillation that can be used to build equivalent norming functionals. We will introduce several basic ones and state the equivalence results from [**346**] and [**345**]. Later we introduce others adapted to particular needs. As we mentioned earlier, there is overhead cost in introducing a variety of norms, but once they are in place they can be quite valuable.

14.2.1. The Spaces and Their Duals.

We have the reproducing formula of Bergman [**288**, Theorem 3.1.3],

$$f(z) = \frac{n!}{\pi^n} \int_{\mathbb{B}_n} \frac{f(w)}{(1-\overline{w}\cdot z)^{n+1}} dV, \qquad f \in L^1(d\lambda_n) \cap \mathrm{Hol}(\mathbb{B}_n), \tag{14.14}$$

and the following variants [**288**, Theorem 7.1.2] and Theorem B.6,

$$f(z) = \frac{n!}{\pi^n} \binom{n+s}{n} \int_{\mathbb{B}_n} \frac{(1-|w|^2)^s}{(1-\overline{w}\cdot z)^{s+n+1}} f(w)\, dV, \qquad \mathrm{Re}\, s > -1, \tag{14.15}$$

valid for all $f \in \mathrm{Hol}(\mathbb{B}_n)$ for which the integrand is in L^1.

We now recall the invertible "radial" operators $R^{\gamma,t} : \mathrm{Hol}(\mathbb{B}_n) \to \mathrm{Hol}(\mathbb{B}_n)$ given in [**346**]. For $f \in \mathrm{Hol}(\mathbb{B}_n)$ with homogenous expansion $f(z) = \sum_{k=0}^{\infty} f_k(z)$ define

$$R^{\gamma,t} f(z) := \sum_{k=0}^{\infty} \frac{\Gamma(n+1+\gamma)\Gamma(n+1+k+\gamma+t)}{\Gamma(n+1+\gamma+t)\Gamma(n+1+k+\gamma)} f_k(z),$$

provided neither $n+\gamma$ nor $n+\gamma+t$ is a negative integer.

If the inverse of $R^{\gamma,t}$ is denoted $R_{\gamma,t}$, then [**346**, Proposition 1.4] yields

$$R^{\gamma,t}\left(\frac{1}{(1-\overline{w}\cdot z)^{n+1+\gamma}}\right)=\frac{1}{(1-\overline{w}\cdot z)^{n+1+\gamma+t}},\tag{14.16}$$
$$R_{\gamma,t}\left(\frac{1}{(1-\overline{w}\cdot z)^{n+1+\gamma+t}}\right)=\frac{1}{(1-\overline{w}\cdot z)^{n+1+\gamma}},$$

for all $w\in\mathbb{B}_n$. Thus for any γ, $R^{\gamma,t}$ is approximately differentiation of order t. From [**346**, Definitions 6.1 and 6.4] we have that, for m large, the derivatives $R^{\gamma,m}f(z)$ are "L^p norm equivalent" to $\sum_{k=0}^{m-1}\left|\nabla^k f(0)\right|+\nabla^m f(z)$.

PROPOSITION 14.4 ([**346**, Theorems 6.1 and 6.4]). *Suppose that $0<p<\infty$, $n+\gamma$ is not a negative integer, and $f\in\mathrm{Hol}(\mathbb{B}_n)$. Then the following four conditions are equivalent:*

$$(1-|z|^2)^m\nabla^m f(z)\in L^p(d\lambda_n)\ \textit{for}\ \text{some}\ m>\frac{n}{p},m\in\mathbb{N},$$
$$(1-|z|^2)^m\nabla^m f(z)\in L^p(d\lambda_n)\ \textit{for}\ \text{all}\ m>\frac{n}{p},m\in\mathbb{N},$$
$$(1-|z|^2)^m R^{\gamma,m}f(z)\in L^p(d\lambda_n)\ \textit{for}\ \text{some}\ m>\frac{n}{p},m+n+\gamma\notin-\mathbb{N},$$
$$(1-|z|^2)^m R^{\gamma,m}f(z)\in L^p(d\lambda_n)\ \textit{for}\ \text{all}\ m>\frac{n}{p},m+n+\gamma\notin-\mathbb{N}.$$

Moreover, we have for $1<p<\infty$, for all $m_1,m_2>n/p$, $m_1+n+\gamma\notin-\mathbb{N}$, $m_2\in\mathbb{N}$,

$$\sum_{k=0}^{m_2-1}\left|\nabla^k f(0)\right|^p+\int_{\mathbb{B}_n}\left|(1-|z|^2)^{m_2}\nabla^{m_2}f(z)\right|^p d\lambda_n\approx\left\|(1-|z|^2)^{m_1}R^{\gamma,m_1}f\right\|_{L^p(\lambda_n)}^p\tag{14.17}$$

and the implied constant depends only on m_1, m_2, n, γ and p.

DEFINITION 14.5. *We define the analytic Besov spaces B_p on the ball $\mathbb{B}_n$ by taking $\gamma=0$ and $m=[n/p]+1$ and setting*

$$B_p=B_p(\mathbb{B}_n):=\left\{f\in\mathrm{Hol}(\mathbb{B}_n):\left\|(1-|z|^2)^m R^{0,m}f\right\|_{L^p(\lambda_n)}<\infty\right\}\tag{14.18}$$

and we norm B_p with any of the equivalent expressions in (14.17).

REMARK 14.6. *In particular this leads to a canonical definition of the norming function on $B_p(\mathbb{B}_n)$;*

$$\|f\|_{B_p(\mathbb{B}_n)}^p=\sum_{k=0}^{m-1}\left|\nabla^k f(0)\right|^p+\left\|(1-|z|^2)^m R^{0,m}f\right\|_{L^p(\lambda_n)}^p.$$

In the previous proposition we noted that there are many equivalent norms on B_p and as we go forward we will see many more. At times we will elide those differences and write equality of two expressions when the equality (as opposed to "$\approx$") actually only holds for a specific, unspecified, choice of equivalent norm.

For $\alpha>-1$ and with $d\nu_\alpha$ given by (14.4), we consider the weighted Bergman space $A_\alpha^2=L^2(\mathbb{B}_n,d\nu_\alpha)\cap\mathrm{Hol}(\mathbb{B}_n)$ and we denote its inner product by $\langle\cdot,\cdot\rangle_\alpha$:

$$\langle f,g\rangle_\alpha=\int_{\mathbb{B}_n}f(z)\overline{g(z)}d\nu_\alpha(z),\quad f,g\in A_\alpha^2.$$

We see from (14.15) that $K_w^\alpha(z) = K^\alpha(z,w) = (1 - \overline{w} \cdot z)^{-n-1-\alpha}$ is the reproducing kernel for A_α^2: $f(w) = \langle f, K_w^\alpha \rangle_\alpha$ for $f \in A_\alpha^2$. Because the polynomials are dense in A_α^p, $1 < p < \infty$, and we have the boundedness result Theorem B.6, this reproducing formula holds as well for $f \in A_\alpha^p$, and we will use it as a basis for a reproducing formula for the Besov spaces.

Notation: For the rest of this chapter and for the next chapter we will use the following notation, for n, α, and p fixed and $p' = (p-1)/p$ we set

$$\begin{aligned} \mathbf{s} &:= (n+1+\alpha)/p, \\ \mathbf{s}' &:= (n+1+\alpha)/p', \text{ and hence} \\ \mathbf{s} + \mathbf{s}' &= n+1+\alpha. \end{aligned} \tag{14.19}$$

Note that $\mathbf{s}$ and $\mathbf{s}'$ *are* ***not*** *conjugate indices.*

[**346**, Corollary 6.5] states that $R^{\gamma,\mathbf{s}}$ is a bounded invertible operator from B_p onto A_α^p, provided that neither $n+\gamma$ nor $n+\gamma+\mathbf{s}$ is a negative integer. It turns out to be convenient to take $\gamma = \alpha - \mathbf{s}$ here (with this choice we can explicitly compute certain derivatives and $B_{p'}$ norms of our reproducing kernels - see (14.23) and (14.24) below), and thus we single out the special operators

$$\mathcal{R}_t^\alpha := R^{\alpha-t,t}.$$

Note that the operators $\mathcal{R}_t^\alpha$ and their inverses $(\mathcal{R}_t^\alpha)^{-1} = R_{\alpha-t,t}$ are self-adjoint with respect to $\langle \cdot, \cdot \rangle_\alpha$. To see this note that the monomials are orthogonal with respect to $\langle \cdot, \cdot \rangle_\alpha$ (see [**346**, Equations (1.21) and (1.23)]), and the action of the operators is to multiply each monomial in the power series of f by a positive constant. The next definition is motivated by the fact that $\mathcal{R}_{\mathbf{s}}^\alpha$ is a bounded invertible operator from B_p onto A_α^p, and that $\mathcal{R}_{\mathbf{s}'}^\alpha$ is a bounded invertible operator from $B_{p'}$ onto $A_\alpha^{p'}$, provided that neither $n+\alpha$, $n+\alpha-\mathbf{s}$ nor $n+\alpha-\mathbf{s}'$ is a negative integer. In particular this proviso holds for $\alpha > -1$.

DEFINITION 14.7. *For $\alpha > -1$ and $1 < p < \infty$, we define a pairing $\langle \cdot, \cdot \rangle_{\alpha,p}$ for B_p and $B_{p'}$ using $\langle \cdot, \cdot \rangle_\alpha$ as follows:*

$$\begin{aligned} \langle f, g \rangle_{\alpha,p} &:= \langle \mathcal{R}_{\mathbf{s}}^\alpha f, \mathcal{R}_{\mathbf{s}'}^\alpha g \rangle_\alpha = \int_{\mathbb{B}_n} \mathcal{R}_{\mathbf{s}}^\alpha f(z) \overline{\mathcal{R}_{\mathbf{s}'}^\alpha g(z)} d\nu_\alpha(z) \\ &= \int_{\mathbb{B}_n} \left((1-|z|^2)^{\mathbf{s}} \mathcal{R}_{\mathbf{s}}^\alpha f(z) \right) \overline{\left((1-|z|^2)^{\mathbf{s}'} \mathcal{R}_{\mathbf{s}'}^\alpha g(z) \right)} d\lambda_n(z). \end{aligned}$$

By Hölder's inequality we have

$$\left| \langle f, g \rangle_{\alpha,p} \right| \le \|f\|_{B_p} \|g\|_{B_{p'}}.$$

By [**346**, Theorem 2.12], we also have that every continuous linear functional Λ on B_p is given by $\Lambda f = \langle f, g \rangle_{\alpha,p}$ for a unique $g \in B_{p'}$ satisfying

$$\|g\|_{B_{p'}} = \sup_{\|f\|_{B_p}=1} \left| \langle f, g \rangle_{\alpha,p} \right|. \tag{14.20}$$

Indeed, if $\Lambda \in (B_p)^*$, then $\Lambda \circ (\mathcal{R}_{\mathbf{s}}^\alpha)^{-1} \in (A_\alpha^p)^*$, and by [**346**, Theorem 2.12], there is $G \in A_\alpha^{p'}$ with $\|G\|_{A_\alpha^{p'}} = \|\Lambda\|$ such that $\Lambda \circ (\mathcal{R}_{\mathbf{s}}^\alpha)^{-1} F = \langle F, G \rangle_\alpha$ for all $F \in A_\alpha^p$.

If we set $g = (\mathcal{R}^{\alpha}_{\mathbf{s}'})^{-1} G$, then we have $\|g\|_{B_{p'}} = \|G\|_{A^{p'}_{\alpha}} = \|\Lambda\|$ and with $F = \mathcal{R}^{\alpha}_{\mathbf{s}} f$, we also have

$$\Lambda f = \Lambda \circ (\mathcal{R}^{\alpha}_{\mathbf{s}})^{-1} F = \langle F, G\rangle_{\alpha} = \langle \mathcal{R}^{\alpha}_{\mathbf{s}} f, \mathcal{R}^{\alpha}_{\mathbf{s}'} g\rangle_{\alpha} = \langle f, g\rangle_{\alpha,p}$$

for all $f \in B_p$. Then (14.20) follows from

$$\|g\|_{B_{p'}} = \|\Lambda\| = \sup_{\|f\|_{B_p}=1} |\Lambda(f)| = \sup_{\|f\|_{B_p}=1} \left|\langle f, g\rangle_{\alpha,p}\right|.$$

Note: The Besov space pairing $\langle\cdot,\cdot\rangle$ introduced in [**346**] is given by

$$\langle f, g\rangle = \left\langle R^{0,n+1} f, R^{0,n+1} g\right\rangle_{n+1} = \langle f, g\rangle_{n+1,2},$$

and so coincides with our pairing for the Hilbert space B_2 with the choice $\alpha = n+1$. However, for $p \neq 2$, our pairing $\langle\cdot,\cdot\rangle_{\alpha,p}$ uses the operators $\mathcal{R}^{\alpha}_{t} = R^{\alpha-t,t}$ with $t = \mathbf{s}$ and $t' = \mathbf{s}'$, and so does not coincide with $\langle\cdot,\cdot\rangle$ in [**346**].

With $K^{\alpha}_{w}(z)$ the reproducing kernel for A^2_{α}, we now claim that the kernel

$$k^{\alpha,p}_{w}(z) := (\mathcal{R}^{\alpha}_{\mathbf{s}'})^{-1} (\mathcal{R}^{\alpha}_{\mathbf{s}})^{-1} K^{\alpha}_{w}(z) \tag{14.21}$$

satisfies the following reproducing formula for B_p:

$$f(w) = \langle f, k^{\alpha,p}_{w}\rangle_{\alpha,p} = \int_{\mathbb{B}_n} \mathcal{R}^{\alpha}_{\mathbf{s}} f(z)\, \overline{\mathcal{R}^{\alpha}_{\mathbf{s}'} k^{\alpha,p}_{w}(z)} d\nu_{\alpha}(z), \qquad f \in B_p. \tag{14.22}$$

Indeed, for f a polynomial, we have

$$\begin{aligned} f(w) &= \langle f, K^{\alpha}_{w}\rangle_{\alpha} \\ &= \left\langle (\mathcal{R}^{\alpha}_{\mathbf{s}})^{-1} \mathcal{R}^{\alpha}_{\mathbf{s}} f, K^{\alpha}_{w}\right\rangle_{\alpha} \\ &= \left\langle \mathcal{R}^{\alpha}_{\mathbf{s}} f, (\mathcal{R}^{\alpha}_{\mathbf{s}})^{-1} K^{\alpha}_{w}\right\rangle_{\alpha} \\ &= \left\langle \mathcal{R}^{\alpha}_{\mathbf{s}} f, \mathcal{R}^{\alpha}_{\mathbf{s}'} (\mathcal{R}^{\alpha}_{\mathbf{s}'})^{-1} (\mathcal{R}^{\alpha}_{\mathbf{s}})^{-1} K^{\alpha}_{w}\right\rangle_{\alpha} \\ &= \left\langle f, (\mathcal{R}^{\alpha}_{\mathbf{s}'})^{-1} (\mathcal{R}^{\alpha}_{\mathbf{s}})^{-1} K^{\alpha}_{w}\right\rangle_{\alpha,p}. \end{aligned}$$

We now obtain the claim since the polynomials are dense in B_p and the kernels $k^{\alpha,p}_{w}$ are in $B_{p'}$ for each fixed $w \in \mathbb{B}_n$. Thus we have proved the following duality theorem.

THEOREM 14.8. *Let $1 < p < \infty$ and $\alpha > -1$. Then the dual space of B_p can be identified with $B_{p'}$ under the pairing $\langle\cdot,\cdot\rangle_{\alpha,p}$. The reproducing kernel $k^{\alpha,p}_{w}$ for this pairing is given by* (14.21).

From (14.21) and (14.16) we have

$$\begin{aligned} \mathcal{R}^{\alpha}_{\mathbf{s}'} k^{\alpha,p}_{w}(z) &= (\mathcal{R}^{\alpha}_{\mathbf{s}})^{-1} K^{\alpha}_{w}(z) \\ &= R_{\alpha-\mathbf{s},\mathbf{s}}\left((1-\overline{w}\cdot z)^{-(n+1+\alpha)}\right) \\ &= (1-\overline{w}\cdot z)^{-\mathbf{s}'}. \end{aligned} \tag{14.23}$$

The parameter $\alpha > -1$ appearing in condition (14.23) plays only a limited role and is sometimes suppressed. The next result is an example of this.

LEMMA 14.9. *For $\alpha > -1$ and $1 < p < \infty$, we have*

$$\|k_w^{\alpha,p}\|_{B_{p'}}^{p'} \approx 1 + \log \frac{1}{1-|w|^2} \approx (1 + \beta(0,w)). \tag{14.24}$$

PROOF. Using (14.23) and $m = \mathbf{s}' > n/p'$, we compute that

$$\begin{aligned}
\|k_w^{\alpha,p}\|_{B_{p'}}^{p'} &= \int_{\mathbb{B}_n} \left|(1-|z|^2)^{\mathbf{s}'} \mathcal{R}_{\mathbf{s}'}^{\alpha} k_w^{\alpha,p}(z)\right|^{p'} d\lambda_n(z) \\
&= \int_{\mathbb{B}_n} \left|\frac{1-|z|^2}{1-\overline{w}\cdot z}\right|^{n+1+\alpha} d\lambda_n(z) \\
&= \int_{\mathbb{B}_n} \frac{(1-|z|^2)^{\alpha}}{|1-\overline{w}\cdot z|^{n+1+\alpha}} dV \\
&\approx 1 + \log \frac{1}{1-|w|^2};
\end{aligned}$$

the last estimate by [**346**, Theorem 1.12] or Lemma B.5. □

14.2.1.1. *Going Forward.* We have introduced some basic facts about the spaces $B_p(\mathbb{B}_n)$ including the relationship between a variety of norming functionals (we will see more of those later) and basic duality statements. Going forward we will use analogs of these properties for more general Besov spaces, often without further comment. The proofs we give here, or reference, cover those extensions, *mutatis mutandis;* the primary new feature being additional notation.

There are two main types of extension we will look at. First, beyond the $B_p(\mathbb{B}_n)$ we will also work with related spaces of vector valued functions. It was clear from the start that this would be necessary because our analysis, and even our definitions, involve derivatives of functions of several variables and those derivatives are not scalars.

Second, we will introduce and consider the spaces $B_p^{\sigma}(\mathbb{B}_n)$, $0 \le \sigma \le 1/2$; where $B_p(\mathbb{B}_n) = B_p^0(\mathbb{B}_n)$. This allows to include consideration of the spaces $B_2^{1/2}(\mathbb{B}_n)$, the Drury-Arveson spaces which play a central role in the theory of RKHS with CPP. Very informally $B_p^{\sigma}(\mathbb{B}_n)$ is a space obtained by applying differentiation of order σ to functions in $B_p(\mathbb{B}_n)$. This family of spaces continues to be of interest for $\sigma > 1/2$ but our techniques do not extend to that range. Some results in that range are in [**338**].

We now want to study multipliers and Carleson measures on the Besov spaces. Both definitions are analogs of the single variable case:

DEFINITION 14.10. *Given $1 < p < \infty$, a positive Borel measure μ on the ball $\mathbb{B}_n$ is a B_p-Carleson measure, $\mu \in CM(B_p)$, if*

$$\int_{\mathbb{B}_n} |f(z)|^p \, d\mu(z) \lesssim \|f\|_{B_p}^p, \qquad f \in B_p. \tag{14.25}$$

DEFINITION 14.11. *We say that φ is a (pointwise) multiplier on B_p, $\varphi \in \operatorname{Mult} B_p$, if $\varphi f \in B_p$ for all $f \in B_p$. By the closed graph theorem, this is equivalent to*

$$\|\varphi f\|_{B_p} \lesssim \|f\|_{B_p}, \qquad f \in B_p. \tag{14.26}$$

The relation between Carleson measures and multipliers uses the definition of Carleson measures but not their characterization, and we develop that relationship in the next section. There we also study the Carleson measures, following the path we used in Chapter 5. We introduce spaces of functions on trees which model the Besov spaces and introduce Carleson measures for the tree spaces. We will then prove a discretization result relating the two types of Carleson measures. After that we give some results characterizing both types of Carleson measures, results similar to the one dimensional Theorem 5.17. However we leave open the problem of characterizing the Carleson measures of $B_p(\mathcal{T}_n)$ for $p \geq 2 + 1/(n-1)$.

One aspect of the theory of Besov spaces that we will not develop is the invariance properties of seminorms used in norming the B_p. Earlier we used the fact that the functional $\mathfrak{D}$, which can be used to define the the Dirichlet space, is invariant under the action of the automorphism group of the disk. Some similar facts hold for seminorms defining the B_p, but we will not pursue that topic and only refer to [**259**] and [**346**]. However, to show the usefulness of invariance we use it here to give a simple proof of the global modulus of continuity estimate for functions in $B_p(\mathbb{B}_n)$. The case $n = 1$, $p = 2$ of this result was Proposition 2.14.

LEMMA 14.12. *For $f \in B_p$ and $z, w \in \mathbb{B}_n$,*

$$|f(z) - f(w)| \lesssim \|f\|_{B_p} \beta(z,w)^{1/p'}. \tag{14.27}$$

PROOF. To see (14.27), we first observe that by (14.24),

$$\begin{aligned}
&\|k_w^{\alpha,p} - k_0^{\alpha,p}\|_{B_{p'}} \\
&= \left(\int_{\mathbb{B}_n} \left| (1-|z|^2)^{\mathbf{s}'} \mathcal{R}_{\mathbf{s}'}^{\alpha} \left(k_w^{\alpha,p} - k_0^{\alpha,p} \right)(z) \right|^{p'} d\lambda_n(z) \right)^{1/p'} \\
&= \left(\int_{\mathbb{B}_n} (1-|z|^2)^{\alpha} \left| \frac{1}{(1-\overline{w}\cdot z)^{\mathbf{s}'}} - 1 \right|^{p'} dV \right)^{1/p'} \\
&\lesssim |w| \left(\int_{\mathbb{B}_n} \frac{(1-|z|^2)^{\alpha}}{|1-\overline{w}\cdot z|^{n+1+\alpha}} dV \right)^{1/p'} \lesssim \beta(0,w)^{1/p'}.
\end{aligned} \tag{14.28}$$

By the invariance of B_p and the Bergman metric β, [**259**], [**346**], we then get

$$\begin{aligned}
|f(z) - f(w)| &= |f\circ\varphi_w(\varphi_w(z)) - f\circ\varphi_w(0)| \\
&= \left| \left\langle f\circ\varphi_w, k_{\varphi_w(z)}^{\alpha,p} - k_0^{\alpha,p} \right\rangle_{\alpha,p} \right| \\
&\leq \|f\circ\varphi_w\|_{B_p} \left\| k_{\varphi_w(z)}^{\alpha,p} - k_0^{\alpha,p} \right\|_{B_{p'}} \\
&\lesssim \|f\circ\varphi_w\|_{B_p} \beta(0,\varphi_w(z))^{1/p'} \\
&\approx \|f\|_{B_p} \beta(z,w)^{1/p'}.
\end{aligned}$$

□

14.3. Function Spaces on Trees

In Chapter 5 we introduced and studied the tree Besov space $B_2(\mathcal{T})$, there called the tree Dirichlet space and denoted by $D(\mathcal{T})$. In this chapter and the next two we look at several families of Besov type spaces defined on the $n-$dimensional Bergman tree $\mathcal{T}_n$. In addition to their intrinsic interest these spaces are useful; and at times necessary, for development of our results for spaces of holomorphic functions.

For the moment we just introduce the simplest of those families, the spaces $B_{p,1}(\mathcal{T}_n)$. We make the definition for positive integers n and for the full range $1 < p < \infty$. However the motivation is clearest if we look at the case $p > n$ because in that case, taking note of Proposition 14.4, we can define the Besov space $B_p(\mathbb{B}_n)$ using just first derivatives.

The tree Besov spaces $B_p(\mathcal{T})$, $1 < p < \infty$, are generalizations of $D(\mathcal{T})$. They are defined similarly but using p^{th} power summability of differences in the place of square summability. Those spaces, which are natural companions to the spaces of holomorphic functions $B_p(\mathbb{D}) = B_p(\mathbb{B}_1)$, have their own multiplier algebras, Carleson measures, etc. and their theory closely resembles that of the spaces $B_p(\mathbb{D})$ [**48**], [**41**], [**49**].

We define the spaces $B_{p,1}(\mathcal{T}_n)$ using the same recipe as for the $B_p(\mathcal{T})$ and informed by the same analogies. That is, given $f \in B_p(\mathbb{B}_n)$ we have that the derivative, f', is in an L^p Bergman space of the ball. Modulus of continuity estimates ensure that f' is roughly constant on each of the cubes in the Bergman tree, a set which is, or is in bijection with, the set of vertices of $\mathcal{T}_n$. We use versions of those constants as our model of f' on $\mathcal{T}_n$. The invariant volume, $d\lambda_n$, is roughly the same for each cube and we model it with counting measure on $\mathcal{T}_n$. Thus we expect our model of f' will be in l^p of $\mathcal{T}_n$. Noting that $B_p(\mathbb{B}_n)$ consists of primitives of Bergman space functions we define $B_{p,1}(\mathcal{T}_n)$ to be the space of " summation primitives" of functions in l^p of the tree.

Specifically, the *tree Besov spaces* $B_{p,1}(\mathcal{T}_n)$, are defined as follows. We have $\mathcal{T}_n$, a rooted tree defined in Section 14.1.6, and we continue the notation and language of Chapter 5 for the partial ordering induced by the root, for parents and children, for the predecessor and successor sets, and for the operators I and Δ. For $1 < p < \infty$ we let $B_{p,1}(\mathcal{T}_n)$ be the space of functions F defined on $\mathcal{T}_n$ such that ΔF defined by $\Delta F(\alpha) = F(\alpha) - F(\alpha^-)$ is in $\ell^p(\mathcal{T}_n)$. The space is normed by $\|F\|^p_{B_{p,1}(\mathcal{T}_n)} = |F(o)|^p + \|\Delta F\|^p_{\ell^p(\mathcal{T}_n)}$.

The $B_{p,1}(\mathcal{T}_n)$ are defined using first differences and are discrete analogs of those $B_p(\mathbb{B}_n)$ spaces which are defined using first derivatives. However the definition of $B_p(\mathbb{B}_n)$ requires working with $m > n/p$ derivatives; hence we need to consider $m > 1$ if we work with $n > 1$ and the full range of p. With that in mind we return to these themes in Chapter 16 and introduce spaces $B_{p,m}(\mathcal{T}_n)$ defined using mth order differences. We also go further and introduce more subtle tree spaces $HB_{p,m}(\mathcal{T}_n)$ which allow a more accurate modeling of the relationship between smoothness of functions in $\mathrm{Hol}(\mathbb{B}_n)$ and the nonisotropic complex geometry of $\mathbb{B}_n$.

As we go forward we will see a difference between intrinsic results about spaces on trees and the interactions of tree spaces with spaces of holomorphic functions. For instance, the characterization of Carleson measures for the tree spaces given in (14.37) is intrinsic to the structure of those spaces as is the result on interpolating sequences, Theorem 16.7. However, sometimes when tree spaces are used to study holomorphic functions the interaction between the tree geometry and the ball geometry must be controlled. For instance, in the proof of Corollary 14.21 on monotonicity of Carleson measures both parameters in the tree construction, θ and

λ, must be small (to invoke the atomic decomposition of Besov spaces). In developing the theory of the refined holomorphic tree Besov spaces $HB_p(\mathcal{T}_n)$ of Section 16.3 we will require both λ small and θ large. The structural inequality (16.58), used to define the spaces $HB_p(\mathcal{T}_n)$ and prove the restriction Theorem 16.38, requires both λ small and θ sufficiently large, while the positivity property (16.64) in Lemma 16.35 requires in addition that λ is sufficiently small compared to θ. The theme here is that the Bergman cubes should be chosen very eccentric with their extension θ in the radial direction much larger than their extension λ in the tangential directions.

14.4. Carleson Measures

14.4.1. Discretization of Carleson Measures. Recall that Carleson measures for B_p are defined in (14.25).

A measure μ on $\mathcal{T}_n$ is said to be a Carleson measure for $B_{p,1}(\mathcal{T}_n)$ if for all $F = If$ in $B_{p,1}(\mathcal{T}_n)$ with $f \geq 0$

$$\sum_{\alpha\in\mathcal{T}_n} F(\alpha)^p \mu(\alpha) = \sum_{\alpha\in\mathcal{T}_n} If(\alpha)^p \mu(\alpha) \lesssim \sum_{\alpha\in\mathcal{T}_n} f(\alpha)^p . \tag{14.29}$$

Equivalently, for all F, $\|F\|_{\ell^p(\mu)} \lesssim \|F\|_{B_{p,1}(\mathcal{T}_n)}$.

We now show that μ is a Carleson measure on $B_p(\mathbb{B}_n)$ if and only if an averaged version and a discretized version are Carleson measures. We now define those. Fix a measure μ on $\mathbb{B}_n$. Pick a large positive R and let RK_α denote $B_\beta(c_\alpha, R)$, the Bergman ball with center c_α and radius R. Note that $1K_\alpha$ does not equal K_α. We will have $K_\alpha \subset RK_\alpha$ and as we saw in Lemma 14.3 the RK_α have bounded overlap

$$\sum_{\alpha\in\mathcal{T}_n} \chi_{RK_\alpha} \lesssim C = C(R). \tag{14.30}$$

We denote the μ measure of RK_α by $\mu(\alpha)$ (suppressing R in that notation) and let $\widetilde{\mu_R}$ denote an averaged version of μ given by

$$d\widetilde{\mu_R}(z) := \sum_{\alpha\in\mathcal{T}_n} \frac{\mu(\alpha)}{|RK_\alpha|} \chi_{RK_\alpha}(z)\, d\lambda_n(z), \tag{14.31}$$

and we define the discretized version $\mu_R^\natural$ by

$$d\mu_R^\natural := \sum_{\alpha\in\mathcal{T}_n} \mu(\alpha)\delta_{c_\alpha}. \tag{14.32}$$

Finally, we will denote by $\hat{\mu}$ the associated measure on the tree $\mathcal{T}_n$; that is $\forall\alpha \in \mathcal{T}_n$, $\hat{\mu}(\alpha) = \mu(\alpha) = \mu(RK_\alpha)$.

We will need the following:

THEOREM 14.13. *Let $1 < p < \infty$, $r \leq R < \infty$, let μ be a positive Borel measure on the ball $\mathbb{B}_n$ and $\widetilde{\mu_R}$, $\mu_R^\natural$ be defined as in (14.31) and (14.32) respectively. Then μ is a B_p-Carleson measure on $\mathbb{B}_n$ if and only if $\widetilde{\mu_R}$ is a B_p-Carleson measure on $\mathbb{B}_n$ if and only if $\mu_R^\natural$ is a B_p-Carleson measure on $\mathbb{B}_n$. The implied constants depend only on R, n and p.*

We make no restriction here on the structural constants θ and λ.

PROOF. Testing the definition of Carleson measure on the constant function we see that such measures as in the Theorem assign finite mass to the tree $\mathcal{T}_n$. Hence the bounded overlap condition (14.30) ensures that $\mu(\mathbb{B}_n) \approx \widetilde{\mu_R}(\mathbb{B}_n) \approx \mu_R^\natural(\mathbb{B}_n) \approx \hat{\mu}(\mathcal{T}_n)$. Fix $f \in B_p$ and, for the moment, $R \geq r$; and let $\{w_\alpha\}_{\alpha\in\mathcal{T}_n}$ and $\{\xi_\alpha\}_{\alpha\in\mathcal{T}_n}$ be sequences with $w_\alpha, \xi_\alpha \in \overline{RK_\alpha}$ satisfying

$$|f(w_\alpha)| = \max_{w\in\overline{RK_\alpha}} |f(w)|, \qquad |f(\xi_\alpha)| = \min_{w\in\overline{RK_\alpha}} |f(w)|, \qquad \alpha \in \mathcal{T}_n.$$

By Minkowski's inequality and (14.12),

$$\begin{aligned}
&\left|\left(\sum_{\alpha\in\mathcal{T}_n} |f(w_\alpha)|^p \mu(RK_\alpha)\right)^{1/p} - \left(\sum_{\alpha\in\mathcal{T}_n} |f(\xi_\alpha)|^p \mu(RK_\alpha)\right)^{1/p}\right| \\
&\qquad \leq \left(\sum_{\alpha\in\mathcal{T}_n} |f(w_\alpha) - f(\xi_\alpha)|^p \mu(RK_\alpha)\right)^{1/p} \\
&\qquad \lesssim \left(\sum_{\alpha\in\mathcal{T}_n} \|f\|_{B_p}^p \mu(RK_\alpha)\right)^{1/p} \lesssim \mu(\mathbb{B}_n)^{1/p} \|f\|_{B_p}.
\end{aligned}$$

From the inequalities

$$\begin{aligned}
|f(\xi_\alpha)|^p \mu(RK_\alpha) &\lesssim \int_{RK_\alpha} |f(z)|^p d\mu(z) \lesssim |f(w_\alpha)|^p \mu(RK_\alpha), \\
|f(\xi_\alpha)|^p \mu(RK_\alpha) &\lesssim \int_{RK_\alpha} |f(z)|^p \left(\int_{RK_\alpha} d\mu\right) d\lambda_n(z) \lesssim |f(w_\alpha)|^p \mu(RK_\alpha), \\
|f(\xi_\alpha)|^p \mu(RK_\alpha) &\lesssim \int_{RK_\alpha} |f(z)|^p \left(\int_{RK_\alpha} d\mu\right) \delta_{c_\alpha}(z) \lesssim |f(w_\alpha)|^p \mu(RK_\alpha),
\end{aligned}$$

it follows that if ν is any of the three measures μ, $\widetilde{\mu_R}$, $\mu_R^\natural$, then

$$\sum_{\alpha\in\mathcal{T}_n} |f(\xi_\alpha)|^p \mu(RK_\alpha) \lesssim \|f\|_{L^p(\nu)}^p \lesssim \sum_{\alpha\in\mathcal{T}_n} |f(w_\alpha)|^p \mu(RK_\alpha),$$

and hence if ν_1 and ν_2 denote any two of the three measures μ, $\widetilde{\mu_R}$, $\mu_R^\natural$, then

$$\begin{aligned}
\|f\|_{L^p(\nu_1)}^p &= \left(\|f\|_{L^p(\nu_1)}^p - \|f\|_{L^p(\nu_2)}^p\right) + \|f\|_{L^p(\nu_2)}^p \\
&\lesssim \mu(\mathbb{B}_n) \|f\|_{B_p}^p + \|f\|_{L^p(\nu_2)}^p.
\end{aligned}$$

The conclusion of the theorem now follows immediately. □

14.4.2. Multipliers and Carleson Measures. Recall the definitions of Carleson measures and multipliers for B_p; (14.25) and (14.26). We denote by $H^\infty(\mathbb{B}_n)$ the space of bounded analytic functions on $\mathbb{B}_n$ with the usual supremum norm.

As was true for functions of a single variable (Section 2.4.2), multipliers are bounded and there is an intimate relation between multipliers and Carleson measures.

THEOREM 14.14. *A function φ defined on $\mathbb{B}_n$ is a multiplier on B_p if and only if $\varphi \in H^\infty(\mathbb{B}_n) \cap B_p$ and for some (or every) $m > n/p$*

$$d\mu_\varphi = \left|(1 - |z|^2)^m \nabla^m \varphi(z)\right|^p d\lambda_n(z) \tag{14.33}$$

is a B_p-Carleson measure.

PROOF. Just as in Section 2.4.2, it is straightforward to see that if φ is a multiplier on B_p, then $\varphi \in H^\infty(\mathbb{B}_n) \cap B_p$. First note that, since $1 \in B_p$, the measure μ_φ must be finite which ensures $\varphi \in B_p$. Next, the adjoint M_φ^* of the multiplier operator $M_\varphi f = \varphi f$ is bounded on the dual space $1B_{p'}$, and if e_z is the point evaluation functional on B_p, then

$$\left\langle f, M_\varphi^* e_z \right\rangle_p = \left\langle M_\varphi f, e_z \right\rangle_p = \varphi(z) f(z) = \varphi(z) \left\langle f, e_z \right\rangle_p = \left\langle f, \overline{\varphi(z)} e_z \right\rangle_p, \qquad f \in B_p.$$

This shows that $M_\varphi^* e_z = \overline{\varphi(z)} e_z$. Thus

$$|\varphi(z)| \, \|e_z\|_{B_p'} = \left\| \overline{\varphi(z)} e_z \right\|_{B_p'} = \left\| M_\varphi^* e_z \right\|_{B_p'} \leq \left\| M_\varphi^* \right\| \|e_z\|_{B_p'}$$

which implies that $|\varphi(z)| \leq \left\| M_\varphi^* \right\| = \|M_\varphi\|$ since $\|e_z\|_{B_p'} < \infty$.

The analysis of the Carleson measure condition is complicated slightly by the need to include analysis of derivatives past the first. Fix $1 < p < \infty$ and $m > n/p$. Let $f, \varphi \in B_p$. Then

$$\|\varphi f\|_{B_p} = \sum_{k=0}^{m-1} \left| \nabla^k (\varphi f)(0) \right| + \left(\int_{\mathbb{B}_n} \left| (1-|z|^2)^m \nabla^m (\varphi f)(z) \right|^p d\lambda_n(z) \right)^{1/p},$$

and, for some constants $c_{m,k}$,

$$\nabla^m (\varphi f)(z) = \sum_{k=0}^{m} c_{m,k} \left(\nabla^{m-k} \varphi(z) \right) \left(\nabla^k f(z) \right). \tag{14.34}$$

Hence, using this estimate,

$$\begin{aligned}
&\left(\int_{\mathbb{B}_n} \left| (1-|z|^2)^m \nabla^m (\varphi f)(z) \right|^p d\lambda_n(z) \right)^{1/p} \\
&\lesssim \left(\int_{\mathbb{B}_n} \left| (1-|z|^2)^m \nabla^m \varphi(z) \right|^p |f(z)|^p \, d\lambda_n(z) \right)^{1/p} \\
&+ \sum_{k=1}^{m-1} \left(\int_{\mathbb{B}_n} \left| (1-|z|^2)^{m-k} \nabla^{m-k} \varphi(z) \right|^p \left| (1-|z|^2)^k \nabla^k f(z) \right|^p d\lambda_n(z) \right)^{1/p} \\
&+ \left(\int_{\mathbb{B}_n} |\varphi(z)|^p \left| (1-|z|^2)^m \nabla^m f(z) \right|^p d\lambda_n(z) \right)^{1/p}.
\end{aligned}$$

For $1 \leq k \leq m-1$ set

$$q_k = \frac{m}{m-k}, \quad q_k' = \frac{m}{k}.$$

Noting that $m-k=m/q_k>n/(pq_k)$ and $k=m/q_k'>n/(pq_k')$, apply Holder's inequality for each k to obtain

$$\begin{aligned}\left(\int_{\mathbb{B}_n}\left|(1-|z|^2)^{m-k}\nabla^{m-k}\varphi(z)\right|^p\left|(1-|z|^2)^k\nabla^k f(z)\right|^p d\lambda_n(z)\right)^{1/p}\\ \leq\left(\int_{\mathbb{B}_n}\left|(1-|z|^2)^{m-k}\nabla^{m-k}\varphi(z)\right|^{pq_k}d\lambda_n(z)\right)^{1/(pq_k)}\\ \times\left(\int_{\mathbb{B}_n}\left|(1-|z|^2)^k\nabla^k f(z)\right|^{pq_k'}d\lambda_n(z)\right)^{1/(pq_k')}\\ \leq\|\varphi\|_{B_{pq_k}(\mathbb{B}_n)}\|f\|_{B_{pq_k'}(\mathbb{B}_n)}.\end{aligned}$$

Now the atomic decomposition of Besov spaces, [**346**, Theorem 6.6], implies in particular that the inclusions of the Besov spaces B_p are determined by those of the ℓ^p spaces; that is,

$$B_p\subset B_q,\qquad 0<p<q<\infty,$$

and hence, because $q_k,q_k'>1$, we also have

$$\|\varphi\|_{B_{pq_k}}\|f\|_{B_{pq_k'}}\leq\|\varphi\|_{B_p}\|f\|_{B_p},\qquad 1\leq k\leq m-1.$$

Also,

$$\begin{aligned}\sum_{k=0}^{m-1}\left|\nabla^k(\varphi f)(0)\right|&\lesssim\sum_{k=0}^{m-1}\left|\sum_{j=0}^{k}\left(\nabla^{k-j}\varphi(0)\right)\left(\nabla^j f(0)\right)\right|\\ &\lesssim\left(\sum_{k=0}^{m-1}\left|\nabla^k\varphi(0)\right|\right)\left(\sum_{k=0}^{m-1}\left|\nabla^k f(0)\right|\right)\\ &\leq\|\varphi\|_{B_p}\|f\|_{B_p}.\end{aligned}$$

Combining all of these inequalities and recalling $\mu=d\mu_\varphi$ from (14.33) we obtain

$$\|\varphi f\|_{B_p}\lesssim\left\{\|f\|_{L^p(\mu)}+\|\varphi\|_{B_p}\|f\|_{B_p}+\|\varphi\|_{H^\infty(\mathbb{B}_n)}\|f\|_{B_p}\right\},\tag{14.35}$$

which shows φ is a multiplier. In the other direction, if we rewrite (14.34) as

$$c_{m,0}\left(\nabla^m\varphi(z)\right)f(z)=-\nabla^m(\varphi f)(z)+\sum_{k=1}^{m}c_{m,k}\left(\nabla^{m-k}\varphi(z)\right)\left(\nabla^k f(z)\right),$$

and multiply through by $(1-|z|^2)^m$, the above inequalities yield

$$\|f\|_{L^p(\mu)}\lesssim\left\{\|\varphi f\|_{B_p}+\|\varphi\|_{B_p}\|f\|_{B_p}+\|\varphi\|_{H^\infty(\mathbb{B}_n)}\|f\|_{B_p}\right\}.\tag{14.36}$$

That shows that μ is a Carleson measure if φ is a multiplier on B_p (and hence is also in $H^\infty(\mathbb{B}_n)\cap B_p$). □

14.4.3. Characterizing Carleson Measures. In this section we give generalizations of results of Chapter 5 concerning the relation between Carleson measures for tree spaces and those for spaces of holomorphic functions. We also investigate the role of the *tree conditions*, such as (TC), (14.29) and (14.37) below, in characterizing those measures.

We show that, given a nonnegative measure μ on the ball, $\hat{\mu}$ is a $B_p(\mathcal{T}_n)$-Carleson measure if and only if $\hat{\mu}$ satisfies (14.37). If $1<p<2+1/(n-1)$ then those conditions are equivalent to μ being a B_p-Carleson measure, and for any p knowing $\hat{\mu}$ is a $B_p(\mathcal{T}_n)$-Carleson measure implies μ is a B_p-Carleson measure. However for $2+1/(n-1) \le p < \infty$ we do not know if the conditions are equivalent; that is, in that range we don't know if every B_p-Carleson measure satisfies the tree condition.

PROBLEM 14.15. *Characterize the $B_p(\mathbb{B}_n)$-Carleson measures for $2+1/(n-1) \le p < \infty$.*

THEOREM 14.16. *Suppose $1 < p < \infty$, $1/p + 1/p' = 1$, and that $0 < \lambda, \theta < \infty$ are the structural constants in the construction of $\mathcal{T}_n$ in Section 14.1.6. Let μ be a positive measure on $\mathbb{B}_n$. Then with constants depending only on p, λ, θ, n, conditions (2) and (3) below are equivalent. Condition (3) is sufficient for condition (1), and, provided $1 < p < 2 + 1/(n-1)$, condition (3) is also necessary for condition (1):*

(1) *μ is a B_p-Carleson measure on $\mathbb{B}_n$, i.e. (14.25) holds.*
(2) *$\widehat{\mu} = \{\mu(\alpha)\}_{\alpha \in \mathcal{T}_n}$ is a $B_p(\mathcal{T}_n)$-Carleson measure on the Bergman tree $\mathcal{T}_n$, i.e. (14.29) holds .*
(3) *There is $C < \infty$ such that for $\alpha \in \mathcal{T}_n$*

$$I^*((I^*\mu)^{p'})(\alpha) = \sum_{\beta \ge \alpha} I^*\mu(\beta)^{p'} \lesssim I^*\mu(\alpha) < \infty. \tag{14.37}$$

PROOF. The tree theorem in [**48**] yields the equivalence of conditions (2) and (3) in Theorem 14.16. That proof, for general p, is essentially the same as the proof for the case $p = 2$ given in Chapter 5.

We will consider the necessity and sufficiency of condition (3) for condition (1) separately in the next two subsections. For convenience we prove only the case $\lambda = 1$ and $\theta = \frac{\ln 2}{2}$, so that by Lemma 14.3, the dimension of $\mathcal{T}_n$ is n, and

$$1 - |z|^2 \approx e^{-2\beta(0,z)} \approx e^{-2\theta d(\alpha)} = 2^{-d(\alpha)} \tag{14.38}$$

for $z \in K_\alpha$ by (14.11). The proof of the general case is similar.

Duality: First we dualize the Carleson embedding by computing its adjoint relative to the pairing $\langle \cdot, \cdot \rangle_{\alpha,p}$ introduced above. Let

$$\langle f, g \rangle_\mu = \int_{\mathbb{B}_n} f(z)\overline{g(z)}d\mu(z) \tag{14.39}$$

be the usual pairing between $L^p(\mu)$ and $L^{p'}(\mu)$, and suppose that $1 < p < \infty$, $\alpha > -1$. We claim that for a polynomial $f \in B_p(\mathbb{B}_n)$ and a simple function $g \in L^{p'}(\mu)$, we have

$$\langle f, g \rangle_\mu = \langle f, \Theta g \rangle_{\alpha,p}$$

where $\Theta g \in B_{p'}(\mathbb{B}_n)$ is given by the formula

$$\mathcal{R}^\alpha_{\mathbf{s}'} \Theta g(w) := \int_{\mathbb{B}_n} \left(\frac{1}{1 - \overline{z} \cdot w} \right)^{\mathbf{s}'} g(z)\, d\mu(z).$$

In fact, by (14.22) and (14.23) we have

$$\begin{aligned}\langle f, g\rangle_\mu &= \int_{\mathbb{B}_n} f(z)\overline{g(z)}d\mu(z) = \int_{\mathbb{B}_n} \langle f, k_z^{\alpha,p}\rangle_{\alpha,p}\overline{g(z)}d\mu(z) \\ &= \int_{\mathbb{B}_n}\left\{\int_{\mathbb{B}_n} \mathcal{R}_{\mathbf{s}}^\alpha f(w)\overline{\mathcal{R}_{\mathbf{s}'}^\alpha k_z^{\alpha,p}(w)}d\nu_\alpha(w)\right\}\overline{g(z)}d\mu(z) \\ &= \int_{\mathbb{B}_n} \mathcal{R}_{\mathbf{s}}^\alpha f(w)\overline{\left\{\int_{\mathbb{B}_n}\left(\frac{1}{1-\overline{z}\cdot w}\right)^{\mathbf{s}'} g(z)\,d\mu(z)\right\}}d\nu_\alpha(w).\end{aligned}$$

The Carleson embedding statement

$$\|f\|_{L^p(\mu)} = \left(\int_{\mathbb{B}_n} |f(z)|^p\,d\mu(z)\right)^{1/p} \lesssim \|f\|_{B_p(\mathbb{B}_n)}, \qquad f \in B_p(\mathbb{B}_n),$$

is equivalent to its Banach space dual statement. By the density of polynomials in Besov spaces, the density of simple functions in Lebesgue spaces, and the previous computation; the dual statement is

$$\|\Theta g\|_{B_{p'}(\mathbb{B}_n)} \lesssim \|g\|_{L^{p'}(\mu)}.$$

Since we have

$$\begin{aligned}\|\Theta g\|_{B_{p'}(\mathbb{B}_n)} &= \left(\int_{\mathbb{B}_n}\left|\int_{\mathbb{B}_n}\left(\frac{1}{1-\overline{z}\cdot w}\right)^{\mathbf{s}'} g(z)\,d\mu(z)\right|^{p'} d\nu_\alpha(w)\right)^{1/p'} \\ &= \left(\int_{\mathbb{B}_n}\left|\int_{\mathbb{B}_n}\left(\frac{1-|w|^2}{1-\overline{z}\cdot w}\right)^{\mathbf{s}'} g(z)\,d\mu(z)\right|^{p'} d\lambda_n(w)\right)^{1/p'},\end{aligned}$$

we can restate the dual inequality as

$$\|S_\mu^* g\|_{L^{p'}(\lambda_n)} \lesssim \|g\|_{L^{p'}(\mu)}, \qquad g \in L^{p'}(\mu), \tag{14.40}$$

where the operator S_μ^* is given by

$$S_\mu^* g(w) := \int_{\mathbb{B}_n}\left(\frac{1-|w|^2}{1-\overline{z}\cdot w}\right)^{\mathbf{s}'} g(z)\,d\mu(z). \tag{14.41}$$

REMARK 14.17. *The implication* (1) *implies* (2) *of Theorem* 14.16 *is equivalent to the implication that boundedness of S_μ^* from $L^{p'}(\mu)$ to $L^{p'}(\lambda_n)$ (as in* (14.40)*) implies the boundedness of T_μ^* between the same two spaces, where*

$$T_\mu^* g(w) := \int_{\mathbb{B}_n}\left|\frac{1-|w|^2}{1-\overline{z}\cdot w}\right|^{\mathbf{s}'} g(z)\,d\mu(z); \tag{14.42}$$

that is, the kernel for T_μ^* is the modulus of the kernel for S_μ^*. Roughly speaking, the Carleson embedding implies the tree condition if and only if we can take absolute values inside the operator S_μ^* without destroying the boundedness in (14.40)*. This can be shown using the argument in the next section.*

Necessity in a Small Range: We now give a relatively simple argument which establishes necessity in the small range $1 < p < 1 + 1/(n-1)$. Suppose that

μ is a B_p-Carleson measure on $\mathbb{B}_n$ where $1 < p < 1 + 1/(n-1)$. Choose $\alpha > -1$ so that $\mathbf{s}' = 1$. We then obtain from (14.40) that

$$\begin{aligned}\int_{\mathbb{B}_n}\left|\int_{\mathbb{B}_n}\operatorname{Re}\left(\frac{1-|w|^2}{1-\overline{z}\cdot w}\right)g(z)\,d\mu(z)\right|^{p'}d\lambda_n(w)&\\ \leq \left\|S_\mu^* g\right\|_{L^{p'}(\lambda_n)}^{p'}&\\ \lesssim \int_{\mathbb{B}_n}|g|^{p'}\,d\mu&\end{aligned}$$

for all $g \geq 0$. The tree inequality (14.29) now follows as in the one-dimensional case in [**48**], [**41**]. Indeed, fix $\alpha \in \mathcal{T}_n$ and let $g = \sum_{\alpha\in_n} g(\alpha)\chi_{K_\alpha}$. Here g is constant on K_α with value $g(\alpha)$. Then since

$$\operatorname{Re}\left(\frac{1-|w|^2}{1-\overline{z}\cdot w}\right) \gtrsim 1 > 0, \qquad w \in K_\beta, z \in S_\beta$$

for $\beta \geq \alpha$, and $\operatorname{Re}\left(\frac{1-|w|^2}{1-\overline{z}\cdot w}\right) \geq 0$ otherwise, we obtain

$$\begin{aligned}\left\|I^*(g\mu)\right\|_{L^{p'}(\mathcal{T}_n)}^{p'} &\approx \sum_{\alpha\in\mathcal{T}_n}\left(\sum_{\beta\geq\alpha} g(\alpha)\mu(K_\beta)\right)^{p'}\\ &\lesssim \int_{\mathbb{B}_n}\left(\int_{\mathbb{B}_n}\operatorname{Re}\left(\frac{1-|w|^2}{1-\overline{z}\cdot w}\right)g(z)\,d\mu(z)\right)^{p'}d\lambda_n(w)\\ &\lesssim \int_{\mathbb{B}_n}|g|^{p'}\,d\mu\\ &\approx \sum_{\alpha\in\mathcal{T}_n}|g(\alpha)|^{p'}\mu(\alpha) \approx \|g\|_{L^{p'}(\mathcal{T}_n)}^{p'},\end{aligned}$$

which yields the dual of inequality (14.29).

Unfortunately, this elegant proof breaks down for $p \geq 1 + 1/(n-1)$. For those p, and with the restriction that $\alpha > -1$, we cannot ensure $\theta = \mathbf{s}' \in (0,1)$, thus we no longer have the crucial estimate $\operatorname{Re}\left(\frac{1-|w|^2}{1-\overline{z}\cdot w}\right)^\theta > 0$.

Sufficiency: We will show condition (3) is sufficient for condition (1). Suppose that $\widehat{\mu}$ satisfies the tree inequality (14.29). We want to show that condition (1) holds. We will do that by establishing the dual inequality (14.40). Since $\widehat{\mu} = \widehat{\widetilde{\mu}}$, we now replace μ by $\widetilde{\mu}$ in (14.40) and establish the boundedness for $S_{\widetilde{\mu}}^*$ instead of S_μ^* in (14.40). We will do that by establishing the boundedness of $T_{\widetilde{\mu}}^*$, given by a formula such as (14.42), from $L^{p'}(\widetilde{\mu})$ to $L^{p'}(\lambda_n)$. When that is done, it follows that $\widetilde{\mu}$, and hence also μ, is a B_p-Carleson measure.

We prove a slightly more general result than the boundedness of $T_{\widetilde{\mu}}^*$. □

LEMMA 14.18. *For $0 \leq \sigma \leq 1$, $1 < r < \infty$, $s + \sigma r > n$ and $f \in L^1$ define*

$$\widehat{T}f(w) := \int_{\mathbb{B}_n}\frac{(1-|w|^2)^{s/r}}{|1-w\cdot\overline{z}|^{\sigma+s/r}}f(z)\,dV.$$

Then we have

$$\left\|\widehat{T}f\right\|_{L^r(d\lambda_n)}^r \lesssim \sum_{\alpha\in\mathcal{T}_n}\left[e^{2\theta\sigma d(\alpha)}I^*f(\alpha)\right]^r,$$

where $I^*f(\alpha)=\sum_{\beta\in\mathcal{T}_n:\beta\geq\alpha}\hat{f}(\beta)$ *and* $\hat{f}(\beta)=\int_{K_\beta}|f(z)|\,dV$.

Note that the discretization of $\hat{f}$ of f involves the measure dV, as compared to the discretization of g above that uses $d\lambda_n$.

PROOF OF LEMMA. Let $\theta=\frac{\ln 2}{2}$; that gives the notationally convenient (16.83). We compute for $f\geq 0$,

$$\begin{aligned}\left\|\widehat{T}f\right\|_{L^r(\lambda_n)}^r &= \int_{\mathbb{B}_n}\left(\int_{\mathbb{B}_n}\frac{(1-|w|^2)^{s/r}}{|1-w\cdot\overline{z}|^{\sigma+s/r}}f(z)\,dV\right)^r d\lambda_n(w)\\ &\leq \int_{\mathbb{B}_n}\left(\frac{1}{(1-|w|^2)^\sigma}\sum_{m=0}^{-\log_2(1-|w|)}2^{-m(\sigma+s/r)}\int_{S_{w_m}}f(z)\,dV\right)^r d\lambda_n(w).\end{aligned}$$

Here we are writing for any $w\in\mathbb{B}_n$, $m=0,1,$

$$w_m=2^m(1-|w|)\frac{w}{|w|}.$$

and S_{w_m} for the tent below w_m; for any $\zeta\in\mathbb{B}_n$,

$$S_\zeta=\left\{z:\left|1-z\cdot\frac{\zeta}{|\zeta|}\right|\leq 1-|\zeta|^2\right\}.$$

We now fix a constant μ such that $n/(s+\sigma r)<\mu<1$ and continue the string of inequalities starting with Hölder's inequality:

$$\begin{aligned}\left\|\widehat{T}f\right\|_{L^r(\lambda_n)}^r &\leq \int_{\mathbb{B}_n}\left\{\sum_{m=0}^{-\log_2(1-|w|)}2^{-m(\sigma+s/r)(1-\mu)r'}\right\}^{r/r'}\\ &\quad\times\left\{\sum_{m=0}^{-\log_2(1-|w|)}2^{-m(s+\sigma r)\mu}\left(\frac{1}{(1-|w|^2)^\sigma}\int_{S_{w_m}}f(z)\,dV\right)^r\right\}d\lambda_n(w)\\ &\leq C_{s,r,\sigma,\mu}\sum_{m=0}^{\infty}2^{-m(s+\sigma r)\mu}\int_{\mathbb{B}_n}\left(\frac{1}{(1-|w|^2)^\sigma}\int_{S_{w_m}}f(z)\,dV\right)^r\\ &\quad\times\chi_{[m,\infty)}(-\log_2(1-|w|))\,d\lambda_n(w)\\ &\leq C_{s,r,\sigma,\mu}\sum_{m=0}^{\infty}2^{-m(s+\sigma r)\mu}\sum_{\alpha\in\mathcal{T}_n}\chi_{[m,\infty)}(d(\alpha))\left(2^{\sigma d(\alpha)}\sum_{\beta\in\mathcal{T}_n:\beta\geq P^m(\alpha)}\hat{f}(\beta)\right)^r,\end{aligned}$$

with $C_{s,r,\sigma,\mu} = \left(\sum_{m=0}^{\infty} 2^{-m\sigma+s/r(1-\mu)r'}\right)^{r-1}$, and where $P^m(\alpha)$ denotes the mth predecessor of α. The final term above satisfies

$$\sum_{m=0}^{\infty} 2^{-m(s+\sigma r)\mu} \sum_{\alpha\in\mathcal{T}_n} \chi_{[m,\infty)}(d(\alpha)) \left(2^{\sigma d(\alpha)} \sum_{\beta\in\mathcal{T}_n:\beta\geq P^m(\alpha)} \hat{f}(\beta)\right)^r$$

$$\leq \sum_{m=0}^{\infty} 2^{m(n-(s+\sigma r)\mu)} \sup_{m\geq 0} 2^{-nm} \sum_{\alpha\in\mathcal{T}_n} \chi_{[m,\infty)}(d(\alpha)) \left(2^{\sigma d(\alpha)} \sum_{\beta\in\mathcal{T}_n:\beta\geq P^m(\alpha)} \hat{f}(\beta)\right)^r$$

$$\lesssim C'_{s,n,\sigma,\mu} \sum_{\alpha\in\mathcal{T}_n} \left(2^{\sigma d(\alpha)} \sum_{\beta\in\mathcal{T}_n:\beta\geq\alpha} \hat{f}(\beta)\right)^r = \sum_{\alpha\in\mathcal{T}_n} \left(2^{\sigma d(\alpha)} I^* f(\alpha)\right)^r,$$

with $C'_{s,n,\sigma,\mu} = \sum_{m=0}^{\infty} 2^{m(n-(s+\sigma r)\mu)} < \infty$ since $n/(s+\sigma r) < \mu$, and where we have used the fact that

$$\operatorname{card}\{\alpha \in \mathcal{T}_n : P^m(\alpha) = \gamma\} \lesssim (2^n)^m,$$

which follows from Lemma 14.3, our choice of $\theta = \frac{\ln 2}{2}$, and (14.38). This completes the proof of Lemma 14.18. □

PROOF OF SUFFICIENCY. We now apply Lemma 14.18 with $\sigma = 0$, $s = n + 1 + \alpha$, $r = p'$ and

$$f(z) = g(z) \sum_{\alpha\in\mathcal{T}_n} \mu(K_\alpha)(1-|z|^2)^{-n-1} \chi_{K_\alpha}(z),$$

we obtain, with $\widehat{T}$ as in the lemma,

$$\left\|T^*_{\widetilde{\mu}} g\right\|^{p'}_{L^{p'}(\lambda_n)} = \left\|\widehat{T}(g\widetilde{\mu})\right\|^{p'}_{L^{p'}(\lambda_n)} \lesssim \sum_{\alpha\in\mathcal{T}_n} I^*(g\widetilde{\mu})(\alpha)^{p'},$$

where

$$I^*(g\widetilde{\mu}) = \sum_{\beta\in\mathcal{T}_n:\beta\geq\alpha} \int_{K_\beta} g\,d\widetilde{\mu} = \sum_{\beta\in\mathcal{T}_n:\beta\geq\alpha} \mu(K_\alpha) \int_{K_\beta} g\,d\lambda_n$$

and $g(\alpha) = \int_{K_\alpha} g\,d\lambda_n$. The tree inequality (14.29) holds for $\widetilde{\mu}(\alpha) = \widetilde{\mu}(K_\alpha)$, and this in turn is equivalent to

$$\sum_{\alpha\in\mathcal{T}_n} I^*(g\widetilde{\mu})(\alpha)^{p'} \lesssim \sum_{\alpha\in\mathcal{T}_n} g(\alpha)^{p'} \widetilde{\mu}(\alpha).$$

Finally, since $\widetilde{\mu}$ is constant on K_α, we have

$$g(\alpha)^{p'} \widetilde{\mu}(\alpha) = \mu(K_\alpha)\left(\int_{K_\alpha} g\,d\lambda_n\right)^{p'} \leq \mu(K_\alpha)\int_{K_\alpha} g^{p'}\,d\lambda_n = \int_{K_\alpha} g^{p'}\,d\widetilde{\mu},$$

for all $\alpha \in \mathcal{T}_n$, and hence

$$\sum_{\alpha\in\mathcal{T}_n} g(\alpha)^{p'} \widetilde{\mu}(\alpha) \leq \int_{\mathbb{B}_n} g^{p'}\,d\widetilde{\mu}.$$

Combining these inequalities establishes (14.40) as required.

We have now completed the proof that condition (3) is sufficient for condition (1) in Theorem 14.16.

Necessity in the Extended Range: We now prove necessity for $1 < p < 2 + 1/(n-1)$, but with a more substantial argument than we used earlier for the restricted range.

Suppose that μ is a B_p-Carleson measure on $\mathbb{B}_n$ where $1 < p < 2 + 1/(n-1)$. We have from (14.40) that

$$\int_{\mathbb{B}_n} \left| \int_{\mathbb{B}_n} \left(\frac{1-|w|^2}{1-\overline{z}\cdot w} \right)^{\mathbf{s}'} g(z)\, d\mu(z) \right|^{p'} d\lambda_n(w) \tag{14.43}$$

$$= \left\| S_\mu^* g \right\|_{L^{p'}(\lambda_n)}^{p'} \lesssim \left(\int_{\mathbb{B}_n} |g|^{p'} d\mu \right), \tag{14.44}$$

where the operator S_μ^* is given by (14.41). The left side of (14.43) raised to the power $1/p'$ is

(14.45)

$$\begin{aligned}
&\left(\int_{\mathbb{B}_n} \left| \int_{\mathbb{B}_n} \left(\frac{1-|w|^2}{1-\overline{z}\cdot w} \right)^{\mathbf{s}'} g(z)\, d\mu(z) \right|^{p'} d\lambda_n(w) \right)^{1/p'} \\
&\quad = \sup_{\|F\|_{L^p(\lambda_n)} \le 1} \left| \int_{\mathbb{B}_n} \left\{ \int_{\mathbb{B}_n} \left(\frac{1-|w|^2}{1-\overline{z}\cdot w} \right)^{\mathbf{s}'} g(z)\, d\mu(z) \right\} \overline{F(w)} d\lambda_n(w) \right| \\
&\quad \ge \sup_{f\ge 0} \left| \int_{\mathbb{B}_n} \left\{ \int_{\mathbb{B}_n} \left(\frac{1-|w|^2}{1-\overline{z}\cdot w} \right)^{\mathbf{s}'} g(z)\, d\mu(z) \right\} \frac{\overline{V_\eta f(w)}}{\|V_\eta f\|_{L^p(d\lambda_n)}} d\lambda_n(w) \right|,
\end{aligned}$$

where for $0 < \eta < 1$ we define the operator V_η by

$$\begin{aligned}
V_\eta f(w) &:= \int_{\mathbb{B}_n} (1-|w|^2)^{\mathbf{s}} \left\{ \mathcal{R}_{\mathbf{s}}^{\alpha} \left(\frac{1}{1-\overline{z}\cdot w} \right)^{\eta} \right\} f(z)\, dV \\
&= (1-|w|^2)^{\mathbf{s}} \mathcal{R}_{\mathbf{s}}^{\alpha} \int_{\mathbb{B}_n} \frac{f(z)}{(1-\overline{z}\cdot w)^{\eta}} dV.
\end{aligned}$$

From (14.16) we have

$$\left(\frac{1}{1-\overline{z'}\cdot w} \right)^{\eta} = R_{\eta-n-1,\mathbf{s}} \left(\frac{1}{1-\overline{z'}\cdot w} \right)^{\mathbf{s}+\eta},$$

and so with

$$\begin{aligned}
\widetilde{V_\eta} f(w) &:= \int_{\mathbb{B}_n} \frac{(1-|w|^2)^{\mathbf{s}}}{(1-\overline{z}\cdot w)^{\mathbf{s}+\eta}} f(z)\, dV, \\
U_\eta f(w) &:= \frac{V_\eta f(w)}{(1-|w|^2)^{\mathbf{s}}}, \\
\widetilde{U_\eta} f(w) &:= \frac{\widetilde{V_\eta} f(w)}{(1-|w|^2)^{\mathbf{s}}},
\end{aligned}$$

we obtain

$$U_\eta f(w) = R^{\alpha-\mathbf{s},\mathbf{s}} R_{\eta-n-1,\mathbf{s}} \widetilde{U_\eta} f(w).$$

We now use the fact that $R^{\gamma_1,t}R_{\gamma_2,t}$ is bounded on the Bergman space A_p^α provided $n+\gamma_i$, $n+\gamma_i+t\notin -\mathbb{N}$ ([**346**, Corollary 6.5]) to conclude that

$$\|V_\eta f\|_{L^p(\lambda_n)} = \|U_\eta f\|_{L^p(\nu_\alpha)} = \left\|R^{\alpha-\mathbf{s},\mathbf{s}}R_{\eta-n-1,\mathbf{s}}\widetilde{U_\eta f}\right\|_{L^p(\nu_\alpha)} \lesssim \left\|\widetilde{U_\eta f}\right\|_{L^p(d\nu_\alpha)} \approx \left\|\widetilde{V_\eta f}\right\|_{L^p(\lambda_n)}. \tag{14.46}$$

REMARK 14.19. *Note that in the special case* $p=2$, $S_\mu^* g = \widetilde{T}_0(g\mu)$. *In that case the adaptation of the argument below to* $\eta=0$ *reduces to a familiar* TT^* *argument.*

The final line in (14.45) is

$$\sup_{f\geq 0}\frac{1}{\|V_\eta f\|_{L^p(\lambda_n)}}\left|\int_{\mathbb{B}_n} S_\mu^* g(w)\,\overline{V_\eta f(w)}d\lambda_n(w)\right|. \tag{14.47}$$

To compute this supremum, we note that the integral in (14.47) for $f\geq 0$ is

(14.48)

$$\begin{aligned}&\int_{\mathbb{B}_n} S_\mu^* g(w)\,\overline{V_\eta f(w)}d\lambda_n(w)\\ &= \int_{\mathbb{B}_n}\int_{\mathbb{B}_n}\left\{\int_{\mathbb{B}_n}\left(\frac{1}{1-\overline{z}\cdot w}\right)^{\mathbf{s}'}\left\{\mathcal{R}_{\mathbf{s}}^\alpha\left(\frac{1}{1-\overline{w}\cdot z'}\right)^\eta\right\}d\nu_\alpha(w)\right\}f(z')\,dV(z')g(z)\,d\mu(z)\\ &= \int_{\mathbb{B}_n}\int_{\mathbb{B}_n}\left(\frac{1}{1-\overline{z}\cdot z'}\right)^\eta f(z')\,dV(z')g(z)\,d\mu(z)\end{aligned}$$

since with $\ell_{z'}^\eta(z) = \left(\frac{1}{1-\overline{z'}\cdot z}\right)^\eta$, the integral in braces above is, by (14.23),

$$\begin{aligned}\int_{\mathbb{B}_n}\mathcal{R}_{\mathbf{s}'}^\alpha k_z^{\alpha,p}(w)\,\overline{\mathcal{R}_{\mathbf{s}}^\alpha \ell_{z'}^\eta(w)}d\nu_\alpha(w) &= \overline{\langle \ell_{z'}^\eta, k_z^{\alpha,p}\rangle_{\alpha,p}}\\ &= \overline{\ell_{z'}^\eta(z)} = \left(\frac{1}{1-\overline{z}\cdot z'}\right)^\eta.\end{aligned}$$

We can discretize the last integral in (14.48) by breaking up the integrals over the balls into Bergman cubes K_α and using the fact that $\left(\frac{1}{1-\overline{z}\cdot z'}\right)^\eta$ is essentially constant on products of Bergman cubes. Observe also that by our choice of $\theta = \frac{\ln 2}{2}$ in (14.38),

$$\operatorname{Re}\left(\frac{1}{1-\overline{z}\cdot z'}\right)^\eta \geq c_\eta 2^{\eta d(\alpha\wedge\beta)}, \qquad z\in K_\alpha, z'\in K_\beta,$$

for a positive constant c_η provided $0<\eta<1$. Note that c_η tends to 0 as $\eta\to 1$, so that we can not use $\eta=1$ even though $\operatorname{Re}\frac{1}{1-\overline{z}\cdot z'}>0$ on the ball. We obtain that

$$\begin{aligned}&\left|\int_{\mathbb{B}_n} S_\mu^* g(w)\,\overline{V_\eta f(w)}d\lambda_n(w)\right|\\ &\qquad\geq \left|\operatorname{Re}\int_{\mathbb{B}_n} S_\mu^* g(w)\,\overline{V_\eta f(w)}d\lambda_n(w)\right|\\ &\qquad\geq c_\eta\int_{\mathbb{B}_n}\int_{\mathbb{B}_n}\left\{\operatorname{Re}\left(\frac{1}{1-\overline{z}\cdot z'}\right)^\eta\right\}g(z)\,d\mu(z)\,\overline{f(z')}dV(z')\\ &\qquad\geq c_\eta\sum_{\alpha\in\mathcal{T}_n}\sum_{\beta\in\mathcal{T}_n}2^{\eta d(\alpha\wedge\beta)}g\mu(\alpha)\,f(\beta)\end{aligned}$$

for $f, g \geq 0$ on the ball $\mathbb{B}_n$, and where $f(\beta) = \int_{K_\beta} f(z)\, dV$ and $g\mu(\alpha) = \int_{K_\alpha} g(z)\, d\mu(z)$. We also observe that

$$\begin{aligned}\sum_{\gamma\in\mathcal{T}_n} 2^{\eta d(\gamma)} I^* g\mu(\gamma) I^* f(\gamma) &= \sum_{\gamma\in\mathcal{T}_n} 2^{\eta d(\gamma)} \sum_{\alpha\in\mathcal{T}_n:\alpha\geq\gamma} g\mu(\alpha) \sum_{\beta\in\mathcal{T}_n:\beta\geq\gamma} f(\beta) \\ &= \sum_{\alpha\in\mathcal{T}_n}\sum_{\beta\in\mathcal{T}_n}\left(\sum_{\gamma\in\mathcal{T}_n:\alpha,\beta\geq\gamma} 2^{\eta d(\gamma)}\right) g\mu(\alpha) f(\beta) \\ &\approx \sum_{\alpha\in\mathcal{T}_n}\sum_{\beta\in\mathcal{T}_n} 2^{\eta d(\alpha\wedge\beta)} g\mu(\alpha) f(\beta),\end{aligned}$$

and so altogether we obtain

$$\sup_{f\geq 0} \frac{1}{\|V_\eta f\|_{L^p(d\lambda_n)}} \sum_{\gamma\in\mathcal{T}_n} 2^{\eta d(\gamma)} I^* g\mu(\gamma) I^* f(\gamma) \lesssim \left(\int_{\mathbb{B}_n} |g(z)|^{q'} d\mu(z)\right)^{1/q'}, \tag{14.49}$$

for $g \in L^{q'}(\mu)$. Provided g is nonnegative and essentially constant on the Bergman boxes K_α, we can discretize the final integral above as

$$\int_{\mathbb{B}_n} |g(w)|^{q'} d\mu(w) \approx \sum_{\alpha\in\mathcal{T}_n} g(\alpha)^{q'} \mu(\alpha). \tag{14.50}$$

Then (14.49) becomes

$$\sup_{f\geq 0} \frac{1}{\|V_\eta f\|_{L^p(d\lambda_n)}} \sum_{\gamma\in\mathcal{T}_n} 2^{\eta d(\gamma)} I^* g\mu(\gamma) I^* f(\gamma) \lesssim \left(\sum_{\alpha\in\mathcal{T}_n} g(\alpha)^{q'} \mu(\alpha)\right)^{1/q'}.$$

With $\sigma = \eta$, $\mathbf{s} = n+1+\alpha$ and $r = p$ in Lemma 14.18, we obtain

$$\left\|\widetilde{V_\eta f}\right\|_{L^p(\lambda_n)} \lesssim \left(\sum_{\alpha\in\mathcal{T}_n} \left(2^{\eta d(\alpha)} I^* f(\alpha)\right)^p\right)^{1/p},$$

which together with (14.46), yields the discretized inequality

$$\begin{aligned}&\sum_{\gamma\in\mathcal{T}_n} I^* g\mu(\gamma) 2^{\eta d(\gamma)} I^* f(\gamma) \\ &\qquad \lesssim \left(\sum_{\alpha\in\mathcal{T}_n} g(\alpha)^{q'} \mu(\alpha)\right)^{1/q'} \left(\sum_{\alpha\in\mathcal{T}_n} \left(2^{\eta d(\alpha)} I^* f(\alpha)\right)^p\right)^{1/p},\end{aligned} \tag{14.51}$$

for $f, g \geq 0$ on $\mathcal{T}_n$. If we rewrite the left side of (14.51) as

$$\sum_{\gamma\in\mathcal{T}_n} I^* g\mu(\gamma) 2^{\eta d(\gamma)} I^* f(\gamma) = \sum_{\alpha\in\mathcal{T}_n} g(\alpha)\mu(\alpha) I 2^{\eta d} I^* f(\alpha)$$

and then take the supremum over all g of norm one we obtain the inequality (14.52)

$$\left(\sum_{\alpha\in\mathcal{T}_n} \left[I 2^{\eta d} I^* h(\alpha)\right]^q \mu(\alpha)\right)^{1/q} \lesssim \left(\sum_{\alpha\in\mathcal{T}_n} \left[2^{\eta d(\alpha)} I^* h(\alpha)\right]^p\right)^{1/p}, \quad h \geq 0 \text{ on } \mathcal{T}_n,$$

which is (14.49) tested over f of the form $f = 2^{\eta d} I^* h$, and with $\mu(\alpha) = \int_{K_\alpha} d\mu$ and $\mathcal{T}$ replaced by $\mathcal{T}_n$. Unfortunately it seems difficult to derive the tree condition from such an inequality. So we will continue instead with the bilinear form (14.51).

We may assume without loss of generality that μ has finite support on the tree $\mathcal{T}_n$. Indeed, if we simply restrict μ on the ball to a finite union $\mathcal{F}$ of Carleson boxes, then this restriction $\mu_{\mathcal{F}}$ is a Carleson measure with controlled norm. If we can show that condition (14.51), and thus also (14.52), holds for $\mu_{\mathcal{F}}$ with a constant independent of $\mathcal{F}$, then obviously (14.51) holds for μ with the same constant.

If we are able to find $f \geq 0$ such that

$$2^{\eta d(\gamma)} I^* f(\gamma) = (I^* g\mu(\gamma))^{p'-1}, \quad \gamma \in \mathcal{T}_n, \tag{14.53}$$

then from (14.51) we have

$$\begin{aligned}\sum_{\gamma\in\mathcal{T}_n} (I^* g\mu(\gamma))^{p'} &\lesssim \left(\sum_{\alpha\in\mathcal{T}_n} g(\alpha)^{q'} \mu(\alpha)\right)^{1/q'} \left(\sum_{\alpha\in\mathcal{T}_n} (I^* g\mu(\alpha))^{(p'-1)p}\right)^{1/p} \\ &\approx \left(\sum_{\alpha\in\mathcal{T}_n} g(\alpha)^{q'} \mu(\alpha)\right)^{1/q'} \left(\sum_{\alpha\in\mathcal{T}_n} (I^* g\mu(\alpha))^{p'}\right)^{1/p},\end{aligned}$$

which yields (14.37) as required. So it remains to solve (14.53) for as large a range of p as we can, using $0 < \eta < 1$. To this end, we invoke the following elementary result on a tree $\mathcal{T}$.

Recall that for α a vertex of a tree we write $\mathcal{C}(\alpha)$ for the set of vertices which are children of α. We now extend that notation and will write $\mathcal{C}_\ell(\alpha)$ for the set of grand$^{\ell-1}$-children α_j of α, i.e. those with $\alpha_j > \alpha$ and $d(\alpha_j, \alpha) = \ell$.

LEMMA 14.20. *Given $G \geq 0$ on $\mathcal{T}$ such that $G(o) < \infty$, there is $h \geq 0$ on $\mathcal{T}$ satisfying*

$$I^* h(\alpha) = G(\alpha), \quad \alpha \in \mathcal{T}, \tag{14.54}$$

if and only if both

$$\sum_{d(\alpha)=N} G(\alpha) \to 0 \text{ as } N \to \infty, \tag{14.55}$$

$$G(\alpha) - \sum_{\beta\in\mathcal{C}(\alpha)} G(\beta) \geq 0, \quad \alpha \in \mathcal{T}. \tag{14.56}$$

PROOF OF THE LEMMA. For the necessity, (14.54) and $h \geq 0$ imply

$$0 \leq h(\alpha) = I^* h(\alpha) - \sum_{\beta\in\mathcal{C}(\alpha)} I^* h(\beta) = G(\alpha) - \sum_{\beta\in\mathcal{C}(\alpha)} G(\beta),$$

which is (14.56), while (14.54) and $\sum_{\alpha\in\mathcal{T}} h(\alpha) = I^* h(o) = G(o) < \infty$ with $h \geq 0$ yield

$$\sum_{d(\alpha)=N} G(\alpha) = \sum_{d(\alpha)=N} I^* h(\alpha) = \sum_{\beta\in\mathcal{T}:d(\beta)\geq N} h(\beta) \to 0 \text{ as } N \to \infty,$$

which is (14.55).

Conversely, (14.56) and (14.55) yield (14.54):

$$\begin{aligned}
I^*h(\alpha) &= \sum_{\beta\in\mathcal{T}:\beta\geq\alpha} h(\beta) = \lim_{N\to\infty}\sum_{\beta\geq\alpha,d(\beta)<N} h(\beta)\\
&= \lim_{N\to\infty}\sum_{\beta\geq\alpha,d(\beta)<N}\left\{G(\beta)-\sum_{\gamma\in\mathcal{C}(\beta)}G(\gamma)\right\}\\
&= \lim_{N\to\infty}\left\{G(\alpha)-\sum_{d(\gamma)=N}G(\gamma)\right\}\\
&= G(\alpha).
\end{aligned}$$

□

We now use the lemma to show we can find $f\geq 0$ so that (14.53) is satisfied. Because we could assume μ had finite support condition (14.55) is not a problem. Hence to know there is an f we need to establish the appropriate version of (14.56), namely

$$2^{-\eta d(\gamma)}\left(I^*g\mu(\gamma)\right)^{p'-1}\geq\sum_{\gamma_j\in\mathcal{C}(\gamma)}2^{-\eta d(\gamma_j)}\left(I^*g\mu(\gamma_j)\right)^{p'-1},\qquad \gamma\in\mathcal{T}_n,$$

This inequality is trivial if $p\leq 2$ since then $p'-1\geq 1$ implies

$$\left(I^*g\mu(\gamma)\right)^{p'-1}\geq\left(\sum_{\gamma_j\in\mathcal{C}(\gamma)}I^*g\mu(\gamma_j)\right)^{p'-1}\geq\sum_{\gamma_j\in\mathcal{C}(\gamma)}I^*g\mu(\gamma_j)^{p'-1},$$

and $d(\gamma_j)>d(\gamma)$ for $\gamma_j\in\mathcal{C}(\gamma)$. Now suppose that $p>2$. Since $d(\gamma_j)=d(\gamma)+1$, and j runs from 1 to the maximum branching number N of the Bergman tree $\mathcal{T}_n$, we have using Hölder's inequality with exponents $p-1$ and $(p-1)/(p-2)$,

$$\begin{aligned}
&\sum_{\gamma_j\in\mathcal{C}(\gamma)}2^{-\eta d(\gamma_j)}\left(I^*g\mu(\gamma_j)\right)^{p'-1}\\
&\quad\leq 2^{-\eta d(\gamma)-\eta}\left(\sum_{\gamma_j\in\mathcal{C}(\gamma)}1\right)^{(p-2)/(p-1)}\left(\sum_j\left(I^*g\mu(\gamma_j)\right)^{(p'-1)(p-1)}\right)^{1/(p-1)}\\
&\quad\leq 2^{-\eta d(\gamma)}2^{-\eta}(N)^{(p-2)/(p-1)}\left(\sum_j I^*g\mu(\gamma_j)\right)^{1/(p-1)}\\
&\quad\leq 2^{(\log_2 N)(p-2)/(p-1)-\eta}\left\{2^{-\eta d(\gamma)}\left(I^*g\mu(\gamma_j)\right)^{p'-1}\right\}\\
&\quad\leq 2^{-\eta d(\gamma)}\left(I^*g\mu(\gamma_j)\right)^{p'-1},
\end{aligned}$$

as required provided

$$(\log_2 N)\frac{p-2}{p-1}-\eta\leq 0 \text{ or } p\leq\frac{2(\log_2 N)-\eta}{(\log_2 N)-\eta}=2+\frac{\eta}{(\log_2 N)-\eta}.$$

In order to obtain the full range $1 < p < 2+1/(n-1)$, we will take η sufficiently close to 1, and use the device of splitting the sum $\sum_{\alpha\in\mathcal{T}_n}$ into "sparse" pieces $\sum_{\alpha\in\mathcal{T}_n:d(\alpha)\in\ell\mathbb{N}+m}$ for $0 \leq m < \ell$, where ℓ is chosen so large in Definition 14.2 that

$$\log_2 (N_\ell)^{1/\ell} = \log_2 \left(\sup_{\alpha\in\mathcal{T}} \operatorname{card} \mathcal{C}_l(\alpha) \right)^{1/\ell} \tag{14.57}$$

$$< \frac{p-1}{p-2}. \tag{14.58}$$

This can be done if $n < (p-1)/(p-2)$, or equivalently $p < 2 + 1/(n-1)$, since the dimension $n(\mathcal{T}_n)$ of the tree $\mathcal{T}_n$ is n by Lemma 14.3 when $\theta = \frac{\ln 2}{2}$ (and also $1 - |z|^2 \approx 2^{-d(\alpha)}$ for $z \in K_\alpha$).

With ℓ chosen so that (14.57) holds, we consider solving the equation

$$2^{\eta d(\gamma)} I^* f(\gamma) = (I^* g\mu(\gamma))^{p'-1}, \qquad \gamma \in \mathcal{T}_n, d(\gamma) \in \ell\mathbb{N}, \tag{14.59}$$

for $0 < \eta < 1$ and with $f \geq 0$ on $\mathcal{T}_n$ and supported in

$$\Omega_\ell := \{\gamma \in \mathcal{T}_n : d(\gamma) \in \ell\mathbb{N}\}.$$

By Lemma 14.20 applied to the tree Ω_ℓ, this is equivalent to the inequality

$$2^{-\eta d(\gamma)} (I^* g\mu(\gamma))^{p'-1} \geq \sum_{\gamma_j\in\mathcal{C}_\ell(\gamma)} 2^{-\eta d(\gamma_j)} (I^* g\mu(\gamma_j))^{p'-1}, \qquad \gamma \in \Omega_\ell,$$

where the sum is now taken over the set $\mathcal{C}_\ell(\gamma)$ of grand$^{\ell-1}$-children γ_j of γ, i.e. those with $\gamma_j > \gamma$ and $d(\gamma_j) = \ell$. From Hölder's inequality with exponents $p-1$ and $(p-1)/(p-2)$ again,

$$\begin{aligned}
&\sum_{\gamma_j\in\mathcal{C}_\ell(\gamma)} 2^{-\eta d(\gamma_j)} (I^* g\mu(\gamma_j))^{p'-1} \\
&\quad \leq 2^{-\eta d(\gamma)-\eta\ell} \left(\sum_{\gamma_j\in\mathcal{C}_\ell(\gamma)} 1 \right)^{(p-2)/(p-1)} \left(\sum_j (I^* g\mu(\gamma_j))^{(p'-1)(p-1)} \right)^{1/(p-1)} \\
&\quad \leq 2^{-\eta d(\gamma)} 2^{-\eta\ell} (N_\ell)^{(p-2)/(p-1)} \left(\sum_j I^* g\mu(\gamma_j) \right)^{1/(p-1)} \\
&\quad \leq 2^{(\log_2 N_\ell)\frac{p-2}{p-1}-\eta\ell} \left\{ 2^{-\eta d(\gamma)} (I^* g\mu(\gamma))^{p'-1} \right\} \\
&\quad \leq 2^{-\eta d(\gamma)} (I^* g\mu(\gamma))^{p'-1},
\end{aligned}$$

since $\left(\log_2 (N_\ell)^{1/\ell}\right) \frac{p-2}{p-1} - \eta < 0$ for η sufficiently close to 1 by (14.57).

Thus we have solved (14.59) for $f \geq 0$ on $\mathcal{T}_n$, and supported in Ω_ℓ. We also have, for $\gamma \in \Omega_\ell$.

$$\sum_{\beta>\gamma:1\leq d(\beta,\gamma)\leq\ell} \left[2^{\eta d(\beta)} I^* f(\beta)\right]^p \lesssim \left[2^{\eta d(\gamma)} I^* f(\gamma)\right]^p \approx (I^* g\mu(\gamma))^{p'}. \tag{14.60}$$

Thus using first (14.59) and (14.51), and then (14.60), we have

$$\begin{aligned}\sum_{\gamma\in\Omega_\ell}\left(I^*g\mu(\gamma)\right)^{p'} &\lesssim \left(\sum_{\alpha\in\mathcal{T}_n} g(\alpha)^{p'}\mu(\alpha)\right)^{1/p'}\left(\sum_{\alpha\in\mathcal{T}_n}\left[2^{\eta d(\alpha)}I^*f(\alpha)\right]^p\right)^{1/p}\\ &\lesssim \left(\sum_{\alpha\in\mathcal{T}_n} g(\alpha)^{p'}\mu(\alpha)\right)^{1/p'}\left(\sum_{\gamma\in\Omega_\ell}\left(I^*g\mu(\gamma)\right)^{p'}\right)^{1/p}.\end{aligned}$$

Since $\sum_{\alpha\in\mathcal{T}_n}\left(I^*g\mu(\alpha)\right)^{p'} \leq \sum_\ell\sum_{\gamma\in\Omega_\ell}\left(I^*g\mu(\gamma)\right)^{p'}$, this yields the dual of (14.37) as required. □

Note: If the weight $2^{\eta d(\gamma)}$ in (14.53) were replaced by $2^{nd(\gamma)}$, then the argument above would solve (14.53) for all $1<p<\infty$. However, this would require using $\ell^n_{z'}(z) = \left(\frac{1}{1-\overline{z'}\cdot z}\right)^n$ above, and the real part of $\ell^n_{z'}(z)$ would no longer be positive. We do not know if inequality (14.52) characterizes (14.29) when $p \geq 2+1/(n-1)$.

14.4.4. A monotonicity property of Carleson measures.

Theorem 14.16 yields the following monotonicity property for $B_p(\mathbb{B}_n)$-Carleson measures on $\mathbb{B}_n$.

COROLLARY 14.21. *If μ is a $B_{p_0}(\mathbb{B}_n)$-Carleson measure for some $1<p_0<\infty$, then μ is a $B_p(\mathbb{B}_n)$-Carleson measure for all $1<p<p_0$. Moreover, μ satisfies the tree condition* (14.37) *with*

$$p=\begin{cases} p_0 & \text{if}\quad p_0<2+\frac{1}{n-1}\\ 2+\frac{1}{n-1}-\varepsilon,\ \varepsilon>0 & \text{if}\quad 2+\frac{1}{n-1}\leq p_0<4+\frac{2}{n-1}\\ \frac{p_0}{2} & \text{if}\quad p_0\geq 4+\frac{2}{n-1}\end{cases}.$$

PROOF. For $1<p_0<2+1/(n-1)$, we simply use Theorem 14.16 together with the fact that Carleson measure inequality (14.29) has the corresponding monotonicity property: for $f\geq 0$ and $1<p<p_0$,

$$\begin{aligned}\left(\sum_{\alpha\in\mathcal{T}} If(\alpha)^p\mu(\alpha)\right)^{1/p} &= \left(\sum_{\alpha\in\mathcal{T}}\left[\sum_{\beta\leq\alpha} f(\beta)\right]^{\frac{p}{p_0}p_0}\mu(\alpha)\right)^{1/p}\\ &\leq \left(\sum_{\alpha\in\mathcal{T}}\left[\sum_{\beta\leq\alpha} f(\beta)^{\frac{p}{p_0}}\right]^{p_0}\mu(\alpha)\right)^{1/p}\\ &\lesssim \left(\sum_{\alpha\in\mathcal{T}}\left[f(\beta)^{\frac{p}{p_0}}\right]^{p_0}\right)^{1/p} \approx \left(\sum_{\alpha\in\mathcal{T}} f(\beta)^p\right)^{1/p},\end{aligned}$$

if the tree condition (14.37) holds for the exponent p_0.

For $p_0>2$, we use the atomic decomposition of $B_{p_0}(\mathbb{B}_n)$, see [**346**, Theorem 6.6], Khintchine's inequality for the Rademacher functions $r_k(t)$, and complex interpolation of the Besov spaces $B_{p_0}(\mathbb{B}_n)$, see [**346**, Theorem 6.12]. If the measure μ is $B_{p_0}(\mathbb{B}_n)$-Carleson, and $a>n/p_0'$, then with $f(z)=\sum_{\alpha\in\mathcal{T}_n} f_\alpha\left(\frac{1-|c_\alpha|^2}{1-\overline{c_\alpha}\cdot z}\right)^a$ and

$p = q = p_0$ in (14.25), we have

$$\int_{\mathbb{B}_n} \left| \sum_{\alpha \in \mathcal{T}_n} f_\alpha \left(\frac{1 - |c_\alpha|^2}{1 - \overline{c_\alpha} \cdot z} \right)^a \right|^{p_0} d\mu(z) \lesssim \sum_{\alpha \in \mathcal{T}_n} |f_\alpha|^{p_0}$$

for all $\{f_\alpha\}_{\alpha \in \mathcal{T}_n} \in \ell^{p_0}$ by the atomic decomposition of $B_{p_0}(\mathbb{B}_n)$. Here we use the atomic decomposition in [**346**, Theorem 6.6], which requires that we choose both parameters λ and θ sufficiently small in the construction of the Bergman tree $\mathcal{T}_n$. Using Khinchine's inequality for the Rademacher functions, we obtain

$$\begin{aligned} &\int_{\mathbb{B}_n} \left\{ \sum_{\alpha \in \mathcal{T}_n} \left| f_\alpha \left(\frac{1 - |c_\alpha|^2}{1 - \overline{c_\alpha} \cdot z} \right)^a \right|^2 \right\}^{\frac{p_0}{2}} d\mu(z) \\ &\approx \int_{\mathbb{B}_n} \int_0^1 \left| \sum_{\alpha \in \mathcal{T}_n} r_\alpha(t) f_\alpha \left(\frac{1 - |c_\alpha|^2}{1 - \overline{c_\alpha} \cdot z} \right)^a \right|^{p_0} dt d\mu(z) \\ &\lesssim \int_0^1 \sum_{\alpha \in \mathcal{T}_n} |c_k r_\alpha(t) f_\alpha|^{p_0} dt \approx \sum_{\alpha \in \mathcal{T}_n} |f_\alpha|^{p_0}. \end{aligned}$$

Now we observe that for $z \in K_\beta$,

$$\sum_{\alpha \in \mathcal{T}_n} \left| f_\alpha \left(\frac{1 - |c_\alpha|^2}{1 - \overline{c_\alpha} \cdot z} \right)^a \right|^2 \gtrsim \sum_{\alpha \leq \beta} |f_\alpha|^2 \approx Ig(\beta),$$

where $g(\alpha) = |f_\alpha|^2$. Thus we have the tree inequality (14.37)

$$\sum_{\beta \in \mathcal{T}_n} Ig(\beta)^{\frac{p_0}{2}} \mu(\beta) \lesssim \sum_{\beta \in \mathcal{T}_n} g(\beta)^{\frac{p_0}{2}}, \qquad g \geq 0,$$

and hence by Theorem 14.16, μ satisfies the tree condition (14.29) with exponent $p = p_0/2 > 1$. As above, we conclude that μ is $B_p(\mathbb{B}_n)$-Carleson for all $1 < p < p_0/2$. Since μ is also $B_{p_0}(\mathbb{B}_n)$-Carleson, complex interpolation shows that μ is $B_p(\mathbb{B}_n)$-Carleson for all $1 < p < p_0$. The assertion regarding the tree condition now follows easily, and this completes the proof of Corollary 14.21. □

14.5. Local Analysis of Besov Functions

We will now work in the situation described in Section 14.1.6. That is, we have the tree $\mathcal{T}_n$ and the corresponding decomposition of the ball $\mathbb{B}_n = \cup_\alpha K_\alpha$. Select a particular $\alpha \in \mathcal{T}_n$ and set $a = c_\alpha$.

To work with the spaces $B_p(\mathbb{B}_n)$ we need to work with m^{th} derivatives, $m > n/p$, Hence, in general, we will need to consider $m > 1$. This introduces complications. For functions in $\mathrm{Hol}(\mathbb{B}_n)$, $n > 1$, the radial derivatives satisfy different estimates than the derivatives in the tangential (orthogonal to the radial) directions. This lack of isotropy is a distinctive feature of the higher dimensional theory and we need a convenient formalism for working with it. Higher partial derivatives involve mixtures of these cases and estimation becomes more complicated. Although results such as Proposition 14.4 allow global comparison of the size of various higher order derivative, we will also need local information.

In (14.9) we defined the invariant derivative $\widetilde{\nabla} f$ of $f \in \mathrm{Hol}(\mathbb{B}_n)$ by $\widetilde{\nabla} f(a) = f'(a)\varphi_a'(0)$ where φ_a is the ball automorphism which interchanges 0 and a. This is a good tool for first order derivatives. However, because $\widetilde{\nabla} f(a)$ is not a holomorphic function of a, using it as a first step on a path to studying higher order derivatives would lead to a variety of new complications. Instead we introduce an *almost invariant holomorphic derivative* $\{D_a F\}_{\alpha\in\mathcal{T}_n}$, a collection of locally defined operators indexed by vertices of the tree.

Specifically, on each Bergman cube K_α we construct a differential operator D_a that, on K_α, is close to the invariant gradient $\widetilde{\nabla}$, and which has the additional property that for $f \in \mathrm{Hol}(K_\alpha)$ and $m \geq 1$ we have $D_a^m f \in \mathrm{Hol}(K_\alpha)$. This construct offers a good formalism for dealing with the lack of isotropy. Further, it keeps us inside the holomorphic category and hence offers a straightforward path to higher derivatives. We will see in Proposition (14.26) we can estimate the oscillation of a Besov space function on some K_α using the values $D_\beta^m f$ for K_β near K_α. Also, most importantly for our purposes, we will show in Proposition (14.25) that $\{D_a^m\}$ can be used to define an equivalent norm on the Besov space B_p.

Most of the rest of this chapter is devoted to establishing properties of $\{D_a\}$.

14.5.1. An Almost Invariant Holomorphic Derivative. We write the usual Euclidean gradient as the pair $(\nabla, \overline{\nabla})$ where

$$\nabla = \left(\frac{\partial}{\partial z_1}, \dots, \frac{\partial}{\partial z_n}\right) \text{ and } \overline{\nabla} = \left(\frac{\partial}{\partial \overline{z_1}}, \dots, \frac{\partial}{\partial \overline{z_n}}\right).$$

Recall that for any $a \in \mathbb{B}_n$, P_a and Q_a are given by (14.1). The invariant derivative of $f \in \mathrm{Hol}(\mathbb{B}_n)$ is then given by

$$\begin{aligned}\widetilde{\nabla} f(a) &= (f \circ \varphi_a)'(0) = f'(a)\,\varphi_a'(0)\\ &= -f'(a)\left\{\left(1-|a|^2\right)P_a + \left(1-|a|^2\right)^{1/2} Q_a\right\}\\ &= -\nabla f(a)\left\{\left(1-|a|^2\right)P_a + \left(1-|a|^2\right)^{1/2} Q_a\right\}\end{aligned}$$

which fails to be holomorphic in a. To rectify this, we define, as in [**47**],

$$\begin{aligned}D_a f(z) &:= f'(z)\,\varphi_a'(0)\\ &= -f'(z)\left\{\left(1-|a|^2\right)P_a + \left(1-|a|^2\right)^{1/2} Q_a\right\}\\ &= -\nabla f(z)\left\{\left(1-|a|^2\right)P_a + \left(1-|a|^2\right)^{1/2} Q_a\right\} \qquad (14.61)\end{aligned}$$

which, for a fixed, is a holomorphic function of z.

We denote the full collection of these operators by $\mathfrak{D}$; $\mathfrak{D} := \{D_a\}$.

To see more clearly what is going on we work out an example in detail. It is convenient here to use matrix notation. We view each $w \in \mathbb{B}_n$ as an $n \times 1$ complex matrix, and denote by w^t the $1 \times n$ transpose of w. Using this notation we have that $\nabla_z(\overline{a} \cdot z) = \overline{a}^t$ and thus the derivative of the projection $P_a(z) = \frac{\overline{a}\cdot z}{|a|^2} a$ is the linear operator $P_a = \frac{a\overline{a}^t}{|a|^2} = |a|^{-2}\left[a_i \overline{a_j}\right]_{1\leq i,j\leq n}$.

EXAMPLE 14.22. *In the case $n = 2$ we can calculate the differential operator D_a for $a \neq 0$ more explicitly using the basis $\left\{a, a^\perp\right\}$ of $\mathbb{C}^2$ where $a = \begin{pmatrix} a_1 \\ a_2 \end{pmatrix}$ and*

$a^\perp = \begin{pmatrix} -\overline{a_2} \\ \overline{a_1} \end{pmatrix}$. *For* $w = \mu a + \nu a^\perp$, *we compute the action of the linear operator* $D_a f(z)$ *on* w *as*

$$\begin{aligned} -D_a f(z)\, w &= f'(z) \left\{ (1-|a|^2)\mu a + (1-|a|^2)^{1/2} \nu a^\perp \right\} \\ &= \mu(1-|a|^2) \left(a_1 \frac{\partial}{\partial z_1} f(z) + a_2 \frac{\partial}{\partial z_2} f(z) \right) \\ &\quad + \nu (1-|a|^2)^{1/2} \left(-\overline{a_2} \frac{\partial}{\partial z_1} f(z) + \overline{a_1} \frac{\partial}{\partial z_2} f(z) \right). \end{aligned}$$

Thus in the basis $\{a, a^\perp\}$, *we have*

$$-D_a = \left((1-|a|^2) \left(a_1 \frac{\partial}{\partial z_1} + a_2 \frac{\partial}{\partial z_2} \right), (1-|a|^2)^{1/2} \left(-\overline{a_2} \frac{\partial}{\partial z_1} + \overline{a_1} \frac{\partial}{\partial z_2} \right) \right),$$

where at the point a, $\left(a_1 \frac{\partial}{\partial z_1} + a_2 \frac{\partial}{\partial z_2} \right) f(a) = f'(a)\, a$ *is the complex radial derivative of* f *at* a, *and* $\left(-\overline{a_2} \frac{\partial}{\partial z_1} + \overline{a_1} \frac{\partial}{\partial z_2} \right) f(a) = f'(a)\, a^\perp$ *is the complex tangential derivative of* f *at* a.

We also define the conjugate operator. For any smooth function, $f \in C^\infty(\mathbb{B}_n)$, we define, for each a,

$$\overline{D}_a f(z) = -\overline{\nabla} f(z) \left\{ \left(1-|a|^2\right) P_a + \left(1-|a|^2\right)^{1/2} Q_a \right\},$$

and denote the collection of all of them by $\overline{\mathfrak{D}}$. Notice that these operators, which are not restricted to holomorphic functions, in fact annihilate holomorphic functions.

We can iterate these constructions, in each case regarding a as a fixed parameter and z as the variable. Thus $D_a^2 f(z) = D_a(D_a f(z))$ and we can continue this way to form D_a^m and $\overline{D}_a^m$ for nonnegative integer m.

In general we have that $D_b^m f(z)$ is a symmetric m-linear form. For instance suppose that $m = 2$ and f is holomorphic; viewing $D_a^2 f(z)$ as a symmetric bilinear form, we have that

$$\begin{aligned} D_a^2 f(z) &= (1-|a|^2)^2 P_a f''(z)\, P_a \\ &\quad + (1-|a|^2)^{3/2} \left(P_a f''(z)\, Q_a + Q_a f''(z)\, P_a \right) \\ &\quad + (1-|a|^2) Q_a f''(z)\, Q_a. \end{aligned} \tag{14.62}$$

The next lemma shows that D_a^m and D_b^m are comparable when a and b are close in the Bergman metric. In particular, if $z \in K_a$ then $\widetilde{\nabla} f(z)$ is close to $D_a f(z)$.

LEMMA 14.23. *Let* $a, b \in \mathbb{B}_n$ *satisfy* $\beta(a,b) \le C$. *Then*

$$|D_a^m f(z)| \approx |D_b^m f(z)|,$$

for all $f \in Hol(\mathbb{B}_n)$ *and the implied constants only depend on* C, m, *and* n.

PROOF. The equivalence is easy if a or b is near the origin so we now assume that $|a|, |b| \ge 1/2$. We first note that $b \in B_\beta(a, C)$ implies the estimates

$$\begin{aligned} |P_a(a-b)| + |Q_a(a-b)|^2 &\lesssim (1-|a|^2), \\ (1-|a|^2) &\approx (1-|b|^2), \end{aligned}$$

and since $Q_a a = 0$, this yields

$$|Q_a b| = |Q_a (a - b)| \lesssim (1 - |a|^2)^{1/2},$$

and by symmetry,

$$|Q_b a| \lesssim (1 - |b|^2)^{1/2} \approx (1 - |a|^2)^{1/2}.$$

We can now estimate the modulus of $\varphi_b' (0) a$ by

$$\begin{aligned}|\varphi_b' (0) a| &= \left|(1 - |b|^2) P_b a + (1 - |b|^2)^{1/2} Q_b a\right| \\ &\lesssim (1 - |a|^2) + (1 - |a|^2)^{1/2} |Q_b a| \\ &\lesssim (1 - |a|^2).\end{aligned}$$

Thus we may write

$$\varphi_b' (0) a = (1 - |a|^2) \left(\sigma a + \tau v^\perp\right), \qquad |\sigma|, |\tau| \leq C, \tag{14.63}$$

where $v^\perp$ is a unit vector in $(\mathbb{C}a)^\perp$.

Recall that $\varphi_b(z)$ is a ball automorphism which interchanges 0 and b. We want estimates on $\varphi_b' (0)$. We start with the derivative formula in (14.3). Let $a^\perp$ be any unit vector in $(\mathbb{C}a)^\perp$. We want to know that

$$\begin{aligned}\varphi_b' (0) a^\perp &= (1 - |b|^2) P_b a^\perp + (1 - |b|^2)^{1/2} Q_b a^\perp \\ &= (1 - |a|^2) \mu a + (1 - |a|^2)^{1/2} \nu w^\perp, \qquad |\mu|, |\nu| \lesssim C,\end{aligned} \tag{14.64}$$

where $w^\perp$ is another unit vector in $(\mathbb{C}a)^\perp$. To see this we use

$$P_b a^\perp = P_a P_b a^\perp + Q_a P_b a^\perp$$

and

$$\begin{aligned}Q_b a^\perp &= a^\perp - P_b a^\perp = a^\perp - \frac{\overline{b} \cdot a^\perp}{|b|^2} b \\ &= a^\perp - \frac{\overline{b} \cdot a^\perp}{|b|^2} (P_a b + Q_a b)\end{aligned}$$

to write

$$\begin{aligned}\varphi_b' (0) a^\perp &= (1 - |b|^2) \left(P_a P_b a^\perp + Q_a P_b a^\perp\right) \\ &\quad + (1 - |b|^2)^{1/2} \left(a^\perp - \frac{\overline{b} \cdot a^\perp}{|b|^2} P_a b - \frac{\overline{b} \cdot a^\perp}{|b|^2} Q_a b\right),\end{aligned}$$

so that for (14.64) to hold we must have

$$(1 - |a|^2) \mu a = (1 - |b|^2) P_a P_b a^\perp - (1 - |b|^2)^{1/2} \frac{\overline{b} \cdot a^\perp}{|b|^2} P_a b$$

and

$$(1 - |a|^2)^{1/2} \nu w^\perp = (1 - |b|^2) Q_a P_b a^\perp + (1 - |b|^2)^{1/2} \left(a^\perp - \frac{\overline{b} \cdot a^\perp}{|b|^2} Q_a b\right).$$

Since P_a and Q_a are orthogonal projections onto $\mathbb{C}a$ and $(\mathbb{C}a)^\perp$ respectively, we see that the representation (14.64) holds with the stated bounds upon using

$$\left|\overline{b} \cdot a^\perp\right| = \left|\overline{(b - a)} \cdot a^\perp\right| \lesssim |b - a| \lesssim (1 - |a|^2)^{1/2}.$$

From (14.63) and (14.64) above we obtain in particular that for every unit vector $v \in \mathbb{C}^n$, there is a unit vector $v^\perp \in (\mathbb{C}a)^\perp$ and bounded scalars μ, ν such that

$$\varphi_b'(0)\, v = (1 - |a|^2)\mu a + (1 - |a|^2)^{1/2}\nu v^\perp \tag{14.65}$$

(simply write v as a linear combination of a and a vector perpendicular to a and use (14.63) and (14.64)).

Now suppose that $m = 1$. Then for any unit vector $v \in \mathbb{C}^n$, we have from (14.65) that

$$\begin{aligned}
|D_b f(z)\, v| &= |f'(z)\, \varphi_b'(0)\, v| \\
&= \left|(1 - |a|^2)\mu f'(z)\, a + (1 - |a|^2)^{1/2}\nu f'(z)\, v^\perp\right| \\
&\lesssim (1 - |a|^2)\, |f'(z)\, a| + (1 - |a|^2)^{1/2} \left|f'(z)\, v^\perp\right| \\
&\lesssim |D_a f(z)|,
\end{aligned}$$

since

$$\begin{aligned}
|D_a f(z)| &= \left|f'(z) \left\{(1 - |a|^2)P_a + (1 - |a|^2)^{1/2} Q_a\right\}\right| \\
&\approx \sup_{|v| \le 1} \left|f'(z) \left\{(1 - |a|^2)P_a v + (1 - |a|^2)^{1/2} Q_a v\right\}\right| \\
&\approx (1 - |a|^2)\, |f'(z)\, P_a| + (1 - |a|^2)^{1/2}\, |f'(z)\, Q_a|.
\end{aligned}$$

Now suppose that $m = 2$. Then viewing $D_a^2 f(z)$ as a symmetric bilinear form, we have that

$$\begin{aligned}
\left|D_a^2 f(z)\right| \approx{}& (1 - |a|^2)^2\, |P_a f''(z)\, P_a| \\
&+ (1 - |a|^2)^{3/2} \left(|P_a f''(z)\, Q_a| + |Q_a f''(z)\, P_a|\right) \\
&+ (1 - |a|^2)\, |Q_a f''(z)\, Q_a|.
\end{aligned} \tag{14.66}$$

Indeed, write $\widetilde{P_a} = (1 - |a|^2)P_a$ and $\widetilde{Q_a} = (1 - |a|^2)^{1/2} Q_a$, and note that the set of vectors

$$\mathcal{V} = \{\sigma P_a v + \tau Q_a w : \sigma, \tau \in \mathbb{C}; v, w \in \mathbb{C}^n; |\sigma|, |\tau|, |v|, |w| \le 1\}$$

satisfies

$$\{V \in \mathbb{C}^n : |V| \le 1\} \subset \mathcal{V} \subset \{V \in \mathbb{C}^n : |V| \le 2\}.$$

Then with $V_i = \sigma_i P_a v_i + \tau_i Q_a w_i$ for $i = 1, 2$ we have

$$\widetilde{P_a} V_i + \widetilde{Q_a} V_i = \sigma_i \widetilde{P_a} v_i + \tau_i \widetilde{Q_a} w_i, \qquad i = 1, 2.$$

Hence

$$\begin{aligned}
&\sup_{V_1,V_2\in\mathcal{V}}\left|\left\{\widetilde{P_a}V_1+\widetilde{Q_a}V_1\right\}f''(z)\left\{\widetilde{P_a}V_2+\widetilde{Q_a}V_2\right\}\right| \\
&=\sup_{\substack{|\sigma_i|,|\tau_i|\le 1\\|v_i|,|w_i|\le 1}}\left|\left\{\sigma_1\widetilde{P_a}v_1+\tau_1\widetilde{Q_a}w_1\right\}f''(z)\left\{\sigma_2\widetilde{P_a}v_2+\tau_2\widetilde{Q_a}w_2\right\}\right| \\
&=\sup_{\substack{|\sigma_i|,|\tau_i|\le 1\\|v_i|,|w_i|\le 1}}\left|(\sigma_1,\tau_1)\begin{bmatrix}\widetilde{P_a}v_1f''(z)\widetilde{P_a}v_2 & \widetilde{P_a}v_1f''(z)\widetilde{Q_a}w_2\\ \widetilde{Q_a}w_1f''(z)\widetilde{P_a}v_2 & \widetilde{Q_a}w_1f''(z)\widetilde{Q_a}w_2\end{bmatrix}\begin{pmatrix}\sigma_2\\ \tau_2\end{pmatrix}\right| \\
&\approx\sup_{|v_i|,|w_i|\le 1}\left|\widetilde{P_a}v_1f''(z)\widetilde{P_a}v_2\right|+\left|\widetilde{P_a}v_1f''(z)\widetilde{Q_a}w_2\right| \\
&\qquad\qquad+\left|\widetilde{Q_a}w_1f''(z)\widetilde{P_a}v_2\right|+\left|\widetilde{Q_a}w_1f''(z)\widetilde{Q_a}w_2\right|
\end{aligned}$$

which yields (14.66). Since $v^tD_b^2f(z)w=(\varphi_b'(0)v)^tf''(z)\varphi_b'(0)w$, we must now show that the expression

$$\left|\left((1-|a|^2)\sigma a+(1-|a|^2)^{1/2}\tau v^\perp\right)^tf''(z)\left((1-|a|^2)\mu a+(1-|a|^2)^{1/2}\nu w^\perp\right)\right| \tag{14.67}$$

is at most $\left|D_a^2f(z)\right|$. However, the expression (14.67) is dominated by a constant multiple of

$$\begin{aligned}
&(1-|a|^2)^2\left|P_af''(z)P_a\right|+(1-|a|^2)^{3/2}\left|P_af''(z)Q_a\right| \\
&\quad+(1-|a|^2)^{3/2}\left|Q_af''(z)P_a\right|+(1-|a|^2)\left|Q_af''(z)Q_a\right|,
\end{aligned}$$

which is at most $\left|D_a^2f(z)\right|$ by (14.66). This completes the proof of Lemma 14.23 in cases $m=1,2$.

The case $m>2$ is handled similarly viewing $D_b^mf(z)$ as a symmetric m-linear form on $\mathbb{C}^n\times\cdots\times\mathbb{C}^n$ (m times). Briefly, the argument is this:

$$\begin{aligned}
\left|D_b^mf(z)\{v_j\}_{j=1}^m\right|&\le\left|f^{(m)}(z)\{\varphi_b'(0)v_j\}_{j=1}^m\right| \\
&\le\left|f^{(m)}(z)\left\{(1-|a|^2)\mu_ja+(1-|a|^2)^{1/2}\nu_jw_j^\perp\right\}_{j=1}^m\right| \\
&\lesssim\sum_{\ell=0}^m(1-|a|^2)^{m-\ell}(1-|a|^2)^{\ell/2}\left|f^{(m)}(z)\{u_j\}_{j=1}^m\right|,
\end{aligned}$$

where in the summand corresponding to the index ℓ, $m-\ell$ of the u_j are equal to a and the remaining u_j are equal to $w_j^\perp$. Each summand is dominated by $|D_a^mf(z)|$, thereby completing the proof of Lemma 14.23. □

14.5.2. Equivalence of Semi-norms.

DEFINITION 14.24. *Suppose $1<p<\infty$ and $m\ge 1$. We define a "tree semi-norm" $\|\cdot\|_{B_{p,m}}^*$ by*

$$\|f\|_{B_{p,m}}^*:=\left(\sum_{\alpha\in\mathcal{T}_n}\int_{B_d(c_\alpha,C_2)}\left|D_{c_\alpha}^mf(z)\right|^pd\lambda_n(z)\right)^{1/p}. \tag{14.68}$$

The finiteness of the semi-norms $\|\cdot\|^*_{B_{p,m}}$, and the associated norms, turn out to be independent of $m > 2n/p$. We will obtain this fact as a corollary of the equivalence of the norm in (14.18) with the corresponding norm based on radial derivatives, as in Proposition 14.4. Let $\mathcal{R}_m$ be one of the operators $R^{\gamma,m}$ from Proposition 14.4.

Note that the restriction $m > 2n/p$ is dictated by the fact that $\left|D^m_{c_\alpha} f(z)\right|$ involves the factor $(1-|z|^2)^{m/2}$ times the mth order tangential derivatives of f, and so we must have that $(1-|z|^2)^{mp/2} d\lambda_n(z)$ is a finite measure, i.e. $mp/2-n-1 > -1$.

We will generally not mention it going forward, but if $n = 1$ the restriction $m > 2n/p$ can often be replaced by $m > 1/p$, see Section 16.1.2.

PROPOSITION 14.25. *Let $1 < p < \infty$ and $m > 2n/p$. Then*

$$\begin{aligned} \|f\|^*_{B_{p,m}} + \sum_{j=0}^{m-1} \left|\nabla^j f(0)\right| &:= \left(\sum_{\alpha\in\mathcal{T}_n} \int_{B_d(c_\alpha,C_2)} \left|D^m_{c_\alpha} f(z)\right|^p d\lambda_n(z)\right)^{1/p} + \sum_{j=0}^{m-1} \left|\nabla^j f(0)\right| \\ &\approx \left(\int_{\mathbb{B}_n} \left|(1-|z|^2)^m \mathcal{R}_m f(z)\right|^p d\lambda_n(z)\right)^{1/p}. \end{aligned} \tag{14.69}$$

PROOF. We have

$$|D_a f(z)| = \left|f'(z)\left\{(1-|a|^2)P_a + (1-|a|^2)^{1/2} Q_a\right\}\right| \geq \left|(1-|a|^2) f'(z)\right|, \tag{14.70}$$

and iterating with f replaced by (the components of) $D_a f$ in (14.70), we obtain

$$\left|D^2_a f(z)\right| \geq \left|(1-|a|^2)(D_a f)'(z)\right|.$$

Applying (14.70) once more with f replaced by (the components of) f', we get

$$\left|(1-|a|^2)(D_a f)'(z)\right| = \left|(1-|a|^2) D_a(f')(z)\right| \geq \left|(1-|a|^2)^2 f''(z)\right|,$$

which when combined with the previous inequality yields

$$\left|D^2_a f(z)\right| \geq \left|(1-|a|^2)^2 f''(z)\right|.$$

Continuing by induction we have

$$|D^m_a f(z)| \geq \left|(1-|a|^2)^m f^{(m)}(z)\right|, \qquad m \geq 1. \tag{14.71}$$

Proposition 14.4 and (14.71) now show that

$$\begin{aligned}
&\left(\int_{\mathbb{B}_n} \left|(1-|z|^2)^m \mathcal{R}_m f(z)\right|^p d\lambda_n(z)\right)^{1/p} \\
&\lesssim \left(\int_{\mathbb{B}_n} \left|(1-|z|^2)^m f^{(m)}(z)\right|^p d\lambda_n(z)\right)^{1/p} + \sum_{j=0}^{m-1} \left|\nabla^j f(0)\right| \\
&\lesssim \left(\sum_{\alpha\in\mathcal{T}_n} \int_{B_d(c_\alpha, C_2)} \left|(1-|z|^2)^m f^{(m)}(z)\right|^p d\lambda_n(z)\right)^{1/p} + \sum_{j=0}^{m-1} \left|\nabla^j f(0)\right| \\
&\lesssim \left(\sum_{\alpha\in\mathcal{T}_n} \int_{B_d(c_\alpha, C_2)} \left|(1-|c_\alpha|^2)^m f^{(m)}(z)\right|^p d\lambda_n(z)\right)^{1/p} + \sum_{j=0}^{m-1} \left|\nabla^j f(0)\right| \\
&\lesssim \left(\sum_{\alpha\in\mathcal{T}_n} \int_{B_d(c_\alpha, C_2)} \left|D_a^m f(z)\right|^p d\lambda_n(z)\right)^{1/p} + \sum_{j=0}^{m-1} \left|\nabla^j f(0)\right| \\
&\approx \|f\|_{B_{p,m}}^* + \sum_{j=0}^{m-1} \left|\nabla^j f(0)\right|.
\end{aligned}$$

For the opposite inequality, we employ some of the ideas in the proofs of [**340**, Lemma 3.3 and Theorem 6.11], where the case $m = 1 > 2n/p$ is proved. Suppose $f \in H(\Omega)$ and that the right side of (14.69) is finite. By Proposition 14.4 and [**346**, Theorem 6.7] we have

$$f(z) = \frac{n!}{\pi^n} \int_{\mathbb{B}_n} \frac{g(w)}{(1-\overline{w}\cdot z)^{n+1}} dV, \qquad z \in \mathbb{B}_n, \tag{14.72}$$

for some $g \in L^p(\lambda_n)$ where

$$\|g\|_{L^p(\lambda_n)} \approx \sum_{j=0}^{m-1} \left|\nabla^j f(0)\right| + \left(\int_{\mathbb{B}_n} \left|(1-|z|^2)^m \mathcal{R}_m f(z)\right|^p d\lambda_n(z)\right)^{1/p}. \tag{14.73}$$

Fix $\alpha \in \mathcal{T}_n$ and let $a = c_\alpha \in \mathbb{B}_n$. We claim that
(14.74)

$$|D_a^m f(z)| \leq C_m (1-|a|^2)^{m/2} \int_{\mathbb{B}_n} \frac{|g(w)|}{|1-\overline{w}\cdot z|^{n+1+m/2}} dV, \qquad m \geq 1, z \in B_d(a, C).$$

We now compute $D_a^m f(z)$ for $z \in B_d(a, C)$, beginning with the case $m = 1$. Since

$$\begin{aligned}
D_a(\overline{w}\cdot z) &= (\overline{w}\cdot z)' \varphi_a'(0) = -\overline{w}^t \left\{(1-|a|^2)P_a + (1-|a|^2)^{1/2} Q_a\right\} \\
&= -\overline{\left\{(1-|a|^2)P_a w + (1-|a|^2)^{1/2} Q_a w\right\}}^t,
\end{aligned}$$

we have
(14.75)

$$\begin{aligned} D_a f(z) &= \frac{n!}{\pi^n} \int_{\mathbb{B}_n} D_a (1-\overline{w}\cdot z)^{-(n+1)} g(w) dV \\ &= -\frac{(n+1)!}{\pi^n} \int_{\mathbb{B}_n} (1-\overline{w}\cdot z)^{-(n+2)} D_a(\overline{w}\cdot z) g(w) dV \\ &= \frac{(n+1)!}{\pi^n} \int_{\mathbb{B}_n} (1-\overline{w}\cdot z)^{-(n+2)} \overline{\left\{(1-|a|^2)P_a w + (1-|a|^2)^{1/2} Q_a w\right\}}^t g(w) dV. \end{aligned}$$

Taking absolute values inside, we obtain

$$|D_a f(z)| \lesssim (1-|a|^2)^{1/2} \int_{\mathbb{B}_n} \frac{(1-|a|^2)^{1/2} |P_a w| + |Q_a w|}{|1-\overline{w}\cdot z|^{n+2}} |g(w)| \, dV. \tag{14.76}$$

From the following elementary inequalities

$$\begin{aligned} |Q_a w| &= \left| w - \frac{w\cdot \overline{a}}{|a|^2} a \right| \le |w-a| + |a|^{-1} |1-w\cdot\overline{a}|, \\ |w-a|^2 &= |w|^2 + |a|^2 - 2\,\mathrm{Re}\,(w\cdot\overline{a}) \\ &\le (1-|w|^2) + (1-|a|^2) + |1-w\cdot\overline{a}| \\ &\lesssim |1-w\cdot\overline{a}|, \end{aligned}$$

we obtain that $|Q_a w| \lesssim |1-\overline{w}\cdot a|^{1/2}$. Now using $(1-|a|^2)+|1-\overline{w}\cdot a| \lesssim |1-\overline{w}\cdot z|$ for $z \in B_d(a, C)$, we see that

$$\frac{(1-|a|^2)^{1/2} |P_a w| + |Q_a w|}{|1-\overline{w}\cdot z|^{n+2}} \lesssim \frac{1}{|1-\overline{w}\cdot z|^{n+3/2}}, \qquad z \in B_d(a, C).$$

Plugging this estimate into (14.76) yields

$$|D_a f(z)| \lesssim (1-|a|^2)^{1/2} \int_{\mathbb{B}_n} \frac{|g(w)|}{|1-\overline{w}\cdot z|^{n+3/2}} dV,$$

which is the case $m=1$ of (14.74).

To obtain the case $m=2$ of (14.74), we differentiate (14.75) again to get

$$D_a^2 f(z) = -\frac{(n+2)!}{\pi^n} \int_{\mathbb{B}_n} (1-\overline{w}\cdot z)^{-(n+3)} W\overline{W}^t g(w) \, dV.$$

where we have written $W = \left\{(1-|a|^2)P_a w + (1-|a|^2)^{1/2} Q_a w\right\}$ for convenience. Again taking absolute values inside, we obtain

$$\left|D_a^2 f(z)\right| \lesssim (1-|a|^2) \int_{\mathbb{B}_n} \frac{\left((1-|a|^2)^{1/2} |P_a w| + |Q_a w|\right)^2}{|1-\overline{w}\cdot z|^{n+3}} |g(w)| \, dV.$$

Once again, using $|Q_a w| \lesssim |1-\overline{w}\cdot a|^{1/2}$ and $(1-|a|^2)+|1-\overline{w}\cdot a| \lesssim |1-\overline{w}\cdot z|$ for $z \in B_d(a, C)$, we see that

$$\frac{\left((1-|a|^2)^{1/2} |P_a w| + |Q_a w|\right)^2}{|1-\overline{w}\cdot z|^{n+3}} \le \frac{C}{|1-\overline{w}\cdot z|^{n+2}}, \qquad z \in B_d(a, C),$$

which yields the case $m = 2$ of (14.74). The general case of (14.74) follows by induction on m.

Now with $a = m/2$, $b = 0$ and $t = -n - 1$, our assumption that $m > 2n/p$ yields

$$-pa < t + 1 < p(b + 1).$$

The inequality (14.74) shows that $|D_a^m f(z)| \leq C_m S|g|(z)$ for $z \in B_d(a, C)$, where

$$Sg(z) := \int_{\mathbb{B}_n} \frac{(1 - |a|^2)^{m/2}}{|1 - \overline{w} \cdot z|^{n+1+m/2}} g(w)\, dV$$

is the operator in [**346**, Theorem 2.10], see also Theorem B.6. Thus the finite overlap property of the balls $B_d(c_\alpha, C_2)$ together with [**346**, Theorem 2.10] yield

$$\begin{aligned}
\|f\|^*_{B_{p,m}} &= \left(\sum_{\alpha \in \mathcal{T}_n} \int_{B_d(c_\alpha, C_2)} \left| D^m_{c_\alpha} f(z) \right|^p d\lambda_n(z) \right)^{1/p} \\
&\leq C_m \left(\int_{\mathbb{B}_n} |Sg(z)|^p \, d\lambda_n(z) \right)^{1/p} \\
&\leq C_m \left(\int_{\mathbb{B}_n} |g(z)|^p \, d\lambda_n(z) \right)^{1/p} \\
&\leq C_m \left(\int_{\mathbb{B}_n} \left| (1 - |z|^2)^m \mathcal{R}_m f(z) \right|^p d\lambda_n(z) \right)^{1/p}
\end{aligned}$$

by (14.73). This completes the proof of Proposition 14.25. □

14.5.3. The Oscillation Inequality. We will obtain an oscillation inequality for Besov spaces by transporting Taylor polynomials at the origin to the cubes K_α using the affine map

$$\psi_a(z) = a + \varphi_a'(0)\, z,$$

and its inverse $\psi_a^{-1}(z) = \varphi_a'(a)(z - a)$. However, the map ψ_a takes a Bergman ball $B_\beta(0, r_0) = B_\rho(0, t_0)$ (with $t_0 = \frac{e^{2r_0} - 1}{e^{2r_0} + 1}$) centered at the origin with radius r_0 into a Bergman ball $B_\beta(a, r)$ centered at a with a larger radius $r = Cr_0$ only if r_0, equivalently t_0, is small enough to begin with. In fact $r_0 < \frac{1}{2}$ is required, and this in turn requires that both of the structure parameters λ, θ be sufficiently small. However, we will apply the oscillation inequality in a tree with structure constant θ sufficiently large so that property (16.55) below holds, which is needed to define the holomorphic tree spaces used there. This then requires that after obtaining the oscillation inequality for small parameters, we then use an additional argument to complete the proof when θ is large. But first we will rigorously determine the mapping properties of ψ_a on a pseudohyperbolic ball $B_\rho(0, t_0) = \{z \in \mathbb{B}_n : t_0 > \rho(0, z) = |\varphi_0(z)| = |z|\}$, which coincides with the Euclidean ball of the same radius at the origin.

So now we determine when a point $\psi_a(z)$ with $z \in B_\rho(0, t_0) = B_\beta(0, r_0)$ lies in the Bergman ball $B_\rho(a, t) = B_\beta(a, r)$, i.e. when

$$t^2 > \rho(a, \psi_a(z))^2 = |\varphi_a(\psi_a(z))|^2 = |P_a \varphi_a(\psi_a(z))|^2 + |Q_a \varphi_a(\psi_a(z))|^2,$$

where

$$\begin{aligned}
\psi_a(z) &= a+\varphi_a'(0)\,z = a-\left\{(1-|a|^2)P_a z+(1-|a|^2)^{1/2}Q_a z\right\};\\
P_a\psi_a(z) &= a-(1-|a|^2)P_a z \text{ and } Q_a\psi_a(z)=-(1-|a|^2)^{1/2}Q_a z;\\
\varphi_a(\psi_a(z)) &= \frac{a-P_a\psi_a(z)-(1-|a|^2)^{1/2}Q_a\psi_a(z)}{1-\psi_a(z)\cdot\overline{a}}\\
&= \frac{(1-|a|^2)P_a z+(1-|a|^2)Q_a z}{1-\psi_a(z)\cdot\overline{a}} = \frac{1-|a|^2}{1-\psi_a(z)\cdot\overline{a}}z\\
&= \frac{1-|a|^2}{1-(a+\varphi_a'(0)\,z)\cdot\overline{a}}z\\
&= \frac{1-|a|^2}{1-|a|^2-\left\{(1-|a|^2)P_a z+(1-|a|^2)^{1/2}Q_a z\right\}\cdot\overline{a}}z,
\end{aligned}$$

and where

$$(P_a z)\cdot\overline{a}=\frac{z\cdot\overline{a}}{|a|^2}a\cdot\overline{a}=z\cdot\overline{a} \text{ and } (Q_a z)\cdot\overline{a}=\left(z-\frac{z\cdot\overline{a}}{|a|^2}a\right)\cdot\overline{a}=z\cdot\overline{a}-z\cdot\overline{a}=0.$$

Thus we have

$$\varphi_a(\psi_a(z))=\frac{1-|a|^2}{1-|a|^2-(1-|a|^2)(z\cdot\overline{a})}z=\frac{z}{1-z\cdot\overline{a}},$$

and so $\psi_a(z)\in B_\beta(a,r)$ if and only if

$$\frac{e^{2r}-1}{e^{2r}+1}=t>\rho(a,\psi_a(z))=|\varphi_a(\psi_a(z))|=\frac{|z|}{|1-z\cdot\overline{a}|}.$$

If we are given that $|z|=t_0$ then we need

$$\frac{t_0}{t}<|1-z\cdot\overline{a}|.$$

But since $1\leq|1-z\cdot\overline{a}|+|z\cdot\overline{a}|\leq|1-z\cdot\overline{a}|+|z|\,|a|$, it suffices to take

$$\frac{t_0}{t}<1-|z|\,|a|=1-t_0\,|a|$$

which gives

$$t>\frac{t_0}{1-t_0\,|a|}.$$

However, this is only possible provided $\frac{t_0}{1-t_0|a|}<1$, i.e. $t_0<\frac{1}{1+|a|}$, which prevents $\psi_a(z)\in B_\beta(a,r)$ from happening for all $a\in\mathbb{B}_n$ unless $t_0<\frac{1}{2}$.

Inequality (14.77) below is due to Peloso [**258**].

PROPOSITION 14.26. *Suppose $\mathcal{T}_n$ is a Bergman tree with structural constants $0<\lambda<1$ sufficiently small and $0<\theta<\infty$. For $1<p<\infty$ and $m\geq1$, we have the Taylor polynomial oscillation inequality, where $(z-a_\alpha)\cdot\partial=\sum_{i=1}^n(z_i-(a_\alpha)_i)\frac{\partial}{\partial z_i}$:*

$$\left\{\sum_{\alpha\in\mathcal{T}_n}\left(\sup_{z\in K_\alpha}\left|f(z)-\sum_{k=0}^{m-1}\frac{((z-a_\alpha)\cdot\partial)^k f(a_\alpha)}{k!}\right|\right)^p\right\}^{1/p}\lesssim\|f\|_{B^*_{p,m}}, \tag{14.77}$$

for every sequence of points $\{a_\alpha\}_{\alpha\in\mathcal{T}_n}$ with $a_\alpha \in K_\alpha$, $\alpha \in \mathcal{T}_n$. We also have the local estimate that there is M so that, with K_α^* is the union of the Carleson box K_α and its neighbors at most M boxes away,

(14.78)

$$\sup_{\xi\in K_\alpha}\left|f(\xi)-\sum_{k=0}^{m-1}\frac{((\xi-a_\alpha)\cdot\partial)^k f(a_\alpha)}{k!}\right| \lesssim \left(\int_{K_\alpha^*}\left|D_{c_\alpha}^m f(z)\right|^p d\lambda_n(z)\right)^{1/p}.$$

Here the implied constants depend on the structural constants λ and θ.

Proof. Recall that $K_0 = \{z \in \mathbb{B}_n : \beta(0,z) < \theta\} = B_\rho\left(0, \frac{e^{2\theta}-1}{e^{2\theta}+1}\right)$ is the Euclidean ball of radius $\frac{e^{2\theta}-1}{e^{2\theta}+1}$ centered at the origin. From part 1 of Lemma 14.3, and provided we choose λ, θ sufficiently small and M sufficiently large, we obtain that there are constants A_1, A_2 in $[0,1)$ such that

$$\begin{aligned} B(0,A_1) &\subset K_0 \subset B(0,A_2), \\ \psi_a(B(0,A_1)) &\subset K_\alpha \subset \psi_a(B(0,A_2)) \subset \psi_a(B(0,2A_2)) \subset K_\alpha^*, \end{aligned}$$

for $\alpha \in \mathcal{T}_n$. We first establish the estimate

$$\sup_{z\in B(0,A_2)}|f(z)| \le C_{m,n,p}\left(\int_{K_0^*}|D_0^m f(z)|^p d\lambda_n(z)\right)^{1/p}, \tag{14.79}$$

for all $f \in B_p$ satisfying the initial conditions

$$\nabla^k f(0) = 0, \qquad 0 \le k < m. \tag{14.80}$$

We have by standard methods, for $|z| < A_3$ and $f \in B_p$ satisfying (14.80),

$$\begin{aligned} |f(z)| &\lesssim \int_{B(0,2A_2)} |\nabla^m f(w)|\, dV \\ &\lesssim \int_{K_0^*} |\nabla^m f(w)|\, dV \\ &\lesssim \left(\int_{K_0^*} |\nabla^m f(\xi)|^p d\lambda_n(\xi)\right)^{1/p}, \end{aligned}$$

which establishes (14.79) since $D_0^m = \nabla^m$.

Using the affine maps ψ_a, we now show that (14.79) implies the estimate (14.78). We set $a = c_\alpha$. Since $\psi_a(B(0,A_2)) \supset K_\alpha$, we will apply (14.79) to the function

$$g(z) = (f\circ\psi_a)(z) - \sum_{k=0}^{m-1}(z\cdot\nabla)^k (f\circ\psi_a)(0),$$

which satisfies (14.80) with g in place of f. We also restrict the supremum over z in (14.79) to $\psi_a^{-1}(K_\alpha)$. We compute that

$$(w\cdot\nabla)(f\circ\psi_a)(0) = f'(a)\,\varphi_a'(0)\,w = D_a f(a)\,w = (w\cdot D_a)\,f(a),$$

and more generally

$$(w\cdot\nabla)^k (f\circ\psi_a)(0) = (w\cdot D_a)^k f(a), \qquad 0 \le k \le m-1, w\in\mathbb{C}^n,$$

so that

$$g(z)=f(\psi_a(z))-\sum_{k=0}^{m-1}\frac{(z\cdot D_a)^k f(a)}{k!}.$$

Now using $\psi_a^{-1}(\xi)=\varphi_a'(a)(\xi-a)$ and

$$\begin{aligned}(\varphi_a'(a)(\xi-a)\cdot D_a)f(a)&=f'(a)\varphi_a'(0)\varphi_a'(a)(\xi-a)\\&=f'(a)(\xi-a)\\&=((\xi-a)\cdot\partial)f(a),\end{aligned}$$

we get from (14.79), with g in place of f, that

$$\begin{aligned}&\max_{\xi\in K_\alpha}\left|f(\xi)-\sum_{k=0}^{m-1}\frac{((\xi-a)\cdot\partial)^k f(a)}{k!}\right|^p\\&=\max_{z\in K_0}|g(z)|^p\\&\lesssim\int_{K_0^*}|D_0^m g(z)|^p\,d\lambda_n(z).\end{aligned}$$

Noting that $D_0^m g=D_0^m(f\circ\psi_a)=(D_a^m f)\circ\psi_a$ and $d\lambda_n(\xi)\approx d\lambda_n(z)$, we have

$$\int_{K_0^*}|D_0^m(f\circ\psi_a)(\xi)|^p\,d\lambda_n(\xi)\approx\int_{K_\alpha^*}|D_a^m f(z)|^p\,d\lambda_n(z),$$

which yields the first inequality (14.78) since $a=c_\alpha$.

Now we apply (14.78) to obtain the case $a_\alpha=c_\alpha$ of (14.77) as follows:

$$\begin{aligned}&\sum_{\alpha\in\mathcal{T}_n}\left(\max_{z\in K_\alpha}\left|f(z)-\sum_{k=0}^{m-1}\frac{((z-c_\alpha)\cdot\partial)^k f(a_\alpha)}{k!}\right|\right)^p\\&\lesssim\sum_\alpha\int_{K_\alpha^*}\left|D_{c_\alpha}^m f(z)\right|^p d\lambda_n(z)\lesssim\|f\|_{B_p}^p,\end{aligned}$$

by the finite overlap condition (5) in Lemma 14.3. The general case with $a_\alpha\in K_\alpha$ is similar. This completes the proof of Proposition 14.26 under the assumption that both structural constants λ and θ are sufficiently small.

Before proceeding to the general case $0<\theta<\infty$, we first observe that in the situation above, where both λ and θ are small, the techniques used there also prove the following extension of inequality (14.78):

$$\begin{aligned}(14.81)\quad&\sup_{\xi\in K_\alpha}\left|f^{(\ell)}(\xi)-\sum_{k=0}^{m-\ell-1}\frac{((\xi-a_\alpha)\cdot\partial)^k f^{(\ell)}(a_\alpha)\left[\otimes^\ell(\xi-a_\alpha)\right]}{k!}\right|\\&\qquad\lesssim\left(\int_{K_\alpha^*}\left|D_{c_\alpha}^m f(z)\right|^p d\lambda_n(z)\right)^{1/p},\end{aligned}$$

for all $0\le\ell<m$. Note that $f^{(\ell)}(a_\alpha)$ is an ℓ-tensor here, and so acts on the tensor $\otimes^\ell(\xi-a_\alpha)$. The case $\ell=0$ is what was proved above, and the case $0<\ell<m$ is almost verbatim the same proof.

Now we suppose that $\theta > 1$ is arbitrary. We will make use of Lemma 14.23, which gives the equivalence of almost invariant holomorphic derivatives with distance between base points held under control in the Bergman metric, namely

$$|D_a^m f(z)| \approx |D_b^m f(z)|, \qquad \text{for } a, b \in \mathbb{B}_n \text{ with } \beta(a,b) \leq C_0, \tag{14.82}$$

for all $f \in \mathrm{Hol}(\mathbb{B}_n)$, and where the implied constants only depend on C_0, m, and n. The idea now is to fix $z \in K_\alpha = B_\beta(a,\theta)$ and choose points $\{a = a_1, a_2, ..., a_J, a_{J+1} = z\}$ in the ball along a Euclidean segment with

$$\beta(a_{j+1}, a_j) \approx \theta_0 \ ,$$

where θ_0 is a radius sufficiently small that the argument above proves the oscillation inequality. Then we see that $J \leq C\frac{\theta}{\theta_0}$.

We start with the simplest case $m = 1$. Then we have from (14.82) that

$$\begin{aligned} |f(z) - f(a)| &\leq \sum_{j=1}^{J} |f(a_{j+1}) - f(a_j)| \lesssim \sum_{j=1}^{J} \left| D_{a_j}^1 f(a_j) \right| \\ &\lesssim J \left| D_a^1 f(z) \right| \leq C\frac{\theta}{\theta_0} \left| D_a^1 f(z) \right|, \end{aligned}$$

for $z \in K_\alpha$. The case $m = 1$ is now finished using the mean value property of holomorphic functions to obtain

$$\sup_{z \in K_\alpha} \left| D_a^1 f(z) \right| \lesssim \left(\int_{K_\alpha^*} \left| D_{c_\alpha}^1 f(z) \right|^p d\lambda_n(z) \right)^{1/p} .$$

Now we turn to the case $m = 2$. We write

$$\begin{aligned} & f(z) - \{f(a) + f'(a)(z-a)\} = \{f(a_{J+1}) - f(a_1)\} - f'(a)(a_{J+1} - a_1) \\ = \ & \sum_{j=1}^{J} (f(a_{j+1}) - f(a_j)) - f'(a_1) \sum_{j=1}^{J} (a_{j+1} - a_j) \\ = \ & \sum_{j=1}^{J} (f(a_{j+1}) - f(a_j)) - \sum_{j=1}^{J} f'(a_j)(a_{j+1} - a_j) + \sum_{j=1}^{J} (f'(a_j) - f'(a_1))(a_{j+1} - a_j), \end{aligned}$$

and if we now apply summation by parts to the final sum, we see that it equals

$$\begin{aligned} \sum_{j=1}^{J} (f'(a_j) - f'(a_1))(a_{j+1} - a_j) &= \sum_{j=1}^{J} \sum_{i=1}^{j-1} (f'(a_{i+1}) - f'(a_i))(a_{j+1} - a_j) \\ &= \sum_{i=1}^{J-1} (f'(a_{i+1}) - f'(a_i)) \sum_{j=i+1}^{J} (a_{j+1} - a_j) \\ &= \sum_{i=1}^{J-1} (f'(a_{i+1}) - f'(a_i))(a_{J+1} - a_{i+1}). \end{aligned}$$

Thus we have

$$\begin{aligned} f(z)-\{f(a)+f'(a)(z-a)\} &= \sum_{j=1}^{J}[f(a_{j+1})-\{f(a_j)-f'(a_j)(a_{j+1}-a_j)\}] \\ &\quad+\sum_{i=1}^{J-1}[(f'(a_{i+1})-f'(a_i))(a_{J+1}-a_{i+1})], \end{aligned}$$

where the factors in square brackets satisfy

$$|f(a_{j+1})-\{f(a_j)+f'(a_j)(a_{j+1}-a_j)\}|\lesssim\left|D^2_{a_j}f(a_j)\right|,\qquad \text{for } 1\le j<J,$$

and

$$|(f'(a_{i+1})-f'(a_i))(a_{i+1}-a_i)|\lesssim\left|D^2_{a_i}f(a_i)\right|,\qquad \text{for } 1\le i\le J-1,$$

by what we proved in the first part of the proof, namely (14.81) in the case when $m=2$, and when ℓ is first 0 and then 1. Moreover from (14.82) we have

$$\left|D^2_{a_j}f(a_j)\right|\lesssim\sup_{w\in K_\alpha}\left|D^2_a f(w)\right|,\qquad \text{for } 1\le j<J.$$

Putting all of these estimates together with the fact that $a_{J+1}-a_{i+1}$ is parallel to $a_{i+1}-a_i$, we obtain

$$\begin{aligned} &|f(z)-\{f(a)+f'(a)(z-a)\}| \\ \lesssim &\sum_{j=1}^{J}\sup_{w\in K_\alpha}\left|D^2_a f(w)\right|+\sum_{i=1}^{J-1}\frac{|a_{J+1}-a_{i+1}|}{|a_{i+1}-a_i|}\sup_{w\in K_\alpha}\left|D^2_a f(w)\right|\lesssim C_\theta\sup_{w\in K_\alpha}\left|D^2_a f(w)\right|, \end{aligned}$$

since $J\lesssim\theta$ and $\frac{|a_{J+1}-a_{i+1}|}{|a_{i+1}-a_i|}\lesssim C_\theta$, $1\le j\le J$, for some large constant C_θ if $\beta(a_{j+1},a_j)\approx\theta_0$. The case $m=2$ is again finished using the mean value property of holomorphic functions to obtain

$$\sup_{z\in K_\alpha}\left|D^2_a f(z)\right|\lesssim\left(\int_{K^*_\alpha}\left|D^2_{c_\alpha}f(z)\right|^p d\lambda_n(z)\right)^{1/p}.$$

The cases $m>2$ are similar, only complicated by more algebra involving summation by parts. □

14.6. Notes and Comments

The books by Rudin [**288**] and Zhu [**346**] are basic sources for information about function theory on the ball and function spaces on the ball. The more specialized material in this chapter is adapted from [**47**].

CHAPTER 15

Interpolating Sequences

In this chapter we describe the interpolating sequences for the spaces $B_p(\mathbb{B}_n)$ and $\text{Mult}(B_p(\mathbb{B}_n))$. Although the results have the same form as the earlier results for H^2 and $\mathcal{D}$, the approach here is very different. In the earlier work we used Hilbert space theory to give an abstract existence theorem for interpolating functions; here we give constructive proofs. The proof uses results about function spaces on trees. The development of those results is deferred until the next chapter.

In this chapter we identify the interpolating sequences for the Besov spaces and for their multiplier algebras on the ball. As was the case in Chapter 7 for the Hardy space and the Dirichlet space; the sequences must be separated and an associated measure must be a Carleson measure. However we must take a different path for the proof. We no longer have use of Theorems such as 7.2 and 7.10 which concerned Hilbert spaces of holomorphic functions. Following Böe's lead from [**77**, Theorem 1.1] we obtain a Banach space substitute for some of those results, Theorems 15.1 and 15.2 below. This lets us move to the central issue of demonstrating the existence of interpolating functions. We do that using the constructive techniques pioneered by Marshall-Sundberg [**220**], Böe [**77**] and Bishop [**72**]. Also, in further analogy with results in Chapter 7, we find that the interpolating sequences for each Besov space and its multiplier algebra are the same. However in contrast to Chapter 7 where that fact was obtained in Theorem 7.3 using Hilbert space techniques and was proved before the sequences were characterized, for each Besov space we characterize the sequences independently of each other and then observe that the descriptions are the same.

15.1. The Theorems and Comments

For typographic convenience, throughout this chapter we will set

$$\kappa_j := \kappa_{z_j} = \left\| k_{z_j}^{\alpha,p} \right\|_{B_{p'}}^{-1} \tag{15.1}$$

and hence the normalized reproducing kernels are $\kappa_j k_{z_j}^{\alpha,p}$.

Recall the duality statement and the formula for the reproducing kernel for the Besov spaces. For α, p given, (15.2) states that $(B_p)^*$ and $B_{p'}$ are identified under the pairing, for $f \in B_p$, $g \in B_{p'}$

$$\begin{aligned} \langle f, g \rangle_{\alpha,p} = \langle \mathcal{R}_{\mathbf{s}}^{\alpha} f, \mathcal{R}_{\mathbf{s}'}^{\alpha} g \rangle_{\alpha} &= \int_{\mathbb{B}_n} \mathcal{R}_{\mathbf{s}}^{\alpha} f(z) \overline{\mathcal{R}_{\mathbf{s}'}^{\alpha} g(z)} d\nu_{\alpha}(z) \\ &= \int_{\mathbb{B}_n} \left((1-|z|^2)^{\mathbf{s}} \mathcal{R}_{\mathbf{s}}^{\alpha} f(z)\right) \overline{\left((1-|z|^2)^{\mathbf{s}'} \mathcal{R}_{\mathbf{s}'}^{\alpha} g(z)\right)} d\lambda_n(z), \end{aligned} \tag{15.2}$$

Further, the reproducing kernel for B_p relative to this pairing is given by

$$\begin{aligned} k_w^{\alpha,p}(z) &= \left(\mathcal{R}_{\mathbf{s}'}^{\alpha}\right)^{-1}\left(\mathcal{R}_{\mathbf{s}}^{\alpha}\right)^{-1} K_w^{\alpha}(z) \\ &= \left(\mathcal{R}_{\mathbf{s}'}^{\alpha}\right)^{-1}\left(1-\overline{w}\cdot z\right)^{-\mathbf{s}'}, \end{aligned}$$

where the last formula above follows from (14.23).

We now state our interpolation theorem in two separate statements. For $n=1$ both results are due to Böe [**77**]. In particular, for $n=1, p=2$ we recapture the result in Chapter 7 for the Dirichlet space, but with a different proof. New complications that arise in higher dimensions are discussed in Section 15.1.1 below.

THEOREM 15.1 (Besov Space Interpolation). *Let $1<p<\infty$, $\alpha>-1$ and $k_w^{\alpha,p}(z)$ be the reproducing kernel for B_p relative to the pairing* (15.2). *Let $\mathcal{Z}$ be a sequence in the unit ball $\mathbb{B}_n$. Then the following conditions are equivalent.*

(1) *$\mathcal{Z}$ is an interpolating sequence for B_p. That is*

$$\text{the map } f \mapsto \{\kappa_j f(z_j)\}_{j=1}^{\infty} \text{ takes } B_p \text{ boundedly into and onto } \ell^p. \tag{15.3}$$

(2) *The following norm equivalence holds:*

$$\left\| \sum_{j=1}^{\infty} a_j \kappa_j k_{z_j}^{\alpha,p} \right\|_{B_{p'}} \approx \left(\sum_{j=1}^{\infty} |a_j|^{p'} \right)^{1/p'}. \tag{15.4}$$

(3) *The following separation condition and Carleson embedding hold:*

$$\begin{aligned} &\beta(z_i,0) \lesssim \beta(z_i,z_j), i\neq j \text{ and} \\ &d\mu_{\mathcal{Z},p} = \sum_{j=1}^{\infty} \kappa_j^p \delta_{z_j} \text{ is a } B_p\text{-Carleson measure.} \end{aligned} \tag{15.5}$$

or, in an equivalent form that is sometimes more convenient

$$\begin{aligned} &\beta(z_i,0) \lesssim \beta(z_i,z_j) \text{ and} \\ &\sum_{j=1}^{\infty} \left(1+\log\frac{1}{1-|z_j|^2}\right)^{1-p} \delta_{z_j} \text{ is a } B_p\text{-Carleson measure.} \end{aligned} \tag{15.6}$$

THEOREM 15.2 (Multiplier Interpolation). *With the same hypotheses and notation as the previous theorem, condition* (15.5) *also characterizes multiplier interpolation sequences. That is, the three conditions in Theorem* 15.1 *are also equivalent to the following condition:*

(1) *$\mathcal{Z}=\{z_j\}_{j=1}^{\infty}$ is an interpolating sequence for* $\mathrm{Mult}(B_p)$*; that is,*

$$\text{the map } f \mapsto \{f(z_j)\}_{j=1}^{\infty} \text{ takes } \mathrm{Mult}(B_p) \text{ boundedly into and onto } \ell^{\infty}. \tag{15.7}$$

Also, we have further relations involving the following two conditions.

First, $\left\{k_{z_j}^{\alpha,p}\right\}_{j=1}^{n}$ is an unconditional basic sequence in $B_{p'}$:

$$\left\| \sum_{j=1}^{\infty} b_j k_{z_j}^{\alpha,p} \right\|_{B_{p'}} \lesssim \left\| \sum_{j=1}^{\infty} a_j k_{z_j}^{\alpha,p} \right\|_{B_{p'}}, \quad \text{whenever } |b_j| \leq |a_j|. \tag{15.8}$$

Second, for $\mathcal{Z} = \{z_j\}_{j=1}^{\infty}$ of the particular form $\{z_j\}_{j=1}^{\infty} = \{c_{\alpha_j}\}_{j=1}^{\infty}$ for $\{\alpha_j\}_{j=1}^{\infty}$ a sequence in a Bergman tree $\mathcal{T}_n$, satisfying

$$\beta(z_i, 0) \lesssim \beta(z_i, z_j), i \neq j \text{ and} \tag{15.9}$$
$$\sum_{j=1}^{\infty} (1 + d(\alpha_j, o))^{1-p} \delta_{\alpha_j} \text{ is a } B_p(\mathcal{T}_n) \text{ Carleson measure (14.29).}$$

(The role of the condition that $\mathcal{Z}$ have a "particular form" is discussed just prior to Claim 15.7 below.)

THEOREM 15.3. *Condition* (15.9) *implies condition* (15.7) *implies condition* (15.8).

If $p \in (1, 2 + 1/(n-1))$, *then each of conditions* (15.7) *and* (15.9) *is equivalent to the three conditions in Theorem* 15.1.

If $p \in (1, 1 + 1/(n-1)) \cup [2, \infty)$, *then condition* (15.8) *implies the three conditions in Theorem* 15.1.

REMARK 15.4. *Our proofs show that the interpolations in* (15.7) *and* (15.3) *can be taken to be linear, i.e. there are bounded linear maps* $R : \ell^\infty \to \mathrm{Mult}(B_p)$ *and* $S : \ell^p \to B_p$ *that yield right inverses to the restriction maps in* (15.7) *and* (15.3) *respectively. For multiplier interpolation in dimension* $n = 1$ *Böe has shown* [**77**] *the stronger result that there are functions* $f_k \in \mathrm{Mult}(B_p)$ *with* $f_k(z_j) = \delta_k^j$ *and* $\|\sum_i a_i f_i\|_{\mathrm{Mult}(B_p)} \lesssim \|\{a_i\}\|_\infty$ *(compare with* [**161**, Theorem 2.1, Chapter 7]. *It seems plausible that this extends to* $1 < p < \infty$ *for* $n > 1$, *but we do not know.*

In [**47**, Corollary 3.2] we saw that if a *fixed* measure μ on the ball is a B_{p_0} Carleson measure for some p_0 then it is also a B_p Carleson measure for all $p < p_0$; the condition propagates downward. In contrast, we show in the next proof that if a sequence $\mathcal{Z}$ satisfies (15.5) for a particular p_0 then this fact propagates upward; it also satisfies (15.5) for all $p > p_0$. We just state the consequences of this for interpolating sequences.

COROLLARY 15.5. *Suppose* $1 < p_0 < p < \infty$.

1. *If* $\mathcal{Z}$ *is an interpolating sequence for* B_{p_0}, *i.e. if* (15.3) *holds for* p_0, *then* $\mathcal{Z}$ *is also an interpolating sequence for* B_p.
2. *If* $\mathcal{Z}$ *is an interpolating sequence for* $\mathrm{Mult}(B_p)$, *i.e. if* (15.9) *holds for* p_0, *then* $\mathcal{Z}$ *is also an interpolating sequence for* $\mathrm{Mult}(B_p)$.

PROOF. By the two previous theorems it suffices to show that if for some p_0 $\mu_{\mathcal{Z},p_0}$ is a B_{p_0} Carleson measure then for any $p > p_0$, $\mu_{\mathcal{Z},p}$ is a B_p Carleson measure.

Suppose $\mathcal{Z}$, p_0 are given and $\mu_{\mathcal{Z},p_0}$ is a B_{p_0} Carleson measure. We write that Carleson embedding in the following form,

$$\sum_{j=1}^{\infty} \left| \frac{f(c_{\alpha_j})}{1 + \log \frac{1}{1 - |c_{\alpha_j}|^2}} \right|^{p_0} \left(1 + \log \frac{1}{1 - |c_{\alpha_j}|^2} \right) \lesssim \|f\|_{B_{p_0}}^{p_0},$$

or

$$\|Tf\|_{L^{p_0}(\nu)} \lesssim \|f\|_{B_{p_0}},$$

where

$$Tf(z) = \left(1 + \log \frac{1}{1 - |z|^2}\right)^{-1} f(z),$$

$$d\nu = \sum_{j=1}^{\infty} \left(1 + \log \frac{1}{1 - |c_{\alpha_j}|^2}\right) \delta_{c_{\alpha_j}}.$$

Because T is bounded from the Bloch space B_∞ to $L^\infty(\nu)$, see [**346**, Corollary 3.7], we have by interpolation, [**346**, Theorem 6.12], that for any $p_0 < p < \infty$ T is bounded from B_p to $L^p(\nu)$, which is the statement that it is a B_p Carleson measure. We remark that in the proof of [**346**, Theorem 6.12], [**346**, Theorem 3.4] substitutes for [**346**, Theorem 6.7] when $p_1 = \infty$ in the conclusion

$$B_p = [B_{p_0}, B_{p_1}]_\theta, \qquad \frac{1}{p} = \frac{1-\theta}{p_0} + \frac{\theta}{p_1}.$$

□

It is not apparent to us how to reach the conclusion of the corollary without passing through the equivalence of interpolation with condition (15.5).

15.1.1. The Proof Outline. The proof of the theorems will take the rest of this chapter as well as parts of the next, and various of the arguments will only be valid for certain ranges of p. We begin with an overview.

We will see that the one-dimensional arguments used in [**77**, Theorem 1.1] by Böe to prove that (15.7) implies (15.8), and that (15.3) implies (15.4) implies (15.5) extend to higher dimensions with any choice of pairing and corresponding reproducing kernel in Theorem 15.2. However the one-dimensional proof of the implication (15.8) implies (15.4) does not extend to the full range of p. In the proof of that implication Böe uses what he calls his "curious lemma." That lemma is a bit of analysis which is used in place of certain Hilbert space tools. To use that lemma he needs to know that his reproducing kernel, $k_w(z) = \log \frac{1}{1-\overline{w}z}$, satisfies the positivity property

$$\operatorname{Re} \frac{1}{\overline{w}} k_w'(z) = \operatorname{Re} \frac{1}{1 - \overline{w}z} > 0. \tag{15.10}$$

With the choice $p = (n+1+\alpha)/(n+\alpha) = 1+\alpha/(n+\alpha)$ it follows from (14.23) that the reproducing kernel $k_w^{\alpha,p}(z)$ given in Theorem 14.8 has the analogous positivity property,

$$\operatorname{Re} \mathcal{R}_{\mathbf{s}'}^{\alpha} k_w^{\alpha,p}(z) = \operatorname{Re} \mathcal{R}_1^{\alpha} k_w^{\alpha,p}(z) = \operatorname{Re} \frac{1}{1 - \overline{w} \cdot z} > 0. \tag{15.11}$$

This does in fact lead to a proof that (15.8) implies (15.4), but because we must have $\alpha > -1$, it does so only for the restricted range $1 < p < 1 + 1/(n-1)$. It is the failure of this argument for the remaining values of p, as well as our failure to obtain the necessity of the tree condition (14.37) for the Carleson embedding (14.29) when $p \geq 2 + 1/(n-1)$, that forces us to proceed via a different route.

The series of implications (15.3) implies (15.4) implies (15.5) implies (15.3), as well as the implication (15.5) implies (15.7), do not use our characterization of Carleson measures in Theorem 14.16, relying instead on only the embedding statement which defines Carleson measures (we do however use the simple condition (8.16) in conjunction with the general tree theorem from [**48**]). Also, most of the remaining implications use only the embedding definition of Carleson measure together with one-dimensional techniques. The exception is the final implication, (15.7) implies (15.9), which is the most difficult step in the proof. That argument involves the introduction of a new class of function spaces on trees and the development of the theory of Carleson measures for those spaces.

We refer to the implications that (15.5) implies (15.7) and that (15.5) implies (15.3) as sufficiency implications, and the remaining ones as necessity implications.

We will prove Theorems 15.1 and 15.2 by demonstrating the following implications:

(1) In Section 15.3, *Proof of Multiplier Space Necessity*, we prove that (15.7) implies (15.8), that (15.3) and (15.4) are equivalent, that (15.3) implies (15.5), and that if $p \in (1, 1 + 1/(n-1)) \cup [2, \infty)$, then (15.8) implies (15.5).

(2) In Section 15.4, *Proof of Multiplier Space sufficiency*, we prove that (15.5) implies (15.7).

(3) In Section 15.5, *Proof of Besov Space Interpolation*, we prove that (15.6), which is equivalent to (15.5), implies (15.3).

(4) In Section 15.6, *Completing the Multiplier Interpolation Loop*, we show that if $p \in (1, 2 + 1/(n-1))$, then (15.7) implies (15.9) (this implication already follows for $p \in (1, 1 + 1/(n-1)) \cup [2, 2 + 1/(n-1))$ from the previous ones and Theorem 14.16).

(5) However the necessity is more difficult to establish in the subrange $1 + 1/(n-1) \leq p < 2$ and we postpone that portion of the proof until the next chapter where additional properties of a refined Bergman tree $\mathcal{T}_n$ are developed. That more delicate argument includes Claim 15.7, which shows that we may assume the points z_j occur as the centers c_{α_j} of certain cubes K_{α_j}.

REMARK 15.6. *We do not know if* (15.5) *is sufficient for* (15.9) *in the range* $p \in [2 + 1/(n-1), \infty)$. *The 'geometric' condition* (15.9) *is flexible due to its positivity and simplicity, and would thus be a welcome equivalence for* (15.5). *The only other implication left is whether or not the unconditional basic sequence property* (15.8) *implies multiplier interpolation* (15.7) *in the 'missing' range* $[1 + 1/(n-1), 2 + 1/(n-1))$. *For the range of small* $p \in (1, 1 + 1/(n-1))$, *Böe's curious lemma suffices for this, while for* $p \in [2 + 1/(n-1), \infty)$, *Khintchine's inequality suffices for this.*

15.2. Setting Things up to Use Trees

We will be working with the Besov spaces $B_p = B_p(\mathbb{B}_n)$, $1 < p < \infty$, and we let p' be the dual index; $p^{-1} + p'^{-1} = 1$. We will carry a parameter $\alpha > -1$ that specifies which pairing we are using for the duality between B_p and $B_{p'}$, and will continue to have $\mathbf{s}$ and $\mathbf{s}'$ given by (14.19). The choice of α does not effect, for example, the estimate for the norms of the reproducing kernels in Lemma 14.9. We will choose a specific value of α later to ensure the positivity condition (15.11) for the kernel.

Let $\mathcal{Z} = \{z_j\}_{j=1}^{\infty}$ be a sequence of points in the unit ball $\mathbb{B}_n$. We will prove that one can perform weighted ℓ^p interpolation on the sequence $\mathcal{Z}$ using functions from the Besov space B_p, that is (15.3) holds, if and only if $\mathcal{Z}$ is separated and the measure

$$d\mu = d\mu_{\mathcal{Z}} := \mu_{\mathcal{Z},p} = \sum_{j=1}^{\infty} \left(1 + \log \frac{1}{1 - |z_j|^2}\right)^{1-p} \delta_{z_j} \tag{15.12}$$

is a B_p Carleson measure; that is, if and only if condition (15.5) holds. For $n = 1$ and $p = 2$ this is the interpolation result we presented for the Dirichlet space in Theorem 7.13. However there we wrote the separation condition in the equivalent form $\delta_{\mathcal{D}}(z_i, z_j) \gtrsim 1 > 0$.

Our construction of the Bergman tree $\mathcal{T}_n$ with its associated set of centers $\mathcal{C} = \{c_{\alpha_j}\}$ is relatively rigid. However when studying a separated sequence $\mathcal{Z}$ in the ball it is often convenient to work using a nearby sequence $\mathcal{Z}'$ which is contained in $\mathcal{C}$. The claim below shows how to find an appropriate $\mathcal{C}$ and $\mathcal{Z}'$. Theorem 14.13 allows us to pass between Carleson measure conditions for $d\mu_{\mathcal{Z}}$ and those for $d\mu_{\mathcal{Z}'}$. The claim below also gives the perturbation argument showing why an interpolating sequence result for $\mathcal{Z}$ follows from a similar result for $\mathcal{Z}'$. With these things in mind, we will use without comment the reduction of questions about general $\mathcal{Z}$ to questions about a favorably situated $\mathcal{Z}'$. Here now is the first step toward this reduction, in which we assume both lambda and theta are sufficiently small.

CLAIM 15.7. *We may assume that the points z_j occur as the centers c_{α_j} for a corresponding sequence $\{\alpha_j\}_{j=1}^{\infty}$ in the Bergman tree $\mathcal{T}_n$, provided that in the construction of $\mathcal{T}_n$ we take $\lambda > 0$ sufficiently small and $\theta > 0$ sufficiently small.*

REMARK 15.8. *Being able to obtain this claim requires only the much weaker separation condition $\beta(z_i, z_j) \geq c > 0$.*

PROOF. First, in Theorem 15.1 below that treats Besov space interpolation, we make no appeal whatsoever to Bergman trees, and so no special assumptions are needed there. In Theorem 15.2 that deals with multiplier space interpolation, we only invoke Bergman trees $\mathcal{T}_n = \mathcal{T}_n(\lambda, \theta)$, with parameters $\lambda > 0$ sufficiently small and $\theta > 0$ sufficiently large, in order to prove the necessity of separation and Carleson embedding (15.5) for multiplier space interpolation. So we now suppose that Theorem 15.2 holds when $\mathcal{Z}$ is a subset of such a Bergman tree $\mathcal{T}_n(\lambda, \theta)$, i.e. we assume that $\mathcal{Z}$ satisfies the separation and Carleson embedding condition (15.5) whenever multiplier space interpolation (15.7) holds with such $\mathcal{Z} = \{z_j\}_{j=1}^{\infty} \subset \{c_\alpha\}_{\alpha \in \mathcal{T}_n(\lambda,\theta)} \subset \mathbb{B}_n$, i.e. that the map $f \to \{f(z_j)\}_{j=1}^{\infty}$ takes $\mathrm{Mult}(B_p)$ boundedly into and onto ℓ^∞. We will now apply a Banach space perturbation argument to obtain a 'Bergman tree' $\mathcal{T}_n$ and a subset $\mathcal{W} := \{w_j\}_{j=1}^{\infty} \subset \mathcal{T}_n$ such that multiplier space interpolation (15.7) holds for $\mathcal{W}$. But a complication involving the parameter θ arises, and which requires subtle adjustments to an otherwise standard perturbation argument. Here are the details, beginning with a familiar abstract result for Banach spaces.

Let X and Y be Banach spaces and let $\mathcal{B}(X, Y)$ be the Banach space of bounded linear maps $T : X \to Y$ from X to Y topologized by the operator norm $\|T\|_{X \to Y} := \sup_{x \neq 0} \frac{\|Tx\|}{\|x\|}$. Denote by $\mathcal{O}(X, Y)$ the subset of maps $T \in \mathcal{B}(X, Y)$ that are *onto* Y, i.e. $T(X) = Y$. We recall the well-known result that $T : X \to Y$ is bounded and onto *if and only if* $T^* : Y^* \to X^*$ is bounded and there exists a positive constant c such that

$$\|T^*y^*\|_{X^*} \geq c\,\|y^*\|_{Y^*}, \qquad y^* \in Y^*.$$

Let $c(T^*)$ denote the largest such constant c for a given T^*.

LEMMA 15.9. *The set of onto maps $\mathcal{O}(X,Y)$ is an open subset of $\mathcal{B}(X,Y)$. More precisely, if $S \in \mathcal{B}(X,Y)$ and $T \in \mathcal{O}(X,Y)$ satisfy $\|S-T\|_{X\to Y} < c(T^*)$, then $S \in \mathcal{O}(X,Y)$.*

PROOF. Suppose $\|S-T\|_{X\to Y} < c(T^*)$. Then

$$\|S^* - T^*\|_{Y^*\to X^*} = \|S-T\|_{X\to Y},$$

and so

$$\|S^*y^*\|_{X^*} \geq \|T^*y^*\|_{X^*} - \left\|(S-T)^*\,y^*\right\|_{X^*} \geq \left(c(T^*) - \|S-T\|_{X\to Y}\right)\|y^*\|_{Y^*}.$$

Thus $c(S^*) \geq c(T^*) - \|S-T\|_{X\to Y} > 0$, which shows that $S : X \to Y$ is onto, hence $S \in \mathcal{O}(X,Y)$. □

Now let $T_{\mathcal{Z}} : \mathrm{Mult}(B_p) \to \ell^\infty$ denote the restriction map $f \to \{f(z_j)\}_{j=1}^\infty$, and suppose that $T_{\mathcal{Z}}$ is bounded and onto, i.e. $T_{\mathcal{Z}} \in \mathcal{O}(\mathrm{Mult}(B_p), \ell^\infty)$. First choose $0 < \varepsilon < c(T_{\mathcal{Z}}^*)$, where $c(T_{\mathcal{Z}}^*)$ is the positive constant in the proof of Lemma 15.9 above. Then choose parameters λ and θ in the construction of the Bergman tree $\mathcal{T}_n$, so small that the set of centers $\{c_\alpha\}_{\alpha\in\mathcal{T}}$ is ε-dense in the ball $\mathbb{B}_n$. Then for every $z_j \in \mathcal{Z}$ we can choose $\alpha_j \in \mathcal{T}_n$ such that $\beta\left(z_j, c_{\alpha_j}\right) < \varepsilon$. Taking $\varepsilon > 0$ even smaller if necessary, it follows from the separation of $\mathcal{Z}$ that the cubes α_j so chosen are distinct. Now take $\mathcal{W} = \{w_j\}_{j=1}^\infty$ where $w_j = c_{\alpha_j}$. Then for $f \in \mathrm{Mult}(B_p)$, using (14.27) we have

$$\begin{aligned}\|(T_{\mathcal{Z}} - T_{\mathcal{W}})(f)\|_{\ell^\infty} &= \sup_j |f(z_j) - f(w_j)| \\ &\leq \sup_j \|f\|_{B_p}\,\beta(z_j, w_j)^{1/p'} \\ &\leq \varepsilon^{1/p'}\,\|f\|_{B_p} \leq \varepsilon^{1/p'}\,\|f\|_{\mathrm{Mult}(B_p)},\end{aligned}$$

which implies $\|T_{\mathcal{Z}} - T_{\mathcal{W}}\|_{\mathrm{Mult}(B_p)\to\ell^\infty} \leq \varepsilon^{1/p'} < c(T_{\mathcal{Z}}^*)$. Now Lemma 15.9 above applies to show that $T_{\mathcal{W}}$ is onto, i.e. the map $f \to \left\{f\left(c_{\alpha_j}\right)\right\}_{j=1}^\infty$ takes $\mathrm{Mult}(B_p)$ boundedly into and onto ℓ^∞.

At this point we would like to appeal to the special case of Theorem 15.2 that we are assuming holds, i.e. the case in which the set $\mathcal{Z}$ in the theorem is a subset of a Bergman tree $\mathcal{T}_n := \mathcal{T}_n(\lambda,\theta)$ having parameters $\lambda > 0$ sufficiently small and $\theta > 0$ sufficiently large. But in the first step of our construction of $\mathcal{T}_n(\lambda,\theta)$ above we assumed that $\lambda > 0$ and $\theta > 0$ were *both* sufficiently small. In our second step, we circumvent this difficulty surrounding the parameter θ, by splitting the Bergman tree $\mathcal{T}_n(\lambda,\theta)$ into finitely many subtrees $\mathcal{T}_n^1, \mathcal{T}_n^2, \ldots, \mathcal{T}_n^M$ as follows. Recall that the cubes K_j^N in $\mathcal{T}_n$ were defined by

$$K_j^N := \left\{z \in \mathbb{B}_n : N\theta \leq \beta(0,z) < (N+1)\theta,\ P_{N\theta}z \in Q_j^N\right\}, \qquad N \geq 1, j \geq 1,$$

and that their centers c_j^N were defined by

$$c_j^N := P_{\left(N+\frac{1}{2}\right)\theta} z_j^N.$$

Now we define the subtree $\mathcal{T}_n^k$ to consist of those cubes $\delta_M\left(K_j^N\right)$ where $K_j^N \in \mathcal{T}_n$ such that $N \in M\mathbb{N} + k = \{Mi + k\}_{i=0}^{\infty}$, and where

$$\delta_M\left(K_j^N\right) := \left\{z \in \mathbb{B}_n : \left(N - \frac{M}{2}\right)\theta \leq \beta(0, z) < \left(N + \frac{M}{2} + 1\right)\theta,\ P_{N\theta}z \in Q_j^N\right\}$$

is the Bergman cube K_j^N expanded in the radial direction so that the collection of such sets $\left\{\delta_M\left(K_j^N\right)\right\}_{N \in M\mathbb{N}+k,\ j \geq 1}$ is a pairwise disjoint decomposition of the ball $\mathbb{B}_n$ (provided the cube K^0 is redefined appropriately). Then without loss of generality, we *may view* $\mathcal{T}_n^k$ as a 'Bergman tree' with parameters λ' and θ' where $\lambda' = \lambda$ and $\theta' = M\theta$. Indeed, the subtree $\mathcal{T}_n^M$ is *precisely* the Bergman tree $\mathcal{T}_n(\lambda, M\theta)$ with parameters λ and $M\theta$, while the remaining subtrees $\mathcal{T}_n^k$ simply have the centers of $\mathcal{T}_n^M = \mathcal{T}_n(\lambda, M\theta)$ shifted by $k\theta$ in the radial direction, and the cube K^0 centered at the origin modified appropriately. What we then mean by saying that we *may view* these subtrees $\mathcal{T}_n^k$ as 'Bergman trees', is simply that all of the arguments in this chapter that involve Bergman trees, turn out to hold just as well for these subtrees $\mathcal{T}_n^k$, $1 \leq k \leq M$, *provided* we take both $\lambda > 0$ and $\theta > 0$ sufficiently small and $M > 1$ sufficiently large - something that the reader can easily verify when reading through the chapter.

So now we split our sequence $\mathcal{W} \subset \mathcal{T}_n$ into finitely many sequences $\mathcal{W}^k := \mathcal{W} \cap \mathcal{T}_n^k$ for $1 \leq k \leq M$. Now each $\mathcal{W}^k$ inherits multiplier space interpolation (15.7) from $\mathcal{W}$. Since $\mathcal{W}^k \subset \mathcal{T}_n^k$, and we may view each $\mathcal{T}_n^k$ as a Bergman tree with parameters λ and $M\theta$, we can apply our assumption that Theorem 15.2 holds for such $\mathcal{W}^k$, and deduce that each sequence $\mathcal{W}^k$ satisfies the separation and Carleson embedding condition (15.5). But since $\mathcal{W}$ itself is already separated, it now follows, from adding the measures of Carleson tents, that $\mathcal{W}$ itself satisfies the separation and Carleson embedding condition (15.5). Finally, from the construction of $\mathcal{W} = \{w_j\}_{j=1}^{\infty} = \{c_{\alpha_j}\}_{j=1}^{\infty}$ we have $\beta(z_j, w_j) < \varepsilon$ for each j, and it follows easily that the sequence $\mathcal{Z} = \{z_j\}_{j=1}^{\infty}$ also satisfies the separation and Carleson embedding condition (15.5), with possibly a larger Carleson embedding constant. This finally completes the proof of Claim 15.7. □

Having gotten this far, namely accomplishing a reduction to the special case where $\mathcal{Z} = \{z_j\}_{j=1}^{\infty}$ is a subset of a Bergman tree $\mathcal{T}_n = \mathcal{T}_n(\lambda, \theta)$ with λ small and θ large, one might hope that no further reductions are needed. On the contrary, an even more delicate reduction is needed in the next chapter when considering the holomorphic Besov spaces $HB_{p,m}(\mathcal{T}_n)$ on the Bergman tree $\mathcal{T}_n$, the tool that permits establishing the necessity of separation and Carleson embedding for multiplier interpolation on the ball. The problem here is that we need a certain structural inequality to hold for the complex structure we impose there on $HB_{p,m}(\mathcal{T}_n)$, and this structural inequality requires a further perturbation of the centers of the Bergman trees already constructed above. We refer to these perturbed trees as modified Bergman trees, and defer their construction until they are needed in the next chapter.

We will show that for $1 < p < \infty$ the analog of condition (15.5) on the Bergman tree $\mathcal{T}_n$, condition (15.9) below, a Carleson measure condition and a separation condition, is necessary and sufficient for ℓ^{∞} interpolation on $\mathcal{Z}$ by the multiplier spaces $\mathrm{Mult}(B_p)$.

There are various ways to rewrite the Carleson measure conditions we are considering. First, note that (15.5) and (15.9) are equivalent by Theorem 14.16. Also, recall from Lemma 14.9 that

$$\|k_w^{\alpha,p}\|_{B_{p'}}^{p'} \approx 1 + \log \frac{1}{1-|w|^2} \approx (1+\beta(w,0)),$$

Recalling that d denotes the edge counting distance in the Bergman tree $\mathcal{T}_n$, (14.38) yields

$$d(\alpha_i, o) \approx \beta(c_{\alpha_i}, 0) \approx \log \frac{1}{1-|c_{\alpha_i}|^2}. \tag{15.13}$$

We will sometimes pass between these various conditions without comment.

Also note that the separation condition $\beta(z_i, 0) \lesssim \beta(z_i, z_j)$ on the ball implies the tree separation condition $d(\alpha_i, o) \lesssim d(\alpha_i, \alpha_j)$, but not conversely. The converse will fail for pairs of leaves that are far from the root and whose placements in the disk are very close, but whose branches have split from each other near the root. For instance, in the "tidy case" described in Section 5.2.1 this will happen for pairs of neighboring small cubes on the opposite sides of an edge determined by a (relatively) very large cube. This phenomena, which limits the ability of discrete structures such as dyadic trees to model the continuum effectively, is sometimes called an *edge effect*. Sometimes those limitations can be overcome by combining results from several, or even a continuum, of discrete models. The proof of Proposition 13.7 is a simple example.

15.3. Proof of Multiplier Space Necessity

We begin with the straightforward necessity implications; (15.7) implies (15.8), (15.3) implies (15.4), and (15.4) implies (15.5). For the most part, we follow Böe [**77**], who in turn generalized the Hilbert space arguments in Marshall and Sundberg [**220**]. Many of the arguments in this section are analogous to arguments in Chapter 7.

First, we have that condition (15.8) follows from (15.7) since if we choose $\varphi \in \operatorname{Mult}(B_p)$ so that $b_j = \overline{\varphi(z_j)} a_j$, then $\mathcal{M}_\varphi^*\left(k_{z_j}^{\alpha,p}\right) = \overline{\varphi(z_j)} k_{z_j}^{\alpha,p}$ and so

$$\begin{aligned}
\left\|\sum_{j=1}^\infty b_j k_{z_j}^{\alpha,p}\right\|_{B_{p'}} &= \left\|\mathcal{M}_\varphi^*\left(\sum_{j=1}^\infty a_j k_{z_j}^{\alpha,p}\right)\right\|_{B_{p'}} \\
&\le \|\mathcal{M}_\varphi\| \left\|\sum_{j=1}^\infty a_j k_{z_j}^{\alpha,p}\right\|_{B_{p'}} \\
&\lesssim \sup_j |\varphi(z_j)| \left\|\sum_{j=1}^\infty a_j k_{z_j}^{\alpha,p}\right\|_{B_{p'}}.
\end{aligned}$$

Next we prove the equivalence of (15.3) and (15.4). First, if the map $Tf = \{\kappa_j f(z_j)\}_{j=1}^{\infty}$ in (15.3) maps B_p *into* ℓ^p, then we have

$$\begin{aligned}\left\|\sum_{j=1}^{\infty} a_j \kappa_j k_{z_j}^{\alpha,p}\right\|_{B_{p'}} &= \sup_{\|f\|_{B_p}=1} \left|\left\langle f, \sum_{j=1}^{\infty} a_j \kappa_j k_{z_j}^{\alpha,p}\right\rangle_{\alpha,p}\right| \\ &= \sup_{\|f\|_{B_p}=1} \left|\sum_{j=1}^{\infty} \kappa_j f(z_j)\overline{a_j}\right| \\ &\leq \sup_{\|f\|_{B_p}=1} \left(\sum_{j=1}^{\infty} |\kappa_j f(z_j)|^p\right)^{1/p} \left\|\{a_j\}_{j=1}^{\infty}\right\|_{\ell^{p'}} \\ &\lesssim \left\|\{a_j\}_{j=1}^{\infty}\right\|_{\ell^{p'}}.\end{aligned}$$

If the map T is also *onto*, then its adjoint T^*, given by

$$T^*\left(\{a_j\}_{j=1}^{\infty}\right) = \sum_{j=1}^{\infty} \overline{a_j}\kappa_j k_{z_j}^{\alpha,p},$$

satisfies

$$\left\|T^*\left(\{a_j\}_{j=1}^{\infty}\right)\right\|_{B_{p'}} \gtrsim \left\|\{a_j\}_{j=1}^{\infty}\right\|_{\ell^{p'}},$$

which is the opposite inequality in (15.4), and completes the proof that (15.3) implies (15.4). Conversely, if the inequality $\lesssim$ in (15.4) holds, then

$$\begin{aligned}\left(\sum_{j=1}^{\infty} |\kappa_j f(z_j)|^p\right)^{1/p} &= \sup_{\left\|\{a_j\}_{j=1}^{\infty}\right\|_{\ell^{p'}}=1} \left|\sum_{j=1}^{\infty} \kappa_j f(z_j)\overline{a_j}\right| \\ &= \sup_{\left\|\{a_j\}_{j=1}^{\infty}\right\|_{\ell^{p'}}=1} \left|\sum_{j=1}^{\infty} \left\langle f, a_j\kappa_j k_{z_j}^{\alpha,p}\right\rangle_{\alpha,p}\right| \qquad (15.14) \\ &\leq \sup_{\left\|\{a_j\}_{j=1}^{\infty}\right\|_{\ell^{p'}}=1} \|f\|_{B_p} \left\|\sum_{j=1}^{\infty} a_j\kappa_j k_{z_j}^{\alpha,p}\right\|_{B_{p'}} \lesssim \|f\|_{B_p},\end{aligned}$$

and thus the map T in (15.3) is a bounded map *into*. If the reverse inequality $\gtrsim$ in (15.4) also holds, then

$$\left\|T^*\left(\{a_j\}_{j=1}^{\infty}\right)\right\|_{B_{p'}} = \left\|\sum_{j=1}^{\infty} \overline{a_j}\kappa_j k_{z_j}^{\alpha,p}\right\|_{B_{p'}} \gtrsim \left\|\{a_j\}_{j=1}^{\infty}\right\|_{\ell^{p'}},$$

which shows that T is also *onto*.

Note that, in particular, we have just shown that the inequality $\lesssim$ in (15.4) implies that the map T in (15.3) is bounded. This will be used below.

The implication (15.4) implies (15.5) will now follow if we show that (15.3) implies (15.5). The statement in (15.5) that the measure is a Carleson measure is, by definition, the statement that the map T in (15.3) is bounded, and thus *into*. To obtain the separation condition, fix i and use that T is *onto* to obtain $f \in B_p$

satisfying $f(z_i)=1$ and $f(z_j)=0$ for $i\neq j$. It now follows from the open mapping theorem, the Hölder estimate (14.27), and (15.1) that

$$\begin{aligned}\|f\|_{B_p} &\lesssim \|Tf\|_{\ell^p} \approx \kappa_i |f(z_i)| \\ &\approx \kappa_i |f(z_i)-f(z_j)| \lesssim \|f\|_{B_p} \frac{\beta(z_i,z_j)^{1/p'}}{\beta(z_i,0)^{1/p'}},\end{aligned}$$

for all $i\neq j$.

15.3.1. The Necessity of Separation and Carleson Measure. Now we turn to proving the more difficult necessity implication (15.8) implies (15.5). First we dispose of the straightforward part – namely that the separation condition in (15.5) follows from (15.8). Indeed, by (14.24), (15.8) and (14.27) we have

$$\begin{aligned}(1+\beta(0,z_i))^{1/p'} &\approx \left\|k_{z_i}^{\alpha,p}\right\|_{B_{p'}} \lesssim \left\|k_{z_i}^{\alpha,p}-k_{z_j}^{\alpha,p}\right\|_{B_{p'}} \\ &\approx \sup_{\|f\|_{B_p}=1} \left|\left\langle f, k_{z_i}^{\alpha,p}-k_{z_j}^{\alpha,p}\right\rangle_{\alpha,p}\right| \\ &\approx \sup_{\|f\|_{B_p}=1} |f(z_i)-f(z_j)| \\ &\lesssim \beta(z_i,z_j)^{1/p'}.\end{aligned}$$

It remains to prove that the Carleson embedding follows from (15.8). For this, we show that (15.8) implies (15.4) for both $1<p<1+1/(n-1)$ and $p=2$, and also that (15.8) implies the inequality $\lesssim$ of (15.4) for $p>2$. Once that is done the argument in the last two paragraphs of the previous section allow us to conclude that we have the Carleson embedding.

15.3.2. The Case $1<p<1+1/(n-1)$. Here we prove the implication (15.8) implies (15.4) for the special case $1<p<1+1/(n-1)$. Given $1<p<1+1/(n-1)$, we make the choice $-1<\alpha<\infty$ to satisfy

$$p=\frac{n+1+\alpha}{n+\alpha}, \tag{15.15}$$

which accounts for our restriction $1<p<1+1/(n-1)$. Note that $p'=n+1+\alpha$, so that

$$\begin{aligned}\mathbf{s} &= n+\alpha, \\ \mathbf{s}' &= 1.\end{aligned}$$

Thus in this case we have $\mathcal{R}_{\mathbf{s}}^{\alpha}=\mathcal{R}_{n+\alpha}^{\alpha}$ and $\mathcal{R}_{\mathbf{s}'}^{\alpha}=\mathcal{R}_1^{\alpha}$ where α is as in (15.15), so that

$$\begin{aligned}\langle f,g\rangle_{\alpha,p} &= \left\langle \mathcal{R}_{n+\alpha}^{\alpha}f, \mathcal{R}_1^{\alpha}g\right\rangle_{A_\alpha^2} \\ &= \int_{\mathbb{B}_n} (1-|z|^2)^{n+\alpha}\mathcal{R}_{n+\alpha}^{\alpha}f(z)\,\overline{(1-|z|^2)\mathcal{R}_1^{\alpha}g(z)}\,d\lambda_n(z).\end{aligned}$$

Let $\mathcal{Z}$ be a sequence in the ball $\mathbb{B}_n$. We will need the following two results. The first result is Böe's curious lemma.

LEMMA 15.10 ([**77**, Lemma 3.1]). *If* $\{f_j\}_{j=1}^{\infty}$ *is an unconditional basic sequence of positive functions in* $L^q(\mu)$, $1<q<\infty$, *then*

$$\left\|\sum_{j=1}^{\infty}|a_jf_j|\right\|_{L^q(\mu)}\approx\left\|\sup_{j\geq 1}|a_jf_j|\right\|_{L^q(\mu)}\approx\left(\sum_{j=1}^{\infty}|a_j|^q\,\|f_j\|_{L^q(\mu)}^q\right)^{1/q}.$$

PROOF. For convenience we sketch Böe's proof. Since the f_n are positive and unconditional in $L^q(\mu)$, we have, letting $\{r_j(t)\}_{j=1}^{\infty}$ denote the Rademacher functions,

$$\left\|\sum_{j=1}^{\infty}|a_jf_j|\right\|_{L^q(\mu)}=\left\|\sum_{j=1}^{\infty}|a_j|\,f_j\right\|_{L^q(\mu)}\lesssim\left\|\sum_{j=1}^{\infty}r_j(t)\,a_jf_j\right\|_{L^q(\mu)}$$

for all $t\in[0,1]$ (since $\{f_j\}$ is an unconditional basic sequence and $|a_j|=|r_j(t)\,a_j|$). Now average the qth power of this inequality over $t\in[0,1]$, and use Khintchine's inequality (see e.g. [**317**, Chapter 5]) to obtain

$$\begin{aligned}\left\|\sum_{j=1}^{\infty}|a_jf_j|\right\|_{L^q(\mu)}^q&\lesssim\int_0^1\left\|\sum_{j=1}^{\infty}r_j(t)\,a_jf_j\right\|_{L^q(\mu)}^q dt\\&=\int\int_0^1\left|\sum_{j=1}^{\infty}r_j(t)\,a_jf_j\right|^q dtd\mu\\&\lesssim\int\left\|\{a_jf_j\}_{j=1}^{\infty}\right\|_{\ell^2}^q d\mu.\end{aligned}$$

Since $\left\|\{a_jf_j\}_{j=1}^{\infty}\right\|_{\ell^2}\leq\left\|\{a_jf_j\}_{j=1}^{\infty}\right\|_{\ell^1}^{1/2}\left\|\{a_jf_j\}_{j=1}^{\infty}\right\|_{\ell^\infty}^{1/2}$, we have by the Cauchy-Schwartz inequality,

$$\left\|\sum_{j=1}^{\infty}|a_jf_j|\right\|_{L^q(\mu)}^q\lesssim\left(\int\left\|\{a_jf_j\}_{j=1}^{\infty}\right\|_{\ell^1}^q d\mu\right)^{q/2}\left(\int\left\|\{a_jf_j\}_{j=1}^{\infty}\right\|_{\ell^\infty}^q d\mu\right)^{q/2},$$

which, after dividing through by the first factor on the right, yields the inequality

$$\left\|\sum_{j=1}^{\infty}|a_jf_j|\right\|_{L^q(\mu)}\lesssim\left\|\sup_{j\geq 1}|a_jf_j|\right\|_{L^q(\mu)}.$$

Thus the expressions

$$\left\|\left(\sum_{j=1}^{\infty}|a_jf_j|^r\right)^{\frac{1}{r}}\right\|_{L^q(\mu)}$$

are all comparable for $1\leq r\leq\infty$, and the choice $r=q$ yields the final equivalence in the lemma. □

LEMMA 15.11. *For* $-1<\alpha<\infty$, $1<q<\infty$ *and* $F\in\mathrm{Hol}(\mathbb{B}_n)$ *with* $\mathrm{Im}\,F(0)=0$,

$$\left(\int_{\mathbb{B}_n}|F(z)|^q\,d\nu_\alpha(z)\right)^{1/q}\approx\left(\int_{\mathbb{B}_n}|\mathrm{Re}\,F(z)|^q\,d\nu_\alpha(z)\right)^{1/q}. \tag{15.16}$$

PROOF OF THE LEMMA. The Korànyi-Vagi theorem [**288**, Theorem 6.3.1] shows the equivalence of the left and right hand sides in (15.16) when the measure $d\nu_\alpha(z)$ on the ball $\mathbb{B}_n$ is replaced by surface measure $d\sigma(z)$ on the sphere $\partial\mathbb{B}_n$ (and F is say a polynomial). This immediately yields (15.16) by an integration in polar coordinates. □

Since $p' = n + 1 + \alpha$, we have from (14.23) that

$$\mathcal{R}_1^\alpha k_w^{\alpha,p}(z) = \frac{1}{1 - \overline{w} \cdot z}. \tag{15.17}$$

We will now use the assumption that $\{k_w^{\alpha,p}\}$ is an unconditional basic sequence (15.8). It is a straightforward exercise (which we omit) using the previous Lemma that (15.8) is equivalent to the statement that $\left\{\operatorname{Re} \mathcal{R}_1^\alpha k_{z_j}^{\alpha,p}\right\}$ is an unconditional basic sequence which is equivalent to the fact that the unconditional basic sequence condition holds when the coefficients $\{a_j\}$ and $\{b_j\}$ are taken to be real.

We now compute that

$$\begin{aligned}
\left\| \sum_{j=1}^{\infty} a_j \kappa_j k_{z_j}^{\alpha,p} \right\|_{B_{p'}} &= \left\| (1 - |z|^2) \mathcal{R}_1^\alpha \left(\sum_{j=1}^{\infty} a_j \kappa_j k_{z_j}^{\alpha,p} \right) \right\|_{L^{p'}(\lambda_n)} \\
&= \left\| (1 - |z|^2) \left(\sum_{j=1}^{\infty} a_j \kappa_j \mathcal{R}_1^\alpha k_{z_j}^{\alpha,p} \right) \right\|_{L^{p'}(\lambda_n)} \\
&= \left\| \sum_{j=1}^{\infty} a_j \kappa_j \frac{1}{1 - \overline{z_j} \cdot z} \right\|_{L^{p'}(\nu_\alpha)}
\end{aligned}$$

Taking note of (15.17) we set

$$f_j = \kappa_j \operatorname{Re} \mathcal{R}_1^\alpha k_{z_j}^{\alpha,p}(z) = \kappa_j \operatorname{Re} \frac{1}{1 - \overline{z_j} \cdot z} > 0.$$

Using the two previous lemmas we continue with

$$\begin{aligned}
\left\| \sum_{j=1}^{\infty} a_j \kappa_j \frac{1}{1 - \overline{z_j} \cdot z} \right\|_{L^{p'}(\nu_\alpha)} &\approx \left\| \sum_{j=1}^{\infty} a_j \kappa_j \operatorname{Re} \frac{1}{1 - \overline{z_j} \cdot z} \right\|_{L^{p'}(\nu_\alpha)} \\
&\approx \left(\sum_{j=1}^{\infty} |a_j|^{p'} \|f_j\|_{L^{p'}(\nu_\alpha)}^{p'} \right)^{1/p'} \\
&\approx \left(\sum_{j=1}^{\infty} |a_j|^{p'} \right)^{1/p'},
\end{aligned}$$

since

$$\begin{aligned}\|f_j\|_{L^{p'}(\nu_\alpha)} &= \kappa_j \left\| \operatorname{Re} \frac{1}{1-\overline{z_j}\cdot z} \right\|_{L^{p'}(\nu_\alpha)} \\ &\approx \kappa_j \left\| \frac{1}{1-\overline{z_j}\cdot z} \right\|_{L^{p'}(\nu_\alpha)} \\ &= \kappa_j \left\| (1-|z|^2)\mathcal{R}_1 k_{z_j}^{\alpha,p} \right\|_{L^{p'}(\lambda_n)} \\ &= \kappa_j \kappa_j^{-1} = 1\end{aligned}$$

upon using the previous lemma again. This completes the proof of condition (15.4) in the case $1 < p < 1 + 1/(n-1)$.

15.3.3. The Case $p \geq 2$. (Note that this range of p, together with the range from the previous section, do not fill the range of interest, $1 < p < \infty$. This is discussed in Section 15.1.1.)

Here we show that (15.8) implies the inequality $\lesssim$ in (15.4) for $p > 2$, and also that (15.8) implies (15.4) for $p = 2$. First we claim that the unconditional basis condition (15.8) and Khinchine's inequality yield the inequality

$$\left\| \sum_{j=1}^{\infty} a_j \kappa_j k_{z_j}^{\alpha,p} \right\|_{B_{p'}} \lesssim \left(\sum_{j=1}^{\infty} |a_j|^{p'} \right)^{1/p'}$$

for $p \geq 2$, and with equality in the case $p = 2$. To see this, we compute using first (15.8) and then Khinchine's inequality, that for any $m > n/p'$, we have

$$\begin{aligned}\left\| \sum_{j=1}^{\infty} a_j \kappa_j k_{z_j}^{\alpha,p} \right\|_{B_{p'}}^{p'} &\approx \int_0^1 \left\| \sum_{j=1}^{\infty} a_j \kappa_j k_{z_j}^{\alpha,p} r_j(t) \right\|_{B_{p'}}^{p'} dt \\ &= \int_0^1 \int_{\mathbb{B}_n} \left| \sum_{j=1}^{\infty} a_j \kappa_j (1-|z|^2)^m \mathcal{R}_m^\alpha k_{z_j}^{\alpha,p}(z)\, r_j(t) \right|^{p'} d\lambda_n(z)\, dt \\ &\approx \int_{\mathbb{B}_n} \left(\sum_{j=1}^{\infty} \left| a_j \kappa_j (1-|z|^2)^m \mathcal{R}_m^\alpha k_{z_j}^{\alpha,p}(z)\, r_j(t) \right|^2 \right)^{p'/2} d\lambda_n(z).\end{aligned}$$

Since $p'/2 \leq 1$, we continue with

$$\begin{aligned}\left\| \sum_{j=1}^{\infty} a_j \kappa_j k_{z_j}^{\alpha,p} \right\|_{B_{p'}}^{p'} &\lesssim \int_{\mathbb{B}_n} \sum_{j=1}^{\infty} \left| a_j \kappa_j (1-|z|^2)^m \mathcal{R}_m^\alpha k_{z_j}^{\alpha,p}(z)\, r_j(t) \right|^{p'} d\lambda_n(z) \\ &= \sum_{j=1}^{\infty} |a_j|^{p'} \left\| k_{z_j}^{\alpha,p} \right\|_{B_{p'}}^{-p'} \int_{\mathbb{B}_n} \left| (1-|z|^2)^m \mathcal{R}_m^\alpha k_{z_j}^{\alpha,p}(z) \right|^{p'} d\lambda_n(z) \\ &= \sum_{j=1}^{\infty} |a_j|^{p'},\end{aligned}$$

which is the inequality $\lesssim$ in (15.4). In the case $p = 2$ we have equality, and so then (15.4).

15.4. Proof of Multiplier Space Sufficiency

Here we prove that (15.5) implies (15.7) for $1 < p < \infty$. We generalize the main ideas in Böe's one-dimensional proof to the unit ball $\mathbb{B}_n$.

For $z \in \mathbb{B}_n$ and $\beta < 1$, define the region V_z^β by

$$V_z^\beta := \left\{ w \in \mathbb{B}_n : |1 - \overline{w} \cdot Pz| \leq (1 - |z|)^\beta \right\},$$

where Pz denotes the radial projection of z onto the unit sphere $\partial\mathbb{B}_n$. The intersection of V_z^β with the complex line $\mathbb{C}z$ through z and the origin is

$$\left\{ w \in \mathbb{B}_n \cap \mathbb{C}z : |w - Pz| \leq (1 - |z|)^\beta \right\},$$

and the intersection of V_z^β with the sphere $\partial\mathbb{B}_n$ is an "ellipse" with radius $(1 - |z|)^\beta$ in the radial tangential direction, and radius $(1 - |z|)^{\beta/2}$ in the complex tangential directions. Using arguments in Marshall and Sundberg [**220**], the separation condition in (15.5) implies the following geometric separation conditions.

LEMMA 15.12. *Suppose the separation condition in* (15.5) *holds. Then there are constants* $0 < \beta < 1 < \beta\eta < \eta$ *such that if* $V_{z_i}^\beta \cap V_{z_j}^\beta \neq \phi$ *and* $|z_j| \geq |z_i|$, *then* $z_i \notin V_{z_j}^\beta$ *and*

$$(1 - |z_j|) \leq (1 - |z_i|)^\eta. \tag{15.18}$$

PROOF OF LEMMA. Fix z_i, z_j with $|z_j| \geq |z_i|$, suppose $w \in V_{z_i}^\beta \cap V_{z_j}^\beta$ and set $\omega = |z_j| w$ so that

$$1 - |\omega|^2 = 1 - |z_j|^2 |w|^2 \geq 1 - |z_j|^2.$$

Then the separation condition $\beta(z_i, z_j) \geq c\beta(z_j, 0)$ (We apologize, β is both a positive constant and the Bergman distance.) yields

$$c\beta(0, \omega) \leq c\beta(0, z_j) \leq \beta(z_i, z_j) \leq \beta(z_i, \omega) + \beta(\omega, z_j),$$

and so

$$c\beta(0, \omega) \leq 2\beta(z_k, \omega),$$

where k is either i or j. Now the identity

$$\rho(z, w) = |\varphi_z(w)| = \tanh \beta(z, w),$$

for the pseudohyperbolic metric ρ yields for this k,

$$\begin{aligned}
1 - \rho(\omega, z_k)^2 &= 1 - \tanh^2 \beta(\omega, z_k) \\
&\leq 1 - \tanh^2 (c\beta(\omega, 0)) \\
&= \frac{4}{e^{2c\beta(\omega,0)} + 2 + e^{-2c\beta(\omega,0)}} \\
&\leq \frac{4}{e^{2c\beta(\omega,0)}} = 4\frac{(1 - |\omega|)^c}{(1 + |\omega|)^c} \\
&\leq 4(1 - |\omega|)^c \leq (1 - |\omega|^2)^\lambda,
\end{aligned}$$

for some $0 < \lambda < c$ provided z_i is large enough, which we may assume by discarding finitely many of the points z_i.

Then the final identity in (14.3) gives

$$\frac{(1-|\omega|^2)(1-|z_k|^2)}{|1-\overline{\omega}\cdot z_k|^2}=1-\rho(\omega,z_k)^2\leq(1-|\omega|^2)^\lambda.$$

Now we use that

$$1-\overline{\omega}\cdot z_k=1-|z_k|+|z_k|\left(1-\overline{w}\cdot Pz_k+\overline{w-\omega}\cdot Pz_k\right)$$

implies

$$\begin{aligned}|1-\overline{\omega}\cdot z_k|&\leq(1-|z_k|)+(1-|z|^2)^\beta+(1-|z_j|)\\&\leq 3(1-|z|^2)^\beta,\end{aligned}$$

to obtain

$$\begin{aligned}(1-|\omega|^2)(1-|z|^2)&\leq|1-\overline{\omega}\cdot z_k|^2(1-|\omega|^2)^\lambda\\&\lesssim(1-|z|^2)^{2\beta}(1-|\omega|^2)^\lambda.\end{aligned}$$

Hence with $1<\eta<(2\beta-1)/(1-\lambda)$ and $|z_i|$ sufficiently large, we have

$$\begin{aligned}1-|z_j|^2\leq 1-|\omega|^2&\lesssim(1-|z|^2)^{(2\beta-1)/(1-\lambda)}\\&\leq(1-|z|^2)^\eta\leq(1-|z|^2)^\eta,\end{aligned}$$

which is (15.18). In particular, if $z_i\in V_{z_j}^\beta$, then

$$1-|z_i|\leq|1-\overline{z_i}\cdot Pz_j|\leq(1-|z_j|^2)^\beta\leq\left(1-|z_i|^2\right)^{\beta\eta}\lesssim(1-|z_i|)^{\beta\eta}$$

yields a contradiction for $|z_i|$ sufficiently large if $\beta\eta>1$. This completes the proof of Lemma 15.12. □

We now fix constants β and η as in Lemma 15.12, and write $V_z=V_z^\beta$.

Lemma 15.16 below is the key construction in the sufficiency proof and is motivated by the formula contained in [**346**, Equation (1.35)]

$$R^{s-n,n}\left(\frac{1}{(1-\overline{w}\cdot z)^{1+s}}\right)=\frac{1}{(1-\overline{w}\cdot z)^{n+1+s}},$$

valid for s not a negative integer. The point is that if we define

$$\Gamma_s^n g(z):=\int_{\mathbb{B}_n}\frac{g(w)(1-|w|^2)^s}{(1-\overline{w}\cdot z)^{1+s}}dV\tag{15.19}$$

for a given (not necessarily holomorphic) function g, then with $\varphi(z)=\Gamma_s^n g(z)$,

$$R^{s-n,n}\varphi(z)=R^{s-n,n}\Gamma_s^n g(z)=\int_{\mathbb{B}_n}\frac{g(w)(1-|w|^2)^s}{(1-\overline{w}\cdot z)^{n+1+s}}dV,\tag{15.20}$$

and by the reproducing formula (14.15), valid for $\operatorname{Re}s>-1$, we also have that

$$R^{s-n,n}\varphi(z)=c_{n,s}\int_{\mathbb{B}_n}\frac{R^{s-n,n}\varphi(w)(1-|w|^2)^s}{(1-\overline{w}\cdot z)^{n+1+s}}dV.$$

Thus $R^{s-n,n}\varphi(w)$ behaves essentially like $g(w)$, and this provides flexibility in choosing g so that φ has desirable algebraic multiplier properties on the one hand, while controlling the multiplier norm of φ on the other hand. Now we use the fact that by Theorem 14.14, the multiplier norm is equivalent to the Carleson norm of $\left|(1-|z|^2)^n \nabla^n \varphi(z)\right|^p d\lambda_n(z)$.

REMARK 15.13. *The same operator* Γ *was used in Section* 13.2.2 *for roughly the same purpose; moving from from a precisely controlled function which was not holomorphic to a holomorphic function with analogous properties. At an informal level the use of this operator is both related to, and an alternative to, obtaining a controlled solution to a* $\overline{\partial}$ *equation.*

15.4.1. Transformation of Carleson Measures. At times it is useful to know that a controlled spreading process takes Carleson measures to Carleson measures. We saw a single variable result of that type in Section 10.2.1. A similar result is given in [**77**, Lemma 2.4]. Finally, we note that the inclusion of $1<p<\infty$ in the lemma below is what permits repairing the gap left by the interpolation theorem in [**47**].

LEMMA 15.14 (Spreading Lemma). *Suppose that*

$$\sup_{\zeta\in\mathbb{B}_n}\left|(1-|\zeta|^2)^n g(\zeta)\right| \le C_0, \tag{15.21}$$

and that the measure

$$d\mu(z) = \left|(1-|z|^2)^n g(z)\right|^p d\lambda_n(z)$$

is a Carleson measure for B_p *with norm* C_1. *Then for* $1<p<\infty$ *and* $s>n+\frac{2n}{p}$, *both*

$$\left|(1-|z|^2)^n R^{s-n,n}\Gamma_s^n g(z)\right|^p d\lambda_n(z)$$

and

$$\left|(1-|z|^2)^n \nabla^n \Gamma_s^n g(z)\right|^p d\lambda_n(z)$$

are Carleson measures for B_p *with norms at most a constant multiple of* (C_0+C_1).

We prove Lemma 15.14 using the following extension due to Peloso [**258**] of the characterization of B_p given by [**346**, Theorem 6.28]:

$$\|f\|_{B_p}^p \approx \int_{\mathbb{B}_n}\int_{\mathbb{B}_n}\frac{\left|f(z)-\sum_{\ell=0}^{L}\frac{((z-w)\cdot\nabla)^\ell}{\ell!}f(w)\right|^p}{|1-\overline{w}\cdot z|^{2(n+1)}}dV\,dV, \tag{15.22}$$

provided $p>1$ and $L>2n/p$.

PROOF (OF LEMMA 15.14). We begin with the case $R^{s-n,n}\Gamma_s^n g$. Define $T_s g(z) := R^{s-n,n}\Gamma_s^n g(z)$, and $\widehat{T_s}g(z)$ by

$$\widehat{T_s}g(z) := c_{n,s}\int_{\mathbb{B}_n}\frac{g(w)(1-|w|^2)^s}{|1-\overline{w}\cdot z|^{n+1+s}}dV.$$

Then by (15.20) and [**346**, Theorem 2.10], we have that T_s, and in fact $\widehat{T_s}$, is bounded on $L^p(d\nu_\alpha)$, $\alpha>-1$, if and only if $0<\alpha+1<p(s+1)$. Thus with

$\alpha = np - (n+1)$, we obtain that

$$\begin{aligned}\|T_s(hg)\|_{L^p(\nu_\alpha)} \lesssim \|hg\|_{L^p(\nu_\alpha)} &= \int_{\mathbb{B}_n} |h(z)\, g(z)|^p \,(1-|z|^2)^\alpha dV \\ &= \int_{\mathbb{B}_n} |h(z)|^p \left|(1-|z|^2)^{(\alpha+n+1)/p} g(z)\right|^p d\lambda_n(z) \\ &= \int_{\mathbb{B}_n} |h(z)|^p \left|(1-|z|^2)^n g(z)\right|^p d\lambda_n(z) \lesssim \|h\|_{B_p}\end{aligned}$$

for all $h \in B_p$ provided that $0 < np - n - 1 < p(s+1)$, i.e. $p > 1$ and

$$s > n - \frac{n+1}{p} - 1 = \frac{n}{p'} - \frac{1}{p}.$$

We need the analogous extension of this inequality to the Taylor polynomial

$$P_L h(w,z) := \sum_{\ell=0}^{L} \frac{((z-w)\cdot\nabla)^\ell}{\ell!} h(w).$$

To see this we split the polynomial into its homogeneous terms and estimate each separately:

$$\begin{aligned}&\|T_s[(P_L h(\cdot,z))\, g](z)\|_{L^p(\nu_\alpha)} \\ &= \left\| \int_{\mathbb{B}_n} \left\{ \sum_{\ell=0}^{L} \frac{((z-w)\cdot\nabla)^\ell}{\ell!} h(w) \right\} g(w) \frac{(1-|w|^2)^s}{(1-\overline{w}\cdot z)^{n+1+s}} dV \right\|_{L^p(\nu_\alpha)} \\ &\le \sum_{\ell=0}^{L} \frac{1}{\ell!} \left\| \int_{\mathbb{B}_n} \left\{ ((z-w)\cdot\nabla)^\ell h(w) \right\} g(w) \frac{(1-|w|^2)^s}{(1-\overline{w}\cdot z)^{n+1+s}} dV \right\|_{L^p(\nu_\alpha)} .\end{aligned}$$

For $z, w \in \mathbb{B}_n$ and $m \in \mathbb{N}$, we introduce $\triangle(w,z)$, a symmetric function of z and w which has many equivalent descriptions.

$$\begin{aligned}\triangle(w,z) &= |1-w\overline{z}|^2 - (1-|w|^2)(1-|z|^2) \\ &= (1-|z|^2)\,|w-z|^2 + |\overline{z}(w-z)|^2 \\ &= (1-|w|^2)\,|w-z|^2 + |\overline{w}(w-z)|^2 \\ &= |1-w\overline{z}|^2\, |\varphi_w(z)|^2 \\ &= |1-w\overline{z}|^2\, |\varphi_z(w)|^2 \\ &= \left| P_w(z-w) + \sqrt{1-|w|^2}\, Q_w(z-w) \right|^2 \\ &= \left| P_z(z-w) + \sqrt{1-|z|^2}\, Q_z(z-w) \right|^2 . \\ &= \sigma(w,z)^4 \rho(w,z)^2 .\end{aligned} \tag{15.23}$$

In the last line σ and ρ are the anisotropic metrics on the sphere and ball respectively. They were introduced in Sections 14.1.3 and 14.1.2.

Letting D denote the almost invariant holomorphic derivative, and invoking (17.62) below, see also [**121**, page 526, inequality (5-7)], we have, for $z, w \in \mathbb{B}_n$ and

$m \in \mathbb{N}$,
(15.24)

$$\left| (z-w)^{\alpha} \frac{\partial^m}{\partial w^{\alpha}} F(w) \right| \lesssim \frac{\triangle(w,z)^{m/2}}{(1-|w|^2)^m} |D^m F(w)|, \qquad F \in \operatorname{Hol}(\mathbb{B}_n), m = |\alpha|.$$

We then note that h is holomorphic and hence (15.24) applies to h, so

$$\begin{aligned}
&\left| \int_{\mathbb{B}_n} \left\{ ((z-w)\cdot\nabla)^{\ell} h(w) \right\} g(w) \frac{(1-|w|^2)^s}{(1-\overline{w}\cdot z)^{n+1+s}} dV \right| \\
&\qquad \leq \int_{\mathbb{B}_n} |D^{\ell} h(w)| \, |g(w)| \frac{\triangle(w,z)^{\ell/2} (1-|w|^2)^{s-\ell}}{|1-\overline{w}\cdot z|^{n+1+s}} dV \\
&\qquad = \widehat{T}_{0,s-\ell,\ell} \left(g D^{\ell} h \right)(z),
\end{aligned}$$

where

$$\widehat{T}_{a,b,c} f(z) = \int_{\mathbb{B}_n} \frac{(1-|z|^2)^a (1-|w|^2)^b \triangle(w,z)^{c/2}}{|1-w\overline{z}|^{n+1+a+b+c}} f(w) \, dV.$$

The following lemma gives a boundedness result for $\widehat{T}_{a,b,c}$ (whose proof is provided as Lemma B.10 in Appendix B. The case $c=0$ is [**346**, Theorem 2.10]) and will let us control this last term above.

LEMMA 15.15. *Let $a,b,c,t \in \mathbb{R}$. Then the operator $\widehat{T}_{a,b,c}$, is bounded on $L^p\left(\mathbb{B}_n; (1-|w|^2)^t dV(w)\right)$ if and only if $c > -2n$ and*

$$-pa < t+1 < p(b+1). \tag{15.25}$$

If

$$0 < np - (n+1) + 1 < p(s-\ell+1),$$

and in particular, if $p > 1$ and $s > n + \ell - 1$ then, setting

$$d\mu(w) = \left| (1-|w|^2)^n g(w) \right|^p d\lambda_n(w)$$

we can use this lemma in the estimate

$$\begin{aligned}
&\left\| \int_{\mathbb{B}_n} \left\{ ((z-w)\cdot\nabla)^{\ell} h(w) \right\} g(w) \frac{(1-|w|^2)^s}{(1-\overline{w}\cdot z)^{n+1+s}} dV \right\|_{L^p(\nu_\alpha)}^p \\
&\leq \left\| \widehat{T}_{0,s-\ell,\ell} \left(g D^{\ell} h \right) \right\|_{L^p(d\nu_\alpha)}^p \leq \left\| g D^{\ell} h \right\|_{L^p(d\nu_\alpha)}^p \\
&= \int_{\mathbb{B}_n} \left| D^{\ell} h(w) \right|^p |g(w)|^p (1-|w|^2)^{pm-n-1} dV \\
&= \int_{\mathbb{B}_n} \left| D^{\ell} h(w) \right|^p \left| (1-|w|^2)^m g(w) \right|^p d\lambda_n(w) = \int_{\mathbb{B}_n} \left| D^{\ell} h(w) \right|^p d\mu(w).
\end{aligned}$$

Writing ∇_{rad} and ∇_{tan} for the radial and tangential components of the gradient on the ball, and taking note of the estimates on D in the previous chapter, we have, for all functions h holomorphic in the ball,

$$\int_{K_\alpha} \left|D^\ell h(w)\right| d\lambda_n(w) \lesssim \int_{K_\alpha} \left|\left(1-|c_{K_\alpha}|^2\right)^\ell \nabla^\ell_{\mathrm{rad}} h(w) + \left(1-|c_{K_\alpha}|^2\right)^{\ell/2} \nabla^\ell_{\tan} h(w)\right| d\lambda_n(w) \lesssim \int_{RK_\alpha} |h(w)|\, d\lambda_n(w). \tag{15.26}$$

Here RK_α is the Bergman ball defined at 14.30 with R sufficiently large, say $R = 2C_2$ where C_2 is the constant from (5.1). To obtain this estimate we use the automorphism φ_α to reduce consideration to the set $\varphi_\alpha K_\alpha$ which contains the origin. In that case we can use the Cauchy reproducing formula [**288**]

$$f(z) = \int_{\partial\mathbb{B}_n} \frac{f(w)}{(1-\overline{w}\cdot z)^n} d\sigma(w), \qquad z \in \mathbb{B}_n,$$

rescaled to $B_\beta(0, 2R)$ together with simple estimates to get (15.26).

Then since h is holomorphic, we can continue with

$$\begin{aligned}\int_{\mathbb{B}_n} \left|D^\ell h(w)\right|^p d\mu(w) &\lesssim \sum_{\alpha\in\mathcal{T}_n} \left(\int_{K_\alpha} \left|D^\ell h(w)\right| d\lambda_n(w)\right)^p \left(\int_{K_\alpha} d\mu\right) \\ &\lesssim \sum_{\alpha\in\mathcal{T}_n} \left(\int_{RK_\alpha} |h(w)|\, d\lambda_n(w)\right)^p \left(\int_{K_\alpha} d\mu\right) \\ &\lesssim \sum_{\alpha\in\mathcal{T}_n} \int_{RK_\alpha} |h(w)|^p\, d\lambda_n(w) \left(\int_{RK_\alpha} d\mu\right) \\ &= \int_{\mathbb{B}_n} |h(w)|^p\, d\widetilde{\mu_R}(w) \lesssim \|h\|_{B_p^p},\end{aligned}$$

since Proposition 14.13 shows that $\widetilde{\mu_R}$, defined in (14.31) is a Carleson measure for B_p if μ also has that property. The conclusion of the spreading lemma, Lemma 15.14, is equivalent to the inequality

$$\begin{aligned}\|hT_s g\|_{L^p(\nu_\alpha)} &= \left\{\int_{\mathbb{B}_n} |h(z)|^p \left|(1-|z|^2)^n R^{s-n,n} \Gamma^n_s g(z)\right|^p d\lambda_n(z)\right\}^{1/p} \\ &\lesssim \|h\|_{B_p}, \qquad h \in B_p,\end{aligned}$$

and thus in particular, it suffices to show that

$$\|T_s((P_L h(\cdot,z))g) - hT_s g(z)\|_{L^p(\nu_\alpha)} \lesssim \|h\|_{B_p}, \qquad h \in B_p. \tag{15.27}$$

We have that:

$$\begin{aligned}&T_s((P_L h(\cdot,z))g)(z) - h(z)T_s g(z) \\ &\qquad = c_{n,s} \int_{\mathbb{B}_n} (P_L h(w,z) - h(z)) \frac{g(w)(1-|w|^2)^s}{(1-\overline{w}\cdot z)^{n+1+s}} dV,\end{aligned}$$

and we obtain a pointwise estimate on this using the sup norm estimate (15.21) on g and Hölder's inequality. We find that $|T_s(P_z(h;w)g)(z) - h(z)T_s g(z)|$ is dominated by

$$\left(\int_{\mathbb{B}_n} \frac{|P_L h(w,z) - h(z)|^p}{|1-\overline{w}\cdot z|^{2(n+1)}} dV\right)^{1/p}$$
$$\times \left(\int_{\mathbb{B}_n} \frac{(1-|w|^2)^{(s-n)p'}}{|1-\overline{w}\cdot z|^{[(1-1/p)(n+1)+s]p'}} dV\right)^{1/p'}$$
$$\leq C_0(1-|z|^2)^{-n+(n+1)/p}$$
$$\times \left(\int_{\mathbb{B}_n} \frac{|P_L h(w,z) - h(z)|^p}{|1-\overline{w}\cdot z|^{2(n+1)}} dV\right)^{1/p},$$

by [**346**, Theorem 1.12] provided $(s-n)p' > -1$, i.e. $s > n - 1/p'$.

To obtain the required norm estimate we use (15.22) with $p > 1$ and $L > 2n/p$ and estimate

$$\begin{aligned}
&\|T_s(P_L h(\cdot,z)g) - hT_s g(z)\|^p_{L^p(d\nu_\alpha)} \\
&\lesssim \int_{\mathbb{B}_n} (1-|z|^2)^{-pn+(n+1)} \int_{\mathbb{B}_n} \frac{|P_L h(w,z) - h(z)|^p}{|1-\overline{w}\cdot z|^{2(n+1)}} dV\, d\nu_\alpha(z) \\
&\lesssim \int_{\mathbb{B}_n}\int_{\mathbb{B}_n} \frac{|P_L h(w,z) - h(z)|^p}{|1-\overline{w}\cdot z|^{2(n+1)}} dV dV \\
&\lesssim \int_{\mathbb{B}_n}\int_{\mathbb{B}_n} \frac{\left|h(z) - \sum_{\ell=0}^L \frac{((z-w)\cdot\nabla)^\ell}{\ell!} h(w)\right|^p}{|1-\overline{w}\cdot z|^{2(n+1)}} dV dV \\
&\lesssim \|h\|^p_{B_p},
\end{aligned}$$

which yields (15.27) as required.

The case where we consider $\nabla^n\Gamma_s$ in place of $T_s = R^{s-n,n}\Gamma_s$ is handled just as above using the pointwise estimate

$$|\nabla^n\Gamma^n_s g(z)| \leq C_{s,n} \int_{\mathbb{B}_n} \frac{|g(w)|(1-|w|^2)^s}{|1-\overline{w}\cdot z|^{n+1+s}} dV \lesssim \widehat{T_s}|g|(z). \tag{15.28}$$

□

15.4.2. Multiplier Approximations. The next lemma constructs a holomorphic function that is close to 1 on the Carleson region associated to a point $w \in \mathbb{B}_n$, and decays appropriately away from the Carleson region. We follow Böe's proof in [**77**], which adapts a real-variable argument of Marshall and Sundberg in [**220**] to produce a holomorphic multiplier approximation. Given $\beta < \rho < \alpha < 1$, we will use the cutoff function $c_{\rho,\alpha}$ defined by

$$c_{\rho,\alpha}(\gamma) = \begin{cases} 0 & \text{for } \gamma < \rho \\ \frac{\gamma-\rho}{\alpha-\rho} & \text{for } \rho \leq \gamma \leq \alpha \\ 1 & \text{for } \alpha < \gamma. \end{cases} \tag{15.29}$$

LEMMA 15.16 (analog of [**77**, Lemma 4.1]). *Suppose $s > -1$. There are ρ and α satisfying $\beta < \rho < \alpha < 1$ such that for every $w \in \mathbb{B}_n$, we can find a function g_w*

so that

$$\varphi_w(z) = \Gamma_s g_w(z) = \int_{\mathbb{B}_n} \frac{g_w(\zeta)(1-|\zeta|^2)^s}{(1-\overline{\zeta}\cdot z)^{1+s}} dV$$

satisfies

$$\text{(15.30)} \qquad \begin{cases} \varphi_w(w) &= 1 \\ \varphi_w(z) &= c_{\rho,\alpha}(\gamma_w(z)) + O\left(\left(\log \frac{1}{1-|w|^2}\right)^{-1}\right), & z \in V_w \\ |\varphi_w(z)| &\leq C\left(\log \frac{1}{1-|w|^2}\right)^{1-p}, & z \notin V_w, \end{cases}$$

where $\gamma_w(z)$ *is defined by*

$$|1 - \overline{z} \cdot Pw| = (1-|w|^2)^{\gamma_w(z)}$$

and $c_{\rho,\alpha}$ *is as in* (15.29)*. Furthermore we have the estimate*

$$\text{(15.31)} \qquad \int_{\mathbb{B}_n} \left|(1-|\zeta|^2)^n g_w(\zeta)\right|^p d\lambda_n(\zeta) \lesssim \left(\log \frac{1}{1-|w|^2}\right)^{1-p}.$$

REMARK 15.17. *The proof of Lemma* 15.16 *shows that the third estimate in* (15.30) *can be vastly improved, and also holds for a larger range of* z*; namely there is* $\beta < \beta_1 < \rho$ *such that*

$$|\varphi_w(z)| \lesssim \left(\log \frac{1}{1-|w|^2}\right)^{-1} (1-|w|^2)^{(\rho-\beta_1)(1+s)}, \qquad z \notin V_w^{\beta_1}.$$

This fact will be used in the proof of Lemma 15.19 *below.*

PROOF. Define $g_w(\zeta)$ by

$$\text{(15.32)} \qquad \frac{g_w(\zeta)(1-|\zeta|^2)^s}{(1-\overline{\zeta}\cdot w)^{1+s}} = K\left(\log \frac{1}{1-|w|^2}\right)^{-1} \left|1-\overline{\zeta}\cdot Pw\right|^{-n-1},$$

when ζ lives in the annular sector $\mathcal{S}$ centered at Pw; $\mathcal{S} := \mathcal{A} \cap \mathcal{C}$ where $\mathcal{A}$ is the annulus

$$\text{(15.33)} \qquad \mathcal{A} := \mathcal{A}_w = \left\{\zeta \in \mathbb{B}_n : (1-|w|^2)^\alpha \leq \left|1-\overline{\zeta}\cdot Pw\right| \leq (1-|w|^2)^\rho\right\}$$

and $\mathcal{C}_w$ the cone

$$\mathcal{C}_w := \left\{\zeta \in \mathbb{B}_n : \left|\operatorname{Im}\left(\zeta \cdot \overline{Pw}\right)\right| + \left|\zeta - \left(\zeta\cdot\overline{Pw}\right)Pw\right|^2 \leq c\left(1 - \left|\overline{\zeta}\cdot Pw\right|\right)\right\},$$

where c is a suitably small constant. Define $g_w(\zeta) = 0$ otherwise. The constant K is then choose so that $\varphi_w(w) = 1$, i.e.

$$K = \left(\log \frac{1}{1-|w|^2}\right)\left(\int_{\mathcal{S}} \left|1-\overline{\zeta}\cdot Pw\right|^{-n-1} dV\right)^{-1}.$$

The following observation will be used repeatedly.

REMARK 15.18. *The cone* $\mathcal{C}_w$ *corresponds to the geodesic in the Bergman tree* $\mathcal{T}_n$ *joining the root to the "boundary point"* Pw*. To see this, consider the case* $w = (t, 0, \ldots, 0)$ *and* $\zeta = \left(re^{i\theta}, \zeta'\right)$ *with* $re^{i\theta} = x + iy$*, so that* $\operatorname{Im}\left(\zeta\cdot\overline{Pw}\right) = y$*,* $\zeta - \left(\zeta\cdot\overline{Pw}\right)Pw = (0,\zeta')$ *and* $1 - \left|\overline{\zeta}\cdot Pw\right| = 1 - r$*.*

We can estimate K by

$$K \approx K_{\alpha,\rho,n} = \frac{c_n}{\alpha - \rho} \tag{15.34}$$

since the annular sector

$$\mathcal{E}_a := \left\{ \zeta \in \mathbb{B}_n : a \leq \left| 1 - \overline{\zeta} \cdot Pw \right| \leq 2a \right\} \cap \mathcal{C}$$

is comparable to a Bergman ball of radius one, $\left|1 - \overline{\zeta} \cdot Pw\right|^{-n-1} dV$ is comparable to invariant measure $d\lambda_n(\zeta)$ on $\mathcal{E}_a$, and $\mathcal{S} \approx \bigcup_{j=0}^{J} \mathcal{E}_{2^j(1-|w|^2)^\alpha}$ where

$$J = \log \frac{(1 - |w|^2)^\rho}{(1 - |w|^2)^\alpha} = (\rho - \alpha) \log(1 - |w|^2).$$

Note also that

$$\left|1 - \overline{\zeta} \cdot Pw\right| \approx \left|1 - \overline{\zeta} \cdot w\right| \approx 1 - |\zeta|^2, \qquad \zeta \in \mathcal{S}, \tag{15.35}$$

and so g_w satisfies the estimate

$$|g_w(\zeta)| \lesssim \left(\log \frac{1}{1 - |w|^2} \right)^{-1} \left|1 - \overline{\zeta} \cdot Pw\right|^{-n}, \qquad \zeta \in \mathbb{B}_n. \tag{15.36}$$

Now fix $z \in V_w$ and set

$$\begin{aligned} E_1 &:= \left\{ \zeta \in \mathbb{B}_n : \left|1 - \overline{\zeta} \cdot Pw\right| \leq (1 - |w|^2)^{\gamma_w(z)} \right\}, \\ E_2 &:= \mathbb{B}_n \setminus E_1 = \left\{ \zeta \in \mathbb{B}_n : \left|1 - \overline{\zeta} \cdot Pw\right| > (1 - |w|^2)^{\gamma_w(z)} \right\}. \end{aligned}$$

Thus the common boundary of E_1 and E_2 passes through z. The main contribution to $\varphi_w(z)$ will come from integration over E_2. Thus we write

$$\varphi_w(z) = \int_{E_1} \frac{g_w(\zeta)(1 - |\zeta|^2)^s}{\left(1 - \overline{\zeta} \cdot z\right)^{1+s}} dV + \int_{E_2} \frac{g_w(\zeta)(1 - |\zeta|^2)^s}{\left(1 - \overline{\zeta} \cdot z\right)^{1+s}} dV = I + II.$$

By equations (15.35) and (15.36) and the definition of $\gamma_w(z)$, term I satisfies

$$|I| \lesssim \frac{-1}{\log(1 - |w|^2)} \int_{\left\{(1-|w|^2)^\alpha \leq \left|1 - \overline{\zeta} \cdot Pw\right| \leq \left|1 - \overline{z} \cdot Pw\right|\right\} \cap \mathcal{C}} \left(\frac{1 - |\zeta|^2}{\left|1 - \overline{\zeta} \cdot z\right|} \right)^{1+s} d\lambda_n(\zeta).$$

The integral is at most a constant because

$$\left|1 - \overline{\zeta} \cdot z\right| \approx \left|1 - \overline{z} \cdot Pw\right|, \qquad \zeta \in \mathcal{C} \cap E_1.$$

Thus we have

$$|I| \lesssim \left(\log \frac{1}{1 - |w|^2} \right)^{-1}.$$

We now write

$$\begin{aligned}
II &= \int_{E_2 \cap \mathcal{S}} \frac{g_w(\zeta)(1-|\zeta|^2)^s}{\left(1-\overline{\zeta}\cdot z\right)^{1+s}} dV \\
&= \int_{E_2 \cap \mathcal{S}} \left\{ \frac{g_w(\zeta)(1-|\zeta|^2)^s}{\left(1-\overline{\zeta}\cdot z\right)^{1+s}} - \frac{g_w(\zeta)(1-|\zeta|^2)^s}{\left(1-\overline{\zeta}\cdot w\right)^{1+s}} \right\} dV \\
&\qquad + \int_{E_2 \cap \mathcal{S}} \frac{g_w(\zeta)(1-|\zeta|^2)^s}{\left(1-\overline{\zeta}\cdot w\right)^{1+s}} dV \\
&= III + IV.
\end{aligned}$$

Using (15.34), and that g_w is supported in $\mathcal{S}$, we calculate that

$$\begin{aligned}
IV &\approx \frac{K}{-\log(1-|w|^2)} \int_{\left\{(1-|w|^2)^{\gamma_w(z)} \leq \left|1-\overline{\zeta}\cdot Pw\right| \leq (1-|w|^2)^{\rho}\right\} \cap \mathcal{C}} \left|1-\overline{\zeta}\cdot Pw\right|^{-n-1} dV \\
&\approx \frac{\gamma(z)-\rho}{\alpha-\rho}
\end{aligned}$$

in the case $\rho < \gamma_w(z) < \alpha$. We also have $IV = 0$ in the case $\gamma_w(z) < \rho$, and $IV \approx \log \frac{1}{1-|w|^2}$ in the case $\alpha < \gamma_w(z)$. This gives the estimate

$$IV = c_{\rho,\alpha}(\gamma_w(z)) + O\left(\left(\log \frac{1}{1-|w|^2}\right)^{-1}\right), \qquad z \in V_w.$$

Using

$$\left| \frac{1}{\left(1-\overline{\zeta}\cdot z\right)^{1+s}} - \frac{1}{\left(1-\overline{\zeta}\cdot w\right)^{1+s}} \right| \lesssim \frac{|z-w|}{(1-|\zeta|^2)^{2+s}}$$

together with (15.35) and (15.36), we obtain that

$$\begin{aligned}
|III| &\lesssim \int_{E_2 \cap \mathcal{S}} |g_w(\zeta)| (1-|\zeta|^2)^{-2} |z-w| \, dV \\
&\lesssim |z-w| \left(\log \frac{1}{1-|w|^2}\right)^{-1} \int_{E_2 \cap \mathcal{S}} (1-|\zeta|^2)^{-1} d\lambda_n(\zeta) \\
&\lesssim \left(\log \frac{1}{1-|w|^2}\right)^{-1},
\end{aligned}$$

as required. This completes the proof of the second estimate in (15.30).

We now turn to the third estimate in (15.30). For $z \notin V_w$, we have $\left|1-\overline{\zeta}\cdot z\right| \gtrsim (1-|w|^2)^{\beta}$ for $\zeta \in \mathcal{S}$, and thus

$$\begin{aligned} |\varphi_w(z)| &\leq \left(\log\frac{1}{1-|w|^2}\right)^{-1} \int_{\mathcal{S}} \left(\frac{1-|\zeta|^2}{\left|1-\overline{\zeta}\cdot z\right|}\right)^{1+s} d\lambda_n(\zeta) \\ &\lesssim \left(\log\frac{1}{1-|w|^2}\right)^{-1} (1-|w|^2)^{(\rho-\beta)(1+s)} \\ &\lesssim \left(\log\frac{1}{1-|w|^2}\right)^{1-p}. \end{aligned}$$

Finally, the estimate (15.31) is a calculation using (15.35), (15.36) the definition of the support of g_w. In fact, the left side of (15.31) is at most

$$\int_{\mathcal{S}} \left(\log\frac{1}{1-|w|^2}\right)^{-p} d\lambda_n(\zeta) \lesssim \left(\log\frac{1}{1-|w|^2}\right)^{1-p}.$$

□

The next lemma uses Lemma 15.14 to construct inductively a holomorphic function whose restriction to the sequence $\mathcal{Z}$ approximates an arbitrarily prescribed bounded sequence $\{\xi_j\}_{j=1}^{\infty}$

LEMMA 15.19 (analog of [**77**, Lemma 4.2]). *Suppose $s>-1$, that $\{\xi_j\}_{j=1}^{\infty} \in \ell^{\infty}$ and let $0<\delta<1$. Let φ_j, g_j and γ_j correspond to z_j as in Lemma 15.16 and with the same s. Then there is $\{a_i\}_{i=1}^{\infty} \in \ell^{\infty}$ such that $\varphi=\sum_{i=1}^{\infty} a_i\varphi_i$ satisfies*

$$\left\|\{\xi_j-\varphi(z_j)\}_{j=1}^{\infty}\right\|_{\ell^{\infty}} < \delta \left\|\{\xi_j\}_{j=1}^{\infty}\right\|_{\ell^{\infty}} \tag{15.37}$$

and

$$\|\{a_i\}_{i=1}^{\infty}\|_{\ell^{\infty}}, \|\varphi\|_{H^{\infty}(\mathbb{B}_n)} \lesssim \left\|\{\xi_j\}_{j=1}^{\infty}\right\|_{\ell^{\infty}}. \tag{15.38}$$

REMARK 15.20. *The series $\sum_{i=1}^{\infty} a_i\varphi_i$ in Lemma 15.19 converges absolutely for each $z \in \mathbb{B}_n$. In fact, the proof below will show that (using the estimate* card $\mathcal{G}_\ell \lesssim \beta(0,z_\ell)$ *for the chains $\mathcal{G}_\ell$ defined there)*

$$\sum_{i=1}^{\infty} |\varphi_i(z)| \lesssim \left(1+\log\frac{1}{1-|z|^2}\right), \qquad z \in \mathbb{B}_n.$$

REMARK 15.21. *The construction in the proof below shows that both the sequence $\{a_i\}_{i=1}^{\infty}$ and the function φ depend linearly on the data $\{\xi_j\}_{j=1}^{\infty}$.*

PROOF. We follow the proof in [**77**, Lemma 4.2]. Let $\left\|\{\xi_j\}_{j=1}^{\infty}\right\|_{\ell^{\infty}}=1$. We first choose J so large that

$$\sup_{j\geq J}\left(\log\frac{1}{1-|z_j|}\right)^{-1} + \sum_{j=J}^{\infty}\left(\log\frac{1}{1-|z_j|}\right)^{1-p} < \varepsilon, \tag{15.39}$$

where $\varepsilon>0$ will be determined later. Note that the series above converges by the Carleson embedding in (15.6).

We would like to discard the finitely many points $\{z_j\}_{j=1}^{J-1}$ from consideration and assume that $J=1$ in (15.39), then perform the basic construction, and then, after that, adjust things on the finite point set we had discarded. To see that is possible, it suffices to show that if we adjoin a single point to an interpolating sequence we still have an interpolating sequence. That is, given $Z=\{z_j\}_{j=1}^{\infty}$ an interpolating sequence for B_p, i.e. that (15.3) holds, and given $w\in\mathbb{B}_n\setminus Z$ we want to show that $Z\cup\{w\}$ satisfies (15.3). To do this it suffices to exhibit a function $h\in B_p$ that vanishes on Z, but not at w. In one dimension this is straightforward. Pick an f in B_p which is not the zero function and which vanishes on $Z\setminus\{z_1\}$ but not at z_1. For some nonnegative integer k, $h(z)=(z-z_1)f(z)/(z-w)^k$ will be the required function. We need to show $h\in B_p$ but we do not need a norm estimate and hence that is easy to see.

In higher dimensions we must work a bit harder. Suppose that $Z=\{z_j\}_{j=1}^{\infty}$ is an interpolating sequence for B_p, i.e. that (15.3) holds. Suppose that $w\in\mathbb{B}_n\setminus Z$. In order to show that $Z\cup\{w\}$ satisfies (15.3), it suffices to exhibit a function $h\in B_p$ that vanishes on Z, but not at w. Using the Hahn-Banach theorem and the reflexivity of B_p, this is equivalent to showing that the point evaluation $e_w=k_w^{\alpha,p}$ is *not* in the closure $\mathcal{S}$ of the linear span of the set of point evaluations $\{e_{z_j}\}_{j=1}^{\infty}$ in $B_p'=B_{p'}$. So suppose, in order to derive a contradiction, that $e_w\in\mathcal{S}$. Above, we showed the equivalence of (15.3) for Z and the Riesz condition (15.4) for Z,

$$\left\|\sum_{j=1}^{\infty}a_j\frac{k_{z_j}^{\alpha,p}}{\left\|k_{z_j}^{\alpha,p}\right\|_{B_{p'}}}\right\|_{B_{p'}}\approx\left(\sum_{j=1}^{\infty}|a_j|^{p'}\right)^{\frac{1}{p'}},$$

and consequently we have

$$e_w=\sum_{j=1}^{\infty}a_j\omega_je_{z_j},\tag{15.40}$$

where $\omega_j=1/\left\|e_{z_j}\right\|_{B_{p'}}$ and

$$\sum_{j=1}^{\infty}|a_j|^{p'}<\infty.$$

Now let $f_\ell\in B_p$ satisfy $f_\ell(z_k)=\delta_k^\ell$. Then from (15.40) we have

$$f_\ell(w)=\sum_{j=1}^{\infty}a_j\omega_jf_\ell(z_j)=\omega_\ell a_\ell,$$

for all $\ell\geq 1$, and we will have the desired contradiction, namely $e_w=0$, if we can show $f_\ell(w)=0$ for all $\ell\geq1$. To see this, choose a linear function g_ℓ that vanishes at z_ℓ and takes the value 1 at w. Then $g_\ell\in M_{B_p}$ implies $f_\ell g_\ell\in B_p$, and (15.40) yields

$$f_\ell(w)=(f_\ell g_\ell)(w)=\sum_{j=1}^{\infty}a_j\omega_j(f_\ell g_\ell)(z_j)=a_\ell\omega_\ell g_\ell(z_\ell)=0.$$

We now suppose $J=1$ and go forward. Order the points $\mathcal{Z}$ so that $1-|z_{j+1}|\leq 1-|z_j|$ for $j\geq1$. We now define a "forest structure" on the index set $\mathbb{N}$ by declaring that j is a child of i (or that i is a parent of j) provided that

$$\begin{aligned} & i<j, \\ & V_{z_j} \subset V_{z_i}, \\ & V_{z_j} \subsetneq V_{z_k} \text{ for } i<k<j. \end{aligned} \tag{15.41}$$

Note that a child j chooses the "nearest" parent i if we have competing indices i and i' with $V_{z_j} \subset V_{z_i} \cap V_{z_{i'}}$. We define a partial order associated with this parent-child relationship by declaring that j is a successor of i (or that i is a predecessor of j) if there is a "chain" of indices $\{i = k_1, k_2, \ldots, k_m = j\} \subset \mathbb{N}$ such that $k_{\ell+1}$ is a child of k_ℓ for $1 \leq \ell < m$. Under this partial ordering, $\mathbb{N}$ decomposes into a disjoint union of trees. Thus associated to each index $\ell \in \mathbb{N}$, there is a unique tree containing ℓ and, unless ℓ is the root of the tree, a unique parent $P(\ell)$ of ℓ in that tree. Denote by $\mathcal{G}_\ell$ the unique geodesic joining the root of the tree to ℓ. We will now define the coefficients $\{a_i\}_{i=1}^\infty$ of $\varphi = \sum_{i=1}^\infty a_i \varphi_i$, where φ_i is the function φ_{z_i} in Lemma 15.16 with w there replaced by z_i, by considering separately the indices in each tree of the forest $\mathbb{N}$.

Let $\mathcal{Y}$ be a tree in the forest $\mathbb{N}$ with root k_0. For each $k \in \mathcal{Y} \setminus \{k_0\}$, define $\beta_k \in [0,1]$ by

$$\beta_k := c\left(\gamma_{P(k)}\left(z_k\right)\right),$$

where the functions $c = c_{\rho,\alpha}$ and $\gamma_j = \gamma_{z_j}$ are defined as in the statement of Lemma 15.16 with w there replaced by z_j. Note that by Lemma 15.16 with $w = z_{P(k)}$, we have the estimate

$$\begin{aligned} \varphi_{P(k)}\left(z_k\right) &= c\left(\gamma_{P(k)}\left(z_k\right)\right) + O\left(\left(\log \frac{1}{1-\left|z_{P(k)}\right|^2}\right)^{-1}\right) \\ &= \beta_k + O\left(\left(\log \frac{1}{1-\left|z_{P(k)}\right|^2}\right)^{-1}\right), \end{aligned}$$

which can serve as motivation for the definition of the coefficients given below in (15.43). Indeed, with gross oversimplification, what we want is

$$\begin{aligned} \xi_k = \varphi\left(z_k\right) &\approx a_k \varphi_k\left(z_k\right) + a_{P(k)} \varphi_{P(k)}\left(z_k\right) + \ldots \\ &\approx a_k + a_{P(k)} \beta_k + \ldots, \end{aligned}$$

which leads to (15.43).

We will now define numbers $\{a_k\}_{k \in \mathcal{Y}}$ by induction on the linear ordering in $\mathcal{Y}$ induced from the natural ordering of $\mathbb{N}$, so that

$$\left\{ \begin{array}{lcl} \left|a_k\right| & \leq & 2 \\ \left|\sum_{i \in \mathcal{G}_k \setminus \{k_0\}} \beta_i a_{P(i)}\right| & \leq & 1 \end{array} \right. \tag{15.42}$$

holds for all $k \in \mathcal{Y}$. First define $a_{k_0} = \xi_{k_0}$. Now fix $\ell \in \mathcal{Y} \setminus \{k_0\}$ and assume that a_k has been defined for all $k \in \mathcal{Y}$ for which $k < \ell$ so that (15.42) holds for all $k \in \mathcal{Y}$ for which $k < \ell$. We now define a_ℓ by

$$a_\ell = \xi_\ell - \sum_{i \in \mathcal{G}_\ell \setminus \{k_0\}} \beta_i a_{P(i)}. \tag{15.43}$$

Of crucial importance is the observation that the geodesics $\mathcal{G}_\ell$ and $\mathcal{G}_{P(\ell)}$ are related by

$$\mathcal{G}_{P(\ell)} = \mathcal{G}_\ell \setminus \{\ell\},$$

i.e. if $\mathcal{G}_\ell = [k_0, k_1, \ldots, k_{m-1}, k_m]$ with $k_m = \ell$, then $k_{m-1} = P(\ell)$ and $\mathcal{G}_{P(\ell)} = [k_0, k_1, \ldots, k_{m-1}]$. By the induction assumption and the fact that $P(\ell) \in \mathcal{Y}$ and $P(\ell) < \ell$, we have

$$\left| \sum_{i \in \mathcal{G}_{P(\ell)} \setminus \{k_0\}} \beta_i a_{P(i)} \right| \leq 1.$$

We have from (15.43) and the above that

$$\begin{aligned}
\left| \sum_{i \in \mathcal{G}_\ell \setminus \{k_0\}} \beta_i a_{P(i)} \right| &= \left| \left(\sum_{i \in \mathcal{G}_{P(\ell)} \setminus \{k_0\}} \beta_i a_{P(i)} \right) + \beta_\ell \left(\xi_{P(\ell)} - \sum_{i \in \mathcal{G}_{P(\ell)} \setminus \{k_0\}} \beta_i a_{P(i)} \right) \right| \\
&= \left| \beta_\ell \xi_{P(\ell)} + (1 - \beta_\ell) \sum_{i \in \mathcal{G}_{P(\ell)} \setminus \{k_0\}} \beta_i a_{P(i)} \right| \\
&\leq \beta_\ell \left| \xi_{P(\ell)} \right| + (1 - \beta_\ell) \left| \sum_{i \in \mathcal{G}_{P(\ell)} \setminus \{k_0\}} \beta_i a_{P(i)} \right| \leq 1.
\end{aligned}$$

From this, and (15.43) once more, it immediately follows that $|a_\ell| \leq 2$, which shows that (15.42) holds for $k = \ell$ as well. This completes the inductive definition of the sequence $\{a_k\}_{k \in \mathcal{Y}}$ satisfying (15.42) on the tree $\mathcal{Y}$, and hence defines the entire sequence $\{a_i\}_{i=1}^\infty$.

We now show that both (15.37) and (15.38) hold for the function $\varphi = \sum_{i=1}^\infty a_i \varphi_i$. Fix an index $\ell \in \mathbb{N}$, and with notation as above, let $\mathcal{F}_\ell = \mathbb{N} \setminus \mathcal{G}_\ell$ and write using $\varphi_\ell(z_\ell) = 1$ and (15.43),

$$\begin{aligned}
\varphi(z_\ell) - \xi_\ell &= \sum_{i=1}^\infty a_i \varphi_i(z_\ell) - \xi_\ell \\
&= \left(\sum_{i \in \mathcal{G}_{P(\ell)}} a_i \varphi_i(z_\ell) + a_\ell \varphi_\ell(z_\ell) + \sum_{i \in \mathcal{F}_\ell} a_i \varphi_i(z_\ell) \right) \\
&\qquad - \left(a_\ell + \sum_{i \in \mathcal{G}_\ell \setminus \{k_0\}} \beta_i a_{P(i)} \right) \\
&= \sum_{i \in \mathcal{G}_\ell \setminus \{k_0\}} a_{P(i)} \left(\varphi_{P(i)}(z_\ell) - \beta_i \right) + \sum_{i \in \mathcal{F}_\ell} a_i \varphi_i(z_\ell) \\
&= I + II.
\end{aligned}$$

We now claim that

$$|I| \lesssim \sup_{i \geq 1} \left(\log \frac{1}{1 - |z_i|} \right)^{-1}, \tag{15.44}$$

$$|II| \lesssim \sum_{i=1}^\infty \left(\log \frac{1}{1 - |z_i|} \right)^{1-p}.$$

With this done we obtain from (15.39) (recall that we assume $J=1$ there) that

$$\sup_{j\geq 1}|\varphi(z_j)-\xi_j|\lesssim \varepsilon<\delta$$

provided we choose $\varepsilon>0$ small enough, and this proves (15.37). We note in passing that the proof below will show that the supremum in (15.44) need only be taken over indices i that are a root of a tree in the forest $\mathbb{N}$.

To estimate term I, we begin with

$$\begin{aligned}|I| &= \left|\sum_{i\in\mathcal{G}_\ell\setminus\{k_0\}} a_{P(i)}\left(\varphi_{P(i)}(z_\ell)-\beta_i\right)\right| \\ &\leq \sum_{i\in\mathcal{G}_\ell\setminus\{k_0\}} \left|a_{P(i)}\right|\left|\varphi_{P(i)}(z_\ell)-c\left(\gamma_{P(i)}(z_i)\right)\right| \\ &\leq 2\sum_{i\in\mathcal{G}_\ell\setminus\{k_0\}}\left\{\left|c\left(\gamma_{P(i)}(z_\ell)\right)-c\left(\gamma_{P(i)}(z_i)\right)\right|+C\left(\log\frac{1}{1-\left|z_{P(i)}\right|^2}\right)^{-1}\right\},\end{aligned}$$

where the final inequality follows from Lemma 15.16 since $z_\ell\in V_{z_{P(i)}}$. Now if V_{z_i} has empty intersection with the annulus $\mathcal{A}_{P(i)}$ given in (15.33) with $w=z_{P(i)}$, then both $c\left(\gamma_{P(i)}(z_\ell)\right)$ and $c\left(\gamma_{P(i)}(z_i)\right)$ have the same value, either 0 or 1. Otherwise, since c is Lipschitz continuous with norm $1/(\alpha-\rho)$, we have

$$\begin{aligned}\left|c\left(\gamma_{P(i)}(z_\ell)\right)-c\left(\gamma_{P(i)}(z_i)\right)\right| &\lesssim \left|\gamma_{P(i)}(z_\ell)-\gamma_{P(i)}(z_i)\right| \\ &\approx \left|\frac{-\log\left|1-\overline{z_\ell}\cdot Pz_{P(i)}\right|}{-\log\left(1-\left|z_{P(i)}\right|^2\right)}-\frac{-\log\left|1-\overline{z_i}\cdot Pz_{P(i)}\right|}{-\log\left(1-\left|z_{P(i)}\right|^2\right)}\right| \\ &\lesssim \left(\log\frac{1}{1-\left|z_{P(i)}\right|^2}\right)^{-1},\end{aligned}$$

since

$$1\lesssim\frac{\left|1-\overline{z_\ell}\cdot Pz_{P(i)}\right|}{\left|1-\overline{z_i}\cdot Pz_{P(i)}\right|}\lesssim 1$$

for $z_i,z_\ell\in\mathcal{A}_{P(i)}$. Now if $\mathcal{G}_\ell=[k_0,k_1,\ldots,k_{m-1},k_m]$, then by applying the separation condition repeatedly, we obtain

$$(1-|z_{k_i}|^2)\leq\left(1-|z_{k_0}|^2\right)^{\eta^i},$$

and so combining these estimates we have

$$\begin{aligned}|I| &\lesssim \sum_{i\in\mathcal{G}_\ell\setminus\{k_0\}}\left(\log\frac{1}{1-\left|z_{P(i)}\right|^2}\right)^{-1} \\ &\lesssim \left(\sum_{j=0}^{m-1}\eta^{-j}\right)\left(\log\frac{1}{1-|z_{k_0}|^2}\right)^{-1} \\ &\lesssim \left(\log\frac{1}{1-|z_{k_0}|^2}\right)^{-1}\end{aligned}$$

since $\eta>1$, which shows the first inequality in (15.44).

To estimate term $II = \sum_{i \in \mathcal{F}_\ell} a_i \varphi_i (z_\ell)$ in (15.44), we first note that if $z_\ell \notin V_{z_i}$, then

$$|\varphi_i (z_\ell)| \lesssim \left(\log \frac{1}{1 - |z_i|^2} \right)^{-(p-1)} \tag{15.45}$$

by Lemma 15.16. On the other hand, if $z_\ell \in V_{z_i}$, then $|z_i| < |z_\ell|$, and if $\mathcal{G}_\ell = [k_0, k_1, \ldots, k_{m-1}, k_m]$, then either $|z_i| < |z_{k_0}|$ or there is j such that $\left|z_{k_{j-1}}\right| < |z_i| \leq \left|z_{k_j}\right|$. Note however that equality cannot hold here by Lemma 15.12, and so we actually have $\left|z_{k_{j-1}}\right| < |z_i| < \left|z_{k_j}\right|$. From (15.41) we obtain that no index $m \in (k_{j-1}, k_j)$ satisfies $V_{z_{k_j}} \subset V_{z_m}$. Since $i \notin \mathcal{G}_\ell$, we have $i \in (k_{j-1}, k_j)$ and thus we have both

$$V_{z_{k_j}} \not\subsetneq V_{z_i} \text{ and } \left|z_{k_j}\right| > |z_i| .$$

Now using Lemma 15.12 and $\beta\eta > 1$, we obtain

$$(1 - |z|^2)^\beta \leq \left(1 - |z_i|^2\right)^{\beta\eta} \leq \left(1 - |z_i|^2\right).$$

If we choose $w \in V_{z_{k_j}} \setminus V_{z_i}$, then $w, z_\ell \in V_{z_{k_j}}$ implies $|z_\ell - w| \lesssim (1 - |z|^2)^\beta$ by definition, and $w \notin V_{z_i}$ implies $|1 - \overline{w} \cdot P z_i| \gtrsim \left(1 - |z_i|^2\right)^\beta$. Together with the reverse triangle inequality we thus have

$$\begin{aligned} |1 - \overline{z_\ell} \cdot P z_i| &\geq |1 - \overline{w} \cdot P z_i| - |\overline{z_\ell} \cdot P z_i - \overline{w} \cdot P z_i| \\ &\gtrsim \left(1 - |z_i|^2\right)^\beta - \left(1 - |z_i|^2\right)^{\beta\eta} \\ &\gtrsim \left(1 - |z_i|^2\right)^{\beta_1}, \end{aligned}$$

for some $\beta_1 \in (\beta, \rho)$ (again provided the $|z_i|$ are large enough). Thus estimate (15.45) now follows by Remark 15.17. With this done, we have completed the proof of (15.44), as well as the first estimate in (15.38).

To prove the second estimate $\|\varphi\|_{H^\infty(\mathbb{B}_n)} \lesssim 1$ in (15.38), we fix $z \in \mathbb{B}_n$. If $z \in \bigcup_{k \in \mathbb{N}} V_{z_k}$, let ℓ be such that

$$z \in V_{z_\ell} \text{ and } z \notin V_{z_k} \text{ for any } k > \ell.$$

Then we have using (15.43) that

$$\begin{aligned} \varphi (z) - \xi_\ell &= \sum_{i=1}^{\infty} a_i \varphi_i (z) - \xi_\ell \\ &= \left(\sum_{i \in \mathcal{G}_{P(\ell)}} a_i \varphi_i (z) + a_\ell \varphi_\ell (z) + \sum_{i \in \mathcal{F}_\ell} a_i \varphi_i (z) \right) - \left(a_\ell + \sum_{i \in \mathcal{G}_\ell \setminus \{k_0\}} \beta_i a_{P(i)} \right) \\ &= \left(\sum_{i \in \mathcal{G}_\ell \setminus \{k_0\}} a_{P(i)} \left(\varphi_{P(i)} (z) - \beta_i\right) \right) + a_\ell \left(\varphi_\ell (z) - 1\right) + \sum_{i \in \mathcal{F}_\ell} a_i \varphi_i (z), \end{aligned}$$

and the estimates we proved above for z_ℓ show that with z_ℓ replaced by z, we also have

$$\begin{aligned}\left|\varphi(z)-\xi_\ell\right| &\leq \left|\sum_{i\in\mathcal{G}_\ell\setminus\{k_0\}} a_{P(i)}\left(\varphi_{P(i)}(z)-\beta_i\right)\right| + \left|a_\ell\left(\varphi_\ell(z)-1\right)\right| + \left|\sum_{i\in\mathcal{F}_\ell} a_i\varphi_i(z)\right| \\ &\leq C\left(\log\frac{1}{1-\left|z_{k_0}\right|^2}\right)^{-1} + C + \sum_{i\in\mathcal{F}_\ell}\left(\log\frac{1}{1-\left|z_i\right|^2}\right)^{-(p-1)},\end{aligned}$$

which yields $|\varphi(z)| \lesssim 1$. Finally, if $z \notin \bigcup_{k\in\mathbb{N}} V_{z_k}$, then

$$|\varphi(z)| = \left|\sum_{i=1}^{\infty} a_i\varphi_i(z)\right| \lesssim \sum_{i\in\mathbb{N}}\left(\log\frac{1}{1-\left|z_i\right|^2}\right)^{-(p-1)} \lesssim 1$$

by Lemma 15.16. Easy modifications of these estimates also prove Remark 15.20. At this point we can use a normal families argument to prove that φ is holomorphic. This completes the proof of Lemma 15.19. □

15.4.3. The Proof of Multiplier Interpolation. Using Lemma 15.19, we first complete the proof that (15.9) implies (15.7) for $1 < p < \infty$. Fix $s > -1$, $0 < \delta < 1$ and $\{\xi_j\}_{j=1}^{\infty}$ with $\left\|\{\xi_j\}_{j=1}^{\infty}\right\|_{\ell^\infty} = 1$. Then by Lemma 15.19 there is $f_1 = \sum_{i=1}^{\infty} a_i^1\varphi_i \in H^\infty(\mathbb{B}_n)$ such that $\left\|\{\xi_j - f_1(z_j)\}_{j=1}^{\infty}\right\|_{\ell^\infty} < \delta$ and $\left\|\{a_i^1\}_{i=1}^{\infty}\right\|_{\ell^\infty}, \|f_1\|_{H^\infty(\mathbb{B}_n)} \lesssim 1$ where the implied constant is as in (15.38). Now apply Lemma 15.19 to the sequence $\{\xi_j - f_1(z_j)\}_{j=1}^{\infty}$ to obtain the existence of $f_2 = \sum_{i=1}^{\infty} a_i^2\varphi_i \in H^\infty(\mathbb{B}_n)$ such that $\left\|\{\xi_j - f_1(z_j) - f_2(z_j)\}_{j=1}^{\infty}\right\|_{\ell^\infty} < \delta^2$ and $\left\|\{a_i^2\}_{i=1}^{\infty}\right\|_{\ell^\infty}, \|f_2\|_{H^\infty(\mathbb{B}_n)} \lesssim \delta$ where the implied constant is as in (15.38). Continuing inductively, we obtain $f_m = \sum_{i=1}^{\infty} a_i^m\varphi_i \in H^\infty(\mathbb{B}_n)$ such that

$$\begin{aligned}\left\|\left\{\xi_j - \sum_{i=1}^{m} f_i(z_j)\right\}_{j=1}^{\infty}\right\|_{\ell^\infty} &< \delta^m, \\ \|\{a_i^m\}_{i=1}^{\infty}\|_{\ell^\infty}, \|f_m\|_{H^\infty(\mathbb{B}_n)} &\lesssim \delta^{m-1}.\end{aligned}$$

If we now take $\varphi = \sum_{m=1}^{\infty} f_m$, we have

$$\begin{aligned}\xi_j &= \varphi(z_j), \quad 1 \leq j < \infty, \\ \|\varphi\|_{H^\infty(\mathbb{B}_n)} &\leq C_\delta,\end{aligned} \tag{15.46}$$

as well as $\varphi = \sum_{i=1}^{\infty} a_i\varphi_i$ with $\|\{a_i\}_{i=1}^{\infty}\|_{\ell^\infty} \leq C_\delta$. Recall that the series $\varphi = \sum_{i=1}^{\infty} a_i\varphi_i$ converges absolutely by Remark 15.20, and depends linearly on the data $\{\xi_j\}_{j=1}^{\infty}$ by Remark 15.21 and the linear construction in this paragraph. Thus $\varphi \in H^\infty(\mathbb{B}_n)$ linearly interpolates the values $\{\xi_j\}_{j=1}^{\infty}$ on the sequence $\mathcal{Z}$, and it remains to prove that $\varphi \in \mathrm{Mult}(B_p)$. Recall that our function φ depends on our choice of $s > -1$.

By Theorem 14.14, $\varphi \in \mathrm{Mult}(B_p)$ will follow if we show that

$$\left\|\left|(1-|z|^2)^n\nabla^n\varphi(z)\right|^p d\lambda_n(z)\right\|_{CM(B_p)} \lesssim 1. \tag{15.47}$$

Since

$$\varphi = \sum_{i=1}^{\infty} a_i \varphi_i = \sum_{i=1}^{\infty} a_i \Gamma_s g_i = \Gamma_s g$$

where $g = \sum_{i=1}^{\infty} a_i g_i$ with $\sup_{i \geq 1} |a_i| \leq C_\delta$, (15.47) will follow from Lemma 15.14 for s sufficiently large provided we show that (15.21) holds and that

$$\left|(1-|z|^2)^n g(z)\right|^p d\lambda_n(z)$$

is a Carleson measure. From the definition of g_i in (15.32), and the fact that the supports of the g_i are pairwise disjoint by the separation condition, we see that the estimate in (15.21) holds. The Carleson condition will follow from the next lemma.

LEMMA 15.22. *Suppose* (15.5) *holds. With* $1 < p < \infty$ *and* $g = \sum_{i=1}^{\infty} a_i g_i$ *as above, we have*

$$\left\| \left|(1-|z|^2)^n g(z)\right|^p d\lambda_n(z) \right\|_{CM(B_p)} \lesssim 1. \tag{15.48}$$

PROOF. Inequality (15.48) follows from estimate (15.31) for g_{z_i},

$$\int_{\mathbb{B}_n} \left|(1-|\zeta|^2)^n g_{z_i}(\zeta)\right|^p d\lambda_n(\zeta) \lesssim \left(\log \frac{1}{1-|z_i|^2}\right)^{1-p}, \tag{15.49}$$

as follows. Fix an index i. From Remark 15.18 we see that the support of g_{z_i} is essentially the union of a geodesic segment of Bergman cubes $K_1^i, K_2^i, \ldots, K_{\mathcal{M}_i}^i$ where

$$\mathcal{M}_i \approx (\alpha - \rho) \log \frac{1}{1-|z_i|^2}.$$

Indeed, recall that the support g_{z_i} is contained in the intersection of the cone $\mathcal{C}_{z_i}$ and the annulus $\mathcal{A}_{z_i}$. Now for ζ in the cone $\mathcal{C}_{z_i}$, we have $\left|1-\overline{\zeta} \cdot Pz_i\right| \approx 1-|\zeta|^2$, and thus for ζ in the annulus $\mathcal{A}_{z_i}$ as well, we have approximately

$$\log \frac{1}{1-|\zeta|^2} \in \left(\rho \log \frac{1}{1-|z_i|^2}, \alpha \log \frac{1}{1-|z_i|^2}\right).$$

Thus ζ, which is in the support of g_{z_i}, lies in the union of those cubes in $\mathcal{T}_n$ along the geodesic joining the root to the "boundary point" Pz_i, and having tree distance from the root lying roughly between $\rho\beta(0, z_i)$ and $\alpha\beta(0, z_i)$. Moreover, this segment can be continued to a longer sequence of adjacent Bergman cubes $K_1^i, K_2^i, \ldots, K_{\mathcal{M}_i}^i, \ldots, K_{J_i}^i = K_{z_i}$ connecting the support of g_{z_i} to the cube K_{z_i} containing z_i, and where

$$J_i \approx \log \frac{1}{1-|z_i|^2}. \tag{15.50}$$

Choose $w_j^i \in K_j^i$ for $1 \leq j < J_i$. Then with a large integer $L > 2n/p$, and $P_L h(w, z)$ denoting the Lth Taylor polynomial centered at w,

$$P_L h(w, z) := \sum_{\ell=0}^{L} \frac{((z-w) \cdot \nabla)^{\ell}}{\ell!} h(w),$$

we have for $z \in K_m^i$, $1 \le m \le \mathcal{M}_i$, and $f \in B_p$,

$$f(z) = f(z) - \int_{K_m^i} P_L f\left(w_m^i, \cdot\right) d\lambda_n$$
$$+ \sum_{j=m}^{J_i - 1} \left[\int_{K_j^i} P_L f\left(w_j^i, \cdot\right) d\lambda_n - \int_{K_{j+1}^i} P_L f\left(w_{j+1}^i, \cdot\right) d\lambda_n \right] + \int_{K_{z_i}} P_L f\left(z_i, \cdot\right) d\lambda_n ,$$

and so

$$|f(z)|^p \lesssim \left| f(z) - \int_{K_m^i} P_L f\left(w_m^i, \cdot\right) d\lambda_n \right|^p$$
$$+ (J_i)^{p-1} \sum_{j=1}^{J_i - 1} \left| \int_{K_j^i} P_L f\left(w_j^i, \cdot\right) d\lambda_n - \int_{K_{j+1}^i} P_L f\left(w_{j+1}^i, \cdot\right) d\lambda_n \right|^p$$
$$+ \left| \int_{K_{z_i}} P_L f\left(z_i, \cdot\right) d\lambda_n \right|^p$$
$$\lesssim \left| f(z) - \int_{K_m^i} P_L f\left(w_m^i, \cdot\right) d\lambda_n \right|^p$$
$$+ (J_i)^{p-1} \sum_{j=1}^{J_i} \left(\max_{\zeta \in RK_j^i} \left| \int_{RK_j^i} P_L f(\zeta, \cdot) \, d\lambda_n \right| \right)^p + \left| \int_{K_{z_i}} P_L f\left(z_i, \cdot\right) d\lambda_n \right|^p ,$$

for a sufficiently large R. Using this together with (15.49), (15.50), the inequality

$$\left| f(\zeta) - P_L f\left(w_j^i, \zeta\right) \right| \lesssim \left| D_{RK_j^i}^L f \right|, \qquad \text{for } \zeta \in RK_j^i,$$

and the fact that the supports of the g_{z_i} are pairwise disjoint, we obtain

$$\int_{\mathbb{B}_n} |f(z)|^p \left| (1 - |z|^2)^n g(z) \right|^p d\lambda_n(z)$$
$$\lesssim \sum_i \left| D_{K_m^i}^L f \right|^p + \sum_i \sum_{j=1}^{J_i} \left| D_{RK_j^i}^L f \right|^p + \sum_i \left| \int_{K_{z_i}} P_L f\left(z_i, \cdot\right) d\lambda_n \right|^p \left(\log \frac{1}{1 - |z_i|^2} \right)^{1-p}$$
$$:= I + II + III.$$

Since the expanded cubes $\left\{ RK_j^i \right\}$ have bounded overlap in j and are pairwise disjoint in i by Lemma 15.12, the first two terms on the right are dominated by

$$I + II \le \sum_{\alpha \in \mathcal{T}_n} \left| D_{RK_\alpha}^L f \right|^p \lesssim \|f\|_{B_p}^p$$

by the oscillation estimate, Proposition 14.26, using $L > 2n/p$.

The third term III is dominated by $C \|f\|_{B_p}^p$ since $\left| \int_{K_{z_i}} Pf\left(z_i, \cdot\right) d\lambda_n \right| \lesssim \int_{RK_{z_i}} |f| \, d\lambda_n$ by (15.26), and since we are assuming in (15.5) that $d\mu := \sum_i \left(\log \frac{1}{1 - |z_i|^2} \right)^{1-p} \delta_{z_i}$ is a B_p-Carleson measure. Indeed, Theorem 14.13 shows that the measure

$$d\widetilde{\mu_R}(z) = \sum_{\alpha \in \mathcal{T}_n} \left(\int_{RK_\alpha} d\mu \right) \chi_{RK_\alpha}(z) \, d\lambda_n(z)$$

in (14.31) is a Carleson measure, and so from this and (15.26) we obtain

$$\begin{aligned} III &\lesssim \sum_i \left(\int_{RK_{z_i}} |f|\, d\lambda_n \right)^p \left(\log \frac{1}{1-|z_i|^2} \right)^{1-p} \\ &\lesssim \sum_i \left(\int_{RK_{z_i}} |f|^p\, d\lambda_n \right) \mu\left(RK_{z_i}\right) \leq \left(\int_{\mathbb{B}_n} |f|^p\, d\widetilde{\mu_R} \right) \lesssim \|f\|_{B_p}^p . \end{aligned}$$

This completes the proof of Lemma 15.22, and hence completes the proof that (15.5) implies (15.7). □

15.5. Proof of Besov Space Interpolation

Here we adapt the above arguments to prove that (15.6), or equivalently (15.5), implies (15.3) for $1 < p < \infty$, and with linear interpolation. We note in passing that we already have that (15.6) implies (15.3) for $1 < p \leq 2$. In fact, the Carleson embedding in (15.6) implies that the map

$$Tf = \{\kappa_j f(z_j)\}_{j=1}^{\infty}$$

in (15.3) maps B_p *into* ℓ^p. On the other hand, (15.6) implies (15.7) implies (15.8) implies (this is where we use $p \leq 2$) the inequality $\gtrsim$ of (15.4), which in turn implies that the map T in (15.3) maps B_p *onto* ℓ^p.

Fix $\alpha, s > -1$ and $\{\xi_j\}_{j=1}^{\infty}$ with

$$\left\| \left\{ \frac{\xi_j}{\left\|k_{z_j}^{\alpha,p}\right\|_{B_{p'}}} \right\}_{j=1}^{\infty} \right\|_{\ell^p} = 1.$$

We may suppose that the sequence $\{z_j\}_{j=1}^J$ is finite, and obtain an appropriate estimate independent of $J \geq 1$ (see (15.52) below). Since the sequence $\{\xi_j\}_{j=1}^J$ is bounded (we are not concerned that this bound blows up as $J \to \infty$), the construction in the previous section leading to (15.46) yields $\varphi = \sum_{m=1}^J \varphi_m$ satisfying

$$\xi_j = \varphi(z_j), \qquad 1 \leq j < J,$$

$$\|\varphi\|_{H^\infty(\mathbb{B}_n)} \lesssim \left\| \{\xi_j\}_{j=1}^J \right\|_{\ell^\infty},$$

as well as $\varphi = \sum_{i=1}^J a_i \varphi_i$ with

$$\left\| \{a_i\}_{i=1}^J \right\|_{\ell^\infty} \lesssim \left\| \{\xi_j\}_{j=1}^J \right\|_{\ell^\infty}.$$

We need additional weighted ℓ^p control on the coefficients $\{a_i\}_{i=1}^J$. Let μ be the measure assigning mass $\left(\log \frac{1}{1-|z_j|^2}\right)^{1-p}$ to the point $j \in \{0, 1, 2, \ldots, J\}$. Recall that by (14.24), $\left\|k_{z_j}^{\alpha,p}\right\|_{B_{p'}} \approx \left(\log \frac{1}{1-|z_j|^2}\right)^{1/p'}$ so that we have

$$\left\| \{b_j\}_{j=1}^J \right\|_{\ell^p(\mu)} \approx \left\| \left\{ \frac{b_j}{\left\|k_{z_j}^{\alpha,p}\right\|_{B_{p'}}} \right\}_{j=1}^{J} \right\|_{\ell^p},$$

for any sequence $\{b_j\}_{j=1}^J$.

LEMMA 15.23. *The sequence* $\{a_i\}_{i=1}^J$ *constructed in* (15.46) *using Lemma* 15.19 *satisfies*

$$\left\|\{a_j\}_{j=1}^J\right\|_{\ell^p(\mu)} \lesssim \left\|\{\xi_j\}_{j=1}^J\right\|_{\ell^p(\mu)} \approx 1. \tag{15.51}$$

Once this lemma is established we can easily complete the proof. We do that now and postpone for a moment the proof of the lemma.

We have that $\varphi \in H^\infty(\mathbb{B}_n)$ interpolates the values $\{\xi_j\}_{j=1}^J$ on the sequence $\{z_j\}_{j=1}^J$, and it remains only to prove that $\varphi \in B_p$ with $\|\varphi\|_{B_p} \lesssim 1$ whenever $\left\|\{\xi_j\}_{j=1}^J\right\|_{\ell^p(\mu)} = 1$, independent of $J \geq 1$. Thus we must show that

$$\int_{\mathbb{B}_n} \left|(1-|z|^2)^n \nabla^n \varphi(z)\right|^p d\lambda_n(z) \lesssim 1,$$

independent of $J \geq 1$. Now

$$\varphi = \sum_{i=1}^J a_i \varphi_i = \sum_{i=1}^J a_i \Gamma_s g_i = \Gamma_s g$$

where $g = \sum_{i=1}^J a_i g_i$ with $\left\|\{a_i\}_{i=1}^J\right\|_{\ell^\infty} \leq C_\delta \left\|\{\xi_j\}_{j=1}^J\right\|_{\ell^\infty}$. Moreover,

$$|\nabla^n \Gamma_s g(z)| \leq C'_{s,n} \widehat{T_s}|g|(z)$$

by (15.28) where the operator $\widehat{T_s}$ is given by

$$\widehat{T_s} f(z) = c_{s,n} \int_{\mathbb{B}_n} \frac{f(w)(1-|w|^2)^s}{|1-\overline{w}\cdot z|^{n+1+s}} dV.$$

Thus we must estimate

$$\int_{\mathbb{B}_n} \left|(1-|z|^2)^n \widehat{T_s}|g|(z)\right|^p d\lambda_n(z) = \int_{\mathbb{B}_n} \left|\widehat{T_s}|g|(z)\right|^p d\nu_t(z)$$

where $t = pn - n - 1$. Now by [**346**, Theorem 2.10], $\widehat{T_s}$ is bounded on $L^p(d\nu_t)$ if and only if $0 < t+1 < p(s+1)$, i.e.

$$p > 1 \text{ and } s > \frac{n}{p'} - 1.$$

Thus choosing $s > -1 + n/p'$, we have

$$\begin{aligned}\int_{\mathbb{B}_n} \left|(1-|z|^2)^n \nabla^n \varphi(z)\right|^p d\lambda_n(z) &\lesssim \int_{\mathbb{B}_n} |g(z)|^p d\nu_t(z) \\ &= \int_{\mathbb{B}_n} \left|(1-|z|^2)^n g(z)\right|^p d\lambda_n(z).\end{aligned}$$

Since the supports of the g_i are pairwise disjoint by the separation condition, we obtain from (15.31) and then (15.51) that $g = \sum_{i=1}^{J} a_i g_i$ satisfies

$$\begin{aligned}\int_{\mathbb{B}_n} \left|(1-|z|^2)^n g(z)\right|^p d\lambda_n(z) &= \sum_{j=1}^{J} |a_i|^p \int_{\mathbb{B}_n} \left|(1-|z|^2)^n g_i(z)\right|^p d\lambda_n(z)\\ &\lesssim \sum_{j=1}^{J} |a_i|^p \left(\log \frac{1}{1-|z_j|^2}\right)^{1-p}\\ &\approx \left\|\{a_j\}_{j=1}^{J}\right\|_{\ell^p(\mu)}\\ &\lesssim \left\|\{\xi_j\}_{j=1}^{J}\right\|_{\ell^p(\mu)} \approx 1.\end{aligned}$$

Altogether we have proved that $\varphi \in B_p$ interpolates the values $\{\xi_j\}_{j=1}^{J}$ on the sequence $\{z_j\}_{j=1}^{J}$ with the norm estimate

$$\|\varphi\|_{B_p} \lesssim \left\|\{\xi_j\}_{j=1}^{J}\right\|_{\ell^p(\mu)}, \tag{15.52}$$

where the implied constant is independent of $J \geq 1$ as required. A limiting argument now finishes the proof that (15.6) implies (15.3).

Proof of Lemma 15.23. Recall from the proof of Lemma 15.19 that the approximating sequence $\{a_j\}_{j=1}^{J}$ (not the actual sequence used in (15.51) above, but rather the sequence appearing in the conclusion of Lemma 15.19) is given by

$$a_\ell = \xi_\ell - \sum_{i \in \mathcal{G}_\ell \setminus \{k_0\}} \beta_i a_{A(i)}, \qquad 1 \leq \ell \leq J,$$

where for any given ℓ, the numbers β_i lie in $[0,1]$, and the geodesics $\mathcal{G}_\ell$ and $\mathcal{G}_{A(\ell)}$ lie in a fixed tree $\mathcal{Y}$ of the forest $\{j\}_{j=1}^{J}$, and are related by

$$\mathcal{G}_{A(\ell)} = \mathcal{G}_\ell \setminus \{\ell\},$$

i.e. if $\mathcal{G}_\ell = [k_0, k_1, \ldots, k_{m-1}, k_m]$ with $k_m = \ell$, then $k_{m-1} = A(\ell)$ and $\mathcal{G}_{A(\ell)} = [k_0, k_1, \ldots, k_{m-1}]$. Without loss of generality, we may assume that the forest of indices $\{j\}_{j=1}^{J}$ is actually a single tree $\mathcal{Y}$.

Now fix ℓ. For the moment we pass to more convenient notation and relabel the points $\{z_j\}_{j\in\mathcal{G}_\ell} = \{z_{k_0}, z_{k_1}, \ldots, z_{k_m}\}$ as $\{z_0, z_1, \ldots, z_m\}$, and similarly relabel $\{a_0, a_1, \ldots, a_m\}$, $\{\xi_0, \xi_1, \ldots, \xi_m\}$ and $\{\beta_0, \beta_1, \ldots, \beta_m\}$ so that

$$a_k = \xi_k - \sum_{i=1}^{k} \beta_i a_{i-1}, \qquad 0 \leq k \leq \ell.$$

We also have $\mu(j) = \left(\log \frac{1}{1-|\gamma_j|^2}\right)^{1-p}$ where z_j now denotes the point which, before relabelling, had been denoted z_{k_j} and lies in the ball corresponding to what was then k_j. In other words, we are restricting attention to the geodesic $\mathcal{G}_\ell$ and relabeling sequences so as to conform to the ordering in the geodesic. Now let

$$\omega_k = \sum_{i=1}^{k} \beta_i a_{i-1}$$

for $1 \leq k \leq \ell$ so that

$$a_k = \xi_k - \omega_k, \quad 0 \leq k \leq \ell.$$

We now claim that

$$(15.53) \qquad \begin{pmatrix} \omega_1 \\ \omega_2 \\ \vdots \\ \omega_k \end{pmatrix} = \begin{bmatrix} b_{11} & 0 & \cdots & 0 \\ b_{21} & b_{22} & \cdots & 0 \\ \vdots & \vdots & \ddots & \vdots \\ b_{k1} & b_{k2} & \cdots & b_{kk} \end{bmatrix} \begin{pmatrix} \xi_0 \\ \xi_1 \\ \vdots \\ \xi_{k-1} \end{pmatrix}, \quad 1 \leq k \leq \ell,$$

where $b_{ij} \in [0,1]$ and $b_{ij} = 0$ if $i < j$. We prove this by induction on k. This is evident for $k = 1$ since then $\omega_1 = \beta_1 a_0 = \beta_1 \xi_0$. Assuming its truth for k, we obtain from

$$\omega_{k+1} = \beta_{k+1} a_k + \sum_{i=1}^{k} \beta_i a_{i-1} = \beta_{k+1} (\xi_k - \omega_k) + \omega_k = (1 - \beta_{k+1}) \omega_k + \beta_{k+1} \xi_k,$$

that the vector $\begin{pmatrix} \omega_1 \\ \omega_2 \\ \vdots \\ \omega_k \\ \omega_{k+1} \end{pmatrix}$ is given by

$$\begin{bmatrix} b_{11} & 0 & & 0 & 0 \\ b_{21} & b_{22} & \cdots & 0 & 0 \\ \vdots & \vdots & \ddots & \vdots & \vdots \\ b_{k1} & b_{k2} & \cdots & b_{kk} & 0 \\ (1-\beta_{k+1}) b_{k1} & (1-\beta_{k+1}) b_{k2} & \cdots & (1-\beta_{k+1}) b_{kk} & \beta_{k+1} \end{bmatrix} \begin{pmatrix} \xi_0 \\ \xi_1 \\ \vdots \\ \xi_{k-1} \\ \xi_k \end{pmatrix},$$

which has the desired form, thus proving (15.53). The consequence we need from (15.53) is

$$(15.54) \qquad |a_k| = |\xi_k - \omega_k| \leq |\xi_k| + |\omega_k| \leq \sum_{j=0}^{k} |\xi_j|, \quad 0 \leq k \leq \ell.$$

We now return our attention to the tree $\mathcal{Y}$. For each $\alpha \in \mathcal{Y}$, with corresponding index $j \in \{j\}_{j=1}^{J}$, there are values $a(\alpha) = a_j$ and $\xi(\alpha) = \xi_j$. Define functions $f(\alpha) = |a(\alpha)|$ and $g(\alpha) = |\xi(\alpha)|$ on the tree $\mathcal{Y}$. Note that we are simply relabelling the indices $\{j\}_{j=1}^{J}$ as $\alpha \in \mathcal{Y}$ to emphasize the tree structure of $\mathcal{Y}$ when convenient. The inequality (15.54) says that

$$(15.55) \qquad f(\alpha) \leq I g(\alpha), \quad \alpha \in \mathcal{Y}.$$

Recall also that we are assuming that the measure $d\mu = \sum_{\alpha \in \mathcal{Y}} \left(\log \frac{1}{1-|z_\alpha|^2} \right)^{1-p}$ is a B_p Carleson measure on the ball $\mathbb{B}_n$, where $z_\alpha = z_j \in \mathbb{B}_n$ if α corresponds to j. The only consequence of this that we need here is the simple condition,

$$(15.56) \qquad \beta(0,\alpha)^{p-1} \sum_{j: z_j \in S(\alpha)} \mu(j) \lesssim 1, \quad \alpha \in \mathcal{T}_n.$$

Note that this last inequality refers to the tree $\mathcal{T}_n$ rather than to $\mathcal{Y}$. Using the fact that $\beta(0,\alpha) \approx \log \frac{1}{1-|z_\alpha|^2}$, we obtain from this simple condition that if $S(\alpha) \approx V_{z_k}$, i.e. $\alpha \approx \left[1-\left(1-|z_k|^\beta\right)\right] z_k$, then

$$\begin{aligned}\sum_{\beta:\beta\geq\alpha}\mu(\beta) = \sum_{j:z_j\in S(\alpha)}\mu(j) &\lesssim \beta(0,\alpha)^{1-p} \\ &\lesssim \left(\log\frac{1}{(1-|z|^2)^\beta}\right)^{1-p} \\ &\lesssim \left(\log\frac{1}{1-|z_k|^2}\right)^{1-p} \lesssim \mu(\alpha),\end{aligned}$$

by the definition of the region V_{z_k}. Thus on the tree $\mathcal{Y}$, we have

$$I^*\mu(\alpha) \lesssim \mu(\alpha), \qquad \alpha\in\mathcal{Y}. \tag{15.57}$$

We now claim that (15.57) implies the inequality

$$\sum_{\alpha\in\mathcal{Y}} Ig(\alpha)^p\,\mu(\alpha) \lesssim \sum_{\alpha\in\mathcal{Y}} g(\alpha)^p\,\mu(\alpha), \tag{15.58}$$

which together with (15.55) yields (15.51).

To see (15.58), we use our more general tree theorem for the tree $\mathcal{Y}$ which we referred to after Theorem 5.17. That is,

$$\sum_{\alpha\in\mathcal{Y}} Ig(\alpha)^p\,w(\alpha) \lesssim \sum_{\alpha\in\mathcal{Y}} g(\alpha)^p\,v(\alpha), \qquad g\geq 0,$$

if and only if

$$\sum_{\beta\geq\alpha} I^*w(\beta)^{p'}\,v(\beta)^{1-p'} \lesssim I^*w(\alpha)<\infty, \qquad \alpha\in\mathcal{Y}. \tag{15.59}$$

With $w=v=\mu$, (15.57) yields condition (15.59) as follows:

$$\sum_{\beta\geq\alpha} I^*\mu(\beta)^{p'}\,\mu(\beta)^{1-p'} \lesssim \sum_{\beta\geq\alpha}\mu(\beta)^{p'}\,\mu(\beta)^{1-p'} \approx \sum_{\beta\geq\alpha}\mu(\beta)\approx I^*\mu(\alpha).$$

This finishes the proof that (15.6) implies (15.3) for all $1<p<\infty$. □

REMARK 15.24. *Note that the above proof shows that the map T in* (15.3) *is* onto, *provided both* (15.56) *and the separation condition in* (15.5) *hold. Moreover, the proof shows that separation and condition* (15.59) *with $w=v=\mu$, i.e.*

$$\sum_{z_j\in V_{z_k}}\left[\sum_{z_i\in V_{z_j}}\left(\log\frac{1}{1-|z_i|^2}\right)^{1-p}\right]^{p'}\left(\log\frac{1}{1-|z_j|^2}\right) \tag{15.60}$$

$$\lesssim \sum_{z_j\in V_{z_k}}\left(\log\frac{1}{1-|z_j|^2}\right)^{1-p}<\infty, \tag{15.61}$$

for all k, suffices for the map T in (15.3) *to be onto. In fact, in the presence of the separation condition, the map T in* (15.3) *takes the subspace $(B_p\cap\operatorname{span}\{\varphi_{z_i}\})^-$ boundedly into and onto ℓ^p if and only if* (15.60) *holds. The necessity of* (15.60) *can be obtained by testing the inequality for T with functions of the form $f(z)=\sum a_i\varphi_{z_i}(z)$ and noting that $f(z_j)=\sum a_i\varphi_{z_i}(z_j)\approx\sum_{z_i\leq z_j}a_i$ by Lemma* 15.16 *and the Remark following it. This yields* (15.58), *and hence* (15.60). *While the separation condition is necessary for the map T in* (15.3) *to be* onto, *we do not* *know a characterization of when the map T in* (15.3) *is* onto. *Onto interpolation is discussed in Chapter* 8 *and in* [**42**].

15.6. Completing the Multiplier Interpolation Loop

At this point Theorem 15.1 has been completely proved, and the only assertion remaining to be proved in Theorem 15.2 is that (15.7) implies (15.9) in the range $1 + 1/(n-1) \leq p < 2$. We do that in the next chapter, the main tool being the introduction and study of the spaces $HB_p(\mathcal{T}_n)$, "holomorphic Besov spaces on the Bergman tree $\mathcal{T}_n$" and their multiplier algebras $\operatorname{Mult}(HB_p(\mathcal{T}_n))$. That material, which allows us to complete the proof of Theorem 15.2, seems intrinsically interesting and we present it in some detail. That requires a lengthy digression from the current proof, and so we pause before going to those details to give a brief informal look at the path ahead, to indicate how, with those techniques in hand, we finish the proof.

The spaces we introduce require an elaborate "complex structure" on $\mathcal{T}_n$, and membership in the spaces involves restrictions on the functions which are somewhat similar to, but more complicated than, those placed on martingales. These spaces serve as analogs for and models of the Besov and multiplier spaces $B_p(\mathbb{B}_n)$ and $\operatorname{Mult}(B_p(\mathbb{B}_n))$ and have three properties of particular importance:

(1) The natural restriction map is bounded from $\operatorname{Mult}(B_p)$ to $\operatorname{Mult}(HB_p(\mathcal{T}_n))$.
(2) The reproducing kernels for $HB_p(\mathcal{T}_n)$ satisfy a positivity property, analogous to the property $\operatorname{Re} \overline{w}^{-1} k'_w(z) > 0$ satisfied by the reproducing kernel of $B_p(\mathbb{B}_1)$. That is crucial in proving an analog of Lemma 15.10, Böe's curious lemma.
(3) Carleson measures for $HB_p(\mathcal{T}_n)$ are characterized by the tree condition of the type (14.37).

With these three properties available we can complete the proof that (15.7) implies (15.9). Indeed, property (1) will show that knowing interpolation in $\operatorname{Mult}(B_p(\mathbb{B}_n))$ is possible implies $\operatorname{Mult}(HB_p(\mathcal{T}_n))$ interpolation is possible on the Bergman tree. Property (2) will then show that the atomic measure μ associated with the interpolation sequence is a Carleson measure for $HB_p(\mathcal{T}_n)$. Finally, property (3) will show that μ satisfies the tree condition, and hence, by Theorem 14.16, μ is a B_p-Carleson measure. This will complete the multiplier interpolation loop since we have already shown in Section 15.5, that if μ is a B_p-Carleson measure, then $\operatorname{Mult}(B_p)$ interpolation holds on the ball.

We will see that property (1) follows from Theorem 16.38 if $m > 2n/p$ and the structural constant θ is large enough; property (2) follows from Lemma 16.35 if in addition the structural constant λ is small enough compared to θ; and finally, property (3) follows from Theorem 16.36 if both λ is small enough and θ is large enough.

Here are some of the details. If $\mathcal{Z} \subset \mathbb{B}_n$ interpolates $\operatorname{Mult}(B_p)$, i.e. (15.3) holds, and if we construct the Bergman tree $\mathcal{T}_n$ so that $\{c_\alpha\}_{\alpha \in \mathcal{T}_n}$ contains $\mathcal{Z}$, say with $z_j = c_{\alpha_j}$, then it follows easily from Theorem 16.38 that $\{\alpha_j\}_{j=1}^\infty$ interpolates $\operatorname{Mult}(HB_p(\mathcal{T}_n))$, i.e.

$$f \to \{f(\alpha_j)\}_{j=1}^\infty \text{ maps } \operatorname{Mult}(HB_p(\mathcal{T}_n)) \text{ boundedly into and onto } \ell^\infty. \tag{15.62}$$

Indeed, to see that (15.62) holds, suppose that $\{\xi_j\}_{j=1}^{\infty} \in \ell^{\infty}$. Using (15.3) we can find $\varphi \in \mathrm{Mult}(B_p)$ satisfying

$$\varphi(z_j) = \xi_j, \qquad 1 \leq j < \infty,$$
$$\|\varphi\|_{\mathrm{Mult}(B_p)} \lesssim \left\| \{\xi_j\}_{j=1}^{\infty} \right\|_{\infty}.$$

Now define f on the tree $\mathcal{T}_n$ by

$$f(\alpha) = \varphi(c_\alpha), \qquad \alpha \in \mathcal{T}_n.$$

Then we have

$$f(\alpha_j) = \varphi\left(c_{\alpha_j}\right) = \varphi(z_j) = \xi_j$$

and Theorem 16.38 shows that

$$\|f\|_{\mathrm{Mult}(HB_p(\mathcal{T}_n))} \lesssim \|\varphi\|_{\mathrm{Mult}(B_p)},$$

thus establishing (15.62).

We can now use functional analytic arguments, together with the positivity property (16.64) of the reproducing kernels $k_\alpha^{(m,0)}$, with $m > 2n/p$ in Lemma 16.35, to show that the measure

$$d\mu = \sum_{j=1}^{\infty} \left\| k_{\alpha_j}^{(m,0)} \right\|_{HB_{p'}(\mathcal{T}_n)}^{-p} \delta_{\alpha_j}$$

is a $HB_p(\mathcal{T}_n)$-Carleson measure. Theorem 16.36 then shows that μ satisfies the tree condition (14.37). Finally then, to obtain that

$$d\nu = \sum_{j=1}^{\infty} \left(\log \frac{1}{1 - |\alpha_j|^2} \right)^{1-p} \delta_{\alpha_j}$$

satisfies the tree condition (14.37), we use

$$\|k_\alpha\|_{HB_{p'}(\mathcal{T}_n)}^{p'} \approx \sum_{\gamma \in [o,\alpha]} 1 = d(\alpha) \approx \left(\log \frac{1}{1 - |\alpha|^2} \right),$$

by (16.64).

Here are those arguments in more detail. For convenience in notation, we abbreviate the space $HB_{p'}(\mathcal{T}_n)$ by $HB_{p'}$, the kernel $k_{\alpha_j}^{(m,0)}$ by k_{α_j} and write K_j for the normalized kernel function $\left\|k_{\alpha_j}\right\|_{HB_{p'}}^{-1} k_{\alpha_j}$. From (15.62) we obtain in the usual way that $\left\{k_{\alpha_j}\right\}_{j=1}^{\infty}$ is an unconditional basic sequence in $HB_{p'}$:

$$\left\| \sum_{j=1}^{\infty} b_j k_{\alpha_j} \right\|_{HB_{p'}} \lesssim \left\| \sum_{j=1}^{\infty} a_j k_{\alpha_j} \right\|_{HB_{p'}}, \qquad \text{whenever } |b_j| \leq |a_j|. \tag{15.63}$$

We will now use (16.64),

$$r^{-md(\gamma)} \operatorname{Re}\left(D_m^m k_\alpha^{(m,0)}(\gamma) \wedge \{\otimes^m \gamma\} \right) \approx 1$$
$$\left| D_m^\ell k_\alpha^{(m,t)}(\gamma) \right|_\gamma + \left| \mathbf{Q}_\gamma^{(m,t)} \left(\mathcal{D}_\gamma k_\alpha^{(m,t)} \right) \right|_\gamma \leq \begin{cases} C & \text{for} \quad \gamma \leq \alpha \\ 0 & \qquad\quad \text{otherwise,} \end{cases}$$

and (15.63), along with a modification of the technique of Böe's curious lemma in [**77**, Lemma 3.1], to obtain the following norm equivalence:

$$\left\| \sum_{j=1}^{\infty} a_j K_j \right\|_{HB_{p'}} \approx \left(\sum_{j=1}^{\infty} |a_j|^{p'} \right)^{1/p'} . \tag{15.64}$$

Indeed, we use the ℓ^p variant (16.79) of Definition 16.33 for the scalar function f, namely

$$\begin{aligned} \|f\|_{HB_p(\mathcal{T}_n)}^p &:= \|f\|_{HB_{p,m}(\mathcal{T}_n)}^p \\ &= \sum_{s_k<m} \left| D_{m-s_{k-1}}^{\ell_k} \cdots D_{m-\ell_1}^{\ell_2} D_m^{\ell_1} f(o) \right|^p \\ &\qquad + \sum_{s_k=m} \sum_{\alpha\in\mathcal{T}_n} \left| \mathbf{Q}_\alpha^{(m-s_{k-1},s_{k-1})} \mathcal{D}_\alpha D_{m-s_{k-2}}^{\ell_{k-1}} \cdots D_m^{\ell_1} f \right|_\alpha^p \\ &\qquad + \sum_{s_k=m} \sum_{\alpha\in\mathcal{T}_n} \left| D_{m-s_{k-1}}^{\ell_k} \cdots D_m^{\ell_1} f(\alpha) \right|_\alpha^p \\ &= \left\| \check{D} f \right\|_{\ell^p(\mathcal{T}_n)}^p , \end{aligned}$$

where $\check{D}f$ is defined on the tree in the obvious way and the ℓ^p norm involves the metric $|\cdot|_\alpha$ as usual, together with the Rademacher functions $r_j(t)$ in conjunction with (15.63) to obtain

$$\begin{aligned} \left\| \sum_{j=1}^{\infty} a_j K_j \right\|_{HB_{p'}}^{p'} &\approx \int_0^1 \left\| \sum_{j=1}^{\infty} a_j r_j(t) K_j \right\|_{HB_{p'}}^{p'} dt \\ &\approx \int_0^1 \left\| \sum_{j=1}^{\infty} a_j r_j(t) \check{D} K_j(\gamma) \right\|_{\ell^{p'}(\gamma)}^{p'} dt. \end{aligned}$$

Now Khinchine's inequality holds for finite-dimensional vector spaces in place of scalars. Indeed, if $\mathbf{a}_j = (a_j(i))_{i=1}^N \in \mathbb{C}^N$, then

$$\begin{aligned} \left(\int_0^1 \left| \sum_{j=1}^{\infty} r_j(t) \mathbf{a}_j \right|^q dt \right)^{1/q} &\approx \left(\int_0^1 \left\{ \sum_{i=1}^{N} \left| \sum_{j=1}^{\infty} r_j(t) a_j(i) \right|^q \right\} dt \right)^{1/q} \\ &\approx \sum_{i=1}^{N} \left(\int_0^1 \left\{ \left| \sum_{j=1}^{\infty} r_j(t) a_j(i) \right|^q \right\} dt \right)^{1/q} \\ &\approx \sum_{i=1}^{N} \left(\sum_{j=1}^{\infty} |a_j(i)|^2 \right)^{1/2} \\ &\approx \left(\sum_{j=1}^{\infty} |\mathbf{a}_j|^2 \right)^{1/2} . \end{aligned}$$

Thus with N sufficiently large, we have that $\left\|\sum_{j=1}^{\infty} a_j K_j\right\|_{HB_{p'}}^{p'}$ is dominated by a constant multiple of

$$\left\|\left\{\sum_{j=1}^{\infty}\left|\left|a_j\right|\check{D}K_j\left(\gamma\right)\right|_{\gamma}^{2}\right\}^{1/2}\right\|_{\ell^{p'}(\gamma)}^{p'}.$$

Using (16.64), we can then continue the above with

$$\left\|\sum_{j=1}^{\infty} a_j K_j\right\|_{HB_{p'}}^{p'}$$

$$\lesssim \left\|\left\{\sum_{j=1}^{\infty}\left|\frac{\left|a_j\right|}{\left\|k_{\alpha_j}\right\|_{HB_{p'}}} r^{-md(\gamma)}\operatorname{Re}\left(D_m^m k_{\alpha}^{(m,0)}\left(\gamma\right)\wedge\left\{\otimes^m\gamma\right\}\right)\right|^2\right\}^{1/2}\right\|_{\ell^{p'}(\gamma)}^{p'}.$$

Let

$$\begin{aligned}A_j\left(\gamma\right) &:= \frac{\left|a_j\right|}{\left\|k_{\alpha_j}\right\|_{HB_{p'}}} r^{-md(\gamma)}\operatorname{Re}\left(D_m^m k_{\alpha}\left(\gamma\right)\wedge\left\{\otimes^m\gamma\right\}\right)\\ &= \operatorname{Re}\left\{\frac{\left|a_j\right|}{\left\|k_{\alpha_j}\right\|_{HB_{p'}}} r^{-md(\gamma)} D_m^m k_{\alpha}\left(\gamma\right)\right\}\wedge\left\{\otimes^m\gamma\right\}.\end{aligned}$$

We now use $\left\|\cdot\right\|_{\ell^2}\leq\sqrt{\left\|\cdot\right\|_{\ell^1}\left\|\cdot\right\|_{\ell^\infty}}$ and the nonnegativity of $A_j\left(\gamma\right)$ to obtain

$$\begin{aligned}\left\|\sum_{j=1}^{\infty} a_j K_j\right\|_{HB_{p'}}^{p'} &\lesssim \left\|\left\{\sum_{j=1}^{\infty} A_j\left(\gamma\right)^2\right\}^{1/2}\right\|_{\ell^{p'}(\gamma)}^{p'}\\ &\lesssim \left\|\left\{\left[\sum_{j=1}^{\infty} A_j\left(\gamma\right)\right]\left[\sup_{1\leq j<\infty} A_j\left(\gamma\right)\right]\right\}^{1/2}\right\|_{\ell^{p'}(\gamma)}^{p'}\\ &\lesssim \left\|\sum_{j=1}^{\infty} A_j\left(\gamma\right)\right\|_{\ell^{p'}(\gamma)}^{p'/2}\left\|\sup_{1\leq j<\infty} A_j\left(\gamma\right)\right\|_{\ell^{p'}(\gamma)}^{p'/2}.\end{aligned}$$

Now set $K=\sum_{j=1}^{\infty} K_j$ so that

$$\begin{aligned}\sum_{j=1}^{\infty} A_j\left(\gamma\right) &= \operatorname{Re}\left\{\sum_{j=1}^{\infty}\frac{\left|a_j\right|}{\left\|k_{\alpha_j}\right\|_{HB_{p'}}} r^{-md(\gamma)} D_m^m k_{\alpha_j}\left(\gamma\right)\right\}\wedge\left\{\otimes^m\gamma\right\}\\ &= \operatorname{Re}\left\{r^{-md(\gamma)} D_m^m K\left(\gamma\right)\right\}\wedge\left\{\otimes^m\gamma\right\}.\end{aligned}$$

Using the inequality

$$\left|r^{-md(\gamma)} D_m^m K\left(\gamma\right)\right|^2\leq\left|\check{D}K\right|_{\gamma}^{2},\tag{15.65}$$

and (16.79) again, we obtain

$$\begin{aligned}\left\|\sum_{j=1}^{\infty} A_j(\gamma)\right\|_{\ell^{p'}(\gamma)} &= \left\|\operatorname{Re}\left\{r^{-md(\gamma)} D_m^m K(\gamma)\right\} \wedge \{\otimes^m \gamma\}\right\|_{\ell^{p'}(\gamma)} \\ &\lesssim \left\|\check{D}K\right\|_{\ell^{p'}(\gamma)} \lesssim \|K\|_{HB_{p'}} \\ &\approx \left\|\sum_{j=1}^{\infty} |a_j| K_j\right\|_{HB_{p'}} .\end{aligned}$$

Another application of the unconditional basic sequence property (15.63), shows that this is dominated by a constant multiple of

$$\left\|\sum_{j=1}^{\infty} a_j K_j\right\|_{HB_{p'}} .$$

Altogether we now have

$$\left\|\sum_{j=1}^{\infty} a_j K_j\right\|_{HD_{p'}}^{p'} \lesssim \left\|\sum_{j=1}^{\infty} a_j K_j\right\|_{HD_{p'}}^{p'/2} \left\|\sup_{1\le j<\infty} A_j(\gamma)\right\|_{\ell^{p'}(\gamma)}^{p'/2},$$

which yields

$$\begin{aligned}\left\|\sum_{j=1}^{\infty} a_j K_j\right\|_{HB_{p'}} &\lesssim \left\|\sup_{1\le j<\infty} A_j(\gamma)\right\|_{\ell^{p'}(\gamma)} \lesssim \left\|\left(\sum_{j=1}^{\infty} A_j(\gamma)^{p'}\right)^{1/p'}\right\|_{\ell^{p'}(\gamma)} \\ &\approx \left\{\sum_{j=1}^{\infty}\left(\frac{|a_j|}{\|k_{\alpha_j}\|_{HB_{p'}}}\right)^{p'} \left\|\operatorname{Re}\left\{r^{-md(\gamma)} D_m^m k_{\alpha_j}(\gamma)\right\} \wedge \{\otimes^m \gamma\}\right\|_{\ell^{p'}(\gamma)}^{p'}\right\}^{1/p'} \\ &\lesssim \left\{\sum_{j=1}^{\infty} |a_j|^{p'}\right\}^{1/p'},\end{aligned}$$

since by (15.65) and (16.79) once more,

$$\begin{aligned}&\left\|\operatorname{Re}\left\{r^{-md(\gamma)} D_m^m k_{\alpha_j}(\gamma)\right\} \wedge \{\otimes^m \gamma\}\right\|_{\ell^{p'}(\gamma)} \\ &\qquad \lesssim \left\|\check{D}k_{\alpha_j}(\gamma)\right\|_{\ell^{p'}(\gamma)} \lesssim \left\|k_{\alpha_j}\right\|_{HB_{p'}} .\end{aligned}$$

This completes the proof of the inequality $\lesssim$ in (15.64), and the opposite inequality is standard.

From the inequality $\lesssim$ in (15.64), we obtain in the usual way, see e.g. the relevant part of the proof that (15.4) implies (15.3), that the measure

$$d\mu = \sum_{j=1}^{\infty} \left\|k_{\alpha_j}\right\|_{HB_{p'}}^{-p} \delta_{\alpha_j}$$

is a $HB_p(\mathcal{T}_n)$-Carleson measure, and as shown above, this completes the loop.

15.7. Notes and Comments

An excellent overview of interpolating sequences in a variety of function spaces is given in [**302**].

The approach to interpolating sequences in this chapter is from [**47**]. That work evolves from the constructive approach to interpolating sequences for the Dirichlet space by Marshall-Sundberg [**220**] and Bishop [**72**], and for the single variable Besov spaces by Böe [**77**].

Questions similar to the ones in this chapter have also been studied for other spaces of holomorphic functions on the ball, including the work in [**20**] and [**19**].

CHAPTER 16

Spaces on Trees

In this chapter we present several types of function spaces on the tree $\mathcal{T}_n$.

The first, $B_{p,1}(\mathcal{T}_n)$, defined using the size of the first differences of functions, is a straightforward analog of the dyadic Dirichlet space $\mathcal{D}(\mathcal{T}_n)$. We develop basic properties of $B_{p,1}(\mathcal{T}_n)$ including its relation to the space of holomorphic functions $B_{p,1}(\mathbb{B}_n)$, and we find that, if $p > 2n$, this relationship can be used to complete the proof in the previous chapter. We also define and study spaces $B_{p,m}(\mathcal{T}_n)$, defined using m^{th} order differences, $m > 1$, and we develop some basic function theoretic properties for these spaces, including discussion of duality, reproducing kernels, Carleson measures, multipliers, and interpolating sequences. Although these spaces are interesting, the functions in them do not mimic holomorphic functions adequately for all of our needs.

We then go further; we define and develop the properties of *holomorphic Besov spaces on trees,* $HB_{p,m}(\mathcal{T}_n)$. At the heart of their definition is a splitting of the difference operator analogous to the classical splitting of the derivative of holomorphic functions on the ball into radial and tangential components. The spaces $HB_{p,m}(\mathcal{T}_n)$ are defined using powers of the split difference operator and functions in them mimic the asymmetric smoothness of functions in the ball. In particular the restriction map taking the multiplier space of $B_{p,m}(\mathbb{B}_n)$ to that of $HB_{p,m}(\mathcal{T}_n)$ is bounded; the crucial fact being that Carleson measures for $HB_{p,m}(\mathcal{T}_n)$ are characterized by a condition similar to the tree condition which characterizes Carleson measures for $B_{p,m}(\mathbb{B}_n)$, Using those facts we complete the program described in Section 15.6, thus finishing the proofs of the theorems in the previous chapter.

The need to work with higher order differences in defining the $HB_{p,m}(\mathcal{T}_n)$, $m > 1$, leads to work with tensor valued functions. That program requires adaptation and extension of the earlier ideas, and a substantial amount of linear algebra. We give a detailed description of those notationally complicated developments.

16.1. Besov Spaces on Trees

Recall that in defining the spaces $B_p(\mathbb{B}_n)$ it was always possible, and generally necessary, to use higher derivatives. This ensured, among other things, that the coordinate functions would be in the space. We will see in Example 16.4 that similar difficulties occur with the spaces $B_{p,1}(\mathcal{T}_n)$. Here we get past those difficulties by introducing the spaces $B_{p,m}(\mathcal{T}_n)$ which are based on m^{th} order differences. As one would hope, the space obtained is independent of m if m is large enough. We then define $B_p(\mathcal{T}_n)$, the Besov space of $\mathcal{T}_n$, to be $B_{p,m}(\mathcal{T}_n)$ for any sufficiently large m.

We begin with a discussion of the spaces $B_{p,m}(\mathcal{T})$ for an abstract rooted tree $\mathcal{T}$ with our primary example being the Bergman tree $\mathcal{T}_n$. The root defines a partial ordering on the tree and thus we can define $A\alpha$, the predecessor or immediate ancestor of α, sometimes denoted α^-. Define the difference operator Δ on the tree by

$$\Delta f(\alpha) := f(\alpha) - f(A\alpha), \qquad \alpha \in \mathcal{T}.$$

The right inverse of Δ is the summation operator I which is defined by

$$If(\alpha) := \sum_{o \leq \beta \leq \alpha} f(\beta).$$

Thus ΔI is the identity.

Define the pointwise multiplier operator 2^d (d here denotes distance and not dimension!) on the tree by

$$2^d f(\alpha) := 2^{d(o,\alpha)} f(\alpha), \qquad \alpha \in \mathcal{T}.$$

DEFINITION 16.1. *For $1 < p < \infty$ and $m \geq 0$, define the Besov space $B_{p,m}(\mathcal{T})$ on a tree $\mathcal{T}$ to consist of all sequences $f = \{f(\alpha)\}_{\alpha \in \mathcal{T}}$ such that*

$$\|f\|_{B_{p,m}(\mathcal{T})} := \left(\sum_{\alpha \in \mathcal{T}: d(\alpha) \geq m} \left| \left(2^{-d}\right)^m \left(2^d \Delta\right)^m f(\alpha) \right|^p \right)^{1/p} + \sum_{d(\alpha) \leq m-1} |f(\alpha)| < \infty.$$

For $m = 0$ these are ℓ^p spaces. For $m = 1$ and with $\mathcal{T}$ the Bergman tree in the disk, they are the Besov spaces on trees studied in [**48**]. If also $p = 2$ we have the tree space $B_2(\mathcal{T}_1)$ of Chapters 5 and 8.

Comparing this definition to the norming functionals of $B_p(\mathbb{B}_n)$ given in Proposition 14.4, we see that the term 2^{-d} plays the role of $(1 - |z|^2)$, and $2^d \Delta$ plays the role of gradient, or, equivalently, Δ plays the role of the invariant gradient.

We say that a positive function $\{\rho(\alpha)\}_{\alpha \in \mathcal{T}}$ is a $B_{p,m}(\mathcal{T})$-Carleson measure if

(16.1)
$$\left(\sum_{\alpha \in \mathcal{T}} |f(\alpha)|^p \rho(\alpha) \right)^{1/p} \lesssim \left(\sum_{\alpha \in \mathcal{T}} \left| \left(2^{-d}\right)^m \left(2^d \Delta\right)^m f(\alpha) \right|^p \right)^{1/p} + \sum_{d(\alpha) \leq m-1} |f(\alpha)|,$$

It turns out that the set of Carleson measures on $B_{p,m}(\mathcal{T}_n)$ is the same for all $m \geq 1$.

LEMMA 16.2. *Let $1 < p < \infty$ and $m \geq 1$. Then $\{\rho(\alpha)\}_{\alpha \in \mathcal{T}}$ is a $B_{p,m}(\mathcal{T})$-Carleson measure if and only if $\{\rho(\alpha)\}_{\alpha \in \mathcal{T}}$ is a $B_{p,1}(\mathcal{T})$-Carleson measure, which in turn holds if and only if ρ satisfies a tree condition:*

$$\sum_{\beta \in \mathcal{T}: \beta \geq \alpha} I^* \rho(\beta)^{p'} \leq C^{p'} I^* \rho(\alpha), \qquad \alpha \in \mathcal{T}.$$

PROOF. If we write $g(\alpha) = \left(2^{-d}\right)^m \left(2^d \Delta\right)^m f(\alpha)$ for $d(\alpha) \geq m$ and then invert the operators, we have

(16.2)

$$
\begin{aligned}
f(\alpha) &= \left(I2^{-d}\right)^m \left(2^d\right)^m g(\alpha) \\
&= \sum_{\beta_1 \leq \alpha} 2^{-d(\beta_1)} \sum_{\beta_2 \leq \beta_1} 2^{-d(\beta_2)} \cdots \sum_{\beta_{m-1} \leq \beta_{m-2}} 2^{-d(\beta_{m-1})} \sum_{\gamma \leq \beta_{m-1}} 2^{-d(\gamma)} 2^{md(\gamma)} g(\gamma) \\
&= \sum_{\gamma \leq \beta_{m-1} \leq \cdots \leq \beta_2 \leq \beta_1 \leq \alpha} 2^{-d(\beta_1)-d(\beta_2)-d(\beta_{m-1})} 2^{(m-1)d(\gamma)} g(\gamma) \\
&= C_1 \sum_{\gamma \leq \beta_{m-1} \leq \cdots \leq \beta_2 \leq \alpha} 2^{-2d(\beta_2)-d(\beta_3)\cdots-d(\beta_{m-1})} 2^{(m-1)d(\gamma)} g(\gamma) \\
&= C_2 \sum_{\gamma \leq \beta_{m-1} \leq \cdots \leq \beta_3 \leq \alpha} 2^{-3d(\beta_3)\cdots-d(\beta_{m-1})} 2^{(m-1)d(\gamma)} g(\gamma) \\
&= C_{m-1} \sum_{\gamma \leq \beta_{m-1} \leq \alpha} 2^{-(m-1)d(\beta_{m-1})} 2^{(m-1)d(\gamma)} g(\gamma) \approx Ig(\alpha).
\end{aligned}
$$

The condition (16.1) is equivalent to

$$\left(\sum_{\alpha \in \mathcal{T}} \left| \left(I2^{-d}\right)^m \left(2^d\right)^m g(\alpha) \right|^p \rho(\alpha)\right)^{1/p} \lesssim \left(\sum_{\alpha \in \mathcal{T}} |g(\alpha)|^p\right)^{1/p}.$$

From (16.2) we obtain that this latter inequality holds if and only if

$$\left(\sum_{\alpha \in \mathcal{T}} |Ig(\alpha)|^p \rho(\alpha)\right)^{1/p} \lesssim \left(\sum_{\alpha \in \mathcal{T}} |g(\alpha)|^p\right)^{1/p},$$

which in turn holds if and only if $\{\rho(\alpha)\}_{\alpha \in \mathcal{T}}$ is a $B_{p,1}(\mathcal{T})$-Carleson measure. The final assertion follows from the equivalence the second and third statements in Theorem 14.16. □

In analogy with the various equivalent norms on $B_p(\mathbb{B}_n)$, the norms $\|f\|_{B_{p,m}(\mathcal{T})}$ and $\|f\|_{B_{p,m'}(\mathcal{T})}$ are equivalent for m, m' large enough. Just how large depends on the notion of *upper dimension* $\overline{n}(\mathcal{T})$ of a tree $\mathcal{T}$, defined in Definition 14.2 as

$$\overline{n}(\mathcal{T}) = \limsup_{\ell \to \infty} \log_2 \left(\sup_{\alpha \in \mathcal{T}} \operatorname{card}\{\beta \in \mathcal{T} : \beta \geq \alpha \text{ and } d(\beta) \leq d(\alpha) + \ell\}\right)^{1/\ell}.$$

LEMMA 16.3. *For $m > m'$, we have $\|f\|_{B_{p,m}(\mathcal{T})} \leq \|f\|_{B_{p,m'}(\mathcal{T})}$. If $\overline{n} = \overline{n}(\mathcal{T})$ is the upper dimension of the tree $\mathcal{T}$, then $\|f\|_{B_{p,m}(\mathcal{T})} \approx \|f\|_{B_{p,m'}(\mathcal{T})}$ for $m, m' > \overline{n}/p$.*

PROOF. Let $g(\alpha) = \left(2^{-d}\right)^{m'} \left(2^d \Delta\right)^{m'} f(\alpha)$ so that $f(\alpha) = \left(I2^{-d}\right)^{m'} \left(2^d\right)^{m'} g(\alpha)$ for $d(\alpha) \geq m'$. Then

$$\left(2^{-d}\right)^m \left(2^d \Delta\right)^m f(\alpha) = \left(2^{-d}\right)^m \left(2^d \Delta\right)^m \left(I2^{-d}\right)^{m'} \left(2^d\right)^{m'} g(\alpha).$$

If $m > m'$, then

$$\begin{aligned}\left|\left(2^{-d}\right)^m\left(2^d\Delta\right)^m f(\alpha)\right| &= \left|\left(2^{-d}\right)^m\left(2^d\Delta\right)^{m-m'}\left(2^d\right)^{m'} g(\alpha)\right| \\ &\lesssim \sum_{k=0}^{m-m'}\left|g\left(A^k\alpha\right)\right|,\end{aligned}$$

and so

$$\begin{aligned}\|f\|_{B_{p,m}(\mathcal{T})} &= \left(\sum_{\alpha\in\mathcal{T}}\left|\left(2^{-d}\right)^m\left(2^d\Delta\right)^m f(\alpha)\right|^p\right)^{1/p} \\ &\lesssim \left(\sum_{\alpha\in\mathcal{T}}|g(\alpha)|^p\right)^{1/p} = \|f\|_{B_{p,m'}(\mathcal{T})}.\end{aligned}$$

On the other hand, if $m < m'$, then using (16.2) with m replaced by $m' - m$, we obtain

$$\begin{aligned}\left|\left(2^{-d}\right)^m\left(2^d\Delta\right)^m f(\alpha)\right| &= \left|\left(2^{-d}\right)^m\left(I2^{-d}\right)^{m'-m}\left(2^d\right)^{m'-m}\left(2^d\right)^m g(\alpha)\right| \\ &\lesssim \left(2^{-d}\right)^m I\left(2^d\right)^m|g(\alpha)| \\ &\approx \sum_{\beta\leq\alpha}2^{m[d(\beta)-d(\alpha)]}|g(\beta)| \\ &\approx \sum_{\beta\in\mathcal{T}}K(\alpha,\beta)|g(\beta)|,\end{aligned}$$

where the kernel $K(\alpha,\beta)$ is given by $\chi_{[o,\alpha]}(\beta)\,2^{m[d(\beta)-d(\alpha)]}$. We now show that $K(\alpha,\beta)$ is the kernel of a bounded operator. To do this we apply Schur's test, Theorem B.2, with auxiliary functions $h(\beta) = k(\beta) = 2^{td(\beta)}$. If $m + p't > 0$. then we have

$$\sum_{\beta\in\mathcal{T}}K(\alpha,\beta)\,h(\beta)^{p'} = \sum_{\beta\leq\alpha}2^{(m+p't)d(\beta)}2^{-md(\alpha)} \lesssim h(\alpha)^{p'}.$$

To estimate the complementary sum

$$\sum_{\alpha\in\mathcal{T}}K(\alpha,\beta)\,h(\alpha)^p = \sum_{\alpha\geq\beta}2^{(pt-m)d(\alpha)}2^{md(\beta)},$$

we use the device of splitting the sum $\sum_{\alpha\geq\beta}$ into "sparse" pieces $\sum\limits_{\alpha\in\mathcal{T}:d(\alpha)\in d(\beta)+\ell\mathbb{N}+j}$ for $0 \leq j < \ell$, where ℓ is chosen so large in Definition 14.2 that given $\varepsilon > 0$,

$$\begin{aligned}\log_2(N_\ell) &= \log_2\left(\sup_{\alpha\in\mathcal{T}}\operatorname{card}\{\beta\in\mathcal{T}:\beta\geq\alpha \text{ and } d(\beta) = d(\alpha)+\ell\}\right)^{1/\ell} \\ &< \overline{n}(\mathcal{T}) + \varepsilon.\end{aligned}$$

This is similar to the "sparse" argument used surrounding (14.57) to obtain the necessity of the tree condition for the Carleson embedding on the ball. We then have

$$\sum_{\alpha\in\mathcal{T}}K(\alpha,\beta)\,h(\alpha)^p \lesssim \sum_{k=0}^{\infty}\left(2^{\overline{n}(\mathcal{T})+\varepsilon}\right)^{k\ell}2^{(pt-m)[d(\beta)+k\ell]}2^{md(\beta)} \lesssim h(\beta)^p,$$

provided $\overline{n}(\mathcal{T})+\varepsilon+pt-m<0$. Thus if we can choose $-m/p'<t<(m-\overline{n}(\mathcal{T})-\varepsilon)/p$ for some $\varepsilon>0$, Schur's test shows that

$$\begin{aligned}\|f\|_{B_{p,m}(\mathcal{T})} &= \left(\sum_{\alpha\in\mathcal{T}}\left|\left(2^{-d}\right)^m\left(2^d\Delta\right)^m f(\alpha)\right|^p\right)^{1/p} \\ &\lesssim \left(\sum_{\alpha\in\mathcal{T}}\left|\sum_{\beta\in\mathcal{T}}K(\alpha,\beta)\,|g(\beta)|\right|^p\right)^{1/p} \\ &\lesssim \left(\sum_{\alpha\in\mathcal{T}}|g(\alpha)|^p\right)^{1/p} \approx \|f\|_{B_{p,m'}(\mathcal{T})}\end{aligned}$$

as required. This choice of t is possible if and only if for some $\varepsilon>0$ $m/p'<(m-\overline{n}(\mathcal{T})-\varepsilon)/p$, or $m>\overline{n}(\mathcal{T})/p$. □

With a small abuse of notation the norm $\|f\|_{B_{p,m}(\mathcal{T})}$ will be denoted simply by $\|f\|_{B_p(\mathcal{T})}$ when it is understood that $m>\overline{n}(\mathcal{T})/p$. Otherwise, we will write $B_{p,m}(\mathcal{T})$ to emphasize the dependence on m. We note that in general $B_{p,m}(\mathcal{T})\neq B_{p,m'}(\mathcal{T})$ if $m<m'$ and $m<n/p$.

EXAMPLE 16.4. *In the case $m=1$ and $m'=2$, define f on the Bergman tree $\mathcal{T}_n$, with $\theta=\frac{\ln 2}{2}$ as in* (14.38), *by $f(\alpha)=2^{-d(\alpha)}$. Then $\Delta f(\alpha)=-2^{-d(\alpha)}$ and*

$$2^{2d}\left(2^d\Delta\right)^2 f(\alpha)=2^{-d}\Delta\left(-2^d2^{-d}\right)(\alpha)=0$$

for $d(\alpha)\geq 2$. Thus $f\in B_{p,2}(\mathcal{T}_n)$ for all $1<p<\infty$. On the other hand, using the "sparse" argument again, $f\in B_{p,1}(\mathcal{T}_n)$ if $p>n$ and only if $p\geq n$:

$$\begin{aligned}\|f\|_{B_{p,1}(\mathcal{T}_n)} &= \left(\sum_{\alpha\in\mathcal{T}_n:d(\alpha)\geq 1}|\Delta f(\alpha)|^p\right)^{1/p}+|f(o)| \\ &= \left(\sum_{\alpha\in\mathcal{T}_n}2^{-pd(\alpha)}\right)^{1/p}+|f(o)| \\ &\lesssim \left(\sum_{k=0}^{\infty}2^{(n+\varepsilon)k\ell}2^{-pk\ell}\right)^{1/p}+|f(o)|,\end{aligned}$$

if $\log_2(N_\ell)^{1/l}<n+\varepsilon$, and similarly $\|f\|_{B_{p,1}(\mathcal{T}_n)}\gtrsim\left(\sum_{k=0}^{\infty}2^{(n-\varepsilon)k\ell}2^{-pk\ell}\right)^{1/p}+|f(o)|$ if $\log_2(N_\ell)^{1/l}>n-\varepsilon$, since by Lemma 14.3, *the dimension of the Bergman tree $\mathcal{T}_n$ is n when $\theta=\frac{\ln 2}{2}$. Thus if $n=1$ the two spaces are the same, but not otherwise.*

We can now characterize the pointwise multipliers on $B_{p,m}(\mathcal{T})$.

LEMMA 16.5. *Let $m>\overline{n}(\mathcal{T})/p$. Then $f\in\mathrm{Mult}(HB_p(\mathcal{T}_n))$ if and only if f is bounded and $\left\{\left|\left(2^{-d}\right)^m\left(2^d\Delta\right)^m f(\alpha)\right|^p\right\}_{\alpha\in\mathcal{T}}$ is a $B_{p,m}(\mathcal{T})$-Carleson measure.*

PROOF. We just provide the details for the case $m = 2 > \overline{n}(\mathcal{T})/p$. For the sufficiency, let $a_n = f(\alpha)$, $a_{n-1} = f(A\alpha)$, $a_{n-2} = f(A^2\alpha)$ and similarly $b_n = \varphi(\alpha)$, $b_{n-1} = \varphi(A\alpha)$, $b_{n-2} = \varphi(A^2\alpha)$ for $\varphi \in B_{p,2}(\mathcal{T})$. Then

$$
\begin{aligned}
&2^{-d}\Delta 2^d\Delta(f\varphi)(\alpha) \\
&\quad = 2^{-d(0,\alpha)}\left(\frac{(f\varphi)(\alpha)-(f\varphi)(P\alpha)}{2^{-d(0,\alpha)}} - \frac{(f\varphi)(P\alpha)-(f\varphi)(P^2\alpha)}{2^{-d(0,P\alpha)}}\right) \\
&\quad = (a_nb_n - a_{n-1}b_{n-1}) - \frac{1}{2}(a_{n-1}b_{n-1} - a_{n-2}b_{n-2}) \\
&\quad = a_n\left((b_n - b_{n-1}) - \frac{1}{2}(b_{n-1} - b_{n-2})\right) \\
&\qquad\quad + b_{n-2}\left((a_n - a_{n-1}) - \frac{1}{2}(a_{n-1} - a_{n-2})\right) \\
&\qquad\quad + \frac{3}{2}(a_n - a_{n-1})(b_{n-1} - b_{n-2}) \\
&\quad = f(\alpha)\left(2^{-d}\Delta 2^d\Delta\varphi\right)(\alpha) + \varphi\left(A^2\alpha\right)\left(2^{-d}\Delta 2^d\Delta f\right)(\alpha) \\
&\qquad\quad + \frac{3}{2}(\Delta f)(\alpha)(\Delta\varphi)(A\alpha).
\end{aligned}
$$

Because f is bounded and $\varphi \in B_{p,2}(\mathcal{T})$, we have $f(\alpha)\left(2^{-d}\Delta 2^d\Delta\varphi\right)(\alpha) \in \ell^p$. Appealing to the equivalence of norms in Lemma 16.3 and noting that $1 > \overline{n}(\mathcal{T})/2p$ we have $B_{2p,2}(\mathcal{T}) \approx B_{2p,1}(\mathcal{T}) \supset B_{p,1}(\mathcal{T})$. Also we have the observation that

$$
\left|\frac{3}{2}(\Delta f)(\alpha)(\Delta\varphi)(P\alpha)\right| \leq |(\Delta f)(\alpha)|^2 + |(\Delta\varphi)(P\alpha)|^2 .
$$

The expression on the right, and hence also the expression on the left, is a generic term from an ℓ^p sequence. Thus the ℓ^p norm of $\varphi\left(A^2\alpha\right)\left(2^{-d}\Delta 2^d\Delta f\right)(\alpha)$ is controlled by the $B_{p,2}(\mathcal{T})$ norm of φ, showing that $\left\{\left|2^{-d}\Delta 2^d\Delta f(\alpha)\right|^p\right\}_{\alpha\in\mathcal{T}}$ is a $B_{p,2}(\mathcal{T})$-Carleson measure. The case $m > 2$ is similar.

The necessity of the boundedness of f is standard. One can reverse the above argument to obtain the necessity of the Carleson embedding. □

16.1.1. Interpolating Sequences. Let $\mathcal{A} := \{\alpha_j\}_{j=1}^{\infty}$ be a sequence of points in a tree $\mathcal{T}$ having finite dimension $n = n(\mathcal{T})$. In this section we will describe the equivalence of weighted ℓ^p interpolation for the Besov spaces $B_{p,m}(\mathcal{T})$, $m > n/p$, on the sequence $\mathcal{A}$, ℓ^∞ interpolation for their multiplier spaces $\mathrm{Mult}(B_{p,m}(\mathcal{T}))$ on $\mathcal{A}$, and the separation and Carleson embeddings on the tree, (16.10) below. However we will not give complete proofs; only some proof sketches and discussions.

As a preliminary step we introduce duality pairings and reproducing formulas for $B_{p,m}(\mathcal{T})$ and $B_{p',m}(\mathcal{T})$.

For $m \geq 1$, define the pairing

$$
\langle F, G\rangle_m := \sum_{\alpha\in\mathcal{T}} \left(2^{-d}\right)^m \left(2^d\Delta\right)^m F(\alpha)\,\overline{\left(2^{-d}\right)^m \left(2^d\Delta\right)^m G(\alpha)}.
$$

In the case $m = 1$, $\left(2^{-d}\right)^m\left(2^d\Delta\right)^m = \Delta$ and the reproducing kernel $k_\alpha^1(\beta) = k^1(\alpha,\beta)$ with respect to the pairing

$$
\langle F, G\rangle_1 = \sum_{\alpha\in\mathcal{T}_n} \Delta F(\alpha)\,\overline{\Delta G(\beta)},
$$

for the Besov space $B_{p,1}(\mathcal{T})$ on the tree $\mathcal{T}$ is given by

$$k_\alpha^1(\beta) = \chi_{[0,\alpha]}(\beta)\, d(0,\beta) + \chi_{S(\alpha)}(\beta)\, d(0,\alpha), \qquad \alpha,\beta \in \mathcal{T},$$

where $S(\alpha)$ denotes the successor set of α,sometimes called the " shadow" of α; the tree analog of a Carleson tent: $S(\alpha) = \{\beta \in \mathcal{T}_n : \beta \geq \alpha\}$.

Indeed, we have

$$\Delta k_\alpha^1(\beta) = \chi_{[0,\alpha]}(\beta) \tag{16.3}$$

and so

$$\langle F, k_\alpha^1 \rangle_{\mathcal{T}} = \sum_{\beta\in\mathcal{T}} \Delta F(\beta)\, \overline{\Delta k_\alpha^1(\beta)} = \sum_{0\leq\beta\leq\alpha} \Delta F(\beta) = F(\alpha).$$

The important property here is that the difference kernel, Δk_α^1 in (16.3), is nonnegative and pointwise comparable to $\chi_{[0,\alpha]}$. This is easily generalized to the reproducing kernels k_α^m for all $m \geq 1$:

$$\left(2^{-d}\right)^m \left(2^d\Delta\right)^m k_\alpha^m(\beta) \approx \chi_{[0,\alpha]}(\beta) \geq 0, \qquad \alpha,\beta\in\mathcal{T}. \tag{16.4}$$

For example, when $m = 2$, $\left(2^{-d}\right)^m \left(2^d\Delta\right)^m = 2^{-d}\Delta 2^d \Delta$ and the reproducing kernel $k_\alpha^2(\beta) = k^2(\alpha,\beta)$ with respect to the pairing

$$\langle F, G\rangle_2 = \sum_{\alpha\in\mathcal{T}} 2^{-d}\Delta 2^d \Delta F(\alpha)\, \overline{2^{-d}\Delta 2^d\Delta G(\beta)},$$

for the Besov space $B_{p,2}(\mathcal{T})$ on the tree $\mathcal{T}$ satisfies

$$2^{-d}\Delta 2^d \Delta k_\alpha^2(\gamma) = \chi_{[0,\alpha]}(\gamma) \left\{ \sum_{\beta\in[\gamma,\alpha]} 2^{d(\gamma)-d(\beta)} \right\}, \qquad \alpha,\beta\in\mathcal{T}. \tag{16.5}$$

In fact, with $f = 2^{-d}\Delta 2^d \Delta F$, we have $F = I2^{-d}I2^d f$ and

$$F(\alpha) = I2^{-d}I2^d f(\alpha) = \sum_{\beta\leq\alpha} 2^{-d(\beta)} \sum_{\gamma\leq\beta} 2^{d(\gamma)} f(\gamma) = \sum_{\gamma\leq\beta} \left\{ \sum_{\beta\in[\gamma,\alpha]} 2^{d(\gamma)-d(\beta)} \right\} f(\gamma).$$

Thus we have

$$\begin{aligned} \langle F, k_\alpha^2 \rangle_{B_{p,2}(\mathcal{T})} &= \sum_{\gamma\in\mathcal{T}} 2^{-d}\Delta 2^d\Delta F(\gamma)\, \overline{2^{-d}\Delta 2^d \Delta k_\alpha^2(\gamma)} \\ &= \sum_{0\leq\gamma\leq\alpha} \left\{ \sum_{\beta\in[\gamma,\alpha]} 2^{d(\gamma)-d(\beta)} \right\} f(\gamma) = F(\alpha). \end{aligned}$$

Continuing this analysis leads to the following theorem.

THEOREM 16.6. *Let $1 < p < \infty$ and $m > n/p$ where $n = n(\mathcal{T})$ is the dimension of the tree $\mathcal{T}$. Then the dual space of $B_{p,m}(\mathcal{T})$ can be identified with $B_{p',m}(\mathcal{T})$ under the pairing $\langle\cdot,\cdot\rangle_m$, and the reproducing kernel k_w^m for this pairing satisfies* (16.4).

Now we suppose that $m > n/p, n/p'$ and suppress the dependence of $k_\alpha = k_\alpha^m$ and of $B_p = B_p(\mathcal{T}) = B_{p,m}(\mathcal{T})$ on m and $\mathcal{T}$. Here is the abstract tree analog of Böe's interpolation theorem. Note that by (16.4), we have

$$\|k_\alpha\|_{B_{p'}} = \left\| \left(2^{-d}\right)^m \left(2^d\Delta\right)^m k_\alpha^m \right\|_{\ell^{p'}} \approx d(o,\alpha)^{1/p'}, \qquad \alpha\in\mathcal{T}.$$

We continue to write $\kappa_{\alpha_i} = \left\|k_{\alpha_j}\right\|_{B_{p'}}^{-1}$ and hence, again, the normalized kernels are $\kappa_{\alpha_i} k_{\alpha_j}$.

THEOREM 16.7. *Let $1 < p < \infty$, $m > n/p, n/p'$ and k_α be the reproducing kernel for $B_p(\mathcal{T})$ relative to the pairing $\langle \cdot, \cdot \rangle_m$ given in Theorem 16.6 above. Let $\mathcal{A}$ be a sequence in the tree $\mathcal{T}$. Then the following conditions are equivalent.*

(1) *$\mathcal{A}$ interpolates $\mathrm{Mult}(B_p)$:*

$$\text{The map } f \to \{f(\alpha_j)\}_{j=1}^{\infty} \text{ takes } \mathrm{Mult}(B_p) \text{ boundedly into and onto } \ell^\infty. \tag{16.6}$$

(2) *$\left\{k_{\alpha_j}\right\}_{j=1}^{\infty}$ is an unconditional basic sequence in $B_{p'}$:*

$$\left\|\sum_{j=1}^{\infty} b_j k_{\alpha_j}\right\|_{B_{p'}} \lesssim \left\|\sum_{j=1}^{\infty} a_j k_{\alpha_j}\right\|_{B_{p'}}, \quad \text{whenever } |b_j| \leq |a_j|. \tag{16.7}$$

(3) *The following norm equivalence holds:*

$$\left\|\sum_{j=1}^{\infty} a_j \kappa_{\alpha_j} k_{\alpha_j}\right\|_{B_{p'}} \approx \left(\sum_{j=1}^{\infty} |a_j|^{p'}\right)^{1/p'}. \tag{16.8}$$

(4) *$\mathcal{A}$ interpolates B_p:*

$$\text{The map } f \to \left\{\frac{f(z_j)}{\left\|k_{\alpha_j}\right\|_{B_{p'}}}\right\}_{j=1}^{\infty} \text{ takes } B_p \text{ boundedly into and onto } \ell^p. \tag{16.9}$$

(5) *The following separation condition and Carleson embedding hold:*

$$\begin{aligned} &d(\alpha_i, o) \lesssim d(\alpha_i, \alpha_j) \text{ and} \\ &\sum_{j=1}^{\infty} d(o, \alpha_j)^{1-p} \delta_{\alpha_j} \text{ is a Carleson measure for } B_{p,m}(\mathcal{T}). \end{aligned} \tag{16.10}$$

REMARK 16.8. *The interpolation operators in* (16.6) *and* (16.9) *can be taken to be linear, i.e. there are bounded linear maps $R : \ell^\infty \to \mathrm{Mult}(B_p)$ and $S : \ell^p \to B_p$ that yield right inverses to the restriction maps in* (16.6) *and* (16.9) *respectively.*

We give a brief sketch of those parts of the proof that we will need in the more refined setting of holomorphic Besov spaces on trees. The proofs that (16.6) implies (16.7) implies (16.8) implies (16.9) implies (16.10) follow the corresponding arguments in Böe [**77**]. The details are standard, with the exception of the implication (16.7) implies (16.8), which depends upon ideas from [**77**, Lemma 3.1] together with the crucial property (16.4) that the difference operator $\left(2^{-d}\right)^m \left(2^d \Delta\right)^m$ applied to the reproducing kernel on the tree is nonnegative for all $1 < p < \infty$. Following the pattern used in Lemma 15.10, again denoting the Rademacher functions by $\{r_j\}$, and using first (16.4) and then (16.7), we have

$$\begin{aligned}
\left\|\left\|a_j\left(2^{-d}\right)^m\left(2^d\Delta\right)^m\kappa_{\alpha_j}k_{\alpha_j}\right\|_{\ell^1}\right\|_{\ell^{p'}}^{p'} &= \left\|\sum_{j=1}^{\infty}|a_j|\,\kappa_{\alpha_j}k_{\alpha_j}\right\|_{B_{p'}}^{p'} \\
&\lesssim \int_0^1\left\|\sum_{j=1}^{\infty}|a_j|\,r_j(t)\,\kappa_{\alpha_j}k_{\alpha_j}\right\|_{B_{p'}}^{p'} dt \\
&= \sum_{\alpha\in\mathcal{T}}\int_0^1\left|\sum_{j=1}^{\infty}|a_j|\,r_j(t)\left(2^{-d}\right)^m\left(2^d\Delta\right)^m\kappa_{\alpha_j}k_{\alpha_j}(\alpha)\right|^{p'} dt \\
&\approx \left\|\left\|a_j\left(2^{-d}\right)^m\left(2^d\Delta\right)^m\kappa_{\alpha_j}k_{\alpha_j}\right\|_{\ell^2}\right\|_{\ell^{p'}}^{p'}.
\end{aligned}$$

Continuing to follow Böe, we now use the inequality $\|\cdot\|_{\ell^2}\leq\|\cdot\|_{\ell^1}\|\cdot\|_{\ell^\infty}$ along with the Cauchy-Schwarz inequality to obtain

$$\begin{aligned}
&\left\|\left\|a_j\left(2^{-d}\right)^m\left(2^d\Delta\right)^m\kappa_{\alpha_j}k_{\alpha_j}\right\|_{\ell^1}\right\|_{\ell^{p'}}^{p'} \\
&\quad\lesssim \left\|\left\|a_j\left(2^{-d}\right)^m\left(2^d\Delta\right)^m\kappa_{\alpha_j}k_{\alpha_j}\right\|_{\ell^2}\right\|_{\ell^{p'}}^{p'} \\
&\quad\lesssim \left\|\left\|a_j\left(2^{-d}\right)^m\left(2^d\Delta\right)^m\kappa_{\alpha_j}k_{\alpha_j}\right\|_{\ell^1}\right\|_{\ell^{p'}}^{p'/2} \\
&\qquad\qquad\times\left\|\left\|a_j\left(2^{-d}\right)^m\left(2^d\Delta\right)^m\kappa_{\alpha_j}k_{\alpha_j}\right\|_{\ell^\infty}\right\|_{\ell^{p'}}^{p'/2},
\end{aligned}$$

which yields using (16.7) again,

$$\begin{aligned}
\left\|\sum_{j=1}^{\infty}a_j\kappa_{\alpha_j}k_{\alpha_j}\right\|_{B_{p'}}^{p'} &\approx \left\|\left\|a_j\left(2^{-d}\right)^m\left(2^d\Delta\right)^m\kappa_{\alpha_j}k_{\alpha_j}\right\|_{\ell^1}\right\|_{\ell^{p'}}^{p'} \\
&\approx \left\|\left\|a_j\left(2^{-d}\right)^m\left(2^d\Delta\right)^m\kappa_{\alpha_j}k_{\alpha_j}\right\|_{\ell^\infty}\right\|_{\ell^{p'}}^{p'/2} \\
&\approx \left\|\left\|a_j\left(2^{-d}\right)^m\left(2^d\Delta\right)^m\kappa_{\alpha_j}k_{\alpha_j}\right\|_{\ell^{p'}}\right\|_{\ell^{p'}}^{p'/2} = \left(\sum_{j=1}^{\infty}|a_j|^{p'}\right)^{1/p'}
\end{aligned}$$

since

$$\left\|\left(2^{-d}\right)^m\left(2^d\Delta\right)^m\kappa_{\alpha_j}k_{\alpha_j}\right\|_{\ell^{p'}}=1,\qquad 1\leq j<\infty.$$

We have not yet discussed why the separation condition and Carleson embedding in (16.10) are sufficient for the multiplier interpolation in (16.6). We do not use this result later and hence, rather than give a proof, we only make a few comments.

The result can be proved by following the argument given in the previous chapter for the ball. The proof there is for the range $1<p<1/(n-1)$ however that range is dictated by the range for which we have the positivity condition (16.4). Here, as we just saw, we have the positivity condition (16.4) for $1<p<\infty$ and hence that earlier argument can be used here for all p. (See also [**48**, Theorem 26], where the implication (16.10) implies (16.9) is proved in the case $m = 1$.)

16.1.2. The Restriction Map. Let $p > \widehat{n}$ where

$$\widehat{n} := \begin{cases} 1 & \text{if } n = 1 \\ 2n & \text{if } n > 1. \end{cases}$$

(The odd pattern reflects the fact that the complex tangential directions only appear in dimension $n > 1$.) Note that in this case we can take $m = 1$ in both the definition of B_p on the unit ball $\mathbb{B}_n$ and $B_p(\mathcal{T}_n)$ on the tree $\mathcal{T}_n$. We have from the case $m = 1$ of Proposition 14.26, or from [**346**, Theorem 6.30], the following estimate on the oscillation of a B_p function on Bergman balls of bounded radius.

LEMMA 16.9. *The oscillation inequality*

$$\left\{ \sum_{\alpha \in \mathcal{T}_n} \left(\max_{z_1, z_2 \in K_\alpha} |f(z_1) - f(z_2)| \right)^p \right\}^{1/p} \lesssim \|f\|_{B_p}, \tag{16.11}$$

holds for $p > \widehat{n}$.

The following local version of Lemma 16.9, which is essentially contained in Proposition 14.26, will prove useful in estimating Carleson measure norms.

LEMMA 16.10. *The local oscillation inequality*

$$\left\{ \sum_{\beta \in \mathcal{T}_n : \beta \geq \alpha} \left(\max_{z_1, z_2 \in K_\beta} |f(z_1) - f(z_2)| \right)^p \right\}^{1/p}$$
$$\lesssim \left(\sum_{\beta \in \mathcal{T}_n : \beta \geq \alpha} \int_{B_d(c_\beta, C_2)} \left| \widetilde{\nabla} f(z) \right|^p d\lambda_n(z) \right)^{1/p}, \qquad \alpha \in \mathcal{T}_n,$$

holds for $p > \widehat{n}$.

This local oscillation inequality is not in [**346**]. We will here give a proof of Lemma 16.9 using the invariant derivative and our proof yields Lemma 16.10. These results could, alternatively, be approached using the almost invariant holomorphic derivatives D_a in Proposition 14.26.

PROOF. (of Lemmas 16.9 and 16.10 without using Proposition 14.26) Denote by K_α^* the union of the Carleson box K_α and its neighbors at most M boxes away. From part (1) of Lemma 14.3, and provided we choose M sufficiently large, we obtain that there are constants A_1, A_2 and A_2^* in $[0, 1)$ with $1 - A_2^* = \frac{1}{2}(1 - A_2)$ depending only on n such that

$$B(c_\alpha, A_1) \subset K_\alpha \subset B(c_\alpha, A_2),$$
$$B(c_\alpha, A_2^*) \subset K_\alpha^*,$$

for $\alpha \in \mathcal{T}_n$. We first establish the estimate

$$\max_{z_1, z_2 \in K_0} |f(z_1) - f(z_2)| \lesssim \left(\int_{K_0^*} \left| \widetilde{\nabla} f(z) \right|^p d\lambda_n(z) \right)^{1/p}, \tag{16.12}$$

for the root Carleson box $K_0 = \frac{1}{4}\mathbb{B}_n$. Here $\widetilde{\nabla} f$ is the invariant gradient of f introduced in Section 14.1.4; $\widetilde{\nabla} f(z) = \nabla(f \circ \varphi_z)(0)$. Note that while the vector $\widetilde{\nabla} f(z) = \nabla(f \circ \varphi_z)(0)$ is not invariant under the action of the automorphism group, its length $\left|\widetilde{\nabla} f(z)\right|$ *is* invariant, i.e.

$$\left|\widetilde{\nabla}(f \circ \psi)\right| = \left|\left(\widetilde{\nabla} f\right) \circ \psi\right|, \qquad \psi \in \operatorname{Aut}(\mathbb{B}_n). \tag{16.13}$$

Indeed, given $\psi \in \operatorname{Aut}(\mathbb{B}_n)$ and $z \in \mathbb{B}_n$, let $w = \psi(z)$. Then $\varphi_w \circ \psi \circ \varphi_z(0) = 0$ and so $\varphi_w \circ \psi \circ \varphi_z = U$ is unitary. For any unitary U, the lengths of the (ordinary) gradients of g and $g \circ U$ are the same at the origin, hence

$$\begin{aligned} \left|\widetilde{\nabla}(f \circ \psi)(z)\right| &= |\nabla(f \circ \psi \circ \varphi_z)(0)| = |\nabla(f \circ \varphi_w \circ U)(0)| \\ &= |\nabla(f \circ \varphi_w)(0)| = \left|\left(\widetilde{\nabla} f\right)(w)\right| = \left|\left(\left(\widetilde{\nabla} f\right) \circ \psi\right)(z)\right|. \end{aligned}$$

To obtain (16.12), we use the standard Euclidean inequality,

$$\int_B |f(z) - f(w)|\, dV \lesssim \int_B |z - w|\, |\nabla f(w)|\, dV,$$

valid for continuously differentiable f on a Euclidean ball B, together with Bergman's formula (applied to the derivative) scaled to the ball $\frac{3}{4}\mathbb{B}_n$,

$$f'(w) = c_n \int_{\frac{3}{4}\mathbb{B}_n} \frac{f'(\xi)}{\left(\left(\frac{3}{4}\right)^2 - \overline{\xi} \cdot w\right)^{n+1}} dV(\xi),$$

valid for f holomorphic on $\frac{3}{4}\mathbb{B}_n$, to obtain that for $|z| < \frac{1}{4}$ and $f \in B_p$,

$$\begin{aligned} &\left| f(z) - \frac{1}{\left|\frac{1}{4}\mathbb{B}_n\right|} \int_{\frac{1}{4}\mathbb{B}_n} f(w)\, dV \right| \\ &\quad \leq \frac{1}{\left|\frac{1}{4}\mathbb{B}_n\right|} \int_{\frac{1}{4}\mathbb{B}_n} |f(z) - f(w)|\, dV \\ &\quad \lesssim \int_{\frac{1}{4}\mathbb{B}_n} |z - w|\, |f'(w)|\, dV \\ &\quad \lesssim \int_{\frac{1}{4}\mathbb{B}_n} |z - w| \left\{ \int_{\frac{3}{4}\mathbb{B}_n} |f'(\xi)|\, dV(\xi) \right\} dV \\ &\quad \lesssim \int_{\frac{3}{4}\mathbb{B}_n} |f'(\xi)|\, dV \\ &\quad \lesssim \left(\int_{\frac{3}{4}\mathbb{B}_n} |f'(\xi)|^p\, dV \right)^{1/p}. \end{aligned}$$

This establishes (16.12) since $|f'(\xi)| \approx \left|\widetilde{\nabla} f(\xi)\right|$ and $dV \approx d\lambda_n$ for $\xi \in \frac{3}{4}\mathbb{B}_n$.

Using the holomorphic automorphisms φ_w of the ball, we now show that (16.12) implies the estimate

$$\max_{z_1, z_2 \in K_\alpha} |f(z_1) - f(z_2)| \lesssim \left(\int_{K_\alpha^*} \left|\widetilde{\nabla} f(\xi)\right|^p d\lambda_n(\xi) \right)^{1/p}, \qquad \alpha \in \mathcal{T}_n. \tag{16.14}$$

Recall that c_α denotes the "center" of K_α. Since φ_{c_α} maps T_α to T_0 (at least approximately) with inverse φ_{c_α}, we can apply (16.12) to the function $g = f \circ \varphi_w$ with $w = c_\alpha$ and then use the invariance of $\widetilde{\nabla}$ (see (16.13)) and λ_n together with the change of variable $\xi = \varphi_w(z)$ to obtain

$$\begin{aligned} \max_{z_1, z_2 \in K_\alpha} |f(z_1) - f(z_2)|^p &= \max_{\xi_1, \xi_2 \in K_0} |g(\xi_1) - g(\xi_2)|^p \\ &\lesssim \int_{K_0^*} \left| \widetilde{\nabla} (f \circ \varphi_w)(\xi) \right|^p d\lambda_n(\xi) \\ &\approx \int_{K_0^*} \left| \widetilde{\nabla} f(\varphi_w(\xi)) \right|^p d\lambda_n(\xi) \\ &\approx \int_{K_\alpha^*} \left| \widetilde{\nabla} f(z) \right|^p d\lambda_n(z). \end{aligned}$$

This proves (16.14).

We now apply (16.14) to obtain (16.11) as follows:

$$\begin{aligned} \sum_{\alpha \in \mathcal{T}_n} \left(\max_{z_1, z_2 \in K_\alpha} |f(z_1) - f(z_2)| \right)^p &\lesssim \sum_{\alpha \in \mathcal{T}_n} \int_{K_\alpha^*} \left| \widetilde{\nabla} f(z) \right|^p d\lambda_n(z) \\ &\lesssim \int_{\mathbb{B}_n} \left| \widetilde{\nabla} f(z) \right|^p d\lambda_n(z), \end{aligned}$$

by the finite overlap condition (5) in Lemma 14.3. By [**346**, Theorem 6.11], the above expression is equivalent to the Besov space norm $\|f\|_{B_p}$ for $p > \widehat{n}$. This completes the proof of Lemma 16.9. Lemma 16.10 follows by adding up estimates in (16.14), yielding for each $\alpha \in \mathcal{T}_n$

$$\sum_{\beta \in \mathcal{T}_n : \beta \geq \alpha} \left(\max_{z_1, z_2 \in K_\beta} |f(z_1) - f(z_2)| \right)^p \lesssim \sum_{\beta \in \mathcal{T}_n : \beta \geq \alpha} \int_{K_\beta^*} \left| \widetilde{\nabla} f(z) \right|^p.$$

□

COROLLARY 16.11. *Let $p > \widehat{n}$. Then the restriction map*

$$Tf := \{f(\alpha)\}_{\alpha \in \mathcal{T}_n}, \qquad \text{where } Tf(\alpha) = f(c_\alpha),$$

is bounded from B_p to $B_p(\mathcal{T}_n)$. If, in addition, $n = 1$ and $1 < p < \infty$, then T is also bounded from $\text{Mult}(B_p(\mathbb{B}_1))$ *to* $\text{Mult}(B_p(\mathcal{T}_1))$.

The reason the dimension is restricted to $n = 1$ for the boundedness of the restriction map on the multiplier space, is that this restriction requires that Carleson embeddings be characterized by the tree condition. We have only established this latter result for $p < 2 + 1/(n-1)$, which together with the restriction $p > \widehat{n} = 2n > 2 + 1/(n-1)$ for $n \geq 2$, leaves $n = 1$ as the only possible dimension in which our argument applies. Nevertheless, in the interest of motivating proofs in later sections, we will present our arguments in as general a setting as possible. Later we will prove a version of this restriction result for $n > 1$, but working in the more elaborate context of spaces of holomorphic functions on trees; a context in which the Carleson embedding is characterized by the tree condition.

PROOF. With the tree difference operator $\left(2^{-d}\right)\left(2^{d}\Delta\right)=\Delta$ as in Definition 16.1 above with $m=1$, we have

$$|\Delta f(\alpha)|\leq \max_{z_1,z_2\in K_\alpha^*}|f(z_1)-f(z_2)|, \tag{16.15}$$

where K_α^* is the Bergman ball centered at c_α with radius C. Then from a modification of (16.11) where K_α is replaced by K_α^*, we obtain

$$\begin{aligned}\|f\|_{B_p(\mathcal{T}_n)} &= \left(\sum_{\alpha\in\mathcal{T}_n:d(\alpha)\geq 1}|\Delta f(\alpha)|^p\right)^{1/p}+|f(o)|\\ &\lesssim \left(\sum_{\alpha\in\mathcal{T}_n:d(\alpha)\geq 1}\left|\max_{z_1,z_2\in K_\alpha^*}|f(z_1)-f(z_2)|\right|^p\right)^{1/p}+|f(o)|\\ &\lesssim \|f\|_{B_p}.\end{aligned}$$

To handle the restriction map T on the multiplier space $\mathrm{Mult}(B_p(\mathbb{B}_1))$, we need the analog of Stegenga's characterization of multipliers on the Dirichlet space, $\mathrm{Mult}(B_2(\mathbb{B}_1))$, as bounded functions which generate Carleson measures. Here is the result we need at the moment. We will see more elaborate variations later. □

LEMMA 16.12. *Let $\psi\in H^\infty(\mathbb{B}_n)\cap B_p$ and $p>\widehat{n}$. Then ψ is a multiplier on B_p if and only if*

$$\left|\widetilde{\nabla}\psi(z)\right|^p d\lambda_n(z)$$

is a B_p-Carleson measure on $\mathbb{B}_n$.

PROOF OF THE PREVIOUS LEMMA. The restriction on p allows us to define B_p without using higher gradients. For the first order gradient we have the product rule

$$\begin{aligned}\widetilde{\nabla}(\psi g)(z) &= \nabla\left[(\psi g)\circ\varphi_z\right](0)=\nabla\left[(\psi\circ\varphi_z)(g\circ\varphi_z)\right](0)\\ &= \psi\circ\varphi_z(0)\,\nabla(g\circ\varphi_z)(0)+\nabla(\psi\circ\varphi_z)(0)\,(g\circ\varphi_z)(0)\\ &= \psi(z)\,\widetilde{\nabla}g(z)+\widetilde{\nabla}\psi(z)\,g(z).\end{aligned}$$

The proof now follows the pattern of the proof of Proposition 2.20. □

PROOF OF THE COROLLARY CONTINUED. Now let $f\in\mathrm{Mult}(B_p(\mathbb{B}_1))$ and set

$$d\mu(z)=\sum_{\alpha\in\mathcal{T}_n}\chi_{K_\alpha}(z)\int_{K_\alpha^*}\left|\widetilde{\nabla}f(\zeta)\right|^p d\lambda_1(\zeta).$$

The above lemma shows that μ is a $B_p(\mathbb{B}_1)$-Carleson measure. Define the discretization of μ in the usual way by

$$\mu(\alpha)=\int_{K_\alpha^*}\left|\widetilde{\nabla}f(\zeta)\right|^p d\lambda_1(\zeta).$$

Since $p<2+1/(n-1)$ when $n=1$, it then follows from Theorem 14.16 that $\{\mu(\alpha)\}_{\alpha\in\mathcal{T}_1}$ is a $B_{p,1}(\mathcal{T}_1)$-Carleson measure. Now set

$$\omega(\alpha)=|\Delta f(\alpha)|^p.$$

From (16.15) and a modification of Remark 16.10, we obtain

$$\begin{aligned}
I^*\omega(\alpha) &= \sum_{\beta\in\mathcal{T}_1:\beta\geq\alpha} |\Delta f(\alpha)|^p \\
&\lesssim \sum_{\beta\in\mathcal{T}_1:\beta\geq\alpha} \left(\max_{z_1,z_2\in K_\alpha^*} |f(z_1)-f(z_2)|\right)^p \\
&\lesssim \sum_{\alpha\in\mathcal{T}_1:\beta\geq\alpha} \int_{B_d(c_\beta,C_2)} \left|\widetilde{\nabla} f(z)\right|^p d\lambda_1(z) \\
&\approx I^*\mu(\alpha).
\end{aligned}$$

It now follows that $\{\omega(\alpha)\}_{\alpha\in\mathcal{T}_1}$ is a $B_{p,1}(\mathcal{T}_1)$-Carleson measure with norm bounded by that of the $B_p(\mathbb{B}_1)$-Carleson measure μ. Finally, the tree multiplier characterization in Theorem 16.5, shows that $Tf \in \text{Mult}(B_{p,1}(\mathcal{T}_1)) = \text{Mult}(B_p(\mathcal{T}_1))$ with $\|Tf\|_{\text{Mult}(B_{p,1}(\mathcal{T}_1))} \lesssim \|f\|_{\text{Mult}(B_p(\mathcal{T}_1))}$. □

REMARK 16.13. *Let $p > \widehat{n}$.* Assume *that the restriction map in Corollary* 16.11 *is bounded from* $\text{Mult}(B_p)$ *to* $\text{Mult}(B_p(\mathcal{T}_n))$. *Suppose* $\{z_j\}_{j=1}^\infty \subset \mathbb{B}_n$ *interpolates* $\text{Mult}(B_p)$, *i.e.*

(16.16) *The map* $f \to \{f(z_j)\}_{j=1}^\infty$ *takes* $\text{Mult}(B_p)$ *boundedly into and onto* ℓ^∞.

If we construct the Bergman tree $\mathcal{T}_n$ so that $\{c_\alpha\}_{\alpha\in\mathcal{T}_n}$ contains $\{z_j\}_{j=1}^\infty$, say with $z_j = c_{\alpha_j}$, then it follows that $\mathcal{A}$ interpolates $\text{Mult}(B_p(\mathcal{T}_n))$, *i.e. that* (16.6) *holds with $\mathcal{T}_n$ in place of $\mathcal{T}$:*

(16.17)
The map $f \to \{f(\alpha_j)\}_{j=1}^\infty$ *takes* $\text{Mult}(B_p(\mathcal{T}_n))$ *boundedly into and onto* ℓ^∞.

PROOF OF REMARK. To see that (16.17) holds if (16.16) does, suppose that $\{\xi_j\}_{j=1}^\infty \in \ell^\infty$. Using (15.7) we can find $\varphi \in \text{Mult}(B_p)$ satisfying

$$\varphi(z_j) = \xi_j, \quad 1 \leq j < \infty,$$

$$\|\varphi\|_{\text{Mult}(B_p)} \lesssim \left\|\{\xi_j\}_{j=1}^\infty\right\|_\infty.$$

Now define f on the tree $\mathcal{T}_n$ by

$$f(\alpha) := \varphi(c_\alpha), \quad \alpha \in \mathcal{T}_n.$$

Then we have

$$f(\alpha_j) = \varphi\left(c_{\alpha_j}\right) = \varphi(z_j) = \xi_j$$

and our assumption on the restriction map shows that

$$\|f\|_{\text{Mult}(B_p(\mathcal{T}_n))} \lesssim \|\varphi\|_{\text{Mult}(B_p)},$$

thus completing the proof of (16.17). □

REMARK 16.14. *It would be desirable to extend the conclusion of the previous remark to all $1 < p < \infty$ and $n > 1$, or at least to the case $1 < p < 2 + 1/(n-1)$ where we have the equivalence of Carleson embeddings with the tree condition. The argument above breaks down for higher order differences since the analog of* (16.15) *fails to hold. In fact, the restrictions of linear functions on the ball fail to belong to $B_p(\mathcal{T}_n)$ for $p \leq 2n$, $n \geq 2$ (on the other hand, we showed above that the analogous linear functions, $f(\alpha) = 2^{-d(\alpha)}$ on the tree, belong to $B_{p,2}(\mathcal{T}_n)$ for all $1 < p < \infty$).*

Indeed, if $f(z)=z_1$*, then for most* $\alpha\in\mathcal{T}_n$*, in particular for those* α *at a distance at least* $c>0$ *from the complex line* $\mathbb{C}(1,0,\ldots,0)$*, we have*

$$\sum_{\beta\in\mathcal{C}(\alpha)}|f(\beta)-f(\alpha)|=\sum_{\beta\in\mathcal{C}(\alpha)}|\beta_1-\alpha_1|\approx e^{-d(\alpha)\theta},$$

where $\mathcal{C}(\alpha)$ *denotes the set of children of* α*. By property (4) of Lemma* 14.3*, we have*

$$\operatorname{card}\{\alpha\in\mathcal{T}_n:d(\alpha)=N\}\approx e^{2nN\theta},$$

and thus

$$\begin{aligned}\|f\|^p_{B_p(\mathcal{T}_n)}=\|f\|^p_{B_{p,1}(\mathcal{T}_n)}&\gtrsim\sum_{\alpha\in\mathcal{T}_n}\left(\sum_{\beta\in\mathcal{C}(\alpha)}|f(\beta)-f(\alpha)|\right)^p\\&\gtrsim\sum_{\alpha\in\mathcal{T}_n}\left(e^{-d(\alpha)\theta}\right)^p\approx\sum_{N=1}^{\infty}\sum_{\alpha\in\mathcal{T}_n:d(\alpha)=N}e^{-pN\theta}\\&\approx\sum_{N=1}^{\infty}e^{2nN\theta}e^{-pN\theta}=\infty\end{aligned}$$

if $2n-p\geq 0$.

What is needed now is a definition of Besov space on a tree that involves complex structure sufficient for higher order differences, or "derivatives", to be properly defined. We discuss that next.

16.2. Structured Trees

We will now move to the "holomorphic spaces" $HB_{p,m}(\mathcal{T}_n)$ and $HB_p(\mathcal{T}_n)$ mentioned at the start of the chapter. We begin by giving an alternative description of $B_{p,1}(\mathcal{T}_n)$ that reflects more clearly the relation between the tree and the complex geometry of the ball $\mathbb{B}_n$ in which it sits. Rather than impose a geometry on the tree we will use the complex geometry of the ball to inform the way in which the local differences of functions on the tree are combined to produce a norm. The resulting scheme will then be generalized to higher order differences producing the "holomorphic Besov spaces on trees", $HB_p(\mathcal{T}_n)$, which will meet the needs described in Section 15.6.

Although $\mathcal{T}_n$ is an abstract tree, we view it now as sitting inside the ball $\mathbb{B}_n$. Each $\alpha\in\mathcal{T}_n$ can be viewed as the index point for a small piece of volume V_α, which, in spherical coordinates, is roughly $S_\alpha\times I_\alpha$ where S_α is parameterized by a small patch of surface in $\partial\mathbb{B}_n$ and I_α is a small interval of radial values. The children of α, $\{\alpha_j^+\}$ are obtained, roughly, by changing I_α and bisecting S_α along each of its $2n-1$ (real) dimensions; producing 2^{2n-1} children. Hence a function F defined on $\mathcal{T}_n$ it will have 2^{2n-1} differences $F\left(\alpha_j^+\right)-F(\alpha)$. On the other hand, if F is the restriction to $\mathcal{T}_n$ of some $\mathbf{F}\in\operatorname{Hol}(\mathbb{B}_n)$ then those differences should be be well approximated by the set of numbers $\{\mathbf{F}'(\alpha)\left(\alpha_j^+-\alpha\right)\}$, a set determined by data from $\mathcal{T}_n$ and the $2n$ real parameters from $\mathbf{F}'(\alpha)$. If $n=1$ then $2^{2n-1}=2n$, suggesting it might be straightforward to pass information between holomorphic functions and functions on trees. In fact in some cases it is and we saw instances of that earlier. However if $n\geq 2$ then $2^{2n-1}>2n$ which suggests there might be problems in passing information both ways. In particular, membership of $\mathbf{F}$

in a Besov space is characterized by quantitative statements about the quality of differential approximations, and we would like to see echoes of that in our function spaces on trees. With all this in mind, we introduce additional structure on the set of differences $\left\{F\left(\alpha_j^+\right)-F(\alpha)\right\}$. We do this in several steps. First we introduce a "complex structure" on a tree, and we then define an associated "complex derivative" of a function on the tree. We then build "spaces of holomorphic functions on a tree" defined by quantitative approximation properties based on that derivative.

We begin by introducing a "complex structure" on the trees we consider.

DEFINITION 16.15. *An n-dimensional* complex structure $\mathcal{V}$ *on a tree* $\mathcal{T}$ *is a collection of n-vectors* $\mathcal{V}:=\{\mathbf{v}_\alpha\}_{\alpha\in\mathcal{T}}$, $\mathbf{v}_\alpha\in\mathbb{C}^n$.

We can "immerse" the structured tree $(\mathcal{T},\mathcal{V})$ in $\mathbb{C}^n$ by identifying $\alpha\in\mathcal{T}$ with the point $c(\alpha):=\mathbf{v}_o+\sum_{o<\beta\leq\alpha}\mathbf{v}_\beta\in\mathbb{C}^n$. For example, the standard embedding of the Bergman tree $\mathcal{T}_n$ in the ball arises in this way from the complex structure $\mathcal{V}=\{c_\alpha-c_{A\alpha}\}_{\alpha\in\mathcal{T}_n}$ on $\mathcal{T}_n$. In general however, the map $\alpha\to c(\alpha)$ need not be one-to-one, hence the term "immerse". Additional properties of $\mathcal{V}$ will be required below.

Recall that the (backward) difference operator Δ on functions f mapping the tree $\mathcal{T}$ to $\mathbb{C}$ is given by, for $\alpha\in\mathcal{T}$ and with $A\alpha$ denoting the immediate ancestor (predecessor) of α; $\Delta f(\alpha)=f(\alpha)-f(A\alpha)$. We denote the set of children of $\alpha\in\mathcal{T}$ by $\mathcal{C}(\alpha)$. We now assume that there is an upper bound N for the branching number of the tree, i.e. $\operatorname{card}\mathcal{C}(\alpha)\leq\log N$ for all $\alpha\in\mathcal{T}_n$. In particular this ensures that $\dim(\mathcal{T})\leq\log_2 N$. Let $\left\{\alpha^j\right\}_{j=1}^{\operatorname{card}\mathcal{C}(\alpha)}=\mathcal{C}(\alpha)$ be an enumeration of the children of α. For notational convenience we artificially expand these sets to sets $\left\{\alpha^j\right\}_{j=1}^N$ by setting $\alpha^j=\alpha$ if $\operatorname{card}\mathcal{C}(\alpha)<j\leq N$. For $\alpha\in\mathcal{T}$ we then define the linear map $L_\alpha:\mathbb{C}^n\to\mathbb{C}^N$ by

$$L_\alpha(\mathbf{w}):=\left(\mathbf{w}\cdot\mathbf{v}_{\alpha^j}\right)_{j=1}^N,$$

where, by our convention, $\mathbf{v}_{\alpha^j}=0$ if $\operatorname{card}\mathcal{C}(\alpha)<j\leq N$. We make the further assumption that L_α is *one-to-one* for all $\alpha\in\mathcal{T}$, equivalently that for each α, $\{\mathbf{v}_{\alpha^j}\}_{j=1}^N$ spans $\mathbb{C}^n$.

Given a complex-valued function f on $\mathcal{T}$ we define its *complex derivative* $f'(\alpha)\in\mathbb{C}^n$ as follows. Let $\mathbf{P}_\alpha$ denote orthogonal projection of $\mathbb{C}^N$ onto the range of L_α, and let $\mathbf{Q}_\alpha:=I-\mathbf{P}_\alpha$. Denote by $\mathcal{D}_\alpha f$ the N-vector of (forward) differences of f:

$$\mathcal{D}_\alpha f:=\left(f\left(\alpha^j\right)-f(\alpha)\right)_{j=1}^N=\left(\Delta f\left(\alpha^j\right)\right)_{j=1}^N\in\mathbb{C}^N,$$

where, by our convention, $\Delta f\left(\alpha^j\right)=0$ if $\operatorname{card}\mathcal{C}(\alpha)<j\leq N$. We then define

$$f'(\alpha):=L_\alpha^{-1}\mathbf{P}_\alpha\left(\mathcal{D}_\alpha f\right),$$

so that

$$\left(f\left(\alpha^j\right)-f(\alpha)\right)_{j=1}^N=\left(f'(\alpha)\cdot\mathbf{v}_{\alpha^j}\right)_{j=1}^N+\mathbf{Q}_\alpha\left(\mathcal{D}_\alpha f\right). \tag{16.18}$$

(Thus as we go forward the "prime" will be used for differentiation of holomorphic function as well as the construction just defined applied to functions on structured trees.) In the case of the natural complex structure introduced above on the Bergman tree $\mathcal{T}_n$, we have $\mathbf{v}_{\alpha^j}=\alpha^j-\alpha$ and (16.18) is thus an analog of Taylor's formula of degree one on $\mathcal{T}_n$.

We now make this more explicit for the particular case of the Bergman tree $\mathcal{T}_n$. For the Bergman tree $\mathcal{T}_n$ we can choose $N = Ce^{2n\theta}$, with $\lambda \leq 1$ and $\theta > 1$ to be chosen below, by property (4) of Lemma 14.3. We also define the difference

$$\Delta\alpha = \alpha - A\alpha, \qquad \alpha \in \mathcal{T}_n,$$

where we identify α with the center c_α of the Bergman cube K_α. With our notational convention, the set of differences $\mathcal{D}_\alpha f := \left(\Delta f\left(\alpha^j\right)\right)_{j=1}^N$ is an $N-$ vector of complex numbers; $\mathcal{D}_\alpha f \in \mathbb{C}^N$. We also consider the corresponding family of differences $\left(\Delta\alpha^j\right)_{j=1}^N$, as a vector of points in $\mathbb{C}^n$ of length N, i.e. in $\left(\mathbb{C}^n\right)^N$.

The linear map L_α defined above sends $\mathbf{v} \in \mathbb{C}^n$ to the point

$$L_\alpha \mathbf{v} = \left(\mathbf{v} \cdot \left(\alpha^j - \alpha\right)\right)_{j=1}^N \in \mathbb{C}^N.$$

Note that for the Bergman tree, if $\theta > 1$ is chosen large enough, then $N = Ce^{2n\theta} >> n$ and the collection of n-vectors $\left(\alpha^j - \alpha\right)_{j=1}^N$ has rank n and hence the map L_α is one-to-one. Recall that $\mathbf{P}_\alpha$ is the orthogonal projection of $\mathbb{C}^N$ onto the range of L_α and $\mathbf{Q}_\alpha = I - \mathbf{P}_\alpha$. Because L_α is injective it has a well defined inverse on range($\mathbf{P}_\alpha$) = range(L_α) and we use that inverse to define the "complex derivative." The complex derivative $f'(\alpha)$ of f at the point α is then the unique vector $\mathbf{v}$ such that $L_\alpha \mathbf{v} = \mathbf{P}_\alpha\left(\mathcal{D}_\alpha f\right)$. Thus we have

$$L_\alpha f'(\alpha) = \mathbf{P}_\alpha\left(\mathcal{D}_\alpha f\right) = \left(f'(\alpha) \cdot \left(\alpha^j - \alpha\right)\right)_{j=1}^N.$$

Denote the radial and tangential components of $f'(\alpha)$ by $f'(\alpha) P_\alpha$ and $f'(\alpha) Q_\alpha$ respectively, where $P_\alpha z := \frac{z \cdot \overline{\alpha}}{|\alpha|^2}\alpha$ as in (14.1), and $Q_\alpha := I - P_\alpha$. Here we are viewing $f'(\alpha)$ as belonging to the space $\mathcal{L}\left(\mathbb{C}^n, \mathbb{C}\right)$ of linear maps from $\mathbb{C}^n$ to $\mathbb{C}$, and $\mathbf{P}_\alpha, \mathbf{Q}_\alpha$ as belonging to the corresponding space of linear maps $\mathcal{L}\left(\mathbb{C}^N, \mathcal{L}\left(\mathbb{C}^N, \mathbb{C}\right)\right)$. Thus we have decomposed the difference set $\mathcal{D}_\alpha f$ as

(16.19)

$$\begin{aligned} \mathcal{D}_\alpha f &= \left(\Delta f\left(\alpha^j\right)\right)_{j=1}^N \\ &= \left(f'(\alpha) P_\alpha \cdot \left(\alpha^j - \alpha\right)\right)_{j=1}^N + \left(f'(\alpha) Q_\alpha \cdot \left(\alpha^j - \alpha\right)\right)_{j=1}^N + \mathbf{Q}_\alpha\left(\mathcal{D}_\alpha f\right). \end{aligned}$$

REMARK 16.16. *This is one of the main places where we need to choose* $\theta > 1$ *sufficiently large in the construction of the Bergman tree, here to obtain that* L_α *is one-to-one and thus to define the complex derivative on the Bergman tree. Another place we will use* $\theta > 1$ *sufficiently large is in* (16.20) *below, which is used to efficiently split a derivative on the tree into radial and tangential components.*

We now begin our introduction of the holomorphic Besov spaces on the tree $\mathcal{T}_n$, $HB_{p,k}\left(\mathcal{T}_n\right)$, a construction which uses the complex structure $\mathcal{T}_n$ inherits from $\mathbb{B}_n$. We start with $k = 1$. In that case, but only in that case, we merely obtain an alternative description of a space we have already introduced; $HB_{p,1}\left(\mathcal{T}_n\right)$ turns out to be $B_{p,1}\left(\mathcal{T}_n\right)$ with an equivalent norm.

The Bergman metric on $\mathbb{B}_n$ (the Riemannian metric which induces the Bergman distance β) assigns to the tangent vector $\mathbf{v}$ at the point α in $\mathbb{B}_n$ the length

$$\sqrt{\left(\frac{P_\alpha \mathbf{v}}{\left(1 - |\alpha|^2\right)^2} + \frac{Q_\alpha \mathbf{v}}{1 - |\alpha|^2}\right) \cdot \overline{\mathbf{v}}},$$

or, equivalently, setting

$$B(z) := \frac{1}{n+1} \frac{\partial^2}{\partial \overline{z_i} \partial z_j} \log \frac{1}{(1-|z|^2)^{n+1}}$$

the length of $\mathbf{v}$ is $\sqrt{(B(\alpha)\mathbf{v}) \cdot \overline{\mathbf{v}}}$ - see (14.11) and [**346**, Chapter 1].

The expression

$$\left| e^{-2d(\alpha)\theta} \mathbf{v} P_\alpha + e^{-d(\alpha)\theta} \mathbf{v} Q_\alpha \right|$$

is the tree analog of this length functional. Thus we make the following definition.

DEFINITION 16.17. *For a vector $\mathbf{v} \in \mathbb{C}^n$ and $\alpha \in \mathcal{T}_n$, let*

$$|\mathbf{v}|_\alpha := \left| e^{-2d(\alpha)\theta} \mathbf{v} P_\alpha + e^{-d(\alpha)\theta} \mathbf{v} Q_\alpha \right| = \sqrt{(B(\alpha)\mathbf{v}) \cdot \overline{\mathbf{v}}}.$$

DEFINITION 16.18. *For $1 < p < \infty$, define the* holomorphic *Besov space $HB_{p,1}(\mathcal{T}_n)$ on $\mathcal{T}_n$ to consist of all complex-valued sequences $f = \{f(\alpha)\}_{\alpha \in \mathcal{T}_n}$ such that*

$$\begin{aligned} \|f\|^p_{HB_{p,1}(\mathcal{T}_n)} &:= |f(o)|^p + \sum_{\alpha \in \mathcal{T}_n} \left| e^{-2d(\alpha)\theta} f'(\alpha) P_\alpha + e^{-d(\alpha)\theta} f'(\alpha) Q_\alpha \right|^p \\ &\quad + \sum_{\alpha \in \mathcal{T}_n} |\mathbf{Q}_\alpha \mathcal{D}_\alpha f|^p \\ &= |f(o)|^p + \sum_{\alpha \in \mathcal{T}_n} |f'(\alpha)|^p_\alpha + \sum_{\alpha \in \mathcal{T}_n} |\mathbf{Q}_\alpha \mathcal{D}_\alpha f|^p < \infty. \end{aligned}$$

For $1 < p < \infty$ and θ sufficiently large, the norms $\|\cdot\|_{B_{p,1}(\mathcal{T}_n)}$ and $\|\cdot\|_{HB_{p,1}(\mathcal{T}_n)}$ are equivalent. This follows from the definitions, the following proof, and noting that, for $m = 1$ we have

$$\mathcal{D}_\alpha = \left(2^{-n}\right)^m (2^n \Delta)^m = \Delta.$$

PROPOSITION 16.19. *For θ sufficiently large in the construction of the Bergman tree $\mathcal{T}_n$, and for all $1 < p < \infty$, we have*

$$\left(\sum_{\alpha \in \mathcal{T}_n} |\mathcal{D}_\alpha f|^p \right)^{1/p} \approx \left(\sum_{\alpha \in \mathcal{T}_n} |f'(\alpha)|^p_\alpha \right)^{1/p} + \left(\sum_{\alpha \in \mathcal{T}_n} |\mathbf{Q}_\alpha \mathcal{D}_\alpha f|^p \right)^{1/p}.$$

PROOF. Recalling that $\mathbf{P}_\alpha$ and $\mathbf{Q}_\alpha$ are orthogonal projections on $\mathbb{C}^N$ and that P_α and Q_α are orthogonal projections on $\mathbb{C}^n$, we compute

$$\begin{aligned} |\mathcal{D}_\alpha f|^2 &= |\mathbf{P}_\alpha \mathcal{D}_\alpha f|^2 + |\mathbf{Q}_\alpha \mathcal{D}_\alpha f|^2 \\ &= |L_\alpha f'(\alpha)|^2 + |\mathbf{Q}_\alpha \mathcal{D}_\alpha f|^2 \\ &\approx \left| e^{-2d(\alpha)\theta} f'(\alpha) P_\alpha \right|^2 + \left| e^{-d(\alpha)\theta} f'(\alpha) Q_\alpha \right|^2 + |\mathbf{Q}_\alpha \mathcal{D}_\alpha f|^2 \\ &= |f'(\alpha)|^2_\alpha + |\mathbf{Q}_\alpha \mathcal{D}_\alpha f|^2, \end{aligned}$$

where the third line follows from

$$\begin{aligned} |\mathbf{P}_\alpha \mathcal{D}_\alpha f|^2 &= |L_\alpha f'(\alpha)|^2 \\ &= \left| \left\{ f'(\alpha) P_\alpha \cdot P_\alpha \left(\alpha^j - \alpha\right) + f'(\alpha) Q_\alpha \cdot Q_\alpha \left(\alpha^j - \alpha\right) \right\}_{j=1}^N \right|^2 \\ &\approx \left| e^{-2d(\alpha)\theta} f'(\alpha) P_\alpha \right|^2 + \left| e^{-d(\alpha)\theta} f'(\alpha) Q_\alpha \right|^2. \end{aligned}$$

To see this last equivalence, we note using (14.11) that both

$$\sum_{j=1}^{N}\left|P_\alpha\left(\alpha^j-\alpha\right)\right|\gtrsim e^{-2d(\alpha)\theta}, \tag{16.20}$$

$$\sum_{j=1}^{N}\left|Q_\alpha\left(\alpha^j-\alpha\right)\right|\gtrsim e^{-d(\alpha)\theta}.$$

The first inequality is obvious. The second inequality follows if $\theta>1$ is chosen sufficiently large, since the set of projections onto the sphere $\mathcal{S}_{d(\alpha)\theta}$ of the children $\mathcal{C}(\alpha)$ is $e^{-2\theta}C_2$-dense in the cube $Q_j^{d(\alpha)}=K_\alpha\cap\mathcal{S}_{d(\alpha)\theta}$ corresponding to K_α. This is because, by property (4) of Proposition 14.3, there are roughly $e^{2n\theta}$ children of α whose projections onto the Bergman sphere $S_{d(\alpha)\theta}$ all lie in $Q_j^{d(\alpha)}$. Since the Bergman distance β is preserved by automorphisms, we see upon mapping matters to the origin that these projections are roughly $e^{-2\theta}$-dense in the Bergman distance. With this established for $e^{-2\theta}$ sufficiently small, we now see that the vectors $\left\{Q_\alpha\left(\alpha^j-\alpha\right)\right\}_{j=1}^N$ are sufficiently well distributed to ensure that (16.20) holds uniformly in α. □

16.2.1. A Digression. This general approach can be used to build many types of function spaces on trees, some analogous to spaces of classical analysis. Before going on to our case of primary interest, $HB_{p,m}(\mathcal{T}_n)$, we digress briefly to describe a few examples. Other than the work in [**41**] on martingale variants of tree Besov spaces, these variations have not been developed in any detail.

In the discussions we will use the notation $\|\cdot\|_{B_p(\mathcal{T})}$ in a context dependent manner.

Given a tree $\mathcal{T}$ with branching number bounded by N, we can more generally than above, suppose that we are given an m-dimensional complex vector space W, and for each $\alpha\in\mathcal{T}$, Hilbert space norms $[\cdot]_\alpha$ and $\{\cdot\}_\alpha$ on W and $\mathbb{C}^N$ respectively, and a one-to-one linear map L_α from W to $\mathbb{C}^N$. Then we can define a "derivative" $f'(\alpha)\in W$ by

$$f'(\alpha):=L_\alpha^{-1}\mathbf{P}_\alpha\left(\mathcal{D}_\alpha f\right),$$

where $\mathcal{D}_\alpha f$ is the (forward) difference set defined as above, and $\mathbf{P}_\alpha$ is orthogonal projection onto the range of L_α. With $\mathbf{Q}_\alpha=I-\mathbf{P}_\alpha$, define a norm on $\mathbb{C}^N$ by

$$|w|_\alpha^2:=\left[L_\alpha^{-1}\mathbf{P}_\alpha(w)\right]_\alpha^2+\left\{\mathbf{Q}_\alpha(w)\right\}_\alpha^2.$$

Then we define a Besov space norm $\|f\|_{B_p(\mathcal{T})}$ by

$$\|f\|_{B_p(\mathcal{T})}^p:=|f(o)|^p+\sum_{\alpha\in\mathcal{T}}|f'(\alpha)|_\alpha^p.$$

Rather than develop this abstract viewpoint further, we list some examples.

EXAMPLE 16.20. *In the case $\mathcal{T}=\mathcal{T}_n$ and $W=\mathbb{C}^n$, with L_α defined as above and*

$$[\mathbf{v}]_\alpha=|\mathbf{v}|_\alpha=\left|e^{-2d(\alpha)\theta}\mathbf{v}P_\alpha+e^{-d(\alpha)\theta}\mathbf{v}Q_\alpha\right|,\quad \mathbf{v}\in\mathbb{C}^n,$$

$$\{w\}_\alpha=|w|,\quad w\in\mathbb{C}^N,$$

we obtain that $B_p(\mathcal{T})$ *is the holomorphic Besov space* $HB_{p,1}(\mathcal{T}_n)$ *defined above. If however we take* $W=\mathbb{C}^N$, $L_\alpha=I$ *and the norms* $[\cdot]_\alpha$ *and* $\{\cdot\}_\alpha$ *to be the Euclidean norm on* $\mathbb{C}^N$, *then* $B_p(\mathcal{T})$ *is the abstract space* $B_{p,1}(\mathcal{T})$ *defined earlier.*

EXAMPLE 16.21. *Suppose that* $\mathcal{T}$ *is a homogeneous tree with branching number* N. *Take* W *to be the orthogonal complement in* $\mathbb{C}^N$ *of the one-dimensional subspace* $\mathbb{C}(1,1,\dots,1)$ *generated by* $(1,1,\dots,1)$, *and let* $[\cdot]_\alpha$ *be the restriction of the Euclidean norm to* W. *Let* L_α *be the natural inclusion of* W *into* $\mathbb{C}^N$. *Finally, let* $\{\cdot\}_\alpha$ *be the trivial norm on* $\mathbb{C}^N$ *that is infinite on all nonzero vectors and vanishes on the zero vector. Then* $B_p(\mathcal{T})$ *is the martingale of all* ℓ^p *functions on the tree* $\mathcal{T}$ *satisfying*

$$f(\alpha)=\frac{1}{N}\sum_{\beta\in\mathcal{C}(\alpha)}f(\beta),\qquad \alpha\in\mathcal{T}.$$

Spaces of this type are considered in [**41**].

EXAMPLE 16.22. *Take* $\mathcal{T}=\mathcal{T}_1$, $W=\mathbb{C}$, $L_\alpha v=\left(v\left(\alpha^j-\alpha\right)\right)_{j=1}^2$ *and*

$$\begin{aligned}[v]_\alpha&=\left|e^{-3d(\alpha)\theta}v\right|,\qquad v\in\mathbb{C},\\ \{w\}_\alpha&=|w|,\qquad w\in\mathbb{C}^2.\end{aligned}$$

The resulting space $B_2(\mathcal{T})$ *normed by*

$$\|f\|_{B_2(\mathcal{T})}^2=|f(o)|^2+\sum_{\alpha\in\mathcal{T}_1}\left|e^{-3d(\alpha)\theta}f'(\alpha)\right|^2+\sum_{\alpha\in\mathcal{T}_1}|\mathbf{Q}_\alpha\mathcal{D}_\alpha f|^2$$

is a tree model for the Hardy space

$$H^2(\mathbb{D})=\left\{F\in H(\mathbb{D}):|F(0)|^2+\int_{\mathbb{D}}\left|(1-|z|^2)^{3/2}F'(z)\right|^2d\lambda_2(z)<\infty\right\}$$

on the unit disk. Note that $|\mathcal{D}_\alpha f|\approx\left|e^{-2d(\alpha)\theta}f'(\alpha)\right|+|\mathbf{Q}_\alpha\mathcal{D}_\alpha f|$, *so that the tree model* $B_2(\mathcal{T})$ *for the Hardy space permits the differences of* f *to grow rapidly in the holomorphic direction, and is thus much larger than the abstract space* $B_{2,1}(\mathcal{T}_1)$.

EXAMPLE 16.23. *By including higher order derivatives, this example can be extended to higher dimensions to provide a tree model for the Drury-Arveson space,* H_n^2, *mentioned in Section* 2.5.7, *that consists of all* $F\in\mathrm{Hol}(\mathbb{B}_n)$ *whose radial derivative* $\mathcal{R}^mF$ *satisfies*

$$\begin{aligned}&\int_{\mathbb{B}_n}\left|(1-|z|^2)^m\mathcal{R}^mF(z)\right|^2\frac{dV}{(1-|z|^2)^n}\\&=\int_{\mathbb{B}_n}\left|(1-|z|^2)^{m+1/2}\mathcal{R}^mF(z)\right|^2d\lambda_n(z)<\infty,\end{aligned}$$

where $m=1+\left[\frac{n-1}{2}\right]$. *The Hardy space* $H^2(\mathbb{D})$ *on the disk is the case* $m=n=1$.

16.3. Holomorphic Besov Spaces on Trees

16.3.1. Prelude. We now go beyond the spaces $HB_{p,1}(\mathcal{T}_n)$ and define the *holomorphic Besov spaces on trees*, $HB_{p,m}(\mathcal{T}_n)$ and $HB_p(\mathcal{T}_n)$. (Similarly to before, $HB_p(\mathcal{T}_n)$ is $HB_{p,m}(\mathcal{T}_n)$ for any sufficiently large m.) Although we saw that the spaces $HB_{p,m}(\mathcal{T}_n)$ and $B_{p,m}(\mathcal{T}_n)$ are the same for $m=1$, that is no longer true for larger m.

We have several overlapping reasons for moving to these additional spaces. The first is aesthetic/conceptual. There is a sense in which the spaces $B_p(\mathcal{T}_n)$ fail to fully capture the consequences of smoothness. It is a fundamental fact that for a holomorphic function F in B_p, the differences $F(\beta) - F(\alpha)$ are *related* as β ranges over the children of a fixed α; namely they are close to $F'(\alpha)(\beta - \alpha)$ which is a complex linear function of $(\beta - \alpha)$. Furthermore that closeness can be quantified using the B_p norm. (A similar comment applies to F which are smooth but not holomorphic.) However the local differences of functions in $B_p(\mathcal{T}_n)$ do not, apparently, have such constraints. Thus the functions in those tree Besov spaces can have more local fluctuation, can be rather "floppy", in contrast to the more constrained behavior of smooth functions. We define the space $HB_p(\mathcal{T}_n)$ to consist of functions whose higher order differences resemble the restriction to the tree $\mathcal{T}_n$ of complex multi-linear functions obtained as higher order almost invariant holomorphic derivatives, as defined in Section 14.5.1.

A second reason we pass to the space $HB_p(\mathcal{T}_n)$ is to complete the characterization of interpolating sequences $B_p(\mathbb{B}_n)$. For $n = 1$ the results on interpolating sequences for $B_p(\mathcal{T}_n)$ ([**48**] and Section 16.1.1) are similar to, and independent from, the results for $B_p(\mathbb{B}_n)$ ([**77**] and the previous chapter). For $n > 1$ the theorems are not independent. The results about tree spaces are used in completing the outline in Section 15.6 and thus finishing the proof of the results for $B_p(\mathbb{B}_n)$. For this to succeed we need to pass information back and forth between holomorphic functions on $\mathbb{B}_n$ and functions $\mathcal{T}_n$. A natural approach in one direction is to restrict the holomorphic functions to the tree vertices, and we would like that to be a bounded map between the Besov spaces and also between their multiplier algebras. This is not generally true if we use the $B_p(\mathcal{T}_n)$; but using the oscillation inequality from Section 14.5.3 we find the situation is much better for the $HB_p(\mathcal{T}_n)$. In particular we will see that the restriction map from B_p to $HB_{p,m}(\mathcal{T}_n)$ (as in Corollary 16.11) is bounded if $p > 2n/m$, and that we are free to let m be large. This will let us show that Carleson measures for $HB_{p,m}(\mathcal{T}_n)$ are characterized by the tree condition (14.37), and that is the condition we need to prove the remaining necessity implication for multiplier interpolation on the ball.

Another specific reason for passing to the space $HB_p(\mathbb{B}_n)$ is for (generalized) positivity of the reproducing kernel. Some of the Hilbert space tools we used in Chapter 7 are not available when $p \neq 2$. For $n = 1$, instead Böe uses [**77**, Lemma 3.1] which he refers to as a "rather curious lemma." We will use that lemma and variations on it, referring to it as Böe's "curious lemma".

The curious lemma in [**77**] is a piece of functional analysis for positive functions in L^p spaces using positivity of the real part of the reproducing kernel for $B_p(\mathbb{B}_1)$; we will follow a similar path using a positivity property of the kernel functions for $HB_p(\mathbb{B}_n)$. That lets us go forward to prove the necessity of the tree condition for multiplier interpolation on the ball in the "difficult range" $1 + 1/(n-1) \leq p < 2$.

The spaces $B_{p,1}(\mathcal{T}_n)$ satisfies some of our requirements. In the previous section we showed in Corollary 16.11 that the restriction map is bounded from B_p to $B_{p,1}(\mathcal{T}_n)$ in the range $p > \widehat{n} = \frac{2n}{1}$. The Carleson measures for $B_{p,1}(\mathcal{T}_n)$ are characterized by the tree condition, and the first order difference of the reproducing kernel for $B_{p,1}(\mathcal{T}_n)$ is nonnegative (the analogs of these latter two properties actually hold for *all* $B_{p,m}(\mathcal{T}_n)$ by Lemma 16.2 and 16.4). This demonstrates that the abstract Besov space $B_{p,1}(\mathcal{T}_n)$ has the properties desired of our holomorphic

Besov space for $p > 2n$, and in view of Lemma 16.19, we have

$$HB_{p,1}(\mathcal{T}_n) = B_{p,1}(\mathcal{T}_n). \tag{16.21}$$

However the more general $B_{p,m}(\mathcal{T}_n)$ spaces do *not* have the desired properties; they do not capture the higher order derivatives of holomorphic functions in the ball. Indeed, higher order tree differences vanish on appropriate polynomial functions of $r^{-d(\alpha)}$ on the tree, but *not* on the restrictions to the tree of polynomials on the ball, even though the corresponding derivatives of polynomials do vanish. In particular, recall from Remark 16.14 that linear functions on the ball do not restrict to $B_{p,m}(\mathcal{T}_n)$ for $p \leq 2n$ for any $m \geq 1$. The proof of the restriction theorem given in Corollary 16.11 is not amenable to generalization to higher order differences.

With these comments in mind, we move forward using the pattern in our definition of $HB_{p,1}(\mathcal{T}_n)$. For general m we take the almost invariant holomorphic derivatives $D_a^m f$ used in the seminorms (17.48) for B_p, as the model for the higher order complex tree derivatives $f^{(k)}(\alpha)$ which we use to define $HB_{p,m}(\mathcal{T}_n)$. We begin with the holomorphic Besov space $HB_{p,1}(\mathcal{T}_n)$, already defined in Section 16.2, and identify its reproducing kernels relative to the duality pairing induced by the norm $\|\cdot\|_{HB_{p,1}(\mathcal{T}_n)}$, and show they have the required positivity properties. The function $f^{(k)}(\alpha)$ will be a tensor-valued function on the tree. Hence, in preparation for an inductive definition of the higher order Besov spaces $HB_{p,m}(\mathcal{T}_n)$, we also develop the analogous theory of the Besov spaces $HB_{p,1}^{(k)}(\mathcal{T}_n)$ of k-tensors defined on $\mathcal{T}_n$. To expedite this process, it is convenient to consider first the order zero case, and develop the required tensor apparatus for ℓ^p spaces, the order zero analog of a holomorphic Besov space. Then, having defined $HB_{p,1}(\mathcal{T}_n)$ earlier, we proceed to define $HB_{p,m}(\mathcal{T}_n)$ inductively for $m \geq 2$ and establish the appropriate positivity properties of their reproducing kernels. That will require a careful choice of the structural constants λ and θ in the construction of the Bergman tree $\mathcal{T}_n$, along with an additional modification of the centers of the Bergman balls. Finally, we establish for these spaces the Carleson measure theorem and the restriction theorem, and then complete the proof of the multiplier interpolation loop for $1 < p < 2+1/(n-1)$, that was left open at the end of the previous chapter.

Although the general themes in the following sections are familiar from the previous chapters (and they recur in the next chapter), a substantial amount of technical detail is needed. Some of the details are standard for work on function spaces and sequences: establishing that point evaluations are continuous functionals, obtaining a description of the dual space and bilinear pairing, representing point evaluation functionals in the dual space (i.e. describing the reproducing kernels), etc. Other details are typical to the study of Besov spaces. In particular we show that a variety of measures of local variation can be used to define equivalent norms and we analyze the details of those measures of variation to learn the valid parameter ranges for the equivalence of the norms. Finally, there is a substantial notational burden associated to the interaction of tensor-valued higher derivatives with the anisotropic geometry of the ball.

These sections are rich in details; however, we emphasize that the underlying ideas are more straightforward than the notational intricacies.

16.3.2. Holomorphic Besov Spaces; Orders Zero and One. Recall that for $1 < p < \infty$ we defined the order 1 holomorphic Besov space $HB_{p,1}(\mathcal{T}_n)$ on $\mathcal{T}_n$

in Definition 16.18 to consist of all complex-valued sequences $f = \{f(\alpha)\}_{\alpha\in\mathcal{T}_n}$ such that

$$\|f\|^p_{HB_{p,1}(\mathcal{T}_n)} = |f(o)|^p + \sum_{\alpha\subset\mathcal{T}_n} \left| r^{-d(\alpha)} f'(\alpha) P_\alpha + r^{-d(\alpha)/2} f'(\alpha) Q_\alpha \right|^p + \sum_{\alpha\in\mathcal{T}_n} |\mathbf{Q}_\alpha \mathcal{D}_\alpha f|^p < \infty,$$

where for convenience we have written

$$r = e^{2\theta} \tag{16.22}$$

and hence by (14.11) we have the approximations

$$1 - |\alpha|^2 \approx e^{-2\beta(0,\alpha)} \approx e^{-2\theta d(\alpha)} = r^{-d(\alpha)}. \tag{16.23}$$

In comparing this definition with Definition 16.1 for the real Besov space $B_{p,1}(\mathcal{T}_n)$ on the Bergman tree $\mathcal{T}_n$,

$$\|f\|^p_{B_{p,1}(\mathcal{T}_n)} = |f(o)|^p + \sum_{\alpha\in\mathcal{T}_n} |\mathcal{D}_\alpha f|^p < \infty,$$

recall that the set of differences $\mathcal{D}_\alpha f$ can be written as the linear sum of the two pieces $\mathbf{P}_\alpha \mathcal{D}_\alpha f$ and $\mathbf{Q}_\alpha \mathcal{D}_\alpha f$; the first piece $\mathbf{P}_\alpha \mathcal{D}_\alpha f$ lying in the range of L_α (the holomorphic part), and the second piece $\mathbf{Q}_\alpha \mathcal{D}_\alpha f$ orthogonal to the range of L_α. The first piece $\mathbf{P}_\alpha \mathcal{D}_\alpha f$ can be further decomposed using the identities

$$\begin{aligned} L_\alpha f'(\alpha) &= \mathbf{P}_\alpha \mathcal{D}_\alpha f, \\ f'(\alpha) &= L_\alpha^{-1} \mathbf{P}_\alpha \mathcal{D}_\alpha f, \end{aligned}$$

where the second one follows since L_α is one-to-one and $\mathbf{P}_\alpha$ is orthogonal projection onto the range of L_α. We then decompose $f'(\alpha)$ as $f'(\alpha) P_\alpha + f'(\alpha) Q_\alpha$. Now L_α has the $N \times n$ matrix representation $\left[\alpha^j_k - \alpha_k\right]_{1\le j\le N, 1\le k\le n}$ where $\alpha^j = \left(\alpha^j_k\right)^n_{k=1}$ and $\alpha = (\alpha_k)^n_{k=1}$, and so resembles the anisotropic linear operator

$$\mathbf{R}^{-d(\alpha)} := r^{-d(\alpha)} P_\alpha + r^{-d(\alpha)/2} Q_\alpha,$$

whose action on a vector $v \in \mathcal{L}(\mathbb{C}^n, \mathbb{C})$ is given by $\mathbf{R}^{-d(\alpha)} v := r^{-d(\alpha)} v P_\alpha + r^{-d(\alpha)/2} v Q_\alpha$. Thus we formally have that

$$\begin{aligned} \mathcal{D}_\alpha f &= L_\alpha f'(\alpha) + \mathbf{Q}_\alpha \mathcal{D}_\alpha f \\ &\sim \mathbf{R}^{-d(\alpha)} f'(\alpha) + \mathbf{Q}_\alpha \mathcal{D}_\alpha f. \end{aligned}$$

We actually proved that $|\mathcal{D}_\alpha f| \approx \left|\mathbf{R}^{-d(\alpha)} f'(\alpha)\right| + |\mathbf{Q}_\alpha \mathcal{D}_\alpha f|$ in the course of proving Proposition 16.19, see (16.20) where $\theta > 1$ sufficiently large was required.

Thus locally we measure the holomorphic parts $\mathbf{P}_\alpha \mathcal{D}_\alpha f$ of the differences $\mathcal{D}_\alpha f$ by the Bergman Riemannian metric, where the radial directions $f'(\alpha) P_\alpha$ are weighted by $r^{-d(\alpha)}$, and the tangential directions $f'(\alpha) Q_\alpha$ weighted by $\sqrt{r^{-d(\alpha)}} = r^{-d(\alpha)/2}$. This is analogous to the definition (14.61) of the almost invariant holomorphic derivative D_α given in the ball by

$$D_a f(z) = -f'(z) \left\{ \left(1 - |a|^2\right) P_a + \left(1 - |a|^2\right)^{1/2} Q_a \right\}, \qquad a \in \mathbb{B}_n.$$

We then take the $\ell^p(\mathcal{T}_n)$ norm of these local measures. We measure the "antiholomorphic" parts $\mathbf{Q}_\alpha \mathcal{D}_\alpha f$ of the differences $\mathcal{D} f(\alpha)$ by the $\ell^p(\mathcal{T})$-norm.

16.3.3. Reproducing Kernels. We now consider reproducing kernels associated to these norms. The reproducing kernel k_α for the space $\ell^p(\mathcal{T}_n)$ is trivial: $k_\alpha(\gamma) = \chi_{\{\alpha\}}(\gamma)$, the delta function at α. We must work harder to obtain the reproducing kernel for $HB_{p,1}(\mathcal{T}_n)$. We first observe that we can recover a function $f \in HB_{p,1}(\mathcal{T}_n)$ from its differences

$$\begin{aligned}\mathcal{D}_\alpha f &= \mathbf{P}_\alpha \mathcal{D}_\alpha f + \mathbf{Q}_\alpha \mathcal{D}_\alpha f \\ &= L_\alpha f'(\alpha) + \mathbf{Q}_\alpha \mathcal{D}_\alpha f\end{aligned}$$

and hence also from its derivatives and antiholomorphic differences $\mathbf{Q}_\alpha \mathcal{D}_\alpha f$. In order to give an exact formula, we write

$$\mathbf{P}_\alpha \mathcal{D}_\alpha f = \left\{\mathbf{P}_\alpha \mathcal{D}_\alpha f\left(\alpha^j\right)\right\}_{j=1}^N \text{ and } \mathbf{Q}_\alpha \mathcal{D}_\alpha f = \left\{\mathbf{Q}_\alpha \mathcal{D}_\alpha f\left(\alpha^j\right)\right\}_{j=1}^N,$$

so that $\Delta f\left(\alpha^j\right) = \mathbf{P}_\alpha \mathcal{D}_\alpha f\left(\alpha^j\right) + \mathbf{Q}_\alpha \mathcal{D}_\alpha f\left(\alpha^j\right)$, where $\alpha = A\alpha^j$, the parent of α^j. Thus $\mathbf{P}_\alpha \mathcal{D}_\alpha f\left(\alpha^j\right)$ is the jth component of the N-vector $\mathbf{P}_\alpha \mathcal{D}_\alpha f = L_\alpha f'(\alpha)$, i.e.

$$\mathbf{P}_{A\beta} \mathcal{D}_{A\beta} f(\beta) = f'(A\beta) \cdot (\beta - A\beta),$$

and we have the tree version of the holomorphic Taylor expansion of order 1 at the point $A\beta$;

$$f(\beta) = f(A\beta) + f'(A\beta) \cdot (\beta - A\beta) + \mathbf{Q}_{A\beta} \mathcal{D}_{A\beta} f(\beta).$$

An explicit formula for $f(\alpha)$, when $\alpha \in \mathcal{T}_n$ has geodesic $[o,\alpha] = \{o, \alpha_1, \alpha_2, \ldots, \alpha_m\}$ (note that here α_j is a generic point of the geodesic $[o,\alpha]$, not a generic point of $\mathcal{T}_n$) is now given by

$$\begin{aligned}f(\alpha) = f(\alpha_m) &= \sum_{k=1}^m \Delta f(\alpha_k) + f(o) \\ &= f(o) + \sum_{k=0}^m \mathbf{P}_{\alpha_{k-1}} \mathcal{D}_{\alpha_{k-1}} f(\alpha_k) + \sum_{k=0}^m \mathbf{Q}_{\alpha_{k-1}} \mathcal{D}_{\alpha_{k-1}} f(\alpha_k) \\ &= f(o) + \sum_{k=0}^m f'(\alpha_{k-1}) \cdot (\alpha_k - \alpha_{k-1}) + \sum_{k=0}^m \mathbf{Q}_{\alpha_{k-1}} \mathcal{D}_{\alpha_{k-1}} f(\alpha_k).\end{aligned} \tag{16.24}$$

We can rewrite this as

$$f(\alpha) = f(o) + \sum_{\gamma<\alpha} f'(\gamma) \cdot (\gamma_\alpha - \gamma) + \sum_{\gamma<\alpha} \mathbf{Q}_\gamma \mathcal{D}_\gamma f(\gamma_\alpha), \tag{16.25}$$

where γ_α denotes the child of γ lying on the geodesic $[o,\alpha]$, so that $\gamma_\alpha = \alpha_k$ if $\gamma = \alpha_{k-1}$. Note that an immediate consequence of (16.25) is an inequality in the spirit of Lemma 14.12;

$$|f(\alpha)| \lesssim \|f\|_{HB_{p,1}(\mathcal{T}_n)} d(\alpha)^{1/p'}. \tag{16.26}$$

In fact, using $I = P_\gamma^2 + Q_\gamma^2$ we have the formula

$$f'(\gamma) \cdot (\gamma_\alpha - \gamma) = f'(\gamma) P_\gamma \cdot P_\gamma (\gamma_\alpha - \gamma) + f'(\gamma) Q_\gamma \cdot Q_\gamma (\gamma_\alpha - \gamma), \tag{16.27}$$

and hence the estimate

$$\begin{aligned}|f'(\gamma) \cdot (\gamma_\alpha - \gamma)| &\le |f'(\gamma) P_\gamma \cdot P_\gamma (\gamma_\alpha - \gamma)| + |f'(\gamma) Q_\gamma \cdot Q_\gamma (\gamma_\alpha - \gamma)| \\ &\le |f'(\gamma) P_\gamma| r^{-d(\gamma)} + |f'(\gamma) Q_\gamma| r^{-d(\gamma)/2} \\ &\approx \left| r^{-d(\gamma)} f'(\gamma) P_\gamma + r^{-d(\gamma)/2} f'(\gamma) Q_\gamma \right| =: |f'(\gamma)|_\gamma.\end{aligned}$$

(The expression on the left in the final line is the anisotropic norm of the vector $f'(\gamma)$. That expression shows up often. Here we are foreshadowing the introduction, in (16.41), of the notation for it: $|f'(\gamma)|_\gamma$.) Plugging this estimate into (16.25) and using Holder's inequality yields (16.26).

Now consider the case $p = 2$. Since P_α and Q_α are orthogonal projections with vanishing product $P_\gamma Q_\gamma = P_\gamma (I - P_\gamma) = P_\gamma - P_\gamma^2 = 0$, polarization shows that the inner product $\langle\langle f, g\rangle\rangle_1$ for the Hilbert space $HB_{2,1}(\mathcal{T}_n)$, with norm $\|\cdot\|_{HB_{2,1}(\mathcal{T}_n)}$ as in Definition 16.18, is given by

(16.28)
$$\begin{aligned}\langle\langle f, g\rangle\rangle_1 &= f(o)\overline{g(o)} \\ &+ \sum_{\alpha\in\mathcal{T}_n} \left(r^{-2d(\alpha)} f'(\alpha) P_\alpha \cdot \overline{g'(\alpha) P_\alpha} + r^{-d(\alpha)} f'(\alpha) Q_\alpha \cdot \overline{g'(\alpha) Q_\alpha}\right) \\ &+ \sum_{\alpha\in\mathcal{T}_n} \mathbf{Q}_\alpha \mathcal{D}_\alpha f \cdot \overline{\mathbf{Q}_\alpha \mathcal{D}_\alpha g}.\end{aligned}$$

Note that the dot product in the second line of (16.28) is n-dimensional, while that in the third line is N-dimensional.

By inequality (16.26), point evaluation at $\alpha \in \mathcal{T}_n$ is a continuous linear functional on $HB_{2,1}(\mathcal{T}_n)$. Thus there is a unique $k_\alpha^{(1)} \in HB_{2,1}(\mathcal{T}_n)$ such that $f(\alpha) = \left\langle\left\langle f, k_\alpha^{(1)}\right\rangle\right\rangle_1$, which written out explicitly is

(16.29)
$$\begin{aligned}f(\alpha) &= f(o)\overline{k_\alpha^{(1)}(o)} \\ &+ \sum_{\gamma\in\mathcal{T}_n} \left\{ r^{-2d(\gamma)} f'(\gamma) P_\gamma \cdot \overline{k_\alpha^{(1)\prime}(\gamma) P_\gamma} + r^{-d(\gamma)} f'(\gamma) Q_\gamma \cdot \overline{k_\alpha^{(1)\prime}(\gamma) Q_\gamma}\right\} \\ &+ \sum_{\gamma\in\mathcal{T}_n} \mathbf{Q}_\gamma \mathcal{D}_\gamma f \cdot \overline{\mathbf{Q}_\gamma \mathcal{D}_\gamma k_\alpha^{(1)}}.\end{aligned}$$

Just as for the real Besov spaces $B_{p,1}(\mathcal{T}_n)$, we can construct by hand a function $k_\alpha^{(1)} \in HB_{p,1}(\mathcal{T}_n)$ so that the right hand side of (16.25) matches the expression in (16.29).

Indeed, for $\alpha \in \mathcal{T}_n$ with geodesic $[o, \alpha] = \{o, \alpha_1, \alpha_2, \ldots, \alpha_m = \alpha\}$, we choose

$$k_\alpha^{(1)\prime}(\gamma) = \begin{cases} r^{2d(\gamma)}\overline{P_\gamma(\gamma_\alpha - \gamma)} + r^{d(\gamma)}\overline{Q_\gamma(\gamma_\alpha - \gamma)} & \text{if } \gamma = \alpha_{k-1} \\ 0 & \text{if } \gamma \notin [o, \alpha]. \end{cases} \tag{16.30}$$

We will use the facts that

$$P_\gamma \overline{z} = \overline{zP_\gamma}, \qquad Q_\gamma \overline{z} = \overline{zQ_\gamma}. \tag{16.31}$$

The first of these follows from

$$\left\langle \overline{zP_\gamma}, w\right\rangle = \langle \overline{z}, P_\gamma w\rangle = \left\langle \overline{P_\gamma w}, z\right\rangle = \overline{\langle P_\gamma w, \overline{z}\rangle} = \overline{\langle w, P_\gamma \overline{z}\rangle} = \langle P_\gamma \overline{z}, w\rangle,$$

where $\langle a, b\rangle = a \cdot \overline{b}$ is the usual inner product in $\mathbb{C}^n$. The second is similar.

We use (16.31) to rewrite the terms on the right hand side of (16.30):

$$\begin{aligned}\overline{k_\alpha^{(1)\prime}(\gamma)\,P_\gamma} &= P_\gamma \overline{k_\alpha^{(1)\prime}(\gamma)}\\ &= P_\gamma\left\{r^{2d(\gamma)}P_\gamma(\gamma_\alpha-\gamma)+r^{d(\gamma)}Q_\gamma(\gamma_\alpha-\gamma)\right\}\\ &= r^{2d(\gamma)}P_\gamma(\gamma_\alpha-\gamma),\end{aligned}$$

and

$$\begin{aligned}\overline{k_\alpha^{(1)\prime}(\gamma)\,Q_\gamma} &= Q_\gamma \overline{k_\alpha^{(1)\prime}(\gamma)}\\ &= Q_\gamma\left\{r^{2d(\gamma)}P_\gamma(\gamma_\alpha-\gamma)+r^{d(\gamma)}Q_\gamma(\gamma_\alpha-\gamma)\right\}\\ &= r^{d(\gamma)}Q_\gamma(\gamma_\alpha-\gamma).\end{aligned}$$

Using the identity (16.27), and taking note of (16.30) and of the previous two rewritings we have

$$\begin{aligned}(16.32)\quad & f'(\gamma)\cdot(\gamma_\alpha-\gamma)\\ &= f'(\gamma)P_\gamma\cdot P_\gamma(\gamma_\alpha-\gamma)+f'(\gamma)Q_\gamma\cdot Q_\gamma(\gamma_\alpha-\gamma)\\ &= r^{-2d(\gamma)}f'(\gamma)P_\gamma\cdot r^{2d(\gamma)}P_\gamma(\gamma_\alpha-\gamma)+r^{-d(\gamma)}f'(\gamma)Q_\gamma\cdot r^{d(\gamma)}Q_\gamma(\gamma_\alpha-\gamma)\\ &= r^{-2d(\gamma)}f'(\gamma)P_\gamma\cdot\overline{k_\alpha^{(1)\prime}(\gamma)\,P_\gamma}+r^{-d(\gamma)}f'(\gamma)Q_\gamma\cdot\overline{k_\alpha^{(1)\prime}(\gamma)\,Q_\gamma},\end{aligned}$$

The vectors $\mathbf{Q}_\gamma\mathcal{D}_\gamma k_\alpha^{(1)}$ must be chosen to lie in the orthogonal complement of the range of L_γ. Let $M_\gamma = \operatorname{Range} L_\gamma$ and let $M_\gamma^\perp$ be its orthogonal complement. For $\gamma=\alpha_{k-1}$ we must also have the identity

$$\mathbf{Q}_{\alpha_{k-1}}\mathcal{D}_{\alpha_{k-1}}f\cdot\overline{\mathbf{Q}_{\alpha_{k-1}}\mathcal{D}_{\alpha_{k-1}}k_\alpha^{(1)}}=\mathbf{Q}_{\alpha_{k-1}}\mathcal{D}_{\alpha_{k-1}}f(\alpha_k)$$

or

$$(16.33)\qquad \mathbf{Q}_\gamma\mathcal{D}_\gamma f\cdot\overline{\mathbf{Q}_\gamma\mathcal{D}_\gamma k_\alpha^{(1)}}=\mathbf{Q}_\gamma\mathcal{D}_\gamma f(\gamma_\alpha),\qquad \gamma<\alpha,$$

for all f, which implies that for $\gamma<\alpha$, $\mathbf{Q}_\gamma\mathcal{D}_\gamma k_\alpha^{(1)}$ is the unique vector V_γ in $M_\gamma^\perp$ whose inner product gives rise to the "γ_α-coordinate" linear functional. Recall that the γ_α-coordinate points to the path along which α lies. In fact we have

$$\mathbf{Q}_\gamma\mathcal{D}_\gamma k_\alpha^{(1)}=V_\gamma=\mathbf{Q}_\gamma\mathbf{e}_{\gamma_\alpha}$$

where $\mathbf{e}_{\gamma_\alpha}$ is the coordinate vector in $\mathbb{C}^N$ in the direction of the γ_α coordinate. Moreover we have that

$$\langle w,\mathbf{Q}_\gamma\mathbf{e}_{\gamma_\alpha}\rangle=\langle\mathbf{Q}_\gamma w,\mathbf{e}_{\gamma_\alpha}\rangle=\langle w,\mathbf{e}_{\gamma_\alpha}\rangle$$

if $w\in M_\gamma^\perp$ since $\mathbf{Q}_\gamma$ is a projection onto $M_\gamma^\perp$. For future reference we note that

$$(16.34)\qquad \left|\mathbf{Q}_\gamma\mathcal{D}_\gamma k_\alpha^{(1)}\right|=|\mathbf{Q}_\gamma\mathbf{e}_{\gamma_\alpha}|\le 1,\qquad \gamma\le\alpha,$$

since $\mathbf{Q}_\gamma$ is an orthogonal projection in $\mathbb{C}^N$. Thus we define

$$(16.35)\qquad \mathbf{Q}_\gamma\mathcal{D}k_\alpha^{(1)}(\gamma):=\begin{cases}\mathbf{Q}_\gamma\mathbf{e}_{\gamma_\alpha}, & \text{if}\quad \gamma\in[o,\alpha)\\ 0 & \text{if}\quad \gamma\notin[o,\alpha)\end{cases},$$

so that

$$\sum_{\gamma\in\mathcal{T}_n}\mathbf{Q}_\gamma\mathcal{D}f(\gamma)\cdot\overline{\mathbf{Q}_\gamma\mathcal{D}k_\alpha^{(1)}(\gamma)}=\sum_{k=1}^m\mathbf{Q}_a\mathcal{D}_{\alpha_{k-1}}f(\alpha_k)=\sum_{\gamma<\alpha}\mathbf{Q}_\gamma\mathcal{D}_\gamma f(\gamma_\alpha).$$

Finally, we set

$$k_\alpha^{(1)}(o) = 1 \tag{16.36}$$

so that

$$f(o)\,\overline{k_\alpha^{(1)}(o)} = f(o).$$

Combining these definitions and observations with (16.24) and (16.29) we obtain

$$f(\alpha) = \left\langle\left\langle f, k_\alpha^{(1)}\right\rangle\right\rangle_1.$$

Using the fact that the representation given in (16.25),

$$g(\alpha) = g(o) + \sum_{\gamma<\alpha} g'(\gamma)\cdot(\gamma_\alpha - \gamma) + \sum_{\gamma<\alpha} \mathbf{Q}_\gamma \mathcal{D}_\gamma g(\gamma_\alpha), \tag{16.37}$$

is uniquely determined by the condition that

$$\mathbf{Q}_\gamma \mathcal{D}_\gamma g(\gamma_\alpha) \in M_\gamma^\perp, \qquad \gamma < \alpha,$$

(since $g'(\gamma)\cdot(\gamma_\alpha - \gamma)$ is obviously in M_γ) we thus see that there does indeed exist a unique function $k_\alpha^{(1)}$ satisfying properties (16.30), (16.35) and (16.36). Indeed, if $\beta \in [0, \alpha]$, then with $g = k_\alpha^{(1)}$ in (16.37) we have

(16.38)

$$\begin{aligned}
&k_\alpha^{(1)}(\beta) \\
&= k_\alpha^{(1)}(o) + \sum_{\gamma<\beta} k_\alpha^{(1)\prime}(\gamma)\cdot(\gamma_\beta - \gamma) + \sum_{\gamma<\beta} \mathbf{Q}_\gamma \mathcal{D}_\gamma k_\alpha^{(1)}(\gamma_\beta) \\
&= 1 + \sum_{\gamma<\beta} \left\{ r^{2d(\gamma)}\overline{P_\gamma(\gamma_\alpha - \gamma)} + r^{d(\gamma)}\overline{Q_\gamma(\gamma_\alpha - \gamma)} \right\}\cdot(\gamma_\beta - \gamma) + \sum_{\gamma<\beta} \mathbf{Q}_\gamma \mathbf{e}_{\gamma_\alpha}(\gamma_\beta)
\end{aligned}$$

where $\mathbf{Q}_\gamma \mathbf{e}_{\gamma_\alpha}(\gamma_\beta)$ denotes the γ_β-coordinate of the N-vector $\mathbf{Q}_\gamma \mathbf{e}_{\gamma_\beta}$. If $\beta \neq \alpha$ is at distance exactly one from the geodesic $[o, \alpha)$, then $A\beta \in [o, \alpha)$ and the formula for $k_\alpha^{(1)}(\beta)$ is identical to that above except that the final term in the sum $\sum_{\gamma<\beta} \mathbf{Q}_\gamma \mathbf{e}_{\gamma_\alpha}(\gamma_\beta)$ is now $\mathbf{Q}_{A\beta}\mathbf{e}_{A\beta_\alpha}(\beta)$ instead of $\mathbf{Q}_{A\beta}\mathbf{e}_{A\beta_\alpha}(A\beta_\alpha)$. The function $k_\alpha^{(1)}$ is then determined for all remaining β by the requirement that $k_\alpha^{(1)}$ be constant on all successor sets $S(\gamma)$ with vertex γ at distance exactly one from the geodesic $[o, \alpha)$.

These calculations generalize to yield duality and reproducing kernels for the holomorphic Besov spaces $HB_{p,1}(\mathcal{T}_n)$, $1 < p < \infty$. In fact, Hölder's inequality yields

$$|\langle\langle f, g\rangle\rangle_1| \leq \|f\|_{HB_{p,1}(\mathcal{T}_n)} \|g\|_{HB_{p',1}(\mathcal{T}_n)},$$

and to see that

$$\|f\|_{HB_{p,1}(\mathcal{T}_n)} = \sup_{\|g\|_{HB_{p',1}(\mathcal{T}_n)}=1} |\langle\langle f, g\rangle\rangle_1|, \tag{16.39}$$

we choose G to be the unique function satisfying

$$\begin{aligned}
G(o) &= f(o)\,|f(o)|^{p-2}, \\
G'(\gamma) &= f'(\gamma)\,|f'(\gamma)|^{p-2}, \\
\mathbf{Q}_\gamma \mathcal{D}_\gamma G &= \mathbf{Q}_\gamma \mathcal{D}_\gamma f\,|\mathbf{Q}_\gamma \mathcal{D}_\gamma f|^{p-2}.
\end{aligned}$$

Thus

$$\mathcal{D}G(\alpha) = \{G'(\alpha)(\beta_j - \alpha)\}_{j-1}^{N} + \mathbf{Q}_\alpha \mathcal{D}f(\alpha)\left|\mathbf{Q}_a \mathcal{D}f(\alpha)\right|^{p-2}$$

where $\mathbf{Q}_\alpha \mathcal{D}f(\alpha)\left|\mathbf{Q}_\alpha \mathcal{D}f(\alpha)\right|^{p-2} \in M_\alpha^\perp$ and $\{G'(\alpha)(\beta_j - \alpha)\}_{j-1}^{N} \in M_\alpha$. With this choice we have $\|G\|_{HB_{p',1}(\mathcal{T}_n)} = \|f\|_{HB_{p,1}(\mathcal{T}_n)}^{p-1}$ since $(p-1)p' = p$, as well as

$$\begin{aligned}
\langle\langle f, G\rangle\rangle_1 &= |f(o)|^2 |f(o)|^{p-2} \\
&\quad + \sum_{\alpha\in\mathcal{T}_n : d(\alpha)\geq 1} \left\{|f'(\alpha)P_\alpha|^2 + |f'(\alpha)Q_\alpha|^2\right\}|\Delta f'(\alpha)|^{p-2} \\
&\qquad\qquad + \sum_{\gamma\in\mathcal{T}_n} |\mathbf{Q}_\gamma\mathcal{D}_\gamma f|^2 |\mathbf{Q}_\gamma\mathcal{D}_\gamma f|^{p-2} \\
&= \|f\|_{HB_{p,1}(\mathcal{T}_n)}^p,
\end{aligned}$$

since $|f'(\alpha)P_\alpha|^2 + |f'(\alpha)Q_\alpha|^2 = |f'(\alpha)|^2$. Then taking $g = G/\|G\|_{HB_{p',1}(\mathcal{T}_n)}$ we obtain (16.39). We summarize these results in the following Proposition.

PROPOSITION 16.24. *Let $1 < p < \infty$. Then the dual space of $HB_{p,1}(\mathcal{T}_n)$ can be identified with $HB_{p',1}(\mathcal{T}_n)$ under the pairing $\langle\langle\cdot,\cdot\rangle\rangle_1$ given in* (16.28), *and the reproducing kernel $k_\alpha^{(1)}(\gamma)$ for this pairing is the unique function $k_\alpha^{(1)}$ satisfying* (16.30), (16.35) *and* (16.36), *and given explicitly in* (16.38).

The corresponding formula (16.3) for Δk_α^1, the difference operator applied to the reproducing kernel k_α^1 for the abstract Besov space $B_{p,1}(\mathcal{T}_n)$, consists entirely of nonnegative entries, a feature that plays prominently in deriving the Carleson embedding property from multiplier interpolation using Böe's curious lemma. The terms $k_\alpha^{(1)}(o)$, $k_\alpha^{(1)\prime}(\gamma)$ and $\mathbf{Q}_\gamma\mathcal{D}_\gamma k_\alpha^{(1)}$ arising in the above formula do not consist entirely of nonnegative entries, but the following two properties will serve as a suitable substitute:

$$\begin{aligned}
(16.40)\qquad &\left|r^{-d(\gamma)}k_\alpha^{(1)\prime}(\gamma)P_\gamma + r^{-d(\gamma)/2}k_\alpha^{(1)\prime}(\gamma)Q_\gamma\right| \\
&\qquad\approx r^{-d(\gamma)}\operatorname{Re}\left(\overline{\gamma}\cdot r^{2d(\gamma)}P_\gamma(\gamma_\alpha - \gamma) + \overline{\gamma}\cdot r^{d(\gamma)}Q_\gamma(\gamma_\alpha - \gamma)\right) \\
&\qquad\approx r^{-d(\gamma)}\operatorname{Re}\left(\gamma\cdot k_\alpha^{(1)\prime}(\gamma)\right) \approx 1; \\
&\left|\mathbf{Q}_\gamma\mathcal{D}_\gamma k_\alpha^{(1)}\right| \leq 1.
\end{aligned}$$

Analogs of these properties were used in Section 15.6 to complete the proof of equivalence of multiplier interpolation with the separation and tree conditions when $1 < p < 2 + 1/(n-1)$.

To see the equivalence in the first line of (16.40), we compute that

$$\overline{\gamma}\cdot P_\gamma(\gamma_\alpha - \gamma) = \overline{\gamma}\cdot\frac{\overline{\gamma}\cdot(\gamma_\alpha - \gamma)}{|\gamma|^2}\gamma = \overline{\gamma}\cdot(\gamma_\alpha - \gamma)$$

has real part approximately $r^{-d(\gamma)}$, that

$$|P_\gamma(\gamma_\alpha - \gamma)| \geq cr^{-d(\gamma)},$$

and that

$$|\overline{\gamma}\cdot Q_\gamma(\gamma_\alpha - \gamma)| \leq |Q_\gamma(\gamma_\alpha - \gamma)| \leq |\gamma_\alpha - \gamma| \lesssim r^{-d(\gamma)/2}.$$

Using (16.30), we thus obtain

$$\begin{aligned}
&r^{-d(\gamma)} \operatorname{Re}\left(\gamma \cdot k_{\alpha}^{(1)\prime}(\gamma)\right) \\
&= r^{-d(\gamma)} \operatorname{Re}\left(\overline{\gamma} \cdot \overline{k_{\alpha}^{(1)\prime}(\gamma)}\right) \\
&= r^{-d(\gamma)} \operatorname{Re}\left(\overline{\gamma} \cdot r^{2d(\gamma)} P_{\gamma}(\gamma_{\alpha} - \gamma) + \overline{\gamma} \cdot r^{d(\gamma)} Q_{\gamma}(\gamma_{\alpha} - \gamma)\right) \\
&\approx r^{-d(\gamma)}\left(r^{d(\gamma)} + O\left(r^{d(\gamma)/2}\right)\right) \approx 1.
\end{aligned}$$

Using (16.30) again, we obtain from (16.31) and the above that

$$\begin{aligned}
&\left|r^{-d(\gamma)} k_{\alpha}^{(1)\prime}(\gamma) P_{\gamma} + r^{-d(\gamma)/2} k_{\alpha}^{(1)\prime}(\gamma) Q_{\gamma}\right|^2 \\
&= \left|r^{-d(\gamma)} k_{\alpha}^{(1)\prime}(\gamma) P_{\gamma}\right|^2 + \left|r^{-d(\gamma)/2} k_{\alpha}^{(1)\prime}(\gamma) Q_{\gamma}\right|^2 \\
&= \left|r^{-d(\gamma)} P_{\gamma} \overline{k_{\alpha}^{(1)\prime}(\gamma)}\right|^2 + \left|r^{-d(\gamma)/2} Q_{\gamma} \overline{k_{\alpha}^{(1)\prime}(\gamma)}\right|^2 \\
&= \left|r^{d(\gamma)} P_{\gamma}(\gamma_{\alpha} - \gamma)\right|^2 + \left|r^{d(\gamma)/2} Q_{\gamma}(\gamma_{\alpha} - \gamma)\right|^2 \\
&\gtrsim 1 > 0,
\end{aligned}$$

which completes the proof of the first line in (16.40). The inequality in the second line of (16.40) is (16.34).

16.3.4. An Example. We close this section by computing $k_{\alpha}^{(1)}$ for the simple case of the Bergman tree $\mathcal{T}_1$ in terms of the geometric embedding of $\mathcal{T}_1$ in the unit disk, and then verifying (16.40) in this case. The branching number for the tree $\mathcal{T}_1$ is 2. Fix $\alpha \in \mathcal{T}_1$ with geodesic $[o, \alpha] = \{o, \alpha_1, \alpha_2, \ldots, \alpha_m = \alpha\}$ as above, and let $\gamma = \alpha_{k-1}$ with children $\gamma_1 = \alpha_k$ and γ_2. Then

$$\begin{aligned}
\Delta\gamma_1 &= \gamma_1 - \gamma, \\
\Delta\gamma_2 &= \gamma_2 - \gamma, \\
\mathcal{E}(\gamma) &= \{\Delta\gamma_1, \Delta\gamma_2\},
\end{aligned}$$

and for $v \in \mathbb{C}$,

$$L_{\gamma} v = \{v\Delta\gamma_1, v\Delta\gamma_2\}.$$

Thus

$$\begin{aligned}
M_{\gamma} &= \{v\Delta\gamma_1, v\Delta\gamma_2 : v \in \mathbb{C}\} = \mathbb{C}\{\Delta\gamma_1, \Delta\gamma_2\}, \\
M_{\gamma}^{\perp} &= \mathbb{C}\left\{-\overline{\Delta\gamma_2}, \overline{\Delta\gamma_1}\right\},
\end{aligned}$$

since if $V = v\{\Delta\gamma_1, \Delta\gamma_2\} \in M_{\gamma}$ and $W = w\left\{-\overline{\Delta\gamma_2}, \overline{\Delta\gamma_1}\right\} \in M_{\gamma}^{\perp}$, then

$$\langle V, W\rangle = v\overline{w}\{(\Delta\gamma_1)(-\Delta\gamma_2) + (\Delta\gamma_2)(\Delta\gamma_1)\} = 0.$$

We now claim that

$$\begin{aligned}
\mathbf{Q}_{\gamma}\mathcal{D}_{\gamma}k_{\alpha}^{(1)} &= \frac{\Delta\gamma_2}{|\Delta\gamma_1|^2 + |\Delta\gamma_2|^2}\left\{-\overline{\Delta\gamma_2}, \overline{\Delta\gamma_1}\right\} \\
&= \left\{\frac{|\Delta\gamma_2|^2}{|\Delta\gamma_1|^2 + |\Delta\gamma_2|^2}, \frac{-\overline{\Delta\gamma_1}\Delta\gamma_2}{|\Delta\gamma_1|^2 + |\Delta\gamma_2|^2}\right\}.
\end{aligned}$$

Indeed, we have $\mathbf{Q}_\gamma \mathcal{D}_\gamma k_\alpha^{(1)} \in M_\gamma^\perp$ from the definition, and if

$$W = \{W_1, W_2\} = w\left\{-\overline{\Delta\gamma_2}, \overline{\Delta\gamma_1}\right\} \in M_\gamma^\perp,$$

then the left side of (16.33) is

$$\begin{aligned}&\left\langle W, \mathbf{Q}_\gamma \mathcal{D}_\gamma k_\alpha^{(1)}\right\rangle \\ &= w\left(-\overline{\Delta\gamma_2}\right)\overline{\frac{|\Delta\gamma_2|^2}{|\Delta\gamma_1|^2 + |\Delta\gamma_2|^2}} + w\left(\overline{\Delta\gamma_1}\right)\overline{\frac{-\overline{\Delta\gamma_1}\Delta\gamma_2}{|\Delta\gamma_1|^2 + |\Delta\gamma_2|^2}} \\ &= w\left(-\overline{\Delta\gamma_2}\right) = W_1,\end{aligned}$$

which is the right side of (16.33) since $\gamma_1 = \alpha_k \in [o, \alpha]$.

Since the projection P_γ is the identity in dimension $n = 1$, we also have

$$k_\alpha^{(1)\prime}(\gamma) = r^{2d(\gamma)}\overline{(\gamma_\alpha - \gamma)},$$

and since it is geometrically evident that $\operatorname{Re}\gamma \cdot \overline{\gamma_\alpha - \gamma} \approx |\gamma|\,|\gamma_\alpha - \gamma|$, we now have

$$\operatorname{Re}\left(\gamma \cdot k_\alpha^{(1)\prime}(\gamma)\right) = r^{2d(\gamma)}\operatorname{Re}\gamma \cdot \overline{\gamma_\alpha - \gamma} \approx r^{d(\gamma)} \approx \left|k_\alpha^{(1)\prime}(\gamma)\right|,$$

$$|\mathbf{Q}_\gamma \mathcal{D}_\gamma k_\alpha|^2 \le \frac{|\Delta\gamma_2|^4 + |\Delta\gamma_1|^2|\Delta\gamma_2|^2}{\left(|\Delta\gamma_1|^2|\Delta\gamma_2|^2\right)^2} \le 1,$$

which is (16.40) for the Bergman tree $\mathcal{T}_1$.

Using (16.38), we give an exact formula for $k_\alpha(\beta)$, $\alpha, \beta \in \mathcal{T}_1$. When $\beta \in [o, \alpha]$ we have

$$k_\alpha^{(1)}(\beta) = 1 + \sum_{\gamma<\beta} r^{2d(\gamma)}\overline{(\gamma_\alpha - \gamma)}(\gamma_\beta - \gamma) + \sum_{\gamma<\beta}\frac{\left|\gamma_\beta^\perp - \gamma\right|^2}{|\gamma_\beta - \gamma|^2 + \left|\gamma_\beta^\perp - \gamma\right|^2},$$

where $\gamma_\beta^\perp$ is the child of γ not lying in $[o, \beta]$. The formula for $k_\alpha^{(1)}\left(\beta_\alpha^\perp\right)$, where $\beta \in [o, \alpha)$ and $\beta_\alpha^\perp$ is the child of β not lying in $[o, \alpha]$, is given by

$$\begin{aligned}k_\alpha^{(1)}\left(\beta_\alpha^\perp\right) &= 1 + \sum_{o<\gamma\le\beta} r^{2d(\gamma)}\overline{(\gamma_\alpha - \gamma)}\left(\gamma_{\beta_\alpha^\perp} - \gamma\right) \\ &\quad + \sum_{o<\gamma<\beta}\frac{\left|\gamma_\beta^\perp - \gamma\right|^2}{|\gamma_\beta - \gamma|^2 + \left|\gamma_\beta^\perp - \gamma\right|^2} - \frac{\overline{(\beta_\alpha - \beta)}\left(\beta_\alpha^\perp - \beta\right)}{|\beta_\alpha - \beta|^2 + |\beta_\alpha^\perp - \beta|^2}.\end{aligned}$$

The values of $k_\alpha^{(1)}$ remain constant on the successor sets $S(\alpha)$ and $S\left(\beta_\alpha^\perp\right)$ for $\beta \in [o, \alpha)$, and this completes the evaluation of $k_\alpha^{(1)}(\beta)$ for all $\beta \in \mathcal{T}_1$.

16.3.5. Tensor-valued Functions. In order to introduce higher order derivatives on the tree we will need to use multilinear maps or tensors, and in order to extend our definitions to tensor-valued functions on the Bergman tree, it is convenient to consider first the simplest case of order zero.

DEFINITION 16.25. *Let $1 < p < \infty$. For a $\mathbb{C}$-valued function $f(\alpha)$ defined for $\alpha \in \mathcal{T}_n$, define*

$$\|f\|_{HB_{p,0}(\mathcal{T}_n)} := \|f\|_{\ell^p(\mathcal{T}_n)} = \left(\sum_{\alpha\in\mathcal{T}_n} |f(\alpha)|^p\right)^{1/p}.$$

Define an operator $\mathbf{R}^{-d}$ on $\mathbb{C}^n$-valued functions $\mathbf{v}$ on the tree $\mathcal{T}_n$ by

$$\left(\mathbf{R}^{-d}\mathbf{v}\right)(\alpha) := r^{-d(\alpha)}\mathbf{v}(\alpha) P_\alpha + r^{-d(\alpha)/2}\mathbf{v}(\alpha) Q_\alpha.$$

For vectors $\mathbf{v} \in \mathbb{C}^n$, let

$$|\mathbf{v}|_\alpha := \left|\left(\mathbf{R}^{-d}\mathbf{v}\right)(\alpha)\right| = \left|e^{-2d(\alpha)\theta}\mathbf{v}P_\alpha + e^{-d(\alpha)\theta}\mathbf{v}Q_\alpha\right| \approx \sqrt{(B(\alpha)\mathbf{v})\cdot\overline{\mathbf{v}}},$$

and for a $\mathbb{C}^n$-valued function $\mathbf{v}(\alpha)$ defined for $\alpha \in \mathcal{T}_n$, define

$$\|\mathbf{v}\|_{HB^{(1)}_{p,0}(\mathcal{T}_n)} := \left\||\mathbf{v}|_\alpha\right\|_{\ell^p(\mathcal{T}_n)} = \left(\sum_{\alpha\in\mathcal{T}_n} |\mathbf{v}|^p_\alpha\right)^{1/p}.$$

REMARK 16.26. *Recall from (16.21) that we have $HB_{p,1}(\mathcal{T}_n) = B_{p,1}(\mathcal{T}_n)$. Let $\boldsymbol{Q}\mathcal{D}f(\alpha) = \boldsymbol{Q}_\alpha\mathcal{D}_\alpha f$. Then using Definitions 16.25 and 16.18, we have the following observation that will provide the basis for extending the definition of $HB_{p,0}(\mathcal{T}_n)$ to $HB_{p,m}(\mathcal{T}_n)$ for larger m:*

$$\begin{aligned}\|f\|^p_{HB_{p,1}(\mathcal{T}_n)} &= |f(o)|^p + |f'(o)|^p + \sum_{\alpha\in\mathcal{T}_n} |f'(\alpha)|^p_\alpha + \sum_{\alpha\subset\mathcal{T}_n} |\mathbf{Q}_\alpha\mathcal{D}_\alpha f|^p \\ &= |f(o)|^p + \|f'\|^p_{HB^{(1)}_{p,0}(\mathcal{T}_n)} + \|\mathbf{Q}\mathcal{D}f\|^p_{\ell^p(\mathcal{T}_n)}.\end{aligned}$$

In Definition 16.25 above, we defined the order zero holomorphic Besov space $HB^{(1)}_{p,0}(\mathcal{T}_n)$ on $\mathcal{T}_n$ for $\mathbb{C}^n$-valued functions $\mathbf{v}(\alpha)$ using the anisotropic norm

$$|\mathbf{v}|_\alpha = \left|r^{-d(\alpha)}\mathbf{v}P_\alpha + r^{-d(\alpha)/2}\mathbf{v}Q_\alpha\right|, \tag{16.41}$$

and where $\mathbf{v}(\alpha)$ was interpreted as a covariant tensor of order 1 acting on the "tangent space" $\mathbb{C}^n$ at α. We now wish to extend this definition to order zero holomorphic Besov spaces $HB^{(t)}_{p,0}(\mathcal{T}_n)$ of symmetric covariant tensors of order t, or t-tensors, on $\mathcal{T}_n$. First we review the tensor setup.

Let E_α be the n-dimensional Hilbert space $(\mathbb{C}^n, \langle\cdot,\cdot\rangle_\alpha)$ whose inner product is given by

$$\left\langle v^1, v^2\right\rangle_\alpha = r^{-2d(\alpha)}\left\langle v^1, p\right\rangle\overline{\left\langle v^2, p\right\rangle} + r^{-d(\alpha)}\sum_{i=2}^n \left\langle v^1, q_i\right\rangle\overline{\left\langle v^2, q_i\right\rangle}$$

where $\{p, q_2, \ldots, q_n\}$ is an orthonormal basis of $\mathbb{C}^n$ with $M_\alpha = \operatorname{span}\{p\}$ and $M_\alpha^\perp = \operatorname{span}\{q_2, \ldots, q_n\}$. Thus $p = e^{is}\frac{\alpha}{|\alpha|}$ for some $s \in \mathbb{R}$, and we will take $s = 0$ so that

$$p = \frac{\alpha}{|\alpha|}. \tag{16.42}$$

In terms of the operator $\mathbf{R}^{-d}$ we have

$$\left\langle v^1, v^2\right\rangle_\alpha = \left\langle \mathbf{R}^{-d}v^1, \mathbf{R}^{-d}v^2\right\rangle,$$

where $\langle\cdot,\cdot\rangle$ denotes the usual inner product on $\mathbb{C}^n$. For $t\geq 1$, denote by $\mathcal{E}_\alpha^{(t)}$ the vector space of *symmetric* multilinear maps, or symmetric t-tensors, from the product space $E_\alpha^t = E_\alpha \times \cdots \times E_\alpha$ (t times) to the complex numbers $\mathbb{C}$.

Using the identification of E_α^* with E_α under the *Euclidean* inner product $\langle\cdot,\cdot\rangle$ (not $\langle\cdot,\cdot\rangle_\alpha$), every symmetric t-tensor $\mathbf{A}\in\mathcal{E}_\alpha^{(t)}$ can be written

$$\mathbf{A} = \sum_{1\leq i_1\leq i_2\leq\cdots\leq i_t\leq n} a^{i_1,i_2,\ldots,i_t} e^{i_1}\otimes\cdots\otimes e^{i_t},$$

where e^1 (respectively e^j, $j\geq 2$) is the Euclidean dual vector to $e_1 = p$ (respectively $e_j = q_j$, $j\geq 2$) so that $e^j(e_i) = \langle e_i, e_j\rangle = \delta_j^i$. The vectors e_i depend on α, but we will usually suppress this dependence. We have

$$\mathbf{A}\left[v^1,\ldots,v^t\right] = \sum_{1\leq i_1\leq i_2\leq\cdots\leq i_t\leq n} a^{i_1,i_2,\ldots,i_t}\left\langle v^1, e_{i_1}\right\rangle\times\cdots\times\left\langle v^t, e_{i_t}\right\rangle. \tag{16.43}$$

We now define an inner product $\langle\cdot,\cdot\rangle_\alpha^{(t)}$ on $\mathcal{E}_\alpha^{(t)}$ by

$$\left\langle e^{i_1}\otimes\cdots\otimes e^{i_t}, e^{j_1}\otimes\cdots\otimes e^{j_t}\right\rangle_\alpha^{(t)} := \prod_{k=1}^t\left\langle e^{i_k}, e^{j_k}\right\rangle_\alpha = \prod_{k=1}^t\left\langle \mathbf{R}^{-d(\alpha)}e_{i_k}, \mathbf{R}^{-d(\alpha)}e_{j_k}\right\rangle,$$

and then extend the definition by linearity so that

$$\langle\mathbf{A},\mathbf{B}\rangle_\alpha^{(t)} := \sum_{1\leq i_1\leq i_2\leq\cdots\leq i_t\leq n} a^{i_1,i_2,\ldots,i_t}\overline{b^{i_1,i_2,\ldots,i_t}}\eta_{i_1,i_2,\ldots,i_t}(\alpha) = \sum_i a^i\overline{b^i}\eta_i(\alpha), \tag{16.44}$$

where we write $i=(i_1,\ldots,i_t)$, $j=(j_1,\ldots,j_t)$ and

$$\begin{aligned}\eta_i(\alpha) &= \eta_{i_1,i_2,\ldots,i_t}(\alpha)\\ &= \left\langle e^{i_1}, e^{i_1}\right\rangle_\alpha\times\cdots\times\left\langle e^{i_t}, e^{i_t}\right\rangle_\alpha\\ &= \left\langle \mathbf{R}^{-d(\alpha)}e_{i_1}, \mathbf{R}^{-d(\alpha)}e_{i_1}\right\rangle\times\cdots\times\left\langle \mathbf{R}^{-d(\alpha)}e_{i_t}, \mathbf{R}^{-d(\alpha)}e_{i_t}\right\rangle,\end{aligned} \tag{16.45}$$

and where $\mathbf{A} = \sum_i a^i e^{i_1}\otimes\cdots\otimes e^{i_t}$ and $\mathbf{B} = \sum_i b^i e^{i_1}\otimes\cdots\otimes e^{i_t}$. We extend this inner product to t-tensor-valued functions $\mathbf{A}$ and $\mathbf{B}$ on the Bergman tree $\mathcal{T}_n$ in the obvious way:

$$\langle\langle\mathbf{A},\mathbf{B}\rangle\rangle_0^{(t)} := \sum_{\gamma\in\mathcal{T}_n}\langle\mathbf{A}(\gamma),\mathbf{B}(\gamma)\rangle_\gamma^{(t)}. \tag{16.46}$$

We denote by $|\mathbf{A}|_\alpha$ the norm of $\mathbf{A}$ in the Hilbert space $\mathcal{E}_\alpha^{(t)}$ when there is no confusion regarding the order t of the tensor in question.

EXAMPLE 16.27. *If $F\in\mathrm{Hol}(\mathbb{B}_n)$ and $\alpha\in\mathbb{B}_n$, then $F''(\alpha)\in\mathcal{E}_\alpha^{(2)}$ is a symmetric 2-tensor, or equivalently an $n\times n$ matrix $[F''(\alpha)]$ relative to the basis $\{e_1,\ldots,e_n\}$ satisfying $F''(\alpha)[v\otimes w] = v'[F''(\alpha)]w$, and whose norm squared in $\mathcal{E}_\alpha^{(2)}$ is given* *by*

$$\begin{aligned}
|F''(\alpha)|_\alpha^2 &= \langle F''(\alpha), F''(\alpha)\rangle_\alpha^{(t)} \\
&= \left\langle \sum_{i,j=1}^n \frac{\partial^2 F}{\partial z_i \partial z_j}(\alpha) e^i \otimes e^j, \sum_{k,\ell=1}^n \frac{\partial^? F}{\partial z_k \partial z_\ell}(\alpha) e^k \otimes e^\ell \right\rangle_\alpha^{(t)} \\
&= \sum_{i,j,k,\ell=1}^n \frac{\partial^2 F}{\partial z_i \partial z_j}(\alpha) \overline{\frac{\partial^2 F}{\partial z_k \partial z_\ell}}(\alpha) \left\langle e^i, e^k\right\rangle_\alpha \overline{\left\langle e^j, e^\ell\right\rangle_\alpha} \\
&= \sum_{i,j=1}^n \left|\frac{\partial^2 F}{\partial z_i \partial z_j}(\alpha)\right|^2 \eta_{(i,j)},
\end{aligned}$$

where

$$\eta_{(i,j)} = \begin{cases} r^{-2d(\alpha)} & \textit{if} \quad i=j=1 \\ r^{-\frac{3}{2}d(\alpha)} & \textit{if} \quad i=1, 2\le j\le n \\ r^{-\frac{3}{2}d(\alpha)} & \textit{if} \quad j=1, 2\le i\le n \\ r^{-d(\alpha)} & \textit{if} \quad 2\le i,j\le n. \end{cases}$$

This is of course comparable to the operator norm

$$\sup_{|v|,|w|\le 1} \left|F''(\alpha)\left[\mathbf{R}^{-d}v, \mathbf{R}^{-d}w\right]\right|^2$$

where $\left|F''(\alpha)\left[\mathbf{R}^{-d}v, \mathbf{R}^{-d}w\right]\right|^2$ *is equal to*

$$\begin{aligned}
&\left|r^{-2d(\alpha)}F''(\alpha)[P_\alpha v, P_\alpha w]\right|^2 + \left|r^{-\frac{3}{2}d(\alpha)}F''(\alpha)[P_\alpha v, Q_\alpha w]\right|^2 \\
&+ \left|r^{-\frac{3}{2}d(\alpha)}F''(\alpha)[Q_\alpha v, P_\alpha w]\right|^2 + \left|r^{-d(\alpha)}F''(\alpha)[Q_\alpha v, Q_\alpha w]\right|^2.
\end{aligned}$$

In similar fashion we define the Hilbert space $\mathcal{E}_\alpha^{(s,t)}$ of symmetric (s,t)-tensors that are covariant of order s and contravariant of order t (see for example [**313**, Chapter 4]). Then $\mathcal{E}_\alpha^{(t)} = \mathcal{E}_\alpha^{(t,0)}$ and we will stop referring to tensors as covariant or contravariant. We will see later that reproducing kernels for Besov spaces of t-tensor-valued functions can be interpreted as (t,t)-tensor-valued functions on $\mathcal{T}_n$.

We define the tensor product of a (s_1,t_1)-tensor $\mathbf{B}$ and a (s_2,t_2)-tensor $\mathbf{A}$ to be the (s_1+s_2, t_1+t_2)-tensor $\mathbf{B}\otimes\mathbf{A}$ in the usual way,

$$\begin{aligned}
&\mathbf{B}\otimes\mathbf{A}\left[v^1,\dots,v^{s_1},w^1,\dots,w^{t_1},x^1,\dots,x^{s_2},y^1,\dots,y^{t_2}\right] \\
&\qquad := \mathbf{B}\left[v^1,\dots,v^{s_1},w^1,\dots,w^{t_1}\right]\times\mathbf{A}\left[x^1,\dots,x^{s_2},y^1,\dots,y^{t_2}\right],
\end{aligned}$$

We define the α-*contraction* $\mathbf{B}\wedge_\alpha\mathbf{A}$ of an (s,t)-tensor

$$\mathbf{B} = \sum_{1\le i_1\le\cdots\le i_t\le n} b_{j_1,\dots,j_t}^{i_1,\dots,i_s} e^{i_1}\otimes\cdots\otimes e^{i_s}\otimes e_{j_1}\otimes\cdots\otimes e_{j_t}$$

and a t-tensor

$$\mathbf{A} = \sum_{1\le i_1\le\ldots\le i_t\le n} a^{i_1,i_2,\dots,i_t} e^{i_1}\otimes\cdots\otimes e^{i_t}$$

to be the s-tensor given by

$$\mathbf{B}\wedge_\alpha \mathbf{A} := \sum_{1\le i_1\le\cdots\le i_t\le n} b^{i_1,\ldots,i_s}_{j_1,\ldots,j_t} a^{j_1,\ldots,j_t} \langle e_{j_1}, e_{j_1}\rangle_\alpha \times\cdots\times \langle e_{j_t}, e_{j_t}\rangle_\alpha e^{i_1}\otimes\cdots\otimes e^{i_s}$$
$$= \sum_i b^i_j a^j \eta_j(\alpha) e^{i_1}\otimes\cdots\otimes e^{i_s},$$

where by the summation convention, we also sum over the repeated upper and lower indices $j_1,\ldots,j_t$. The Euclidean contraction $\mathbf{B}\wedge\mathbf{A}$ is given by $\sum_i b^i_j a^j e^{i_1}\otimes\cdots\otimes e^{i_s}$ without the η_j. Thus $\mathbf{B}\wedge\mathbf{A}[v_1,\ldots,v_s]$ is the contraction (trace if $t=1$) of the linear map λ given by

$$\lambda(w_1,\ldots,w_t,u_1,\ldots,u_t) = \mathbf{B}\otimes\mathbf{A}(v_1,\ldots,v_s,w_1,\ldots,w_t,u_1,\ldots,u_t)$$

(see [**313**, pages 4-27]). Note that if we interpret v^j as a $(0,1)$-tensor (contravariant of order 1), then from (16.43) we have

$$\mathbf{A}\left[v^1,\ldots,v^t\right] = \mathbf{A}\wedge\left(v^1\otimes\cdots\otimes v^t\right). \tag{16.47}$$

DEFINITION 16.28. *We define the "inner product" $\langle\mathbf{A},\mathbf{B}\rangle^{(t)}_\alpha$ of a t-tensor $\mathbf{A}$ and a (t,t)-tensor $\mathbf{B}$ to be the t-tensor given by*

$$\langle\mathbf{A},\mathbf{B}\rangle^{(t)}_\alpha := \overline{\mathbf{B}}\wedge_\alpha\mathbf{A}, \tag{16.48}$$

so that

$$\langle\mathbf{A},\mathbf{B}\rangle^{(t)}_\alpha = \sum_i \overline{b^i_j} a^j \eta_j(\alpha) e^{i_1}\otimes\cdots\otimes e^{i_t}. \tag{16.49}$$

We also define an "inner product" $\langle\langle\mathbf{A},\mathbf{B}\rangle\rangle^{(t)}_0$, for a t-tensor-valued function $\mathbf{A}$ and a (t,t)-tensor-valued function $\mathbf{B}$ on the tree $\mathcal{T}_n$, by

$$\langle\langle\mathbf{A},\mathbf{B}\rangle\rangle^{(t)}_0 := \sum_{\gamma\in\mathcal{T}_n}\langle\mathbf{A}(\gamma),\mathbf{B}(\gamma)\rangle^{(t)}_\gamma. \tag{16.50}$$

The use of the same notation $\langle\mathbf{A},\mathbf{B}\rangle^{(t)}_\alpha$ in both (16.44) and (16.48) should not cause confusion as the former is a scalar and the latter a t-tensor, and similarly for the same notation $\langle\langle\mathbf{A},\mathbf{B}\rangle\rangle^{(t)}_0$ in both (16.46) and (16.50).

16.3.5.1. *Order Zero Spaces of t-tensors.* We now define the order zero holomorphic Besov spaces $HB^{(t)}_{p,0}(\mathcal{T}_n)$ of t-tensors on $\mathcal{T}_n$.

DEFINITION 16.29. *For $1<p<\infty$ and $t\in\mathbb{N}$, let $HB^{(t)}_{p,0}(\mathcal{T}_n)$ consist of all t-tensor-valued functions $\mathbf{A}(\alpha)$ defined for $\alpha\in\mathcal{T}_n$ such that the norm*

$$\|\mathbf{A}\|_{HB^{(t)}_{p,0}(\mathcal{T}_n)} := \left\||\mathbf{A}|_\alpha\right\|_{\ell^p(\mathcal{T}_n)} = \left(\sum_{\alpha\in\mathcal{T}_n}|\mathbf{A}(\alpha)|^p_\alpha\right)^{1/p}$$

is finite.

The inner product for the Hilbert space $HB^{(t)}_{2,0}(\mathcal{T}_n)$ is given by (16.46), and the dual space of $HB^{(t)}_{p,0}(\mathcal{T}_n)$ can be identified with $HB^{(t)}_{p',0}(\mathcal{T}_n)$ under the pairing $\langle\langle\cdot,\cdot\rangle\rangle^{(t)}_0$.

LEMMA 16.30. *For $1 < p < \infty$ and $t \in \mathbb{N}$, we have*

$$\left|\langle\langle \mathbf{A}, \mathbf{B}\rangle\rangle_0^{(t)}\right| \leq \|\mathbf{A}\|_{HB_{p,0}^{(t)}(\mathcal{T}_n)} \|\mathbf{B}\|_{HB_{p',0}^{(t)}(\mathcal{T}_n)},$$

$$\|\mathbf{B}\|_{HB_{p',0}^{(t)}(\mathcal{T}_n)} = \sup_{\|\mathbf{A}\|_{HB_{p,0}^{(t)}(\mathcal{T}_n)} \leq 1} \left|\langle\langle \mathbf{A}, \mathbf{B}\rangle\rangle_0^{(t)}\right|.$$

Moreover, in the case $p = 2$, the function $\Lambda_\alpha^{v^1,\dots,v^t}$ defined by

$$\Lambda_\alpha^{v^1,\dots,v^t}\mathbf{A} := \mathbf{A}(\alpha)\left[v^1,\dots,v^t\right]$$

is a continuous linear functional on $HB_{2,0}^{(t)}(\mathcal{T}_n)$ for $\alpha \in \mathcal{T}_n$ and every choice of $v^1,\dots,v^t$. Thus there is a unique $\mathbf{k}_\alpha^{v^1,\dots,v^t}$ in the Hilbert space $HB_{2,0}^{(t)}(\mathcal{T}_n)$ such that

$$\left\langle\left\langle \mathbf{A}, \mathbf{k}_\alpha^{v^1,\dots,v^t}\right\rangle\right\rangle_0^{(t)} = \Lambda_{v^1,\dots,v^t}\mathbf{A} = \mathbf{A}(\alpha)\left[v^1,\dots,v^t\right]. \tag{16.51}$$

By this uniqueness, the function that sends $v^1,\dots,v^t$ to the t-tensor $\mathbf{k}_\alpha^{v^1,\dots,v^t}(\gamma)$ is multi-conjugate linear in $v^1,\dots,v^t$, and so there is a unique (t,t)-tensor $\mathbf{k}_\alpha^{(0,t)}$ such that

$$\mathbf{k}_\alpha^{v^1,\dots,v^t}(\gamma)\left[w^1,\dots,w^t\right] = \mathbf{k}_\alpha^{(0,t)}(\gamma)\left[w^1,\dots,w^t,\overline{v^1},\dots,\overline{v^t}\right],$$

which by (16.47) is

$$\mathbf{k}_\alpha^{v^1,\dots,v^t}(\gamma) = \mathbf{k}_\alpha^{(0,t)}(\gamma) \wedge \overline{v^1 \otimes \cdots \otimes v^t}.$$

We refer to this (t,t)-tensor-valued function $\mathbf{k}_\alpha^{(0,t)}$ as the reproducing kernel for the holomorphic Besov space of t-tensors $HB_{p,0}^{(t)}(\mathcal{T}_n)$ relative to the pairing $\langle\langle\cdot,\cdot\rangle\rangle_0^{(t)}$ since by (16.47), (16.48) and (16.50), we have

$$\begin{aligned}
\left\langle\left\langle \mathbf{A}, \mathbf{k}_\alpha^{(0,t)}\right\rangle\right\rangle_0^{(t)}\left[v^1,\dots,v^t\right] &= \sum_{\gamma\in\mathcal{T}_n}\left\langle \mathbf{A}(\gamma), \mathbf{k}_\alpha^{(0,t)}(\gamma)\right\rangle_\gamma^{(t)}\left[v^1,\dots,v^t\right] \\
&= \sum_{\gamma\in\mathcal{T}_n}\overline{\mathbf{k}_\alpha^{(0,t)}(\gamma)} \wedge_\gamma \mathbf{A}(\gamma)\left[v^1,\dots,v^t\right] \\
&= \sum_{\gamma\in\mathcal{T}_n}\overline{\mathbf{k}_\alpha^{(0,t)}(\gamma)} \wedge_\gamma \mathbf{A}(\gamma) \wedge v^1 \otimes \cdots \otimes v^t \\
&= \sum_{\gamma\in\mathcal{T}_n}\overline{\mathbf{k}_\alpha^{(0,t)}(\gamma)} \wedge v^1 \otimes \cdots \otimes v^t \wedge_\gamma \mathbf{A}(\gamma)
\end{aligned}$$

since $\wedge$ and $\wedge_\gamma$ "commute", as they act on different sets of variables. We then continue with

$$\begin{aligned}
\left\langle\left\langle \mathbf{A}, \mathbf{k}_\alpha^{(0,t)}\right\rangle\right\rangle_0^{(t)}\left[v^1,\dots,v^t\right] &= \sum_{\gamma\in\mathcal{T}_n}\left\langle \mathbf{A}(\gamma), \mathbf{k}_\alpha^{(0,t)}(\gamma) \wedge_\gamma \overline{v^1 \otimes \cdots \otimes v^t}\right\rangle_\gamma^{(t)} \\
&= \sum_{\gamma\in\mathcal{T}_n}\left\langle \mathbf{A}(\gamma), \mathbf{k}_\alpha^{v^1,\dots,v^t}(\gamma)\right\rangle_\gamma^{(t)} \\
&= \left\langle\left\langle \mathbf{A}, \mathbf{k}_\alpha^{v^1,\dots,v^t}\right\rangle\right\rangle_0^{(t)} \\
&= \mathbf{A}(\alpha)\left[v^1,\dots,v^t\right],
\end{aligned}$$

by (16.51). Thus we have shown that

$$\mathbf{A}(\alpha) = \left\langle \left\langle \mathbf{A}, \mathbf{k}_\alpha^{(0,t)} \right\rangle \right\rangle_0^{(t)}, \quad \alpha \in \mathcal{T}_n. \tag{16.52}$$

The reproducing kernel $\mathbf{k}_\alpha^{(0,t)}$ is in fact given by the (t,t)-tensor

$$\sum_i \eta_i(\alpha)^{-1} e^{i_1} \otimes \cdots \otimes e^{i_t} \otimes e_{i_1} \otimes \cdots \otimes e_{i_t}$$

with $\eta_i(\alpha)$ as in (16.45) times the delta function at α, i.e.

(16.53)

$$\begin{aligned}\mathbf{k}_\alpha^{(0,t)}(\gamma) &= \chi_{\{\alpha\}}(\gamma) \sum_i \eta_i(\alpha)^{-1} \left\{e^{i_1} \otimes \cdots \otimes e^{i_t}\right\} \otimes \left\{e_{i_1} \otimes \cdots \otimes e_{i_t}\right\} \\ &= \chi_{\{\alpha\}}(\gamma) \sum_i \left\{\mathbf{R}^{d(\alpha)} e^{i_1} \otimes \cdots \otimes \mathbf{R}^{d(\alpha)} e^{i_t}\right\} \otimes \left\{\mathbf{R}^{d(\alpha)} e_{i_1} \otimes \cdots \otimes \mathbf{R}^{d(\alpha)} e_{i_t}\right\}.\end{aligned}$$

Indeed, by (16.48) and (16.49),

$$\begin{aligned}&\sum_{\gamma \in \mathcal{T}_n} \left\langle \mathbf{A}(\gamma), \mathbf{k}_\alpha^{(0,t)}(\gamma) \right\rangle_\gamma^{(t)} \\ &= \mathbf{k}_\alpha^{(0,t)}(\alpha) \wedge_\alpha \mathbf{A}(\alpha) \\ &= \left\{\sum_i \eta_i(\alpha)^{-1} \left\{e^{i_1} \otimes \cdots \otimes e^{i_t}\right\} \otimes \left\{e_{i_1} \otimes \cdots \otimes e_{i_t}\right\}\right\} \wedge_\alpha \left\{\sum_i a^i e^{i_1} \otimes \cdots \otimes e^{i_t}\right\}, \\ &= \sum_i \eta_i(\alpha)^{-1} a^i \eta_i(\alpha) e^{i_1} \otimes \cdots \otimes e^{i_t} = \mathbf{A}(\alpha).\end{aligned}$$

16.3.5.2. *Order One Spaces of t-tensors.* We next turn to defining the order one holomorphic Besov spaces $HB_{p,1}^{(t)}(\mathcal{T}_n)$ of t-tensors on $\mathcal{T}_n$. In Remark 16.26, we have already defined the scalar case $HB_{p,1}(\mathcal{T}_n)$ using the norm (to the pth power)

$$\|f\|_{HB_{p,1}(\mathcal{T}_n)}^p = |f(o)|^p + \|f'\|_{HB_{p,0}^{(1)}(\mathcal{T}_n)}^p + \|\mathbf{Q}\mathcal{D}f\|_{\ell^p(\mathcal{T}_n)}^p.$$

In order to replace f with a tensor, we first need to define the complex derivative $\mathbf{A}'(\alpha)$ of a t-tensor-valued function $\mathbf{A}(\alpha)$ on the tree $\mathcal{T}_n$. The derivative $\mathbf{A}'$ will be a $(t+1)$-tensor-valued function on the tree defined in the same spirit as f'.

Define the forward difference $\mathcal{D}_\alpha \mathbf{A}$ of a t-tensor-valued function $\mathbf{A}(\alpha)$ in the obvious way. Define the linear map $L_\alpha^{(t)}$ from the space of $(t+1)$-tensors $\mathcal{E}_\alpha^{(t+1)}$ to the space $\left(\mathcal{E}_\alpha^{(t)}\right)^N$ by sending $\mathbf{v} \in \mathcal{E}_\alpha^{(t+1)}$ to

$$L_\alpha^{(t)} \mathbf{v} := \left(\mathbf{v} \wedge \left(\alpha^j - \alpha\right)\right)_{j=1}^N \in \left(\mathcal{E}_\alpha^{(t)}\right)^N,$$

where $\mathbf{v} \wedge \left(\alpha^j - \alpha\right)$ denotes the t-tensor obtained by contracting the $(t+1)$-tensor $\mathbf{v}$ with the vector $\left(\alpha^j - \alpha\right)$ viewed as a $(0,1)$-tensor or contravariant 1 tensor, i.e.

$$\mathbf{v} \wedge \left(\alpha^j - \alpha\right)\left[v^1, \ldots, v^t\right] = \mathbf{v}\left[\alpha^j - \alpha, v^1, \ldots, v^t\right],$$

since if $\mathbf{v} = \sum_{i,k} v_k^i e_k \otimes \left\{e^{i_1} \otimes \cdots \otimes e^{i_t}\right\}$, then both sides of the above equation equal

$$\sum_{i,k} v_k^i \left\langle \alpha^j - \alpha, e_k \right\rangle \left\langle v^1, e^{i_1} \right\rangle \times \cdots \times \left\langle v^t, e^{i_t} \right\rangle.$$

In Section 16.3.12 below, we will show that with θ sufficiently large in the construction of the Bergman tree in Section 14.1.6, and then modifying the centers, we can make the map $L_\alpha^{(t)}$ one-to-one for all $0 \leq t \leq M-1$ for any finite M - we only need to take $M = 2n$ for our purposes in this proof. Let $\mathbf{P}_\alpha^{(t)}$ be the orthogonal projection of $\left(\mathcal{E}_\alpha^{(t)}\right)^N$ onto the range $M_\alpha^{(t)}$ of $L_\alpha^{(t)}$ with respect to the inner product on the product Hilbert space $\mathcal{E}_\alpha^{(t)} \times \cdots \times \mathcal{E}_\alpha^{(t)}$ (N times). Let $\mathbf{Q}_a^{(t)} = I - \mathbf{P}_\alpha^{(t)}$ be orthogonal projection onto $\left(M_\alpha^{(t)}\right)^\perp$. The complex derivative $\mathbf{A}'(\alpha)$ of $\mathbf{A}$ at the point α is then the unique $(t+1)$-tensor $\mathbf{v}$ such that

$$L_\alpha \mathbf{v} = \mathbf{P}_\alpha^{(t)}\left(\mathcal{D}_\alpha \mathbf{A}\right).$$

Thus we have

$$L_\alpha^{(t)} \mathbf{A}'(\alpha) = \mathbf{P}_\alpha^{(t)}\left(\mathcal{D}_\alpha \mathbf{A}\right) = \left(\mathbf{A}'(\alpha) \wedge \left(\alpha^j - \alpha\right)\right)_{j=1}^N,$$

and the first order Taylor formula for t-tensor-valued functions

$$\mathcal{D}_\alpha \mathbf{A} = \left(\mathbf{A}'(\alpha) \wedge \left(\alpha^j - \alpha\right)\right)_{j=1}^N + \mathbf{Q}_\alpha^{(t)}\left(\mathcal{D}_\alpha \mathbf{A}\right).$$

We can now define the order one holomorphic Besov space $HB_{p,1}^{(t)}(\mathcal{T}_n)$ of t-tensors on $\mathcal{T}_n$. First we define

$$\left\|\mathbf{Q}^{(t)} \mathcal{D}\mathbf{A}\right\|_{\ell^p(\mathcal{T}_n)}^p := \sum_{\alpha \in \mathcal{T}_n} \left|\mathbf{Q}_\alpha^{(t)}\left(\mathcal{D}_\alpha \mathbf{A}\right)\right|_\alpha^p$$

where $\left|\left(\mathbf{v}^1, \ldots, \mathbf{v}^N\right)\right|_\alpha^2 = \sum_{j=1}^N \left|\mathbf{v}^j\right|_\alpha^2$ for $\left(\mathbf{v}^1, \ldots, \mathbf{v}^N\right) \in \left(\mathcal{E}_\alpha^{(t)}\right)^N$. Note that at the root o, $|\cdot|_o$ is always a Euclidean norm.

Definition 16.31. *For $1 < p < \infty$ and $0 \leq t \leq M-1$, let $HB_{p,1}^{(t)}(\mathcal{T}_n)$ consist of all t-tensor-valued functions $\mathbf{A}$ on the tree $\mathcal{T}_n$ such that the norm (to the pth power)*

$$\|\mathbf{A}\|_{HB_{p,1}^{(t)}(\mathcal{T}_n)}^p := |\mathbf{A}(o)|^p + \|\mathbf{A}'\|_{HB_{p,0}^{(t+1)}(\mathcal{T}_n)}^p + \left\|\mathbf{Q}^{(t)} \mathcal{D}\mathbf{A}\right\|_{\ell^p(\mathcal{T}_n)}^p$$

is finite.

As in Proposition 16.24, we can obtain the duality of $HB_{p,1}^{(t)}(\mathcal{T}_n)$ and $HB_{p',1}^{(t)}(\mathcal{T}_n)$ relative to the inner product $\langle\langle \cdot, \cdot \rangle\rangle_1^{(t)}$ for the Hilbert space $HB_{2,1}^{(t)}(\mathcal{T}_n)$:

$$\begin{aligned} \langle\langle \mathbf{A}, \mathbf{B} \rangle\rangle_1^{(t)} &= \mathbf{A}(o)\overline{\mathbf{B}(o)} + \sum_{\alpha \in \mathcal{T}_n} \langle \mathbf{A}', \mathbf{B}' \rangle_\alpha^{(t+1)} + \sum_{\alpha \in \mathcal{T}_n} \mathbf{Q}_\alpha \mathcal{D}_\alpha \mathbf{A} \cdot \overline{\mathbf{Q}_\alpha \mathcal{D}_\alpha \mathbf{B}} \\ &= \mathbf{A}(o)\overline{\mathbf{B}(o)} + \langle\langle \mathbf{A}', \mathbf{B}' \rangle\rangle_0^{(t+1)} + \sum_{\alpha \in \mathcal{T}_n} \mathbf{Q}_\alpha \mathcal{D}_\alpha \mathbf{A} \cdot \overline{\mathbf{Q}_\alpha \mathcal{D}_\alpha \mathbf{B}}. \end{aligned} \tag{16.54}$$

Lemma 16.32. *For $1 < p < \infty$ and $t \in \mathbb{N}$, we have*

$$\left|\langle\langle \mathbf{A}, \mathbf{B} \rangle\rangle_1^{(t)}\right| \leq \|\mathbf{A}\|_{HB_{p,1}^{(t)}(\mathcal{T}_n)} \|\mathbf{B}\|_{HB_{p',1}^{(t)}(\mathcal{T}_n)},$$

$$\|\mathbf{B}\|_{HB_{p',1}^{(t)}(\mathcal{T}_n)} = \sup_{\|\mathbf{A}\|_{HB_{p,1}^{(t)}(\mathcal{T}_n)} \leq 1} \left|\langle\langle \mathbf{A}, \mathbf{B} \rangle\rangle_1^{(t)}\right|.$$

Combining the arguments for the order one space $HB_{p,1}(\mathcal{T}_n)$ in Section 16.2 with the arguments above for the order zero space $HB_{p,0}^{(t)}(\mathcal{T}_n)$ of t-tensors, we can show there is a unique reproducing kernel $\mathbf{k}_\alpha^{(1,t)}$ for the holomorphic Besov space of t-tensors $HB_{p,1}^{(t)}(\mathcal{T}_n)$ relative to this pairing. Again, $\mathbf{k}_\alpha^{(1,t)}$ is a (t,t)-tensor-valued function on the tree satisfying

$$\mathbf{A}(\alpha) = \left\langle\left\langle \mathbf{A}, \mathbf{k}_\alpha^{(1,t)} \right\rangle\right\rangle_1^{(t)}, \qquad \alpha \in \mathcal{T}_n,$$

for $\mathbf{A} \in HB_{p,1}^{(t)}(\mathcal{T}_n)$, where just as in Definition 16.28, the notation $\langle\langle \mathbf{A}, \mathbf{B} \rangle\rangle_1^{(t)}$ represents a scalar if $\mathbf{B}$ is a t-tensor, and a t-tensor if $\mathbf{B}$ is a (t,t)-tensor.

16.3.6. Higher Order Holomorphic Besov Spaces. We cannot simply define the mth order holomorphic Besov space $HB_{p,m}^{(t)}(\mathcal{T}_n)$ of t-tensor-valued functions inductively to consist of all f such that $f' \in HB_{p,m-1}^{(t+1)}(\mathcal{T}_n)$. Besides the question of how to handle the error term $\left\{\mathbf{Q}_\alpha^{(t)}(\mathcal{D}_\alpha \mathbf{A})\right\}_{\alpha \in \mathcal{T}_n}$, the restriction theorem from B_p to $HB_{p,2}(\mathcal{T}_n)$ will fail because $\|f'\|_{HB_{p,1}^{(1)}(\mathcal{T}_n)}$ will not in general be controlled by $\|F\|_{B_p}$ when f is the restriction of $F \in B_p$ to the Bergman tree $\mathcal{T}_n$. The problem arises when we minimize the distance from $\mathcal{D}_\alpha f$ to M_α by letting $f'(\alpha) = L_\alpha^{-1} P_\alpha \mathcal{D}_\alpha f$. The resulting vector $f'(\alpha)$ is within order 2, but not within order 3, of the restriction $F'(\alpha)$ of F' to the tree. In order to circumvent this difficulty, we will simultaneously define the first, second and through to the mth order derivatives $f'(\alpha), f''(\alpha), \dots, f^{(m)}(\alpha)$ at α using a single orthogonal projection onto the range of an appropriate generalization of the operator L_α. We will also need to define holomorphic Besov spaces of t-tensor-valued functions as well, in order to implement an inductive definition. Recall that we defined a variant $L_\alpha^{(t)}$ of $L_\alpha = L_\alpha^{(0)}$ in order to define the complex derivative of a t-tensor-valued function on the Bergman tree. We will now use the notation $L_\alpha^{(1,t)}$ to denote these operators, and use the notation $L_\alpha^{(m,t)}$ to define a corresponding linear operator that will allow us to simultaneously define first through m^{th} order complex derivatives of a t-tensor-valued function on the Bergman tree. Here is the setup for the scalar case, $t = 0$.

We define the operator $L_\alpha^{(m,0)}$ as the linear map taking the point $\mathbf{x} = (\mathbf{v}^1, \dots, \mathbf{v}^m)$ in the product space $\mathcal{E}_\alpha^{(1)} \times \mathcal{E}_\alpha^{(2)} \times \cdots \times \mathcal{E}_\alpha^{(m)}$ to the point $L_\alpha^{(m,0)}\mathbf{x}$ given by

$$\begin{aligned} L_\alpha^{(m,0)}\mathbf{x} &:= \left\{ \sum_{\ell=1}^m \frac{1}{\ell!} \mathbf{v}^\ell \wedge \left\{ \otimes^\ell (\alpha^j - \alpha) \right\} \right\}_{j=1}^N \\ &= \left\{ \sum_{\ell=1}^m \frac{1}{\ell!} \mathbf{v}^\ell \left\lfloor \alpha^j - \alpha, \dots, \alpha^j - \alpha \right\rceil \right\}_{j=1}^N \in \mathbb{C}^N, \end{aligned}$$

where the second equality follows from (16.47). Here $\otimes^\ell (\alpha^j - \alpha)$ denotes the $(0,\ell)$-tensor

$$(\alpha^j - \alpha) \otimes \cdots \otimes (\alpha^j - \alpha),$$

where $(\alpha^j - \alpha)$ is repeated ℓ times. Let $M_\alpha^{(m,0)} = \operatorname{Range} L_\alpha^{(m,0)}$ and denote by $\left(M_\alpha^{(m,0)}\right)^\perp$ the orthogonal complement of $M_\alpha^{(m,0)}$ in $\mathbb{C}^N$. Let $\mathbf{P}_\alpha^{(m,0)}$ denote orthogonal projection of $\mathbb{C}^N$ onto $M_\alpha^{(m,0)}$, let $\mathbf{Q}_\alpha^{(m,0)} = I - \mathbf{P}_\alpha^{(m,0)}$ be orthogonal projection of $\mathbb{C}^N$ onto $\left(M_\alpha^{(m,0)}\right)^\perp$, and define as usual

$$\mathcal{D}_\alpha f = \left(f(\alpha^j) - f(\alpha)\right)_{j=1}^N \in \mathbb{C}^N.$$

At this point we need to know that $L_\alpha^{(m,0)}$ is one-to-one. In fact, in order to prove the boundedness of the restriction map later, we will need the following inequality

$$c_m \sum_{\ell=1}^m \left|\mathbf{v}^\ell\right|_\alpha \le \left|L_\alpha^{(m,0)}\left(\mathbf{v}^1, \ldots, \mathbf{v}^m\right)\right| \le C_m \sum_{\ell=1}^m \left|\mathbf{v}^\ell\right|_\alpha, \tag{16.55}$$

for all $1 \le m \le M$, uniformly for $\alpha \in \mathcal{T}_n$ (we actually only need $M = 2n$ for our purposes). This is established in Section 16.3.12 below using a careful reconstruction of the Bergman tree $\mathcal{T}_n$ with parameter θ sufficiently large depending on M.

Assuming (16.55) for the moment, we see that $L_\alpha^{(m,0)}$ is one-to-one. If we now define the derivatives $\left(D_m f(\alpha), D_m^2 f(\alpha), \ldots, D_m^m f(\alpha)\right)$ by

$$L_\alpha^{(m,0)}\left(D_m f(\alpha), D_m^2 f(\alpha), \ldots, D_m^m f(\alpha)\right) := \mathbf{P}_\alpha^{(m,0)} \mathcal{D}_\alpha f,$$

then we have that

$$\left|\mathbf{Q}_\alpha^{(m,0)} \mathcal{D}_\alpha f\right|^2 = \left|\mathcal{D}_\alpha f - \mathbf{P}_\alpha^{(m,0)} \mathcal{D}_\alpha f\right|^2$$

is given by

$$\sum_{j=1}^N \left| f(\alpha^j) - \left\{ f(\alpha) + \sum_{\ell=1}^m \frac{1}{\ell!} D_m^\ell f(\alpha) \wedge \left\{\otimes^\ell (\alpha^j - \alpha)\right\} \right\} \right|^2,$$

and is the minimum value of $\left|\mathcal{D}_\alpha f - L_\alpha^{(m,0)} \mathbf{x}\right|^2$ over $\mathbf{x} = \left(\mathbf{v}^1, \ldots, \mathbf{v}^m\right)$ in the product space $\mathcal{E}_\alpha^{(1)} \times \mathcal{E}_\alpha^{(2)} \times \cdots \times \mathcal{E}_\alpha^{(m)}$, where $\left|\mathcal{D}_\alpha f - L_\alpha^{(m,0)} \mathbf{x}\right|^2$ is given by

$$\sum_{j=1}^N \left| f(\alpha^j) - \left\{ f(\alpha) + \sum_{\ell=1}^m \frac{1}{\ell!} \mathbf{v}^\ell(\alpha) \wedge \left\{\otimes^\ell (\alpha^j - \alpha)\right\} \right\} \right|^2.$$

Thus if we write

$$\mathbf{Q}_\alpha^{(m,0)} \mathcal{D}_\alpha f = \left\{ \mathbf{Q}_\alpha^{(m,0)} \mathcal{D}_\alpha f(\alpha^j) \right\}_{j=1}^N,$$

we have the following m^{th} order Taylor expansion of f at $A\alpha$

$$f(\alpha) = f(A\alpha) + \sum_{\ell=1}^m \frac{1}{\ell!} D_m^\ell f(A\alpha) \wedge \left\{\otimes^\ell (\alpha - A\alpha)\right\} + \mathbf{Q}_{A\alpha}^{(m,0)} \mathcal{D}_{A\alpha} f(\alpha). \tag{16.56}$$

The remainder term $\mathbf{Q}_\alpha^{(m,0)} \mathcal{D}_\alpha f$ satisfies the minimizing property

$$\left|\mathbf{Q}_\alpha^{(m,0)} \mathcal{D}_\alpha f\right| = \min_{\mathbf{x} \in \mathcal{E}^{(1)} \times \cdots \times \mathcal{E}^{(m)}} \left|\mathcal{D}_\alpha f - L_\alpha^{(m,0)} \mathbf{x}\right|. \tag{16.57}$$

Note that the complex derivatives $D_m^\ell f(\alpha)$ defined here depend on both the order m and the degree of differentiation ℓ, and are generally different for different m. Also notice that the operator D_m^ℓ is <u>not</u> the lth power of an operator D_m. (However it is closely related to the lth power of a differentiation operator, see Section 16.3.9.)

We now extend the definition of mth order derivatives to t-tensor-valued functions A on the tree. Define the operator $L_\alpha^{(m,t)}$ as the linear map taking the point $\mathbf{x} = (\mathbf{v}^1, \ldots, \mathbf{v}^m)$ in the product space $\mathcal{E}_\alpha^{(t+1)} \times \mathcal{E}_\alpha^{(t+2)} \times \cdots \times \mathcal{E}_\alpha^{(t+m)}$ to the point $L_\alpha^{(m,t)}\mathbf{x}$ given by

$$L_\alpha^{(m,t)}\mathbf{x} := \left\{\sum_{\ell=1}^{m} \frac{1}{\ell!}\mathbf{v}^\ell \wedge \left\{\otimes^\ell\left(\alpha^j - \alpha\right)\right\}\right\}_{j=1}^{N} \in \left(\mathcal{E}_\alpha^{(t)}\right)^N$$

where $\mathbf{v}\wedge\left\{\otimes^\ell\left(\alpha^j - \alpha\right)\right\}$ denotes the t-tensor obtained by contracting the $(t+\ell)$-tensor $\mathbf{v}$ with the $(0,\ell)$-tensor $\otimes^\ell\left(\alpha^j - \alpha\right)$, i.e.

$$\mathbf{v}\wedge\left(\otimes^\ell\left(\alpha^j - \alpha\right)\right)\left[v^1, \ldots, v^t\right] = \mathbf{v}\left[\alpha^j - \alpha, \ldots, \alpha^j - \alpha, v^1, \ldots, v^t\right].$$

Let $M_\alpha^{(m,t)} = \operatorname{Range} L_\alpha^{(m,t)}$ and denote by $\left(M_\alpha^{(m,t)}\right)^\perp$ the orthogonal complement of $M_\alpha^{(m,t)}$ in $\left(\mathcal{E}_\alpha^{(t)}\right)^N$. Let $\mathbf{P}_\alpha^{(m,t)}$ denote orthogonal projection of $\left(\mathcal{E}_\alpha^{(t)}\right)^N$ onto $M_\alpha^{(m,t)}$ with respect to the inner product on the product Hilbert space $\mathcal{E}_\alpha^{(t)} \times \cdots \times \mathcal{E}_\alpha^{(t)}$ (N times), and let $\mathbf{Q}_\alpha^{(m,t)} = I - \mathbf{P}_\alpha^{(m,t)}$ be orthogonal projection of $\left(\mathcal{E}_\alpha^{(t)}\right)^N$ onto $\left(M_\alpha^{(m,t)}\right)^\perp$.

In Section 16.3.12 below, we also demonstrate the extension to t-tensors of inequality (16.55) above, namely that the map $L_\alpha^{(m,t)}$ satisfies the structural inequalities

$$c_m \sum_{\ell=1}^{m} \left|\mathbf{v}^\ell\right|_\alpha \le \left|L_\alpha^{(m,t)}\left(\mathbf{v}^1, \ldots, \mathbf{v}^m\right)\right| \le C_m \sum_{\ell=1}^{m} \left|\mathbf{v}^\ell\right|_\alpha \tag{16.58}$$

for all $0 \le m + t \le M$, uniformly for $\alpha \in \mathcal{T}_n$. This shows in particular that $L_\alpha^{(m,t)}$ is one-to-one. Suppose now that $\mathbf{A}$ is a t-tensor-valued function on the tree $\mathcal{T}_n$. We define the m-tuple $\left(D_m\mathbf{A}(\alpha), D_m^2\mathbf{A}(\alpha), \ldots, D_m^m\mathbf{A}(\alpha)\right)$ of derivatives up to order m of $\mathbf{A}$ so that

$$L_\alpha^{(m,t)}\left(D_m\mathbf{A}(\alpha), D_m^2\mathbf{A}(\alpha), \ldots, D_m^m\mathbf{A}(\alpha)\right) := \mathbf{P}_\alpha^{(m,t)}\mathcal{D}_\alpha\mathbf{A}.$$

As in the case $t = 0$, the remainder term $\mathbf{Q}_\alpha^{(m,t)}\mathcal{D}_\alpha\mathbf{A}$ satisfies the minimizing property

$$\left|\mathbf{Q}_\alpha^{(m,t)}\mathcal{D}_\alpha\mathbf{A}\right|_\alpha = \min_{\mathbf{x}\in\mathcal{E}_\alpha^{(1)}\times\cdots\times\mathcal{E}_\alpha^{(m)}} \left|\mathcal{D}_\alpha\mathbf{A} - L_\alpha^{(m,t)}\mathbf{x}\right|_\alpha. \tag{16.59}$$

We now define by induction on m the order m holomorphic Besov space $HB_{p,m}^{(t)}(\mathcal{T}_n)$ of t-tensors on $\mathcal{T}_n$. As usual, we use

$$\left\|\mathbf{Q}^{(m,t)}\mathcal{D}\mathbf{A}\right\|_{\ell^p(\mathcal{T}_n)}^p = \sum_{\alpha\in\mathcal{T}_n} \left|\mathbf{Q}_\alpha^{(m,t)}\mathcal{D}_\alpha\mathbf{A}\right|_\alpha^p,$$

where as above, we define

$$|w|_\alpha^2 = \sum_{j=1}^{N} |w_j|_\alpha^2, \qquad w = (w_1, \ldots, w_N) \in \left(\mathcal{E}_\alpha^{(t)}\right)^N.$$

DEFINITION 16.33. *For $1 < p < \infty$ and $0 \le m + t \le M$, let $HB_{p,m}^{(t)}(\mathcal{T}_n)$ consist of all t-tensor-valued functions $\mathbf{A}$ on the tree $\mathcal{T}_n$ such that the norm (to the pth power)*

$$\|\mathbf{A}\|_{HB_{p,m}^{(t)}(\mathcal{T}_n)}^p := |\mathbf{A}(o)|^p + \sum_{\ell=1}^{m} \left\|D_m^\ell \mathbf{A}\right\|_{HB_{p,m-\ell}^{(t+\ell)}(\mathcal{T}_n)}^p + \left\|\mathbf{Q}^{(m,t)} \mathcal{D}\mathbf{A}\right\|_{\ell^p(\mathcal{T}_n)}^p$$

is finite. We write simply $HB_{p,m}(\mathcal{T}_n)$ for the scalar case $t = 0$.

16.3.7. Reproducing Kernels and the Positivity Property. In this section we establish the key positivity property of reproducing kernels that will permit us to use the technique of Böe's curious lemma. It is this property that yields the fruit of our labour in developing the theory of holomorphic Besov spaces on trees. Let $p = 2$. Then the inner product corresponding to the Hilbert space norm $\|\mathbf{A}\|_{HB_{2,m}^{(t)}(\mathcal{T}_n)}$ defined on t-tensor-valued functions $\mathbf{A}$ on the tree $\mathcal{T}_n$ is defined by induction on m by

$$\begin{aligned}\langle\langle \mathbf{A}, \mathbf{B}\rangle\rangle_m^{(t)} &:= \mathbf{A}(o) \cdot \overline{\mathbf{B}(o)} + \sum_{\ell=1}^{m} \langle\langle D_m^\ell \mathbf{A}, D_m^\ell \mathbf{B}\rangle\rangle_{m-\ell}^{(t+\ell)} \\ &\quad + \sum_{\alpha \in \mathcal{T}_n} \left\langle \mathbf{Q}_\alpha^{(m,t)} (\mathcal{D}_\alpha \mathbf{A}), \mathbf{Q}_\alpha^{(m,t)} (\mathcal{D}_\alpha \mathbf{B})\right\rangle_\alpha,\end{aligned}$$

where the cases $m = 0$ and $m = 1$ are defined in (16.46) and (16.54) respectively. We have the following duality relation.

PROPOSITION 16.34. *For $1 < p < \infty$ and $t \in \mathbb{N}$, we have*

$$\left|\langle\langle \mathbf{A}, \mathbf{B}\rangle\rangle_m^{(t)}\right| \le \|\mathbf{A}\|_{HB_{p,m}^{(t)}(\mathcal{T}_n)} \|\mathbf{B}\|_{HB_{p',m}^{(t)}(\mathcal{T}_n)},$$

$$\|\mathbf{B}\|_{HB_{p',m}^{(t)}(\mathcal{T}_n)} = \sup_{\|\mathbf{A}\|_{HB_{p,m}^{(t)}(\mathcal{T}_n)} \le 1} \left|\langle\langle \mathbf{A}, \mathbf{B}\rangle\rangle_m^{(t)}\right|.$$

Denote by $\mathbf{k}_\alpha^{(m,t)}$ the reproducing kernel for $\alpha \in \mathcal{T}_n$ relative to the inner product $\langle\langle \cdot, \cdot\rangle\rangle_m^{(t)}$, which exists by a modification of the argument in the case $m = 0$ immediately following Lemma 16.30 in Section 16.3.5. Then with the usual notation convention regarding inner products as in Definition 16.28, we have

$$\begin{aligned}\mathbf{A}(\alpha) &= \left\langle\left\langle \mathbf{A}, \mathbf{k}_\alpha^{(m,t)}\right\rangle\right\rangle_m^{(t)} \\ &= \mathbf{A}(o) \cdot \overline{\mathbf{k}_\alpha^{(m,t)}(o)} + \sum_{\ell=1}^{m} \left\langle\left\langle D_m^\ell \mathbf{A}, D_m^\ell \mathbf{k}_\alpha^{(m,t)}\right\rangle\right\rangle_{m-\ell}^{(t+\ell)} \\ &\quad + \sum_{\alpha \in \mathcal{T}_n} \left\langle \mathbf{Q}_\alpha^{(m,t)} (\mathcal{D}_\alpha \mathbf{A}), \mathbf{Q}_\alpha^{(m,t)} \left(\mathcal{D}_\alpha \mathbf{k}_\alpha^{(m,t)}\right)\right\rangle_\alpha.\end{aligned} \tag{16.60}$$

We can also recover $\mathbf{A}(\alpha)$ from $\mathbf{A}(o)$ together with the data

$$\mathbf{A}'(\gamma), \mathbf{A}''(\gamma), \ldots, \mathbf{A}^{(m)}(\gamma)$$

and the remainder term $\mathbf{Q}_\gamma^{(m,t)}\mathcal{D}_\gamma\mathbf{A}$ for $\gamma\in[o,\alpha]$. In fact with $[o,\alpha]=\{o,\alpha_1,\ldots,\alpha_M=\alpha\}$ we have

$$\begin{aligned}
\mathbf{A}(\alpha)&=\mathbf{A}(o)+\sum_{j=1}^{M}\left[\mathbf{A}(\alpha_j)-\mathbf{A}(\alpha_{j-1})\right] \\
&=\mathbf{A}(o)+\sum_{j=1}^{M}D_m\mathbf{A}(\alpha_{j-1})\wedge(\alpha_j-\alpha_{j-1}) \\
&\quad+\sum_{j=1}^{M}\frac{1}{2}D_m^2\mathbf{A}(\alpha_{j-1})\wedge\{(\alpha_j-\alpha_{j-1})\otimes(\alpha_j-\alpha_{j-1})\} \\
&\quad\vdots \\
&\quad+\sum_{j=1}^{M}\frac{1}{m!}D_m^m\mathbf{A}(\alpha_{j-1})\wedge\{\otimes^m(\alpha_j-\alpha_{j-1})\} \\
&\quad+\sum_{j=1}^{m}\mathbf{Q}_\alpha\mathcal{D}_{\alpha_{j-1}}\mathbf{A}(\alpha_j).
\end{aligned}\tag{16.61}$$

Using the reproducing kernels $\mathbf{k}_{A\beta}^{(m-\ell,t+\ell)}$ to recover $D_m^\ell\mathbf{A}(A\beta)$, we can rewrite this as

$$\begin{aligned}
\mathbf{A}(\alpha)&=\mathbf{A}(o)+\sum_{\ell=1}^{m}\sum_{o<\beta\leq\alpha}\frac{1}{\ell!}D_m^\ell\mathbf{A}(A\beta)\wedge\left\{\otimes^\ell(\beta-A\beta)\right\} \\
&\quad+\sum_{o<\beta\leq\alpha}\mathbf{Q}_{A\beta}^{(m,t)}\mathcal{D}_{A\beta}\mathbf{A}(\beta) \\
&=\mathbf{A}(o)+\sum_{\ell=1}^{m}\sum_{o<\beta\leq\alpha}\frac{1}{\ell!}\left\langle\left\langle D_m^\ell\mathbf{A},\mathbf{k}_{A\beta}^{(m-\ell,t+\ell)}\right\rangle\right\rangle_{m-\ell}^{(t+\ell)}\wedge\left\{\otimes^\ell(\beta-A\beta)\right\} \\
&\quad+\sum_{o<\beta\leq\alpha}\mathbf{Q}_{A\beta}^{(m,t)}\mathcal{D}_{A\beta}\mathbf{A}(\beta),
\end{aligned}$$

and since $\wedge$ and $\wedge_{A\beta}$ "commute", we have

$$\begin{aligned}
&\left\langle D_m^\ell\mathbf{A}(\gamma),\mathbf{k}_{A\beta}^{(m-\ell,t+\ell)}(\gamma)\right\rangle_\gamma^{(t+\ell)}\wedge\left\{\otimes^\ell(\beta-A\beta)\right\} \\
&=\overline{\mathbf{k}_{A\beta}^{(m-\ell,t+\ell)}(\gamma)}\wedge_\gamma D_m^\ell\mathbf{A}(\gamma)\wedge\left\{\otimes^\ell(\beta-A\beta)\right\} \\
&=\overline{\mathbf{k}_{A\beta}^{(m-\ell,t+\ell)}(\gamma)}\wedge\left\{\otimes^\ell(\beta-A\beta)\right\}\wedge_\gamma D_m^\ell\mathbf{A}(\gamma) \\
&=\left\langle D_m^\ell\mathbf{A}(\gamma),\mathbf{k}_{A\beta}^{(m-\ell,t+\ell)}(\gamma)\wedge\left\{\otimes^\ell\overline{\beta-A\beta}\right\}\right\rangle_\gamma^{(t+\ell)},
\end{aligned}$$

and so

$$\begin{aligned}
\mathbf{A}(\alpha)&=\mathbf{A}(o)+\sum_{\ell=1}^{m}\left\langle\left\langle D_m^\ell\mathbf{A},\sum_{o<\beta\leq\alpha}\frac{1}{\ell!}\mathbf{k}_{A\beta}^{(m-\ell,t+\ell)}\wedge\left(\otimes^\ell\overline{\beta-A\beta}\right)\right\rangle\right\rangle_{m-\ell}^{(t+\ell)} \\
&\quad+\sum_{o<\beta\leq\alpha}\mathbf{Q}_{A\beta}^{(m,t)}\mathcal{D}_{A\beta}\mathbf{A}(\beta).
\end{aligned}\tag{16.62}$$

By uniqueness of the representation formula (16.61) subject to the condition that

$$\mathbf{Q}_{\alpha_{j-1}}^{(m,t)}\mathcal{D}_{\alpha_{j-1}}\mathbf{A}\left(\alpha_j\right)\in M_{\alpha_{j-1}}^{\perp},\qquad 1\leq j\leq m,$$

(this uses that $L_{\alpha_{j-1}}^{(m,t)}$ is one-to-one as well as the uniqueness of the orthogonal decompositions into $M_{\alpha_{j-1}}$ and $M_{\alpha_{j-1}}^{\perp}$; see the discussion surrounding (16.37)), we see upon comparing (16.60) and (16.62) that we have the recursion formula

$$D_m^{\ell}\mathbf{k}_{\alpha}^{(m,t)}=\sum_{o<\beta\leq\alpha}\frac{1}{\ell!}\mathbf{k}_{A\beta}^{(m-\ell,t+\ell)}\wedge\left(\otimes^{\ell}\overline{\beta-A\beta}\right),\qquad 1\leq\ell\leq m,\tag{16.63}$$

as well as

$$\mathbf{A}\left(o\right)=\mathbf{A}\left(o\right)\cdot\overline{\mathbf{k}_{\alpha}^{(m,t)}\left(o\right)},$$
$$\sum_{o<\beta\leq\alpha}\mathbf{Q}_{A\beta}^{(m,t)}\mathcal{D}_{A\beta}\mathbf{A}\left(\beta\right)=\sum_{\alpha\in\mathcal{T}_n}\left\langle\mathbf{Q}_{\alpha}^{(m,t)}\left(\mathcal{D}_{\alpha}\mathbf{A}\right),\mathbf{Q}_{\alpha}^{(m,t)}\left(\mathcal{D}_{\alpha}\mathbf{k}_{\alpha}^{(m,t)}\right)\right\rangle_{\alpha}.$$

Note that the left side (16.63) is a tensor of order $2t+\ell$, while that of the right side has order $2\left(t+\ell\right)-\ell$, the same order. We now use the recursion formula in (16.63) to establish by induction the following positivity property for derivatives of the reproducing kernels $\mathbf{k}_{\alpha}^{(m,t)}$.

LEMMA 16.35. *Let $0\leq m+t\leq M$. Then provided we choose λ small enough relative to θ in the construction of the Bergman tree, we have for all $\alpha,\gamma\in\mathcal{T}_n$,*

$$r^{-md(\gamma)}\operatorname{Re}\left(D_m^m k_{\alpha}^{(m,0)}\left(\gamma\right)\wedge\left\{\otimes^m\gamma\right\}\right)\approx 1\tag{16.64}$$
$$\left|D_m^{\ell}k_{\alpha}^{(m,t)}\left(\gamma\right)\right|_{\gamma}+\left|\mathbf{Q}_{\gamma}^{(m,t)}\left(\mathcal{D}_{\gamma}k_{\alpha}^{(m,t)}\right)\right|_{\gamma}\leq\begin{cases}C & \text{for}\quad \gamma\leq\alpha\\ 0 & \text{otherwise,}\end{cases}$$

where $D_m^m k_{\alpha}^{(m,0)}\left(\gamma\right)\wedge\left\{\otimes^m\gamma\right\}=D_m^m k_{\alpha}^{(m,0)}\left(\gamma\right)\left[\gamma,\ldots,\gamma\right]$.

PROOF. Using induction with (16.53) and the recursion formula (16.63), we see that $D_m^{\ell}\mathbf{k}_{\alpha}^{(m,t)}$ is supported in the geodesic $[o,\alpha]$ for $\ell\geq 1$. The case $m=0$ of (16.64) is trivial from (16.53), and for $m=1$ and $t=0$ (16.64) has been established in (16.40) above. Now consider the case $m=1$ and $t\geq 1$. Then from (16.63) we have

$$D_m\mathbf{k}_{\alpha}^{(1,t)}\left(\gamma\right)=\sum_{o<\beta\leq\alpha}\mathbf{k}_{A\beta}^{(0,t+1)}\left(\gamma\right)\wedge\overline{\beta-A\beta}.$$

Note that each side of the above equation is a $(2t+1)$-tensor. By (16.53), we have that $\mathbf{k}_{A\beta}^{(0,t+1)}\left(\gamma\right)$ is a $2\left(t+1\right)$-tensor that vanishes if $\gamma\neq A\beta$, and so

$$D_m\mathbf{k}_{\alpha}^{(1,t)}\left(\gamma\right)=\mathbf{k}_{\gamma}^{(0,t+1)}\left(\gamma\right)\wedge\overline{\gamma_{\alpha}-\gamma}.$$

Recall also from (16.53) that

$$\mathbf{k}_{\gamma}^{(0,t)}\left(\gamma\right)=\sum_i\left\{\mathbf{R}^{d(\gamma)}e^{i_1}\otimes\cdots\otimes\mathbf{R}^{d(\gamma)}e^{i_t}\right\}\otimes\left\{\mathbf{R}^{d(\gamma)}e_{i_1}\otimes\cdots\otimes\mathbf{R}^{d(\gamma)}e_{i_t}\right\}.$$

We now easily obtain the second line in (16.64) for $m=1$ and $t\geq 0$.

For $m = \ell \geq 2$ and $t = 0$, the recursion formula (16.63) yields

$$D_m^m \mathbf{k}_\alpha^{(m,0)}(\gamma) = \sum_{o<\beta\leq\alpha} \frac{1}{m!} \mathbf{k}_{A\beta}^{(0,m)}(\gamma) \wedge \{\otimes^m \overline{\beta - A\beta}\}$$
$$= \frac{1}{m!} \mathbf{k}_\gamma^{(0,m)}(\gamma) \wedge \{\otimes^m \overline{\gamma_\alpha - \gamma}\},$$

and so

$$\begin{aligned} &D_m^m \mathbf{k}_\alpha^{(m,0)}(\gamma) \wedge \{\otimes^m \gamma\} \\ &= \frac{1}{m!} \mathbf{k}_\gamma^{(0,m)}(\gamma) \wedge \{\otimes^m \overline{\gamma_\alpha - \gamma}\} \wedge \{\otimes^m \gamma\} \\ &= \frac{1}{m!} \sum_i \left\langle \mathbf{R}^{d(\gamma)} e^{i_1}, \gamma_\alpha - \gamma \right\rangle \times \cdots \times \left\langle \mathbf{R}^{d(\gamma)} e^{i_m}, \gamma_\alpha - \gamma \right\rangle \\ &\quad \times \left\langle \mathbf{R}^{d(\gamma)} e_{i_1}, \gamma \right\rangle \times \cdots \times \left\langle \mathbf{R}^{d(\gamma)} e_{i_m}, \gamma \right\rangle. \end{aligned}$$

Now we recall from (16.42) that $e_1 = \frac{\gamma}{|\gamma|}$ and that e_j is orthogonal to γ for $j \geq 2$. Thus $\left\langle \mathbf{R}^{d(\gamma)} e_{i_k}, \gamma \right\rangle = r^{d(\gamma)} |\gamma| \, \delta_{i_k}^1$ and so

$$\begin{aligned} D_m^m \mathbf{k}_\alpha^{(m,0)}(\gamma) \wedge \{\otimes^m \gamma\} &= \frac{1}{m!} \left\langle \mathbf{R}^{d(\gamma)} e^1, \gamma_\alpha - \gamma \right\rangle^m \left(r^{d(\gamma)} |\gamma| \right)^m \\ &= \frac{r^{2md(\gamma)}}{m!} \langle \gamma, \gamma_\alpha - \gamma \rangle^m \\ &= \frac{r^{2md(\gamma)}}{m!} \langle \gamma, P_\gamma \gamma_\alpha - \gamma \rangle^m. \end{aligned}$$

Let $\varepsilon > 0$ be given. The vector $P_\gamma \gamma_\alpha$ lies in $P_\gamma K_{\gamma_\alpha}$, and from the construction of the Bergman tree $\mathcal{T}_n$ *with λ chosen sufficiently small relative to θ*, we see that

$$|\arg \langle \gamma, P_\gamma \gamma_\alpha - \gamma \rangle| < \varepsilon,$$

as well as

$$\operatorname{Re} \langle \gamma, P_\gamma \gamma_\alpha - \gamma \rangle = \operatorname{Re} \langle \gamma, \gamma_\alpha - \gamma \rangle \gtrsim r^{-d(\gamma)}.$$

It follows that

$$\operatorname{Re} \left(\langle \gamma, P_\gamma \gamma_\alpha - \gamma \rangle^m \right) \gtrsim r^{-md(\gamma)} (1 - m\varepsilon) \gtrsim r^{-md(\gamma)}.$$

This proves the first line in (16.64) for $0 \leq m \leq M$.

The proof of the second line in (16.64) can easily be completed by induction as follows. By the recursion formula (16.63), we have

$$\begin{aligned} \left| D_m^\ell \mathbf{k}_\alpha^{(m,t)}(\gamma) \right|_\gamma &= \left| \sum_{o<\beta\leq\alpha} \frac{1}{\ell!} \mathbf{k}_{A\beta}^{(m-\ell,t+\ell)}(\gamma) \wedge \{\otimes^\ell \overline{\beta - A\beta}\} \right|_\gamma \\ &\leq \sum_{o<\beta\leq\alpha} \frac{1}{\ell!} \left| \mathbf{k}_{A\beta}^{(m-\ell,t+\ell)}(\gamma) \wedge \{\otimes^\ell \overline{\beta - A\beta}\} \right|_\gamma, \end{aligned}$$

which is bounded by a constant using the induction assumption with $m - \ell < m$, together with the fact that $\beta > A\beta \geq \gamma$ if $\mathbf{k}_{A\beta}^{(m-\ell,t+\ell)}(\gamma)$ is nonzero. The boundedness of $\left| \mathbf{Q}_\gamma^{(m,t)} \left(\mathcal{D}_\gamma \mathbf{k}_\alpha^{(m,t)} \right) \right|_\gamma$ follows from (16.59), and this completes the proof of Lemma 16.35. $\square$

16.3.8. Carleson Measures. Here we characterize Carleson measures on the holomorphic Besov space $HB_{p,m}(\mathcal{T}_n)$.

THEOREM 16.36. *Let $1 < p < \infty$ and $1 \leq m \leq M$. Then there are λ and θ in the construction of the Bergman tree, sufficiently small and large respectively, such that μ is a $HB_{p,m}(\mathcal{T}_n)$-Carleson measure, i.e.*

$$\left(\sum_{\alpha\in\mathcal{T}_n} |f(\alpha)|^p \mu(\alpha)\right)^{1/p} \lesssim \|f\|_{HB_{p,m}(\mathcal{T}_n)}, \tag{16.65}$$

if and only if the tree condition (14.37) *holds, i.e.*

$$\sum_{\beta\geq\alpha}\left(\sum_{\gamma\geq\beta}\mu(\gamma)\right)^{p'} \lesssim \sum_{\beta\geq\alpha}\mu(\beta) < \infty, \qquad \alpha\in\mathcal{T}_n. \tag{16.66}$$

PROOF. We first show that (16.66) implies (16.65). To see this, note that by Definition 16.33 with $t = 0$ and (16.61) with f in place of $\mathbf{A}$, we have

$$\begin{aligned}
|f(\alpha)| &\leq |f(o)| + \sum_{\ell=1}^{m}\sum_{o<\beta\leq\alpha}\frac{1}{\ell!}\left|D_m^\ell f(A\beta)\wedge\left\{\otimes^\ell(\beta - A\beta)\right\}\right| \\
&\quad + \left|\sum_{o<\beta\leq\alpha}\mathbf{Q}_{A\beta}^{(m,t)}\mathcal{D}_{A\beta}f(\beta)\right| \\
&\lesssim |f(o)| + \sum_{\ell=1}^{m}\sum_{o<\beta\leq\alpha}\frac{1}{\ell!}\left|D_m^\ell f(A\beta)\right|_{A\beta} + \left|\sum_{o<\beta\leq\alpha}\mathbf{Q}_{A\beta}^{(m,t)}\mathcal{D}_{A\beta}f(\beta)\right| \\
&\lesssim |f(o)| + Ig(\alpha),
\end{aligned}$$

where

$$g(\beta) = \sum_{\ell=1}^{m}\left|D_m^\ell f(A\beta)\right|_{A\beta} + \left|\mathbf{Q}_{A\beta}\mathcal{D}_{A\beta}f(\beta)\right|. \tag{16.67}$$

Using this, together with our assumption (16.66) and Theorem 14.16, we have

$$\begin{aligned}
\left(\sum_{\alpha\in\mathcal{T}_n}|f(\alpha)|^p\mu(\alpha)\right)^{1/p} &\lesssim \left\{|f(o)| + \left(\sum_{\alpha\in\mathcal{T}_n}|Ig(\alpha)|^p\mu(\alpha)\right)^{1/p}\right\} \\
&\lesssim \left\{|f(o)| + \left(\sum_{\alpha\in\mathcal{T}_n}|g(\alpha)|^p\right)^{1/p}\right\} \\
&\lesssim \|f\|_{HB_{p,m}(\mathcal{T}_n)},
\end{aligned}$$

which is (16.65). Note that we do not use the full size of the $HB_{p,m}(\mathcal{T}_n)$ norm in the last line.

REMARK 16.37. *In deriving* (16.67) *we used the Bergman norm $\left|D_m^\ell f(A\beta)\right|_{A\beta}$ to give a natural pointwise estimate of $\left|D_m^\ell f(A\beta)\wedge\left\{\otimes^\ell(\beta - A\beta)\right\}\right|$, That norm is basic to the analogy between $HB_{p,m}(\mathcal{T}_n)$ and the holomorphic spaces $\|\cdot\|_{B_{p,m}}^*$ of Definition 14.24 which are defined using almost invariant holomorphic derivatives.*

The Bergman norm is also central in the growth estimate (16.26) *which ensures that the point evaluations are continuous linear functionals on* $HB_{p,m}(\mathcal{T}_n)$.

It is not known if it is possible to develop a satisfactory theory of holomorphic Besov spaces on Bergman trees using the somewhat simpler scaled Euclidean norm $r^{-\ell d(A\beta)}\left|D_m^\ell f(A\beta)\right|$.

Conversely, we show that the dual of (16.65) implies (16.66). Let $\mathbf{k}_\alpha^{(m,0)}$ be the (scalar-valued) reproducing kernel for $HB_{p,m}(\mathcal{T}_n)$. Using

$$\begin{aligned}\sum_{\alpha\in\mathcal{T}_n} f(\alpha)\overline{g(\alpha)}\mu(\alpha) &= \sum_{\alpha\in\mathcal{T}_n}\left\langle\left\langle f,\mathbf{k}_\alpha^{(m,0)}\right\rangle\right\rangle_m^{(0)}\overline{g(\alpha)}\mu(\alpha)\\ &= \left\langle\left\langle f,\sum_{\alpha\in\mathcal{T}_n} g(\alpha)\mu(\alpha)\mathbf{k}_\alpha^{(m,0)}\right\rangle\right\rangle_m^{(0)}\end{aligned}$$

and Proposition 16.34 we can obtain the inequality dual to (16.65):

$$\begin{aligned}\left\|\sum_{\alpha\in\mathcal{T}_n} g(\alpha)\mu(\alpha)\mathbf{k}_\alpha^{(m,0)}\right\|_{HB_{p',m}(\mathcal{T}_n)} &= \sup_{\|f\|_{HB_{p,m}(\mathcal{T}_n)}\le 1}\left|\left\langle\left\langle f,\sum_{\alpha\in\mathcal{T}_n} g(\alpha)\mu(\alpha)k_\alpha\right\rangle\right\rangle_m^{(0)}\right|\\ &= \sup_{\|f\|_{HB_{p,m}(\mathcal{T}_n)}\le 1}\left|\sum_{\alpha\in\mathcal{T}_n} f(\alpha)\overline{g(\alpha)}\mu(\alpha)\right|\\ &\le \sup_{\|f\|_{HB_{p,m}(\mathcal{T}_n)}\le 1}\|f\|_{L^p(\mu)}\|g\|_{L^{p'}(\mu)}\\ &\lesssim \|g\|_{L^{p'}(\mu)}.\end{aligned}$$

By Definition 16.33 with $t=0$, this implies in particular that

(16.68)

$$\begin{aligned}\left\|D_m^m\left(\sum_{\alpha\in\mathcal{T}_n} g(\alpha)\mu(\alpha)\mathbf{k}_\alpha^{(m,0)}\right)\right\|_{HB_{p',0}^{(m)}(\mathcal{T}_n)} &\le \left\|\sum_{\alpha\in\mathcal{T}_n} g(\alpha)\mu(\alpha)\mathbf{k}_\alpha^{(m,0)}\right\|_{HB_{p',m}(\mathcal{T}_n)}\\ &\lesssim \|g\|_{L^{p'}(\mu)}.\end{aligned}$$

Note that, as always, that we may assume μ has finite support.

We now restrict g to be nonnegative in (16.68). From the first part of (16.64) in Lemma 16.35, and the fact that the support of $\mathbf{k}_\alpha^{(m,0)}$ is contained in the geodesic

$[o, \alpha)$, we obtain that

$$\begin{aligned}
&\|I^* g\mu\|_{\ell^{p'}(\gamma)} \\
&= \left\| \sum_{\alpha \in \mathcal{T}_n : \alpha \geq \gamma} g(\alpha)\,\mu(\alpha) \right\|_{\ell^{p'}(\gamma)} \\
&\lesssim \left\| \sum_{\alpha \in \mathcal{T}_n} g(\alpha)\,\mu(\alpha)\, r^{-md(\gamma)} \operatorname{Re}\left(D_m^m k_\alpha^{(m,0)}(\gamma) \wedge \{\otimes^m \gamma\} \right) \right\|_{\ell^{p'}(\gamma)} \\
&\approx \left\| r^{-md(\gamma)} \operatorname{Re}\left(\sum_{\alpha \in \mathcal{T}_n} g(\alpha)\,\mu(\alpha)\, D_m^m k_\alpha^{(m,0)}(\gamma) \wedge \{\otimes^m \gamma\} \right) \right\|_{\ell^{p'}(\gamma)} \\
&\lesssim \left\| r^{-md(\gamma)} D_m^m \left(\sum_{\alpha \in \mathcal{T}_n} g(\alpha)\,\mu(\alpha)\, k_\alpha^{(m,0)}(\gamma) \right) \right\|_{\ell^{p'}(\gamma)} \\
&\lesssim \left\| D_m^m \left(\sum_{\alpha \in \mathcal{T}_n} g(\alpha)\,\mu(\alpha)\, k_\alpha^{(m,0)}(\gamma) \right) \right\|_{HB_{p',0}^{(m)}(\mathcal{T}_n)} \lesssim \|g\|_{L^{p'}(\mu)}
\end{aligned}$$

for all $g \geq 0$ by (16.68). The choice $g = \chi_{S(\alpha)}$ now yields (16.66) as required. □

16.3.9. Restricting Holomorphic Functions. In the special case where f arises as the restriction $f = TF = \{F(c_\alpha)\}_{\alpha \in \mathcal{T}_n}$ of a holomorphic function $F \in B_p$, $m > 2n/p$, then $D_m^\ell f(\alpha) \approx F^{(\ell)}(c_\alpha)$ for $1 \leq \ell \leq m$, and using Taylor's formula we will see that $\mathbf{Q}_\alpha \mathcal{D}_\alpha f$ is controlled by $F^{(m+1)}$. Similarly, $D_m^\ell f(\alpha) = F^{(\ell)}(\alpha) + \left\{ D_m^\ell f(\alpha) - F^{(\ell)}(\alpha) \right\}$, and we will show that each term in this sum is also controlled by $F^{(m+1)}$. In this way we will obtain the following Besov space restriction theorem, as well as the corresponding multiplier space restriction theorem. Recall that we need the structural constant theta sufficiently large in the construction of the Bergman tree in order to define the holomorphic tree space.

THEOREM 16.38. *Let $m > 2n/p$. Then provided θ is chosen large enough in the construction of the Bergman tree $\mathcal{T}_n$, the restriction map*

$$TF = \{F(\alpha)\}_{\alpha \in \mathcal{T}_n}, \qquad \textit{where } TF(\alpha) = F(c_\alpha),$$

is bounded from B_p to $HB_{p,m}(\mathcal{T}_n)$, and if in addition $p < 2 + 1/(n-1)$, then T is also bounded from $\operatorname{Mult}(B_p)$ *to* $\operatorname{Mult} HB_{p,m}(\mathcal{T}_n)$.

The proof of the first assertion in Theorem 16.38 will be given immediately below. The more difficult second assertion is proved in Section 16.3.10, and will require the lengthy proof of Proposition 16.41, characterizing the pointwise multipliers of $HB_{p,m}(\mathcal{T}_n)$.

PROOF. We first prove a stronger assertion for Besov spaces by induction on m. To state this stronger version, we need to recast the seminorms $\|\cdot\|^*_{B_{p,m}(\mathbb{B}_n)}$ we introduced for Besov spaces on the ball in Section 14.24, in the language of the discrete tensors we are using on the Bergman tree. This is easily accomplished by observing that for $f \in \operatorname{Hol}(\mathbb{B}_n)$, and $z, \alpha \in \mathbb{B}_n$, we have

$$(16.69) \qquad \left| D_\alpha^\ell f(z) \right| = \left| f^{(\ell)}(z) \right|_\alpha .$$

Indeed,

$$\begin{aligned}
|D_\alpha f(z)|^2 &= \left| f'(z) \left\{ \left(1-|\alpha|^2\right) P_\alpha + \left(1-|\alpha|^2\right)^{1/2} Q_\alpha \right\} \right|^2 \\
&= \left| f'(z) \left\{ r^{-d(\alpha)} P_\alpha + r^{-d(\alpha)/2} Q_\alpha \right\} \right|^2 \\
&= \left| \sum_i \frac{\partial f}{\partial z_i}(z) e^i \left\{ r^{-d(\alpha)} P_\alpha + r^{-d(\alpha)/2} Q_\alpha \right\} \right|^2 \\
&= \sum_i \left| \frac{\partial f}{\partial z_i}(z) \right|^2 \left\langle \mathbf{R}^{-d(\alpha)} e^i, \mathbf{R}^{-d(\alpha)} e^i \right\rangle \\
&= |f'(z)|_\alpha^2,
\end{aligned}$$

and the general case $\ell \geq 1$ can be verified by expanding each of the ℓ-tensors $D_\alpha^\ell f(z)$ and $f^{(\ell)}(z)$ as a sum of the basis tensors $e^{i_1} \otimes \cdots \otimes e^{i_\ell}$, and then using definitions. Then by Proposition 14.25 we have

$$\|f\|_{B_p} \approx \sum_{j=0}^{m-1} \left|\nabla^j f(0)\right| + \left(\sum_{\alpha \in \mathcal{T}_n} \int_{K_\alpha} \left| f^{(m)}(z) \right|_\alpha^p d\lambda_n(z) \right)^{1/p},$$

provided $m > 2n/p$. □

To fully exploit this realization of the Besov space norm, we now define the Besov space $B_{p,m}^{(t)}(\mathbb{B}_n)$ of t-tensors on the ball.

DEFINITION 16.39. *The space $B_{p,m}^{(t)}(\mathbb{B}_n)$ consists of all holomorphic t-tensor-valued functions $\mathbf{A}$ on the ball $\mathbb{B}_n$ such that the norm*

$$\|\mathbf{A}\|_{B_{p,m}^{(t)}(\mathbb{B}_n)} := \sum_{j=0}^{m-1} \left|\nabla^j \mathbf{A}(0)\right| + \left(\sum_{\alpha \in \mathcal{T}_n} \int_{K_\alpha} \left| \mathbf{A}^{(m)}(z) \right|_\alpha^p d\lambda_n(z) \right)^{1/p}$$

is finite.

Note that $\mathbf{A}^{(m)}(z)$ is a $(t+m)$-tensor and $\left|\mathbf{A}^{(m)}(z)\right|_\alpha$ is its norm in the Hilbert space $\mathcal{E}_\alpha^{(t+m)}$ given by $\sqrt{\left\langle \mathbf{A}^{(m)}(z), \mathbf{A}^{(m)}(z) \right\rangle_\alpha^{(t+m)}}$. If $\mathbf{A}$ is a polynomial, then

$$\left|\mathbf{A}^{(m)}(z)\right| \leq C_{\mathbf{A},m,t}(1-|z|^2)^{(t+m)/2} \qquad z \in K_\alpha$$

and we thus see that the space $B_{p,m}^{(t)}(\mathbb{B}_n)$ contains all polynomials if $p(t+m)/2 - n - 1 > -1$, or $m > 2n/p$. One can in fact show that the above norms are equivalent for different m, m' if both $t+m$ and $t+m'$ are greater than $2n/p$, but we will not need this fact. We will use however the trivial inequality

$$\left\|\mathbf{A}^{(\ell)}\right\|_{B_{p,m-\ell}^{(t+\ell)}(\mathbb{B}_n)} \leq \|\mathbf{A}\|_{B_{p,m}^{(t+m)}(\mathbb{B}_n)}, \qquad t+m > 2n/p, \tag{16.70}$$

for holomorphic t-tensor-valued functions, which uses only the definitions and the identity $\left(\mathbf{A}^{(\ell)}\right)^{(k)} = \mathbf{A}^{(\ell+k)}$.

We now extend the definition of the restriction map T to t-tensor-valued functions $\mathbf{A}$ on the ball in the natural way. Continuing to identify $\alpha \in \mathcal{T}_n$ with $c_\alpha \in \mathbb{B}_n$ we have

$$T\mathbf{A} = \{\mathbf{A}(\alpha)\}_{\alpha \in \mathcal{T}_n} = \{\mathbf{A}(c_\alpha)\}_{c_\alpha \in \mathbb{B}_n}.$$

The stronger assertion we will prove by induction on m is:

$$\|T\mathbf{A}\|_{HB^{(t)}_{p,m}(\mathcal{T}_n)} \lesssim \|\mathbf{A}\|_{B^{(t)}_{p,m}(\mathbb{B}_n)}, \qquad t+m>2n/p. \tag{16.71}$$

The case $t=0$ is the required Besov space restriction theorem.

We begin the induction with the case $m=0$ of (16.71), i.e.

$$\sum_{\alpha\in\mathcal{T}_n} |\mathbf{A}(\alpha)|_\alpha^p \lesssim \sum_{\alpha\in\mathcal{T}_n} \int_{K_\alpha} |\mathbf{A}(z)|_\alpha^p \, d\lambda_n(z),$$

follows immediately from the mean value equality for holomorphic functions and tensors,

$$\mathbf{A}(\alpha) = \int_{\varphi_\alpha(K_0)} \mathbf{A}(z)\, d\lambda_n(z), \tag{16.72}$$

followed by an application of Minkowski's inequality with the norm $|\cdot|_\alpha$, and then observing that $\varphi_\alpha(K_0) \approx K_\alpha$. Equation (16.72) is in turn a consequence of the mean value equality $\mathbf{A}(0) = \int_{B(0,1/2)} \mathbf{A}(z)\, d\lambda_n(z)$ and the invariance of the measure λ_n.

Now fix $0<m,t\le M$ with $t+m>2n/p$ and make the induction assumption that (16.71) holds for $m'<m$ and $0\le t'\le M$ satisfying $t'+m'>2n/p$. By Definition 16.33 we have, with $\mathbf{a}=T\mathbf{A}$,

$$\|\mathbf{a}\|^p_{HB^{(t)}_{p,m}(\mathcal{T}_n)} = |\mathbf{a}(o)|^p + \sum_{\ell=1}^m \left\|D^\ell_m \mathbf{a}\right\|^p_{HB^{(t+\ell)}_{p,m-\ell}(\mathcal{T}_n)} + \left\|\mathbf{Q}^{(m,t)}\mathcal{D}\mathbf{a}\right\|^p_{\ell^p(\mathcal{T}_n)}. \tag{16.73}$$

To estimate the term $\left\|D^\ell_m \mathbf{a}\right\|^p_{HB^{(t+\ell)}_{p,m-\ell}(\mathcal{T}_n)}$ in (16.73) we write

$$D^\ell_m\mathbf{a} = T\left(\mathbf{A}^{(\ell)}\right) + \left(D^\ell_m\mathbf{a} - T\left(\mathbf{A}^{(\ell)}\right)\right),$$

and estimate the terms

$$\left\|T\left(\mathbf{A}^{(\ell)}\right)\right\|^p_{HB^{(t+\ell)}_{p,m-\ell}(\mathcal{T}_n)} \tag{16.74}$$

and

$$\left\|D^\ell_m\mathbf{a} - T\left(\mathbf{A}^{(\ell)}\right)\right\|^p_{HB^{(t+\ell)}_{p,m-\ell}(\mathcal{T}_n)} \tag{16.75}$$

separately. The first term (16.74) is easy using the induction assumption:

$$\left\|T\left(\mathbf{A}^{(\ell)}\right)\right\|^p_{HB^{(t+\ell)}_{p,m-\ell}(\mathcal{T}_n)} \lesssim \left\|\mathbf{A}^{(\ell)}\right\|^p_{B^{(t+\ell)}_{p,m-\ell}(\mathbb{B}_n)},$$

since $(t+\ell)+(m-\ell)=t+m>2n/p$, and then (16.70) yields

$$\left\|T\left(\mathbf{A}^{(\ell)}\right)\right\|^p_{HB^{(t+\ell)}_{p,m-\ell}(\mathcal{T}_n)} \lesssim \|\mathbf{A}\|^p_{B^{(t)}_{p,m}(\mathbb{B}_n)}.$$

To handle the second term (16.75), let $\mathbf{a}^{(\ell)} = T\left(\mathbf{A}^{(\ell)}\right)$ denote the restriction of the holomorphic $(t+\ell)$-tensor $\mathbf{A}^{(\ell)}$ to the tree, and use the structure inequality (16.58) with $\mathbf{v}^\ell = D^\ell_m\mathbf{a} - \mathbf{a}^{(\ell)}$, to obtain

$$\begin{aligned}\sum_{\ell=1}^m \left|D^\ell_m\mathbf{a} - \mathbf{a}^{(\ell)}\right|_\alpha &\lesssim \left|L^{(m,t)}_\alpha\left(D^1_m\mathbf{a},\dots,D^m_m\mathbf{a}\right) - L^{(m,t)}_\alpha\left(\mathbf{a}^{(1)},\dots,\mathbf{a}^{(m)}\right)\right|_\alpha \\ &= \left|\mathbf{P}^{(m,t)}_\alpha \mathcal{D}_\alpha\mathbf{A} - \mathsf{U}^{(m,t)}_\alpha\right|_\alpha = \left|\mathsf{V}^{(m,t)}_\alpha\right|_\alpha\end{aligned}$$

where the vectors $\mathsf{U}_\alpha^{(m,t)}$ and $\mathsf{V}_\alpha^{(m,t)}$ in $\left(\mathcal{E}_\alpha^{(t)}\right)^N$ are defined by

$$\begin{aligned}\mathsf{U}_\alpha^{(m,t)} &= L_\alpha^{(m,t)}\left(\mathbf{a}^{(1)},\dots,\mathbf{a}^{(m)}\right) = L_\alpha^{(m,t)}\left(\mathbf{A}^{(1)},\dots,\mathbf{A}^{(m)}\right),\\ \mathsf{V}_\alpha^{(m,t)} &= \mathcal{D}_\alpha\mathbf{a} - \mathsf{U}_\alpha^{(m,t)} = \mathcal{D}_\alpha\mathbf{A} - \mathsf{U}_\alpha^{(m,t)}.\end{aligned}$$

By a generalization of the local oscillation inequality (14.78) in Proposition 14.26, with the scalar f replaced by a t-tensor $\mathbf{A}$, and using θ large, we have the estimate

(16.76)

$$\begin{aligned}\left|\mathsf{V}_\alpha^{(m,t)}\right|_\alpha &= \left|\left\{\mathbf{A}\left(\alpha^j\right)-\mathbf{A}\left(\alpha\right)-\sum_{\ell=1}^m D_m^\ell\mathbf{A}\left(\alpha\right)\wedge\otimes^\ell\left(\alpha^j-\alpha\right)\right\}_{j=1}^N\right|_\alpha\\ &\lesssim\left(\int_{K_\alpha^*}\left|\mathbf{A}^{m+1}\left(z\right)\right|_\alpha^p d\lambda_n\left(z\right)\right)^{1/p}\lesssim\left(\int_{K_\alpha^{**}}\left|\mathbf{A}^{m}\left(z\right)\right|_\alpha^p d\lambda_n\left(z\right)\right)^{1/p}.\end{aligned}$$

Let $\ell_{(t)}^p\left(\mathcal{T}_n\right)$ be the space of t-tensor-valued functions $\mathbf{A}$ on the tree with

$$\|\mathbf{A}\|_{\ell_{(t)}^p(\mathcal{T}_n)} = \left(\sum_{\alpha\in\mathcal{T}_n}\left|\mathbf{A}\left(\alpha\right)\right|_\alpha^p\right)^{1/p}<\infty.$$

We can now estimate the second term, (16.75) above, using the embedding

$$\ell_{(t+\ell)}^p\left(\mathcal{T}_n\right)\subset HB_{p,m-\ell}^{(t+\ell)}\left(\mathcal{T}_n\right),$$

to obtain

$$\begin{aligned}\left\|D_m^\ell\mathbf{a}-T\left(\mathbf{A}^{(\ell)}\right)\right\|_{HB_{p,m-\ell}^{(t+\ell)}(\mathcal{T}_n)}^p &\lesssim\left\|\left|D_m^\ell\mathbf{a}-T\left(\mathbf{A}^{(\ell)}\right)\right|_\alpha\right\|_{\ell_{(t+\ell)}^p(\mathcal{T}_n)}^p\\ &\lesssim\left\|\left|\mathsf{V}_\alpha^{(m,t)}\right|_\alpha\right\|_{\ell^p(\mathcal{T}_n)}^p\\ &\lesssim\left(\sum_{\alpha\in\mathcal{T}_n}\int_{K_\alpha^*}\left|\mathbf{A}^m\left(z\right)\right|_\alpha^p d\lambda_n\left(z\right)\right)^{1/p}\\ &\lesssim\|\mathbf{A}\|_{B_{p,m}^{(t+m)}(\mathbb{B}_n)}^p.\end{aligned}\tag{16.77}$$

To estimate the term $\left\|\mathbf{Q}^{(m,t)}\mathcal{D}\mathbf{A}\right\|_{\ell^p(\mathcal{T}_n)}^p$ in (16.73), we note that the minimizing property (16.59) yields

$$\begin{aligned}\left|\mathbf{Q}_a\mathcal{D}_\alpha\mathbf{A}\right|_\alpha &= \min_{\mathbf{x}\in\mathcal{E}_\alpha^{(1)}\times\cdots\times\mathcal{E}_\alpha^{(m)}}\left|\mathcal{D}_\alpha\mathbf{A}-L_\alpha^{(m,t)}\mathbf{x}\right|_\alpha\\ &\le\left|\mathcal{D}_\alpha\mathbf{A}-\mathsf{U}_\alpha^{(m,t)}\right|_\alpha=\left|\mathsf{V}_\alpha^{(m,t)}\right|_\alpha.\end{aligned}\tag{16.78}$$

Thus by (16.76) we can bound $\left|\mathbf{Q}_a\mathcal{D}_\alpha\mathbf{A}\right|_\alpha$ by $C\left(\int_{K_\alpha^*}\left|\mathbf{A}^m\left(z\right)\right|_\alpha^p d\lambda_n\left(z\right)\right)^{1/p}$, and so

$$\left\|\mathbf{Q}^{(m,t)}\mathcal{D}\mathbf{A}\right\|_{\ell^p(\mathcal{T}_n)}^p\lesssim\left\|\left|\mathsf{V}_\alpha^{(m,t)}\right|_\alpha\right\|_{\ell^p(\mathcal{T}_n)}^p\lesssim\|\mathbf{A}\|_{B_{p,m}^{(t+m)}(\mathbb{B}_n)}^p,$$

by (16.77). This completes the proof of the first assertion in Theorem 16.38.

16.3.10. Restricting Besov Space Multipliers. We now turn to proving that T is bounded from the ball multiplier space $\text{Mult}(B_p)$ to the tree multiplier space $\text{Mult}(HB_{p,m}(\mathcal{T}_n))$ for $2n/m < p < 2 + 1/(n-1)$. First we record a variant of the ball multiplier Theorem 14.14 using the derivatives $D^m_{c_\alpha}$ in place of ∇^m.

LEMMA 16.40. *Let $\varphi \in H^\infty(\mathbb{B}_n) \cap B_p$ and $m > 2n/p$. Then φ is a multiplier on B_p if and only if*

$$\sum_{\alpha \in \mathcal{T}_n} \chi_{K_\alpha}(z) \left\{ \int_{K_\alpha} \left| D^m_{c_\alpha} \varphi(\zeta) \right|^p d\lambda_n(\zeta) \right\} d\lambda_n(z)$$

is a B_p-Carleson measure on $\mathbb{B}_n$.

PROOF. Since the operators $D^m_{c_\alpha}$ satisfy the same product rule as ∇^m, and can be used in place of ∇^m in the seminorm for B_p when $m > 2n/p$ by Lemma 17.37, the proof of Theorem 14.14 applies almost verbatim. □

Next we prove the analog of the tree multiplier Lemma 16.5 for the holomorphic Besov space $HB_{p,m}(\mathcal{T}_n)$. We only need the sufficiency statement in the sequel. In order to state the lemma, we begin by expressing the Besov space norm $HB_{p,m}(\mathcal{T}_n)$ as an ℓ^p norm of appropriate quantities. We have

$$\begin{aligned} \|f\|^p_{HB_{p,m}(\mathcal{T}_n)} &= |f(o)|^p + \sum_{\alpha \in \mathcal{T}_n} |\mathbf{Q}_\alpha \mathcal{D}_\alpha f|^p + \sum_{\ell_1=1}^{m} \left\| D^{\ell_1}_m f(\alpha) \right\|^p_{HB^{(\ell_1)}_{p,m-\ell_1}(\mathcal{T}_n)} \\ &= |f(o)|^p + \sum_{\alpha \in \mathcal{T}_n} |\mathbf{Q}_\alpha \mathcal{D}_\alpha f|^p + \sum_{\alpha \in \mathcal{T}_n} |D^m_m f(\alpha)|^p_\alpha \\ &\quad + \sum_{\ell_1=1}^{m-1} \left\| D^{\ell_1}_m f(\alpha) \right\|^p_{HB^{(\ell_1)}_{p,m-\ell_1}(\mathcal{T}_n)}. \end{aligned}$$

Now write each term $\left\| D^{\ell_1}_m f(\alpha) \right\|^p_{HB^{(\ell_1)}_{p,m-\ell_1}(\mathcal{T}_n)}$ as

$$\begin{aligned} \left\| D^{\ell_1}_m f(\alpha) \right\|^p_{HB^{(\ell_1)}_{p,m-\ell_1}(\mathcal{T}_n)} &= \left| D^{\ell_1}_m f(o) \right| + \sum_{\alpha \in \mathcal{T}_n} \left| \mathbf{Q}_\alpha \mathcal{D}_\alpha D^{\ell_1}_m f \right|^p \\ &\quad + \sum_{\alpha \in \mathcal{T}_n} \left| D^{m-\ell_1}_{m-\ell_1} D^{\ell_1}_m f(\alpha) \right|^p_\alpha \\ &\quad + \sum_{\ell_2=1}^{m-\ell_1-1} \left\| D^{\ell_2}_{m-\ell_1} D^{\ell_1}_m f(\alpha) \right\|^p_{HB^{(\ell_1)}_{p,m-\ell_1}(\mathcal{T}_n)}, \end{aligned}$$

to get

$$\begin{aligned}\|f\|_{HB_{p,m}(\mathcal{T}_n)}^p &= |f(o)|^p + \sum_{\ell_1=1}^{m-1} \left|D_m^{\ell_1} f(o)\right| \\ &+ \sum_{\alpha\in\mathcal{T}_n} |\mathbf{Q}_\alpha \mathcal{D}_\alpha f|^p + \sum_{\ell_1=1}^{m-1} \sum_{\alpha\in\mathcal{T}_n} |\mathbf{Q}_\alpha \mathcal{D}_\alpha D_m^{\ell_1} f|^p \\ &+ \sum_{\alpha\in\mathcal{T}_n} |D_m^m f(\alpha)|_\alpha^p + \sum_{\ell_1=1}^{m-1} \sum_{\alpha\in\mathcal{T}_n} \left|D_{m-\ell_1}^{m-\ell_1} D_m^{\ell_1} f(\alpha)\right|_\alpha^p \\ &+ \sum_{\ell_1=1}^{m-1} \sum_{\ell_2=1}^{m-\ell_1-1} \left\|D_{m-\ell_1}^{\ell_2} D_m^{\ell_1} f(\alpha)\right\|_{HB_{p,m-\ell_1}^{(\ell_1)}(\mathcal{T}_n)}^p .\end{aligned}$$

Continuing in this way we arrive at the desired formula (let $s_k = \ell_1 + \cdots + \ell_k$):

$$\begin{aligned}\|f\|_{HB_{p,m}(\mathcal{T}_n)}^p &= \sum_{s_k<m} \left|D_{m-s_{k-1}}^{\ell_k} \cdots D_{m-\ell_1}^{\ell_2} D_m^{\ell_1} f(o)\right|^p \\ &+ \sum_{s_k=m} \sum_{\alpha\in\mathcal{T}_n} \left|\mathbf{Q}_\alpha \mathcal{D}_\alpha D_{m-s_{k-2}}^{\ell_{k-1}} \cdots D_m^{\ell_1} f\right|_\alpha^p \\ &+ \sum_{s_k=m} \sum_{\alpha\in\mathcal{T}_n} \left|D_{m-s_{k-1}}^{\ell_k} \cdots D_m^{\ell_1} f(\alpha)\right|_\alpha^p .\end{aligned} \tag{16.79}$$

PROPOSITION 16.41. *Let $m > 2n/p$. Then $f \in \mathrm{Mult}(HB_{p,m}(\mathcal{T}_n))$ if and only if f is bounded and $\{\omega(\alpha)\}_{\alpha\in\mathcal{T}}$ is a $HB_{p,m}(\mathcal{T}_n)$-Carleson measure, where*

$$\omega(\alpha) := \sum_{s_k=m} \left|D_{m-s_{k-1}}^{\ell_k} \cdots D_m^{\ell_1} f(\alpha)\right|_\alpha^p + \sum_{s_k=m} \left|\mathbf{Q}_a \mathcal{D}_\alpha D_{m-s_{k-2}}^{\ell_{k-1}} \cdots D_m^{\ell_1} f\right|_\alpha^p . \tag{16.80}$$

For the proof of this characterization of the pointwise multipliers of $HB_{p,m}(\mathcal{T}_n)$, we will need the embeddings

$$\begin{aligned} HB_{q,2}(\mathcal{T}_n) &\subset HB_{r,2}(\mathcal{T}_n), \qquad 1<q<r<\infty \\ HB_{p,m}(\mathcal{T}_n) &= HB_{p,m'}(\mathcal{T}_n), \qquad m,m' > 2n/p. \end{aligned} \tag{16.81}$$

The embedding $HB_{q,2}(\mathcal{T}_n) \subset HB_{r,2}(\mathcal{T}_n)$ for $q<r$ is automatic from Definition 16.18 and the embeddings of ℓ^q spaces. The equality $HB_{p,m}(\mathcal{T}_n) = HB_{p,m'}(\mathcal{T}_n)$ for $m,m' > 2n/p$ is established in the following lemma.

LEMMA 16.42. *$HB_{p,m}(\mathcal{T}_n) \subset HB_{p,m+1}(\mathcal{T}_n)$ for $m \geq 0$, $1<p<\infty$; and $HB_{p,m}(\mathcal{T}_n) = HB_{p,m'}(\mathcal{T}_n)$ for $m,m' > 2n/p$.*

PROOF. We first show by induction on m that

$$\|\mathbf{A}\|_{HB_{p,m+1}^{(t)}(\mathcal{T}_n)} \lesssim \|\mathbf{A}\|_{HB_{p,m}^{(t)}(\mathcal{T}_n)}, \qquad 1<p<\infty, m\geq 0,$$

for all t-tensor-valued functions $\mathbf{A}$ on the tree $\mathcal{T}_n$. We have from (16.58) that

$$\begin{aligned}\|\mathbf{A}\|^p_{HB^{(t)}_{p,1}(\mathcal{T}_n)} &= |\mathbf{A}(o)|^p + \|D_1\mathbf{A}\|^p_{HB_{p,0}(\mathcal{T}_n)} + \left\|\mathbf{Q}^{(1,t)}\mathcal{D}\mathbf{A}\right\|^p_{\ell^p(\mathcal{T}_n)} \\ &\approx |\mathbf{A}(o)|^p + \left\|\mathbf{P}^{(1,t)}\mathcal{D}\mathbf{A}\right\|^p_{\ell^p(\mathcal{T}_n)} + \left\|\mathbf{Q}^{(1,t)}\mathcal{D}\mathbf{A}\right\|^p_{\ell^p(\mathcal{T}_n)} \\ &\lesssim |\mathbf{A}(o)|^p + \|\mathcal{D}\mathbf{A}\|^p_{\ell^p(\mathcal{T}_n)} \\ &\lesssim |\mathbf{A}(o)|^p + \|\mathbf{A}\|^p_{\ell^p(\mathcal{T}_n)} \approx \|\mathbf{A}\|^p_{HB^{(t)}_{p,0}(\mathcal{T}_n)}.\end{aligned}$$

The significant inequality above is $\|\mathcal{D}\mathbf{A}\|^p_{\ell^p(\mathcal{T}_n)} \lesssim \|\mathbf{A}\|^p_{\ell^p(\mathcal{T}_n)}$, which cannot be reversed for general functions $\mathbf{A}$. This establishes the case $m = 0$ of the induction. Now from the case $m = 0$ applied to the $(t+1)$-tensor $D_2\mathbf{A}$ we have

$$\begin{aligned}\|\mathbf{A}\|^p_{HB^{(t)}_{p,2}(\mathcal{T}_n)} &= |\mathbf{A}(o)|^p + \|D_2\mathbf{A}\|^p_{HB_{p,1}(\mathcal{T}_n)} + \left\|D_2^2\mathbf{A}\right\|^p_{HB_{p,0}(\mathcal{T}_n)} + \left\|\mathbf{Q}^{(2,t)}\mathcal{D}\mathbf{A}\right\|^p_{\ell^p(\mathcal{T}_n)} \\ &\leq |\mathbf{A}(o)|^p + \|D_2\mathbf{A}\|^p_{HB_{p,0}(\mathcal{T}_n)} + \left\|D_2^2\mathbf{A}\right\|^p_{HB_{p,0}(\mathcal{T}_n)} + \left\|\mathbf{Q}^{(2,t)}\mathcal{D}\mathbf{A}\right\|^p_{\ell^p(\mathcal{T}_n)} \\ &\approx |\mathbf{A}(o)|^p + \left\|\mathbf{P}^{(2,t)}\mathcal{D}\mathbf{A}\right\|^p_{\ell^p(\mathcal{T}_n)} + \left\|\mathbf{Q}^{(2,t)}\mathcal{D}\mathbf{A}\right\|^p_{\ell^p(\mathcal{T}_n)} \\ &\approx |\mathbf{A}(o)|^p + \|\mathcal{D}\mathbf{A}\|^p_{\ell^p(\mathcal{T}_n)} \\ &\approx |\mathbf{A}(o)|^p + \left\|\mathbf{P}^{(1,t)}\mathcal{D}\mathbf{A}\right\|^p_{\ell^p(\mathcal{T}_n)} + \left\|\mathbf{Q}^{(1,t)}\mathcal{D}\mathbf{A}\right\|^p_{\ell^p(\mathcal{T}_n)} = \|\mathbf{A}\|^p_{HB^{(t)}_{p,1}(\mathcal{T}_n)},\end{aligned}$$

which is the case $m = 1$. The general case is similar.

This proof of the opposite inequality,

$$\|f\|_{HB_{p,m}(\mathcal{T}_n)} \lesssim \|f\|_{HB_{p,m+1}(\mathcal{T}_n)}, \qquad m > 2n/p, \tag{16.82}$$

is similar to the proof of the analogous Theorem 14.4 on the ball - see [**346**, Theorem 6.1]. To illustrate we prove only the case $m = 1$ of (16.82), the case $m = 0$ being trivial. We must show that $\|f\|_{HB_{p,1}(\mathcal{T}_n)} \lesssim \|f\|_{HB_{p,2}(\mathcal{T}_n)}$ provided $p > 2n$. Recall that

$$\begin{aligned}\|f\|_{HB_{p,1}(\mathcal{T}_n)} &\approx |f(o)| + \|D_1 f\|_{HB_{p,0}(\mathcal{T}_n)} + \left\|\mathbf{Q}^{(1,0)}\mathcal{D}f\right\|_{\ell^p(\mathcal{T}_n)}, \\ \|f\|_{HB_{p,2}(\mathcal{T}_n)} &\approx |f(o)| + \|D_2 f\|_{HB_{p,1}(\mathcal{T}_n)} + \left\|D_2^2 f\right\|_{HB_{p,0}(\mathcal{T}_n)} + \left\|\mathbf{Q}^{(2,0)}\mathcal{D}f\right\|_{\ell^p(\mathcal{T}_n)}.\end{aligned}$$

Now we have

$$\begin{aligned}&\|D_1 f\|_{HB_{p,0}(\mathcal{T}_n)} + \left\|\mathbf{Q}^{(1,0)}\mathcal{D}f\right\|_{\ell^p(\mathcal{T}_n)} \\ &\approx \left\|\mathbf{P}^{(1,0)}\mathcal{D}f\right\|_{\ell^p(\mathcal{T}_n)} + \left\|\mathbf{Q}^{(1,0)}\mathcal{D}f\right\|_{\ell^p(\mathcal{T}_n)} \\ &\approx \|\mathcal{D}f\|_{\ell^p(\mathcal{T}_n)} \\ &\approx \left\|\mathbf{P}^{(2,t)}\mathcal{D}f\right\|_{\ell^p(\mathcal{T}_n)} + \left\|\mathbf{Q}^{(2,t)}\mathcal{D}f\right\|_{\ell^p(\mathcal{T}_n)} \\ &\approx \|D_2 f\|_{HB_{p,0}(\mathcal{T}_n)} + \left\|D_2^2 f\right\|_{HB_{p,0}(\mathcal{T}_n)} + \left\|\mathbf{Q}^{(2,0)}\mathcal{D}f\right\|_{\ell^p(\mathcal{T}_n)},\end{aligned}$$

and thus it suffices to show that $\|D_2 f\|_{HB_{p,0}(\mathcal{T}_n)} \lesssim \|D_2 f\|_{HB_{p,1}(\mathcal{T}_n)}$, or

$$\begin{aligned}\sum_{\alpha\in\mathcal{T}_n} |D_2 f(\alpha)|_\alpha^p &\lesssim \left(|D_2 f(o)|^p + \sum_{\alpha\in\mathcal{T}_n} |D_1 D_2 f|_\alpha^p + \sum_{\alpha\in\mathcal{T}_n} |\mathbf{Q}_\alpha \mathcal{D}_\alpha D_2 f|_\alpha^p \right) \\ &\approx |D_2 f(o)|^p + \sum_{\alpha\in\mathcal{T}_n} \left|\mathbf{P}_\alpha^{(1,0)} \mathcal{D}_\alpha D_2 f\right|_\alpha^p + \sum_{\alpha\in\mathcal{T}_n} |\mathbf{Q}_\alpha \mathcal{D}_\alpha D_2 f|_\alpha^p \\ &\approx |D_2 f(o)|^p + \sum_{\alpha\in\mathcal{T}_n} |\mathcal{D}_\alpha D_2 f|_\alpha^p,\end{aligned}$$

i.e.

$$\sum_{\alpha\in\mathcal{T}_n} |D_2 f(\alpha)|_\alpha^p \lesssim \left(|D_2 f(o)|^p + \sum_{\alpha\in\mathcal{T}_n} |\Delta D_2 f(\alpha)|_\alpha^p \right) \tag{16.83}$$

(Recall that by convention $\Delta D_2 f(o) = D_2 f(o)$.)

Now using (16.22), (16.23) and the operator

$$\left(\mathbf{R}^{\pm d}\right)(\alpha) = r^{\pm d(\alpha)} P_\alpha + r^{\pm d(\alpha)/2} Q_\alpha$$

as in Definition 16.25, we have

$$\begin{aligned}\left(\mathbf{R}^{-d} f'\right)(\alpha) &= f'(\alpha)\left(\mathbf{R}^{-d}\right)(\alpha) \\ &= \left(\sum_{o\le\gamma\le\alpha} \Delta f'(\gamma)\right)\left(\mathbf{R}^{-d}\right)(\alpha) \\ &= \sum_{o\le\gamma\le\alpha} \left(\mathbf{R}^{-d}\Delta f'\right)(\gamma)\left(\mathbf{R}^{d}\right)(\gamma)\left(\mathbf{R}^{-d}\right)(\alpha),\end{aligned}$$

and thus the estimate

$$\begin{aligned}|D_2 f(\alpha)|_\alpha &= \left|\left(\mathbf{R}^{-d} D_2 f\right)(\alpha)\right| \\ &\le \sum_{o\le\gamma\le\alpha} \left|\left(\mathbf{R}^{-d}\Delta D_2 f\right)(\gamma)\right| \left|\left(\mathbf{R}^{d}\right)(\gamma)\left(\mathbf{R}^{-d}\right)(\alpha)\right| \\ &= \sum_{o\le\gamma\le\alpha} \left|\left(\mathbf{R}^{d}\right)(\gamma)\left(\mathbf{R}^{-d}\right)(\alpha)\right| |\Delta D_2 f(\gamma)|_\gamma .\end{aligned} \tag{16.84}$$

We now claim that

$$\left|\left(\mathbf{R}^{d}\right)(\gamma)\left(\mathbf{R}^{-d}\right)(\alpha)\right| \lesssim r^{-(d(\alpha)-d(\gamma))/2}. \tag{16.85}$$

To see this we expand the product of operators as

$$\begin{aligned}&\left(\mathbf{R}^{d}\right)(\gamma)\left(\mathbf{R}^{-d}\right)(\alpha) \\ &= \left(r^{d(\gamma)} P_\gamma + r^{d(\gamma)/2} Q_\gamma\right)\left(r^{-d(\alpha)} P_\alpha + r^{-d(\alpha)/2} Q_\alpha\right) \\ &= r^{-(d(\alpha)-d(\gamma))/2}\left\{ r^{-(d(\alpha)-d(\gamma))/2} P_\gamma P_\alpha + r^{d(\gamma)/2} P_\gamma Q_\alpha \right. \\ &\quad \left. + r^{-d(\alpha)/2} Q_\gamma P_\alpha + Q_\gamma Q_\alpha \right\},\end{aligned}$$

which reduces the proof of the claim to

$$|P_\gamma Q_\alpha| \lesssim r^{-d(\gamma)/2}.$$

For this it is enough to show that the adjoint $(P_\gamma Q_\alpha)^* = Q_\alpha^* P_\gamma^* = Q_\alpha P_\gamma$ has norm bounded by $Cr^{-d(\gamma)/2}$. However, for a unit vector $\mathbf{v}$, $P_\gamma \mathbf{v} = \lambda\gamma$ where λ is a complex number of modulus at most one, and so

$$\begin{aligned}|Q_\alpha P_\gamma \mathbf{v}| &= |Q_\alpha(\lambda\gamma)| = |\lambda\gamma - P_\alpha(\lambda\gamma)| \\ &= |\lambda| \left|(\gamma-\alpha) - \frac{((\gamma-\alpha)\cdot\overline{\alpha})\,\alpha}{|\alpha|^2}\right| \\ &\leq 2|\gamma-\alpha| \lesssim \left(1-|\gamma|^2\right)^{1/2} \approx r^{-d(\gamma)/2},\end{aligned}$$

by (16.23).

Combining (16.84) and (16.85) we obtain

$$|D_2 f(\alpha)|_\alpha \lesssim \sum_{o\leq\gamma\leq\alpha} r^{-(d(\alpha)-d(\gamma))/2} |\Delta D_2 f(\gamma)|_\gamma .$$

We write this as

$$|D_2 f(\alpha)|_\alpha \lesssim \sum_{\beta\in\mathcal{T}} K(\alpha,\beta) |\Delta D_2 f(\beta)|_\beta$$

where the kernel $K(\alpha,\beta)$ is given by $\chi_{[o,\alpha]}(\beta)\, r^{[d(\beta)-d(\alpha)]/2}$. We now apply Schur's test , Theorem B.2, with auxiliary functions $k(\beta) = h(\beta) = r^{td(\beta)}$. We have

$$\sum_{\beta\in\mathcal{T}} K(\alpha,\beta)\, h(\beta)^{p'} = \sum_{\beta\leq\alpha} r^{\left(1/2+p't\right)d(\beta)} r^{-d(\alpha)/2} \lesssim h(\alpha)^{p'}$$

provided $1/2 + p't > 0$. We also have using the "sparse" argument of Lemma 16.3, that for $\varepsilon > 0$,

$$\begin{aligned}\sum_{\alpha\in\mathcal{T}} K(\alpha,\beta)\, h(\alpha)^p &= \sum_{\alpha\geq\beta} r^{(pt-1/2)d(\alpha)} r^{d(\beta)/2} \\ &\leq C_\ell \sum_{k=0}^{\infty} \left(r^{n+\varepsilon}\right)^{k\ell} r^{(pt-1/2)(d(\beta)+k\ell)} r^{d(\beta)/2} \leq C_\ell h(\beta)^p\end{aligned}$$

provided $n + \varepsilon + pt - 1/2 < 0$, where ℓ is chosen so large in Definition 14.2 that $2^{(N_\ell)^{1/\ell}} < r^{n+\varepsilon}$. Now since $p > 2n$, we can choose $-1/2p' < t < (1/2 - n - \varepsilon)/p$ for some $\varepsilon > 0$, and then Schur's test shows that

$$\begin{aligned}\left(\sum_{\alpha\in\mathcal{T}} |D_2 f(\alpha)|_\alpha^p\right)^{1/p} &\lesssim \left(\sum_{\alpha\in\mathcal{T}} \left|\sum_{\beta\in\mathcal{T}} K(\alpha,\beta) |\Delta D_2 f(\beta)|_\beta\right|^p\right)^{1/p} \\ &\lesssim \left(\sum_{\beta\in\mathcal{T}} |\Delta D_2 f(\beta)|_\beta^p\right)^{1/p},\end{aligned}$$

which is (16.83) as required. This completes the proof of the case $m = 1$ of (16.82). Again the general case is similar and the details are left to the reader. The proof of Lemma 16.42 is complete. □

PROOF OF PROPOSITION 16.41. We now return to the proof of Proposition 16.41. We first prove the sufficiency assertion. Let f be bounded and suppose that $\{\omega(\alpha)\}_{\alpha\subset\mathcal{T}}$ is a $HB_{p,m}(\mathcal{T}_n)$-Carleson measure where ω is as in (16.80). Then if $g \in HB_{p,m}(\mathcal{T}_n)$, we must show that $fg \in HB_{p,m}(\mathcal{T}_n)$ with norm control

$$\|fg\|_{HB_{p,m}(\mathcal{T}_n)} \lesssim \|g\|_{HB_{p,m}(\mathcal{T}_n)},$$

where the implied constant depends on the Carleson norm $\|\omega\|_{Carleson}$ of $\{\omega(\alpha)\}_{\alpha\in\mathcal{T}}$. We have

$$\begin{aligned}(16.86)\qquad \|fg\|^p_{HB_{p,m}(\mathcal{T}_n)} &= \sum_{s_k<m}\left|D^{\ell_k}_{m-s_{k-1}}\cdots D^{\ell_2}_{m-\ell_1}D^{\ell_1}_m(fg)(o)\right|^p \\ &\quad + \sum_{s_k=m}\sum_{\alpha\in\mathcal{T}_n}\left|\mathbf{Q}_\alpha\mathcal{D}_\alpha D^{\ell_{k-1}}_{m-s_{k-2}}\cdots D^{\ell_1}_m(fg)\right|^p_\alpha \\ &\quad + \sum_{s_k=m}\sum_{\alpha\in\mathcal{T}_n}\left|D^{\ell_k}_{m-s_{k-1}}\cdots D^{\ell_1}_m(fg)(\alpha)\right|^p_\alpha .\end{aligned}$$

Consider first the case that $s_1 = m$ in (16.86), so that $\ell_1 = m$. We must show that

$$(16.87)\quad \sum_{\alpha\in\mathcal{T}_n}\left|\mathbf{Q}^{(m,0)}_\alpha\mathcal{D}_\alpha(fg)\right|^p_\alpha + \sum_{\alpha\in\mathcal{T}_n}\left|D^m_m(fg)(\alpha)\right|^p_\alpha \lesssim \|\omega\|^p_{Carleson}\|g\|^p_{HB_{p,m}(\mathcal{T}_n)}.$$

To see this we write

$$\begin{aligned}\mathcal{D}_\alpha(fg) &= \{(fg)(\alpha_j)-(fg)(\alpha)\}^N_{j=1} \\ &= f(\alpha)\{\Delta g(\alpha_j)\}^N_{j=1} + g(\alpha)\{\Delta f(\alpha_j)\}^N_{j=1} \\ &\quad + \{\Delta f(\alpha_j)\Delta g(\alpha_j)\}^N_{j=1}.\end{aligned}$$

We now use the Taylor expansion (16.56)

$$f(\alpha) = f(A\alpha) + \sum_{\ell=1}^m \frac{1}{\ell!}D^\ell_m f(A\alpha)\wedge\left\{\otimes^\ell(\alpha - A\alpha)\right\} + \mathbf{Q}^{(m,0)}_{A\alpha}\mathcal{D}_{A\alpha}f(\alpha).$$

We use this to replace the first occurrence of Δf and of Δg and obtain

$$\begin{aligned}\mathcal{D}_\alpha(fg) &= f(\alpha)\left\{\sum_{\ell=1}^m\frac{1}{\ell!}D^\ell_m g(\alpha)\wedge\left\{\otimes^\ell(\alpha_j-\alpha)\right\}\right\}^N_{j=1} \\ &\quad + g(\alpha)\left\{\sum_{\ell=1}^m\frac{1}{\ell!}D^\ell_m f(\alpha)\wedge\left\{\otimes^\ell(\alpha_j-\alpha)\right\}\right\}^N_{j=1} \\ &\quad + f(\alpha)\mathbf{Q}^{(m,0)}_\alpha\mathcal{D}_\alpha g + g(\alpha)\mathbf{Q}^{(m,0)}_\alpha\mathcal{D}_\alpha f \\ &\quad + \{\Delta f(\alpha_j)\Delta g(\alpha_j)\}^N_{j=1}.\end{aligned}$$

When we use (16.56) for the product $\Delta f\Delta g$ and expand that product, the resulting collection of terms which do not involve the operator $\mathbf{Q}^{(m,0)}_\alpha$ can be simplified. The simplification for the jth component is

$$\begin{aligned}
&\sum_{k=1}^{m}\sum_{\ell=1}^{m}\frac{1}{k!\ell!}\left(D_m^k f(\alpha)\wedge\left\{\otimes^k(\alpha_j-\alpha)\right\}\right)\left(D_m^\ell g(\alpha)\wedge\left\{\otimes^\ell(\alpha_j-\alpha)\right\}\right)\\
&\quad=\sum_{k=1}^{m}\sum_{\ell=1}^{m}\frac{1}{k!\ell!}\left(D_m^k f(\alpha)\otimes D_m^\ell g(\alpha)\right)\otimes^{k+\ell}(\alpha_j-\alpha)\\
&\quad=\sum_{r=1}^{m}\left[\sum_{\ell=1}^{r}\binom{r}{\ell}\left(D_m^{r-\ell}f(\alpha)\otimes D_m^\ell g(\alpha)\right)\right]\otimes^r(\alpha_j-\alpha).
\end{aligned}$$

Altogether we have a jth component

$$\begin{aligned}
\mathcal{D}_\alpha(fg)(\alpha_j)&=\sum_{r=0}^{m}\left[\sum_{\ell=0}^{r}\binom{r}{\ell}\left(D_m^{r-\ell}f(\alpha)\otimes D_m^\ell g(\alpha)\right)\right]\otimes^r(\alpha_j-\alpha)\\
&\quad+\left(\sum_{\ell=1}^{m}\frac{1}{\ell!}D_m^\ell g(\alpha)\wedge\left\{\otimes^\ell(\alpha_j-\alpha)\right\}\right)\left(\mathbf{Q}_\alpha^{(m,0)}\mathcal{D}_\alpha f(\alpha_j)\right)\\
&\quad+\left(\sum_{\ell=1}^{m}\frac{1}{\ell!}D_m^\ell f(\alpha)\wedge\left\{\otimes^\ell(\alpha_j-\alpha)\right\}\right)\left(\mathbf{Q}_\alpha^{(m,0)}\mathcal{D}_\alpha g(\alpha_j)\right)\\
&\quad+\left(\mathbf{Q}_\alpha^{(m,0)}\mathcal{D}_\alpha f(\alpha_j)\right)\left(\mathbf{Q}_\alpha^{(m,0)}\mathcal{D}_\alpha g(\alpha_j)\right).
\end{aligned}\tag{16.88}$$

Recall the notation $|\mathbf{A}|_\alpha=\sqrt{\langle\mathbf{A},\mathbf{A}\rangle_\alpha^{(t)}}$ introduced in (16.44). The first term on the right side of (16.88) lies in M_α,. Hence when we apply $\mathbf{Q}_\alpha$, which is the orthogonal projections onto $M_\alpha^\perp$, we can proceed with

$$\begin{aligned}
\left|\mathbf{Q}_\alpha^{(m,0)}\mathcal{D}_\alpha(fg)\right|&\leq\sum_{\ell=1}^{m}\left|D_m^\ell f(\alpha)\right|_\alpha\left|\mathbf{Q}_\alpha^{(m,0)}\mathcal{D}_\alpha g\right|\\
&\quad+\sum_{\ell=1}^{m}\left|D_m^\ell g(\alpha)\right|_\alpha\left|\mathbf{Q}_\alpha^{(m,0)}\mathcal{D}_\alpha f\right|\\
&\quad+\left|\mathbf{Q}_\alpha^{(m,0)}\mathcal{D}_\alpha f\right|\left|\mathbf{Q}_\alpha^{(m,0)}\mathcal{D}_\alpha g\right|.
\end{aligned}\tag{16.89}$$

As in the proof of the corresponding multiplier characterization on the ball, Theorem 14.14, we set

$$q_\ell=\frac{m}{m-\ell},\ q_\ell'=\frac{m}{\ell},\qquad 1\leq\ell\leq m-1,\tag{16.90}$$

and continue by estimating the ℓ^p norm of the first terms on the right side of (16.89) by

$$\begin{aligned}
\sum_{\alpha\in\mathcal{T}_n}\left|D_m^\ell f(\alpha)\right|_\alpha^p\left|\mathbf{Q}_\alpha^{(m,0)}\mathcal{D}_\alpha g\right|^p&\leq\left\{\sum_{\alpha\in\mathcal{T}_n}\left|D_m^\ell f(\alpha)\right|_\alpha^{pq_\ell'}\right\}^{1/q_\ell'}\left\{\sum_{\alpha\in\mathcal{T}_n}\left|\mathbf{Q}_\alpha^{(m,0)}\mathcal{D}_\alpha g\right|_\alpha^{pq_\ell}\right\}^{1/q_l}\\
&\leq\|f\|_{HB_{pq_\ell',\ell}(\mathcal{T}_n)}^p\|g\|_{HB_{p,m}(\mathcal{T}_n)}^p\\
&\leq\|f\|_{HB_{pq_\ell',m}(\mathcal{T}_n)}^p\|g\|_{HB_{p,m}(\mathcal{T}_n)}^p\\
&\leq\|f\|_{HB_{p,m}(\mathcal{T}_n)}^p\|g\|_{HB_{p,m}(\mathcal{T}_n)}^p,
\end{aligned}$$

where we have used first that

$$\left\{\sum_{\alpha\in\mathcal{T}_n}\left|D_m^\ell f(\alpha)\right|_\alpha^{pq_\ell'}\right\}^{1/p}\lesssim\|f\|_{HB_{pq_\ell',\ell}(\mathcal{T}_n)}$$

for $\ell>2n/p=2n/pq_\ell'$ and $m\geq\ell$ (the case $m=\ell$ is by definition, and the case $m>\ell$ follows easily), and then (16.81). The ℓ^p norms of the remaining terms in (16.89) are handled similarly and thus we have obtained

$$\sum_{\alpha\in\mathcal{T}_n}\left|\mathbf{Q}_\alpha^{(m,0)}\mathcal{D}_\alpha(fg)\right|^p\lesssim\|f\|_{HB_{p,m}(\mathcal{T}_n)}^p\|g\|_{HB_{p,m}(\mathcal{T}_n)}^p \tag{16.91}$$

$$\lesssim\|\omega\|_{Carleson}^p\|g\|_{HB_{p,m}(\mathcal{T}_n)}^p,$$

which is the first half of (16.87).

The other half of (16.87) requires that

$$\sum_{\alpha\in\mathcal{T}_n}\left|D_m^m(fg)(\alpha)\right|_\alpha^p\lesssim\|\omega\|_{Carleson}^p\|g\|_{HB_{p,m}(\mathcal{T}_n)}^p. \tag{16.92}$$

To prove this we write

$$D_m^m(fg)(\alpha)=\sum_{\ell=0}^m\binom{m}{l}\left(D_m^{m-\ell}f(\alpha)\otimes D_m^\ell g(\alpha)\right) \tag{16.93}$$

$$+\left\{D_m^m(fg)(\alpha)-\sum_{\ell=0}^m\binom{m}{l}\left(D_m^{m-\ell}f(\alpha)\otimes D_m^\ell g(\alpha)\right)\right\},$$

where the first term is what one expects from Leibniz's rule, and the second term is the error. With q_ℓ and q_ℓ' as in (16.90), we estimate the first term on the right side of (16.93) by

(16.94)

$$\sum_{\alpha\in\mathcal{T}_n}\left|D_m^{m-\ell}f(\alpha)\otimes D_m^\ell g(\alpha)\right|_\alpha^p\leq\sum_{\alpha\in\mathcal{T}_n}\left|D_m^{m-\ell}f(\alpha)\right|_\alpha^p\left|D_m^\ell g(\alpha)\right|_\alpha^p$$

$$\leq\left\{\sum_{\alpha\in\mathcal{T}_n}\left|D_m^{m-\ell}f(\alpha)\right|_\alpha^{pq_\ell}\right\}^{1/q_l}\left\{\sum_{\alpha\in\mathcal{T}_n}\left|D_m^\ell g(\alpha)\right|_\alpha^{pq_\ell'}\right\}^{1/q_l'}$$

$$\leq\|f\|_{HB_{pq_\ell,m-\ell}(\mathcal{T}_n)}^p\|g\|_{HB_{pq_\ell',\ell}(\mathcal{T}_n)}^p$$

$$\leq\|f\|_{HB_{p,m}(\mathcal{T}_n)}^p\|g\|_{HB_{p,m}(\mathcal{T}_n)}^p,$$

again since $m-\ell>\frac{2n(m-\ell)}{pm}=\frac{2n}{pq_\ell}$ and $\ell>\frac{2n\ell}{pm}=\frac{2n}{pq_\ell'}$. To estimate the second term on the right side of (16.93), we use (16.55) to obtain

$$\left|D_m^m(fg)(\alpha)-\sum_{\ell=0}^m\binom{m}{l}\left(D_m^{m-\ell}f(\alpha)\otimes D_m^\ell g(\alpha)\right)\right|_\alpha$$

$$\lesssim\left|L_\alpha^{(m,0)}\left\{D_m^r(fg)(\alpha)-\sum_{\ell=0}^r\binom{r}{l}\left(D_m^{r-\ell}f(\alpha)\otimes D_m^\ell g(\alpha)\right)\right\}_{r=1}^m\right|.$$

We now compute that

$$L_\alpha^{(m,0)}\left\{D_m^r(fg)(\alpha)\right\}_{r=1}^m=\mathbf{P}_\alpha^{(m,0)}\mathcal{D}_\alpha(fg)$$

and that the jth component of $L_{\alpha}^{(m,0)}\left\{\sum_{\ell=0}^{r}\binom{r}{l}\left(D_m^{r-\ell}f(\alpha)\otimes D_m^{\ell}g(\alpha)\right)\right\}_{r=1}^{m}$ is

$$\begin{aligned}
&\sum_{r-1}^{m}\sum_{\ell-0}^{r}\frac{1}{(r-\ell)!\ell!}\left(D_m^{r-\ell}f(\alpha)\otimes D_m^{\ell}g(\alpha)\right)\otimes^r(\alpha_j-\alpha)\\
&\quad=\sum_{r=1}^{m}\sum_{\ell=0}^{r}\frac{1}{(r-\ell)!\ell!}\left(D_m^{r-\ell}f(\alpha)\otimes^{r-\ell}(\alpha_j-\alpha)\right)\left(D_m^{\ell}g(\alpha)\otimes^{\ell}(\alpha_j-\alpha)\right)\\
&\quad=\left(\sum_{\ell=0}^{m}\frac{1}{\ell!}D_m^{\ell}f(\alpha)\wedge\left\{\otimes^{\ell}(\alpha_j-\alpha)\right\}\right)\left(\sum_{\ell=0}^{m}\frac{1}{\ell!}D_m^{\ell}g(\alpha)\wedge\left\{\otimes^{\ell}(\alpha_j-\alpha)\right\}\right)\\
&\qquad-f(\alpha)g(\alpha)\\
&\quad=\left(f(\alpha)+\mathbf{P}_{\alpha}^{(m,0)}\mathcal{D}_{\alpha}f(\alpha_j)\right)\left(g(\alpha)+\mathbf{P}_{\alpha}^{(m,0)}\mathcal{D}_{\alpha}g(\alpha_j)\right)-f(\alpha)g(\alpha).
\end{aligned}$$

Thus the jth component of

$$L_{\alpha}^{(m,0)}\left\{D_m^r(fg)(\alpha)-\sum_{\ell=0}^{r}\binom{r}{l}\left(D_m^{r-\ell}f(\alpha)\otimes D_m^{\ell}g(\alpha)\right)\right\}_{r=1}^{m}\tag{16.95}$$

is

$$\begin{aligned}
&\mathbf{P}_{\alpha}^{(m,0)}\mathcal{D}_{\alpha}(fg)(\alpha_j)-g(\alpha)\mathbf{P}_{\alpha}^{(m,0)}\mathcal{D}_{\alpha}f(\alpha_j)-f(\alpha)\mathbf{P}_{\alpha}^{(m,0)}\mathcal{D}_{\alpha}g(\alpha_j)\\
&\qquad-\mathbf{P}_{\alpha}^{(m,0)}\mathcal{D}_{\alpha}f(\alpha_j)\mathbf{P}_{\alpha}^{(m,0)}\mathcal{D}_{\alpha}g(\alpha_j)\\
&\quad=\Delta(fg)(\alpha_j)-\mathbf{Q}_{\alpha}^{(m,0)}\mathcal{D}_{\alpha}(fg)(\alpha_j)-g(\alpha)\left(\Delta f(\alpha_j)-\mathbf{Q}_{\alpha}^{(m,0)}\mathcal{D}_{\alpha}f(\alpha_j)\right)\\
&\qquad-f(\alpha)\left(\Delta g(\alpha_j)-\mathbf{Q}_{\alpha}^{(m0)}\mathcal{D}_{\alpha}g(\alpha_j)\right)\\
&\qquad-\left(\Delta f(\alpha_j)-\mathbf{Q}_{\alpha}^{(m,0)}\mathcal{D}_{\alpha}f(\alpha_j)\right)\left(\Delta g(\alpha_j)-\mathbf{Q}_{\alpha}^{(m,0)}\mathcal{D}_{\alpha}g(\alpha_j)\right),
\end{aligned}$$

and now, noting that all products not involving a projection $\mathbf{Q}_{\alpha}^{(m,0)}$ cancel, we obtain that the jth component of (16.95) is

$$\begin{aligned}
&-\mathbf{Q}_{\alpha}^{(m,0)}\mathcal{D}_{\alpha}(fg)(\alpha_j)+g(\alpha)\mathbf{Q}_{\alpha}^{(m,0)}\mathcal{D}_{\alpha}f(\alpha_j)+f(\alpha)\mathbf{Q}_{\alpha}^{(m,0)}\Delta g(\alpha_j)\\
&\qquad+\Delta f(\alpha_j)\mathbf{Q}_{\alpha}^{(m,0)}\mathcal{D}_{\alpha}g(\alpha_j)+\Delta g(\alpha_j)\mathbf{Q}_{\alpha}^{(m,0)}\mathcal{D}_{\alpha}f(\alpha_j)\\
&\qquad-\mathbf{Q}_{\alpha}^{(m,0)}\mathcal{D}_{\alpha}f(\alpha_j)\mathbf{Q}_{\alpha}^{(m,0)}\mathcal{D}_{\alpha}g(\alpha_j),
\end{aligned}$$

or simply

$$\begin{aligned}
&f(\alpha_j)_{\alpha}^{(m,0)}\mathcal{D}_{\alpha}g(\alpha_j)+g(\alpha_j)\mathbf{Q}_{\alpha}^{(m,0)}\mathcal{D}_{\alpha}f(\alpha_j)\\
&\qquad-\mathbf{Q}_{\alpha}^{(m,0)}\mathcal{D}_{\alpha}(fg)(\alpha_j)-\mathbf{Q}_{\alpha}^{(m,0)}\mathcal{D}_{\alpha}f(\alpha_j)\mathbf{Q}_{\alpha}^{(m,0)}\mathcal{D}_{\alpha}g(\alpha_j)\\
&\quad=I(\alpha_j)+II(\alpha_j)+III(\alpha_j)+IV(\alpha_j).
\end{aligned}$$

To handle the sum

$$\sum_{\alpha\in\mathcal{T}_n}|I(\alpha_j)|^p=\sum_{\alpha\in\mathcal{T}_n}|f(\alpha_j)|^p\left|\mathbf{Q}_{\alpha}^{(m,0)}\mathcal{D}_{\alpha}g(\alpha_j)\right|^p,$$

we simply use the boundedness of f together with the definition of $\|g\|_{HB_{p,m}(\mathcal{T}_n)}^p$. To handle the sum

$$\sum_{\alpha\in\mathcal{T}_n}|II(\alpha_j)|^p=\sum_{\alpha\in\mathcal{T}_n}|g(\alpha_j)|^p\left|\mathbf{Q}_{\alpha}^{(m,0)}\mathcal{D}_{\alpha}f(\alpha_j)\right|^p,$$

we use the inequality

$$\left|\mathbf{Q}_{\alpha}^{(m,0)}\mathcal{D}_{\alpha}f\left(\alpha_j\right)\right|^p \le \omega\left(\alpha\right)$$

together with our assumption that $\{\omega(\alpha)\}_{\alpha\in\mathcal{T}_n}$ is a $HB_{p,m}(\mathcal{T}_n)$-Carleson measure. The sum $\sum_{\alpha\in\mathcal{T}_n}|III(\alpha_j)|^p$ is controlled by (16.91), and the final sum is easy:

$$\begin{aligned}\sum_{\alpha\in\mathcal{T}_n}|IV(\alpha_j)|^p &\lesssim \left(\sum_{\alpha\in\mathcal{T}_n}\left|\mathbf{Q}_{\alpha}^{(m,0)}\mathcal{D}_{\alpha}f\right|^{2p}\right)^{1/2}\left(\sum_{\alpha\in\mathcal{T}_n}\left|\mathbf{Q}_{\alpha}^{(m,0)}\mathcal{D}_{\alpha}g\right|^{2p}\right)^{1/2}\\ &\lesssim \|f\|_{HB_{2p,m}(\mathcal{T}_n)}^p\|g\|_{HB_{2p,m}(\mathcal{T}_n)}^p\\ &\lesssim \|f\|_{HB_{p,m}(\mathcal{T}_n)}^p\|g\|_{HB_{p,m}(\mathcal{T}_n)}^p.\end{aligned}$$

This completes the proof of (16.87), which is the case $s_1 = 1$ in (16.86).

The remaining cases $s_k = \ell_1 + \cdots + \ell_k = m$ with $k > 1$ are handled similarly, using repeated application of Hölder's inequality on products of tensors as in (16.94), and using the structural inequality (16.58) to estimate the error in the Leibniz formula, as in (16.93). The details are routine but long, and are left to the interested reader. The normalizing terms

$$\sum_{s_k<m}\left|D_{m-s_{k-1}}^{\ell_k}\cdots D_{m-\ell_1}^{\ell_2}D_m^{\ell_1}(fg)(o)\right|^p$$

in (16.86) are handled easily, just as the corresponding terms

$$\sum_{k=0}^{m-1}\left|\nabla^k(\varphi f)(0)\right|$$

in the proof of Theorem 14.14 on the ball. This completes the proof of the sufficiency assertion in Proposition 16.41.

For the necessity, a standard argument using the boundedness of the adjoint of the multiplication operator shows that $\|f\|_\infty \lesssim \|f\|_{\mathrm{Mult}(HB_{p,m}(\mathcal{T}_n))}$. The arguments above can then be reversed to show that $\{\omega(\alpha)\}_{\alpha\in\mathcal{T}_n}$ is a $HB_{p,m}(\mathcal{T}_n)$-Carleson measure. This completes the proof of Proposition 16.41. □

Now we return to the proof of the multiplier restriction in Theorem 16.38. Let $F \in \mathrm{Mult}(B_p)$ and set

$$d\mu(z) = \sum_{\alpha\in\mathcal{T}_n}\chi_{K_\alpha}(z)\left\{\int_{K_\alpha^*}\left|D_{c_\alpha}^m F(\zeta)\right|^p d\lambda_n(\zeta)\right\}d\lambda_n(z).$$

Since $p > 2n/m$ by hypothesis, Lemma 16.40 shows that μ is a B_p-Carleson measure. Define the discretization of μ in the usual way by

$$\mu(\alpha) = \int_{K_\alpha^*}\left|D_{c_\alpha}^m F(\zeta)\right|^p d\lambda_n(\zeta).$$

Since $p < 2 + 1/(n-1)$, Theorem 14.16 shows that $\{\mu(\alpha)\}_{\alpha\in\mathcal{T}_n}$ satisfies the tree condition (16.66). Set

$$\omega(\alpha) = \sum_{s_k=m}\left|D_{m-s_{k-1}}^{\ell_k}\cdots D_m^{\ell_1}f(\alpha)\right|_\alpha^p + \sum_{s_k=m}\left|\mathbf{Q}_\alpha^{(m-s_{k-1},s_{k-1})}\mathcal{D}_\alpha D_{m-s_{k-2}}^{\ell_{k-1}}\cdots D_m^{\ell_1}f\right|_\alpha^p,$$

as in (16.80) in the proof of Lemma 16.41. It will follow from Lemma 16.41 that the restriction $f = TF$ lies in Mult$(HB_{p,m}(\mathcal{T}_n))$ if we can show that $\{\omega(\alpha)\}_{\alpha\in\mathcal{T}_n}$ is a $HB_{p,m}(\mathcal{T}_n)$-Carleson measure.

To this end, we invoke the following local version of (16.71):

$$\sum_{s_k=m}\left|D^{\ell_k}_{m-s_{k-1}}\cdots D^{\ell_1}_m\mathbf{A}(\alpha)\right|^p_\alpha+\sum_{s_k=m}\left|\mathbf{Q}^{(m-s_{k-1},t+s_{k-1})}_\alpha\mathcal{D}_\alpha D^{\ell_{k-1}}_{m-s_{k-2}}\cdots D^{\ell_1}_m\mathbf{A}\right|^p_\alpha$$
$$\lesssim\int_{K^{**}_\alpha}\left|\mathbf{A}^{(m)}(z)\right|^p_\alpha d\lambda_n(z),$$

for all holomorphic t-tensor-valued functions $\mathbf{A}$ on the Bergman tree. This can be proved by an induction similar to the proof of (16.71), and we omit the details. From this we obtain that

$$\begin{aligned}I^*\omega(\alpha)&=\sum_{\beta\in\mathcal{T}_n:\beta\geq\alpha}\omega(\beta)\\&\lesssim\sum_{\beta\in\mathcal{T}_n:\beta\geq\alpha}\int_{K^{**}_\beta}\left|D^m_{c_\beta}F(z)\right|^p d\lambda_n(z)\\&\approx I^*\mu(\alpha).\end{aligned}$$

It now follows that $\{(\mu+\omega)(\alpha)\}_{\alpha\in\mathcal{T}_n}$ satisfies the tree condition (16.66), hence is a $HB_{p,m}(\mathcal{T}_n)$-Carleson measure by Theorem 16.36. This finally yields that $\{\omega(\alpha)\}_{\alpha\in\mathcal{T}_n}$ is a $HB_{p,m}(\mathcal{T}_n)$-Carleson measure with norm bounded by that of the B_p-Carleson measure μ. At long last, this completes the proof of Theorem 16.38.

16.3.11. Embeddings and Isomorphisms. Here we make some observations regarding the restriction map, showing in particular that it is always an embedding, but never an isomorphism.

First, if a holomorphic function f in some $B_p(\mathbb{B}_n)$ vanishes on $\mathcal{T}_n$, then its admissible limits f^* on $\partial\mathbb{B}_n$ are zero almost everywhere with respect to surface measure. Because $B_p(\mathbb{B}_n)$ is contained in the Nevanlinna class this insures that f vanishes identically on the ball ([**288**]: Theorem 5.6.4). This shows that $\mathcal{T}_n$ is not a zero set for either B_p, or for Mult(B_p). Hence the restriction map T is one-to-one from B_p to $HB_{p,m}(\mathcal{T}_n)$, as well as from Mult(B_p) to Mult$(HB_{p,m}(\mathcal{T}_n))$.

Second, we observe that Lemma 16.42 shows that $\ell^p(\mathcal{T}_n)=HB_{p,0}(\mathcal{T}_n)$ embeds continuously into $HB_{p,m}(\mathcal{T}_n)$. As a consequence, the restriction map T *cannot* map B_p *onto* $HB_{p,m}(\mathcal{T}_n)$. Indeed, if T is onto $HB_{p,m}(\mathcal{T}_n)$, and $\alpha\in\mathcal{T}_n$, then there is $F\in B_p$ such that $F(\beta)=0$ for all $\beta\in\mathcal{T}_n\setminus\{\alpha\}$, and $F(\alpha)=1$. Thus $F(z)$ is not identically zero, and hence neither is $G(z)=F(z)(z_1-\alpha_1)$ where $\alpha=(\alpha_1,\ldots,\alpha_n)$. But $G\in B_p$ by Theorem 14.14, and this implies that $\mathcal{T}_n$ is a zero set for B_p, a contradiction. The same argument also shows that T *cannot* map Mult(B_p) *onto* Mult$(HB_{p,m}(\mathcal{T}_n))$.

16.3.12. The Modified Bergman Tree. In this section, we construct a Bergman tree $\mathcal{T}_n$ that satisfies the structural inequality (16.58) for all $0\leq m+t\leq M$, where M is chosen so large that $M>2n/p$ for all $1<p<\infty$, e.g. $M=2n$ will do. Our construction will also have the property that given a sequence $Z=\{z_j\}_{j=1}^\infty$ in the ball satisfying the separation condition in (15.5),

$$\beta(z_i,0)\lesssim\beta(z_i,z_j),\qquad 1\leq i\neq j<\infty,$$

we can arrange to have $Z \subset \mathcal{T}_n$. This property is crucial for the arguments in Chapter 15 on interpolating sequences. Moreover, we recall from Claim 15.7 above, that we may assume $Z = \{z_j\}_{j=1}^{\infty}$ is already a subset of a Bergman tree $\mathcal{T}_n = \mathcal{T}_n(\lambda, \theta)$ with structural constants $0 < \lambda < 1$ small and $1 < \theta < \infty$ large.

We now recall (16.58): for $0 \leq m + t \leq M$, we have

$$c_m \sum_{\ell=1}^{m} \left|\mathbf{v}^\ell\right|_\alpha \leq \left|L_\alpha^{(m,t)}\left(\mathbf{v}^1, \ldots, \mathbf{v}^m\right)\right| \leq C_m \sum_{\ell=1}^{m} \left|\mathbf{v}^\ell\right|_\alpha, \tag{16.96}$$

where

$$L_\alpha^{(m,t)}\mathbf{x} = \left\{\sum_{\ell=1}^{m} \frac{1}{\ell!}\mathbf{v}^\ell \wedge \left\{\otimes^\ell \left(\alpha^j - \alpha\right)\right\}\right\}_{j=1}^{N} \in \left(\mathcal{E}_\alpha^{(t)}\right)^N,$$

and $\mathbf{x} = \left(\mathbf{v}^1, \ldots, \mathbf{v}^m\right) \in \mathcal{E}_\alpha^{(t+1)} \times \mathcal{E}_\alpha^{(t+2)} \times \cdots \times \mathcal{E}_\alpha^{(t+m)}$. The main point here is that the constants c_m and C_m in (16.96) are independent of $\alpha \in \mathcal{T}_n$. It is not hard to see that, given a construction of the Bergman tree $\mathcal{T}_n = \{\alpha\}_{\alpha \in \mathcal{T}_n}$, we can perturb the centers $\gamma = c_\gamma$ slightly so that the expression $\left|L_\alpha^{(m,t)}\mathbf{x}\right|_\alpha$ vanishes only when $\mathbf{x}$ vanishes. However, we need a uniform version, and to see this we will use the equivalence of norms on a finite dimensional vector space together with certain affine maps T_α constructed below, that take a prototypical set of points not included in the zero set of any nontrivial polynomial of degree at most M, into a set of candidates for the centers of the children of K_α in a Bergman tree.

For convenience in notation, we will prove only the case $t = 0 \leq m \leq M$ as stated in (16.55) above,

$$c_m \sum_{\ell=1}^{m} \left|\mathbf{v}^\ell\right|_\alpha \leq \left|L_\alpha^{(m,0)}\left(\mathbf{v}^1, \ldots, \mathbf{v}^m\right)\right| \leq C_m \sum_{\ell=1}^{m} \left|\mathbf{v}^\ell\right|_\alpha, \tag{16.97}$$

where

$$L_\alpha^{(m,0)}\mathbf{x} = \left\{\sum_{\ell=1}^{m} \frac{1}{\ell!}\mathbf{v}^\ell \wedge \left\{\otimes^\ell \left(\alpha^j - \alpha\right)\right\}\right\}_{j=1}^{N}, \qquad \mathbf{x} = \left(\mathbf{v}^1, \ldots, \mathbf{v}^m\right).$$

It suffices to take $m = M$.

We begin with the construction of the affine maps T_α. Let Π denote the *real* $(2n-1)$-dimensional vector space perpendicular to $e_1 = (1, 0, \ldots, 0) \in \mathbb{B}_n$, i.e.

$$\Pi = \{z \in \mathbb{C}^n : \operatorname{Re} z_1 = 0\},$$

and we let D denote the Euclidean ball of radius $1/2$ centered at the origin in Π,

$$D = \left\{z \in \Pi ; |z| < \frac{1}{2}\right\}.$$

We also suppose that θ is large enough that $D \subset K_0$.We now construct the affine maps T_α to have the property that, if z_j^N lies on the sphere $\mathcal{S}_{N\theta}$ as in the construction of the Bergman tree $\mathcal{T}_n$ in Chapter 14.1.6, and if $\alpha = c_j^N \in \mathcal{S}_{\left(N+\frac{1}{2}\right)\theta}$ is the center of the associated cube $K_j^N = K_\alpha$, then the affine map T_α takes the $(2n-1)$-dimensional disk D to a disk $T_\alpha D$ that lies in the *tangent plane* Γ to the sphere $\mathcal{S}_{(N+\frac{3}{2})\theta}$ at the radial projection $\alpha^* \equiv P_{(N+\frac{3}{2})\theta}\alpha$ of α onto that sphere $\mathcal{S}_{(N+1)\theta}$. Moreover, we require that the radial projection $P_{(N+\frac{3}{2})\theta}(T_\alpha D)$ of the disk $T_\alpha D$ onto the sphere $\mathcal{S}_{(N+\frac{3}{2})\theta}$ is contained within the projection $P_{(N+\frac{3}{2})\theta}K_\alpha$ of the cube K_α onto the same sphere $\mathcal{S}_{(N+\frac{3}{2})\theta}$, and has approximately the same shape and size. Of course this latter projection $P_{(N+\frac{3}{2})\theta}(T_\alpha D)$ contains the centers c_i^{N+1} of the children of K_α in the Bergman tree. In order to achieve these properties, we note that the ball automorphism φ_α satisfies $B_\beta(\alpha,\lambda) = \varphi_\alpha(B_\beta(0,\lambda))$, and that the affine map $\psi_\alpha(z) = \alpha + \varphi'_\alpha(0)z$ takes $B_\beta(0,\lambda)$ to $B_\beta(\alpha,\lambda)$ with little distortion, and furthermore that distortion goes to zero as $\lambda \searrow 0$. Now we translate the map ψ_α by $P_{(N+\frac{3}{2})\theta}\alpha - \alpha$ where $P_{(N+\frac{3}{2})\theta}\alpha$ is the radial projection of α to the sphere$\mathcal{S}_{(N+\frac{3}{2})\theta}$, and we note that the image of the $(2n-1)$-D under this translation composed with ψ_α, which we call T_α, lies in the tangent plane Γ. Moreover,

$$P_{(N+\frac{3}{2})\theta}T_\alpha(D) \subset P_{(N+\frac{3}{2})\theta}K_\alpha\ ,$$

and $T_\alpha(D)$ has roughly the same size and shape as $P_{(N+\frac{3}{2})\theta}K_\alpha$. This completes our construction of the affine maps T_α.

We now turn to the construction of a prototypical set of points not included in the zero set of any nontrivial polynomial of degree at most M with controlled coefficients. Let $\mathcal{E} = \{z_j\}_{j=1}^J$ be a maximal ρ-separated subset of D in the Bergman metric, i.e.

$$\begin{aligned} \beta(z_i,z_j) &\geq \rho, &&\text{for } i \neq j,\\ \beta(z,\mathcal{E}) &< \rho, &&\text{for } z \in D, \end{aligned}$$

where $0 < \rho < \lambda$ will be chosen later. Note that we can arrange to have J as large as we wish by taking ρ sufficiently small. We now claim that for ρ sufficiently small, the finite set $\mathcal{E} = \{z_j\}_{j=1}^J$ is *not* contained in the zero set of any nontrivial polynomial F on $\mathbb{C}^n$ of degree at most M. To see this, we argue by contradiction. Suppose that for every $\delta > 0$, there is a finite set F_δ of points that are δ-dense in the set D, and a nontrivial polynomial P_δ of degree at most M such that P_δ vanishes on the set F_δ. We can normalize P_δ so that $\int_{\mathbb{B}_n}|P_\delta| = 1$ and then the fact that the P_δ are polynomials of degree at most M shows that they form a normal family of holomorphic functions on the ball. Thus we can extract a sequence of such polynomials P_δ that converge uniformly in the ball $\mathbb{B}_n$ to a holomorphic function P such that both $\int_{\mathbb{B}_n}|P| = 1$ and P vanishes on the set D, clearly a contradiction (for $z' = (z_2,...,z_n)$ with $|z'| < \frac{1}{2}$ the holomorphic function of one variable $P^{z'}(z_1) = P(z_1,z')$ vanishes on the intersection of D with a small disk centered at the origin, hence vanishes identically).

Now define $\mathcal{V}^{(M)}(\mathbb{C}^n)$ to be the vector space of polynomials on $\mathbb{C}^n$ of degree at most M that vanish at the origin. Then for all $\zeta \in \mathbb{C}^n$ the expression

$$\|F\|_\zeta = \left(\sum_{j=1}^J \left|F\left(z'_j - \zeta\right)\right|^2\right)^{1/2}$$

is a norm on $\mathcal{V}^{(M)}(\mathbb{C}^n)$. Another norm on $\mathcal{V}^{(M)}(\mathbb{C}^n)$ is given by the 'coefficient norm' $\|F\| \equiv |\mathbf{x}|_\alpha$, where $\mathbf{x} = (\mathbf{v}^1, \ldots, \mathbf{v}^M)$ is the unique element such that $F(z) = \sum_{\ell=1}^M \frac{1}{\ell!} \mathbf{v}^\ell \wedge \{\otimes^\ell z\}$. Since the vector space $\mathcal{V}^{(M)}(\mathbb{C}^n)$ is finite dimensional, these norms are all equivalent, and uniformly so for ζ in any compact subset K of $\mathbb{C}^n$. Such a K will be fixed below.

To begin our construction of a modified Bergman tree, we will assume that a perturbation has already been performed on the first $(N = 1)$ generation of centers c_j^1 in the construction of the Bergman tree $\mathcal{T}_n$, as given in Chapter 15 above, so that $\{c_j^1\}_j$ is not contained in the zero set of any nontrivial complex polynomial of degree at most M. This will require that the structural constant theta be chosen sufficiently large that the number of children, which is roughly $e^{2n\theta}$, is greater than a large multiple of J where J is the number of points in the prototypical set constructed above, and that the centers of the children are chosen so as to include an affine image of the prototypical set.

We now transport these points and norms to K_γ, where γ is a child of α, by induction on N where $\alpha = c_j^N$ is in the Nth generation of the construction in Chapter 15. So let $N \geq 1$ and $\alpha = c_j^N$ so that $d(\alpha) = N$. We will perturb certain of the children of α as follows. We pick a unitary map U_α that takes e_1 to $\frac{\alpha}{|\alpha|}$ and then use the map $T_\alpha U_\alpha$, where T_α is the affine map constructed above, to transport points and norms to $\cup_{\gamma \in \mathcal{C}(\alpha)} K_\gamma$. The points $\mathcal{E}' = \{z_j'\}_{j=1}^J$ are taken by the affine map $T_\alpha U_\alpha$ to a set of points

$$\mathcal{E}_\alpha' = \{\alpha^{j\prime}\}_{j=1}^J = \{T_\alpha U_\alpha z_j'\}_{j=1}^J .$$

Now project the set of points $\mathcal{E}_\alpha'$ onto the sphere $\mathcal{S}_{(N+1)\theta}$ to obtain the set $P_{(N+1)\theta}\mathcal{E}_\alpha' = \{P_{(N+1)\theta}\alpha^{j\prime}\}_{j=1}^J$. We note that, by slightly perturbing the points, we may arrange matters so that the sets of points $P_{(N+1)\theta}\mathcal{E}_\alpha'$ and $P_{(N+1)\theta}\mathcal{E}_\gamma'$ are well separated from each other for $\alpha \neq \gamma$, $N = d(\alpha) = d(\gamma)$ provided θ is chosen large enough. We now redefine the points $\{z_j^{N+1}\}_{j=1}^J$ and unit cubes $\{Q_j^{N+1}\}_{j=1}^J$ in $\mathcal{S}_{(N+1)\theta}$ satisfying (14.12) in Section 14.1.6. We start with the points $E_{N+1} = \cup_{\alpha \in \mathcal{T}_n : d(\alpha) = N} P_{(N+1)\theta}\mathcal{E}_\alpha$. They are $(1 - \sqrt{r\varepsilon})\,\rho$-separated where $r = e^{2\theta}$ as in (16.22). Now extend the collection E_{N+1} to E_{N+1}^* by adding those original points z_j^{N+1} that are at distance at least λ from the set E_{N+1}. The resulting collection $E_{N+1}^* = \{x_i'\}$ satisfies

$$\begin{aligned} \beta(x_i', x_j') &\geq (1 - \sqrt{r\varepsilon})\,\rho, \quad && i \neq j, \\ \beta(x, E_N^*) &< (1 + \sqrt{r\varepsilon})\,\lambda, \quad && x \in \mathcal{S}_{(d+1)\theta}, \end{aligned}$$

Now we proceed with the construction of the cubes Q_j^{N+1} as in Lemma 14.1, and then construct the new cubes K_j^{N+1} with new centers given by $\mathcal{E}_\alpha'$ instead of by the construction in Section 14.1.6 that placed them on the sphere $\mathcal{S}_{(d+\frac{3}{2})\theta}$- the new centers are now lying in the image $T_\alpha D$ of D under the affine map T_α, and where $T_\alpha D$ is tangent to $\mathcal{S}_{(d+\frac{3}{2})\theta}$ at the point α^*, defined as the radial projection of α onto $\mathcal{S}_{(d+\frac{3}{2})\theta}$. Finally, we repeat this construction inductively for $N \geq 1$ to obtain a perturbed Bergman tree.

We have

$$T_\alpha U_\alpha z = \alpha^* + \varphi'_\alpha(0) U_\alpha z = \alpha^* + S_\alpha z$$

where $S_\alpha = \varphi'_{\alpha^*}(0) U_\alpha$ is linear and invertible, and so for $F(z) = \sum_{\ell=1}^M \frac{1}{\ell!} \mathbf{v}^\ell \wedge \{\otimes^\ell z\}$ and $\mathbf{x} = (\mathbf{v}^1, \ldots, \mathbf{v}^M)$,

$$\left|L_\alpha^{(M,0)}\mathbf{x}\right|^2 = \left|\left\{\sum_{\ell=1}^M \frac{1}{\ell!}\mathbf{v}^\ell \wedge \{\otimes^\ell (\alpha^j - \alpha)\}\right\}_{j=1}^N\right|^2 \tag{16.98}$$
$$\geq \sum_{j=1}^J \left|\sum_{\ell=1}^M \frac{1}{\ell!}\mathbf{v}^\ell \wedge \{\otimes^\ell (\alpha^{j'} - \alpha)\}\right|^2$$
$$= \sum_{j=1}^J \left|\sum_{\ell=1}^M \frac{1}{\ell!}\mathbf{v}^\ell \wedge \{\otimes^\ell S_\alpha S_\alpha^{-1} (\alpha^{j'} - \alpha)\}\right|^2$$
$$= \sum_{j=1}^J \left|\sum_{\ell=1}^M \frac{1}{\ell!}S_\alpha\mathbf{v}^\ell \wedge \{\otimes^\ell [z'_j - S_\alpha^{-1}(\alpha - \alpha^*)]\}\right|^2$$
$$= \|F\|_\zeta^2 \approx \sum_{\ell=1}^M \left|S_\alpha \mathbf{v}^\ell\right|^2 \approx \sum_{\ell=1}^M \left|\mathbf{v}^\ell\right|_\alpha^2,$$

since the perturbed children $\{\alpha^{j'}\}_{j=1}^J$ are a subset of $\mathcal{C}(\alpha)$, and since $\zeta = S_\alpha^{-1}(\alpha - \alpha^*)$ lies uniformly in a sufficiently large compact set K, independent of $\alpha \in \mathcal{T}_n$. Thus we have proved the left-hand inequality of (16.97) for the new centers. The right hand inequality is trivial (and not used anyway).

Finally, we can adapt this construction so that $\mathcal{T}_n$ contains a given sequence $\mathcal{Z} = \{z_j\}_{j=1}^\infty$ in the ball satisfying the separation condition

$$\beta(z_i, 0) \lesssim \beta(z_i, z_j), \quad 1 \leq i \neq j < \infty$$

apart from perhaps a bounded number of points z_j near the origin, which won't affect our application of the holomorphic Bergman trees to completing the multiplier interpolation loop in Section 15.6. Recall that we have already used Lemma 15.7 to assume that $\mathcal{Z} \subset \mathcal{T}_n = \mathcal{T}_n(\lambda, \theta)$ with λ small and θ large. To see the adaptation, let $\mathcal{U}_\alpha = \cup_{j=1}^J K_{\alpha^{j'}}$ be the union of the cubes $K_{\alpha^{j'}}$ corresponding to a set of perturbed points $\mathcal{E}_\alpha$ constructed above. Then the separation condition implies the points z_i are so well separated that no more than one of them occurs in any $\mathcal{U}_\alpha$, and in fact those $\mathcal{U}_\alpha$ that contain a point z_i are themselves pairwise disjoint and well separated. Thus it suffices to choose the model set $\mathcal{E} = \{z'_j\}_{j=1}^J$ to have the somewhat stronger property that, even after the removal of a bounded number of points z'_i (where the bound is independent of J), the resulting set $\mathcal{E}'$ is still not contained in the zero set of any nontrivial complex polynomial of degree at most M. Then if a point z_i from Z lies in the cube K_α, we simply replace α by z_i, and if necessary, modify at most C of the neighboring points so as not to lie too close to z_i.

DEFINITION 16.43. *We define $HB_p(\mathcal{T}_n)$ to be any of the spaces $HB_{p,m}(\mathcal{T}_n)$ with $m > 2n/p$. Lemma* 16.42 *shows that these spaces are identical, and the above* *construction of the Bergman tree* $\mathcal{T}_n$ *shows that we can use the same tree* $\mathcal{T}_n$ *for all Besov spaces* $HB_{p,m}(\mathcal{T}_n)$ *with* $p > 1$ *if we choose* $M = 2n$.

16.4. Notes and Comments

The material in this chapter on trees is from [**47**] which builds on the ideas from [**48**].

CHAPTER 17

Corona Theorems for Besov Spaces in $\mathbb{C}^n$

In this chapter we prove corona type theorems for the Besov spaces $B_p^\sigma(\mathbb{B}_n)$, $1 < p < \infty$, $\sigma \geq 0$.

Those theorems are statements about the solvability of equations in function space. For functions of one variable such theorems are often proved by explicitly constructing solution functions, generally by solving $\overline{\partial}-$equations with function space estimates. As we follow that path here new difficulties arise. The Koszul complex for algebraically reducing corona problems to solving systems of $\overline{\partial}$-equations is more intricate. We deal with that by building on the ideas of Andersson and Carlsson [**29**]. Also, the solution operators of Charpentier [**106**] which we use for those $\overline{\partial}$-equations have homogeneities different from those commonly encountered and techniques must be introduced to obtain the required estimates. Our approach to that is facilitated by using the almost invariant holomorphic derivative of Chapter 14.

Some of the Hilbert spaces we consider have the CPP. For those we can access the Toeplitz corona theorem and hence also obtain a classical corona theorem for their multiplier algebras, $\operatorname{Mult}(B_2^\sigma(\mathbb{B}_n))$, $0 \leq \sigma \leq 1/2$, $n = 1, 2, \ldots$. Special cases of that result include Carleson's celebrated corona theorem [**89**] for H^∞ as well as multiplier corona theorems for the Dirichlet spaces and the Drury-Arveson spaces.

The material in this chapter is less self contained than that of previous chapters.

17.1. Single Variable Corona Questions

We begin with an informal discussion of corona theorems and some of the ideas used in proving them. Additional background is in [**132**], [**194**], [**161**], [**193**], [**245**], and [**294**, Chapter 3]; more recent work and further references are in [**130**].

It is natural to ask how aspects of the theory of the algebra of polynomials extend to algebras of analytic functions. For instance, a set of polynomials $\{p_i\}_{i=1}^N$ generates the full polynomial algebra, equivalently $1 \in [p_i]$, the ideal generated by $\{p_i\}$, if and only if the polynomials have no common zeros. The analogous question makes sense for the algebra of entire functions and was answered by Helmer in 1940 [**176**].

THEOREM 17.1. *Given entire functions* $\{g_i\}_{i=1}^N$, *a necessary and sufficient condition that there are entire functions* $\{f_i\}_{i=1}^N$ *such that*

$$\sum_{j=1}^{N} g_j(z) f_j(z) = 1, \quad z \in \mathbb{C}, \tag{17.1}$$

is that

$$\sum_{i=1}^{N} |g_i(z)|^2 > 0, \quad z \in \mathbb{C}. \tag{17.2}$$

Of course the condition (17.2) is just the condition that the $\{g_i\}$ do not vanish simultaneously. This formulation emphasizes the analogy with the condition (17.4) below; a condition which insures, in a uniform way, that the $\{g_i\}$ are not simultaneously small. The quadratic formulation will be a convenient one when we pass to the case $N = \infty$.

17.1.1. The Corona Theorem for H^∞. It is natural to ask if an analog of the previous theorem holds for $H^\infty = H^\infty(\mathbb{D})$, the algebra of bounded analytic function on the disk. Beyond its intrinsic interest, that question came to prominence because it is equivalent to an interesting question about the Banach algebra structure of H^∞.

The space H^∞ is a commutative Banach algebra. Its spectrum, or maximal ideal space, $\mathfrak{M}$, is a compact Hausdorff space consisting of generalized point evaluations. The maximal ideals are the kernels of those functionals. In particular, associated to each z in $\mathbb{D}$ is the evaluation functional, $f \to f(z)$, and its associated kernel. This association produces a copy $\mathfrak{D}$ of $\mathbb{D}$ inside $\mathfrak{M}$. We know $\mathbb{D}$ is not compact, so $\mathfrak{D}$ cannot be all of $\mathfrak{M}$. Other points of $\mathfrak{M}$ can be obtained as generalized limits of the point evaluations from $\mathfrak{D}$, and the question arose if this was the entire story. Is $\mathfrak{D}$ dense in $\mathfrak{M}$?

The solar corona is a ball of hot gas that is a part of the sun. It extends far beyond the visible disk of the sun and is ordinarily hidden from sight by the brightness of the main disk. However it is visible when the main disk is blocked from view during a solar eclipse. By analogy, the "corona" of $\mathfrak{M}$ is the name given to the part of $\mathfrak{M}$ far beyond the discernible disk $\mathfrak{D}$ and not easily visible. The corona is the complement of the closure of $\mathfrak{D}$; $\mathfrak{M} \setminus \overline{\mathfrak{D}}$. The corona conjecture was that there is a corona, that $\mathfrak{M} \setminus \overline{\mathfrak{D}}$ is not empty.

It is an exercise in Banach algebra theory to show that giving a negative answer to the corona conjecture is equivalent to proving the following analog of Theorem 17.1 [**294**, Chapter 3].

THEOREM 17.2 (The Corona Theorem, Carleson [**89**]). *Given* $\{g_i\}_{i=1}^N \subset H^\infty$, *a necessary and sufficient condition that there are* $\{f_i\}_{i=1}^N \subset H^\infty$ *such that*

$$\sum_{j=1}^{N} g_j(z) f_j(z) = 1, \quad z \in \mathbb{D}, \tag{17.3}$$

is that there exist $\delta > 0$ *such that*

$$\inf_{z\in\mathbb{D}} \sum_{i=1}^{N} |g_i(z)|^2 \geq \delta. \tag{17.4}$$

This states that there is no corona. There are no hidden evaluation functionals and the only obstruction to the closed ideal $[g_i]_{i=1}^N$ being the entire algebra is the obvious one, the one precluded by (17.4).

EXERCISE 17.3. *Show that the condition* (17.4) *is necessary for* (17.3).

17.1.2. A Proof Scheme for Corona Theorems. Suppose we are given $\{g_i\}$ which satisfies (17.4) and want to find $\{f_i\}$ so that (17.3) is satisfied. One path would be to first find $\{F_i\}$ which satisfy all the required properties except holomorphy and then adjust or modify these functions as needed.

Here is an outline of that approach for $N = 2$ functions. Write $|g|^2 = |g_1|^2 + |g_2|^2$. For $i = 1, 2$ set

$$F_i = \frac{\overline{g}_i}{|g|^2}.$$

We then have $g_1F_1 + g_2F_2 = 1$. Also, the F_i are bounded and satisfy estimates of the form

$$|\nabla F_i(z)| \lesssim (1 - |z|^2)^{-1}$$

which is the type of smoothness possessed by bounded holomorphic functions.

EXERCISE 17.4. *Check the estimates for* F_i.

Our goal now is to modify F_1 and F_2 into functions which still satisfy the equation, are bounded, and are also holomorphic. Recall that a function h is holomorphic exactly if $\overline{\partial} h = 0$. Hence, we can finish the proof by modifying F_1 and F_2 as follows. Find $b_{1,2}$ and $b_{2,1}$ which are bounded and satisfy

$$\overline{\partial} b_{ij} = F_i \overline{\partial} F_j \tag{17.5}$$

and then set

$$f_1 = g_1 + (b_{12} - b_{21})g_2$$
$$f_2 = g_2 + (b_{21} - b_{12})g_1.$$

It is straightforward to check that if we can find the required b_{ij} then this would give an acceptable choice of F_i and the proof would be complete.

EXERCISE 17.5. *Check the claim in the previous sentence.*

To use this proof scheme for $N > 2$ we need to identify an algebraic structure, generalizing the equations for b_{ij} and F_i, so that the scheme would produce a formal solution. For functions of one variable this algebraic step is only slightly complicated. When we work with functions of several variables, the algebraic problems are more substantial and we will rely on a sophisticated algebraic organizing tool, the Koszul complex. We will discuss those algebraic details later in this chapter.

In any case, we also need to know that we could find b's which not only solve equations such as (17.5) but which are also in an appropriate function space. That will ensure that the functions F_i we construct are multipliers and thus our formal solution will solve the problem. Often, as in the example just presented, the equation of interest is in the form

$$\overline{\partial} b = \mu \tag{17.6}$$

with b the unknown function and with data, μ, which is a type of Carleson measure. For functions of one variable it is not hard to give a solution to the equation. The function

$$F(z) = \frac{1}{2\pi i} \int_{\mathbb{D}} \frac{d\mu(\zeta)}{\zeta - z} dA(\zeta) \tag{17.7}$$

satisfies $\overline{\partial} F = \mu$ in the sense of distributions.

EXERCISE 17.6. *Prove the claim about* $\overline{\partial} F$.

However we will want solutions in particular function spaces such as a multiplier algebra, and that requires function space estimates for our solution.

There has been a great deal of work on solving equations such as (17.6) with estimates. One powerful tool for doing this is a non-linear solution operator introduced by Jones [**186**] and developed further in [**243**] and [**343**].

THEOREM 17.7. *Let* μ *be a complex measure on* $\mathbb{D}$ *and suppose* $|\mu|$ *is a* H^2*-Carleson measure. Set* $\sigma = \mu / \|\mu\|_{CM(H^2)}$ *and define the integration kernel* K *by*

$$K(\sigma, z, \zeta) := \frac{2i}{\pi} \frac{1 - |\zeta|^2}{(z - \zeta)(1 - \overline{\zeta} z)} \exp\left\{ \int_{|\omega| \geq |\zeta|} \left(-\frac{1 + \overline{\omega} z}{1 - \overline{\omega} z} + \frac{1 + \overline{\omega} \zeta}{1 - \overline{\omega} \zeta} \right) d\sigma(\omega) \right\}.$$

For S *defined by*

$$S(\mu)(z) = \int_{\mathbb{D}} K(\sigma, z, \zeta) \, d\mu(\zeta) \tag{17.8}$$

we have

$$S(\mu) \in L^1_{loc}(\mathbb{D}),$$
$$\overline{\partial} S(\mu) = \mu \text{ in the sense of distributions,}$$
$$\|S(\mu)\|_{L^\infty(\mathbb{T})} \lesssim \|\mu\|_{CM(H^2)}.$$

Jones' theorem can be used to complete the proof outlined above and prove a corona theorem. There are also extensions and variations which can be used with the Dirichlet space [**343**] and with Besov spaces [**40**], and those versions can also be used to prove corona theorems.

We want to at least mention a path we did not take earlier. Methods for solving $\overline{\partial}$-equations with function space estimates, for example the previous theorem and its variations, can be used to study interpolating sequences, [**186**], [**343**], [**294**, Chapter 4], [**40**]. We did not do that in Chapter 7 because we were focusing on Hilbert space methods. In the previous two chapters, when we studied interpolating sequences for functions on the ball we also did not use results about solving $\overline{\partial}$-equations. However there we made fundamental use of an operator Γ,and we noted in the remark at the end of Section 15.4 that using Γ was closely related to an approach based on solving $\overline{\partial}$-equations.

17.1.3. The Baby Corona Theorem for H^2.

The corona theorem asks for existence of functions in H^∞. Experience has shown that it is often more difficult to find a bounded holomorphic function satisfying certain algebraic conditions than to find a function in some other function space which satisfies the same conditions. Hence, it is at least plausible that the following *baby corona theorem* could be easier to establish.

THEOREM 17.8 (Baby Corona Theorem). *Given* $\{g_i\}_{i=1}^N \subset H^\infty$, *a necessary and sufficient condition that, for each* $h \in H^2$ *there are* $\{f_i\}_{i=1}^N \subset H^2$ *such that*

$$\sum_{j=1}^{N} g_j(z) f_j(z) - h(z), \quad z \in \mathbb{D}, \tag{17.9}$$

is that there exist $\delta > 0$ *such that*

$$\inf_{z\in\mathbb{D}} \sum_{i=1}^{N} |g_i(z)|^2 > \delta. \tag{17.10}$$

EXERCISE 17.9. *Show that if Theorem* 17.2 *is known then the hard implication in the previous theorem,* (17.10) *implies* (17.9), *holds.*

17.1.4. The Toeplitz Corona Theorem. We will call Theorem 17.2 and its analogs *multiplier corona theorems;* that is, if X is a Banach space of holomorphic functions on a domain Ω in $\mathbb{C}^n$ and Mult (X) is its multiplier algebra, then we will say the multiplier corona theorem holds for X if Theorem 17.2, as modified for X, is true. The modifications are the obvious ones; H^∞ is replaced by Mult (X), $\mathbb{D}$ is replaced by Ω. The Banach algebra argument which works for H^∞ also works in this general case to show that such a theorem is equivalent to the absence of a corona in the maximal ideal space of the Banach algebra Mult(X).

A baby corona theorem is a consequence of, and often easier to prove than, a multiplier corona theorem. Furthermore in some cases there is also a reverse implication; a baby corona theorem implies a multiplier corona theorem. We mentioned that for a RKHS with a CPP there is a very close interrelation between the function theory of the Hilbert space and the function theory of the associated multiplier algebra. The following *Toeplitz corona theorem* which relates the baby corona theorem to the multiplier corona theorem is a striking example of this. The theorem was proved for H^2 and H^∞ in [**56**] and [**298**]. The general version is due to Ball, Trent and Vinnikov [**60**] (see also Ambrozie and Timotin [**23**] and [**4**, Theorem 8.57]).

THEOREM 17.10 (Toeplitz Corona Theorem). *Suppose* Ω *is an open set in* $\mathbb{C}^n$ *and* $H \subset \mathrm{Hol}(\Omega)$ *is a RKHS with a CPK. The multiplier corona theorem holds for* X *if and only if the baby corona theorem holds for* X.

See also Theorem 17.12 where a more quantitative version is given. In this chapter we will prove the baby corona theorem for a family of Besov spaces on the complex ball. For the spaces in that family that are RKHS with a CPK we are then able to invoke this Toeplitz corona theorem (actually, a slightly more elaborate version stated in a moment) and obtain a multiplier corona theorem.

17.2. Corona Theorems on the Ball

We will obtain a baby corona theorem for the Besov spaces $B_p^\sigma(\mathbb{B}_n;\mathbb{C}^k)$, $0 \leq \sigma < \infty$, $1 < p < \infty$, of $\mathbb{C}^k$-valued functions defined on the $n-$ball $\mathbb{B}_n$. For $p = 2$, $0 \leq \sigma \leq 1/2$ these are Hilbert spaces with the complete Pick property. In those cases we have access to the Toeplitz corona theorem and hence in those cases our baby corona theorem implies a classical multiplier corona theorem. In fact we obtain results for $N-$tuples of multipliers between two such spaces. The family of spaces is less well understood when $\sigma > 1/2$; however see [**318**].

Although we will work in a rather general context the reader may want to focus on special cases. The parameter value $\sigma = 0$ corresponds to the classical Besov spaces, the ones we considered in previous chapters. Most of the main ideas of the proofs are already present in that special case, but some of the details about boundedness of operators are more delicate at the endpoint $\sigma = 1/2$. Most of the core ideas are of the proof are also present in the special cases of Hilbert spaces of scalar-valued functions, $p = 2$, $k = 1$; and for multipliers of a space into itself.

When restricted to functions of one variable, $n = 1$, the parts of the proof involving the Koszul complex can be simplified a great deal and so can the computation of the kernels for the solution operators. However the boundedness estimates for the solution operators still carry some of the basic ideas of the general proof. For $\sigma = 1/2$ and $n = 1$ we obtain an alternative proof of Carleson's corona theorem, however the estimates it provides are not as sharp as those of the standard proof.

The theorem involves $N-$tuples of functions. If $N = 1$ the result is trivial. If $N = 2$ there is a substantial simplification in the algebraic analysis associated with the Koszul complex; that is discussed briefly below. The case $N = 3$ captures the ideas of the general case.

17.2.1. Definitions and Theorems. We define the spaces $B_p^\sigma(\mathbb{B}_n), 1 < p < \infty$, $0 \leq \sigma < \infty$, to be the spaces of holomorphic functions on $\mathbb{B}_n$ for which

(17.11)
$$\|f\|_{B_p^\sigma} := \sum_{k=0}^{m-1} \left|\nabla^k f(0)\right| + \left(\int_{\mathbb{B}_n} \left|(1-|z|^2)^{m+\sigma}\nabla^m f(z)\right|^p d\lambda_n(z)\right)^{1/p} < \infty,$$

for some $m > n/p - \sigma$. The spaces with $\sigma = 0$ are the Besov spaces of previous chapters and many of the results for these spaces follow the same pattern as the results for those spaces and for the same reasons. For instance, as before, the right side is finite for some $m > n/p - \sigma$ (i.e. if m is large enough to ensure that the integral is finite when f is a polynomial) if and only if it is finite for all $m > n/p - \sigma$ in which case the associated norming functionals are equivalent; see Proposition 14.4 (see also [**63**], [**345**]).

The spaces $B_2^\sigma(\mathbb{B}_n)$ are all RKHS and many carry additional names. $B_2^{1/2}(\mathbb{B}_1)$ is the one variable Hardy space, H^2, a space of functions in the disk that can also be studied as a space of boundary value functions on the circle. The space $B_2^{n/2}(\mathbb{B}_n)$ is the $n-$dimensional Hardy space, commonly denoted $H^2(\partial\mathbb{B}_n)$. By passing to boundary values $H^2(\partial\mathbb{B}_n)$ can also be viewed as a closed subspace of $L^2(\partial\mathbb{B}_n)$. The space $B_2^1(\mathbb{B}_1)$ is the Bergman space, A^2, space of holomorphic functions that are square integrable (with respect to area measure) on the disk. The spaces $B_2^{(n+1)/2}(\mathbb{B}_n)$ are the higher dimensional Bergman spaces, sometimes denoted $A^2(\mathbb{B}_n)$, sometimes $L_a^2(\mathbb{B}_n)$. They consist of holomorphic functions on the ball that are square integrable with respect to volume measure. The spaces $B_2^{1/2}(\mathbb{B}_n)$ are RKHS with the kernel functions $k_z(w) = (1 - \overline{z}w)^{-1}$. If $n = 1$ this is the classical Hardy space. The space $B_2^{1/2}(\mathbb{B}_n)$, $n > 1$ is not the n-dimensional Hardy space; it is often called the Drury-Arveson space and sometimes denoted DA_n, sometimes H_n^2. The space $B_2^0(\mathbb{B}_1)$ is the classical Dirichlet space, $\mathcal{D}$. The spaces $B_2^0(\mathbb{B}_n)$ are sometimes called higher dimensional Dirichlet spaces.

We will denote by $B_p^\sigma\left(\mathbb{B}_n; \ell^2\right)$ the corresponding spaces of Hilbert space valued functions, equivalently the space of sequences of scalar-valued functions. Taking into account that we use $|\cdot|$ for both the modulus of a complex number and the norm of an element in ℓ^2 we find that the formula for the norm $B_p^\sigma\left(\mathbb{B}_n;\ell^2\right)$ has the exact same appearance as the formula given above for the norm in $B_p^\sigma(\mathbb{B}_n)$. The space $B_p^\sigma\left(\mathbb{B}_n;\ell^2\right)$ consists of all $f=(f_i)_{i=1}^\infty\subset \mathrm{Hol}(\mathbb{B}_n)$ such that

(17.12)

$$\|f\|_{B_p^\sigma(\mathbb{B}_n;\ell^2)}:=\sum_{k=0}^{m-1}\left|\nabla^k f(0)\right|+\left(\int_{\mathbb{B}_n}\left|(1-|z|^2)^{m+\sigma}\nabla^m f(z)\right|^p d\lambda_n(z)\right)^{1/p}<\infty,$$

for some $m>n/p-\sigma$. Again, the right side is finite for some $m>n/p-\sigma$ if and only if it is finite for all $m>n/p-\sigma$ in which case the associated norming functionals are equivalent. We will continue to write $B_p^\sigma(\mathbb{B}_n)$ for the scalar-valued space.

We will also consider multipliers between spaces of vector-valued functions. We start with the abstract situation. Given a Banach space of functions X and given $N\in\mathbb{N}$ we let $\oplus^N X$ be the vector space of $N-$tuples of elements of X, normed by, for $h=(h_j)_{j=1}^N\in\oplus^N X$,

$$\|h\|_{\oplus^N X}=\left(\sum_{j=1}^N\|h_j\|_X^2\right)^{1/2}. \tag{17.13}$$

We now define multipliers from X to $\oplus^N X$. For $f=(f_\alpha)_{\alpha=1}^N\in\oplus^N X$ the multiplication operator $\mathbb{M}_f$ which sends $h\in X$ to $\mathbb{M}_f h=(f_\alpha h)_{\alpha=1}^N$ is a densely defined map of X to $\oplus^N X$. If it is bounded we say f is a multiplier of X to $\oplus^N X$. We denote the space of all multipliers by $\mathrm{Mult}\left(X,\oplus^N X\right)$ and we norm that space by

$$\|f\|_{\mathrm{Mult}(X,\oplus^N X)}=\|\mathbb{M}_f\|_{X\to\oplus^N X}=\sup_{\|h\|_X\le 1}\|\mathbb{M}_f h\|_{\oplus^N X}.$$

We are comfortable with the choice $N=\infty$ in the previous formulas and we will be particularly interested in the multipliers from $B_p^\sigma(\mathbb{B}_n)$ to $B_p^\sigma\left(\mathbb{B}_n;\ell^2\right)$. Given $g\in B_p^\sigma\left(\mathbb{B}_n;\ell^2\right)$ the multiplication operator $\mathbb{M}_g$ defined by $(\mathbb{M}_g f)(z)=f(z)g(z)$ is a densely defined map from $B_p^\sigma(\mathbb{B}_n)$ to $B_p^\sigma\left(\mathbb{B}_n;\ell^2\right)$. When $\mathbb{M}_g$ extends to a bounded map we say that g is a vector-valued multiplier, $g\in\mathrm{Mult}\left(B_p^\sigma(\mathbb{B}_n),B_p^\sigma\left(\mathbb{B}_n;\ell^2\right)\right)$, and define the multiplier norm of g to be the operator norm of $\mathbb{M}_g$;

$$\|g\|_{\mathrm{Mult}\left(B_p^\sigma(\mathbb{B}_n),B_p^\sigma(\mathbb{B}_n;\ell^2)\right)}=\|\mathbb{M}_g\|_{B_p^\sigma(\mathbb{B}_n)\to B_p^\sigma(\mathbb{B}_n;\ell^2)}.$$

However, that being said, when it seems safe we may abbreviate things; for example writing $\|g\|_{\mathrm{Mult}}$, or $\|\mathbb{M}_g\|$.

This space of vector-valued multipliers, $\mathrm{Mult}\left(X,\oplus^N X\right)$, is very similar to the multiplier algebra $\mathrm{Mult}(X)$; but when considering analogies between the two spaces it should be kept in mind that the first space is *not* an algebra if $N>1$. Here is the baby corona theorem for the ℓ^2-valued Banach spaces $B_p^\sigma\left(\mathbb{B}_n;\ell^2\right)$, $\sigma\ge 0$, $1<p<\infty$ ([**121**]).

THEOREM 17.11 (Baby Corona Theorem). *Let $\delta>0$, $\sigma\ge 0$ and $1<p<\infty$. Then there is a constant $C_{n,\sigma,p,\delta}$ such that given*

$$g=(g_i)_{i=1}^\infty\in\mathrm{Mult}\left(B_p^\sigma(\mathbb{B}_n),B_p^\sigma\left(\mathbb{B}_n;\ell^2\right)\right)$$

which satisfies

$$\|\mathbb{M}_g\|_{B_p^\sigma(\mathbb{B}_n)\to B_p^\sigma(\mathbb{B}_n;\ell^2)} \leq 1,$$

$$\sum_{j=1}^{\infty} |g_j(z)|^2 \geq \delta^2 > 0, \qquad z \in \mathbb{B}_n, \tag{17.14}$$

there is, for each $h \in B_p^\sigma(\mathbb{B}_n)$, *a vector-valued function* $f \in B_p^\sigma(\mathbb{B}_n;\ell^2)$ *satisfying*

$$\|f\|_{B_p^\sigma(\mathbb{B}_n;\ell^2)} \lesssim \|h\|_{B_p^\sigma(\mathbb{B}_n)}, \tag{17.15}$$

$$\sum_{j=1}^{\infty} g_j(z) f_j(z) = h(z), \qquad z \in \mathbb{B}_n. \tag{17.16}$$

For some Hilbert spaces we also have the Toeplitz corona theorem. Let X be a Hilbert space of holomorphic functions in an open set Ω in $\mathbb{C}^n$.

THEOREM 17.12 (Toeplitz Corona Theorem, [**60**], [**23**], [**4**, Theorem 8.57]). *Suppose X is a reproducing kernel Hilbert space with an* irreducible complete Pick kernel *as described in Section* 6.2. *Let $\delta > 0$ and $N \in \mathbb{N}$. The functions $g_1, \ldots, g_N \in$* Mult(X) *satisfy the following "baby corona property"; for every $h \in X$, there are $f_1, \ldots, f_N \in X$ such that*

$$\|f_1\|_X^2 + \cdots + \|f_N\|_X^2 \leq \frac{1}{\delta}\|h\|_X^2, \tag{17.17}$$
$$g_1(z) f_1(z) + \cdots + g_N(z) f_N(z) = h(z), \qquad z \in \Omega,$$

if and only if $g_1, \ldots, g_N \in$ Mult(X) *satisfy the following "multiplier corona property"; there are $\varphi_1, \ldots, \varphi_N \in$* Mult$(X)$ *such that*

$$\|\varphi\|_{Mult(X,\oplus^N X)} \leq 1, \tag{17.18}$$
$$g_1(z)\varphi_1(z) + \cdots + g_N(z)\varphi_N(z) = \sqrt{\delta}, \qquad z \in \Omega.$$

This formulation with $\sqrt{\delta}$ is traditional.

For $0 \leq \sigma \leq 1/2$ the spaces $X = B_2^\sigma(\mathbb{B}_n)$ are RKHS with CPP. That follows from the multivariate version of Quiggen's theorem, Theorem 6.16; see also [**47**]. Hence, for those spaces we can combine the previous two theorems and obtain the following.

COROLLARY 17.13 (Multiplier Corona Theorem). *For the algebras* $\mathcal{A} =$ Mult$(B_2^\sigma(\mathbb{B}_n))$, $0 \leq \sigma \leq 1/2$, *the corona theorem is true; that is,* (17.14) *holds if and only if* (17.15).

This covers the multiplier algebras of the $n-$dimensional Drury-Arveson space, Mult$(B_2^{1/2}(\mathbb{B}_n))$, and of the $n-$dimensional Dirichlet space, Mult$(B_2^0(\mathbb{B}_n))$, For $\sigma = 1/2$, $n = 1$, the algebra is $H^\infty(\mathbb{D})$ and the result is Carleson's corona theorem. The particular case $\sigma = 0, n = 1$ gives a corona theorem for the multiplier algebra of the classical Dirichlet space, a result first proved by Tolokonnikov [**320**].

This corollary actually only requires the baby corona theorem for $\mathbb{C}^N$-valued functions with N finite. However the general statement for Hilbert space valued functions opens the way to further developments. For instance, there is a richer, more complicated, story for operator-valued multipliers between spaces of Hilbert space valued functions. We will just mention one result. Combining the baby corona theorem, the Toeplitz corona theorem, and the work of Trent and Zhang on matricial corona theorems [**329**], [**328**]. We have a semi-infinite matricial corona theorem.

COROLLARY 17.14 (Matricial Corona Theorem). *Let $0 \leq \sigma < 1/2$. Let $\mathcal{H}_1$ be a finite m-dimensional Hilbert space and let $\mathcal{H}_2$ be an infinite dimensional Hilbert space. Suppose that $G \in \mathrm{Mult}(B_2^\sigma(\mathbb{B}_n)(\mathcal{H}_1 \to \mathcal{H}_2))$ satisfies $\delta^2 I_m \leq G^*(z)G(z) \leq I_m$. Then there is $F \in \mathrm{Mult}_{B_2^\sigma(\mathbb{B}_n)(\mathcal{H}_2 \to \mathcal{H}_1)}$ such that*

$$F(z)G(z) = I_m,$$
$$\|F\|_{\mathrm{Mult}(B_2^\sigma(\mathbb{B}_n)(\mathcal{H}_2 \to \mathcal{H}_1))} \leq C_{\sigma,n,\delta,m}.$$

Surprisingly, in a certain sense the story ends here. S. Treil has shown if we permit both $\mathcal{H}_1$ and $\mathcal{H}_2$ to have infinite dimension then this corollary can fail, even for the classical Hardy space on the disk, $H^2 = B_2^{1/2}(\mathbb{B}_1)$ [**323**].

17.2.2. Plan for the Proof. As we said, most of the work in this chapter is in proving the baby corona theorem. Fundamental work on that problem was done in a 2000 paper by J. M. Ortega and J. Fabrega [**248**]. They obtained results for the multiplier algebras $\mathrm{Mult}(B_2^\sigma(\mathbb{B}_n))$ with $0 \leq \sigma < 1/2$ (but not $\sigma = 1/2$). They considered $\{g_i\}_{i=1}^N$ with $N = 2$. That case is already highly nontrivial, note that in [**89**] Carleson uses his new description of the interpolating sequences for H^∞ to resolve the $N = 2$ case of the corona conjecture for H^∞. With the restriction to $N = 2$ the antisymmetric 2-tensor Ω_1^2 of $(0,1)$-forms which arises in the Koszul complex below is, in fact, a 2×2 antisymmetric matrix, one of the form

$$\begin{pmatrix} 0 & \alpha \\ -\overline{\alpha} & 0 \end{pmatrix}.$$

That simple form makes it easy to see that Ω_1^2 is $\overline{\partial}$-closed and hence to move forward in the proof. Substantial new issues arise if this simplification is not present. Nevertheless, the work of Ortega and Fabrega was influential in shaping the basic groundwork and approach in the following sections.

When going past $N = 2$ functions it turns out to be just as easy to treat the case $N = \infty$. This includes the results for ℓ_N^2-valued functions and, in fact, gives all those results with constants independent of N.

We are given an infinite vector of multipliers

$$g = (g_i)_{i=1}^\infty \in \mathrm{Mult}\left(B_p^\sigma(\mathbb{B}_n), B_p^\sigma(\mathbb{B}_n; \ell^2)\right)$$

that satisfy

$$\|g\|_{\mathrm{Mult}\left(B_p^\sigma(\mathbb{B}_n), B_p^\sigma(\mathbb{B}_n;\ell^2)\right)} = \|\mathbb{M}_g\|_{B_p^\sigma(\mathbb{B}_n) \to B_p^\sigma(\mathbb{B}_n;\ell^2)} \leq 1$$
$$\inf_{\mathbb{B}_n} |g|^2 \geq \delta^2 > 0.$$

We are given an element $h \in B_p^\sigma(\mathbb{B}_n)$. We wish to find $f = (f_i)_{i=1}^\infty \in B_p^\sigma(\mathbb{B}_n; \ell^2)$ such that

(1) $\mathrm{Mult}_g f = g \cdot f = h$,
(2) $\overline{\partial} f = 0$,
(3) $\|f\|_{B_p^\sigma(\mathbb{B}_n;\ell^2)} \lesssim \|h\|_{B_p^\sigma(\mathbb{B}_n)}$.

A first approximation to a solution is

$$f = \frac{\overline{g}}{|g|^2}h.$$

That function obviously satisfies (1) and it can be shown to satisfy differential estimates similar to (3), but in general it fails to satisfy (2).

To rectify this we use the Koszul complex in Section 17.4. To help adapt it to our needs we use the remarkable factorization of the Koszul complex due to Andersson and Carlsson [**29**]. Those tools show how to use solutions to $\overline{\partial}$ problems on forms of bidegree $(0, q)$, $1 \leq q \leq n$, to produce a correction term $\Lambda_g \Gamma_0^2$ so that

$$f = \frac{\overline{g}}{|g|^2}h - \Lambda_g \Gamma_0^2$$

now satisfies (1) and (2).

The correction term $\Lambda_g \Gamma_0^2$ is obtained by solving $\overline{\partial}$-equations and to ensure that (3) holds we must have good estimates for the solution operator. In one dimension the formula of Jones, Theorem 17.7, can be used to find solutions which are readily seen to satisfy the needed estimates. However the situation in higher dimension is more complicated. In Section 17.3 we explicitly calculate Charpentier's solution operators for $\overline{\partial}$-equations and use them in solving the equations that arise from the Koszul complex. Those computations are lengthy but the solution operators obtained are remarkably simple in form. Moreover the operators are superbly adapted for obtaining estimates in real-variable analogs of the Besov-Sobolev spaces in the ball. In particular, the kernels $K(w, z)$ of these solution operators involve expressions such as

$$\frac{(1 - w\overline{z})^{\alpha} (1 - |w|^2)^{\beta}}{\triangle (w, z)^{\gamma}}, \tag{17.19}$$

see, for example, (17.24). Here $\triangle (w, z)$ is a quantity related to the complex geometry and the distance between w and z; it has many equivalent descriptions (15.23).

In Section 17.3.2 we recall the clever integration by parts formulas of Ortega and Fabrega involving the left side of (17.20) below, and extend them to the Charpentier solution operators for higher degree forms. When we differentiate (17.19) the power of $\triangle (w, z)$ in the denominator can increase but the integration by parts in Lemma 17.29 below will temper this singularity on the diagonal. On the other hand the radial integration by parts in Corollary 17.32 below will temper singularities on the boundary of the ball.

In Section 17.4.2 we introduce $\Lambda_{p,m}^{\sigma}(\mathbb{B}_n)$, real-variable analogs of the Besov-Sobolev spaces $B_p^{\sigma}(\mathbb{B}_n)$ along with ℓ^2-valued variants. Just as with the spaces $B_p^{\sigma}(\mathbb{B}_n)$ and $HB_p(\mathcal{T}_n)$, the definitions reflect the complex geometry of the ball, a geometry that is reflected in the solution kernels in (17.19). This will be done using the almost invariant holomorphic derivative, D_a, which was introduced in

Section 14.5.1

$$
\begin{aligned}
D_a f(z) &:= f'(z)\,\varphi_a'(0) \\
&= -f'(z)\left\{\left(1-|a|^2\right)P_a+\left(1\quad |a|^2\right)^{1/2}Q_a\right\} \\
&= -\nabla f(z)\left\{\left(1-|a|^2\right)P_a+\left(1-|a|^2\right)^{1/2}Q_a\right\}
\end{aligned}
$$

which, for a fixed, is a holomorphic function of z. Another variation on this theme, denoted D^m will be defined in Definition 17.38 later on.

That operator will give us information when applied to functions which are not necessarily holomorphic.

Several crucial inequalities are then developed to facilitate showing the boundedness of the Charpentier solution operators, most notably

$$
\left|\left(\overline{z-w}\right)^{\alpha}\frac{\partial^m}{\partial\overline{w}^{\alpha}}F(w)\right|\lesssim\triangle(w,z)^{m/2}\left|(1-|w|^2)^{-m}\overline{D}^m F(w)\right|, \tag{17.20}
$$

for $F\in H^\infty\left(\mathbb{B}_n;\ell^2\right)$, which controls the product of Euclidean lengths with Euclidean derivatives on the left, in terms of the product of the smaller length $\sqrt{\triangle(w,z)}$ and the larger derivative $(1-|w|^2)^{-1}\overline{D}$ on the right. (We caution the reader that our definition of $\overline{D}^m$ is *not* simply the composition of m copies of $\overline{D}$ – see Definition 17.38 below.)

We then use Schur's test, Lemma B.10 to establish the boundedness of positive operators with kernels related to (17.19). Those results turn out to be extremely well suited for dealing with the ameliorated solution operators of Charpentier.

Finally in Section 17.5 we put these pieces together to prove theorem 17.11.

17.2.3. Related Results. Relatively little is known about multiplier algebras of other spaces of holomorphic functions of several complex variables.

Of course the scale of Besov spaces is not the only one for which these questions make sense. The Hardy spaces on the ball, $H^p(\mathbb{B}_n)$, $1<p<\infty$, have also been considered. Those spaces all have $H^\infty(\mathbb{B}_n)$ as their multiplier algebra. An early and very influential paper by Varopolous [**335**] gave results in the spirit of a corona theorem for $N=2$ functions but only obtained BMO estimates for the function $\{f_1,f_2\}$, not boundedness. This was taken slightly further by Costea, Sawyer and Wick in [**123**] where it was shown that it is possible to obtain BMO estimates when there are an infinite number of generators. The actual corona question remains open for the algebra $H^\infty(\mathbb{B}_n)$.

Also, baby corona theorems and variants are known for some of the Hardy spaces on the ball, including work by Andersson and Carlsson [**30**], [**28**], Amar [**22**], and Krantz and Li [**198**].

There are also results for the Hardy spaces on the polydisk. In [**325**] Treil and Wick obtain the H^p (baby) corona theorem on the polydisk $\mathbb{D}^n$ (see also Lin [**208**] and Trent [**327**]). However the Hardy space on the polydisk, $H^2(\mathbb{D}^n)$, fails to have the CPP, and the more complicated version of the Toeplitz corona theorem which holds for $H^2(\mathbb{D}^n)$ is not directly applicable. As a result the corona theorem for the algebra $H^\infty(\mathbb{D}^n)$ on the polydisk remains open for $n\geq 2$. Finally, even the baby corona problems, apart from that for H^p, remain open on the polydisk.

17.3. Charpentier's Solution Kernels

In [**106**, Theorem I.4], Charpentier proves the following formula which, when given a $\overline{\partial}$-closed $(0, q+1)$-form f, will let us solve $\overline{\partial}_z u = f$.

THEOREM 17.15. *For $q \geq 0$ and all forms $f(\xi) \in C^1\left(\overline{\mathbb{B}_n}\right)$ of degree $(0, q+1)$, we have for $z \in \mathbb{B}_n$:*

$$f(z) = C_q \int_{\mathbb{B}_n} \overline{\partial} f(\xi) \wedge \mathcal{C}_n^{0,q+1}(\xi, z) + c_q \overline{\partial}_z \left(\int_{\mathbb{B}_n} f(\xi) \wedge \mathcal{C}_n^{0,q}(\xi, z) \right).$$

Here $\mathcal{C}_n^{0,q}(\xi, z)$, defined in Definition 17.17 below, is a $(n, n-q-1)$-form in ξ on the ball and a $(0, q)$-form in z on the ball. Using Theorem 17.15 we can solve $\overline{\partial}_z u = f$ as follows. Set

$$u(z) := c_q \int_{\mathbb{B}_n} f(\xi) \wedge \mathcal{C}_n^{0,q}(\xi, z)$$

Taking $\overline{\partial}_z$ of this we see from Theorem 17.15 and $\overline{\partial} f = 0$ that

$$\overline{\partial}_z u = c_q \overline{\partial}_z \left(\int_{\mathbb{B}_n} f(\xi) \wedge \mathcal{C}_n^{0,q}(\xi, z) \right) = f(z).$$

It is essential for our proof to explicitly compute the kernels $\mathcal{C}_n^{0,q}$ when $0 \leq q \leq n-1$. The case $q = 0$ is given in [**106**] and we briefly recall the setup. Denote by $\triangle : \mathbb{C}^n \times \mathbb{C}^n \to [0, \infty)$ the map

$$\triangle(w, z) := |1 - w\overline{z}|^2 - (1 - |w|^2)(1 - |z|^2)$$

and recall from (15.23) that this quantity is a measure of the distance between w and z and it can be rewritten in a number of ways, some with quite different appearance.

To compute the kernels $\mathcal{C}_n^{0,q}$ we start with the closed Cauchy-Leray form (see [**288**, Equation 16.4.5])

$$\mu(\xi, w, z) := \frac{1}{(\xi(w-z))^n} \sum_{i=1}^{n} (-1)^{i-1} \xi_i \left[\wedge_{j \neq i} d\xi_j\right] \wedge_{i=1}^{n} d(w_i - z_i).$$

One then lifts the form μ via a section s to give a closed form on $\mathbb{C}^n \times \mathbb{C}^n$. Namely, for $s : \mathbb{C}^n \times \mathbb{C}^n \to \mathbb{C}^n$ one defines,

$$s^* \mu(w, z) := \frac{1}{(s(w, z)(w-z))^n} \sum_{i=1}^{n} (-1)^{i-1} s_i(w, z) \left[\wedge_{j \neq i} ds_j\right] \wedge_{i=1}^{n} d(w_i - z_i).$$

Now we select s to be the following section used by Charpentier:

$$s(w, z) := \overline{w}(1 - w\overline{z}) - \overline{z}(1 - |w|^2). \tag{17.21}$$

Simple computations [**121**] demonstrate that

$$s(w, z)(w - z) = \triangle(w, z). \tag{17.22}$$

DEFINITION 17.16. *We define the Cauchy kernel on $\mathbb{B}_n \times \mathbb{B}_n$ to be*

$$\mathcal{C}_n(w, z) := s^* \mu(w, z) \tag{17.23}$$

for the section s given in (17.21) above.

DEFINITION 17.17. *For $0 \le p \le n$ and $0 \le q \le n-1$ we let $\mathcal{C}_n^{p,q}$ be the component of $\mathcal{C}_n(w,z)$ that has bidegree (p,q) in z and bidegree $(n-p,n-q-1)$ in w.*

Thus if η is a $(p,q+1)$-form in w, then $\mathcal{C}_n^{p,q}\wedge\eta$ is a (p,q)-form in z and a multiple of the volume form in w. We now prepare to give explicit formulas for Charpentier's solution kernels $\mathcal{C}_n^{0,q}(w,z)$. First we introduce some notation.

NOTATION 17.18. *Let $\omega_n(z)=\bigwedge_{j=1}^n dz_j$. For n a positive integer and $0\le q\le n-1$ let P_n^q denote the collection of all permutations ν on $\{1,\dots,n\}$ that map to $\{i_\nu,J_\nu,L_\nu\}$ where J_ν is an increasing multi-index with card$(J_\nu)=n-q-1$ and L_ν is an increasing multi-index with card$(L_\nu)=q$. Let $\epsilon_\nu=\operatorname{sgn}(\nu)\in\{-1,1\}$ denote the signature of the permutation ν.*

Note that the number of increasing multi-indices of length $n-q-1$ is $\binom{n}{q+1}$, while the number of increasing multi-indices of length q is $\binom{n}{q}$. Since we are only allowed certain combinations of J_ν and L_ν (they must have disjoint intersection and they must be increasing multi-indices), it is straightforward to see that the total number of permutations in P_n^q that we are considering is $(n-q)\binom{n}{q}$.

From Øvrelid [**250**] we obtain that Charpentier's kernel takes the (abstract) form

$$\mathcal{C}_n^{0,q}(w,z)=\frac{1}{\triangle(w,z)^n}\sum_{\nu\in P_n^q}\operatorname{sgn}(\nu)\,s_{i_\nu}\bigwedge_{j\in J_\nu}\overline{\partial}_w s_j\bigwedge_{l\in L_\nu}\overline{\partial}_z s_l\wedge\omega_n(w).$$

Fundamental for us will be the explicit formula for Charpentier's kernel given in the next theorem. We are informed by Part 2 of Proposition I.1 in [**106**] that $\mathcal{C}_n^{p,q}(w,z)=0$ for $w\in\partial\mathbb{B}_n$, and this serves as a guiding principle in the proof below. It is convenient to isolate the following factor common to all summands in the formula:

$$\Phi_n^q(w,z):=\frac{(1-w\overline{z})^{n-1-q}(1-|w|^2)^q}{\triangle(w,z)^n},\qquad 0\le q\le n-1. \tag{17.24}$$

THEOREM 17.19. *Let n be a positive integer and suppose that $0\le q\le n-1$. Then*

$$\mathcal{C}_n^{0,q}(w,z)=\sum_{\nu\in P_n^q}(-1)^q\,\Phi_n^q(w,z)\;sign(\nu)\left(\overline{w_{i_\nu}}-\overline{z_{i_\nu}}\right)\bigwedge_{j\in J_\nu}d\overline{w_j}\bigwedge_{l\in L_\nu}d\overline{z_l}\bigwedge\omega_n(w). \tag{17.25}$$

REMARK 17.20. *We can rewrite the formula for $\mathcal{C}_n^{0,q}(w,z)$ in (17.25) as*

$$\mathcal{C}_n^{0,q}(w,z)=\Phi_n^q(w,z)\sum_{|J|=q}\sum_{k\notin J}(-1)^{\mu(k,J)}\left(\overline{z_k}-\overline{w_k}\right)d\overline{z}^J\wedge d\overline{w}^{(J\cup\{k\})^c}\wedge\omega_n(w), \tag{17.26}$$

where $J\cup\{k\}$ here denotes the increasing multi-index obtained by rearranging the integers $\{k,j_1,\dots,j_q\}$ as

$$J\cup\{k\}=\left\{j_1,\dots,j_{\mu(k,J)-1},k,j_{\mu(k,J)},\dots,j_q\right\}.$$

Thus k occupies the $\mu(k,J)^{th}$ position when $J\cup\{k\}$ is arranged to be increasing. The notation $(J\cup\{k\})^c$ refers to the multi-index obtained by rearranging the integers of $\{1,2,\dots,n\}\setminus(J\cup\{k\})$ in increasing order. To see (17.26), we note that in

(17.25) *the permutation* ν *takes the* n*-tuple* $(1, 2, \ldots, n)$ *to* (i_ν, J_ν, L_ν). *In* (17.26) *the* n*-tuple* $(k, (J \cup \{k\})^c, J)$ *corresponds to* (i_ν, J_ν, L_ν), *and so* $\operatorname{sgn}(\nu)$ *becomes in* (17.26) *the signature of the permutation that takes* $(1, 2, \ldots, n)$ *to* $(k, (J \cup \{k\})^c, J)$. *This in turn equals* $(-1)^{\mu(k,J)}$ *with* $\mu(k, J)$ *as above.*

We observe at this point that the functional coefficient in the summands in (17.25) has the form

$$(-1)^q \Phi_n^q(w, z)\left(\overline{w_{i_\nu}} - \overline{z_{i_\nu}}\right) = (-1)^q \frac{(1 - w\overline{z})^{n-q-1}(1 - |w|^2)^q}{\triangle(w, z)^n}\left(\overline{w_{i_\nu}} - \overline{z_{i_\nu}}\right).$$

Hence, relative to the invariant measure and the Bergman metric, this kernel has a singularity on the diagonal similar to that of a fractional integral operator of order one. Finally, we will adopt the usual convention of writing

$$\mathcal{C}_n^{0,q} f(z) = \int_{\mathbb{B}_n} f(w) \wedge \mathcal{C}_n^{0,q}(w, z),$$

when we wish to view $\mathcal{C}_n^{0,q}$ as an operator taking $(0, q+1)$-forms f in w to $(0, q)$-forms $\mathcal{C}_n^{0,q} f$ in z.

PROOF OF THEOREM 17.19. The proof is a lengthy computation in exterior differential calculus with more exterior algebra bookkeeping than calculus. See [**121**] for a proof of a different, more geometric flavor

For typographic convenience for dealing with some of the long formulas in this section we introduce the abbreviations:

$$\Omega = (1 - w\overline{z})$$
$$\Psi = (1 - |w|^2)$$

In the computation of the Cauchy kernel $\mathcal{C}_n(w, z)$, we need to compute the full exterior derivative of the section $s(w, z)$. Straightforward computations show that

$$\partial_w s_i(w, z) = \sum_{j=1}^n \left(\overline{z}_i \overline{w}_j - \overline{w}_i \overline{z}_j\right) dw_j, \tag{17.27}$$

$$\begin{aligned}\overline{\partial}_w s_i(w, z) &= \Omega d\overline{w}_i + \sum_{j=1}^n w_j \overline{z}_i d\overline{w}_j \\ &= \Omega d\overline{w}_i + \overline{z}_i \overline{\partial}_w |w|^2,\end{aligned} \tag{17.28}$$

$$\begin{aligned}\overline{\partial}_z s_i(w, z) &= -\Omega d\overline{z}_i - \sum_{j=1}^n \overline{w}_i w_j d\overline{z}_j \\ &= -\Psi d\overline{z}_i - \overline{w}_i \overline{\partial}_z (w\overline{z}),\end{aligned} \tag{17.29}$$

$$\partial_z s_i(w, z) = 0.$$

We also have the following representations of s_k, again following by simple computation. Recall from Notation 17.18 that $\{1, 2, \ldots, n\} = \{i_\nu\} \cup J_\nu \cup L_\nu$ where J_ν and L_ν are increasing multi-indices of lengths $n - q - 1$ and q. We will use the following with $k = i_\nu$,

$$\begin{aligned} s_k &= (\overline{w}_k - \overline{z}_k) + \sum_{l \neq k} w_l(\overline{w}_l \overline{z}_k - \overline{w}_k \overline{z}_l) \\ &= (\overline{w}_k - \overline{z}_k) + \sum_{j \in J_\nu} w_j(\overline{w}_j \overline{z}_k - \overline{w}_k \overline{z}_j) + \sum_{l \in L_\nu} w_l(\overline{w}_l \overline{z}_k - \overline{w}_k \overline{z}_l) \\ &= (\overline{w}_k - \overline{z}_k) + \overline{z}_k \sum_{j \in J_\nu} |w_j|^2 - \overline{w}_k \sum_{j \in J_\nu} w_j \overline{z}_j + \overline{z}_k \sum_{l \in L_\nu} |w_l|^2 - \overline{w}_k \sum_{l \in L_\nu} w_l \overline{z}_l. \end{aligned}$$

REMARK 17.21. *Since $A \wedge A = 0$ for any form, we have in particular that $\overline{\partial_w} |w|^2 \wedge \overline{\partial_w} |w|^2 = 0$ and $\overline{\partial_z} (w\overline{z}) \wedge \overline{\partial_z} (w\overline{z}) = 0$.*

Using this remark we next compute $\bigwedge_{j \in J_\nu} \overline{\partial}_w s_j$. We identify J_ν as $j_1 < j_2 < \cdots < j_{n-q-1}$ and define a map $\imath(j_r) = r$, namely $\imath$ says where j_r occurs in the multi-index. We will frequently abuse notation and simply write $\imath(j)$. Because $\overline{\partial}_w |w|^2 \wedge \overline{\partial}_w |w|^2 = 0$ it is easy to conclude that we cannot have any term in $\overline{\partial}_w |w|^2$ of degree greater than one when expanding the wedge product of the $\overline{\partial_w} s_j$. So we continue,

$$\begin{aligned} \bigwedge_{j \in J_\nu} \overline{\partial}_w s_j &= \bigwedge_{j \in J_\nu} \left\{ \Omega d\overline{w}_j + \overline{z}_j \overline{\partial}_w |w|^2 \right\} \\ &= \Omega^{n-q-1} \bigwedge_{j \in J_\nu} d\overline{w}_j \\ &\quad + \Omega^{n-q-2} \sum_{j \in J_\nu} (-1)^{\imath(j)-1} \overline{z}_j \overline{\partial}_w |w|^2 \wedge \bigwedge_{j' \in J_\nu \setminus \{j\}} d\overline{w}_{j'} \\ &= \Omega^{n-q-2} \left(\Omega + \sum_{j \in J_\nu} w_j \overline{z}_j \right) \bigwedge_{j \in J_\nu} d\overline{w}_j \\ &\quad + \Omega^{n-q-2} \sum_{j \in J_\nu} (-1)^{\imath(j)-1} \overline{z}_j \sum_{k \in L_\nu \cup \{i_\nu\}} w_k d\overline{w}_k \bigwedge_{j' \in J_\nu \setminus \{j\}} d\overline{w}_{j'}. \end{aligned}$$

The last line follows by direct computation using

$$\overline{\partial}_w |w|^2 = \sum_{j \in J_\nu} w_j d\overline{w}_j + \sum_{k \in L_\nu \cup \{i_\nu\}} w_k d\overline{w}_k.$$

A similar computation yields that

$$\begin{aligned}\bigwedge_{l\in L_\nu}\overline{\partial}_z s_l &= (-1)^q\bigwedge_{l\in L_\nu}\left\{\Psi d\overline{z}_l+\overline{w}_l\overline{\partial}_z(w\overline{z})\right\}\\ &=(-1)^q\left(\Psi^q\bigwedge_{l\in L_\nu}d\overline{z}_l\right)\\ &\quad+(-1)^q\Psi^{q-1}\sum_{l\in L_\nu}(-1)^{\imath(l)-1}\overline{w}_l\overline{\partial}_z(w\overline{z})\wedge\bigwedge_{l'\in L_\nu\setminus\{l\}}d\overline{z}_{l'}\\ &=(-1)^q\Psi^{q-1}\left(\Psi+\sum_{l\in L_\nu}|w_l|^2\right)\bigwedge_{l\in L_\nu}d\overline{z}_l\\ &\quad+(-1)^q\Psi^{q-1}\sum_{l\in L_\nu}(-1)^{\imath(l)-1}\overline{w}_l\sum_{k\in J_\nu\cup\{i_\nu\}}w_kd\overline{z}_k\bigwedge_{l'\in L_\nu\setminus\{l\}}d\overline{z}_{l'}.\end{aligned}$$

An important remark at this point is that the multi-index J_ν or L_ν can only appear in the first term of the last line above. The terms after the plus sign have multi-indices that are related to J_ν and L_ν, but differ by one element. This fact will play a role later.

Combining things, we see that

$$\bigwedge_{j\in J_\nu}\overline{\partial}_w s_j\bigwedge_{l\in L_\nu}\overline{\partial}_z s_j=(-1)^q\Omega^{n-q-2}\Psi^{q-1}\left(I_\nu+II_\nu+III_\nu+IV_\nu\right),$$

where

$$\begin{aligned}I_\nu&=\left(\Omega+\sum_{j\in J_\nu}w_j\overline{z}_j\right)\left(\Psi+\sum_{l\in L_\nu}|w_l|^2\right)\bigwedge_{j\in J_\nu}d\overline{w}_j\bigwedge_{l\in L_\nu}d\overline{z}_l,\\ II_\nu&=\left(\Omega+\sum_{j\in J_\nu}w_j\overline{z}_j\right)\bigwedge_{j\in J_\nu}d\overline{w}_j\\ &\qquad\times\left(\sum_{l\in L_\nu}(-1)^{\imath(l)-1}\overline{w}_l\sum_{k\in J_\nu\cup\{i_\nu\}}w_kd\overline{z}_k\bigwedge_{l'\in L_\nu\setminus\{l\}}d\overline{z}_{l'}\right)\\ III_\nu&=\left(\sum_{j\in J_\nu}(-1)^{\imath(j)-1}\overline{z}_j\sum_{k\in L_\nu\cup\{i_\nu\}}w_kd\overline{w}_k\bigwedge_{j'\in J_\nu\setminus\{j\}}d\overline{w}_{j'}\right)\\ &\qquad\times\left(1-|w|^2+\sum_{l\in L_\nu}|w_l|^2\right)\bigwedge_{l\in L_\nu}d\overline{z}_l,\\ IV_\nu&=\left(\sum_{j\in J_\nu}(-1)^{\imath(j)-1}\overline{z}_j\sum_{k\in L_\nu\cup\{i_\nu\}}w_kd\overline{w}_k\bigwedge_{j'\in J_\nu\setminus\{j\}}d\overline{w}_{j'}\right)\\ &\qquad\times\left(\sum_{l\in L_\nu}(-1)^{\imath(l)-1}\overline{w}_l\sum_{k\in J_\nu\cup\{i_\nu\}}w_kd\overline{z}_k\bigwedge_{l'\in L_\nu\setminus\{l\}}d\overline{z}_{l'}\right).\end{aligned}$$

To aid in the computation of the kernel $\mathcal{C}_n^{0,q}(w,z)$ we introduce a bit more notation/ For $1 \le k \le n$ we let $P_n^q(k) = \{\nu \in P_n^q : \nu(1) = i_\nu = k\}$. This divides the set P_n^q into n classes with $\binom{n-1}{q}$ elements. For each k we set

$$I(k) = \sum_{\nu \in P_n^q(k)} \epsilon_\nu I_\nu \quad II(k) - \sum_{\nu \in P_n^q(k)} \epsilon_\nu II_\nu$$
$$III(k) = \sum_{\nu \in P_n^q(k)} \epsilon_\nu III_\nu \quad IV(k) = \sum_{\nu \in P_n^q(k)} \epsilon_\nu IV_\nu.$$

and

$$C(k) = I(k) + II(k) + III(k) + IV(k).$$

Using this notation and that introduced Notation 17.18, together with the previous computations, we are reduced to the calculation of

$$\begin{aligned}
\mathcal{C}_n^{0,q}(w,z) &= \frac{1}{\triangle(w,z)^n} \sum_{\nu \in P_n^q} \epsilon_\nu s_{i_\nu} \bigwedge_{j \in J_\nu} \bar{\partial}_w s_j \bigwedge_{l \in L_\nu} \bar{\partial}_z s_l \wedge \omega(w) \\
&= \frac{(-1)^q \Omega^{n-q-2} \Psi^{q-1}}{\triangle(w,z)^n} \sum_{k=1}^n s_k \sum_{\nu \in P_n^q(k)} \epsilon_\nu (I_\nu + II_\nu + III_\nu + IV_\nu) \\
&= \frac{(-1)^q \Omega^{n-q-2} \Psi^{q-1}}{\triangle(w,z)^n} \sum_{k=1}^n s_k (I(k) + II(k) + III(k) + IV(k)) \\
&= \frac{(-1)^q \Omega^{n-q-2} \Psi^{q-1}}{\triangle(w,z)^n} \sum_{k=1}^n s_k C(k).
\end{aligned}$$

For a fixed $\tau \in P_n^q$ we will compute the coefficient of $\bigwedge_{j \in J_\tau} d\overline{w}_j \bigwedge_{l \in L_\tau} d\overline{z}_l$. We will ignore the functional coefficient in front of the sum since it only needs to be taken into consideration at the final stage. We will show that for this fixed τ the sum on k of $s_k C(k)$ can be replaced by $\epsilon_\tau \Omega \Psi (\overline{w}_{i_\tau} - \overline{z}_{i_\tau}) \bigwedge_{j \in J_\tau} d\overline{w}_j \bigwedge_{l \in L_\tau} d\overline{z}_l$. There will also be other terms that appear in this expression that arise from multi-indices J and I that are not disjoint. Using the computations below it can be seen that these terms actually vanish and hence provide no contribution for $\mathcal{C}_n^{0,q}(w,z)$. Since τ is an arbitrary element of P_n^q this will then complete the computation of the kernel.

Note that when $k = i_\tau$ then we have the following contributions. It is easy to see that $II(i_\tau) = III(i_\tau) = 0$. It is also easy to see that

$$\begin{aligned}
I(i_\tau) &= \epsilon_\tau \left(\Omega + \sum_{j \in J_\tau} w_j \overline{z}_j \right) \left(\Psi + \sum_{l \in L_\tau} |w_l|^2 \right) \bigwedge_{j \in J_\tau} d\overline{w}_j \bigwedge_{l \in L_\tau} d\overline{z}_l \\
&= \epsilon_\tau \Omega \Psi \bigwedge_{j \in J_\tau} d\overline{w}_j \bigwedge_{l \in L_\tau} d\overline{z}_l \\
&\quad + \left(\Omega \sum_{l \in L_\tau} |w_l|^2 + \Psi \sum_{j \in J_\tau} w_j \overline{z}_j + \sum_{l \in L_\tau} |w_l|^2 \sum_{j \in J_\tau} w_j \overline{z}_j \right) \bigwedge_{j \in J_\tau} d\overline{w}_j \bigwedge_{l \in L_\tau} d\overline{z}_l.
\end{aligned}$$

We also receive a contribution from term $IV(i_\tau)$ is this case. This happens by interchanging an index in the multi-index J_τ with one in L_τ. Namely, we consider the permutations $\nu : \{1, \dots, n\} \to \{i_\tau, (J_\tau \setminus \{j\}) \cup \{l\}, (L_\tau \setminus \{l\}) \cup \{j\}\}$. This permutation contributes the term $\overline{z}_l w_l \overline{w}_j w_j$. After summing over all these possible permutations, we arrive at the simplified formula,

$$IV(i_\tau) = -\epsilon_\tau \left(\sum_{j\in J_\tau} |w_j|^2\right)\left(\sum_{l\in L_\tau} w_l\overline{z}_l\right) \bigwedge_{j\in J_\tau} d\overline{w}_j \bigwedge_{l\in L_\tau} d\overline{z}_l.$$

Collecting all these terms, when $k = i_\tau$ we have that the coefficient of $\epsilon_\tau \bigwedge_{j\in J_\tau} d\overline{w}_j \bigwedge_{l\in L_\tau} d\overline{z}_l$ is:

$$\begin{aligned} C(i_\tau) = \Omega\Psi &+ (\Omega + \sum_{j\in J_\tau} w_j\overline{z}_j) \sum_{l\in L_\tau} |w_l|^2 \\ &+ (\Psi + \sum_{l\in L_\tau} |w_l|^2) \sum_{j\in J_\tau} w_j\overline{z}_j \\ &- \sum_{l\in L_\tau} |w_l|^2 \sum_{j\in J_\tau} w_j\overline{z}_j - \sum_{j\in J_\tau} |w_j|^2 \sum_{l\in L_\tau} w_l\overline{z}_l. \end{aligned}$$

We next note that when $k \neq i_\tau$ it is still possible to have terms which contribute to the coefficient of $\bigwedge_{j\in J_\tau} d\overline{w}_j \bigwedge_{l\in L_\tau} d\overline{z}_l$. To see this we further split the conditions on k into the situations where $k \in J_\tau$ and $k \in L_\tau$. First, observe in this situation that if $k \neq i_\tau$ then term $I(k)$ can never contribute. So all contributions must come from terms $II(k)$, $III(k)$, and $IV(k)$. In these terms it is possible to obtain the term $\bigwedge_{j\in J_\tau} d\overline{w}_j \bigwedge_{l\in L_\tau} d\overline{z}_l$ by replacing some index in ν. These terms arise when ν and τ differ from each other by a single index, or when replacing an index with i_τ.

Next, observe that when $k \in L_\tau$ there exists a unique $\nu \in P_n^q(k)$ such that $J_\nu = J_\tau$. Namely, we have that $\nu : \{1, \ldots, n\} \to \{k, J_\tau, (L_\tau \setminus \{k\}) \cup i_\tau\}$. Here, we used that $i_\nu = k$. Terms of this type will contribute to term $II(k)$ but will give no contribution to term $III(k)$. However, they will give a contribution to term $IV(k)$.

Similarly, when $k \in J_\tau$ there will exist a unique $\mu \in P_n^q(k)$ with $L_\mu = L_\tau$. This happens with $\mu : \{1, \ldots, n\} \to \{k, (J_\tau \setminus \{k\}) \cup i_\tau, L_\tau\}$. Here we used that $i_\mu = k$. Again, we get a contribution to term $III(k)$ and $IV(k)$ and they give no contribution to the term $II(k)$.

Using these observations when $k \in L_\tau$ we arrive at the following for $I(k)$, $II(k)$, $III(k)$, and $IV(k)$:

$$\begin{aligned} I(k) &= 0 \\ II(k) &= -\epsilon_\tau \left(\Omega + \sum_{j\in J_\tau} w_j\overline{z}_j\right) \overline{w}_{i_\tau} w_k \bigwedge_{j\in J_\tau} d\overline{w}_j \bigwedge_{l\in L_\tau} d\overline{z}_l \\ III(k) &= 0 \\ IV(k) &= \epsilon_\tau \overline{z}_{i_\tau} w_k \left(\sum_{j\in J_\tau} |w_j|^2\right) \bigwedge_{j\in J_\tau} d\overline{w}_j \bigwedge_{l\in L_\tau} d\overline{z}_l. \end{aligned}$$

Similarly, when $k \in J_\tau$ we arrive at the following for $I(k)$, $II(k)$, $III(k)$, and $IV(k)$:

$$\begin{aligned} I(k) &= 0 \\ II(k) &= 0 \\ III(k) &= -\epsilon_\tau \left(\Psi + \sum_{l \in L_\tau} |w_l|^2 \right) \overline{z}_{i_\tau} w_k \bigwedge_{j \in J_\tau} d\overline{w}_j \bigwedge_{l \in L_\tau} d\overline{z}_l \\ IV(k) &= \epsilon_\tau \overline{w}_{i_\tau} w_k \left(\sum_{l \in L_\tau} w_l \overline{z}_l \right) \bigwedge_{j \in J_\tau} d\overline{w}_j \bigwedge_{l \in L_\tau} d\overline{z}_l . \end{aligned}$$

Collecting these terms, we see the following for the coefficient of $\epsilon_\tau \bigwedge_{j \in J_\tau} d\overline{w}_j \bigwedge_{l \in L_\tau} d\overline{z}_l$:

$$C(k) = -w_k \left(\overline{z}_{i_\tau} \left(\Psi + \textstyle\sum_{l \in L_\tau} |w_l|^2 \right) - \overline{w}_{i_\tau} \left(\textstyle\sum_{l \in L_\tau} w_l \overline{z}_l \right) \right) \qquad \forall k \in J_\tau,$$

$$C(k) = -w_k \left(\overline{w}_{i_\tau} \left(\Omega + \textstyle\sum_{j \in J_\tau} w_j \overline{z}_j \right) - \overline{z}_{i_\tau} \left(\textstyle\sum_{j \in J_\tau} |w_j|^2 \right) \right) \qquad \forall k \in L_\tau.$$

This then implies that the *total* coefficient of $\epsilon_\tau \bigwedge_{j \in J_\tau} d\overline{w}_j \bigwedge_{l \in L_\tau} d\overline{z}_l$ is given by

$$s_{i_\tau} C(i_\tau) + \sum_{k \in J_\tau} s_k C(k) + \sum_{k \in L_\tau} s_k C(k).$$

From this point the remainder of the proof of the Theorem 17.19 is lengthy but straightforward algebra. The term $s_{i_\tau} C(i_\tau)$ will contribute the term $(1 - w\overline{z})(1 - |w|^2)(\overline{w}_{i_\tau} - \overline{z}_{i_\tau})$ and a remainder term. The remainder term will cancel with the terms $\sum_{k \neq i_\tau} s_k C(k)$. The remaining details of the proof are in [**122**, pages 105-107], the electronic appendix of [**121**]. □

17.3.1. Ameliorated Kernels. We now wish to define right inverses with improved behavior at the boundary. We consider the case when the right side f of the $\overline{\partial}$-equation is a $(p, q+1)$-form in $\mathbb{B}_n$.

As usual for a positive integer $s > n$ we will " project" the formula $\overline{\partial}\mathcal{C}_s^{p,q} f = f$ in $\mathbb{B}_s$ for a $\overline{\partial}$-closed form f in $\mathbb{B}_s$ to a formula $\overline{\partial}\mathcal{C}_{n,s}^{p,q} f = f$ in $\mathbb{B}_n$ for a $\overline{\partial}$-closed form f in $\mathbb{B}_n$. To accomplish this we define *ameliorated* operators $\mathcal{C}_{n,s}^{p,q}$ by

$$\mathcal{C}_{n,s}^{p,q} := \mathsf{R}_n \mathcal{C}_s^{p,q} \mathsf{E}_s,$$

where for $n < s$, E_s (R_n) is the extension (restriction) operator that takes forms $\Omega = \sum_{I,J} \eta_{I,J} dw^I \wedge d\overline{w}^J$ in $\mathbb{B}_n$ ($\mathbb{B}^s$) and extends (restricts) them to $\mathbb{B}^s$ ($\mathbb{B}_n$) by

$$\begin{aligned} \mathsf{E}_s \left(\sum_{I,J} \eta_{I,J} dw^I \wedge d\overline{w}^J \right) &:= \sum_{I,J} (\eta_{I,J} \circ R) \, dw^I \wedge d\overline{w}^J, \\ \mathsf{R}_n \left(\sum \eta_{I,J} dw^I \wedge d\overline{w}^J \right) &:= \sum_{I,J \subset \{1,2,\ldots,n\}} (\eta_{I,J} \circ E) \, dw^I \wedge d\overline{w}^J . \end{aligned} \tag{17.30}$$

Here R is the natural orthogonal projection from $\mathbb{C}^s$ to $\mathbb{C}^n$ and E is the natural embedding of $\mathbb{C}^n$ into $\mathbb{C}^s$. In other words, we extend a form by taking the coefficients to be constant in the extra variables, and we restrict a form by discarding all wedge products of differentials involving the extra variables and restricting the coefficients accordingly.

This use of extension and restriction operators lets us obtain new reproducing kernels with different, sometimes more convenient, behavior as the variables approach each other and the boundary, the ameliorated kernels. A model case for this approach is the development in [**288**, Chapter 7] of the kernels described in Corollary B.7.

For $s > n$ we observe that the operator $\mathcal{C}_{n,s}^{p,q}$ has integral kernel

$$\mathcal{C}_{n,s}^{p,q}(w,z) := \int_{\sqrt{1-|w|^2}\mathbb{B}^{s-n}} \mathcal{C}_s^{p,q}((w,w'),(z,0))\,dV(w'), \qquad z,w \in \mathbb{B}_n, \tag{17.31}$$

where $\mathbb{B}_{s-n}$ denotes the unit ball in $\mathbb{C}^{s-n}$ with respect to the orthogonal decomposition $\mathbb{C}^s = \mathbb{C}^n \oplus \mathbb{C}^{s-n}$, and dV denotes Lebesgue measure. If $f(w)$ is a $\overline{\partial}$-closed form on $\mathbb{B}_n$ then $f(w,w') = f(w)$ is a $\overline{\partial}$-closed form on $\mathbb{B}^s$ and we have for $z \in \mathbb{B}_n$,

$$\begin{aligned} f(z) = f(z,0) &= \overline{\partial} \int_{\mathbb{B}^s} \mathcal{C}_s^{p,q}((w,w'),(z,0))\,f(w)\,dV(w)\,dV(w') \\ &= \overline{\partial} \int_{\mathbb{B}_n} \left(\int_{\sqrt{1-|w|^2}\mathbb{B}^{s-n}} \mathcal{C}_s^{p,q}((w,w'),(z,0))\,dV(w') \right) f(w)\,dV(w) \\ &= \overline{\partial} \int_{\mathbb{B}_n} \mathcal{C}_{n,s}^{p,q}(w,z)\,f(w)\,dV(w). \end{aligned}$$

We have proved the following:

THEOREM 17.22. *For all $s > n$ and $\overline{\partial}$-closed forms f in $\mathbb{B}_n$, we have*

$$\overline{\partial}\mathcal{C}_{n,s}^{p,q} f = f \text{ in } \mathbb{B}_n.$$

We will use only the case $p = 0$ of this theorem and from now on we restrict our attention to this case. The operators $\mathcal{C}_{n,s}^{0,0}$ have been computed in [**248**] and are given by

$$\mathcal{C}_{n,s}^{0,0} f(z) = \int_{\mathbb{B}_n} \sum_{j=0}^{n-1} c_{n,j,s} \frac{(1-|w|^2)^{s-n+j}(1-|z|^2)^j}{(1-\overline{w}z)^{s-n+j}(1-w\overline{z})^j} \mathcal{C}_n^{0,0}(w,z) \wedge f(w), \tag{17.32}$$

where

$$\begin{aligned} \mathcal{C}_n^{0,0}(w,z) = c_0 &\frac{(1-w\overline{z})^{n-1}}{\left\{|1-w\overline{z}|^2 - (1-|w|^2)(1-|z|^2)\right\}^n} \\ &\times \sum_{j=1}^n (-1)^{j-1}(\overline{w_j} - \overline{z_j}) \bigwedge_{k\neq j} d\overline{w_k} \bigwedge_{\ell=1}^n dw_\ell. \end{aligned} \tag{17.33}$$

A similar result holds for the operators $\mathcal{C}_{n,s}^{0,q}$. Define

$$\begin{aligned}
\Phi_{n,s}^{q}(w,z) &= \Phi_{n}^{q}(w,z)\left(\frac{1-|w|^{2}}{1-\overline{w}z}\right)^{s-n}\sum_{j=0}^{n-q-1}c_{j,n,s}\left(\frac{\left(1-|w|^{2}\right)\left(1-|z|^{2}\right)}{|1-w\overline{z}|^{2}}\right)^{j}\\
&= \frac{(1-w\overline{z})^{n-1-q}\left(1-|w|^{2}\right)^{q}}{\triangle(w,z)^{n}}\left(\frac{1-|w|^{2}}{1-\overline{w}z}\right)^{s-n}\\
&\quad\times\sum_{j=0}^{n-q-1}c_{j,n,s}\left(\frac{\left(1-|w|^{2}\right)\left(1-|z|^{2}\right)}{|1-w\overline{z}|^{2}}\right)^{j}\\
&= \sum_{j=0}^{n-q-1}c_{j,n,s}\frac{(1-w\overline{z})^{n-1-q-j}\left(1-|w|^{2}\right)^{s-n+q+j}\left(1-|z|^{2}\right)^{j}}{(1-\overline{w}z)^{s-n+j}\triangle(w,z)^{n}}.
\end{aligned}$$

Note that the numerator and denominator are *balanced* in the sense that the sum of the exponents in the denominator minus the corresponding sum in the numerator (counting $\triangle(w,z)$ double) is $s+n+j-(s+j-1)=n+1$, the exponent of the invariant measure of the ball $\mathbb{B}_n$. This is the pattern that shows up in the bounded operators T_{abc} in Lemma B.10 in the Appendix.

THEOREM 17.23. *Suppose that $s>n$ and $0\leq q\leq n-1$. Then we have*

$$\begin{aligned}
\mathcal{C}_{n,s}^{0,q}(w,z) &= \mathcal{C}_{n}^{0,q}(w,z)\left(\frac{1-|w|^{2}}{1-\overline{w}z}\right)^{s-n}\sum_{j=0}^{n-q-1}c_{j,n,s}\left(\frac{\left(1-|w|^{2}\right)\left(1-|z|^{2}\right)}{|1-w\overline{z}|^{2}}\right)^{j}\\
&= \Phi_{n,s}^{q}(w,z)\sum_{|J|=q}\sum_{k\notin J}(-1)^{\mu(k,J)}\left(\overline{z_k}-\overline{w_k}\right)d\overline{z}^{J}\wedge d\overline{w}^{(J\cup\{k\})^{c}}\wedge\omega_n(w).
\end{aligned}$$

PROOF. For $s>n$ recall that the kernels of the ameliorated operators $\mathcal{C}_{n,s}^{0,q}$ are given in (17.31). For ease of notation, we will set $k=s-n$, so we have $\mathbb{C}^s=\mathbb{C}^n\oplus\mathbb{C}^k$. Suppose that $0\leq q\leq n-1$. Recall from (17.25) that

$$\begin{aligned}
\mathcal{C}_{s}^{0,q}(w,z) &= (-1)^{q}\frac{(1-w\overline{z})^{s-q-1}\left(1-|w|^{2}\right)^{q}}{\triangle(w,z)^{s}}\\
&\quad\times\sum_{\nu\in P_s^q}\operatorname{sgn}(\nu)\left(\overline{w_{i_\nu}}-\overline{z_{i_\nu}}\right)\bigwedge_{j\in J_\nu}d\overline{w_j}\bigwedge_{l\in L_\nu}d\overline{z_l}\bigwedge\omega_s(w)\\
&= \sum_{\nu\in P_s^q}F_{s,i_\nu}^{q}(w,z)\bigwedge_{j\in J_\nu}d\overline{w_j}\bigwedge_{l\in L_\nu}d\overline{z_l}\bigwedge\omega_s(w).
\end{aligned}$$

where

$$\begin{aligned}
F_{s,i_\nu}^{q}(w,z) &= \Phi_{s}^{q}(w,z)\left(\overline{w_{i_\nu}}-\overline{z_{i_\nu}}\right)\\
&= \frac{(1-w\overline{z})^{s-q-1}\left(1-|w|^{2}\right)^{q}}{\triangle(w,z)^{s}}\left(\overline{w_{i_\nu}}-\overline{z_{i_\nu}}\right).
\end{aligned}$$

To compute the ameliorations of these kernels, we need only focus on the functional coefficient $F_{s,i_\nu}^{q}(w,z)$ of the kernel. First note that the ameliorated kernel can only give a contribution in the variables when $1\leq i_\nu\leq n$, since when $n+1\leq i_\nu\leq s$ the functional kernel becomes radial in certain variables and thus reduces to zero upon integration.

Next, for any $1 \leq i \leq n$ the corresponding functional coefficient $\digamma^q_{s,i}(w,z)$ has amelioration $\digamma^q_{n,s,i}(w,z)$ given by

$$\begin{aligned}\digamma^q_{n,s,i}(w,z) &= \int_{\sqrt{1-|w|^2}\mathbb{B}^{s-n}} \digamma^q_{s,i}((w,w'),(z,0))\, dV(w') \\ &= \int_{\sqrt{1-|w|^2}\mathbb{B}_k} \frac{(1-w\overline{z})^{s-q-1}(1-|w|^2-|w'|^2)^q(\overline{z}_i-\overline{w}_i)}{\triangle((w,w'),(z,0))^s} dV(w') \\ &= (\overline{z}_i-\overline{w}_i)(1-w\overline{z})^{s-q-1}\int_{\sqrt{1-|w|^2}\mathbb{B}_k} \frac{(1-|w|^2-|w'|^2)^q}{\triangle((w,w'),(z,0))^s} dV(w').\end{aligned}$$

Theorem 17.23 is a thus a consequence of the following elementary lemma, which we will use again in Section 17.3.2 when we are considering integration by parts.

LEMMA 17.24. *We have*

$$\begin{aligned}&(1-w\overline{z})^{s-q-1}\int_{\sqrt{1-|w|^2}\mathbb{B}^{s-n}} \frac{(1-|w|^2-|w'|^2)^q}{\triangle((w,w'),(z,0))^s} dV(w') \\ &= \frac{\pi^{s-n}}{(s-n)!}\Phi^q_n(w,z)\left(\frac{1-|w|^2}{1-\overline{w}z}\right)^{s-n}\sum_{j=0}^{n-q-1} c_{j,n,s}\left(\frac{(1-|w|^2)(1-|z|^2)}{|1-w\overline{z}|^2}\right)^j.\end{aligned}$$

□

17.3.2. Integration by Parts. We begin with an integration by parts formula involving a covariant derivative in [**249**, Lemma 2.1] that reduces the singularity of the solution kernel on the diagonal at the expense of differentiating the form. However, in order to develop a generalization to higher order forms, we replace the covariant derivative with a differential operator $\overline{\mathcal{Z}_{z,w}}$ which we now introduce.

Recall Charpentier's explicit solution $\mathcal{C}^{0,0}_n\eta$ to the $\overline{\partial}$ equation $\overline{\partial}\mathcal{C}^{0,0}_n\eta = \eta$ in the ball $\mathbb{B}_n$ when η is a $\overline{\partial}$-closed $(0,1)$-form with coefficients in $C\left(\overline{\mathbb{B}_n}\right)$. For $(w,z) \in \mathbb{B}_n \times \mathbb{B}_n$ and recalling that $\triangle(w,z)$ is defined in (15.23) and noting (17.33) the kernel is given by

$$\mathcal{C}^{0,0}_n(w,z) = c_0\frac{(1-w\overline{z})^{n-1}}{\triangle(w,z)^n}\sum_{j=1}^n(-1)^{j-1}(\overline{w_j}-\overline{z_j})\bigwedge_{k\neq j} d\overline{w_k}\bigwedge_{\ell=1}^n dw_\ell.$$

Define the Cauchy operator $\mathcal{S}_n$ on $\partial\mathbb{B}_n \times \mathbb{B}_n$ by its kernel

$$\mathcal{S}_n(\zeta,z) = c_1\frac{1}{\left(1-\overline{\zeta}z\right)^n}d\sigma(\zeta), \qquad (\zeta,z)\in\partial\mathbb{B}_n\times\mathbb{B}_n.$$

Let $\eta = \sum_{j=1}^n \eta_j d\overline{w_j}$ be a $(0,1)$-form with smooth coefficients. Let $\overline{\mathcal{Z}} = \overline{\mathcal{Z}_{z,w}}$ be the vector field acting in the variable $w = (w_1,\ldots,w_n)$ and parameterized by $z = (z_1,\ldots,z_n)$ given by

$$\overline{\mathcal{Z}} = \overline{\mathcal{Z}_{z,w}} = \sum_{j=1}^n(\overline{w_j}-\overline{z_j})\frac{\partial}{\partial\overline{w_j}}. \tag{17.34}$$

It will usually be understood from the context what the acting variable w and the parameter variable z are in $\overline{\mathcal{Z}_{z,w}}$ and we will then omit the subscripts and simply write $\overline{\mathcal{Z}}$ for $\overline{\mathcal{Z}_{z,w}}$.

DEFINITION 17.25. *For $m \geq 0$, define the mth order derivative $\overline{\mathcal{Z}}^m \eta$ of a $(0,1)$-form $\eta = \sum_{k=1}^n \eta_k(w)\, d\overline{w_k}$ to be the $(0,1)$-form obtained by componentwise differentiation holding monomials in $\overline{w} - \overline{z}$ fixed:*
(17.35)

$$\overline{\mathcal{Z}}^m \eta(w) := \sum_{k=1}^n \left(\overline{\mathcal{Z}}^m \eta_k\right)(w)\, d\overline{w_k} = \sum_{k=1}^n \left\{ \sum_{|\alpha|=m} \left(\overline{w-z}\right)^\alpha \frac{\partial^m \eta_k}{\partial \overline{w}^\alpha}(w) \right\} d\overline{w_k}.$$

LEMMA 17.26 ([**249**, Lemma 2.1]). *For all $m \geq 0$ and smooth $(0,1)$-forms $\eta = \sum_{k=1}^n \eta_k(w)\, d\overline{w_k}$, we have the formula,*

$$\begin{aligned} \mathcal{C}_n^{0,0}\eta(z) &= \int_{\mathbb{B}_n} \mathcal{C}_n^{0,0}(w,z) \wedge \eta(w) \\ &= \sum_{j=0}^{m-1} c_j \int_{\partial\mathbb{B}_n} \mathcal{S}_n(w,z) \left(\overline{\mathcal{Z}}^j \eta\right)\left[\overline{\mathcal{Z}}\right](w)\, d\sigma(w) \\ &\quad + c_m \int_{\mathbb{B}_n} \mathcal{C}_n^{0,0}(w,z) \wedge \overline{\mathcal{Z}}^m \eta(w). \end{aligned} \tag{17.36}$$

Here the $(0,1)$-form $\overline{\mathcal{Z}}^j \eta$ acts on the vector field $\overline{\mathcal{Z}}$ in the usual way:

$$\left(\overline{\mathcal{Z}}^j \eta\right)\left[\overline{\mathcal{Z}}\right] = \left(\sum_{k=1}^n \overline{\mathcal{Z}}^j \eta_k(w)\, d\overline{w_k}\right)\left(\sum_{i=1}^n \left(\overline{w_i} - \overline{z_i}\right) \frac{\partial}{\partial \overline{w_i}}\right) = \sum_{k=1}^n \left(\overline{w_k} - \overline{z_k}\right) \overline{\mathcal{Z}}^j \eta_k(w).$$

We can also rewrite the final integral in (17.36) as

$$\int_{\mathbb{B}_n} \mathcal{C}_n^{0,0}(w,z) \wedge \overline{\mathcal{Z}}^m \eta(w) = \int_{\mathbb{B}_n} \Phi_n^0(w,z) \left(\overline{\mathcal{Z}}^m \eta\right)\left[\overline{\mathcal{Z}}\right](w)\, dV(w).$$

Lemma 17.26 is proved by following verbatim the proof of [**249**, Lemma 2.1].

We now extend this lemma to $(0,q+1)$-forms. Let

$$\eta = \sum_{|I|=q+1} \eta_I(w)\, d\overline{w}^I$$

be a $(0,q+1)$-form with smooth coefficients. Given a $(0,q+1)$-form $\eta = \sum_{|I|=q+1} \eta_I d\overline{w}^I$ and an increasing sequence J of length $|J| = q$, we define the interior product $\eta \lrcorner d\overline{w}^J$ of η and $d\overline{w}^J$ by

$$\eta \lrcorner d\overline{w}^J := \sum_{|I|=q+1} \eta_I d\overline{w}^I \lrcorner d\overline{w}^J = \sum_{k \notin J} (-1)^{\mu(k,J)} \eta_{J\cup\{k\}} d\overline{w_k}, \tag{17.37}$$

since $d\overline{w}^I \lrcorner d\overline{w}^J = (-1)^{\mu(k,J)} d\overline{w_k}$ if $k \in I \setminus J$ is the $\mu(k,J)^{th}$ index in I, and 0 otherwise. Recall the vector field $\overline{\mathcal{Z}}$ defined in (17.34). The key relation between $\eta \lrcorner d\overline{w}^J$ and the vector field $\overline{\mathcal{Z}}$ is

$$\begin{aligned} \left(\eta \lrcorner d\overline{w}^J\right)\left(\overline{\mathcal{Z}}\right) &= \left(\sum_{k=1}^n (-1)^{\mu(k,J)} \eta_{J\cup\{k\}} d\overline{w_k}\right)\left(\sum_{j=1}^n \left(\overline{w_j} - \overline{z_j}\right) \frac{\partial}{\partial \overline{w_j}}\right) \\ &= \sum_{k=1}^n \left(\overline{w_k} - \overline{z_k}\right)(-1)^{\mu(k,J)} \eta_{J\cup\{k\}}. \end{aligned} \tag{17.38}$$

We now define an mth order derivative $\overline{\mathcal{D}}^m \eta$ of a $(0, q+1)$-form η using the interior product. In the case $q = 0$ we will have $\overline{\mathcal{D}}^m \eta = \left(\overline{\mathcal{Z}}^m \eta\right) \left[\overline{\mathcal{Z}}\right]$ for a $(0,1)$-form η.

REMARK 17.27. *We are motivated by the fact that the Charpentier kernel $\mathcal{C}_n^{0,q}(w,z)$ takes $(0,q+1)$-forms in w to $(0,q)$-forms in z. Thus in order to express the solution operator $\mathcal{C}_n^{0,q}$ in terms of a volume integral rather than the integration of a form in w and z, our definition of $\overline{\mathcal{D}}^m \eta$, even when $m = 0$, must include an appropriate exchange of w-differentials for z-differentials.*

DEFINITION 17.28. *Let $m \geq 0$. For a $(0,q+1)$-form $\eta = \sum_{|I|=q+1} \eta_I d\overline{w}^I$ in the variable w, define the $(0,q)$-form $\overline{\mathcal{D}}^m \eta$ in the variable z by*

$$\overline{\mathcal{D}}^m \eta(w) := \sum_{|J|=q} \overline{\mathcal{Z}}^m \left(\eta \lrcorner d\overline{w}^J\right) \left[\overline{\mathcal{Z}}\right](w)\, d\overline{z}^J.$$

Again it is usually understood what the acting and parameter variables are in $\overline{\mathcal{D}}^m$ but we will write $\overline{\mathcal{D}_{z,w}}^m \eta(w)$ when this may not be the case. Note that for a $(0,q+1)$-form $\eta = \sum_{|I|=q+1} \eta_I d\overline{w}^I$, we have

$$\eta = \sum_{|J|=q} \left(\eta \lrcorner d\overline{w}^J\right) \wedge d\overline{w}^J,$$

and using (17.35) the above definition yields

$$\begin{aligned}(17.39)\quad & \overline{\mathcal{D}}^m \eta(w) \\ &= \sum_{|J|=q} \overline{\mathcal{Z}}^m \left(\eta \lrcorner d\overline{w}^J\right) \left[\overline{\mathcal{Z}}\right](w)\, d\overline{z}^J \\ &= \sum_{|J|=q} \sum_{k=1}^{n} \left(\overline{w_k} - \overline{z_k}\right)(-1)^{\mu(k,J)} \left(\overline{\mathcal{Z}}^m \eta_{J \cup \{k\}}\right)(w)\, d\overline{z}^J \\ &= \sum_{|J|=q} \sum_{k=1}^{n} \left(\overline{w_k} - \overline{z_k}\right)(-1)^{\mu(k,J)} \left\{ \sum_{|\alpha|=m} \left(\overline{w-z}\right)^\alpha \frac{\partial^m \eta_{J\cup\{k\}}}{\partial \overline{w}^\alpha}(w) \right\} d\overline{z}^J.\end{aligned}$$

Thus the effect of $\overline{\mathcal{D}}^m$ on a basis element $\eta_I d\overline{w}^I$ is to replace a differential $d\overline{w_k}$ from $d\overline{w}^I$ ($I = J \cup \{k\}$) with the factor $(-1)^{\mu(k,J)}\left(\overline{w_k} - \overline{z_k}\right)$ (and this is accomplished by acting a $(0,1)$-form on $\overline{\mathcal{Z}}$), replace the remaining differential $d\overline{w}^J$ with $d\overline{z}^J$, and then to apply the differential operator $\overline{\mathcal{Z}}^m$ to the coefficient η_I. We will refer to the factor $\left(\overline{w_k} - \overline{z_k}\right)$ introduced above as a rogue factor since it is not associated with a derivative $\frac{\partial}{\partial \overline{w_k}}$ in the way that $\left(\overline{w-z}\right)^\alpha$ is associated with $\frac{\partial^{|\alpha|}}{\partial \overline{w}^{|\alpha|}}$. The point of this distinction will be explained in Section 17.5 on estimates for solution operators.

The following lemma expresses $\mathcal{C}_n^{0,q}\eta(z)$ in terms of integrals involving $\overline{\mathcal{D}}^j \eta$ for $0 \leq j \leq m$. Note that the overall effect is to reduce the singularity of the kernel on the diagonal by m factors of $\sqrt{\triangle(w,z)}$, at the cost of increasing by m the number of derivatives hitting the form η. Recall from (17.24) that

$$\Phi_n^\ell(w,z) := \frac{(1 - w\overline{z})^{n-1-\ell}\left(1 - |w|^2\right)^\ell}{\triangle(w,z)^n}.$$

We define the operator Φ_n^ℓ on forms η by

$$\Phi_n^\ell \eta(z) := \int_{\mathbb{B}_n} \Phi_n^\ell(w,z)\,\eta(w)\,dV(w).$$

LEMMA 17.29. *Let $q \geq 0$. For all $m \geq 0$ we have the formula,*

$$\mathcal{C}_n^{0,q}\eta(z) = \sum_{k=0}^{m-1} c_k \mathcal{S}_n\left(\overline{\mathcal{D}}^k \eta\right)(z) + \sum_{\ell=0}^{q} c_\ell \Phi_n^\ell \left(\overline{\mathcal{D}}^m \eta\right)(z). \tag{17.40}$$

We noted that Lemma 17.26 is proved by following the proof of [**249**, Lemma 2.1]. The proof of this lemma follows that same path together with the algebra that reduces matters to $(0,1)$-forms.

17.3.3. The Radial Derivative.

Recall the radial derivative from (14.10):

$$R = \sum_{j=1}^{n} w_j \frac{\partial}{\partial w_j}.$$

The following useful radial integration by parts lemma is essentially Lemma 2.2 of [**249**].

LEMMA 17.30. *Let $b > -1$. For $\Psi \in C\left(\overline{\mathbb{B}_n}\right) \cap C^\infty(\mathbb{B}_n)$ we have*

$$\int_{\mathbb{B}_n} (1-|w|^2)^b \Psi(w)\,dV(w)$$
$$= \int_{\mathbb{B}_n} (1-|w|^2)^{b+1} \left(\frac{n+b+1}{b+1} I + \frac{1}{b+1} R\right) \Psi(w)\,dV(w).$$

REMARK 17.31. *Typically the above lemma is applied with*

$$\Psi(w) = \frac{1}{(1-\overline{w}z)^s}\psi(w,z)$$

where z is a parameter in the ball $\mathbb{B}_n$. In that case,because $\frac{1}{(1-\overline{w}z)^s}$ is antiholomorphic in w,

$$R\Psi(w) = \frac{1}{(1-\overline{w}z)^s} R\psi(w,z).$$

We will also need to iterate Lemma 17.30, and for this purpose it is convenient to introduce for $m \geq 1$ the notation

$$R_b = R_{b,n} = \frac{n+b+1}{b+1} I + \frac{1}{b+1} R,$$
$$R_b^m = R_{b+m-1} R_{b+m-2} \cdots R_b = \prod_{k=1}^{m} R_{b+m-k}.$$

COROLLARY 17.32. *Let $b > -1$. For $\Psi \in C\left(\overline{\mathbb{B}_n}\right) \cap C^\infty(\mathbb{B}_n)$ we have*

$$\int_{\mathbb{B}_n} (1-|w|^2)^b \Psi(w)\,dV(w)$$
$$= \int_{\mathbb{B}_n} (1-|w|^2)^{b+m} R_b^m \Psi(w)\,dV(w).$$

REMARK 17.33. *It is common when working with Besov spaces to "trade" radial differentiation for multiplication by powers of* $1-|w|^2$, *roughly the distance to the boundary. This corollary is an instance of that. We will sometimes write this identity as*

$$\int_{\mathbb{B}_n} F(w)\,dV(w) = \int_{\mathbb{B}_n} \mathfrak{R}_b^m F(w)\,dV(w)$$

where

$$\mathfrak{R}_b^m := (1-|w|^2)^{b+m} R_b^m (1-|w|^2)^{-b}, \tag{17.41}$$

and provided that $\Psi(w) = (1-|w|^2)^{-b} F(w)$ *lies in* $C\left(\overline{\mathbb{B}_n}\right) \cap C^\infty(\mathbb{B}_n)$.

17.3.4. Integration by Parts in Ameliorated Kernels. We must now extend Lemma 17.29 and Corollary 17.32 to the ameliorated kernels $\mathcal{C}_{n,s}^{0,q}$ given by

$$\mathcal{C}_{n,s}^{0,q} = \mathsf{R}_n \mathcal{C}_s^{0,q} \mathsf{E}_s,$$

where E is defined in (17.30). Since Corollary 17.32 already applies to general functions $\Psi(w)$, we need only consider an extension of Lemma 17.29. Our procedure for doing this is to apply Lemma 17.29 to $\mathcal{C}_s^{0,q}$ in s dimensions, and then integrate out the additional variables using Lemma 17.24.

LEMMA 17.34. *Suppose that* $s > n$ *and* $0 \le q \le n-1$. *For all* $m \ge 0$ *and smooth* $(0, q+1)$*-forms* η *in* $\overline{\mathbb{B}_n}$ *we have the formula,*

$$\mathcal{C}_{n,s}^{0,q}\eta(z) = \sum_{k=0}^{m-1} c'_{k,n,s} \mathcal{S}_{n,s}\left(\overline{\mathcal{D}}^k \eta\right)\left[\overline{\mathcal{Z}}\right](z) + \sum_{\ell=0}^{q} c_{\ell,n,s} \Phi_{n,s}^\ell \left(\overline{\mathcal{D}}^m \eta\right)(z),$$

where the ameliorated operators $\mathcal{S}_{n,s}$ *and* $\Phi_{n,s}^\ell$ *have kernels given by,*

$$\mathcal{S}_{n,s}(w,z) = c_{n,s} \frac{\left(1-|w|^2\right)^{s-n-1}}{(1-\overline{w}z)^s} = c_{n,s}\left(\frac{1-|w|^2}{1-\overline{w}z}\right)^{s-n-1} \frac{1}{(1-\overline{w}z)^{n+1}},$$

$$\Phi_{n,s}^\ell(w,z) = \Phi_n^\ell(w,z)\left(\frac{1-|w|^2}{1-\overline{w}z}\right)^{s-n} \sum_{j=0}^{n-\ell-1} c_{j,n,s}\left(\frac{\left(1-|w|^2\right)\left(1-|z|^2\right)}{|1-w\overline{z}|^2}\right)^j.$$

PROOF. Recall that for a smooth $(0,q+1)$-form $\eta(w) = \sum_{|I|=q+1} \eta_I d\overline{w}^I$ in $\overline{\mathbb{B}_n}$, the $(0,q)$-form $\overline{\mathcal{D}^m}\mathsf{E}_s\eta$ is given by

$$\begin{aligned}
\overline{\mathcal{D}^m}\mathsf{E}_s\eta(w) &= \sum_{|J|=q} \overline{\mathcal{D}^m}\left(\eta \lrcorner d\overline{w}^J\right) d\overline{z}^J = \sum_{|J|=q} \overline{\mathcal{D}^m}\left(\sum_{k\notin J} (-1)^{\mu(k,J)} \eta_{J\cup\{k\}}(w)\, d\overline{w_k}\right) d\overline{z}^J \\
&= \sum_{|J|=q} \overline{\mathcal{D}^m}\left(\sum_{k\notin J} (-1)^{\mu(k,J)} \eta_{J\cup\{k\}}(w)\, d\overline{w_k}\right) d\overline{z}^J \\
&= \sum_{|J|=q}\sum_{k\notin J} (-1)^{\mu(k,J)} \left(\sum_{|\alpha|=m} \overline{(w_k-z_k)(w-z)^\alpha} \frac{\partial^m}{\partial \overline{w}^\alpha} \eta_{J\cup\{k\}}(w)\right),
\end{aligned}$$

where $J\cup\{k\}$ is a multi-index with entries in $\mathfrak{I}_n := \{1,2,\ldots,n\}$ since the coefficient η_I vanishes if I is not contained in $\mathfrak{I}_n$. Moreover, the multi-index α lies in $(\mathfrak{I}_n)^m$ since the coefficients η_I are constant in the variable $w' = (w_{n+1},\ldots,w_s)$. Thus

$$\overline{\mathcal{D}^m_{(z,0),(w,w')}}\mathsf{E}_s\eta = \overline{\mathcal{D}^m_{z,w}}\eta = \overline{\mathcal{D}^m}\eta,$$

and we compute that

$$\begin{aligned}
&\mathsf{R}_n\Phi_s^\ell\left(\overline{\mathcal{D}^m_{(z,0),(w,w')}}\mathsf{E}_s\eta\right)(z)\\
&=\Phi_s^\ell\left(\overline{\mathcal{D}^m}\eta\right)((z,0))\\
&=\sum_{|J|=q}\sum_{k\in\mathfrak{I}_n\setminus J}(-1)^{\mu(k,J)}\sum_{|\alpha|=m}\Phi_s^\ell\left(\overline{(w_k-z_k)(w-z)^\alpha}\frac{\partial^m}{\partial\overline{w}^\alpha}\eta_{J\cup\{k\}}((w,w'))\right)((z,0)),
\end{aligned}$$

where $J\cup\{k\}\subset\mathfrak{I}_n$ and $\alpha\in(\mathfrak{I}_n)^m$ and

$$\begin{aligned}
&\Phi_s^\ell\left(\overline{(w_k-z_k)(w-z)^\alpha}\frac{\partial^m}{\partial\overline{w}^\alpha}\eta_{J\cup\{k\}}(w)\right)((z,0))\\
&=\int_{\mathbb{B}^s}\frac{(1-w\overline{z})^{s-1-\ell}\left(1-|w|^2-|w'|^2\right)^\ell}{\triangle((w,w'),(z,0))^s}\\
&\qquad\times\overline{(w_k-z_k)(w-z)^\alpha}\frac{\partial^m}{\partial\overline{w}^\alpha}\eta_{J\cup\{k\}}(w)\,dV((w,w'))\\
&=\int_{\mathbb{B}_n}\left\{(1-w\overline{z})^{s-\ell-1}\int_{\mathbb{B}^{s-n}}\frac{\left(1-|w|^2-|w'|^2\right)^\ell}{\triangle((w,w'),(z,0))^s}dV(w')\right\}\\
&\qquad\qquad\times\overline{(w_k-z_k)(w-z)^\alpha}\frac{\partial^m}{\partial\overline{w}^\alpha}\eta_{J\cup\{k\}}(w)\,dV(w).
\end{aligned}$$

By Lemma 17.24 the term in braces above equals

$$\frac{\pi^{s-n}}{(s-n)!}\Phi_n^\ell(w,z)\left(\frac{1-|w|^2}{1-\overline{w}z}\right)^{s-n}\sum_{j=0}^{n-\ell-1}c_{j,n,s}\left(\frac{\left(1-|w|^2\right)\left(1-|z|^2\right)}{|1-w\overline{z}|^2}\right)^j,$$

and now performing the sum $\sum_{|J|=q}\sum_{k\in\mathfrak{I}_n\setminus J}(-1)^{\mu(k,J)}\sum_{|\alpha|=m}$ yields

$$\mathsf{R}_n\Phi_s^\ell\left(\overline{\mathcal{D}^m_{(z,0)}}\mathsf{E}_s\eta\right)(z)=\Phi_s^\ell\left(\overline{\mathcal{D}^m_z}\eta\right)((z,0))=\Phi_{n,s}^\ell\left(\overline{\mathcal{D}^m_z}\eta\right)(z).\tag{17.42}$$

An easy calculation using [**288**, page 14, equation (1)] shows that

$$\mathsf{R}_n\mathcal{S}_s\left(\mathsf{E}_s\overline{\mathcal{D}^k_z}\eta\right)((z,0))=\mathcal{S}_s\left(\overline{\mathcal{D}^k_z}\eta\right)((z,0))=\mathcal{S}_{n,s}\left(\overline{\mathcal{D}^k_z}\eta\right)(z),\tag{17.43}$$

and now the conclusion of Lemma 17.34 follows from (17.42), (17.43), the definition $\mathcal{C}^{0,q}_{n,s}=\mathsf{R}_n\mathcal{C}^{0,q}_s\mathsf{E}_s$, and Lemma 17.29. □

17.4. The Koszul Complex

In Section 17.1.2 we saw how to reduce proving a simple version of a corona theorem to a problem of solving certain $\overline{\partial}$-equations. The algebraic work in formulating those equations could have been cast in the language of the Koszul complex; but, because we were in one dimension and only dealing with two functions, finding the equations was elementary and so we did not need to introduce algebraic machinery. The algebraic work when dealing with n functions is less straightforward; it can be seen in [**161**].

We will use the Koszul complex here to help formulate a set of $\overline{\partial}$-equations, the solving of which will let us complete our proof. However now, working in higher dimensions, the algebra is more substantial. That algebra is presented in [**208**] for the finite dimensional setting and a more detailed treatment in that setting is in [**294**, Section 5.5.3]. Here we just review the basics. There is an interesting discussion of the Koszul complex in [**55**].

Suppose now that $g = (g_j)_{j=1}^{\infty}$ satisfies $|g|^2 = \sum_{j=1}^{\infty} |g_j|^2 \geq \delta^2 > 0$, let

$$\Omega_0^1 = \frac{\overline{g}}{|g|^2} = \left(\frac{\overline{g_j}}{|g|^2}\right)_{j=1}^{\infty} = \left(\Omega_0^1(j)\right)_{j=1}^{\infty},$$

which we view as a 1-tensor (in $\ell^2 = \mathbb{C}^\infty$) of $(0,0)$-forms with components $\Omega_0^1(j) = \frac{\overline{g_j}}{|g|^2}$. Fix h holomorphic as in Theorem 17.11. Then $f = \Omega_0^1 h$ satisfies $\mathrm{Mult}_g f = f \cdot g = h$, but in general fails to be holomorphic. The Koszul complex provides a scheme which we now recall for solving a sequence of $\overline{\partial}$-equations that result in a correction term $\Lambda_g \Gamma_0^2$ that when subtracted from f above yields a holomorphic solution to (17.16).

The 1-tensor of $(0,1)$-forms $\overline{\partial}\Omega_0 = \left(\overline{\partial}\frac{\overline{g_j}}{|g|^2}\right)_{j=1}^{\infty} = \left(\overline{\partial}\Omega_0^1(j)\right)_{j=1}^{\infty}$ is given by

$$\overline{\partial}\Omega_0^1(j) = \overline{\partial}\frac{\overline{g_j}}{|g|^2} = \frac{|g|^2\,\overline{\partial g_j} - \overline{g_j}\,\overline{\partial}|g|^2}{|g|^4} = \frac{1}{|g|^4}\sum_{k=1}^{\infty} g_k \overline{(g_k \partial g_j - g_j \partial g_k)}.$$

Defining the antisymmetric 2-tensor Ω_1^2 of $(0,1)$-forms by

$$\Omega_1^2 = \left[\Omega_1^2(j,k)\right]_{j,k=1}^{\infty} = \left[\frac{\overline{\{g_k \partial g_j - g_j \partial g_k\}}}{|g|^4}\right]_{j,k=1}^{\infty}$$

we can continue with

$$\overline{\partial}\Omega_0^1 = \Lambda_g \Omega_1^2 = \left[\sum_{k=1}^{\infty} \Omega_1^2(j,k)\, g_k\right]_{j=1}^{\infty}.$$

Here $\Lambda_g \Omega_1^2$ denotes its contraction by the vector g in the second index.

Repeating this process and by induction yields

$$\overline{\partial}\Omega_q^{q+1} = \Lambda_g \Omega_{q+1}^{q+2}, \qquad 0 \leq q \leq n,$$

where Ω_q^{q+1} is an alternating $(q+1)$-tensor of $(0,q)$-forms. Recall that h is holomorphic. When $q = n$ we have that $\Omega_n^{n+1} h$ is $\overline{\partial}$-closed and this allows us to solve a chain of $\overline{\partial}$-equations

$$\overline{\partial}\Gamma_{q-2}^q = \Omega_{q-1}^q h - \Lambda_g \Gamma_{q-1}^{q+1},$$

for alternating q-tensors Γ_{q-2}^q of $(0,q-2)$-forms, using the ameliorated Charpentier solution operators $\mathcal{C}_{n,s}^{0,q}$ defined in (17.31) above (note that our notation suppresses the dependence of Γ on h). With the convention that $\Gamma_n^{n+2} = 0$ we have

$$\overline{\partial}\left(\Omega_q^{q+1} h - \Lambda_g \Gamma_q^{q+2}\right) = 0, \qquad 0 \leq q \leq n, \tag{17.44}$$
$$\overline{\partial}\Gamma_{q-1}^{q+1} = \Omega_q^{q+1} h - \Lambda_g \Gamma_q^{q+2}, \qquad 1 \leq q \leq n.$$

Now

$$f := \Omega_0^1 h - \Lambda_g \Gamma_0^2$$

is holomorphic by the first line in (17.44) with $q = 0$, and since Γ_0^2 is antisymmetric, we compute that $\Lambda_g \Gamma_0^2 \cdot g = \Gamma_0^2(g,g) = 0$ and

$$\mathrm{Mult}_g f = f \cdot g = \Omega_0^1 h \cdot g - \Lambda_g \Gamma_0^2 \cdot g = h - 0 = h.$$

Thus $f = (f_i)_{i=1}^{\infty}$ is a vector of holomorphic functions satisfying (17.16). The estimate (17.15) is the subject of the rest of this chapter.

17.4.1. Wedge products and factorization of the Koszul complex. Here we record the factorization of the Koszul complex in Andersson and Carlsson [**29**]. To describe the factorization we introduce an exterior algebra structure on $\ell^2 = \mathbb{C}^{\infty}$. Let $\{e_1, e_2, \ldots\}$ be the usual basis in $\mathbb{C}^{\infty}$, and for an increasing multi-index $I = (i_1, \ldots, i_\ell)$ of integers in $\mathbb{N}$, define

$$e_I = e_{i_1} \wedge e_{i_2} \wedge \cdots \wedge e_{i_\ell},$$

where we use $\wedge$ to denote the wedge product in the exterior algebra $\Lambda^*(\mathbb{C}^{\infty})$ of $\mathbb{C}^{\infty}$, as well as for the wedge product on forms in $\mathbb{C}^n$. Note that $\{e_I : |I| = r\}$ is a basis for the alternating r-tensors on $\mathbb{C}^{\infty}$.

If $f = \sum_{|I|=r} f_I e_I$ is an alternating r-tensor on $\mathbb{C}^{\infty}$ with values that are $(0,k)$-forms in $\mathbb{C}^n$, which may be viewed as a member of the exterior algebra of $\mathbb{C}^{\infty} \otimes \mathbb{C}^n$, and if $g = \sum_{|J|=s} g_J e_J$ is an alternating s-tensor on $\mathbb{C}^{\infty}$ with values that are $(0,\ell)$-forms in $\mathbb{C}^n$, then as in [**29**] we define the wedge product $f \wedge g$ in the exterior algebra of $\mathbb{C}^{\infty} \otimes \mathbb{C}^n$ to be the alternating $(r+s)$-tensor on $\mathbb{C}^{\infty}$ with values that are $(0, k+\ell)$-forms in $\mathbb{C}^n$ given by

$$\begin{aligned} f \wedge g &= \left(\sum_{|I|=r} f_I e_I\right) \wedge \left(\sum_{|J|=s} g_J e_J\right) \\ &= \sum_{|I|=r,|J|=s} (f_I \wedge g_J)(e_I \wedge e_J) \\ &= \sum_{|K|=r+s} \left(\pm \sum_{I+J=K} f_I \wedge g_J\right) e_K. \end{aligned} \tag{17.45}$$

Note that we simply write the exterior product of an element from $\Lambda^*(\mathbb{C}^{\infty})$ with an element from $\Lambda^*(\mathbb{C}^n)$ as juxtaposition, without writing an explicit wedge symbol. This should cause no confusion since the basis we use in $\Lambda^*(\mathbb{C}^{\infty})$ is $\{e_i\}_{i=1}^{\infty}$, while the basis we use in $\Lambda^*(\mathbb{C}^n)$ is $\{dz_j, d\widehat{z_j}\}_{j=1}^n$, quite different in both appearance and interpretation.

In terms of this notation we then have the following factorization of Andersson and Carlsson in [**29**, Theorem 3.1]:

$$\Omega_0^1 \wedge \bigwedge_{i=1}^{\ell} \widetilde{\Omega_0^1} = \left(\sum_{k_0=1}^{\infty} \frac{\overline{g_{k_0}}}{|g|^2} e_{k_0}\right) \wedge \bigwedge_{i=1}^{\ell} \left(\sum_{k_i=1}^{\infty} \frac{\overline{\partial g_{k_i}}}{|g|^2} e_{k_i}\right) = -\frac{1}{\ell+1} \Omega_\ell^{\ell+1}, \tag{17.46}$$

where

$$\Omega_0^1 = \left(\frac{\overline{g_i}}{|g|^2}\right)_{i=1}^{\infty} \text{ and } \widetilde{\Omega_0^1} = \left(\frac{\overline{\partial g_i}}{|g|^2}\right)_{i=1}^{\infty}.$$

The factorization in [**29**] is proved in the finite dimensional case, but this extends to the infinite dimensional case by continuity. Since the ℓ^2-norm is quasi-multiplicative on wedge products by [**29**, Lemma 5.1] we have

$$\left|\Omega_\ell^{\ell+1}\right|_\ell^2 \le C_\ell \left|\Omega_0^1\right|^2 \left|\widetilde{\Omega_0^1}\right|^{2\ell}, \qquad 0 \le \ell \le n, \tag{17.47}$$

where the constant C_ℓ depends only on the number of factors ℓ in the wedge product, and *not* on the underlying dimension of the vector space (which is infinite for $\ell^2 = \mathbb{C}^\infty$).

17.4.2. Real Variable Analogs of Besov Spaces. In order to handle the operators arising from integration by parts formulas below, we will need yet additional equivalent norms on $B^\sigma_{p,m}(\mathbb{B}_n;\ell^2)$ and we will need to deal with functions f on $\mathbb{B}_n$ that are not necessarily holomorphic. In particular we will want to consider tensor products such as

$$\widetilde{\Omega^1_0} \otimes \widetilde{\Omega^1_0} = \left(\sum_{i=1}^{\infty} \frac{\overline{\partial g_i}}{|g|^2} e_i\right) \otimes \left(\sum_{j=1}^{\infty} \frac{\overline{\partial g_j}}{|g|^2} e_j\right) = \sum_{i,j=1}^{\infty} \frac{\overline{\partial g_i} \otimes \overline{\partial g_j}}{|g|^4} e_i \otimes e_j,$$

which are not holomorphic, as well as more general expressions such as $\mathcal{X}^\alpha \widetilde{\Omega^1_0} \otimes \mathcal{X}^\beta \widetilde{\Omega^1_0}$ where $\mathcal{X}^m$ denotes the vector derivative which we define in Definition 17.39 below. We will do this using the almost invariant holomorphic derivative and other notation and constructions from Section 14.5. We will also take advantage of the fact the ℓ^2-norm is *multiplicative* on tensor products.

DEFINITION 17.35 (see [**47**]). *Suppose $\sigma \geq 0$, $1 < p < \infty$ and $m \geq 1$. We define a "tree semi-norm" $\|\cdot\|^*_{B^\sigma_{p,m}(\mathbb{B}_n;\ell^2)}$ by*

$$\|f\|^*_{B^\sigma_{p,m}(\mathbb{B}_n;\ell^2)} := \left(\sum_{\alpha \in \mathcal{T}_n} \int_{B_d(c_\alpha, C_2)} \left|(1-|z|^2)^\sigma D^m_{c_\alpha} f(z)\right|^p d\lambda_n(z)\right)^{1/p}. \tag{17.48}$$

The results from Section 14.2.1 also continue to hold after making the natural adjustments. In particular we still have the operators $R^{\gamma,t}: \mathrm{Hol}(\mathbb{B}_n) \to \mathrm{Hol}(\mathbb{B}_n)$. We can then, if m is a large enough integer, build seminorms using the pth power integrals of the functionals $(1-|z|^2)^m \nabla^m f(z)$ or $(1-|z|^2)^m R^{\gamma,m} f(z)$ where, as before, we require that

$$n+\gamma, m+n+\gamma \notin -\mathbb{N}.$$

The parameter σ shows up in the choice of measure $(1-|z|^2)^{\sigma p} d\lambda_n$ where before we used $d\lambda_n$. We now require that m be large enough so that $(1-|z|^2)^{(m+\sigma)p} d\lambda_n$ is a finite measure. We then have a complete analog of Proposition 14.4: if one of the seminorms is finite then they are all finite and are equivalent to each other.

The extensions of the proof of Proposition 14.4 to $\sigma \geq 0$ and ℓ^2-valued f is routine. In summary we have, with $\psi(z) = 1-|z|^2$, we have for $1 < p < \infty$,

$$\begin{aligned}
&\left\|\psi^{m_1+\sigma} R^{\gamma,m_1} f\right\|_{L^p(d\lambda_n;\ell^2)} \\
&\lesssim \sum_{k=0}^{m_2-1} \left|\nabla^k f(0)\right| + \left(\int_{\mathbb{B}_n} \left|(1-|z|^2)^{m_2+\sigma} \nabla^{m_2} f(z)\right|^p d\lambda_n(z)\right)^{1/p} \\
&\lesssim \left\|\psi^{m_1+\sigma} R^{\gamma,m_1} f\right\|_{L^p(d\lambda_n;\ell^2)}
\end{aligned}$$

for all $1 < p < \infty_1, m_1, m_2 \in \mathbb{N}, m_2 > n/p - \sigma$, $m_1 + n + \gamma \notin -\mathbb{N}$, and where the implied constant depends only on σ, m_1, m_2, n, γ and p.

There is one further equivalent norm involving the radial derivative

$$Rf(z) = z \cdot \nabla f(z) = \sum_{j=1}^{n} z_j \frac{\partial f}{\partial z_j}(z), \tag{17.49}$$

and its iterates $R^k = R \circ R \circ \cdots \circ R$ (k times). See [**63**].

PROPOSITION 17.36. *Suppose that $\sigma \geq 0$, $0 < p < \infty$ and $f \in \mathrm{Hol}\left(\mathbb{B}_n; \ell^2\right)$. Then*

$$\begin{aligned} &\sum_{k=0}^{m_1} \left(\int_{\mathbb{B}_n} \left| (1-|z|^2)^{m_1+\sigma} R^k f(z) \right|^p d\lambda_n(z) \right)^{1/p} \\ &\approx \sum_{k=0}^{m_2-1} \left| \nabla^k f(0) \right| + \left(\int_{\mathbb{B}_n} \left| (1-|z|^2)^{m_2+\sigma} \nabla^{m_2} f(z) \right|^p d\lambda_n(z) \right)^{1/p} \end{aligned} \tag{17.50}$$

for all $m_1, m_2 > n/p - \sigma$, $m_1 + n + \gamma \notin -\mathbb{N}$, $m_2 \in \mathbb{N}$, and where the constants in the equivalence depend only on σ, m_1, m_2, n and p.

The seminorms $\|\cdot\|^*_{B^\sigma_{p,m}(\mathbb{B}_n;\ell^2)}$ turn out to be independent of $m > 2(n/p - \sigma)$. We will obtain this fact as a corollary of the equivalence of the standard norm in (17.50) with the corresponding norm in Proposition 14.4 using the " radial" derivative $R^{0,m}$. Note that the restriction $m > 2(n/p-\sigma)$ is dictated by the fact that $\left|D^m_{v_i} f(z)\right|$ involves the factor $(1-|z|^2)^{m/2}$ times m^{th} order tangential derivatives of f, and so we must have that $(1-|z|^2)^{(m/2+\sigma)p} d\lambda_n(z)$ is a finite measure, i.e. $(m/2+\sigma)p - n - 1 > -1$. The scalar case $\sigma = 0$ of the following lemma is [**47**, Lemma 6.4].

LEMMA 17.37. *Let $1 < p < \infty$, $\sigma \geq 0$ and $m > 2(n/p - \sigma)$. Denote by $B_\beta(c, C)$ the ball center c radius C in the Bergman metric β. Then for $f \in \mathrm{Hol}(\mathbb{B}_n)$,*

(17.51)

$$\begin{aligned} &\|f\|^*_{B^\sigma_{p,m}(\mathbb{B}_n;\ell^2)} + \sum_{j=0}^{m-1} \left| \nabla^j f(0) \right| \\ &= \left(\sum_{\alpha \in \mathcal{T}_n} \int_{B_\beta(c_\alpha, C_2)} \left| (1-|z|^2)^\sigma D^m_{c_\alpha} f(z) \right|^p d\lambda_n(z) \right)^{1/p} + \sum_{j=0}^{m-1} \left| \nabla^j f(0) \right| \\ &\approx \left(\int_{\mathbb{B}_n} \left| (1-|z|^2)^{m+\sigma} R^{\sigma,m} f(z) \right|^p d\lambda_n(z) \right)^{1/p} + \sum_{j=0}^{m-1} \left| \nabla^j f(0) \right| \\ &= \|f\|_{B^\sigma_{p,m}(\mathbb{B}_n;\ell^2)} . \end{aligned}$$

It is a straightforward exercise to adapt the proof in [**47**] to the case $\sigma \geq 0$ and ℓ^2-valued f.

In Chapter 14 we introduced the operators D_a and $\overline{D}_a$ as well as their higher powers. We will use those operators for, among other things, building new operators, $\mathcal{X}^n$, $\mathcal{Y}^n$. The new operators are related to powers of radial differentiation. The $\mathcal{X}^n$ interact well with the generalized integration by parts formulas below. The $\mathcal{Y}^n$ are well suited for studying certain functions that arise in our work with the Koszul complex.

DEFINITION 17.38. *For $m \in \mathbb{N}$ and $f \in C^\infty\left(\mathbb{B}_n;\ell^2\right)$ we define $\Theta^m f(a,z) = D_a^m f(z)$ for $a,z \in \mathbb{B}_n$, and then set*

$$D^m f(z) = \Theta^m f(z,z) = D_z^m f(z), \qquad z \in \mathbb{B}_n.$$

Note that in this definition, we iterate the operator D_z holding z fixed, and then evaluate the result at the same z. If we combine Lemmas 14.23 and 17.37 we obtain that for $f \in \mathrm{Hol}(\mathbb{B}_n)$,

$$\|f\|_{B^\sigma_{p,m}(\mathbb{B}_n;\ell^2)} \approx \sum_{j=0}^{m-1} \left|\nabla^j f(0)\right| + \left(\int_{\mathbb{B}_n} \left|(1-|z|^2)^\sigma D^m f(z)\right|^p d\lambda_n(z)\right)^{1/p}.$$

DEFINITION 17.39. *We denote by $\mathcal{X}^m$ the vector of all differential operators of the form $X_1X_2\cdots X_m$ where each X_i is either $1-|z|^2$ times the identity operator I, the operator $\overline{D}$, or the operator $(1-|z|^2)R$. Just as in Definition 17.38, we calculate the products $X_1X_2\cdots X_m$ by composing $\overline{D_a}$ and $(1-|a|^2)R$ and then setting $a=z$ at the end. Note that $\overline{D_a}$ and $(1-|a|^2)R$ commute since the first is an antiholomorphic derivative and the coefficient z in $R = z\cdot\nabla$ is holomorphic. Similarly we denote by $\mathcal{Y}^m$ the corresponding products of $(1-|z|^2)I$, D (instead of $\overline{D}$) and $(1-|z|^2)R$.*

In the iterated derivative $\mathcal{X}^m$ we are differentiating only with the *antiholomorphic* derivative $\overline{D}$ or the *holomorphic* derivative R. When f is holomorphic, we thus have $\mathcal{X}^m f \sim \left\{(1-|z|^2)^m R^k f\right\}_{k=0}^m$. The reason we allow $1-|z|^2$ times the identity I to occur in $\mathcal{X}^m$ is that this produces a norm (as opposed to just a seminorm) without including the term $\sum_{k=0}^{m-1}\left|\nabla^k f(0)\right|$. We define the norm $\|\cdot\|_{B^\sigma_{p,m}(\mathbb{B}_n;\ell^2)}$ for *smooth* f on the ball $\mathbb{B}_n$ by

$$\|f\|_{B^\sigma_{p,m}(\mathbb{B}_n;\ell^2)} := \left(\sum_{k=0}^m \int_{\mathbb{B}_n} \left|(1-|z|^2)^{m+\sigma} R^k f(z)\right|^p d\lambda_n(z)\right)^{1/p},$$

and note that provided $m+\sigma > n/p$, this provides an equivalent norm for the Besov space $B_p^\sigma\left(\mathbb{B}_n;\ell^2\right)$ of holomorphic functions on $\mathbb{B}_n$. These considerations motivate the following two definitions of a *real-variable* analog of the norm $\|\cdot\|_{B^\sigma_{p,m}(\mathbb{B}_n;\ell^2)}$.

DEFINITION 17.40. *We define the* norms $\|\cdot\|_{\Lambda^\sigma_{p,m}(\mathbb{B}_n;\ell^2)}$ *and* $\|\cdot\|_{\Phi^\sigma_{p,m}(\mathbb{B}_n;\ell^2)}$ *for $f = (f_i)_{i=1}^\infty$ smooth on the ball $\mathbb{B}_n$ by*

$$\|f\|_{\Lambda^\sigma_{p,m}(\mathbb{B}_n;\ell^2)} := \left(\int_{\mathbb{B}_n} \left|(1-|z|^2)^\sigma \mathcal{X}^m f(z)\right|^p d\lambda_n(z)\right)^{1/p}, \tag{17.52}$$

$$\|f\|_{\Phi^\sigma_{p,m}(\mathbb{B}_n;\ell^2)} := \left(\int_{\mathbb{B}_n} \left|(1-|z|^2)^\sigma \mathcal{Y}^m f(z)\right|^p d\lambda_n(z)\right)^{1/p}.$$

It is *not* true that either of the norms $\|\cdot\|_{\Lambda^\sigma_{p,m}(\mathbb{B}_n;\ell^2)}$ or $\|\cdot\|_{\Phi^\sigma_{p,m}(\mathbb{B}_n;\ell^2)}$ are independent of m for large m when acting on smooth functions. Indeed, a smooth function f could artificially be given a highly oscillating $(m+1)^{st}$ derivative, so that the $\Lambda^\sigma_{p,m+1}\left(\mathbb{B}_n;\ell^2\right)$ norm of f is infinite while the $\Lambda^\sigma_{p,m}\left(\mathbb{B}_n;\ell^2\right)$ norm is finite. However, Lemmas 14.23 and 17.37 show the equivalence of norms when restricted to holomorphic vector functions:

LEMMA 17.41. *Let $1 < p < \infty$, $\sigma \geq 0$ and $m > 2(n/p - \sigma)$. Then for f a holomorphic vector function we have*

$$\|f\|_{B^\sigma_{p,m}(\mathbb{B}_n;\ell^2)} \approx \|f\|_{\Lambda^\sigma_{p,m}(\mathbb{B}_n;\ell^2)} \approx \|f\|_{\Phi^\sigma_{p,m}(\mathbb{B}_n;\ell^2)}. \tag{17.53}$$

We need some information about multipliers. We begin with the standard arguments showing that the pointwise multipliers in $M_{B^\sigma_p(\mathbb{B}_n)\to B^\sigma_p(\mathbb{B}_n;\ell^2)}$ are bounded.

LEMMA 17.42.

$$M_{B^\sigma_p(\mathbb{B}_n)\to B^\sigma_p(\mathbb{B}_n;\ell^2)} \subset H^\infty\left(\mathbb{B}_n;\ell^2\right) \cap B^\sigma_p\left(\mathbb{B}_n;\ell^2\right). \tag{17.54}$$

PROOF. The function $1 \in B^\sigma_p(\mathbb{B}_n)$. Hence if $\varphi \in M_{B^\sigma_p(\mathbb{B}_n)\to B^\sigma_p(\mathbb{B}_n;\ell^2)}$, then $\varphi \in B^\sigma_p\left(\mathbb{B}_n;\ell^2\right)$.

If e_z denotes point evaluation at $z \in \mathbb{B}_n$, $x \in \ell^2$ and $f \in B^\sigma_p(\mathbb{B}_n)$, then the calculation

$$\begin{aligned}
\left\langle f, \mathbb{M}^*_\varphi (xe_z)\right\rangle_{B^\sigma_p(\mathbb{B}_n)} &= \left\langle \mathbb{M}_\varphi f, xe_z\right\rangle_{B^\sigma_p(\mathbb{B}_n;\ell^2)} = \sum_{i=1}^\infty \left\langle \varphi_i f, x_i e_z\right\rangle_{B^\sigma_p(\mathbb{B}_n)} \\
&= \sum_{i=1}^\infty \overline{x_i}\varphi_i(z) f(z) = \sum_{i=1}^\infty \overline{x_i}\varphi_i(z)\left\langle f, e_z\right\rangle_{B^\sigma_p(\mathbb{B}_n)} \\
&= \sum_{i=1}^\infty \left\langle f, \overline{\varphi_i(z)} x_i e_z\right\rangle_{B^\sigma_p(\mathbb{B}_n)} = \left\langle f, \left\langle x, \varphi(z)\right\rangle_{\ell^2} e_z\right\rangle_{B^\sigma_p(\mathbb{B}_n)},
\end{aligned}$$

shows that

$$\mathbb{M}^*_\varphi (xe_z) = \left\langle x, \varphi(z)\right\rangle_{\ell^2} e_z.$$

This yields

$$\begin{aligned}
\left|\left\langle x, \varphi(z)\right\rangle_{\ell^2}\right| \|e_z\|_{B^\sigma_p(\mathbb{B}_n)^*} &= \left\|\mathbb{M}^*_\varphi (xe_z)\right\|_{B^\sigma_p(\mathbb{B}_n)^*} \\
&\leq \left\|\mathbb{M}^*_\varphi\right\|_{B^\sigma_p(\mathbb{B}_n;\ell^2)^*\to B^\sigma_p(\mathbb{B}_n)^*} \|xe_z\|_{B^\sigma_p(\mathbb{B}_n;\ell^2)^*} \\
&= \|\mathbb{M}_\varphi\|_{B^\sigma_p(\mathbb{B}_n)\to B^\sigma_p(\mathbb{B}_n;\ell^2)} |x| \|e_z\|_{B^\sigma_p(\mathbb{B}_n)^*},
\end{aligned}$$

which gives

$$|\varphi(z)| = \sup_{x\neq 0} \frac{\left|\left\langle x, \varphi(z)\right\rangle_{\ell^2}\right|}{|x|} \leq \|\mathbb{M}_\varphi\|_{B^\sigma_p(\mathbb{B}_n)\to B^\sigma_p(\mathbb{B}_n;\ell^2)}, \qquad z \in \mathbb{B}_n,$$

and completes the proof of Lemma 17.42. □

In estimating the holomorphic function g in the Koszul complex we will use the following multilinear inequality whose scalar version is, after translating notation, [**249**, Theorem 3.5].

PROPOSITION 17.43. *Suppose that $1 < p < \infty$, $0 \leq \sigma < \infty$, $M \geq 1$, $m > 2(n/p - \sigma)$ and $\alpha = (\alpha_0, \ldots, \alpha_M) \in \mathbb{Z}^{M+1}_+$ with $|\alpha| = m$. For $g \in M_{B^\sigma_p(\mathbb{B}_n)\to B^\sigma_p(\mathbb{B}_n;\ell^2)}$ and $h \in B^\sigma_p(\mathbb{B}_n)$ we have,*

$$\begin{aligned}
&\int_{\mathbb{B}_n} (1-|z|^2)^{p\sigma} \left|\left(\mathcal{Y}^{\alpha_1} g\right)(z)\right|^p \cdots \left|\left(\mathcal{Y}^{\alpha_M} g\right)(z)\right|^p \left|\left(\mathcal{Y}^{\alpha_0} h\right)(z)\right|^p d\lambda_n(z) \\
&\qquad \lesssim \|\mathbb{M}_g\|^{Mp}_{B^\sigma_p(\mathbb{B}_n)\to B^\sigma_p(\mathbb{B}_n;\ell^2)} \|h\|^p_{B^\sigma_p(\mathbb{B}_n)}.
\end{aligned}$$

REMARK 17.44. *The inequalities for $M=1$ in Proposition 17.43 actually characterize multipliers g in the sense that a function $g \in B_p^\sigma(\mathbb{B}_n;\ell^2) \cap H^\infty(\mathbb{B}_n;\ell^2)$ is in $M_{B_p^\sigma(\mathbb{B}_n)\to B_p^\sigma(\mathbb{B}_n;\ell^2)}$* if and only if *the inequalities with $M=1$ in Proposition 17.43 hold. This follows from noting that each term in the Leibniz expansion of $\mathcal{Y}^m(gh)$ occurs on the left side of the display above with $M=1$.*

Proposition 17.43 is proved by adapting the proof of Ortega and Fabrega in [**249**, Theorem 3.5] to ℓ^2-valued functions. This argument uses the complex interpolation theorem of Beatrous [**63**] and Ligocka [**207**], which extends to Hilbert space valued functions with the same proof. In order to apply this extension we will need the following operator norm inequality.

If $\varphi \in M_{B_p^\sigma(\mathbb{B}_n)\to B_p^\sigma(\mathbb{B}_n;\ell^2)}$ and $f=\sum_{|I|=\kappa} f_I e_I \in B_p^\sigma(\mathbb{B}_n;\otimes^{\kappa-1}\ell^2)$, we define

$$\mathbb{M}_\varphi f = \varphi \otimes f = \varphi \otimes \left(\sum_{|I|=\kappa-1} f_I e_I\right) = \sum_{|I|=\kappa-1} (\varphi f_I)\otimes e_I,$$

where $I=(i_1,\dots,i_{\kappa-1})\in\mathbb{N}^{\kappa-1}$ and $e_I = e_{i_1}\otimes\cdots\otimes e_{i_{\kappa-1}}$.

LEMMA 17.45. *Suppose that $\sigma \geq 0$, $1<p<\infty$ and $\kappa \geq 1$. Then there is a constant $C_{n,\sigma,p,\kappa}$ such that*

$$\|\mathbb{M}_g\|_{B_p^\sigma(\mathbb{B}_n;\otimes^{\kappa-1}\ell^2)\to B_p^\sigma(\mathbb{B}_n;\otimes^{\kappa}\ell^2)} \lesssim \|\mathbb{M}_g\|_{B_p^\sigma(\mathbb{B}_n)\to B_p^\sigma(\mathbb{B}_n;\ell^2)}. \tag{17.55}$$

In the case $p=2$ we have equality:

$$\|\mathbb{M}_\varphi\|_{B_2^\sigma(\mathbb{B}_n;\otimes^{\kappa-1}\ell^2)\to B_2^\sigma(\mathbb{B}_n;\otimes^{\kappa}\ell^2)} = \|\mathbb{M}_\varphi\|_{B_2^\sigma(\mathbb{B}_n)\to B_2^\sigma(\mathbb{B}_n;\ell^2)}. \tag{17.56}$$

It turns out that in order to prove (17.55) for $p\neq 2$ we will need the case $M=1$ of Proposition 17.43. Fortunately, the case $M=1$ does not require inequality (17.55), thus avoiding circularity.

PROOF OF PROPOSITION 17.43 AND LEMMA 17.45. We begin with the proof of the case $M=1$ of Proposition 17.43. We will show that for $m=\ell+k$,

$$\int_{\mathbb{B}_n} \left|(1-|z|^2)^\sigma \left(\mathcal{Y}^\ell g\right)\left(\mathcal{Y}^k h\right)\right|^p d\lambda_n(z) \lesssim \|\mathbb{M}_g\|^p_{B_p^\sigma(\mathbb{B}_n)\to B_p^\sigma(\mathbb{B}_n;\ell^2)} \|h\|^p_{B_p^\sigma(\mathbb{B}_n)}. \tag{17.57}$$

Following the proof of [**249**, Theorem 3.1] we use the Leibnitz formula in " divergence form",

$$\left(\mathcal{Y}^\ell g\right)\left(\mathcal{Y}^k h\right) = \sum_{\alpha=0}^{\ell} c_\alpha^\ell \mathcal{Y}^\alpha\left(g\mathcal{Y}^{k+\ell-\alpha}h\right). \tag{17.58}$$

Next we note that for $s>n/p$, $B_p^s(\mathbb{B}_n;\ell^2)$ is a Bergman space, hence $M_{B_p^s(\mathbb{B}_n)\to B_p^s(\mathbb{B}_n;\ell^2)} = H^\infty(\mathbb{B}_n,\ell^2)$. Thus using (17.54) we have for $s>n/p$,

$$g \in M_{B_p^\sigma(\mathbb{B}_n)\to B_p^\sigma(\mathbb{B}_n;\ell^2)} \cap H^\infty(\mathbb{B}_n;\ell^2) = M_{B_p^\sigma(\mathbb{B}_n)\to B_p^\sigma(\mathbb{B}_n;\ell^2)} \cap M_{B_p^s(\mathbb{B}_n)\to B_p^s(\mathbb{B}_n;\ell^2)}.$$

Then, still following the argument in [**249**], we use the complex interpolation theorem of Beatrous [**63**] and Ligocka [**207**] (they prove only the scalar-valued version but the Hilbert space valued version has the same proof),

$$\left(B_p^\sigma(\mathbb{B}_n), B_p^{n/p+\varepsilon}(\mathbb{B}_n)\right)_\theta = B_p^{(1-\theta)\sigma+\theta(n/p+\varepsilon)}(\mathbb{B}_n), \qquad 0\leq\theta\leq 1,$$

$$\left(B_p^\sigma(\mathbb{B}_n;\ell^2), B_p^{n/p+\varepsilon}(\mathbb{B}_n;\ell^2)\right)_\theta = B_p^{(1-\theta)\sigma+\theta(n/p+\varepsilon)}(\mathbb{B}_n;\ell^2), \qquad 0\leq\theta\leq 1,$$

to conclude that $g\in M_{B_p^s(\mathbb{B}_n)\to B_p^s(\mathbb{B}_n;\ell^2)}$ for all $s\geq\sigma$, and with multiplier norm $\|\mathbb{M}_g\|_{B_p^s(\mathbb{B}_n)\to B_p^s(\mathbb{B}_n;\ell^2)}$ bounded by $\|\mathbb{M}_g\|_{B_p^\sigma(\mathbb{B}_n)\to B_p^\sigma(\mathbb{B}_n;\ell^2)}$. Recall now that

$$\|h\|^p_{B_p^\sigma(\mathbb{B}_n)} \approx \int_{\mathbb{B}_n}\left|(1-|z|^2)^\sigma\mathcal{Y}^m h(z)\right|^p d\lambda_n(z),$$

and similarly for $\|f\|^p_{B_p^\sigma(\mathbb{B}_n;\ell^2)}$, provided m satisfies

$$\left(\sigma+\frac{m}{2}\right)p>n, \tag{17.59}$$

where $m/2$ appears in the inequality since the derivatives D that can appear in $\mathcal{Y}^m$ only contribute $(1-|z|^2)^{1/2}$ to the power of $1-|z|^2$ in the integral (see Section 14.5.1).

REMARK 17.46. *At this point we recall the convention established in Definitions* 17.38 *and* 17.39 *that the factors of* $1-|z|^2$ *that are embedded in the notation for the derivative* $\mathcal{Y}^\alpha$ *are treated as constants relative to the actual differentiations. In the calculations below, we will adopt* **the same convention** *for the factors* $(1-|z|^2)^s$ *that we introduce into the integrals. Alternatively, the reader may wish to write out all the derivatives explicitly with the appropriate power of* $1-|z|^2$ *set aside as is done in the proof of* [**249**, Theorem 3.5].

So we have, keeping in mind the previous Remark,

$$\begin{aligned}&\int_{\mathbb{B}_n}\left|(1-|z|^2)^\sigma\mathcal{Y}^\alpha\left(g(z)\,\mathcal{Y}^{k+\ell-\alpha}h(z)\right)\right|^p d\lambda_n\\ &\quad=\int_{\mathbb{B}_n}\left|(1-|z|^2)^s\mathcal{Y}^\alpha\left\{g(z)(1-|z|^2)^{\sigma-s}\mathcal{Y}^{k+\ell-\alpha}h(z)\right\}\right|^p d\lambda_n\\ &\quad=\left\|g(z)(1-|z|^2)^{\sigma-s}\mathcal{Y}^{k+\ell-\alpha}h\right\|^p_{B^s_{p,\alpha}(\mathbb{B}_n;\ell^2)}.\end{aligned}$$

Here the function

$$H(z)=(1-|z|^2)^{\sigma-s}\mathcal{Y}^{k+\ell-\alpha}h(z)$$

is *not* holomorphic, but recall that we have also defined the norm $\|\cdot\|_{B^s_{p,\alpha}(\mathbb{B}_n;\ell^2)}$ on all smooth functions. Now we would like to apply a multiplier property of g, and for this we must be acting on a Besov-Sobolev space of *holomorphic* functions, since that is what we get from the complex interpolation of Beatrous and Ligocka. Precisely, we get that $\mathbb{M}_g$ is a bounded operator from $B_p^s(\mathbb{B}_n)$ to $B_p^s(\mathbb{B}_n;\ell^2)$ for *all* $s\geq\sigma$.

Now we express $\mathcal{Y}^{k+\ell-\alpha}h(z)$ as a sum of terms that are products of a power of $1-|z|^2$ and a derivative $R^iL^jh(z)$ where $i+j=k+\ell-\alpha$ and R is the radial derivative and L denotes a complex tangential derivative $\frac{\partial}{\partial z_q}-\overline{z_q}R$ as in [**249**]. However, the operators R^iL^j have different weights in the sense that the power of $1-|z|^2$ that is associated with R^iL^j is $(1-|z|^2)^{i+j/2}$, i.e.

$$\mathcal{Y}^{k+\ell-\alpha}h(z)=\sum_{i,j}(1-|z|^2)^{i+j/2}R^iL^jh(z).$$

It turns out that to handle the term $(1-|z|^2)^{i+j/2}R^iL^jh(z)$ we will use that g is a multiplier on $B_p^s(\mathbb{B}_n)$ with

$$s=\sigma+i+\frac{j}{2},$$

an exponent that depends on $i+j/2$ and *not* on $i+j=k+\ell-\alpha$. Indeed we have, using the convention noted in Remark 17.46, that

$$\begin{aligned}
&\left\|g(z)(1-|z|^2)^{\sigma-s}(1-|z|^2)^{i+j/2}R^iL^jh(z)\right\|^p_{B^s_{p,\alpha}(\mathbb{B}_n;\ell^2)}\\
&\quad=\int_{\mathbb{B}_n}\left|(1-|z|^2)^s\mathcal{Y}^\alpha\left\{g(z)(1-|z|^2)^{\sigma-s}(1-|z|^2)^{i+j/2}R^iL^jh(z)\right\}\right|^p d\lambda_n\\
&\quad=\int_{\mathbb{B}_n}\left|(1-|z|^2)^{\sigma+i+j/2}\mathcal{Y}^\alpha\left\{g(z)R^iL^jh(z)\right\}\right|^p d\lambda_n\\
&\quad=\left\|g(z)R^iL^jh(z)\right\|^p_{B^s_{p,\alpha}(\mathbb{B}_n;\ell^2)}.
\end{aligned}$$

Now the function $g(z)R^iL^jh(z)$ is holomorphic and $s=\sigma+i+j/2\geq\sigma$ so that we can use that g is a multiplier on $B_p^s(\mathbb{B}_n)=B^s_{p,\alpha}(\mathbb{B}_n)$ (this latter equality holds because $\left(s+\frac{\alpha}{2}\right)p>n$ by (17.59)). The result is that

$$\begin{aligned}
&\left\|g(z)R^iL^jh(z)\right\|^p_{B^s_p(\mathbb{B}_n;\ell^2)}\\
&\leq\|\mathbb{M}_g\|^p_{B^s_p(\mathbb{B}_n)\to B^s_p(\mathbb{B}_n;\ell^2)}\left\|R^iL^jh(z)\right\|^p_{B^s_{p,\alpha}(\mathbb{B}_n)}\\
&\leq\|\mathbb{M}_g\|^p_{B^s_p(\mathbb{B}_n)\to B^s_p(\mathbb{B}_n;\ell^2)}\int_{\mathbb{B}_n}\left|(1-|z|^2)^{\sigma+i+j/2}\mathcal{Y}^\alpha R^iL^jh(z)\right|^p d\lambda_n\\
&=\|\mathbb{M}_g\|^p_{B^s_p(\mathbb{B}_n)\to B^s_p(\mathbb{B}_n;\ell^2)}\int_{\mathbb{B}_n}\left|(1-|z|^2)^\sigma\mathcal{Y}^\alpha\left[(1-|z|^2)R\right]^i\left[\sqrt{1-|z|^2}L\right]^jh(z)\right|^p d\lambda_n\\
&\leq\|\mathbb{M}_g\|^p_{B^s_p(\mathbb{B}_n)\to B^s_p(\mathbb{B}_n;\ell^2)}\int_{\mathbb{B}_n}\left|(1-|z|^2)^\sigma\mathcal{Y}^{\alpha+i+j}h(z)\right|^p d\lambda_n\\
&=\|\mathbb{M}_g\|^p_{B^s_p(\mathbb{B}_n)\to B^s_p(\mathbb{B}_n;\ell^2)}\int_{\mathbb{B}_n}\left|(1-|z|^2)^\sigma\mathcal{Y}^m h(z)\right|^p d\lambda_n\\
&\leq\|\mathbb{M}_g\|^p_{B^\sigma_p(\mathbb{B}_n)\to B^\sigma_p(\mathbb{B}_n;\ell^2)}\|h\|^p_{B^\sigma_p(\mathbb{B}_n)},
\end{aligned}$$

and the case $M=1$ of Proposition 17.43 is proved.

Now we turn to the proof of the operator norm inequality (17.55) in Lemma 17.45. The case $p = 2$ is particularly easy:

$$\begin{aligned}
\|\mathbb{M}_\varphi f\|^2_{B_2^\sigma(\mathbb{B}_n;\otimes^\kappa \ell^2)} &= \int_{\mathbb{B}_n} (1-|z|^2)^{2\sigma} \sum_{|I|=\kappa-1} |\mathcal{Y}^m(\varphi f_I)|^2 \, d\lambda_n \\
&= \sum_{|I|=\kappa-1} \|\mathbb{M}_\varphi f_I\|^2_{B_2^\sigma(\mathbb{B}_n;\ell^2)} \\
&\le \|\mathbb{M}_\varphi\|^2_{B_2^\sigma(\mathbb{B}_n)\to B_2^\sigma(\mathbb{B}_n;\ell^2)} \sum_{|I|=\kappa-1} \|f_I\|^2_{B_2^\sigma(\mathbb{B}_n)} \\
&= \|\mathbb{M}_\varphi\|^2_{B_2^\sigma(\mathbb{B}_n)\to B_2^\sigma(\mathbb{B}_n;\ell^2)} \int_{\mathbb{B}_n} (1-|z|^2)^{2\sigma} \sum_{|I|=\kappa-1} |\mathcal{Y}^m f_I|^2 \, d\lambda_n \\
&= \|\mathbb{M}_\varphi\|^2_{B_2^\sigma(\mathbb{B}_n)\to B_2^\sigma(\mathbb{B}_n;\ell^2)} \|f\|^2_{B_2^\sigma(\mathbb{B}_n;\otimes^{\kappa-1}\ell^2)},
\end{aligned}$$

and from this we easily obtain (17.56).

For $p \ne 2$ it suffices to show that

$$\|\mathbb{M}_\varphi\|_{B_p^\sigma(\mathbb{B}_n;\mathbb{C}^\nu)\to B_p^\sigma(\mathbb{B}_n;\mathbb{C}^\mu\otimes\mathbb{C}^\nu)} \lesssim \|\mathbb{M}_\varphi\|_{B_p^\sigma(\mathbb{B}_n)\to B_p^\sigma(\mathbb{B}_n;\mathbb{C}^\mu)} \tag{17.60}$$

for all $\mu, \nu \ge 1$ where the implied constant is independent of μ, ν. This holds because both ℓ^2 and $\otimes^{\kappa-1}\ell^2$ are separable Hilbert spaces and so can be appropriately approximated by $\mathbb{C}^\mu$ and $\mathbb{C}^\nu$ respectively. For each $z \in \mathbb{B}_n$ we will view $\varphi(z) \in \mathbb{C}^\mu$ as a column vector and $f(z) \in \mathbb{C}^\nu$ as a row vector so that $(\mathbb{M}_\varphi f)(z)$ is the rank one $\mu \times \nu$ matrix

$$(\mathbb{M}_\varphi f)(z) = \begin{bmatrix} (\varphi_1 f_1)(z) & \cdots & (\varphi_1 f_\nu)(z) \\ \vdots & \ddots & \vdots \\ (\varphi_\mu f_1)(z) & \cdots & (\varphi_\mu f_\nu)(z) \end{bmatrix} = \varphi(z) \odot f(z),$$

where we have inserted the symbol $\odot$ simply to remind the reader that this is *not* the dot product $\varphi(z) \cdot f(z) = f(z)\varphi(z)$ of the vectors $\varphi(z)$ and $f(z)$.

Now we consider a single component X^m of the vector differential operator $\mathcal{Y}^m$ for some $m > 2(n/p - \sigma)$, which can be chosen independent of μ and ν. The main point in the proof of the lemma is that the matrix $X^m(\mathbb{M}_\varphi f)(z)$ has rank at most $m+1$ independent of μ and ν. Indeed, the Leibnitz formula yields

$$X^m(\mathbb{M}_\varphi f)(z) = X^m(\varphi(z) \odot f(z)) = \sum_{\ell=0}^m c_{\ell,m} X^{m-\ell}\varphi(z) \odot X^\ell f(z),$$

where each matrix $X^{m-\ell}\varphi(z) \odot X^\ell f(z)$ is rank one. Thus on those terms, where the Hilbert Schmidt norm is multiplicative, we have

$$\left|X^{m-\ell}\varphi(z) \odot X^\ell f(z)\right| = \left|X^{m-\ell}\varphi(z)\right| \left|X^\ell f(z)\right|.$$

(Recall that we are using $|\cdot|$ to denote a general Hilbert space norm; here the Hilbert Schmidt norm.)

Momentarily fix $0 \le \ell \le m$ and define

$$\begin{aligned}
T^\ell h(z) &= X^{m-\ell}\varphi(z) h(z), \qquad h(z) \in \mathbb{C}, \\
T^\ell g(z) &= X^{m-\ell}\varphi(z) \odot g(z), \qquad g(z) \in \mathbb{C}^\nu.
\end{aligned}$$

For $x \in \partial\mathbb{B}_\mu$, which we view as a row vector, define

$$T_x^\ell g(z) = x T^\ell g(z) = x\left(X^{m-\ell}\varphi\right)(z) \odot g(z).$$

Now choose $x(z) \in \partial\mathbb{B}_\mu$ such that $x(z)\left(X^{m-\ell}\varphi\right)(z) = \left|X^{m-\ell}\varphi(z)\right|$ so that

$$T^\ell_{x(z)} g(z) = x(z)\left(X^{m-\ell}\varphi\right)(z) \odot g(z) = \left|X^{m-\ell}\varphi(z)\right| g(z),$$

and hence

$$\begin{aligned}\left|T^\ell_{x(z)}\left(X^\ell f\right)(z)\right| &= \left|X^{m-\ell}\varphi(z)\right|\left|X^\ell f(z)\right| \\ &= \left|X^{m-\ell}\varphi(z) \odot X^\ell f(z)\right| = \left|T^\ell\left(X^\ell f\right)(z)\right|.\end{aligned}$$

Now we follow the standard argument found on, for example in [**316**, page 451]. For $y \in \partial\mathbb{B}_\nu$, which we view as a column vector, and $g(z) \in \mathbb{C}^\nu$ define the scalars

$$\begin{aligned} g_y(z) &= g(z)\, y, \\ \left(T^\ell_{x(z)} g\right)_y(z) &= T^\ell_{x(z)} g(z)\, y = x(z)\left(X^{m-\ell}\varphi\right)(z) \odot g(z)\, y,\end{aligned}$$

and note that

$$T^\ell_{x(z)}\left(X^\ell f\right)(z)\, y = x(z)\left(X^{m-\ell}\varphi\right)(z) \odot \left(X^\ell f\right)(z)\, y = T^\ell_{x(z)}\left(X^\ell f\right)_y(z).$$

Thus we have with $d\sigma_\nu$ surface measure on $\partial\mathbb{B}_\nu$,

$$\begin{aligned}&\int_{\partial\mathbb{B}_\nu}\left|T^\ell_{x(z)}\left(X^\ell f\right)(z)\, y\right|^p d\sigma_\nu(y) \\ &\qquad = \left|T^\ell_{x(z)}\left(X^\ell f\right)(z)\right|^p \int_{\partial\mathbb{B}_\nu}\left|\frac{T^\ell_{x(z)}\left(X^\ell f\right)(z)}{\left|T^\ell_{x(z)}\left(X^\ell f\right)(z)\right|}\cdot y\right|^p d\sigma_\nu(y),\end{aligned}$$

as well as

$$\int_{\partial\mathbb{B}_\nu}\left|\left(X^\ell f\right)_y(z)\right|^p d\sigma_\nu(y) = \left|X^\ell f(z)\right|^p \int_{\partial\mathbb{B}_\nu}\left|\frac{X^\ell f(z)}{\left|X^\ell f(z)\right|}\cdot y\right|^p d\sigma_\nu(y).$$

The crucial observation now is that

$$\int_{\partial\mathbb{B}_\nu}\left|\frac{T^\ell_{x(z)}\left(X^\ell f\right)(z)}{\left|T^\ell_{x(z)}\left(X^\ell f\right)(z)\right|}\cdot y\right|^p d\sigma_\nu(y) = \int_{\partial\mathbb{B}_\nu}\left|\frac{X^\ell f(z)}{\left|X^\ell f(z)\right|}\cdot y\right|^p d\sigma_\nu(y) = \gamma_{p,\nu}$$

is *independent* of the row vector in $\partial\mathbb{B}_\nu$ that is dotted with y. Thus we have

$$\begin{aligned}\left|T^\ell\left(X^\ell f\right)(z)\right|^p = \left|T^\ell_{x(z)}\left(X^\ell f\right)(z)\right|^p &= \frac{1}{\gamma_{p,\nu}}\int_{\partial\mathbb{B}_\nu}\left|T^\ell_{x(z)}\left(X^\ell f\right)(z)\, y\right|^p d\sigma_\nu(y), \\ \left|X^\ell f(z)\right|^p &= \frac{1}{\gamma_{p,\nu}}\int_{\partial\mathbb{B}_\nu}\left|\left(X^\ell f\right)_y(z)\right|^p d\sigma_\nu(y).\end{aligned}$$

So, with $d\omega_{p\sigma}(z) = (1-|z|^2)^{p\sigma} d\lambda_n(z)$, we conclude that

$$\begin{aligned}
&\int_{\mathbb{B}_n} |X^m(\mathbb{M}_\varphi f)|^p \, d\omega_{p\sigma}(z) \\
&\lesssim \sum_{\ell=0}^{m} \int_{\mathbb{B}_n} \left|T^\ell\left(X^\ell f\right)(z)\right|^p d\omega_{p\sigma}(z) \\
&\approx \sum_{\ell=0}^{m} \frac{1}{\gamma_{p,\nu}} \int_{\partial\mathbb{B}_\nu} \int_{\mathbb{B}_n} \left|x(z)\left(X^{m-\ell}\varphi\right)(z)\left(X^\ell f_y\right)(z)\right|^p d\omega_{p\sigma}(z)\, d\sigma_\nu(y) \\
&\lesssim \sum_{\ell=0}^{m} \frac{1}{\gamma_{p,\nu}} \int_{\partial\mathbb{B}_\nu} \int_{\mathbb{B}_n} \left|\left(X^{m-\ell}\varphi\right)(z)\left(X^\ell f\right)_y(z)\right|^p d\omega_{p\sigma}(z)\, d\sigma_\nu(y) \\
&\lesssim \sum_{\ell=0}^{m} \frac{1}{\gamma_{p,\nu}} \int_{\partial\mathbb{B}_\nu} \|\mathbb{M}_\varphi\|^p_{B_p^\sigma(\mathbb{B}_n)\to B_p^\sigma(\mathbb{B}_n;\mathbb{C}^\mu)} \int_{\mathbb{B}_n} \left|(\mathcal{X}^m f)_y(z)\right|^p d\omega_{p\sigma}(z)\, d\sigma_\nu(y)
\end{aligned}$$

by the case $M=1$ of Proposition 17.43, where ℓ^2 there is replaced by $\mathbb{C}^\nu$, g by φ and h by f_y. Now we use the equality

$$\int_{\partial\mathbb{B}_\nu} \left|(\mathcal{X}^m f)_y(z)\right|^p d\sigma_\nu(y) = \gamma_{p,\nu} |\mathcal{X}^m f(z)|^p$$

to obtain

$$\begin{aligned}
\int_{\mathbb{B}_n} |X^m(\mathbb{M}_\varphi f)|^p \, d\omega_{p\sigma}(z) &\lesssim \|\mathbb{M}_\varphi\|^p_{B_p^\sigma(\mathbb{B}_n)\to B_p^\sigma(\mathbb{B}_n;\mathbb{C}^\mu)} \int_{\mathbb{B}_n} |\mathcal{X}^m f(z)|^p \, d\omega_{p\sigma}(z) \\
&\lesssim \|\mathbb{M}_\varphi\|^p_{B_p^\sigma(\mathbb{B}_n)\to B_p^\sigma(\mathbb{B}_n;\mathbb{C}^\mu)} \|f\|^p_{B_p^\sigma(\mathbb{B}_n;\mathbb{C}^\mu)}.
\end{aligned}$$

Since m depends only on n, σ and p, this completes the proof of (17.60), and hence that of Lemma 17.45.

Finally we return to complete the proof of Proposition 17.43. We have already proved the case $M=1$. Now we sketch a proof of the case $M=2$ using the multiplier norm inequality (17.55) with $\kappa=2$. By multiplicativity of $|\cdot|$ on tensors, it suffices to show that for $m=\ell_1+\ell_2+k$,

$$\begin{aligned}
\int_{\mathbb{B}_n} \left|(1-|z|^2)^\sigma \left(\mathcal{Y}^{\ell_1}g\right)\otimes\left(\mathcal{Y}^{\ell_2}g\right)\left(\mathcal{Y}^k h\right)\right|^p d\lambda_n(z) \qquad (17.61)\\
\lesssim \|\mathbb{M}_g\|^{2p}_{B_p^\sigma(\mathbb{B}_n)\to B_p^\sigma(\mathbb{B}_n;\ell^2)} \|h\|^p_{B_p^\sigma(\mathbb{B}_n)}.
\end{aligned}$$

Using (17.58) on tensors we obtain

$$\begin{aligned}
\left(\mathcal{Y}^{\ell_1}g\right)\otimes\left(\mathcal{Y}^{\ell_2}g\right)\left(\mathcal{Y}^k h\right) &= \left(\mathcal{Y}^{\ell_1}g\right)\otimes\left\{\sum_{\alpha=0}^{\ell_2} c_\alpha^{\ell_2}\mathcal{Y}^\alpha\left(g\mathcal{Y}^{k+\ell_2-\alpha}h\right)\right\} \\
&= \sum_{\alpha=0}^{\ell_2} c_\alpha^{\ell_2}\left(\mathcal{Y}^{\ell_1}g\right)\otimes\left[\mathcal{Y}^\alpha\left(g\mathcal{Y}^{k+\ell_2-\alpha}h\right)\right] \\
&= \sum_{\alpha=0}^{\ell_2} c_\alpha^{\ell_2}\left\{\sum_{\beta=0}^{\ell_1} c_\beta^{\ell_1}\mathcal{Y}^\beta\left(g\otimes\mathcal{Y}^{\alpha+\ell_1-\beta}\left(g\mathcal{Y}^{k+\ell_2-\alpha}h\right)\right)\right\}.
\end{aligned}$$

We first use the Hilbert space valued interpolation theorem together with the case $\kappa=2$ of Lemma 17.45 to show that $g \in M_{B_p^{s_1}(\mathbb{B}_n;\ell^2)\to B_p^{s_1}(\mathbb{B}_n;\ell^2\otimes\ell^2)}$ and $g \in$

$M_{B_p^{s_2}(\mathbb{B}_n)\to B_p^{s_2}(\mathbb{B}_n;\ell^2)}$ for appropriate values of s_1 and s_2. We will just write down the case of $\mathcal{Y}=(1-|z|^2)R$. The extension to the general case, which includes the operators $(1-|z|^2)I$ and D, is straightforward. Keeping in mind Remark 17.46, we obtain

$$\begin{aligned}
&\left\|g(z)\otimes(1-|z|^2)^{\sigma-s_1}\mathcal{Y}^{\alpha+\ell_1-\beta}\left(g\mathcal{Y}^{k+\ell_2-\alpha}h\right)\right\|^p_{B_p^{s_1}(\mathbb{B}_n;\ell^2\otimes\ell^2)}\\
&\le \|\mathbb{M}_g\|^p_{B_p^{s_1}(\mathbb{B}_n;\ell^2)\to B_p^{s_1}(\mathbb{B}_n;\ell^2\otimes\ell^2)}\left\|(1-|z|^2)^{\sigma-s_1}\mathcal{Y}^{\alpha+\ell_1-\beta}\left(g\mathcal{Y}^{k+\ell_2-\alpha}h\right)\right\|^p_{B_p^{s_1}(\mathbb{B}_n;\ell^2)}\\
&= \|\mathbb{M}_g\|^p_{B_p^{s_1}(\mathbb{B}_n;\ell^2)\to B_p^{s_1}(\mathbb{B}_n;\ell^2\otimes\ell^2)}\\
&\qquad\times\int_{\mathbb{B}_n}\left|(1-|z|^2)^{s_1}\mathcal{Y}^{\beta}(1-|z|^2)^{\sigma-s_1}\mathcal{Y}^{\alpha+\ell_1-\beta}\left(g\mathcal{Y}^{k+\ell_2-\alpha}h\right)\right|^p d\lambda_n,
\end{aligned}$$

which by (17.55) is at most

$$\begin{aligned}
&\|\mathbb{M}_g\|^p_{B_p^{\sigma}(\mathbb{B}_n)\to B_p^{\sigma}(\mathbb{B}_n;\ell^2)}\\
&\qquad\times\int_{\mathbb{B}_n}\left|(1-|z|^2)^{s_2}\mathcal{Y}^{\alpha+\ell_1}\left(g(1-|z|^2)^{\sigma-s_2}\mathcal{Y}^{k+\ell_2-\alpha}h\right)\right|^p d\lambda_n\\
&\approx \|\mathbb{M}_g\|^p_{B_p^{\sigma}(\mathbb{B}_n)\to B_p^{\sigma}(\mathbb{B}_n;\ell^2)}\left\|g(1-|z|^2)^{\sigma-s_2}\mathcal{Y}^{k+\ell_2-\alpha}h\right\|^p_{B_p^{s_2}(\mathbb{B}_n;\ell^2)}\\
&\lesssim \|\mathbb{M}_g\|^p_{B_p^{\sigma}(\mathbb{B}_n)\to B_p^{\sigma}(\mathbb{B}_n;\ell^2)}\\
&\qquad\times\|\mathbb{M}_g\|^p_{B_p^{s_2}(\mathbb{B}_n)\to B_p^{s_2}(\mathbb{B}_n;\ell^2)}\left\|(1-|z|^2)^{\sigma-s_2}\mathcal{Y}^{k+\ell_2-\alpha}h\right\|^p_{B_p^{s_2}(\mathbb{B}_n)}\\
&\lesssim \|\mathbb{M}_g\|^{2p}_{B_p^{\sigma}(\mathbb{B}_n)\to B_p^{\sigma}(\mathbb{B}_n;\ell^2)}\|h\|^p_{B_p^{\sigma}(\mathbb{B}_n)}.
\end{aligned}$$

Summing up over α and β gives (17.61).

Repeating this procedure for $M\ge 3$ and using (17.55) with $\kappa=M$ finishes the proof of Proposition 17.43. □

17.4.3. Three Crucial Inequalities. In order to establish appropriate inequalities for the Charpentier solution operators, we will need to control terms of the form $\left(\overline{z-w}\right)^{\alpha}\frac{\partial^m}{\partial\overline{w}^{\alpha}}F(w)$, $\overline{D^m_{(z)}}\triangle(w,z)$, $\overline{D^m_{(z)}}\left\{(1-w\overline{z})^k\right\}$ and $R^m_{(z)}\left\{(1-\overline{w}z)^k\right\}$ inside the integral for T as given in the integration by parts formula in Lemma 17.29 above. Here we are using the subscript (z) in parentheses to indicate the variable being differentiated. This is to avoid confusion with the notation D_a introduced in Section 14.5.1. We collect the necessary estimates in the following proposition.

PROPOSITION 17.47. *For $z,w\in\mathbb{B}_n$ and $m\in\mathbb{N}$, we have the following three crucial sets of estimates:*

(17.62)

$$\left|\left(\overline{z-w}\right)^{\alpha}\frac{\partial^m}{\partial\overline{w}^{\alpha}}F(w)\right|\lesssim\left(\frac{\sqrt{\triangle(w,z)}}{1-|w|^2}\right)^m\left|\overline{D}^m F(w)\right|,\qquad F\in C^{\infty}(\mathbb{B}_n),\, m=|\alpha|.$$

$$\text{(17.63)}\qquad \left|D_{(z)}\triangle(w,z)\right|\lesssim\left\{(1-|z|^2)\triangle(w,z)^{1/2}+\triangle(w,z)\right\},$$

$$\left|(1-|z|^2)R_{(z)}\triangle(w,z)\right|\lesssim(1-|z|^2)\sqrt{\triangle(w,z)},$$

$$\left|D^m_{(z)}\left\{(1-\overline{w}z)^k\right\}\right| \lesssim |1-\overline{w}z|^k\left(\frac{1-|z|^2}{|1-\overline{w}z|}\right)^{m/2}, \tag{17.64}$$

$$\left|(1-|z|^2)^m R^m_{(z)}\left\{(1-\overline{w}z)^k\right\}\right| \lesssim |1-\overline{w}z|^k\left(\frac{1-|z|^2}{|1-\overline{w}z|}\right)^{m}.$$

PROOF. To prove (17.62) we view D_a as a differentiation operator in the variable w so that

$$D_a = -\nabla_w\left\{\left(1-|a|^2\right)P_a + \sqrt{1-|a|^2}Q_a\right\}.$$

A basic calculation is then:

$$\begin{aligned}
&(1-\overline{a}z)\,\varphi_a(z)\cdot(D_a)^t\\
&= \left\{P_a(z-a) + \sqrt{1-|a|^2}Q_a(z-a)\right\}\\
&\qquad\cdot\left\{\left(1-|a|^2\right)P_a\nabla_w + \sqrt{1-|a|^2}Q_a\nabla_w\right\}\\
&= P_a(z-a)\left(1-|a|^2\right)P_a\nabla_w\\
&\qquad + \sqrt{1-|a|^2}Q_a(z-a)\sqrt{1-|a|^2}Q_a\nabla_w\\
&= \left(1-|a|^2\right)(z-a)\cdot\nabla_w.
\end{aligned}$$

From this we conclude the inequality

$$\begin{aligned}
\left|(z_i-a_i)\frac{\partial}{\partial w_i}F(w)\right| &\le |(z-a)\cdot\nabla F(w)|\\
&\le \left|\frac{1-\overline{a}z}{1-|a|^2}\varphi_a(z)\right||D_aF(w)|\\
&= \frac{\sqrt{\triangle(a,z)}}{1-|a|^2}|D_aF(w)|,
\end{aligned}$$

as well as its conjugate

$$\left|\overline{(z_i-a_i)}\frac{\partial}{\partial\overline{w_i}}F(w)\right| \lesssim \frac{\sqrt{\triangle(a,z)}}{1-|a|^2}\left|\overline{D_a}F(w)\right|.$$

Moreover, we can iterate this inequality to obtain

$$\left|\left(\overline{z-a}\right)^\alpha\frac{\partial^m}{\partial\overline{w}^\alpha}F(w)\right| \lesssim \left(\frac{\sqrt{\triangle(a,z)}}{1-|a|^2}\right)^m\left|\left(\overline{D_a}\right)^mF(w)\right|,$$

for a multi-index of length m. With $a=w$ this becomes the first estimate (17.62).

To see the second estimate (17.63), recall from (14.61) that

$$D_af(z) = -\left\{\left(1-|a|^2\right)P_a\nabla f(z) + \left(1-|a|^2\right)^{1/2}Q_a\nabla f(z)\right\}.$$

We let $a = z$. By the unitary invariance of

$$\triangle(w,z) = |1 - \overline{w}z|^2 - (1 - |z|^2)(1 - |w|^2),$$

we may assume that $z = (|z|, 0, \ldots, 0)$. Then we have

$$\begin{aligned}
\frac{\partial}{\partial z_j} \triangle(w,z) &= \frac{\partial}{\partial z_j} \left\{ (1 - \overline{w}z)(1 - \overline{z}w) - (1 - \overline{z}z)(1 - |w|^2) \right\} \\
&= -\overline{w_j}(1 - \overline{z}w) + \overline{z_j}(1 - |w|^2) \\
&= (\overline{z_j} - \overline{w_j}) + \overline{w_j}(\overline{z}w) - \overline{z_j}|w|^2 \\
&= (\overline{z_j} - \overline{w_j})(1 - |z|^2) + \overline{z_j}|z|^2 - \overline{w_j}|z|^2 + \overline{w_j}(\overline{z}w) - \overline{z_j}|w|^2 \\
&= (\overline{z_j} - \overline{w_j})(1 - |z|^2) + \overline{z_j}\left(|z|^2 - |w|^2\right) + \overline{w_j}(\overline{z}(w - z)).
\end{aligned}$$

Now $Q_z \nabla f = \left(0, \frac{\partial f}{\partial z_2}, \ldots, \frac{\partial f}{\partial z_n}\right)$ and thus a typical term in $Q_z \nabla \triangle$ is $\frac{\partial}{\partial z_j} \triangle(w,z)$ with $j \geq 2$. From $z = (|z|, 0, \ldots, 0)$ and $j \geq 2$ we have $z_j = 0$ and so

$$\frac{\partial}{\partial z_j} \triangle(w,z) = (\overline{z_j} - \overline{w_j})(1 - |z|^2) - (\overline{z_j} - \overline{w_j})(\overline{z}(w - z)), \qquad j \geq 2.$$

Now (15.23) lets us rewrite $\triangle$,

$$\triangle(w,z) = (1 - |z|^2)|w - z|^2 + |\overline{z}(w - z)|^2, \tag{17.65}$$

which together with the above shows that

(17.66)

$$\begin{aligned}
\sqrt{1 - |z|^2}\, |Q_z \nabla \triangle(w,z)| &\lesssim |z - w|(1 - |z|^2)^{\frac{3}{2}} + \sqrt{1 - |z|^2}\, |z - w|\, |\overline{z}(w - z)| \\
&\lesssim (1 - |z|^2) \triangle(w,z)^{1/2} + \triangle(w,z).
\end{aligned}$$

As for $P_z \nabla D = \left(\frac{\partial f}{\partial z_1}, 0, \ldots, 0\right)$ we use (17.65) to obtain

$$\begin{aligned}
|P_z \nabla \triangle(w,z)| &= \left|(\overline{z_1} - \overline{w_1})(1 - |z|^2) + \overline{z_1}\left(|z|^2 - |w|^2\right) + \overline{w_1}\overline{z}(w - z)\right| \\
&\leq |z - w|(1 - |z|^2) + \left||z|^2 - |w|^2\right| + |\overline{z}(w - z)| \\
&\lesssim \sqrt{\triangle(w,z)} + 2\,||z| - |w||.
\end{aligned}$$

However,

$$\begin{aligned}
\triangle(w,z) &\geq (1 - |w|\,|z|)^2 - (1 - |z|^2)(1 - |w|^2) \\
&= 1 - 2|w|\,|z| + |w|^2|z|^2 - \left\{1 - |z|^2 - |w|^2 + |z|^2|w|^2\right\} \\
&= |z|^2 + |w|^2 - 2|w|\,|z| = (|z| - |w|)^2
\end{aligned}$$

and so altogether we have the estimate

$$|P_z \nabla \triangle(w,z)| \lesssim \sqrt{\triangle(w,z)}. \tag{17.67}$$

Combining (17.66) and (17.67) with the definition (14.61) completes the proof of the first line in (17.63). The second line in (17.63) follows from (17.67) since $R_{(z)} = P_z \nabla$.

To prove the third estimate (17.64) we compute:

$$\begin{aligned}
D_{(z)}\left(1-\overline{w}z\right)^{k} &= k\left(1-\overline{w}z\right)^{k-1}D_{(z)}\left(1-\overline{w}z\right)\\
&= k\left(1-\overline{w}z\right)^{k-1}\left\{(1-|z|^{2})P_{z}\nabla+\sqrt{1-|z|^{2}}Q_{z}\nabla\right\}\left(1-\overline{w}z\right)\\
&= -k\left(1-\overline{w}z\right)^{k-1}\left\{(1-|z|^{2})P_{z}\overline{w}+\sqrt{1-|z|^{2}}Q_{z}\overline{w}\right\};\\
R_{(z)}\left(1-\overline{w}z\right)^{k} &= k\left(1-\overline{w}z\right)^{k-1}\left(-\overline{w}z\right).
\end{aligned}$$

Since $|w|^{2}+|a|^{2}\leq 2$ we have

$$\begin{aligned}
\left|Q_{z}\overline{w}\right|^{2} &= \left|Q_{z}\left(\overline{w}-\overline{z}\right)\right|^{2}\leq\left|\overline{w}-\overline{z}\right|^{2},\\
&= |w|^{2}+|z|^{2}-2\operatorname{Re}\left(w\overline{z}\right)\\
&\leq 2\operatorname{Re}\left(1-w\overline{z}\right)\leq 2\left|1-w\overline{z}\right|,
\end{aligned}$$

which yields (17.64). □

17.5. Operator Estimates

Often, when obtaining solutions to $\overline{\partial}$-equations in function spaces, the crucial estimates involve singular integral operators. In contrast, the approach here leads to operators which can be majorized by bounded operators with positive integration kernels. Those can be studied with tools based on Schur's lemma . In particular we will make heavy use of the following which is Lemma B.10 from Appendix B.

LEMMA 17.48. *Let $a,b,c,t\in\mathbb{R}$. Then the operator*

$$T_{a,b,c}f(z)=\int_{\mathbb{B}_{n}}\frac{(1-|z|^{2})^{a}(1-|w|^{2})^{b}\left(\sqrt{\triangle(w,z)}\right)^{c}}{\left|1-w\overline{z}\right|^{n+1+a+b+c}}f(w)\,dV(w)$$

is bounded on $L^{p}\left(\mathbb{B}_{n};(1-|w|^{2})^{t}dV(w)\right)$ if and only if $c>-2n$ and

$$-pa<t+1<p(b+1). \tag{17.68}$$

Noting that $|1-w\overline{z}|\leq 2$ we also have an immediate corollary of the lemma

REMARK 17.49. *If a,b,c,p,t are as in the previous Lemma, and if $d>0$ then the operator*

$$T_{a,b,c,d}f(z)=\int_{\mathbb{B}_{n}}\frac{(1-|z|^{2})^{a}(1-|w|^{2})^{b}\left(\sqrt{\triangle(w,z)}\right)^{c}}{\left|1-w\overline{z}\right|^{n+1+a+b+c+d}}f(w)\,dV(w)$$

is bounded on $L^{p}\left(\mathbb{B}_{n};(1-|w|^{2})^{t}dV(w)\right)$.

We must show that $f=\Omega_{0}^{1}h-\Lambda_{g}\Gamma_{0}^{2}\in B_{p}^{\sigma}\left(\mathbb{B}_{n};\ell^{2}\right)$ where Γ_{0}^{2} is an antisymmetric 2-tensor of $(0,0)$-forms that solves

$$\overline{\partial}\Gamma_{0}^{2}=\Omega_{1}^{2}h-\Lambda_{g}\Gamma_{1}^{3},$$

and inductively where Γ_{q}^{q+2} is an alternating $(q+2)$-tensor of $(0,q)$-forms that solves

$$\overline{\partial}\Gamma_{q}^{q+2}=\Omega_{q+1}^{q+2}h-\Lambda_{g}\Gamma_{q+1}^{q+3},$$

up to $q = n-1$ (since $\Gamma_n^{n+2} = 0$ and the $(0,n)$-form Ω_n^{n+1} is $\overline{\partial}$-closed). Using the Charpentier solution operators $\mathcal{C}_{n,s}^{0,q}$ on $(0,q+1)$-forms we then get

$$\begin{aligned}
f &= \Omega_0^1 h - \Lambda_g \Gamma_0^2 \\
&= \Omega_0^1 h - \Lambda_g \mathcal{C}_{n,s_1}^{0,0} \left(\Omega_1^2 h - \Lambda_g \Gamma_1^3\right) \\
&= \Omega_0^1 h - \Lambda_g \mathcal{C}_{n,s_1}^{0,0} \left(\Omega_1^2 h - \Lambda_g \mathcal{C}_{n,s_2}^{0,1} \left(\Omega_2^3 h - \Lambda_g \Gamma_2^4\right)\right) \\
&\vdots \\
&= \Omega_0^1 h - \Lambda_g \mathcal{C}_{n,s_1}^{0,0} \Omega_1^2 h + \Lambda_g \mathcal{C}_{n,s_1}^{0,0} \Lambda_g \mathcal{C}_{n,s_2}^{0,1} \Omega_2^3 h - \Lambda_g \mathcal{C}_{n,s_1}^{0,0} \Lambda_g \mathcal{C}_{n,s_2}^{0,1} \Lambda_g \mathcal{C}_{n,s_3}^{0,2} \Omega_3^4 h - \cdots \\
&\qquad \cdots + (-1)^n \Lambda_g \mathcal{C}_{n,s_1}^{0,0} \cdots \Lambda_g \mathcal{C}_{n,s_n}^{0,n-1} \Omega_n^{n+1} h \\
&:= \mathcal{F}^0 + \mathcal{F}^1 + \cdots + \mathcal{F}^n.
\end{aligned}$$

The goal is to establish

$$\|f\|_{B_p^\sigma(\mathbb{B}_n;\ell^2)} \lesssim C_{n,\sigma,p,\delta}(g) \|h\|_{B_p^\sigma(\mathbb{B}_n)},$$

which we accomplish by showing that

$$\|\mathcal{F}^\mu\|_{B_{p,m_1}^\sigma(\mathbb{B}_n;\ell^2)} \lesssim C_{n,\sigma,p,\delta}(g) \|h\|_{\Lambda_{p,m_\mu}^\sigma(\mathbb{B}_n)}, \qquad 0 \le \mu \le n, \tag{17.69}$$

for a choice of integers m_μ satisfying

$$\frac{n}{p} - \sigma < m_1 < m_2 < \cdots < m_\ell < \cdots < m_n.$$

Recall that in (17.52) we defined both of the norms $\|F\|_{B_{p,m_\mu}^\sigma(\mathbb{B}_n;\ell^2)}$ and $\|F\|_{\Lambda_{p,m_\mu}^\sigma(\mathbb{B}_n;\ell^2)}$ for smooth vector functions F in the ball $\mathbb{B}_n$.

We now use the norms $\|\cdot\|_{\Lambda_{p,m}^\sigma(\mathbb{B}_n;\ell^2)}$ to estimate the composition of Charpentier solution operators in each function

$$\mathcal{F}^\mu = \Lambda_g \mathcal{C}_{n,s_1}^{0,0} \cdots \Lambda_g \mathcal{C}_{n,s_\mu}^{0,\mu-1} \Omega_\mu^{\mu+1} h.$$

Specifically we will use the specialized variants of the seminorms given by

$$\|F\|_{\Lambda_{p,m',m''}^\sigma(\mathbb{B}_n;\ell^2)}^p = \int_{\mathbb{B}_n} \left|(1-|z|^2)^\sigma \left\{(1-|z|^2)^{m'} R^{m'}\right\} \overline{D}^{m''} F(z)\right|^p d\lambda_n(z),$$

where we take m'' derivatives in $\overline{D}$ followed by m' derivatives in the invariant radial operator $(1-|z|^2)R$. Recall the vector $\mathcal{X}^m$ from Definition 17.39. Its components are differential operators of the form $X_1 X_2 \cdots X_m$ where each X_i is either I, $\overline{D}$, or $(1-|z|^2)R$, and where, by definition, $1-|z|^2$ is held constant in composing operators. It will also be convenient at times to use the notation

$$\mathcal{R}^m := (1-|z|^2)^m \left(R^k\right)_{k=0}^m. \tag{17.70}$$

Note that the components of $\mathcal{R}^m$ are simply the entries of $\mathcal{X}^m$ in which none of the operators $\overline{D}$ appear. We will make extensive use the multilinear estimate in Proposition 17.43.

Let us fix our attention on the function $\mathcal{F}^{\mu} = \mathcal{F}_0^{\mu}$ and write

$$\mathcal{F}_0^{\mu} = \Lambda_g \mathcal{C}_{n,s_1}^{0,0} \left\{ \Lambda_g \mathcal{C}_{n,s_2}^{0,1} \cdots \Lambda_g \mathcal{C}_{n,s_\mu}^{0,\mu-1} \Omega_{\mu}^{\mu+1} h \right\} = \Lambda_g \mathcal{C}_{n,s_1}^{0,0} \left\{ \mathcal{F}_1^{\mu} \right\},$$
$$\mathcal{F}_1^{\mu} = \Lambda_g \mathcal{C}_{n,s_2}^{0,1} \left\{ \Lambda_g \mathcal{C}_{n,s_3}^{0,2} \cdots \Lambda_g \mathcal{C}_{n,s_\mu}^{0,\mu-1} \Omega_{\mu}^{\mu+1} h \right\} = \Lambda_g \mathcal{C}_{n,s_2}^{0,1} \left\{ \mathcal{F}_2^{\mu} \right\},$$
$$\mathcal{F}_q^{\mu} = \Lambda_g \mathcal{C}_{n,s_{q+1}}^{0,q} \left\{ \mathcal{F}_{q+1}^{\mu} \right\}, \qquad etc,$$

where $\mathcal{F}_q^{\mu}$ is a $(0,q)$-form. We now perform the integration by parts in Lemma 17.34 in each iterated Charpentier operator $\mathcal{F}_q^{\mu} = \Lambda_g \mathcal{C}_{n,s_{q+1}}^{0,q} \left\{ \mathcal{F}_{q+1}^{\mu} \right\}$ to obtain

$$\begin{aligned} \mathcal{F}_q^{\mu} &= \Lambda_g \mathcal{C}_{n,s_{q+1}}^{0,q} \mathcal{F}_{q+1}^{\mu} \\ &= \sum_{j=0}^{m'_{q+1}-1} c'_{j,n,s_{q+1}} \Lambda_g \mathcal{S}_{n,s_{q+1}} \left(\overline{\mathcal{D}}^j \mathcal{F}_{q+1}^{\mu} \right)(z) \\ &\quad + \sum_{\ell=0}^{\mu} c_{\ell,n,s_{q+1}} \Lambda_g \Phi_{n,s_{q+1}}^{\ell} \left(\overline{\mathcal{D}}^{m'_{q+1}} \mathcal{F}_{q+1}^{\mu} \right)(z). \end{aligned} \tag{17.71}$$

Now we compose these formulas for $\mathcal{F}_k^{\mu}$ to obtain an expression for $\mathcal{F}^{\mu}$ that is a complicated sum of compositions of the individual operators in (17.71) above. For now we will concentrate on the main terms $\Lambda_g \Phi_{n,s_{k+1}}^{\mu} \left(\overline{\mathcal{D}}^{m'_{k+1}} \mathcal{F}_{k+1}^{\mu} \right)$ that arise in the second sum above when $\ell = \mu$. We will see that the same considerations apply to any of the other terms in (17.71) . Recall from Lemma 17.34 that the "boundary" operators $\mathcal{S}_{n,s_{q+1}}$ are projections of operators on $\partial \mathbb{B}^s$ to the ball $\mathbb{B}_n$ and have (balanced) kernels even simpler than those of the operators $\Phi_{n,s_{q+1}}^{\ell}$. The composition of these main terms is

$$\begin{aligned} &\left(\Lambda_g \Phi_{n,s_1}^{\mu} \overline{\mathcal{D}}^{m'_1} \right) \mathcal{F}_1^{\mu} \\ &= \left(\Lambda_g \Phi_{n,s_1}^{\mu} \overline{\mathcal{D}}^{m'_1} \right) \left(\Lambda_g \Phi_{n,s_2}^{\mu} \overline{\mathcal{D}}^{m'_2} \right) \mathcal{F}_2^{\mu} \\ &= \left(\Lambda_g \Phi_{n,s_1}^{\mu} \overline{\mathcal{D}}^{m'_1} \right) \left(\Lambda_g \Phi_{n,s_2}^{\mu} \overline{\mathcal{D}}^{m'_2} \right) \cdots \left(\Lambda_g \Phi_{n,s_\mu}^{\mu} \overline{\mathcal{D}}^{m'_\mu} \right) \Omega_{\mu}^{\mu+1} h. \end{aligned} \tag{17.72}$$

At this point we would like to take absolute values inside all of these integrals and use the crucial inequalities in Proposition 17.47 to obtain a composition of positive operators of the type considered in Lemma 17.48. However, there is a difficulty in using the crucial inequality (17.62) to estimate the derivative $\overline{\mathcal{D}}^m$ on $(0, q+1)$-forms η given by (17.39):

$$\overline{\mathcal{D}}^m \eta(z) = \sum_{|J|=q} \sum_{k \notin J} \sum_{|\alpha|=m} (-1)^{\mu(k,J)} \overline{(w_k - z_k)(w-z)^{\alpha}} \frac{\partial^m}{\partial \overline{w}^{\alpha}} \eta_{J \cup \{k\}}(w).$$

The problem is that the factor $\overline{(w_k - z_k)}$ has no derivative $\frac{\partial}{\partial \overline{w_k}}$ naturally associated with it, as do the other factors in $\overline{(w-z)^{\alpha}}$. We refer to the factor $\overline{(w_k - z_k)}$ as a *rogue* factor, as it requires special treatment in order to apply (17.62). Informally, in the first estimate in Proposition 17.47 we control a power of the operator $\overline{(w_k - z_k)} \frac{\partial}{\partial \overline{w_k}}$ with the same power of $\sqrt{\triangle(w,z)} \overline{D}$. This holds in spite of the fact that the operator $\sqrt{\triangle(w,z)}$ is much much *smaller* $(w_k - z_k)$. The compensating fact is that $\overline{D}$ is much much *larger* than $\frac{\partial}{\partial \overline{w_k}}$.

We now describe how to circumvent this difficulty in the composition of operators in (17.72). Let us write each $\overline{\mathcal{D}}^{m'_{q+1}}\mathcal{F}^{\mu}_{q+1}$ as

$$\sum_{|J|=q}\sum_{k\notin J}\sum_{|\alpha|=m'_{q+1}}(-1)^{\mu(k,J)}\overline{(w_k-z_k)}\overline{(w-z)}^{\alpha}\frac{\partial^m}{\partial\overline{w}^{\alpha}}\left(\mathcal{F}^{\mu}_{q+1}\right)_{J\cup\{k\}}(w),$$

where $\left(\mathcal{F}^{\mu}_{q+1}\right)_{J\cup\{k\}}$ is the coefficient of the form $\mathcal{F}^{\mu}_{q+1}$ with differential $d\overline{w}^{J\cup\{k\}}$. We now replace each of these sums with just one of the summands, say

$$\overline{(w_k-z_k)}\overline{(w-z)}^{\alpha}\frac{\partial^m}{\partial\overline{w}^{\alpha}}\left(\mathcal{F}^{\mu}_{q+1}\right)_{J\cup\{k\}}(w). \tag{17.73}$$

Here the factor $\overline{(w_k-z_k)}$ is a *rogue* factor, not associated with a corresponding derivative $\frac{\partial}{\partial\overline{w}_k}$. We will refer to k as the *rogue* index associated with the rogue factor when it is not convenient to explicitly display the variables.

The key fact in treating the rogue factor $\overline{(w_k-z_k)}$ is that its presence in (17.73) means that the coefficient $\left(\mathcal{F}^{\mu}_{q+1}\right)_I$ of the form $\mathcal{F}^{\mu}_{q+1}$ that multiplies it *must* have k in the multi-index I. Since $\mathcal{F}^{\mu}_{q+1}=\Lambda_g\mathcal{C}^{0,q+1}_{n,s_{q+2}}\left\{\mathcal{F}^{\mu}_{q+2}\right\}$, the form of the ameliorated Charpentier kernel $\mathcal{C}^{0,q+1}_{n,s_{q+2}}$ in Theorem 17.23 shows that the coefficients of $\mathcal{C}^{0,q+1}_{n,s_{q+2}}(w,z)$ that multiply the rogue factor *must* have the differential $d\overline{z_k}$ in them. In turn, this means that the differential $d\overline{w_k}$ must be *missing* in the coefficient of $\mathcal{C}^{0,q+1}_{n,s_{q+2}}(w,z)$, and hence finally that the coefficients $\left(\mathcal{F}^{\mu}_{q+2}\right)_H$ with multi-index H that survive the wedge products in the integration *must* have $k\in H$. Furthermore, this observation can be repeated. We now use those facts to derive an important consequence.

Returning to (17.72), each summand in $\overline{\mathcal{D}}^{m'_{q+1}}\mathcal{F}^{\mu}_{q+1}$ has a rogue factor with associated rogue index k_{q+1}. Thus the function in (17.72) is a sum of composite operators of the form

$$\begin{aligned}\left(\Lambda_g\Phi^{\mu}_{n,s_1}\overline{(w_{k_1}-z_{k_1})}\overline{\mathcal{Z}}^{m'_1}\right)&\circ\left(\Lambda_g\Phi^{\mu}_{n,s_2}\overline{(w_{k_2}-z_{k_2})}\overline{\mathcal{Z}}^{m'_2}\right)_{I_1}\circ\\ &\cdots\circ\left(\Lambda_g\Phi^{\nu}_{n,s_\nu}\overline{(w_{k_\nu}-z_{k_\nu})}\overline{\mathcal{Z}}^{m'_\nu}\right)_{I_{\nu-1}}\circ\\ &\cdots\circ\left(\Lambda_g\Phi^{\mu-1}_{n,s_\mu}\overline{(w_{k_\mu}-z_{k_\mu})}\overline{\mathcal{Z}}^{m'_\mu}\right)_{I_{\mu-1}}\circ\left(\Omega^{\mu+1}_{\mu}h\right)_{I_\mu}.\end{aligned}$$

Here the subscript I_ν on the form $\Lambda_g\Phi^{\nu}_{n,s_\nu}\overline{(w_{k_\nu}-z_{k_\nu})}\overline{\mathcal{Z}}^{m'_\nu}$ indicates that we are composing with the component of $\Lambda_g\Phi^{\nu}_{n,s_\nu}\overline{(w_{k_\nu}-z_{k_\nu})}\overline{\mathcal{Z}}^{m'_\nu}$ corresponding to the multi-index $I_{\nu-1}$, i.e. the component with the differential $d\overline{z}^{I_{\nu-1}}$. The notation will become exceedingly unwieldy if we attempt to identify the different variables associated with each of the iterated integrals, so we refrain from this in general. The considerations of the previous paragraph now show that we must have $\{k_1\}=I_1$, $\{k_2\}\cup I_1=I_2$ and more generally

$$\{k_\nu\}\cup I_{\nu-1}=I_\nu,\qquad 1<\nu\le\mu.$$

In particular we see that the associated rogue indices $k_1,k_2,\dots k_\mu$ are all distinct and that as sets

$$\{k_1,k_2,\dots,k_\mu\}=I_\mu.$$

If we denote by ζ the variable in the final form $\Omega_\mu^{\mu+1}h$, we can thus write each rogue factor $\overline{(w_{k_\nu} - z_{k_\nu})}$ as

$$\overline{(w_{k_\nu} - z_{k_\nu})} = \overline{(w_{k_\nu} - \zeta_{k_\nu})} - \overline{(z_{k_\nu} - \zeta_{k_\nu})},$$

and since $k_\nu \in I_\mu$, there is a factor of the form $\frac{\partial}{\partial \overline{\zeta_{k_\nu}}} \frac{\partial^{|\beta|} g_i}{\partial \overline{\zeta}^\beta}$ in each summand of the component $\left(\Omega_\mu^{\mu+1}h\right)_{I_\mu}$ of $\Omega_\mu^{\mu+1}h$. So we are able to associate the rogue factor $\overline{(w_{k_\nu} - z_{k_\nu})}$ with derivatives of g as follows:

$$\left\{\overline{(w_{k_\nu} - \zeta_{k_\nu})} \frac{\partial}{\partial \overline{\zeta_{k_\nu}}}\right\} \frac{\partial^{|\beta|} g_i}{\partial \overline{\zeta}^\beta} - \left\{\overline{(z_{k_\nu} - \zeta_{k_\nu})} \frac{\partial}{\partial \overline{\zeta_{k_\nu}}}\right\} \frac{\partial^{|\beta|} g_i}{\partial \overline{\zeta}^\beta}. \tag{17.74}$$

Thus it is possible to

(1) apply the radial integration by parts in Corollary 17.32,
(2) then take absolute values and ℓ^2-norms inside all the integrals,
(3) and then apply the crucial inequalities in Proposition 17.47.

One of the difficulties remaining after this is that we are now left with additional factors of the form

$$\frac{\sqrt{\triangle(w,\zeta)}}{1-|w|^2}, \frac{\sqrt{\triangle(z,\zeta)}}{1-|z|^2}$$

that result from an application of (17.62) to the derivatives in (17.74). These factors are still rogue in the sense that the variable pairs occurring in them, namely (w,ζ) and (z,ζ), do not consist of consecutive variables in the iterated integrals of (17.72). This is rectified by using the fact that $d(w,z) = \sqrt{\triangle(w,z)}$ is a quasimetric; that is, there is a K so that for all p,q,r, $d(p,r) \leq K(d(p,q)+d(q,r))$. That fact follows from the identity

$$\sqrt{\triangle(w,z)} = |1-w\overline{z}|\,|\varphi_z(w)| = \delta(w,z)^2 \rho(w,z),$$

where $\rho(w,z) = |\varphi_z(w)|$ is the invariant pseudohyperbolic metric on the ball [**346**, Corollary 1.22] and where $\delta(w,z) = |1-w\overline{z}|^{1/2}$ satisfies the triangle inequality on the ball [**288**, Proposition 5.1.2]. Using the quasi-subadditivity of $d(w,z)$ we can, with some care, redistribute appropriate factors back to the iterated integrals where they can be favorably estimated using Lemma 17.48. It is simplest to illustrate this procedure in specific cases, so we defer further discussion of this point until we treat in detail the cases $\mu = 0,1,2$ below. We again emphasize that all of the above observations regarding rogue factors in (17.72) apply equally well to the rogue factors in the other terms $\Phi^\ell_{n,s_{q+1}}\left(\overline{\mathcal{D}}^{m'_q}\mathcal{F}^\mu_{q+1}\right)(z)$ in (17.71), as well as to the boundary terms $\mathcal{S}_{n,s_{q+1}}\left(\overline{\mathcal{D}}^j\mathcal{F}^\mu_{q+1}\right)(z)$ in (17.71).

The other difficulty remaining is that in order to obtain a favorable estimate using Lemma 17.48 for the iterated integrals resulting from the numbered items above, it is necessary to generate additional powers of $(1-|z|^2)$ (we are using z as a generic variable in the iterated integrals here). This is accomplished by applying the radial integrations by parts in Corollary 17.32 to the *previous* iterated integral. Of course this is not possible for the first of the iterated integrals; however in that case, because our candidate f from the Koszul complex is holomorphic, we are only applying the radial derivative R. As a result, we see from (17.63) that $(1-|z|^2)R$, unlike D, generates positive powers of $1-|z|^2$ even when acting on $\triangle(w,z)$. This procedure is also best illustrated in specific cases and will be treated in the next section.

So ignoring these technical issues for the moment, the integrals that result from taking absolute values and ℓ^2-norms inside (17.72) are now estimated using Lemma 17.48 and Remark 17.49. Note that we only use *scalar-valued* Schur estimates (as opposed to those in Section B.5) since all the integrals to which Lemma 17.48 and Remark 17.49 are applied have positive integrands. Here is the rough idea. Suppose that $\{T_1, T_2, \ldots, T_\mu\}$ is a collection of Charpentier solution operators and that for a sequence of large integers $\{m_1', m_1'', m_2', m_2'' \ldots, m_{\mu+1}', m_{\mu+1}''\}$, we have the inequalities

$$\|T_j F\|_{\Lambda^\sigma_{p,m_j',m_j''}(\mathbb{B}_n;\ell^2)} \lesssim \|F\|_{\Lambda^\sigma_{p,m_{j+1}',m_{j+1}''}(\mathbb{B}_n;\ell^2)}, \qquad 1 \le j \le \ell+1, \tag{17.75}$$

for the class of smooth functions F that arise as TG for some Charpentier solution operator T and some smooth G. Then we can estimate

$$\|T_1 \circ T_2 \circ \cdots \circ T_\mu \Omega\|_{B^\sigma_{p,m}(\mathbb{B}_n;\ell^2)}$$

by

$$\begin{aligned}
&\|T_1 \circ T_2 \circ \cdots \circ T_\ell \Omega\|_{\Lambda^\sigma_{p,m_1',m_1''}(\mathbb{B}_n;\ell^2)} \\
&\qquad \le C_1 \|T_2 \circ \cdots \circ T_\ell \Omega\|_{\Lambda^\sigma_{p,m_2',m_2''}(\mathbb{B}_n;\ell^2)} \\
&\qquad \le C_1 C_2 \|T_3 \circ \cdots \circ T_\ell \Omega\|_{\Lambda^\sigma_{p,m_3',m_3''}(\mathbb{B}_n;\ell^2)} \\
&\qquad \le C_1 C_2 \cdots C_\ell \|\Omega\|_{\Lambda^\sigma_{p,m_{\ell+1}',m_{\ell+1}''}(\mathbb{B}_n;\ell^2)}.
\end{aligned}$$

Finally we will show that if Ω is one of the forms Ω_q^{q+1} in the Koszul complex, then

$$\|\Omega\|_{\Lambda^\sigma_{p,m_{\ell+1}',m_{\ell+1}''}(\mathbb{B}_n;\ell^2)} \le \|\Omega\|_{\Lambda^\sigma_{p,m_{\ell+1}'+m_{\ell+1}''}(\mathbb{B}_n;\ell^2)} \lesssim \|h\|_{B^\sigma_{p,m}(\mathbb{B}_n)},$$

and so altogether this proves that

$$\|f\|_{B^\sigma_p(\mathbb{B}_n;\ell^2)} \lesssim \|h\|_{B^\sigma_{p,m}(\mathbb{B}_n)}.$$

We now make some brief comments on how to obtain the inequalities in (17.75). Complete details will be given in the cases $\mu = 0, 1, 2$ below, and the general case $0 \le \mu \le n$ is no different than these three cases. We note that from (17.25) the kernel of $\mathcal{C}_n^{0,q}$ typically looks like a sum of terms

$$\frac{(1 - w\overline{z})^{n-1-q} (1 - |w|^2)^q}{\triangle (w, z)^n} (\overline{z_j} - \overline{w_j}) \tag{17.76}$$

times a wedge product of differentials in which the differential $d\overline{w_j}$ is missing. We again emphasize that the rogue factor $(\overline{z_j} - \overline{w_j})$ cannot simply be estimated by $|\overline{z_j} - \overline{w_j}|$ as the formula (15.23) shows that

$$\sqrt{\triangle} (w, z) = \left| P_z (z - w) + \sqrt{1 - |z|^2} Q_z (z - w) \right|$$

can be much smaller than $|z - w|$. As we mentioned above, it is possible to exploit the fact that any surviving term in the form $\Omega_\mu^{\mu+1}$ must then involve the derivative $\frac{\partial}{\partial \overline{w_j}}$ hitting a component of g. This permits us to absorb part of the complex tangential component of $z - w$ into the almost invariant derivative D which is larger than the usual gradient in the complex tangential directions. This results in a good estimate for the rogue factor $(\overline{z_j} - \overline{w_j})$ in (17.76) based on the smaller quantity $\sqrt{\triangle(w,z)}$. We have already integrated by parts to write (17.76) as (recall that the factors $\overline{z_j} - \overline{w_j}$ are already incorporated into $\overline{\mathcal{D}_z^m}\eta(w)$)

$$\int_{\mathbb{B}_n} \frac{(1-w\overline{z})^{n-1-q}(1-|w|^2)^q}{\triangle(w,z)^n}\overline{\mathcal{D}^m}\eta(w)\,dV(w),$$

plus boundary terms which we ignore for the moment. Then we use the three crucial inequalities (17.62), (17.63) and (17.64);

$$\begin{aligned}
\left|(\overline{z_j}-\overline{w_j})\overline{\mathcal{D}_{z,w}^m}\Omega_\ell^{\ell+1}(w)\right| &\leq \left(\frac{\sqrt{\triangle(w,z)}}{1-|w|^2}\right)^{m+1}\left|\overline{D^m}\widehat{\Omega_\ell^{\ell+1}}(w)\right|,\\
\left|D_{(z)}\triangle(w,z)\right| &\lesssim (1-|z|^2)\triangle(w,z)^{1/2}+\triangle(w,z),\\
\left|(1-|z|^2)R_{(z)}\triangle(w,z)\right| &\lesssim (1-|z|^2)\triangle(w,z)^{1/2},\\
\left|D_{(z)}^m\left\{(1-\overline{w}z)^k\right\}\right| &\lesssim |1-wz|^k\left(\frac{1-|z|^2}{|1-\overline{w}z|}\right)^{m/2}\\
\left|(1-|z|^2)^m R_{(z)}^m\left\{(1-\overline{w}z)^k\right\}\right| &\lesssim |1-\overline{w}z|^k\left(\frac{1-|z|^2}{|1-\overline{w}z|}\right)^m,
\end{aligned}$$

to help show that the resulting iterated kernels can be factored (after accounting for all rogue factors $\overline{z_j} - \overline{w_j}$) into operators that satisfy the hypotheses of Lemma 17.48 or Remark 17.49 above.

DEFINITION 17.50. *The expression* $\widehat{\Omega_\ell^{\ell+1}}$ *denotes the form* $\Omega_\ell^{\ell+1}$ *but with every occurrence of the derivative* $\frac{\partial}{\partial \overline{w_j}}$ *replaced by the derivative* $\overline{D_j}$.

Recall that each summand of $\Omega_\ell^{\ell+1}$ includes a product of exactly ℓ distinct derivatives $\frac{\partial}{\partial \overline{w_j}}$ applied to components of g. Thus the entries of $\overline{D^m}\widehat{\Omega_\ell^{\ell+1}}(w)$ consist of $m+\ell$ derivatives distributed among components of g. Using the factorization of $\Omega_\ell^{\ell+1}$ in (17.46), we obtain the corresponding factorization for $\widehat{\Omega_\ell^{\ell+1}}$:

$$\Omega_0^1 \wedge \bigwedge_{i=1}^{\ell} \widehat{\Omega_0^1} = -\frac{1}{\ell+1}\widehat{\Omega_\ell^{\ell+1}}, \tag{17.77}$$

where $\Omega_0^1 = \left(\frac{\overline{g_i}}{|g|^2}\right)_{i=1}^\infty$ and $\widehat{\Omega_0^1} = \left(\frac{\overline{Dg_i}}{|g|^2}\right)_{i=1}^\infty$.

It is important for this purpose of using Lemma 17.48 and Remark 17.49 to first apply the integration by parts Lemma 17.29 to temper the singularity due to negative powers of $\triangle(w,z)$, and to use the integration by parts Corollary 17.32 to infuse enough powers of $(1-|w|^2)$ for use in the subsequent iterated integral.

Finally it follows from Lemma 17.37, Proposition 17.36 and Proposition 17.43 together with the factorization (17.46) that

$$\left\|(1-|z|^2)^\sigma \mathcal{X}^m \widehat{\Omega_\mu^{\mu+1}} h(z)\right\|_{L^p(\lambda_n;\ell^2)} \lesssim \|\mathbb{M}_g\|_{B_p^\sigma(\mathbb{B}_n)\to B_p^\sigma(\mathbb{B}_n;\ell^2)}^{m+\mu}\|h\|_{B_p^\sigma(\mathbb{B}_n)}. \tag{17.78}$$

We defer the proof of (17.78) until Section 17.5.1.1 when further calculations are available.

REMARK 17.51. *At this point we observe from* (17.69) *that the exponent* $m+\mu$ *in* (17.78) *is at most* m_n+n, *and thus we may take* $\kappa=m_n+n$. *We leave it to the interested reader to estimate the size of* m_n.

Taking into account all of the above, the conclusion is that with $\kappa=m_n+n$,

$$\|f\|_{B_p^\sigma(\mathbb{B}_n;\ell^2)} \lesssim \|\mathbb{M}_g\|^\kappa_{B_p^\sigma(\mathbb{B}_n)\to B_p^\sigma(\mathbb{B}_n;\ell^2)} \|h\|_{B_p^\sigma(\mathbb{B}_n)}.$$

As the arguments described above are rather complicated we illustrate them by considering the three cases $\mu=0,1,2$ in complete detail in the next section before proceeding to the general case.

17.5.1. Estimates in Special Cases. Here we prove the estimates (17.69) for $\mu=0,1,2$. We first give an informal preview and then provide the details.

Recall that

$$\begin{aligned}\mathcal{F}^0 &= \Omega_0^1 h,\\ \mathcal{F}^1 &= \Lambda_g \mathcal{C}^{0,0}_{n,s_1}\Omega_1^2 h,\\ \mathcal{F}^2 &= \Lambda_g \mathcal{C}^{0,0}_{n,s_1}\Lambda_g \mathcal{C}^{0,1}_{n,s_2}\Omega_2^3 h.\end{aligned}$$

To obtain the estimate for $\mathcal{F}^0$ we use the multilinear inequality in Proposition 17.43.

In estimating $\mathcal{F}^1$ we confront for the first time a rogue factor $\overline{z_k-w_k}$ that we must associate with a derivative $\frac{\partial}{\partial \overline{w_k}}$ occurring in each surviving summand of the kth component of the form Ω_1^2. After applying the integration by parts formula in 17.34 as in [**249**], we use the crucial inequalities in Proposition 17.47 and the Schur type operator estimates in Lemma 17.48 with $c=0$ to obtain the desired estimates. Finally we must also deal with the boundary terms in the integration by parts formula for ameliorated Charpentier kernels in Lemma 17.34. This requires using the radial derivative integration by parts formula in Corollary 17.32 as in [**249**], and also requires dealing with the corresponding rogue factors.

The final trick in the proof arises in estimating $\mathcal{F}^2$. This time there are two iterated integrals each with a rogue factor. The problematic rogue factor $\overline{z_k-\zeta_k}$ occurs in the *first* of the iterated integrals since there is *no* derivative $\frac{\partial}{\partial \overline{\zeta_k}}$ hitting the second iterated integral with which to associate the rogue factor $\overline{z_k-\zeta_k}$. Instead we decompose the factor as $\overline{z_k-w_k}-\overline{\zeta_k-w_k}$ and associate each of these summands with a derivative $\frac{\partial}{\partial \overline{w_k}}$ already occurring in Ω_2^3. Then we can apply the crucial inequality (17.62) and use the fact that $\sqrt{\triangle(w,z)}$ is quasisubadditive to redistribute the estimates appropriately. As a result of this redistribution we are forced to use Lemma 17.48 with $c=\pm 1$ this time as well as $c=0$. In applying the Schur type estimates in Lemma 17.48 to the *second* iterated integral, we require a sufficiently large power of $(1-|w|^2)$ to be carried over from the first iterated integral. To ensure this we again use the radial derivative integration by parts formula in Corollary 17.32.

The estimate (17.69) for general μ involves no new ideas. There are now μ rogue terms and we need to apply Lemma 17.48 with $c=0,\pm 1,\ldots,\pm(\mu-1)$. With this noted the arguments needed are those used above in the cases $\mu=0,1,2$.

17.5.1.1. *The Estimate for* $\mathcal{F}^0$. We begin with the estimate

$$\left\|\mathcal{F}^0\right\|_{B^\sigma_{p,m}(\mathbb{B}_n;\ell^2)} = \left\|\Omega^1_0 h\right\|_{B^\sigma_{p,m}(\mathbb{B}_n;\ell^2)} \lesssim \|\mathbb{M}_g\|^m_{B^\sigma_p(\mathbb{B}_n)\to B^\sigma_p(\mathbb{B}_n;\ell^2)} \|h\|_{B^\sigma_{p,m}(\mathbb{B}_n)},$$

for $m+\sigma > n/p$. However, for later use we prove instead the more general estimate with $\mathcal{X}$ in place of R, except that m must then be chosen twice as large to insure $m > 2(n/p - \sigma)$:

(17.79)

$$\int_{\mathbb{B}_n} \left|(1-|z|^2)^\sigma \mathcal{X}^m \left(\Omega^1_0 h\right)(z)\right|^p d\lambda_n(z) \lesssim \|\mathbb{M}_g\|^{mp}_{B^\sigma_p(\mathbb{B}_n)\to B^\sigma_p(\mathbb{B}_n;\ell^2)} \|h\|^p_{B^\sigma_p(\mathbb{B}_n)},$$

Recall that $\mathcal{X}^m$ is the differential operator of order m given in Definition 17.39 that is adapted to the complex geometry of the unit ball $\mathbb{B}_n$. It will be in estimating iterated Charpentier integrals below that the derivatives R^m and $\overline{\mathcal{D}^m}$ will arise from integration by parts in the previous iterated integral, and this will require estimates using $\mathcal{X}^m$.

By Leibnitz' rule for $\mathcal{X}^m$ we have

$$\mathcal{X}^m\left(\Omega^1_0 h\right) = \sum_{k=0}^m c_k \left(\mathcal{X}^k \Omega^1_0\right)\left(\mathcal{X}^{m-k} h\right),$$

and

$$\mathcal{X}^k\left(\Omega^1_0\right) = \mathcal{X}^k\left(\frac{\overline{g}}{|g|^2}\right) = \sum_{\ell=0}^k c_\ell \left(\mathcal{X}^{k-\ell}\overline{g}\right)\left(\mathcal{X}^\ell |g|^{-2}\right). \tag{17.80}$$

It suffices to prove

$$\int_{\mathbb{B}_n} \left|(1-|z|^2)^\sigma \left(\sum_{k=0}^m \sum_{\ell=0}^k c_k c_\ell \left(\mathcal{X}^{k-\ell}\overline{g}\right)\left(\mathcal{X}^\ell |g|^{-2}\right)\left(\mathcal{X}^{m-k} h\right)\right)\right|^p d\lambda_n$$
$$\lesssim \|\mathbb{M}_g\|^{mp}_{B^\sigma_p(\mathbb{B}_n)\to B^\sigma_p(\mathbb{B}_n;\ell^2)} \|h\|^p_{B^\sigma_p(\mathbb{B}_n)},$$

and hence

$$\int_{\mathbb{B}_n} (1-|z|^2)^{p\sigma} \left|\mathcal{X}^{k-\ell}\overline{g}\right|^p \left|\mathcal{X}^\ell |g|^{-2}\right|^p \left|\mathcal{X}^{m-k} h\right|^p d\lambda_n \tag{17.81}$$
$$\lesssim \|\mathbb{M}_g\|^{mp}_{B^\sigma_p(\mathbb{B}_n)\to B^\sigma_p(\mathbb{B}_n;\ell^2)} \|h\|^p_{B^\sigma_p(\mathbb{B}_n)},$$

for each fixed $0 \le \ell \le k \le m$.

Now we can profitably estimate both $\left|\mathcal{X}^{m-k}h\right|$ and $\left|\mathcal{X}^{k-\ell}\overline{g}\right|$ as they are, but we must be more careful with $\left|\mathcal{X}^\ell |g|^{-2}\right|$. In the case $\ell = 1$, we assume for convenience that $\mathcal{X}$ annihilates g_i (if not it will annihilate $\overline{g_i}$ unless $\mathcal{X} = I$, and the estimates are similar) and obtain,

$$\left|\mathcal{X}|g|^{-2}\right|^2 = \left|-|g|^{-4}\sum_{i=1}^\infty g_i \mathcal{X}\overline{g_i}\right|^2$$
$$\le |g|^{-8}\left(\sum_{i=1}^\infty |g_i|^2\right)\left(\sum_{i=1}^\infty \left|\mathcal{X}\overline{g_i}\right|^2\right) \le |g|^{-6}\sum_{i=1}^\infty \left|\mathcal{X}\overline{g_i}\right|^2.$$

Similarly when $\ell = 2$,

$$\left|\mathcal{X}^2 |g|^{-2}\right|^2 = \left| - |g|^{-4} \sum_{i=1}^{\infty} g_i \mathcal{X}^2 \overline{g_i} + 2 |g|^{-6} \sum_{i \neq j} \left(g_i \mathcal{X} \overline{g_i}\right) \left(g_j \mathcal{X} \overline{g_j}\right) \right|^2$$

$$\leq 2 |g|^{-6} \sum_{i=1}^{\infty} \left|\mathcal{X}^2 \overline{g_i}\right|^2 + 4 |g|^{-8} \left(\sum_{i=1}^{\infty} \left|\mathcal{X} \overline{g_i}\right|^2 \right)^2,$$

and the general case is

$$(17.82) \qquad \begin{aligned} &\left|\mathcal{X}^\ell |g|^{-2}\right|^2 \\ &\leq C_\ell |g|^{-6} \sum_{i=1}^{\infty} \left|\mathcal{X}^\ell \overline{g_i}\right|^2 + C_{\ell-1} |g|^{-8} \left(\sum_{i=1}^{\infty} \left|\mathcal{X}^{\ell-1} \overline{g_i}\right|^2 \right) \left(\sum_{i=1}^{\infty} \left|\mathcal{X} \overline{g_i}\right|^2 \right) \\ &\qquad + \cdots + C_0 |g|^{-4-2\ell} \left(\sum_{i=1}^{\infty} \left|\mathcal{X} \overline{g_i}\right|^2 \right)^\ell \\ &= \sum_{\substack{1 \leq \alpha_1 \leq \alpha_2 \leq \cdots \leq \alpha_M \\ \alpha_1 + \alpha_2 + \cdots + \alpha_M = \ell}} c_\alpha |g|^{-4-2\ell} \prod_{m=1}^{M} \left(\sum_{i=1}^{\infty} \left|\mathcal{X}^{\alpha_m} \overline{g_i}\right|^2 \right). \end{aligned}$$

We can ignore the powers of $|g|$ since $|g|$ is bounded above and below by Lemma 17.42 and the hypotheses of Theorem 17.11. Fixing α we see that the left side of (17.81) is thus dominated by a multiple of

$$\int_{\mathbb{B}_n} (1 - |z|^2)^{p\sigma} \left|\mathcal{X}^{k-\ell} \overline{g}\right|^p \left|\mathcal{Y}^{m-k} h\right|^p \left(\prod_{j=1}^{M} \left|\mathcal{X}^{\alpha_j} \overline{g}\right|^p \right) d\lambda_n.$$

Since $\left|\mathcal{X}^{k-\ell} \overline{g}\right|^2 = \sum_{i=1}^{\infty} \left|\mathcal{X}^{k-\ell} \overline{g_i}\right|^2$ and $k - \ell$ could vanish (unlike the exponents α_ℓ which are positive), we see that altogether after renumbering, it suffices to prove (17.83)

$$\int_{\mathbb{B}_n} (1 - |z|^2)^{p\sigma} \left|\mathcal{Y}^{\alpha_1} h\right|^p \left|\mathcal{Y}^{\alpha_2} g\right|^p \cdots \left|\mathcal{Y}^{\alpha_M} g\right|^p d\lambda_n \lesssim \|\mathbb{M}_g\|_{B_p^\sigma(\mathbb{B}_n) \to B_p^\sigma(\mathbb{B}_n;\ell^2)}^{Mp} \|h\|_{B_p^\sigma(\mathbb{B}_n)}^p$$

for each fixed $\alpha = (\alpha_1, \alpha_2, \ldots, \alpha_M)$ with $M \geq 2$, $|\alpha| = m$ and at most one of $\alpha_2, \ldots, \alpha_M$ is zero. We have used here that $\left|\overline{Dg}\right| = |Dg|$. Now Proposition 17.43 yields 17.83 for each $0 \leq k \leq m$ and $|\alpha| = m - k$. Summing these estimates completes the proof of (17.79).

We can now prove the more general inequality (17.78). Using the factorization (17.46) of $\widehat{\Omega_\mu^{\mu+1}}$ together with the Leibnitz formula gives

$$\begin{aligned}\mathcal{X}^m\left(\widehat{\Omega_\mu^{\mu+1}}h\right) &= \mathcal{X}^m\left(\Omega_0^1\wedge\left(\widehat{\Omega_0^1}\right)^\mu h\right)\\ &= \sum_{\alpha\in\mathbb{Z}_+^{\mu+2}:|\alpha|=m}\left(\mathcal{X}^{\alpha_0}\Omega_0^1\right)\wedge\bigwedge_{j=1}^{\mu}\left(\mathcal{X}^{\alpha_j}\widehat{\Omega_0^1}\right)\left(\mathcal{X}^{\alpha_{\mu+1}}h\right)\\ &= \sum_{\alpha\in\mathbb{Z}_+^{\mu+2}:|\alpha|=m}\left(\left(\mathcal{X}^{\alpha_0}\Omega_0^1\right)\wedge\bigwedge_{j=1}^{\mu}\left(\mathcal{X}^{\alpha_j+1}\Omega_0^1\right)\right)\left(\mathcal{X}^{\alpha_{\mu+1}}h\right),\end{aligned}$$

where we have used that $\widehat{\Omega_0^1}$ already has an $\mathcal{X}$ derivative in each summand, and so $\mathcal{X}^{\alpha_j}\widehat{\Omega_0^1}$ can be written as $\mathcal{X}^{\alpha_j+1}\Omega_0^1$. Now use (17.80) and (17.82) to see that $\left|\mathcal{X}^m\left(\widehat{\Omega_\mu^{\mu+1}}h\right)\right|$ is controlled by a tensor product of at most $m+\mu$ factors, and then apply Proposition 17.43 as above to complete the proof of (17.78).

17.5.1.2. *The Estimate for* $\mathcal{F}^1$. The estimate in (17.69) with $\mu=1$ will follow from (17.78) and the estimate

$$\begin{aligned}\left\|(1-|z|^2)^\sigma\mathcal{Y}^{m_1}\left(\Lambda_g\mathcal{C}_{n,s}^{0,0}\Omega_1^2h\right)\right\|_{L^p(\lambda_n)}^p \\ \lesssim\int_{\mathbb{B}_n}\left|(1-|z|^2)^\sigma\mathcal{X}^{m_2}\left(\widehat{\Omega_1^2}h\right)(z)\right|^p d\lambda_n(z),\end{aligned}\tag{17.84}$$

where as in Definition 17.50, we define $\widehat{\Omega_1^2}$ to be Ω_1^2 with ∂ replaced by D throughout:

$$\widehat{\Omega_1^2}:=\sum_{j,k=1}^{N}\frac{\overline{\{g_kDg_j-g_jDg_k\}}}{|g|^4}e_j\wedge e_k,$$

and where $Dh=\sum_{k=1}^n(D_kh)\,dz_k$ and D_k is the kth component of D. We are using here the following observation regarding the interior product $\Omega_1^2 h\lrcorner d\overline{w_k}$:

(17.85) For each summand of $\Omega_1^2h\lrcorner d\overline{w_k}$, there is a unique $1\le i\le N$ so that $\dfrac{\partial g_i}{\partial w_k}$ occurs as a factor in the summand.

We rewrite (17.84) as

$$\begin{aligned}\left\|(1-|z|^2)^\sigma\mathcal{R}^{m_1''}D^{m_1'}\left(\Lambda_g\mathcal{C}_{n,s}^{0,0}\Omega_1^2h\right)\right\|_{L^p(\lambda_n)}^p \\ \lesssim\int_{\mathbb{B}_n}\left|(1-|z|^2)^\sigma\mathcal{R}^{m_2''}\overline{D^{m_2'}}\left(\widehat{\Omega_1^2}h\right)(z)\right|^p d\lambda_n(z),\end{aligned}\tag{17.86}$$

where $\mathcal{R}^m=(1-|z|^2)^m\left(R^k\right)_{k=0}^m$ as in (17.70). As mentioned above, we only need to prove the case $m_1''=0$ since (17.69) only requires that we estimate $\left\|\mathcal{F}^1\right\|_{B_{p,m}^\sigma(\mathbb{B}_n)}$. However, when considering the estimate for $\mathcal{F}^2$ in (17.69) we will no longer have the luxury of using the norm $\|\cdot\|_{B_{p,m}^\sigma(\mathbb{B}_n)}$ in the second iterated integral occurring there, and so we will consider the more general case now in preparation for what comes later. As we will see however, it is necessary to choose m_1' sufficiently large in order to obtain (17.86). It is useful to recall that the operator $(1-|z|^2)R$ is "smaller" than $\overline{D}$ in the sense that

$$\overline{D} = (1-|z|^2)P_z\overline{\nabla} + \sqrt{1-|z|^2}Q_z\overline{\nabla},$$
$$(1-|z|^2)R = (1-|z|^2)P_z\nabla.$$

To prove (17.86) we will ignore the contraction Λ_g since if derivatives hit g in the contraction, the estimates are similar if not easier. Note also that $|\Lambda_g F| \leq |g|\,|F|$ for the contraction $\Lambda_g F$ of any tensor F.

We will also initially suppose that $m_1''=0$ and later take m_1'' sufficiently large. Now we apply Lemma 17.34 to $\mathcal{C}_{n,s}^{0,0}\Omega_1^2 h$ and obtain

$$\begin{aligned}(17.87)\qquad \mathcal{C}_{n,s}^{0,0}\Omega_1^2 h\,(z) &= c_0\mathcal{C}_{n,s}^{0,0}\left(\overline{\mathcal{D}}^{m_2'}\Omega_1^2 h\right)(z) + \text{boundary terms}\\ &= \int_{\mathbb{B}_n}\Phi_{n,s}^0(w,z)\,\overline{\mathcal{D}}^{m_2'}\left(\Omega_1^2 h\right)dV(w) + \text{ boundary terms.}\end{aligned}$$

A typical term above looks like

$$(17.88)\qquad \int_{\mathbb{B}_n}\left(\frac{1-|w|^2}{1-\overline{w}z}\right)^{s-n}\frac{(1-w\overline{z})^{n-1}}{\triangle(w,z)^n}\overline{\mathcal{D}}^{m_2'}\left(\Omega_1^2 h\right)dV(w)$$

where we are discarding the sum of (balanced) factors $\left(\frac{(1-|w|^2)(1-|z|^2)}{|1-w\overline{z}|^2}\right)^j$ for $1\leq j\leq n-1$ in Lemma 17.34 that turn out to only help with the estimates. This can be seen from (17.64) and its trivial counterpart

$$\left|D_{(z)}^m\left\{(1-|z|^2)^k\right\}\right| + \left|(1-|z|^2)^m R_{(z)}^m\left\{(1-|z|^2)^k\right\}\right| \lesssim (1-|z|^2)^k.$$

Recall from the general discussion above that in the integral (17.88) there are rogue factors $\overline{z_k-w_k}$ in $\overline{\mathcal{D}}^{m_2'}\left(\Omega_1^2 h\right)(w)$ that must be associated with a $\frac{\partial}{\partial\overline{w_k}}$ derivative that hits some factor of each summand in the kth component $\Omega_1^2\lrcorner d\overline{w_k}$ of $\Omega_1^2 \approx \overline{\{g_i\partial g_j - g_j\partial g_i\}}$. Thus we can apply (17.62) to the components of $\Omega_1^2 h\,(z)$ to obtain

$$\begin{aligned}\left|\overline{\mathcal{D}}^{m_2'}\Omega_1^2 h\,(z)\right| &\approx \left|\sum_{k=1}^n\sum_{|\alpha|=m_2'}^n(\overline{w_k}-\overline{z_k})\,(\overline{w-z})^\alpha\frac{\partial^{m_2'}}{\partial\overline{w}^\alpha}\left(\Omega_1^2 h\lrcorner d\overline{w_k}\right)\right|\\ &\lesssim \left(\frac{\sqrt{\triangle(w,z)}}{1-|w|^2}\right)^{m_2'+1}\left|\overline{D^{m_2'}}\left(\widehat{\Omega_1^2 h}\right)(w)\right|.\end{aligned}$$

Thus we get

$$\begin{aligned}(17.89)\qquad &(1-|z|^2)^\sigma\left|D^{m_1'}\mathcal{C}_{n,s}^{0,0}\Omega_1^2 h\,(z)\right|\\ &\leq \int_{\mathbb{B}_n}(1-|z|^2)^\sigma\left|D_{(z)}^{m_1'}\left\{\frac{(1-|w|^2)^{s-n}(1-w\overline{z})^{n-1}}{(1-\overline{w}z)^{s-n}\triangle(w,z)^n}\right\}\right|\\ &\qquad\times\left(\frac{\sqrt{\triangle(w,z)}}{1-|w|^2}\right)^{m_2'+1}\left|\overline{D^{m_2'}}\left(\widehat{\Omega_1^2 h}\right)(w)\right|dV(w)\\ &:= S_{m_1',m_2'}^s f(z),\end{aligned}$$

where

$$f(w) = (1-|w|^2)^\sigma \left|\overline{D^{m_2'}} \left(\widehat{\Omega_1^2 h}\right)(w)\right|. \tag{17.90}$$

Now we iterate the estimate (17.63),

$$\left|D_{(z)} \triangle (w,z)\right| \lesssim (1-|z|^2) \triangle (w,z)^{1/2} + \triangle (w,z),$$

to obtain

$$\begin{aligned} &\left| D_{(z)}^{m_1'} \left\{ \frac{(1-|w|^2)^{s-n} (1-w\overline{z})^{n-1}}{(1-\overline{w}z)^{s-n} \triangle (w,z)^n} \right\} \right| \\ &\qquad \leq \frac{(1-|z|^2)^{m_1'} (1-|w|^2)^{s-n} \triangle (w,z)^{m_1'/2}}{|1-w\overline{z}|^{s-2n+1} \triangle (w,z)^{n+m_1'}} \\ &\qquad\qquad + \cdots + \frac{(1-|w|^2)^{s-n}}{|1-w\overline{z}|^{s-2n+1} \triangle (w,z)^n} + OK, \end{aligned} \tag{17.91}$$

where the terms in OK are obtained when some of the derivatives D hit the factor $\frac{1}{(1-\overline{w}z)^{s-n}}$ or factors $D \triangle (w,z)$ already in the numerator. Leaving the OK terms for later, we combine all the estimates above to get that if we plug the first term on the right in (17.91) into the left side of (17.86), then the result is dominated by

$$\begin{aligned} &S_{m_1',m_2'}^s f(z) \\ &= \int_{\mathbb{B}_n} \frac{(1-|z|^2)^{m_1'+\sigma} (1-|w|^2)^{s-n-m_2'-1-\sigma} \triangle (w,z)^{(m_1'+m_2'+1)/2}}{|1-w\overline{z}|^{s-2n+1} \triangle (w,z)^{n+m_1'}} f(w)\, dV(w) \\ &= \int_{\mathbb{B}_n} \frac{(1-|z|^2)^{m_1'+\sigma} (1-|w|^2)^{s-n-1-m_2'-\sigma}}{|1-w\overline{z}|^{s-2n+1}} \sqrt{\triangle (w,z)}^{m_2'-m_1'-2n+1} f(w)\, dV(w). \end{aligned}$$

Now for convenience choose $m_2' = m_1' + 2n - 1$ so that the factor of $\sqrt{\triangle (w,z)}$ disappears. We then get

$$(1-|z|^2)^\sigma \left|D^{m_1'} \mathcal{C}_{n,s}^{0,0} \Omega_1^2 h(z)\right| \tag{17.92}$$

$$\leq \int_{\mathbb{B}_n} \frac{(1-|z|^2)^{m_1'+\sigma} (1-|w|^2)^{s-3n-m_1'-\sigma}}{|1-w\overline{z}|^{s-2n+1}} f(w)\, dV(w). \tag{17.93}$$

Lemma 17.48 shows that the operator

$$T_{a,b,0} f(z) = \int_{\mathbb{B}_n} \frac{(1-|z|^2)^a (1-|w|^2)^b}{|1-w\overline{z}|^{n+1+a+b}} f(w)\, dV(w)$$

is bounded on $L^p\left(\mathbb{B}_n; (1-|w|^2)^t dV(w)\right)$ if and only if

$$-pa < t+1 < p(b+1).$$

We apply this lemma with $t = -n-1$, $a = m_1' + \sigma$ and $b = s - 3n - m_1' - \sigma$. Note that the sum of the exponents in the numerator and denominator of (17.92) are equal if we write the integral in terms of invariant measure $d\lambda_n(w) = (1-|w|^2)^{-n-1} dV(w)$. We conclude that $S_{m_1',m_2'}^s$ is bounded on $L^p(d\lambda_n)$ provided T is, and that this latter happens if and only if

$$-p(m_1'+\sigma) < -n < p(s-3n+1-m_1'-\sigma).$$

This requires $m_1' + \sigma > n/p$ and $s > 3n - 1 + m_1' + \sigma - n/p$.

REMARK 17.52. *Suppose instead that we choose m_2' above to be a positive integer satisfying $c = m_2' - m_1' - 2n + 1 > -2n$. Then we would be dealing with the operator $T_{a,b,c}$ where $a = m_1' + \sigma$ and*

$$b = s - n - 1 - m_2' - \sigma = s - 3n - c - m_1' - \sigma.$$

By Lemma 17.48, $T_{a,b,c}$ is bounded on $L^p(d\lambda_n)$ if and only if

$$-p(m_1' + \sigma) < -n < p(s - 3n + 1 - c - m_1' - \sigma),$$

i.e. $m_1' + \sigma > n/p$ and $s > c + 3n - 1 + m_1' + \sigma - n/p$. Thus we can use any *value of $c > -2n$ provided we choose $m_2' \geq m_1'$ and s large enough.*

Now we turn to the second displayed term on the right side of (17.91) which leads to the operator $T_{a,b,0}$ with $a = \sigma$, $b = s - 3n - \sigma$. This time we will *not* in general have the required boundedness condition $\sigma > n/p$. It is for this reason that we must return to (17.86) and insist that m_1'' be chosen sufficiently large that $m_1'' + \sigma > n/p$. For convenience we let $m_1' = 0$ for now. Indeed, it follows from the second line in the crucial inequality (17.63) that the second displayed term on the right side of (17.91) /is

$$\frac{(1-|z|^2)^{m_1''}(1-|w|^2)^{s-n} \triangle(w,z)^{m_1''/2}}{|1-w\overline{z}|^{s-2n+1} \triangle(w,z)^{n+m_1''}} + \text{better terms}.$$

Using this expression and choosing $m_2' = m_1'' + 2n - 1$ so that the term $\sqrt{\triangle(w,z)}$ disappears from the ensuing integral, we obtain the following analog of (17.92):

$$\begin{aligned}(1-|z|^2)^{\sigma}(1-|z|^2)^{m_1''} &\left|\mathcal{R}^{m_1''}\mathcal{C}_{n,s}^{0,0}\Omega_1^2 h(z)\right| \\ &\leq \int_{\mathbb{B}_n} \frac{(1-|z|^2)^{m_1''+\sigma}(1-|w|^2)^{s-3n-m_1''-\sigma}}{|1-w\overline{z}|^{s-2n+1}} f(w)\,dV(w).\end{aligned}$$

The corresponding operator $T_{a,b,0}$ has $a = m_1'' + \sigma$ and $b = s - 3n - m_1'' - \sigma$ and is bounded on $L^p(\lambda_n)$ when $-p(m_1'' + \sigma) < -n < p(s - 3n + 1 - m_1'' - \sigma)$. Thus there is no unnecessary restriction on σ if m_1'' and s are chosen appropriately large. Note that the only difference between this operator $T_{a,b,0}$ and the previous one is that m_1' has been replaced by m_1''.

The above arguments are easily modified to handle the general case of (17.86) provided $m_1'' + \sigma > n/p$ and s is chosen sufficiently large.

Now we return to consider the OK terms in (17.91). For this we use the inequality (17.64):

$$\left|D_{(z)}^m\left\{(1-\overline{w}z)^k\right\}\right| \lesssim |1-\overline{w}z|^k \left(\frac{1-|z|^2}{|1-\overline{w}z|}\right)^{m/2}.$$

We ignore the derivative $(1-|z|^2)R$ as the second line in (17.64) shows that it satisfies a better estimate. We also write m_1 and m_2 in place of m_1' and m_2' now. As a result, one of the extremal OK terms in (17.91) is

$$\frac{(1-|z|^2)^{m_1/2}(1-|w|^2)^{s-n}}{|1-w\overline{z}|^{s-2n+1+m_1/2} \triangle(w,z)^n},$$

which when combined with the other estimates leads to the integral operator

$$\int_{\mathbb{B}_n} \frac{(1-|z|^2)^{m_1/2+\sigma}(1-|w|^2)^{s-n-1-m_2-\sigma}}{|1-w\overline{z}|^{s-2n+1+m_1/2}} \sqrt{\triangle(w,z)}^{m_2-2n-1} f(w)\, dV(w).$$

This is $T_{a,b,c}$ with $a = m_1/2+\sigma$, $b = s-n-1-m_2-\sigma$ and $c = m_2-2n-1$. This is bounded on $L^p(\lambda_n)$ provided $m_2 \geq 2$ and

$$-p\left(\frac{m_1}{2}+\sigma\right) < -n < p(s-n-m_2-\sigma),$$

i.e. $m_1/2+\sigma > n/p$ and $s > n+m_2+\sigma-n/p$. The intermediate OK terms are handled similarly. Note that the crux of the matter is that all of the positive operators have the form $T_{a,b,c}$, and moreover, if s and the $m's$ are chosen appropriately large, then $T_{a,b,c}$ is bounded on $L^p(\lambda_n)$.

17.5.1.3. *Boundary Terms for $\mathcal{F}^1$.* Now we turn to estimating the boundary terms in (17.87). Considering upon Lemma 17.34 we see that a typical term we need to consider is, with $0 \leq k \leq m-1$,

$$\mathcal{S}_{n,s}\left(\overline{\mathcal{D}}^k\left(\Omega_1^2 h\right)\right)\left[\overline{\mathcal{Z}}\right](z) = \int_{\mathbb{B}_n} \frac{(1-|w|^2)^{s-n-1}}{(1-\overline{w}z)^s} \overline{\mathcal{D}}^k\left(\Omega_1^2 h\right)\left[\overline{\mathcal{Z}}\right](w)\, dV(w). \tag{17.94}$$

We now apply the operator $(1-|z|^2)^{m_1+\sigma} R^{m_1}$ to the integral in the right side of (17.94) and using Proposition 17.47 we obtain that the absolute value of the result is dominated by

$$\begin{aligned}
&\int_{\mathbb{B}_n} \frac{(1-|z|^2)^{m_1+\sigma}(1-|w|^2)^{s-n-1}}{|1-\overline{w}z|^{s+m_1}} \left(\frac{\sqrt{\triangle(w,z)}}{1-|w|^2}\right)^{k+1} \left|\overline{D}^k\left(\widehat{\Omega_1^2 h}\right)\right| dV(w) \\
&= \int_{\mathbb{B}_n} \frac{(1-|z|^2)^{m_1+\sigma}(1-|w|^2)^{s-n-2-k-\sigma}\sqrt{\triangle(w,z)}^{k+1}}{|1-\overline{w}z|^{s+m_1}} \\
&\qquad\qquad \times \left|(1-|w|^2)^{\sigma}\overline{D}^k\left(\widehat{\Omega_1^2 h}\right)(w)\right| dV(w).
\end{aligned}$$

The operator in question here is $T_{a,b,c}$ with $a = m_1+\sigma$, $b = s-n-2-k-\sigma$ and $c = k+1$ since

$$a+b+c+n+1 = s+m_1.$$

Lemma 17.48 applies to prove the desired boundedness on $L^p(\lambda_n)$ provided $m_1 + \sigma > n/p$.

However, if k fails to satisfy $k+1 > 2(n/p-\sigma)$, then the derivative $D^{k+1}\Omega$ cannot be used to control the norm $\|\Omega\|_{B_p^\sigma(\mathbb{B}_n)}$. To compensate for a small k, we must then apply Corollary 17.32 to the right side of (17.94) (which for fixed z is in $C\left(\overline{\mathbb{B}_n}\right) \cap C^\infty(\mathbb{B}_n)$) before differentiating and taking absolute values inside the integral. This then leads to operators of the form

$$\begin{aligned}
(1-|z|^2)^{m_1+\sigma} R^{m_1} &\int_{\mathbb{B}_n} \frac{(1-|w|^2)^{s-n-1}}{(1-\overline{w}z)^s} \\
&\qquad \times (1-|w|^2)^m R^m \left[\overline{\mathcal{D}}^k\left(\Omega_1^2 h\right)(w)\right] dV(w)
\end{aligned}$$

which are dominated by

$$\int_{\mathbb{B}_n} \frac{(1-|z|^2)^{m_1+\sigma}(1-|w|^2)^{s-n-1}}{|1-\overline{w}z|^{s+m_1}} \times \left(\frac{\sqrt{\triangle(w,z)}}{1-|w|^2}\right)^{k+1} \left|\mathcal{R}^m\overline{\mathcal{D}}^k\left(\widehat{\Omega_1^2 h}\right)(w)\right| dV(w),$$

which is

$$\int_{\mathbb{B}_n} \frac{(1-|z|^2)^{m_1+\sigma}(1-|w|^2)^{s-n-2-k-\sigma}\sqrt{\triangle(w,z)}^{k+1}}{|1-\overline{w}z|^{s+m_1}} \times \left|(1-|w|^2)^{\sigma}\mathcal{R}^m\overline{\mathcal{D}}^k\left(\widehat{\Omega_1^2 h}\right)(w)\right| dV(w).$$

This latter operator is $T_{a,b,c}H(z)$ with

$$a=m_1+\sigma, b=s-n-2-k-\sigma, c=k+1$$

and $H(w)=\left|(1-|w|^2)^{\sigma}R_{b'}^m\overline{D}^k\left(\widehat{\Omega_1^2 h}\right)(w)\right|$. Note that for $m>2(n/p-\sigma)$ we do now have $\|H\|_{L^p(\lambda_n)}\approx\left\|\widehat{\Omega_1^2 h}\right\|_{B_p^\sigma(\mathbb{B}_n)}$. The operator here is the same as that above and so Lemma 17.48 applies to prove the desired boundedness on $L^p(\lambda_n)$.

17.5.1.4. *The Estimate for $\mathcal{F}^2$.* Our next task is to obtain the estimate (17.69) for $\mu=2$, and for this we will show that

$$\int_{\mathbb{B}_n}\left|(1-|z|^2)^{m_1+\sigma}R^{m_1}\Lambda_g\mathcal{C}_{n,s_1}^{0,0}\Lambda_g\mathcal{C}_{n,s_2}^{0,1}\Omega_2^3\right|^p d\lambda_n(z) \lesssim \int_{\mathbb{B}_n}\left|(1-|z|^2)^{\sigma}(1-|z|^2)^{m_3''}R^{m_3''}\overline{\mathcal{D}}^{m_3'}\left(\widehat{\Omega_2^3 h}\right)(z)\right|^p d\lambda_n(z). \tag{17.95}$$

Unlike the previous argument we will have to deal with a rogue term $(\overline{z_2}-\overline{\xi_2})$ this time where there is no derivative $\frac{\partial}{\partial\overline{\xi_2}}$ to associate to the factor $(\overline{z_2}-\overline{\xi_2})$. Again we ignore the contractions Λ_g. Then we use Lemma 17.34 to perform integration by parts m_2' times in the first iterated integral and m_3' times in the second iterated integral. We also use Corollary 17.32 to perform integration by parts in the *radial* derivative m_2'' times in the *first* iterated integral (for fixed z, $\mathcal{C}_{n,s_2}^{0,1}\Omega_2^3\in C\left(\overline{\mathbb{B}_n}\right)\cap C^\infty(\mathbb{B}_n)$ by standard estimates [**106**]), so that the additional factor $(1-|\xi|^2)^{m_2''}$ can be used crucially in the second iterated integral, and also m_3'' times in the *second* iterated integral for use in acting on Ω_2^3.

Recall from Lemma 17.34 that

$$\mathcal{C}_{n,s}^{0,q}\eta(z) = \text{boundary terms (depending on } m) + \sum_{\ell=0}^{q}\int_{\mathbb{B}_n}\frac{(1-w\overline{z})^{n-1-\ell}(1-|w|^2)^{\ell}}{\triangle(w,z)^n}\left(\frac{1-|w|^2}{1-\overline{w}z}\right)^{s-n} \times\left(\sum_{j=0}^{n-\ell-1}c_{j,\ell,n,s}\left[\frac{(1-|w|^2)(1-|z|^2)}{|1-w\overline{z}|^2}\right]^j\right)\overline{\mathcal{D}}^m\eta(z).$$

Recall also that $\overline{\mathcal{D}}^m$ already has the rogue terms built in, as can be seen from (17.39). Now we use the right side above with $q=\ell=j=0$ to substitute for $\mathcal{C}_{n,s_1}^{0,0}$, and the right side above with $q = \ell = 1$ and $j = 0$ to substitute for $\mathcal{C}^{0,1}_{n,s_2}$. Then a typical part of the resulting kernel of the operator $\mathcal{C}^{0,0}_{n,s_1}\mathcal{C}^{0,1}_{n,s_2}\Omega^3_2(z)$ is

$$\begin{aligned}(17.96)\quad &\int_{\mathbb{B}_n} \frac{(1-\xi\overline{z})^{n-1}}{\triangle(\xi,z)^n}\left(\frac{1-|\xi|^2}{1-\overline{\xi}z}\right)^{s_1-n}\left(\overline{z_2}-\xi_2\right)\\ &\quad\times(1-|\xi|^2)^{m_2'}R^{m_2'}\overline{\mathcal{D}}^{m_2''}\int_{\mathbb{B}_n}\frac{\left(1-w\overline{\xi}\right)^{n-2}(1-|w|^2)}{\triangle(w,\xi)^n}\left(\frac{1-|w|^2}{1-\overline{w}\xi}\right)^{s_2-n}\\ &\quad\times\left(\overline{w_1}-\overline{\xi_1}\right)(1-|w|^2)^{m_3'}R^{m_3'}\overline{\mathcal{D}}^{m_3''}\left(\Omega^3_2h\right)(w)\,dV(w)\,dV(\xi),\end{aligned}$$

where we have arbitrarily chosen $\left(\overline{z_2}-\overline{\xi_2}\right)$ and $\left(\overline{w_1}-\overline{\xi_1}\right)$ as the rogue factors.

REMARK 17.53. *It is important to note that the differential operators* $\overline{\mathcal{D}^{m_2}_{\zeta}}$ *are conjugate in the variable* z *and hence* vanish *on the kernels of the boundary terms* $\mathcal{S}_{n,s}\left(\overline{\mathcal{D}}^k\Omega^3_2h\right)(z)$ *in the integration by parts formula* (17.40) *associated to the Charpentier solution operator* $\mathcal{C}^{0,1}_{n,s_2}$ *since these kernels are* holomorphic. *As a result the operator* $\overline{\mathcal{D}}^{m_2'}$ *hits only the factor* $\overline{\mathcal{D}}^k\Omega^3_2h$ *and a typical term is*

$$\overline{(z_i-\zeta_i)}\frac{\partial}{\partial\overline{z_i}}\left\{\overline{(w_i-z_i)}\Omega^3_2h\right\}=-\overline{(z_i-\zeta_i)}\Omega^3_2h,$$

where the derivative $\frac{\partial}{\partial\overline{w_i}}$ *must occur in each surviving term in* Ω^3_2h. *This term is then handled like the rogue terms.*

Now we recall the factorization (17.46) with $\ell = 2$,

$$\Omega^3_2=-4\Omega^1_0\wedge\widetilde{\Omega^1_0}\wedge\widetilde{\Omega^1_0},$$

and that $\Omega^3_2(w)$ must have both derivatives $\frac{\partial g}{\partial\overline{w_1}}$ and $\frac{\partial g}{\partial\overline{w_2}}$ occurring in it, one surviving in each of the factors $\widetilde{\Omega^1_0}$, along with other harmless powers of g that we ignore. Thus we may replace $\widetilde{\Omega^1_0}\wedge\widetilde{\Omega^1_0}$ with $\frac{\partial}{\partial\overline{w_2}}\Omega^1_0\wedge\frac{\partial}{\partial\overline{w_1}}\Omega^1_0$. If we use

$$\overline{z_2}-\overline{\xi_2}=\left(\overline{z_2}-\overline{w_2}\right)-\left(\overline{\xi_2}-\overline{w_2}\right),$$

we can write the above iterated integral as a difference, $A-B$, with

$$\begin{aligned}A=&\int_{\mathbb{B}_n}\frac{(1-\xi\overline{z})^{n-1}}{\triangle(\xi,z)^n}\left(\frac{1-|\xi|^2}{1-\overline{\xi}z}\right)^{s_1-n}\\ &\times\int_{\mathbb{B}_n}(1-|\xi|^2)^{m_2''}R^{m_2''}\overline{\mathcal{D}}^{m_2'}\left(\frac{\left(1-w\overline{\xi}\right)^{n-2}(1-|w|^2)}{\triangle(w,\xi)^n}\left(\frac{1-|w|^2}{1-\overline{w}\xi}\right)^{s_2-n}\right)\\ &\times\left[(1-|w|^2)^{m_3''}R^{m_3''}\left(\overline{\xi_2}-\overline{w_2}\right)\frac{\partial}{\partial\overline{w_2}}\overline{\mathcal{D}}^{m_3'-\ell}\Omega^1_0\right]\\ &\qquad\wedge\left[(1-|w|^2)^{m_3''}R^{m_3''}\left(\overline{\xi_1}-\overline{w_1}\right)\frac{\partial}{\partial\overline{w_1}}\overline{\mathcal{D}}^{\ell}\Omega^1_0\right]dV(w)\,dV(\xi)\end{aligned}$$

$$
\begin{aligned}
B = & \int\limits_{\mathbb{B}_n} \frac{(1-\xi\overline{z})^{n-1}}{\triangle(\xi,z)^n} \left(\frac{1-|\xi|^2}{1-\overline{\xi}z}\right)^{s_1-n} \\
& \times \int\limits_{\mathbb{B}_n} (1-|\xi|^2)^{m_2''} R^{m_2''} \overline{\mathcal{D}}^{m_2'} \left(\frac{(1-w\overline{\xi})^{n-2}(1-|w|^2)}{\triangle(w,\xi)^n} \left(\frac{1-|w|^2}{1-\overline{w}\xi}\right)^{s_2-n} \right) \\
& \times \left[(1-|w|^2)^{m_3''} R^{m_3''} \left(\overline{z_2}-\overline{w_2}\right) \frac{\partial}{\partial \overline{w_2}} \overline{\mathcal{D}}^{m_3'-\ell} \Omega_0^1 \right] \\
& \qquad \wedge \left[(1-|w|^2)^{m_3''} R^{m_3''} \left(\overline{\xi_1}-\overline{w_1}\right) \frac{\partial}{\partial \overline{w_1}} \overline{\mathcal{D}}^{\ell} \Omega_0^1 \right] dV(w)\, dV(\xi).
\end{aligned}
$$

Here we have temporarily ignored the wedge products with terms that do not include derivatives of g, as these terms are bounded and hence harmless.

Now we apply $(1-|z|^2)^\sigma (1-|z|^2)^{m_1''} R^{m_1''} D^{m_1'}$ to these operators. Using the crucial inequalities in Proposition 17.47 together with the factorization (17.77) with $\ell = 2$,

$$
\widehat{\Omega_2^3} = -4\Omega_0^1 \wedge \widehat{\Omega_0^1} \wedge \widehat{\Omega_0^1},
$$

the result of this application on the first integral is then dominated by

(17.97)

$$
\begin{aligned}
& \int_{\mathbb{B}_n} \frac{(1-|z|^2)^\sigma |1-\xi\overline{z}|^{n-1}}{\triangle(\xi,z)^{m_1'+m_1''+n}} \left[(1-|z|^2)\sqrt{\triangle(\xi,z)}\right]^{m_1''} \\
& \quad \times \left(\left[(1-|z|^2)\sqrt{\triangle(\xi,z)}\right]^{m_1'} + \triangle(\xi,z)^{m_1'} \right) \left| \frac{1-|\xi|^2}{1-\xi\overline{z}} \right|^{s_1-n} \\
& \quad \times \int_{\mathbb{B}_n} \frac{(1-|\xi|^2)^{m_2''} \left|1-w\overline{\xi}\right|^{n-2} (1-|w|^2)}{\triangle(w,\xi)^{m_2'+m_2''+n}} \left(\frac{\sqrt{\triangle(\xi,z)}}{1-|\xi|^2} \right)^{m_2'} \\
& \quad \times \left[(1-|\xi|^2)\sqrt{\triangle(w,\xi)}\right]^{m_2''} \left(\left[(1-|\xi|^2)\sqrt{\triangle(w,\xi)}\right]^{m_2'} + \triangle(w,\xi)^{m_2'} \right) \\
& \quad \times \left| \frac{1-|w|^2}{1-w\overline{\xi}} \right|^{s_2-n} \left(\frac{\sqrt{\triangle(w,\xi)}}{1-|w|^2} \right)^{m_3'} \left(\frac{\sqrt{\triangle(w,\xi)}}{1-|w|^2} \right)^2 \\
& \quad \times \left| (1-|w|^2)^{m_3''} R^{m_3''} \overline{D^{m_3'}} \left(\widehat{\Omega_2^3} h\right)(w) \right| dV(w)\, dV(\xi),
\end{aligned}
$$

and the result of this application on the second integral is dominated by

$$
\begin{aligned}
(17.98)\quad &\int_{\mathbb{B}_n} \frac{(1-|z|^2)^\sigma |1-\xi\overline{z}|^{n-1}}{\triangle(\xi,z)^{m_1'+m_1''+2}} \left[(1-|z|^2)\sqrt{\triangle(\xi,z)}\right]^{m_1''} \\
&\times \left(\left[(1-|z|^2)\sqrt{\triangle(\xi,z)}\right]^{m_1'} + \triangle(\xi,z)^{m_1'}\right) \left|\frac{1-|\xi|^2}{1-\xi\overline{z}}\right|^{s_1-n} \\
&\times \int_{\mathbb{B}_n} \frac{(1-|\xi|^2)^{m_2''}|1-w\overline{\xi}|^{n-2}(1-|w|^2)}{\triangle(w,\xi)^{m_2'+m_2''+n}} \left(\frac{\sqrt{\triangle(\xi,z)}}{1-|\xi|^2}\right)^{m_2'} \\
&\times \left[(1-|\xi|^2)\sqrt{\triangle(w,\xi)}\right]^{m_2''} \left(\left[(1-|\xi|^2)\sqrt{\triangle(w,\xi)}\right]^{m} + \triangle(w,\xi)^{m_2'}\right) \\
&\times \left|\frac{1-|w|^2}{1-w\overline{\xi}}\right|^{s_2-n} \left(\frac{\sqrt{\triangle(w,\xi)}}{1-|w|^2}\right)^{m_3'} \left(\frac{\sqrt{\triangle(w,z)}}{1-|w|^2}\right) \left(\frac{\sqrt{\triangle(w,\xi)}}{1-|w|^2}\right) \\
&\times \left|(1-|w|^2)^{m_3''} R^{m_3''}\overline{D^{m_3'}}\left(\widehat{\Omega_2^3}h\right)(w)\right| dV(w)\, dV(\xi).
\end{aligned}
$$

The only difference between these two iterated integrals is that one of the factors $\frac{\sqrt{\triangle(w,\xi)}}{1-|w|^2}$ that occur in the first is replaced by the factor $\frac{\sqrt{\triangle(w,z)}}{1-|w|^2}$ in the second. Note that the ignored wedge products have now been reinstated in $\widehat{\Omega_2^3}$.

We can realize the iterated integral in (17.97), as the composition of two operators of the form treated previously. One factor is the operator

$$
\begin{aligned}
(17.99)\quad &\int_{\mathbb{B}_n} \frac{(1-|z|^2)^\sigma |1-\xi\overline{z}|^{n-1}}{\triangle(\xi,z)^{m_1'+m_1''+n}} \left[(1-|z|^2)\sqrt{\triangle(\xi,z)}\right]^{m_1''} \\
&\times \left(\left[(1-|z|^2)\sqrt{\triangle(\xi,z)}\right]^{m_1'} + \triangle(\xi,z)^{m_1'}\right) \\
&\times \left(\frac{\sqrt{\triangle(\xi,z)}}{1-|\xi|^2}\right)^{m_2'} \left|\frac{1-|\xi|^2}{1-\xi\overline{z}}\right|^{s_1-n} (1-|\xi|^2)^{-\sigma} F(\xi)\, dV(\xi),
\end{aligned}
$$

and the other factor is the operator

(17.100)

$$
\begin{aligned}
F(\xi) = &\int_{\mathbb{B}_n} \frac{(1-|\xi|^2)^\sigma |1-w\overline{\xi}|^{n-2}(1-|w|^2)}{\triangle(w,\xi)^{m_2'+m_2''+n}} \left[(1-|\xi|^2)\sqrt{\triangle(w,\xi)}\right]^{m_2''} \\
&\times \left(\left[(1-|\xi|^2)\sqrt{\triangle(w,\xi)}\right]^{m_2'} + \triangle(w,\xi)^{m_2'}\right) \left|\frac{1-|w|^2}{1-w\overline{\xi}}\right|^{s_2-n} \\
&\times \left(\frac{\sqrt{\triangle(w,\xi)}}{1-|w|^2}\right)^{m_3'+2} (1-|w|^2)^{-\sigma} f(w)\, dV(w),
\end{aligned}
$$

where $f(w) = (1-|w|^2)^\sigma \left|(1-|w|^2)^{m_3''} R^{m_3''}\overline{D^{m_3'}}\left(\widehat{\Omega_2^3}h\right)(w)\right|$. We now show how Lemma 17.48 applies to obtain the appropriate boundedness.

We will in fact compare the corresponding kernels to that in (17.92). We use the sum in the middle line of (17.99) to write that integral operator as the sum of two operators. The second, which includes the factor $\triangle(\xi,z)^{m_1'}$, is simplified by the choice $m_2' = m_1'' + 2n$ which lets the factor $\triangle(\xi,z)$ disappear:

$$\begin{aligned}&\frac{(1-|z|^2)^{\sigma+m_1''}(1-|\xi|^2)^{s_1-n-m_2'-\sigma}}{|1-\xi\bar{z}|^{s_1-2n+1}\triangle(\xi,z)^{m_1'+m_1''+n-(m_1''+2m_1'+m_2')/2}}\\ &\qquad=\frac{(1-|z|^2)^{\sigma+m_1''}(1-|\xi|^2)^{s_1-3n-m_1''-\sigma}}{|1-\xi\bar{z}|^{s_1-2n+1}},\end{aligned}\tag{17.101}$$

This is exactly the same as the kernel of the operator in (17.93) in the previous alternative argument but with m_1'' in place of m_1' there. When we consider the other summand, the one including $\left[(1-|z|^2)\sqrt{\triangle(\xi,z)}\right]^{m_1'}$, we obtain the kernel in (17.93) but with $m_1''+m_1'$ in place of m_1''.

We then do an analogous splitting of the operator in (17.100). For the summand which includes the factor $\triangle(w,\xi)^{m_2'}$ we select $m_3' = m_2'' + 2n - 2$ and use

$$\frac{(1-|\xi|^2)^{m_2''+\sigma}\left(1-|w|^2\right)^{1+s_2-n-m_3'-2-\sigma}}{\left|1-w\bar{\xi}\right|^{s_2-2n+2}\triangle(w,\xi)^{m_2'+m_2''+n-(m_1''+2m_1'+m_2'+2)/2}}\tag{17.102}$$

$$=\frac{(1-|\xi|^2)^{m_2''+\sigma}\left(1-|w|^2\right)^{s_2-3n+1-m_2''-\sigma}}{\left|1-w\bar{\xi}\right|^{s_2-2n+2}}.\tag{17.103}$$

Thus we again obtain the kernel of an operator which is bounded on $L^p(d\lambda_n)$ for m_2'' and s_2 sufficiently large. Finally, for the operator coming from the splitting of (17.100) which contains the factor $\left[(1-|z|^2)\sqrt{\triangle(\xi,z)}\right]^{m_2'}$ we note that it is the same as the kernel in (17.102) but with $m_2''+m_2'$ in place of m_2''.

Note: It is here in choosing m_2'' large that we are using the full force of Corollary 17.32 to perform integration by parts in the *radial* derivative m_2'' times in the first iterated integral.

To handle the iterated integral in (17.98) we must first deal with the rogue factor $\sqrt{\triangle(w,z)}$ whose variable pair (w,z) doesn't match that of either of the denominators $\triangle(\xi,z)$ or $\triangle(w,\xi)$. For this we use the fact that

$$\sqrt{\triangle(w,z)} = |1-w\bar{z}|\,|\varphi_z(w)| = \sigma(w,z)^2\,\rho(w,z),$$

where $\rho(w,z) = |\varphi_z(w)|$ and $\sigma(w,z) = |1-w\bar{z}|^{1/2}$ are the anisotropic metrics on the ball and sphere respectively mentioned in Sections 14.1.2 and 14.1.3. Using the triangle inequality for those metrics we find the estimate

$$\begin{aligned}\sqrt{\triangle(w,z)} &\leq 2\left[\sigma(\xi,z)^2+\sigma(w,\xi)^2\right](|\varphi_z(\xi)|+|\varphi_\xi(w)|)\\ &= 2\left(1+\frac{|1-w\bar{\xi}|}{|1-\xi\bar{z}|}\right)\sqrt{\triangle(\xi,z)}+2\left(1+\frac{|1-\xi\bar{z}|}{|1-w\bar{\xi}|}\right)\sqrt{\triangle(w,\xi)}.\end{aligned}$$

Thus we can write

$$(17.104)\quad \frac{\sqrt{\triangle(w,z)}}{1-|w|^2}$$
$$\lesssim \frac{1-|\xi|^2}{1-|w|^2}\frac{\sqrt{\triangle(\xi,z)}}{1-|\xi|^2}+\frac{|1-w\overline{\xi}|}{1-|w|^2}\frac{1-|\xi|^2}{|1-\xi\overline{z}|}\frac{\sqrt{\triangle(\xi,z)}}{1-|\xi|^2}$$
$$+\frac{\sqrt{\triangle(w,\xi)}}{1-|w|^2}+\frac{|1-\xi\overline{z}|}{1-|\xi|^2}\frac{1-|\xi|^2}{|1-w\overline{\xi}|}\frac{\sqrt{\triangle(w,\xi)}}{1-|w|^2}.$$

All of the terms on the right hand side of (17.104) are of an appropriate form to distribute throughout the iterated integral, and again Lemma 17.48 applies to obtain the appropriate boundedness.

For example, the final two terms on the right side of (17.104) that involve $\frac{\sqrt{\triangle(w,\xi)}}{1-|w|^2}$ are handled in the same way as the operator in (17.97) by taking $m_3'=m_2''+2n-2$ and $m_2'=m_1''+2n$, and taking s_1 and s_2 large as required by the extra factors $\frac{|1-\xi\overline{z}|}{1-|\xi|^2}\frac{1-|\xi|^2}{|1-w\overline{\xi}|}$. With these choices the first two terms on the right side of (17.104) that involve $\frac{\sqrt{\triangle(\xi,z)}}{1-|\xi|^2}$ are then handled using Lemma 17.48 with $c=\pm1$ as follows.

If we substitute the first term $\frac{1-|\xi|^2}{1-|w|^2}\frac{\sqrt{\triangle(\xi,z)}}{1-|\xi|^2}$ on the right in (17.104) for the factor $\frac{\sqrt{\triangle(w,z)}}{1-|w|^2}$ in (17.98) we get a composition of two operators as in (17.99) and (17.100) but with the kernel in (17.99) multiplied by $\frac{\sqrt{\triangle(\xi,z)}}{1-|\xi|^2}$ and the kernel in (17.100) multiplied by $\frac{1-|\xi|^2}{1-|w|^2}$ and divided by $\frac{\sqrt{\triangle(w,\xi)}}{1-|w|^2}$. If we consider the summand $\triangle(\xi,z)^{m_1'}$ in the middle line of (17.99), and with the choice $m_2'=m_1''+2n$ already made, the first operator then has kernel

$$\frac{\sqrt{\triangle(\xi,z)}}{1-|\xi|^2}\frac{(1-|z|^2)^{\sigma+m_1''}(1-|\xi|^2)^{s_1-3n-m_1''-\sigma}}{|1-\xi\overline{z}|^{s_1-2n+1}}$$
$$=\frac{(1-|z|^2)^{m_1''+\sigma}(1-|\xi|^2)^{s_1-m_1''-3n-1-\sigma}\sqrt{\triangle(\xi,z)}}{|1-\xi\overline{z}|^{s_1-2n+1}},$$

and hence is of the form $T_{a,b,c}$ with

$$a=m_1''+\sigma,$$
$$b=s_1-3n-1-m_1''-\sigma,$$
$$c=1,$$

since $a+b+c+n+1=s_1-n-1$. Now we apply Lemma 17.48 to conclude that this operator is bounded on $L^p(\lambda_n)$ if and only if

$$-p(m_1''+\sigma)<-n<p(s_1-3n-m_1''-\sigma),$$

i.e. $m_1''+\sigma>n/p$ and $s_1>m_1''+\sigma+3n-n/p$.

If we consider the summand $\triangle(w,\xi)^{m_2'}$ in the middle line of (17.100), and with the choice $m_3' = m_2'' + 2n - 2$ already made, the second operator has kernel

$$\frac{1-|\xi|^2}{1-|w|^2} \times \left(\frac{\sqrt{\triangle(w,\xi)}}{1-|w|^2}\right)^{-1} \times \frac{(1-|\xi|^2)^{m_2''+\sigma}\left(1-|w|^2\right)^{s_2-3n+1-m_2''-\sigma}}{\left|1-w\overline{\xi}\right|^{s_2-2n+2}}$$
$$= \frac{(1-|\xi|^2)^{m_2''+\sigma+1}\left(1-|w|^2\right)^{s_2-3n+1-m_2''-\sigma}\sqrt{\triangle(w,\xi)}^{-1}}{\left|1-w\overline{\xi}\right|^{s_2-2n+2}},$$

and hence is of the form $T_{a,b,c}$ with

$$\begin{aligned} a &= m_2'' + \sigma + 1, \\ b &= s_2 - 3n + 1 - m_2'' - \sigma, \\ c &= -1. \end{aligned}$$

This operator is bounded on $L^p(\lambda_n)$ if and only if

$$-p\left(m_2'' + \sigma + 1\right) < -n < p\left(s_2 - 3n + 2 - m_2'' - \sigma\right),$$

i.e. $m_2'' + \sigma > n/p - 1$ and $s_2 > m_2'' + \sigma + 3n - 2 - n/p$.

If we now substitute the second term $\frac{|1-w\overline{\xi}|}{1-|w|^2}\frac{1-|\xi|^2}{|1-\xi\overline{z}|}\frac{\sqrt{\triangle(\xi,z)}}{1-|\xi|^2}$ on the right in (17.104) for the factor $\frac{\sqrt{\triangle(w,z)}}{1-|w|^2}$ in (17.98) we similarly get a composition of two operators that are each bounded on $L^p(\lambda_n)$ for m_i and s_i chosen large enough.

17.5.1.5. *Boundary Terms for $\mathcal{F}^2$.* Now we must address in $\mathcal{F}^2$ the boundary terms that arise in the integration by parts formula (17.40). Suppose the first operator $\mathcal{C}_{n,s_1}^{0,0}$ is replaced by a boundary term, but not the second. We proceed by applying Corollary 17.32 to the boundary term. Since the differential operator $(1-|z|^2)^{m_1+\sigma}R^{m_1}$ hits only the kernel of the boundary term, we can apply Remark 17.49 to the first iterated integral and Lemma 17.48 to the second iterated integral in the manner indicated in the above arguments. If the second operator $\mathcal{C}_{n,s_2}^{0,1}$ is replaced by a boundary term, then as mentioned in Remark 17.53, the operators $\overline{D}^{m_2}$ hit only the factors $\overline{\mathcal{D}}^{m_3}$, and this produces rogue terms that are handled as above. If the first operator $\mathcal{C}_{n,s_1}^{0,0}$ was also replaced by a boundary term, then in addition we would have radial derivatives R^m hitting the second boundary term. Since radial derivatives are holomorphic, they hit only the holomorphic kernel and not the antiholomorphic factors in $\overline{\mathcal{D}}^{m_3}$, and so these terms can also be handled as above.

17.5.2. The Estimates for General $\mathcal{F}^\mu$. In view of inequality (17.78), it suffices to establish the following inequality:

$$(17.105)\quad \|\mathcal{F}^\mu\|_{B_p^\sigma(\mathbb{B}_n)}^p$$
$$= \int_{\mathbb{B}_n}\left|(1-|z|^2)^{m_1+\sigma}R^{m_1}\Lambda_g\mathcal{C}_{n,s_1}^{0,0}\cdots\Lambda_g\mathcal{C}_{n,s_\mu}^{0,\mu-1}\Omega_\mu^{\mu+1}h\right|^p d\lambda_n(z)$$
$$\lesssim \int_{\mathbb{B}_n}\left|(1-|z|^2)^\sigma\mathcal{X}^{m_\mu}\left(\widehat{\Omega_\mu^{\mu+1}h}\right)(z)\right|^p d\lambda_n(z).$$

Recall that the absolute value $|F|$ of an element F in the exterior algebra is the square root of the sum of the squares of the coefficients of F in the standard basis.

The case $\mu > 2$ involves no new ideas, and is merely complicated by straightforward algebra. The reason is that the solution operator $\Lambda_g \mathcal{C}^{0,0}_{n,s_1} \cdots \Lambda_g \mathcal{C}^{0,\mu-1}_{n,s_\mu}$ acts *separately* in each entry of the form $\Omega^{\mu+1}_\mu h$, an element of the exterior algebra of $\mathbb{C}^\infty \otimes \mathbb{C}^n$ which we view as an alternating ℓ^2-tensor of $(0,\mu)$ forms in $\mathbb{C}^n$. These operators decompose as a sum of simpler operators with the basic property that their kernels are *identical*, except that the rogue factors in each kernel differ according to the entry. Nevertheless, there are always exactly μ distinct rogue factors in each kernel and after splitting, the μ rogue factors can be associated in one-to-one fashion with each of the $\frac{\partial}{\partial w_j}$ derivatives in the corresponding entry of

$$\Omega^{\mu+1}_\mu h = -(\mu+1)\left(\sum_{k_0=1}^{\infty} \frac{\overline{g_{k_0}}}{|g|^2} e_{k_0}\right) \wedge \bigwedge_{i=1}^{\mu} \left(\sum_{k_i=1}^{\infty} \frac{\overline{\partial g_{k_i}}}{|g|^2} e_{k_i}\right) h.$$

After applying the crucial inequalities, this effectively results in replacing each derivative $\frac{\partial}{\partial w_j}$ by the derivative $\overline{D_j}$, and consequently we can write the resulting form as $\widehat{\Omega^{\mu+1}_\mu} h$.

This completes our proof of Theorem 17.11.

17.6. Notes and Comments

Influential early work on corona results for spaces of functions of several variables was done by Varopolous [**335**]. More recent work, which contributed substantially to the approach in this chapter, was done by Charpentier [**106**], by Andersson and Carlsson [**29**], and Ortega and Fabrega [**249**], [**248**]. The material in this chapter is based on [**121**]. Related Corona-type results can also be found in [**21, 22, 26, 27, 153, 286, 319, 321, 322**].

APPENDIX A

Some Functional Analysis

Here we collect some specific results from functional analysis which are used in the text, as well as some related results.

A.1. Well Defined Everywhere

It is sometimes said that a linear operator that is well defined everywhere is continuous. To make that precise the phrase "well defined everywhere" needs to be interpreted and placed in context. One possible thread begins with the automatic continuity of *symmetric* linear operators on a *Hilbert* space. But for this we first need the Closed Graph Theorem.

A.1.1. The Closed Graph Theorem. If X is any topological space and Y is a Hausdorff space, then every continuous map $f : X \to Y$ has a closed graph (exercise: prove this). A statement that gives conditions under which the converse holds is referred to as a "closed graph theorem." Here is an elementary example. Suppose that X and Y are metric spaces and Y is compact. If the graph of f is closed in $X \times Y$ then f is continuous. Indeed, for metric spaces it is enough to show that every sequence $\{x_n\}_{n=1}^{\infty}$ in X converging to a point $x \in X$ has a subsequence $\{x_{n_k}\}_{k=1}^{\infty}$ such that $f(x_{n_k}) \to f(x)$ as $k \to \infty$. However, since Y is compact, $\{f(x_n)\}_{n=1}^{\infty}$ has a convergent subsequence, say $f(x_{n_k}) \to y \in K$ as $k \to \infty$. Thus (x, y) is a limit point of the graph $G = \{(x, f(x)) : x \in X\}$, and since G is assumed closed, we have $(x, y) \in G$, i.e. $y = f(x)$. The next theorem gives the same conclusion for a *linear* map from one Banach space to another. Note that linearity is needed here since $f : \mathbb{R} \to \mathbb{R}$ by $f(x) = \begin{cases} 1/x & \text{if } x \neq 0 \\ 0 & \text{if } x = 0 \end{cases}$ has a closed graph, but is not continuous at the origin.

THEOREM A.1 (Closed Graph Theorem). *Suppose that X and Y are Banach spaces and $\Lambda : X \to Y$ is linear. If the graph $G = \{(x, \Lambda(x)) : x \in X\}$ is closed in $X \times Y$, then Λ is continuous.*

PROOF. The product $X \times Y$ is a Banach space with the norm $\|(x, y)\| = \|x\|_X + \|y\|_Y$. Since Λ is linear and the graph G of Λ is closed, G is also a Banach space. Now the projection $\pi_1 : X \times Y \to X$ by $(x, y) \to x$ is a continuous linear map from the Banach space G *onto* the Banach space X, and the open mapping theorem A.14 thus implies that π_1 is an open map. However, π_1 is clearly one-to-one and so the inverse map $\pi_1^{-1} : X \to G$ exists and is continuous. But then the composition $\pi_2 \circ \pi_1^{-1} : X \to Y$ is also continuous where $\pi_2 : X \times Y \to Y$ by $(x, y) \to y$. We are done since $\pi_2 \circ \pi_1^{-1} = \Lambda$. □

A.1.2. The Hellinger and Toeplitz Theorem.

THEOREM A.2 (Hellinger and Toeplitz). *Suppose that T is a linear operator on a Hilbert space H satisfying $\langle Tx, y \rangle = \langle x, Ty \rangle$ for all $x, y \in H$. Then T is continuous.*

PROOF. It is enough to show that T has a closed graph G. So let (x, z) be a limit point of G. Then there is a sequence $\{x_n\}_{n=1}^{\infty} \subset X$ such that $x_n \to x$ and $Tx_n \to z$. For every $y \in H$ the symmetry hypothesis now shows that

$$\langle T(x_n - x), y \rangle = \langle x_n - x, Ty \rangle \to 0$$

as $n \to \infty$. But we also have

$$\langle T(x_n - x), y \rangle = \langle Tx_n, y \rangle - \langle Tx, y \rangle \to \langle z, y \rangle - \langle Tx, y \rangle$$

as $n \to \infty$. Thus $\langle z - Tx, y \rangle = 0$ for all $y \in H$ and so $z = Tx$, which shows that $(x, z) \in G$. □

This theorem extends to a linear map $T : A \to B$ between different Banach spaces if symmetry is replaced with weak convergence. More precisely, the same proof gives the following result.

COROLLARY A.3 (of the proof). *Suppose that $T : A \to B$ is a linear operator from one Banach space A to another Banach space B. If $Ta_n \to 0$ weakly in B whenever $a_n \to 0$ strongly in A, then T is continuous.*

We can further relax the assumption of weak convergence provided B is a Banach function space on a set X, in which case we only need to assume continuity of point evaluations composed with T.

DEFINITION A.4. *A Banach space B is a Banach function space on a set X if*

(1) *The elements of B can be identified with a linear space of functions $f : X \to \mathbb{C}$ for which point evaluations (i.e. the functionals $f \to f(x)$ for $x \in X$) are continuous on B.*
(2) *If under the above identification we have $f, g \in B$ and $f(x) = g(x)$ for all $x \in X$, then $f = g$ as elements of B.*

COROLLARY A.5 (of the proof). *Suppose that $T : A \to B$ is a linear map from a Banach space A to a Banach function space B on a set X with the following weak continuity property:* ***Continuous Point Evaluations****: For each $x \in X$, the functional $a \to Ta(x)$ is a continuous function of $a \in A$. Then T is continuous.*

PROOF. We repeat the above proof in the present context. It is enough to show that T has a closed graph G. So let (a, b) be a limit point of G. Then there is a sequence $\{a_n\}_{n=1}^{\infty} \subset A$ such that $a_n \to a$ in A and $Ta_n \to b$ in B. For every $x \in X$ property (1) of Definition A.4 shows that

$$T(a_n - a)(x) \to 0$$

as $n \to \infty$. But we also have for every $x \in X$ that

$$T(a_n - a)(x) = Ta_n(x) - Ta(x) \to b(x) - Ta(x) = (b - Ta)(x)$$

as $n \to \infty$. Thus $(b - Ta)(x) = 0$ for all $x \in X$ and so $b = Ta$ by property (2) of Definition A.4. This shows that $(a, b) \in G$, and hence that G is closed. □

There is a weakening of the notion of 'Banach function space on a set X' under which the above Corollary persists, namely that of a 'μ-Banach space on a set X' for a positive measure μ on X.

Let $(X, \mathcal{M}, \mu)$ be a positive measure space. The space $L^1_{\text{loc}}(\mu)$ consists of those equivalence classes of functions that are locally integrable with respect to μ, and where two functions are equivalent if the set where they disagree has μ-measure zero. As usual, we ignore the distinction between functions and their equivalence classes. If E is a set of positive μ-measure on X, we define the linear functional Λ_E of 'evaluation at E' on $L^1_{\text{loc}}(\mu)$ by

$$\Lambda_E f = \frac{1}{\mu(E)} \int_E f(x)\, d\mu(x), \qquad f \in L^1_{\text{loc}}(\mu).$$

We refer to such linear functionals Λ_E as *mass evaluations* on X. (The factor $\frac{1}{\mu(E)}$ is included for cosmetic reasons only and could be omitted - it plays no role below.)

Mass evaluations are a generalization of point evaluations in the following sense. In the special case where μ is counting measure on X, i.e. $(X, \mathcal{M}, \mu) = (X, \mathcal{P}(X), c)$ where $c(E) = \text{card}(E)$ for all $E \in \mathcal{P}(X)$, then the mass evaluation $\Lambda_{\{x\}}$ for the singleton $\{x\}$ coincides with the point evaluation at x.

DEFINITION A.6. *Let $(X, \mathcal{M}, \mu)$ be a positive measure space. A Banach space B is a μ-Banach function space on X if*

(1) *The elements of B can be identified with a linear subspace of (equivalence classes of) functions $f : X \to \mathbb{C}$ in $L^1_{\text{loc}}(\mu)$ for which mass evaluations (i.e. the linear functionals $f \to \Lambda_E f$ for $\mu(E) > 0$) are continuous on B.*
(2) *If, under the above identification, we have $f, g \in B$ and $f(x) = g(x)$ for μ-almost every $x \in X$, then $f = g$ as elements of B.*

THEOREM A.7. *Let $(X, \mathcal{M}, \mu)$ be a positive measure space. Suppose that $T : A \to B$ is a linear map from a Banach space A to a μ-Banach function space B on X with the following weak continuity property:*

Continuous Averages. *For each set $E \subset X$ of positive μ-measure, the averaging functional $a \to \frac{1}{\mu(E)} \int_E Ta\,d\mu$ is a continuous function of $a \in A$.*

Then T is continuous.

PROOF. We again repeat the above proof in the present context. It is enough to show that T has a closed graph G. So let (a, b) be a limit point of G. Then there is a sequence $\{a_n\}_{n=1}^\infty \subset A$ such that $a_n \to a$ in A and $Ta_n \to b$ in B. For every $E \subset X$ of positive μ-measure, property (1) of Definition A.6 shows that

$$\int_E T(a_n - a)\, d\mu \to 0$$

as $n \to \infty$. But we also have for every $E \subset X$ of positive μ-measure that

$$\int_E T(a_n - a)\, d\mu = \int_E Ta_n d\mu - \int_E Tad\mu \to \int_E bd\mu - \int_E Tad\mu = \int_E (b - Ta)\, d\mu$$

as $n \to \infty$. Thus $\int_E (b - Ta)\, d\mu = 0$ for all $E \subset X$ of positive μ-measure and so $b = Ta$ by property (2) of Definition A.4. This shows that $(a, b) \in G$, and hence that G is closed. Now the Closed Graph Theorem A.1 shows that T is continuous. □

A.2. Hilbert Spaces and Surjective Maps

There are variations of the previous principles for maps $T : A \to L^1_\mu$ that are *onto,* instead of into, a μ-Banach function space B on a set X. For convenience we present the case where A is a Hilbert space, and complementation is automatic. We say that a Hilbert space H is a μ-Hilbert function space on a set X if when considered as a Banach space, it is a μ-Banach function space on X. Here are the preliminaries we need regarding complementation in a Hilbert space.

The next easy theorem lies at the heart of the great success of Hilbert spaces in analysis.

THEOREM A.8. *Suppose E is a nonempty closed convex subset of a Hilbert space H. Then E contains a unique element x of minimal norm, i.e. $\|x\| = \inf_{y\in E} \|y\|$.*

PROOF. Let $d = \inf_{y\in E} \|y\|$, which is finite since E is *nonempty.* Pick $\{x_n\}_{n=1}^\infty \subset E$ with $\|x_n\| \to d$ as $n \to \infty$. Because E is *convex,* each $(x_m+x_n)/2 \in E$ and hence has norm at least d. The parallelogram law now yields

$$\begin{aligned}\left\|\frac{x_m - x_n}{2}\right\|^2 &= \frac{\|x_m\|^2 + \|x_n\|^2}{2} - \left\|\frac{x_m + x_n}{2}\right\|^2 \\ &\leq \frac{\|x_m\|^2 + \|x_n\|^2}{2} - d^2 \\ &\to \frac{d^2 + d^2}{2} - d^2 = 0\end{aligned}$$

as $m, n \to \infty$. Thus $\{x_n\}_{n=1}^\infty$ is Cauchy and since H is complete and E *closed,* $x = \lim_{n\to\infty} x_n \in E$. Since $\|\cdot\|$ is continuous, we have $\|x\| = d$. If $x' \in E$ also satisfies $\|x'\| = d$, then using the parallelogram law as above yields $\left\|\frac{x-x'}{2}\right\|^2 = \frac{\|x\|^2+\|x'\|^2}{2} - \left\|\frac{x+x'}{2}\right\|^2 \leq 0$, hence $x = x'$. □

Let H be a Hilbert space. We say that x and y in H are perpendicular, written $x \perp y$, if $\langle x, y\rangle = 0$. We say subsets E and F of H are perpendicular, written $E \perp F$, if $\langle x, y\rangle = 0$ for all $x \in E$ and $y \in F$. Finally, we define

$$E^\perp = \{y \in H : \langle x, y\rangle = 0 \text{ for all } x \in E\}.$$

The next theorem uses Theorem A.8 to establish an orthogonal decomposition of H relative to any closed subspace M of a Hilbert space H.

THEOREM A.9. *Suppose that M is a closed subspace of a Hilbert space H. Then*

$$H = M \oplus M^\perp,$$

which means that M and $M^\perp$ are closed subspaces of H whose intersection is the smallest subspace $\{0\}$, and whose span is the largest subspace H.

PROOF. $M^\perp$ is a subspace since $\langle x, y\rangle$ is linear in x, and is closed by the Cauchy-Schwarz inequality. The fact that $\langle x, x\rangle = 0 \Longleftrightarrow x = 0$ gives $M \cap M^\perp = \{0\}$. Finally, to show $M + M^\perp = H$, let $x \in H$ and set $E = x - M$, a nonempty closed convex set. Thus there is a unique element $m^\perp \in x - M$ of minimal norm having the form $x - m$ with $m \in M$. Thus for all $z \in M$ and $\lambda \in \mathbb{C}$,

$$\|m^\perp\| \leq \|m^\perp + \lambda z\|.$$

For each $z \in M$, the fact that the expression on the right has a minimum at $\lambda = 0$ implies that $\langle z, m^\perp \rangle = 0$. This yields $m^\perp \in M^\perp$. Thus $x = m + m^\perp \in M + M^\perp$. $\square$

COROLLARY A.10. $\left(M^\perp\right)^\perp = M$.

PROOF. $M \subset \left(M^\perp\right)^\perp$ is obvious, and since $M \oplus M^\perp = H = M^\perp \oplus \left(M^\perp\right)^\perp$, we cannot have that M is a proper subset of $\left(M^\perp\right)^\perp$. $\square$

Now we can give our *onto* variant of the extension of the Hellinger and Toeplitz theorem.

THEOREM A.11. *Let $(X, \mathcal{M}, \mu)$ be a positive measure space. Suppose that $T: H \to B$ is a linear map from a μ-Hilbert function space H on X into $L^1_{\mathrm{loc}}(\mu)$ that is* onto *a μ-Banach function space B on X, and has the following weak continuity property:*

***Continuous Averages**. For each set $E \subset X$ of positive μ-measure, the averaging functional $h \mapsto \frac{1}{\mu(E)} \int_E Th\,d\mu$ is a continuous function of $h \in H$.*

Then there is a positive constant C such that for every $b \in B$, there is $h \in H$ satisfying

(1) $Th = b$,
(2) $\|h\|_H \leq C \|b\|_B$.

PROOF. Let $M = \{h \in H : \Lambda_E h = 0 \text{ for all } E \text{ with } \mu(E) > 0\}$. Then M is closed in H since mass evaluations are continuous on H, and we have $H = M \oplus M^\perp$ by Theorem A.9. Let $P^\perp$ denote projection of H onto $M^\perp$. Then if $b \in B$ and an element $h_b \in H$ satisfies $Th_b = b$, we have that $TP^\perp h_b = b$ with $P^\perp h_b$ the *unique* such element in $M^\perp$. Thus

$$\Gamma b \equiv P^\perp h_b$$

defines a linear map from B to $M^\perp$ with $T\Gamma b = b$. Now we claim that Γ has closed graph. To see this, suppose $(b_k, \Gamma b_k) \to \left(b, m^\perp\right)$ in $B \times M^\perp$, i.e. $b_k \to b$ in B and $\Gamma b_k \to m^\perp$ in $M^\perp$, so that both

$$\int_E b_k d\mu \to \int_E b d\mu \text{ and } \int_E \Gamma b_k d\mu \to \int_E m^\perp d\mu$$

as $k \to \infty$ for each subset E of positive μ-measure. However

$$\int_E b_k d\mu = \int_E \left(TP^\perp h_{b_k}\right) d\mu = \int_E (T\Gamma b_k)\, d\mu$$

then shows that $\int_E Tm^\perp d\mu = \int_E b d\mu$ for all subsets E of positive μ-measure, and hence that $Tm^\perp = b$. By uniqueness of solutions to this equation in $M^\perp$ we conclude that $\Gamma b = m^\perp$, and hence that $\left(b, m^\perp\right)$ is in the graph of Γ. The Closed Graph Theorem A.1 now implies that Γ is continuous. Thus we may take $h = \Gamma b$ to obtain the conclusion of the theorem. $\square$

A.3. Surjective iff Adjoint is Bounded Below

We will need the Hahn-Banach Theorem for this.

THEOREM A.12 (Hahn-Banach). *Let M be a proper subspace of a normed linear space X. If $f \in M^*$, then there is $F \in X^*$ such that $F\mid_M = f$ and $\|F\|_{X^*} = \|f\|_{M^*}$. In the event that X is separable, the axiom of choice is not needed.*

PROOF. Suppose first that the scalar field is $\mathbb{R}$. Without loss of generality $\|f\|_{M^*} = 1$. In the event that X is separable, let $S = \{s_j\}_{j=1}^{\infty}$ be a dense subset of X. Choose $x_0 \in X \setminus M$, where x_0 is the first s_j in $X \setminus M$ if X is separable, and let

$$M_0 = \operatorname{span}\{M, x_0\} = \{x + \lambda x_0 : x \in M, \lambda \in \mathbb{R}\}.$$

Define f_0 on M_0 by $f_0(x + \lambda x_0) = f(x) + \lambda\alpha$ where $\alpha \in \mathbb{R}$ will be chosen below so that f_0 is a norm-preserving extension of f to M_0. Since f_0 is a linear extension of f for any $\alpha \in \mathbb{R}$, we need only require that $|f(x) + \lambda\alpha| \leq \|x + \lambda x_0\|$ for $x \in M, \lambda \in \mathbb{R}$, and if we write $y = \frac{x}{\lambda}$, this is equivalent to

$$-\|y + x_0\| \leq f(y) + \alpha \leq \|y + x_0\|, \qquad y \in M,$$

and with $z = -y$, to

$$f(z) - \|z - x_0\| \leq \alpha \leq f(z) + \|z - x_0\|, \qquad z \in M. \tag{A.1}$$

However, the supremum of the left side and the infimum of the right side satisfy

$$\sup_{z\in M} f(z) - \|z - x_0\| \leq \inf_{w\in M} f(w) + \|w - x_0\|$$

since

$$f(z) - f(w) = f(z - w) \leq \|z - w\| \leq \|z - x_0\| + \|w - x_0\|.$$

This shows that there exists $\alpha \in \mathbb{R}$ such that (A.1) holds, and completes the proof that f_0 is a norm-preserving extension of f to M_0.

In the event that X is separable, we continue to add points x_n such that x_n is the first s_j not in span $\left(M \cup \{x_k\}_{k=0}^{n-1}\right)$, the linear span of M and $\{x_0, x_1, \ldots, x_{n-1}\}$. In this way we obtain a norm preserving extension f' of f to $M' = \operatorname{span}(M \cup \{x_n\}_{n=0}^{\infty})$, and since $M' \supset S$ by construction, we can extend f' to all of X by continuity.

In the general case we use the Hausdorff Maximality Theorem. Let

$$\mathcal{P} = \left\{(M', f') : M \subset M' \subset X, f' \in (M')^*, f'\mid_M = f, \|f'\|_{(M')^*} = 1\right\},$$

where M' denotes a subspace of X. Define a partial order on $\mathcal{P}$ by declaring $(M', f') \preceq (M'', f'')$ if $M' \subset M''$ and $f''\mid_{M'} = f'$. The Hausdorff Maximality Theorem implies the existence of a maximal totally ordered subset Ω of $\mathcal{P}$. We now define

$$\widetilde{M} = \cup\left\{M' : (M', f') \in \Omega \text{ for some } f' \in (M')^*\right\}.$$

Clearly $\widetilde{M}$ is a subspace of X and if we define $\widetilde{f}$ on $\widetilde{M}$ by

$$\widetilde{f}(x) = f'(x) \text{ if } x \in M' \text{ and } (M', f') \in \Omega,$$

then $\widetilde{f}$ is a well-defined linear functional on $\widetilde{M}$ with norm 1, and in fact $\left(\widetilde{M}, \widetilde{f}\right) \in \Omega$. Now we must have $\widetilde{M} = X$ since otherwise we could adjoin a point in $X \setminus \widetilde{M}$ and extend $\widetilde{f}$ by the argument above, contradicting the maximality of Ω. This completes the proof when the scalar field is $\mathbb{R}$.

Now suppose the scalar field is $\mathbb{C}$. Set $u = \operatorname{Re} f$. We have just proved that there is a real-linear $U \in X^*$ such that $U|_M = u$ and $\|U\|^* = 1$. However, there is a unique complex-linear functional Λ whose real part is U:

$$\Lambda x = Ux - iU(ix), \qquad x \in X.$$

Then $\Lambda|_M = f$ since if $\Phi = \Lambda|_M - f$, then

$$|\Phi(m)| = \Phi\left(\frac{|\Phi m|}{\Phi m} m\right) = \operatorname{Re}\Phi\left(\frac{|\Phi m|}{\Phi m} m\right) = 0$$

for $m \in M$. Moreover,

$$|\Lambda x| = \Lambda\left(\frac{|\Lambda x|}{\Lambda x} x\right) = U\left(\frac{|\Lambda x|}{\Lambda x} x\right) \leq \left\|\frac{|\Lambda x|}{\Lambda x} x\right\| = \|x\|, \qquad x \in X.$$

□

Now we obtain a separation result for disjoint closed balanced convex sets A and B, one of which is compact - the case A is a singleton is especially useful. A set E is balanced if $\alpha E \subset E$ for all $\alpha \in \mathbb{C}$ with $|\alpha| \leq 1$.

COROLLARY A.13. *Suppose that A and B are disjoint closed convex balanced subsets of a Banach space X. If A is compact, there are $\Lambda \in X^*$ and $\beta < \gamma$ such that*

$$\operatorname{Re}\Lambda x \leq \beta < \gamma \leq \operatorname{Re}\Lambda y$$

for all $x \in A$ and $y \in B$. In the special case $A = \{x_0\}$ is a singleton, there is $\lambda \in X^*$ such that

$$|\lambda x| \leq 1 < |\lambda x_0|, \qquad x \in B. \tag{A.2}$$

PROOF. It suffices to prove the case of real scalars. Let $r > 0$ be such that $A_r = A + B(0,r)$ and B are disjoint. Let p be the Minkowski functional of $\Gamma = A_r - B + x_0$, i.e.

$$p(x) = \inf\{t > 0 : t^{-1}x \in \Gamma\}.$$

Since Γ is a convex balanced neighborhood of the origin, straightforward computations show that p is a seminorm on X. Fix $a_0 \in A_r$ and $b_0 \in B$ and set $x_0 = b_0 - a_0$. Then $x_0 \notin \Gamma$ since $A_r \cap B = \phi$, and we thus have $p(x_0) \geq 1$. Define $f(tx_0) = t$ on the subspace M generated by x_0. By Remark Hahn-Banach there is a linear functional Λ with $\Lambda x_0 = 1$ and $|\Lambda x| \leq p(x)$ for all $x \in X$. Since Γ is a neighborhood of the origin, we have $\Lambda \in X^*$. Now for $a \in A$ and $b \in B$ we have

$$\Lambda a - \Lambda b + 1 = \Lambda(a - b + x_0) \leq p(a - b + x_0) < 1$$

since Γ is open. We conclude that $\Lambda a < \Lambda b$, i.e. ΛA_r and ΛB are disjoint convex subsets of $\mathbb{R}$ with ΛA_r to the left of ΛB. Now A_r is open and Λ is an open map, and thus ΛA_r is open. If we set $\gamma = \sup \Lambda A_r$, we obtain

$$\operatorname{Re}\Lambda x < \gamma \leq \Lambda y, \qquad x \in A_r, y \in B.$$

Now we observe that ΛA is a compact subset of the open set ΛA_r to get $\beta < \gamma$ such that

$$\operatorname{Re}\Lambda x \leq \beta < \gamma \leq \Lambda y, \qquad x \in A, y \in B.$$

In the case when $A = \{x_0\}$, the above conclusion shows there is $\Lambda \in X^*$ such that $\Lambda x_0 = \rho e^{i\theta}$ lies outside the closure K of ΛB. Since B is balanced, so is K, and thus there is $0 < s < \rho$ so that $|z| \leq s$ for all $z \in K$. It follows that $\lambda = s^{-1}e^{-i\theta}\Lambda$ satisfies (A.2). □

A.3.1. The Open Mapping Theorem. Now we turn to open maps. A map $f : X \to Y$ where X, Y are topological spaces is *open* if $f(G)$ is open in Y for every G open in X. A famous "open mapping theorem" is that a holomorphic function f on a connected open subset Ω of the complex plane is open if it is not constant. If we consider continuous linear maps $\Lambda : X \to Y$ where X, Y are Banach spaces, then Λ is open if it is onto. (Of course some such assumptions are necessary. The inclusion of the $x-$axis into the plane is an isometric map of $\mathbb{R}^1$ into $\mathbb{R}^2$ which is certainly not open.) Note that for a linear map $\Lambda : X \to Y$ from one normed linear space X to another Y, Λ is open if and only if $\Lambda(B_X(0,1)) \supset B_Y(0,r)$ for some $r > 0$.

THEOREM A.14 (Open Mapping Theorem). *Suppose X, Y are Banach spaces and $\Lambda : X \to Y$ is bounded and onto. Then Λ is an open map.*

PROOF. Since Λ is onto we have $Y = \cup_{k=1}^{\infty} \Lambda\left(kB_X\left(0,\frac{1}{4}\right)\right)$, and thus by Baire's Theorem, one of the sets $\overline{\Lambda\left(kB_X\left(0,\frac{1}{4}\right)\right)} = k\overline{\Lambda\left(B_X\left(0,\frac{1}{4}\right)\right)}$ must have nonempty interior, and hence so must $\overline{\Lambda\left(B_X\left(0,\frac{1}{4}\right)\right)}$, say

$$B_Y(y_0,r) \subset \overline{\Lambda\left(B_X\left(0,\frac{1}{4}\right)\right)}.$$

Then we have

$$\begin{aligned} \overline{\Lambda\left(B_X\left(0,\frac{1}{2}\right)\right)} &\supset \overline{\Lambda\left(B_X\left(0,\frac{1}{4}\right)\right) - \Lambda\left(B_X\left(0,\frac{1}{4}\right)\right)} \\ &\supset \overline{\Lambda\left(B_X\left(0,\frac{1}{4}\right)\right)} - \overline{\Lambda\left(B_X\left(0,\frac{1}{4}\right)\right)} \\ &\supset B_Y(y_0,r) - B_Y(y_0,r) \\ &\supset B_Y(0,r). \end{aligned} \tag{A.3}$$

It remains only to prove that $\overline{\Lambda\left(B_X\left(0,\frac{1}{2}\right)\right)} \subset \Lambda(B_X(0,1))$. For this, fix $y_1 \in \overline{\Lambda\left(B_X\left(0,\frac{1}{2}\right)\right)}$. Now the argument above shows that $\overline{\Lambda\left(B_X\left(0,\frac{1}{4}\right)\right)}$ contains an open ball $B_Y(0,r_1)$ about the origin as well. There is $x_1 \in B_X\left(0,\frac{1}{2}\right)$ such that $\Lambda x_1 \in \Lambda\left(B_X\left(0,\frac{1}{2}\right)\right)$ satisfies $\|\Lambda x_1 - y_1\|_Y < r_1$. Then we have

$$\Lambda x_1 \in B_Y(y_1,r_1) \subset \left\{ y_1 - \overline{\Lambda\left(B_X\left(0,\frac{1}{4}\right)\right)} \right\}.$$

Now define

$$y_2 = y_1 - \Lambda x_1 \in \overline{\Lambda\left(B_X\left(0,\frac{1}{4}\right)\right)}.$$

We can repeat this procedure inductively to obtain sequences $\{x_n\}_{n=1}^{\infty} \subset X$ and $\{y_n\}_{n=1}^{\infty} \subset Y$ satisfying

$$\begin{aligned} x_n &\in B_X\left(0,\frac{1}{2^n}\right), \\ y_n &\in \overline{\Lambda\left(B_X\left(0,\frac{1}{2^n}\right)\right)}, \\ y_{n+1} &= y_n - \Lambda x_n, \end{aligned}$$

for all $n \geq 1$. Then $x = \lim_{m\to\infty} \sum_{n=1}^{m} x_n \in B_X(0,1)$ since $\|x\| \leq \sum_{n=1}^{\infty} \|x_n\| < \sum_{n=1}^{\infty} \frac{1}{2^n} = 1$, and since $\|y_n\| \leq \|\Lambda\| 2^{-n}$,

$$\Lambda x = \lim_{m\to\infty} \sum_{n=1}^{m} \Lambda x_n = \lim_{m\to\infty} \sum_{n=1}^{m} (y_n - y_{n+1}) = y_1 - \lim_{m\to\infty} y_{m+1} = y_1.$$

□

We can combine the Hahn-Banach Theorem with the open mapping theorem to obtain a characterization of when a bounded map T is onto in terms of its adjoint T^*.

Corollary A.15. *Suppose $T : X \to Y$ is bounded. Then T is onto Y if and only if there is $\delta > 0$ such that*

$$\|T^* y^*\| \geq \delta \|y^*\|, \qquad y^* \in Y^*.$$

Proof: If T is onto, the open mapping theorem A.14 implies that there is $\delta > 0$ such that $T(U) \supset \delta V$ where U and V are the unit balls in X and Y respectively. Then

$$\|T^* y^*\| = \sup_{x\in U} |\langle x, T^* y^*\rangle| = \sup_{x\in U} |\langle Tx, y^*\rangle| \geq \sup_{y\in\delta V} |\langle y, y^*\rangle| = \delta \|y^*\|.$$

Conversely, if $\|T^* y^*\| \geq \delta \|y^*\|$ for all $y^* \in Y^*$, then a standard argument using the Hahn-Banach Theorem shows that $\overline{T(U)} \supset \delta V$. Indeed, if $y_0 \notin \overline{T(U)}$, then by Corollary A.13 there is y^* such that $|\langle y_0, y^*\rangle| > 1$ and $|\langle y, y^*\rangle| \leq 1$ for all $y \in \overline{T(U)}$. Thus for $x \in U$ we have

$$|\langle x, T^* y^*\rangle| = |\langle Tx, y^*\rangle| \leq 1,$$

and so $\|T^* y^*\| \leq 1$. Then from our hypothesis we have

$$\delta < \delta |\langle y_0, y^*\rangle| \leq \delta \|y_0\| \|y^*\| \leq \|y_0\| \|T^* y^*\| \leq \|y_0\|,$$

which shows that $y \in \overline{T(U)}$ if $\|y\| \leq \delta$.

Now from a variation on an argument used in the proof of the open mapping theorem in the previous section, we get that $T(U) \supset \delta V$, which implies that T is onto Y. Indeed, for $y_1 \in V$ choose $\varepsilon_n > 0$ such that $\sum_{n=1}^{\infty} \varepsilon_n < 1 - \|y_1\|$. Assuming y_n has been chosen we choose x_n and y_{n+1} as follows:

$$\begin{aligned} &y_{n+1} = y_n - Tx_n \text{ where} \\ &\|x_n\| \leq \|y_n\| \text{ and } \|y_n - Tx_n\| < \varepsilon_n. \end{aligned}$$

Then $x = \sum_{n=1}^{\infty} x_n \in U$ (standard exercise) and

$$Tx = \lim_{N\to\infty} \sum_{n=1}^{N} Tx_n = \lim_{N\to\infty} \sum_{n=1}^{N} (y_n - y_{n+1}) = y_1 - \lim_{N\to\infty} y_{N+1} = y_1,$$

since $\|y_{N+1}\| < \varepsilon_N \to 0$ as $N \to \infty$.

A.3.2. Mazur's Lemma. Let V be a normed linear space and $\{u_j\}$ be a sequence converging weakly to u^*. Then there exists a sequence of convex combinations of the u_j which converges in norm to u^*. In order to prove this theorem, we first use the separation result in Corollary A.13 to show the equality of weak and strong closures of convex subsets in normed linear spaces.

LEMMA A.16. *Suppose E is a convex subset of a normed linear space V. Then the weak closure $\overline{E}^{\text{weak}}$ is equal to the original closure $\overline{E}$.*

PROOF. Clearly $\overline{E} \subset \overline{E}^{\text{weak}}$ since every strongly closed superset of E is also weakly closed. To obtain the opposite conclusion, pick $v_0 \in V \setminus \overline{E}$. Then corollary A.13 shows that there is a continuous linear functional $\Lambda \in V^*$ and a real number $\gamma \in \mathbb{R}$ such that

$$\operatorname{Re} \Lambda v_0 < \gamma < \operatorname{Re} \Lambda v, \qquad \text{for all } v \in \overline{E}.$$

The set $\{v : \operatorname{Re} \Lambda v < \gamma\}$ is thus a weak neighborhood of the point v_0 that does not intersect E. It follows that $v_0 \notin \overline{E}^{\text{weak}}$, which shows that $\overline{E}^{\text{weak}} \subset \overline{E}$, and completes the proof that $\overline{E}^{\text{weak}} = \overline{E}$. □

Here now is Mazur's theorem.

THEOREM A.17. *Suppose X is a normed linear space. If $\{x_n\}_{n=1}^{\infty}$ is a sequence in X that converges weakly to some $x \in X$, then there is a sequence $\{y_i\}_{i=1}^{\infty}$ in X such that:*

(1) *Each y_i is a convex combination of finitely many of the x_n.*
(2) *$y_i \to x$ strongly as $i \to \infty$.*

PROOF. Let H be the convex hull of the set of elements $\{x_n : n \geq 1\}$, and let K denote the weak closure of H. Then $x \in K$, and by the above lemma, x lies in the strong closure of H as well. Since a normed linear space is a metric space, it follows that there is a *sequence* $\{y_i\}_{i=1}^{\infty}$ in H that converges strongly to x. □

A.4. Convergence in Spaces of Functions

At times it is useful to know that the pointwise limit of a norm bounded sequence of functions in a function space is also in the space. The following version of a result of Brown and Shields is a useful tool for showing that.

PROPOSITION A.18 (Brown and Shields, [**86**]). *Suppose $f \in \mathcal{D}$ and $[f]$ is the closed shift invariant subspace generated by f. The following are equivalent for $g \in \mathcal{D}$:*

(1) *$g \in [f]$; that is, g is in the norm closure of $\{pf : p \text{ a polynomial}\}$*
(2) *g is in the weak closure of $\{pf : p \text{ a polynomial}\}$*
(3) *There is a sequence $\{g_n\} \subset [f]$ which is norm bounded (i.e. $\{\|g_n\|\}$ is bounded) and such that for all $z \in \mathbb{D}$, $g_n(z) \to g(z)$.*

PROOF. The first two are equivalent because the norm closed subspaces of a Hilbert space are the same as the weakly closed subspaces by Lemma A.16. The first statement clearly implies the third. If the third condition holds then $\{g_n\}$ must have a weakly convergent subsequence. The convergence of $\{g_n(z)\}$ ensures that the weak limit must be g, which establishes the second condition, and then, as noted the weak closure of the subspace is the norm closure. □

APPENDIX B

Schur's Test

It is a classical result of Schur [**299**] that a matrix of positive entries with uniformly bounded row sums and column sums gives a bounded operator on the Hilbert space ℓ^2. Extensions of that result apply to operators defined on Lebesgue spaces by non-negative matrices or non-negative integration kernels. We use several versions of this result in the text. For the reader's convenience we collect them here along with some additional comments.

B.1. Introduction and Motivation

Schur's test is a fundamental tool to prove that integral operators with non-negative kernels are bounded. The utility of Schur's test is that it allows one to test the action of the operator on just a single function (or maybe a pair, or small family of functions), and knowing the behavior in this particular case, one deduces the operator is bounded on all functions. However, in fairness there is an issue with this, a suitable Schur function is nearly impossible to guess. Guessing the correct function, in many situations of interest, is usually influenced by the surrounding geometry.

As simple motivation we consider the following computation for proving a matrix is bounded. Let $A = (a_{ij}) : \ell^2 \to \ell^2$ and we want some *sufficient* conditions to show that the operator is bounded. We proceed under the simple hypothesis that

$$\sup_j \sum_i |a_{ij}| \leq C_1 \quad \text{and} \quad \sup_i \sum_j |a_{ij}| \leq C_2.$$

Under this very simple hypothesis we can deduce the A is a bounded operator. Indeed,

$$\begin{aligned}
|\langle Ax, y\rangle_{\ell^2}| &= \left|\sum_i \sum_j a_{ij} x_j \overline{y}_i\right| \leq \sum_i \sum_j |a_{ij}|^{1/2} |a_{ij}|^{1/2} |x_j|\,|y_i| \\
&\leq \sum_i \left(\sum_j |a_{ij}|\right)^{1/2} \left(\sum_j |a_{ij}|\,|x_j|^2\right)^{1/2} |y_i| \\
&\leq \sqrt{C_2}\left(\sum_i |y_i|^2\right)^{1/2} \left(\sum_i \sum_j |a_{ij}|\,|x_j|^2\right)^{1/2}
\end{aligned}$$

$$\leq \sqrt{C_2}\,\|y\|_{\ell^2}\left(\sum_j\sum_i |x_j|^2\,|a_{ij}|\right)^{1/2}$$

$$\leq \sqrt{C_1C_2}\,\|y\|_{\ell^2}\left(\sum_j |x_j|^2\right)^{1/2} = \sqrt{C_1C_2}\,\|y\|_{\ell^2}\,\|x\|_{\ell^2}\,.$$

B.2. Schur's Test

The proof strategy used above is flexible enough to be modified to yield much more useful results. In the remainder of this appendix, we explore some extensions of this idea and applications that are relevant for the book.

THEOREM B.1 (Simple Schur's Test). *Let (X,μ) be a measure space and $H(x,y)$ be a non-negative symmetric kernel. Define*

$$Tf(x) = \int_X H(x,y)\,f(y)\,d\mu(y)\,.$$

Then T is bounded on $L^2(X,\mu)$ if and only if there is a positive function $h \in L^2(X,\mu)$ such that

$$Th(x) = \int_X H(x,y)\,h(y)\,d\mu(y) \leq Ah(x)\,, \qquad \mu - a.e.x \in X.$$

Moreover, if A is the least such constant, then $\left\|T : L^2(X,\mu) \to L^2(X,\mu)\right\| = A$.

PROOF. We have

$$\begin{aligned}
&\int_X |Tf(x)|^2\,d\mu(x)\\
&\leq \int_X \left(\int_X H(x,y)\,h(y)\,d\mu(y)\right)\left(\int_X H(x,y)\,\frac{f(y)^2}{h(y)}\,d\mu(y)\right)d\mu(x)\\
&\leq A\int_X \left(\int_X H(x,y)\,h(x)\,d\mu(x)\right)\frac{f(y)^2}{h(y)}\,d\mu(y)\\
&\leq A^2\int_X h(y)\,\frac{f(y)^2}{h(y)}\,d\mu(y) = A^2\int_X f(y)^2\,d\mu(y)\,.
\end{aligned}$$

Conversely, if T is bounded, choose $f > 0$ in $L^2(\mu)$, $\gamma > 1$, and define

$$h(x) = \sum_{n=0}^{\infty}\left(\frac{1}{\gamma\,\|T\|}\right)^n T^n f(x)\,, \qquad x \in X.$$

Then

$$\|h\|_{L^2(\mu)} \leq \sum_{n=0}^{\infty}\left(\frac{1}{\gamma\,\|T\|}\right)^n \|T\|^n\,\|f\|_{L^2(\mu)} < \infty,$$

and

$$\begin{aligned} Th(x) &= \sum_{n=0}^{\infty} \left(\frac{1}{\gamma \|T\|} \right)^n T^{n+1} f(x) \\ &= \gamma \|T\| \sum_{n=1}^{\infty} \left(\frac{1}{\gamma \|T\|} \right)^n T^n f(x) \\ &= \gamma \|T\| (h(x) - f(x)) \\ &\le \gamma \|T\| h(x) . \end{aligned}$$

□

The most general (linear) version of Schur's test is the following result. In some applications, one doesn't have a symmetric kernel $H(x, y)$, yet it is still positive and one would like a simple test to verify the boundedness of the operator. This then requires that one test on two different functions as opposed to just a single function. Nevertheless, one obtains a very simple test by which to verify the boundedness.

THEOREM B.2 (Linear Schur's Test). *Let (X, μ) and (Y, ν) be measure spaces and $H(x, y)$ be a non-negative measurable function on $X \times Y$. Define, initially for non-negative functions f,*

$$T(f)(x) = \int_Y H(x, y) f(y)\, d\nu(y), \quad x \in X.$$

For $1 < p < \infty$, suppose there are positive functions h and k on X and Y respectively such that

$$T(k^{p'})(x) = \int_Y H(x, y) k(y)^{p'}\, d\nu(y) \le (Ah(x))^{p'},$$

for μ-a.e. $x \in X$, and

$$T^*(h^p)(y) = \int_X H(x, y) h(x)^p\, d\mu(x) \le (Bk(y))^p,$$

for ν-a.e.$y \in Y$. Then T is a bounded linear transformation from $L^p(Y, \nu)$ to $L^p(X, \mu)$ and $\|T : L^p(Y, \nu) \to L^p(X, \mu)\| \le AB$.

PROOF. The idea is to simply use the definition of the operator T, multiply and divide by one of the functions appearing in the testing conditions appearing in the hypothesis, and then use Fubini (which is permissible since we are dealing with non-negative quantities). Using this approach, we have

$$\begin{aligned} &\int_X |Tf(x)|^p\, d\mu(x) \\ &\le \int_X \left(\int_Y H(x, y) k(y)^{p'}\, d\nu(y) \right)^{p/p'} \left(\int_Y H(x, y) \left(\frac{f(y)}{k(y)} \right)^p d\nu(y) \right) d\mu(x) \\ &\le A^p \int_Y \left(\int_X H(x, y) h(x)^p\, d\mu(x) \right) \left(\frac{f(y)}{k(y)} \right)^p d\nu(y) \\ &\le A^p B^p \int_Y k(y)^p \left(\frac{f(y)}{k(y)} \right)^p d\nu(y) \\ &= (AB)^p \int_Y f(y)^p\, d\nu(y) . \end{aligned}$$

This clearly proves that $\|T : L^p(Y, \nu) \to L^p(X, \mu)\| \le AB$ as claimed. □

We present a slight modification of Schur's Lemma that allows one to deduce a range of boundedness of the operator by knowing the behavior on test functions for a certain range of values. This is a slight modification of a result in [**142**].

THEOREM B.3. *Let* (X,μ) *be a measure space and* $H(x,y)$ *be a non-negative measurable function on* $X\times X$. *Define, initially for non-negative functions* f,

$$T(f)(x)=\int_X H(x,y)\,f(y)\,d\mu(y),\ \ x\in X.$$

Suppose that there exists a positive auxiliary function h *on* X *and numbers* $0<a<b$ *such that for all* $\epsilon\in[a,b)$ *the following estimates hold:*

$$T(h^{-\epsilon})(x)=\int_X H(x,y)\,h^{-\epsilon}(y)\,d\mu(y)\lesssim h^{-\epsilon}(x),$$

for μ-*a.e.* $x\in X$, *and*

$$T(h^{-\epsilon})(y)=\int_X H(x,y)\,h^{-\epsilon}(x)\,d\mu(x)\lesssim h^{-\epsilon}(y),$$

for μ-*a.e.*$y\in X$. *Then* T *is a bounded linear transformation from* $L^p(X,\mu)$ *to* $L^p(X,\mu)$ *for all* $p\in\left(\frac{a+b}{b},\frac{a+b}{a}\right)$.

PROOF. The proof is the same strategy as that appearing in Theorem B.2. Without loss of generality, suppose that $p'\le 2\le p$. Choose $s\in[a,b)$ so that $s\frac{p}{p'}\in[a,b)$.

$$\begin{aligned}
|Tg(z)|^p &\le \left(\int_X |g(w)|\,H(w,z)d\mu(w)\right)^p\\
&\le \left(\int_X |g(w)|^p\,H(w,z)h(w)^{\frac{sp}{p'}}d\mu(w)\right)\left(\int_X H(w,z)h(w)^{-s}d\mu(w)\right)^{\frac{p}{p'}}\\
&\le \left(\int_X |g(w)|^p\,H(w,z)h(w)^{\frac{sp}{p'}}d\mu(w)\right)h(z)^{-\frac{sp}{p'}}.
\end{aligned}$$

Then integrate this in the z variable to have:

$$\begin{aligned}
\|Tg(z)\|^p_{L^p(X,\mu)} &\le \int_X\int_X |g(w)|^p\,H(w,z)h(w)^{\frac{sp}{p'}}d\mu(w)h(z)^{-\frac{sp}{p'}}d\mu(z)\\
&= \int_X |g(w)|^p\left(\int_X H(w,z)h(z)^{-\frac{sp}{p'}}d\mu(z)\right)h(w)^{\frac{sp}{p'}}d\mu(z)\\
&\lesssim \int_X |g(w)|^p\,h(w)^{-\frac{sp}{p'}}h(w)^{\frac{sp}{p'}}d\mu(w)=\|g\|^p_{L^p(X,\mu)}.
\end{aligned}$$

The condition that $s\in[a,b)$ and $s\frac{p}{p'}\in[a,b)$ is equivalent to $a\frac{p}{p'}=a(p-1)<b$, which upon rearrangement gives $p<\frac{a+b}{a}$.

Repeat the computations and argument above to deduce

$$\|Tg\|_{L^{p'}(X,\mu)}\lesssim\|g\|_{L^{p'}(X,\mu)}$$

provided that $s\in[a,b)$ and $s\frac{p'}{p}\in[a,b)$. But, this condition on s is equivalent to $b\frac{p'}{p}=b(p'-1)>a$, which gives $p'>\frac{a+b}{b}$. However since $p'\le 2\le p$, we then have that $p\in\left(\frac{a+b}{b},\frac{a+b}{a}\right)$. □

REMARK B.4. *Although we will not use the fact, it is interesting to note that the converse also holds, if the linear operator T in Theorem* B.2 *is bounded then there must be functions which satisfies the testing conditions. And, additionally, there is a third equivalence that can also play a role in applications of Schur's Lemma.*

The third condition that is equivalent is both of the conditions in Theorem B.2 *is the following: For all $C > \|T : L^p(Y,\nu) \to L^p(X,\mu)\|$ there exists a non-negative finite measurable function l on (X,μ) such that*

$$T^*\left((Tl)^{p-1}\right) \leq C^p l^{p-1}$$

for μ-almost every $x \in X$. If we have this function l, then define $h = (Tl)^{\frac{p-1}{p}}$, and $k = Cl^{\frac{p-1}{p}}$. This implies that $k^{p'} = C^{p'} l$ and $h^{p'} = Tl$. Then we have by the hypothesis, $T^*(h^p) = T^*\left((Tl)^{p-1}\right) \leq C^p l^{p-1} = k^p$. And, we also have that $T(k^{p'}) = C^{p'} T(l) = C^{p'} h^{p'}$. We thus have functions h and k and constants $A = C$ and $B = 1$ so that the hypotheses in Theorem B.2 *hold.*

It then would suffice to prove that if T is bounded between $L^p(Y,\nu)$ and $L^p(X,\mu)$ then we can find the function l used above. We outline the details, and point the interested reader to [**182**, Theorem 4] *for full details. Without loss, we assume that the norm of T is at most* 1. *Define*

$$S(f) = \left(T^*\left(Tf\right)^{p-1}\right)^{\frac{1}{p-1}}.$$

Then we have that $\|S\|_{L^p(Y,\nu)} \leq 1$ when $\|f\|_{L^p(Y,\nu)} \leq 1$, if $0 \leq f_1 \leq f_2$ then $Sf_1 \leq sF_2$ and if $f_n \nearrow f$ in $L^p(Y,\nu)$ then $Sf_n \nearrow Sf$ ν-almost everywhere. Let $f_1 \in L^p(Y,\nu)$ be positive almost everywhere with $\|f_1\|_{L^p(Y,\nu)} \leq \frac{C-1}{C}$. For $n > 1$ define $f_n = f_1 + \frac{1}{C} Sf_{n-1}$. A simple inductive argument gives that $f_n \leq f_{n+1}$ and that $\|f_n\|_{L^p(Y,\nu)} \leq 1$. This further implies that we can find a function $l \in L^p(Y,\nu)$ such that $f_n \nearrow l$ ν-almost everywhere and $\|l\|_{L^p(Y,\nu)} \leq 1$. Since $Sf_n \nearrow Sl$ ν-almost everywhere we have that $l = f_1 + \frac{1}{C} Sl$, which implies that $Sl \leq Cl$. Using the definition of S we arrive at the alternate condition mentioned.

For matrices with positive entries the previous theorem is due to Schur [**299**]. The extension to integral operators and the consideration of the converse goes back to work of Gagliardo [**160**].

B.3. Application of the Linear Schur Estimate

The well-known application of Schur's Lemma is to prove that the Bergman projection is bounded on $L^2(\mathbb{B}_n)$. Recall that the Bergman projection is given by

$$\mathbb{P}(f)(z) = \int_{\mathbb{B}_n} \frac{f(w)}{(1 - \overline{w}z)^{n+1}} \, dV(w)$$

and we have the Bergman space of square integrable analytic functions on the $\mathbb{B}_n$. One would like to know that $\mathbb{P}$ is bounded on L^2, or more generally L^p. This turns out to be a simple application of Schur's Lemma. The only "challenge" is to guess the form of the function to test on. But, the symmetry of the problem suggests that we try $k(z) = h(z) = \left(1 - |z|^2\right)^{\varepsilon}$ for an appropriate choice of ε. To see that this has a chance of working we recall the following well-known lemma which is important in this application. See [**346**, Theorem 1.12] for the proof, which is essentially a computation with power series.

LEMMA B.5 (Rudin-Forelli Estimates [**154**]). *Suppose that c is real and that $t > -1$. The for $z \in \mathbb{B}_n$ we have*

$$\int_{\mathbb{B}_n} \frac{(1-|w|^2)^t}{|1-\overline{w}z|^{n+1+t+c}} dV \approx \begin{cases} C_t & \text{if} \quad c<0,\ t>-1 \\ -C_t \log(1-|z|^2) & \text{if} \quad c=0, \\ C_t (1-|z|^2)^{-c} & \text{if} \quad c>0. \end{cases} \tag{B.1}$$

THEOREM B.6 (Boundedness of the Bergman Projection). *Set*

$$T_{a,b} f(z) = (1-|z|^2)^a \int_{\mathbb{B}_n} \frac{(1-|w|^2)^b}{(1-\overline{w}z)^{n+1+a+b}} f(w)\, dV(w),$$

$$S_{a,b} f(z) = (1-|z|^2)^a \int_{\mathbb{B}_n} \frac{(1-|w|^2)^b}{|1-\overline{w}z|^{n+1+a+b}} f(w)\, dV(w).$$

Suppose $t \in \mathbb{R}$ and $1 \le p < \infty$. Then $T_{a,b}$ is bounded on $L^p\left(\mathbb{B}_n, (1-|z|^2)^t\, dV\right)$ if and only if $S_{a,b}$ is bounded on $L^p\left(\mathbb{B}_n, (1-|z|^2)^t\, dV\right)$ if and only if

$$-pa < t+1 < p(b+1). \tag{B.2}$$

PROOF. We outline this now, but leave the details to the reader. The complete details can be found in [**346**, Theorem 2.10]. Clearly, if $S_{a,b}$ is bounded, then $T_{a,b}$ is bounded. Setting $h_\varepsilon(z) = (1-|z|^2)^\varepsilon$ for ε chosen so that $\varepsilon + b > -1$ lets one deduce through a straightforward computation that $T_{a,b} h_\varepsilon(z) = C_\varepsilon h_a(z)$. Using that $T_{a,b}$ is bounded on the spaces in question lets one deduce the condition that $t+1 > pa$. The condition that $t+1 < p(b+1)$ is deduced by doing the same computations but for $T_{a,b}^*$. It remains to show that the conditions on a, b, p, t imply that $S_{a,b}$ is bounded. But, the conditions on a, b, p and t, coupled with the Forelli-Rudin estimates easily let one deduce that the hypotheses in Schur's Lemma holds, and hence $S_{a,b}$ is bounded. □

For $t > -1$ $T_{0,t}$ is the self adjoint projection of $L^2\left(\mathbb{B}_n, (1-|z|^2)^t\, dV\right)$ onto its subspace of holomorphic functions (the "weighted Bergman space"). Hence, for $1 < p < \infty$ $T_{0,t}$ is a bounded projection of $L^p\left(\mathbb{B}_n, (1-|z|^2)^t\, dV\right)$ onto its subspace of holomorphic functions. In particular the operator reproduces holomorphic functions;

COROLLARY B.7. *For $t > -1$ and $f \in L^p\left(\mathbb{B}_n, (1-|z|^2)^t\, dV\right)$, $T_{0,t} f = f$.*

B.4. Bilinear Schur Estimates

We next show how to obtain a bilinear version of Schur's Test. The proof is essentially the same as above. This can be extended to the multi-linear setting. A further, related multi-linear extension can be found in [**166**].

THEOREM B.8 (Bilinear Schur Lemma). *Let (X, μ), (Y, ν) and (Z, ω) be measure spaces and $H(x, y, z)$ be a non-negative measurable function on $X \times Y \times Z$. Define, initially for non-negative functions f, g,*

$$T(f,g)(x) = \int_{Y \times Z} H(x,y,z)\, f(y)\, d\nu(y)\, g(z)\, d\omega(z), \quad x \in X,$$

For $1 < p < \infty$, suppose there are positive functions h, k and m on X, Y and Z respectively such that

$$\int_{Y\times Z} H(x,y,z)\, k(y)^{p'} m(z)^{p'}\, d\nu(y)\, d\omega(z) \leq (Ah(x))^{p'},$$

for μ-a.e. $x \in X$, and

$$\int_X H(x,y,z)\, h(x)^p\, d\mu(x) \leq (Bk(y)\, m(z))^p,$$

for $\nu\times\omega$-a.e. $(y,z) \in Y\times Z$. Then T is bounded from $L^p(Y;\nu)\times L^p(Z,\omega)$ to $L^p(X;\mu)$ and $\|T : L^p(Y,\nu)\times L^p(Z,\omega) \to L^p(X,\mu)\| \leq AB$.

PROOF. The proof is essentially the same as the linear case. We have

$$\begin{aligned}
&\int_X |Tf(x)|^p\, d\mu(x) \\
&\leq \int_X \left(\int_{Y\times Z} H(x,y,z)\, k(y)^{p'} m(z)^{p'}\, d\nu(y)\, d\omega(z)\right)^{p/p'} \\
&\times \left(\int_{Y\times Z} H(x,y,z) \left(\frac{f(y)}{k(y)}\right)^p d\nu(y) \left(\frac{g(z)}{m(z)}\right)^p d\omega(z)\right) d\mu(x) \\
&\leq A^p \int_{Y\times Z} \left(\int_X H(x,y,z)\, h(x)^p\, d\mu(x)\right) \left(\frac{f(y)}{k(y)}\right)^p d\nu(y) \left(\frac{g(z)}{m(z)}\right)^p d\omega(z) \\
&\leq A^p B^p \int_{Y\times Z} k(y)^p m(z)^p \left(\frac{f(y)}{k(y)}\right)^p d\nu(y) \left(\frac{g(z)}{m(z)}\right)^p d\omega(z) \\
&= (AB)^p \int_Y f(y)^p\, d\nu(y) \int_Z g(z)^p\, d\omega(z).
\end{aligned}$$

□

B.5. Schur's Test for Matrix Valued Kernels

We now give a version of Schur's test for operators with matrix valued kernels acting on vector-valued functions. To do this we need to introduce an appropriate notion of "positivity" for the kernels. The following result is from [**189**].

THEOREM B.9 (Matrix-Valued Kernel Schur Test). *Let T be a linear operator taking $\mathbb{C}^d$-valued functions on a measure space (X,μ) to $\mathbb{C}^d$-valued functions on a measure space (Y,ν). Suppose that the action of T is given by:*

$$Tf(y) = \int_X K(y,x) f(x)\, d\mu(x)$$

where $K(y,x)$ is a $d\times d$ matrix-valued kernel. Suppose that it is possible to write K as a product, $K(y,x) = K_1(y,x)K_2(y,x)$, in such a way that

$$\int_X K_1(y,x)K_1^*(y,x)\, d\mu(x) \leq C_1 \operatorname{Id}_d \quad \nu \text{ a.e. } y\in Y$$

$$\int_Y K_2^*(y,x)K_2(y,x)\, d\nu(y) \leq C_2 \operatorname{Id}_d \quad \mu \text{ a.e. } x\in X$$

with the inequality understood in the sense of operators. Then $T : L^2(X,\mu,\mathbb{C}^d) \to L^2(Y,\nu,\mathbb{C}^d)$ with norm at most $C_1^{1/2}C_2^{1/2}$.

A good exercise is to check that the motivating example at the beginning of this section is a special case of this theorem.

PROOF. It suffices to bound the following:

$$\begin{aligned}
\langle Tf, g\rangle_{L^2(Y,\nu,\mathbb{C}^d)} &= \int_Y \langle Tf(y)g(y)\rangle_{\mathbb{C}^d}\, d\nu(y) \\
&= \int_Y\int_X \langle K(y,x)f(x), g(y)\rangle_{\mathbb{C}^d}\, d\mu(x)\, d\nu(y) \\
&= \int_Y\int_X \langle K_1(y,x)K_2(y,x)f(x), g(y)\rangle_{\mathbb{C}^d}\, d\mu(x)\, d\nu(y) \\
&= \int_Y\int_X \langle K_2(y,x)f(x), K_1^*(y,x)g(y)\rangle_{\mathbb{C}^d}\, d\mu(x)\, d\nu(y).
\end{aligned}$$

Estimating we then have

$$\begin{aligned}
\left|\langle Tf, g\rangle_{L^2(Y,\nu,\mathbb{C}^d)}\right| &\le \int_Y\int_X |\langle K_2(y,x)f(x), K_1^*(y,x)g(y)\rangle_{\mathbb{C}^d}|\, d\mu(x)\, d\nu(y) \\
&\le \int_Y\int_X \|K_2(y,x)f(x)\|_{\mathbb{C}^d}\, \|K_1^*(y,x)g(y)\|_{\mathbb{C}^d}\, d\mu(x)\, d\nu(y) \\
&\le \left(\int_Y\int_X \|K_2(y,x)f(x)\|^2_{\mathbb{C}^d}\, d\mu(x)\, d\nu(y)\right)^{1/2} \\
&\qquad \times \left(\int_Y\int_X \|K_1^*(y,x)g(y)\|^2_{\mathbb{C}^d}\, d\mu(x)\, d\nu(y)\right)^{1/2}.
\end{aligned}$$

It suffices to control each of these terms. Note that

$$\begin{aligned}
&\int_Y\int_X \|K_2(y,x)f(x)\|^2_{\mathbb{C}^d}\, d\mu(x)\, d\nu(y) \\
&\quad = \int_Y\int_X \langle K_2(y,x)f(x), K_2(y,x)f(x)\rangle_{\mathbb{C}^d}\, d\mu(x)\, d\nu(y) \\
&\quad = \int_X \left\langle \int_Y K_2^*(y,x)K_2(y,x)\, d\nu(y) f(x), f(x)\right\rangle_{\mathbb{C}^d} d\mu(x) \\
&\quad \le C_2 \int_X \langle f(x), f(x)\rangle_{\mathbb{C}^d}\, d\mu(x) = C_2\, \|f\|^2_{L^2(X,\mu,\mathbb{C}^d)},
\end{aligned}$$

hence $\left(\int_Y\int_X \|K_2(y,x)f(x)\|^2_{\mathbb{C}^d}\, d\mu(x)\, d\nu(y)\right)^{1/2} \le C_2^{1/2}\, \|f\|_{L^2(X,\mu,\mathbb{C}^d)}$. Similarly, we have that

$$\begin{aligned}
&\int_Y\int_X \|K_1^*(y,x)g(y)\|^2_{\mathbb{C}^d}\, d\mu(x)\, d\nu(y) \\
&\quad - \int_Y\int_X \langle K_1^*(y,x)g(y), K_1^*(y,x)g(y)\rangle_{\mathbb{C}^d}\, d\mu(x)\, d\nu(y) \\
&\quad = \int_Y \left\langle \int_X K_1(y,x)K_1^*(y,x)\, d\mu(x)g(y), g(y)\right\rangle_{\mathbb{C}^d} d\nu(y) \\
&\quad \le C_1 \int_X \langle g(y), g(y)\rangle_{\mathbb{C}^d}\, d\nu(y) = C_1\, \|g\|^2_{L^2(Y,\nu,\mathbb{C}^d)},
\end{aligned}$$

giving $\left(\int_Y \int_X \|K_1^*(y,x)g(y)\|_{\mathbb{C}^d}^2 \, d\mu(x)\, d\nu(y)\right)^{1/2} \leq C_1^{1/2} \|g\|_{L^2(Y,\nu,\mathbb{C}^d)}$. All together this provides:

$$\left|\langle Tf, g\rangle_{L^2(Y,\nu,\mathbb{C}^d)}\right| \leq C_1^{1/2} C_2^{1/2} \|f\|_{L^2(X,\mu,\mathbb{C}^d)} \|g\|_{L^2(Y,\nu,\mathbb{C}^d)}$$

which by duality yields

$$\left\|T : L^2(X,\mu,\mathbb{C}^d) \to L^2(Y,\nu,\mathbb{C}^d)\right\| \leq C_1^{1/2} C_2^{1/2}.$$

□

B.6. Applications of Schur's Lemma Used in the Text

B.6.1. Schur's Test and Solutions to $\overline{\partial}$-problems. Here we characterize boundedness of the positive operators that arise as majorants of solution operators to the $\overline{\partial}$-equation on the unit ball. The case $c = 0$ of the following lemma is [**346**, Theorem 2.10] and was sketched above in Theorem B.6.

LEMMA B.10. *Let $a, b, c, t \in \mathbb{R}$. Then the operator*

$$T_{a,b,c} f(z) = \int_{\mathbb{B}_n} \frac{\left(1-|z|^2\right)^a \left(1-|w|^2\right)^b \left(\sqrt{\triangle(w,z)}\right)^c}{|1-w\overline{z}|^{n+1+a+b+c}} f(w)\, dV(w)$$

is bounded on $L^p\left(\mathbb{B}_n, \left(1-|w|^2\right)^t dV(w)\right)$ if and only if $c > -2n$ and

$$-pa < t+1 < p(b+1). \tag{B.3}$$

REMARK B.11. *Recall from* (15.23) *that $\triangle(w,z)$ is described by many equivalent formulas and that it has a rich set of transformation properties. In this proof we use $\triangle(w,z) = |1-w\overline{z}|^2 |\varphi_w(z)|^2$.*

PROOF. We sketch the proof for the case $c \neq 0$ when $p = 2$ and $t = -n-1$. Let $\psi_\varepsilon(\zeta) = \left(1-|\zeta|^2\right)^\varepsilon$.

We compute conditions on a, b, c and ε such that we have

$$T_{a,b,c}\psi_\varepsilon(z) \lesssim \psi_\varepsilon(z) \text{ and } T^*_{a,b,c}\psi_\varepsilon(w) \lesssim \psi_\varepsilon(w), \qquad z, w \in \mathbb{B}_n,$$

where $T^*_{a,b,c}$ denotes the adjoint relative to $L^2(\mathbb{B}_n, \lambda_n)$. For this we take $\varepsilon \in \mathbb{R}$ and compute

$$T_{a,b,c}\psi_\varepsilon(z) = \int_{\mathbb{B}_n} \frac{\left(1-|z|^2\right)^a \left(1-|w|^2\right)^{n+1+b+\varepsilon} |\varphi_z(w)|^c}{|1-w\overline{z}|^{n+1+a+b}} d\lambda_n(w).$$

Note that the integral defining $T_{a,b,c}\psi_\varepsilon(z)$ is finite if and only if $\varepsilon > -b-1$. Now in this integral make the change of variable $w = \varphi_z(\zeta)$ and use that λ_n is invariant to obtain

$$T_{a,b,c}\psi_\varepsilon(z) = \int_{\mathbb{B}_n} \frac{(1-|z|^2)^a \left(1-|\varphi_z(\zeta)|^2\right)^{n+1+b+\varepsilon}}{\left|1-\overline{\varphi_z(\zeta)}z\right|^{n+1+a+b}} \frac{|\zeta|^c}{(1-|\zeta|^2)^{n+1}} dV(\zeta).$$

Plugging the identities

$$1-\varphi_z(\zeta)\overline{z}=1-\langle\varphi_z(\zeta),\varphi_z(0)\rangle=\frac{1-|z|^2}{1-\zeta\overline{z}},$$

$$1-|\varphi_z(\zeta)|^2=1-\langle\varphi_z(\zeta),\varphi_z(\zeta)\rangle=\frac{\left(1-|z|^2\right)\left(1-|\zeta|^2\right)}{|1-\zeta\overline{z}|^2},$$

into the formula for $T_{a,b,c}\psi_\varepsilon(z)$ we obtain

$$T_{a,b,c}\psi_\varepsilon(z)=\psi_\varepsilon(z)\int_{\mathbb{B}_n}\frac{\left(1-|\zeta|^2\right)^{b+\varepsilon}|\zeta|^c}{|1-\zeta\overline{z}|^{n+1+b-a+2\varepsilon}}dV(\zeta).$$

Then from Lemma B.5 we obtain that

$$\sup_{z\in\mathbb{B}_n}\int_{\mathbb{B}_n}\frac{\left(1-|\zeta|^2\right)^{\alpha}}{|1-\zeta\overline{z}|^{\beta}}dV(\zeta)<\infty$$

if and only if $\beta-\alpha<n+1$. Provided $c>-2n$ it is now easy to see that we also have

$$\sup_{z\in\mathbb{B}_n}\int_{\mathbb{B}_n}\frac{\left(1-|\zeta|^2\right)^{\alpha}|\zeta|^c}{|1-\zeta\overline{z}|^{\beta}}dV(\zeta)<\infty$$

if and only if $\beta-\alpha<n+1$. It now follows from the above that

$$T_{a,b,c}\psi_\varepsilon(z)\lesssim\psi_\varepsilon(z),\qquad z\in\mathbb{B}_n,$$

if and only if

$$-b-1<\varepsilon<a.$$

Arguing as above and provided $c>-2n$, we obtain

$$T^*_{a,b,c}\psi_\varepsilon(w)\lesssim\psi_\varepsilon(w),\qquad w\in\mathbb{B}_n,$$

if and only if

$$-a+n<\varepsilon<b+n+1.$$

Altogether then there is $\varepsilon\in\mathbb{R}$ such that $h=\sqrt{\psi_\varepsilon}$ is a Schur function for $T_{a,b,c}$ on $L^2(\mathbb{B}_n,\lambda_n)$ in Schur's Test if and only if

$$\max\{-a+n,-b-1\}<\min\{a,b+n+1\}.$$

This is equivalent to $-2a<-n<2(b+1)$, which is (B.3) in the case $p=2,t=-n-1$. This completes the proof (in this case) that (B.3) implies the boundedness of $T_{a,b,c}$ on $L^2(\mathbb{B}_n,\lambda_n)$. The converse is easy - see for example the argument for the case $c=0$ of Theorem B.6. □

Because $|1-w\overline{z}|\leq 2$ this lemma has the immediate consequence.

COROLLARY B.12. *The operator*

$$T_{a,b,c,d}f(z)=\int_{\mathbb{B}_n}\frac{\left(1-|z|^2\right)^a\left(1-|w|^2\right)^b\left(\sqrt{\triangle(w,z)}\right)^c}{|1-w\overline{z}|^{n+1+a+b+c+d}}f(w)\,dV(w)$$

is bounded on $L^p\left(\mathbb{B}_n,\left(1-|w|^2\right)^t dV(w)\right)$ *if* $c>-2n$, $d\leq 0$ *and* (B.3) *holds.*

B.6.2. Application of the Bilinear Schur Estimate. When studying the bilinear Hankel form on the Dirichlet space, we are required to estimate certain bilinear forms. This leads to an application of Theorem B.8. Recall that we are working with the dyadic tree on the disk, and need to show that if $\mathcal{A}, \mathcal{B} \subset \mathcal{T}$ are well separated then a certain bilinear operator mapping on $\ell^2(\mathcal{A}) \times \ell^2(\mathcal{B})$ maps boundedly into $L^2(\mathbb{D})$.

LEMMA B.13. *Suppose $\mathcal{A}$ and $\mathcal{B}$ are subsets of T, $h \in \ell^2(\mathcal{A})$ and $k \in \ell^2(\mathcal{B})$, and $1/2 < \alpha < 1$. Suppose further that $\mathcal{A}$ and $\mathcal{B}$ satisfy the separation condition: $\forall \kappa \in \mathcal{A}$, $\gamma \in \mathcal{B}$ we have*

$$|\kappa - \gamma| \geq (1 - |\gamma|^2)^{\alpha}. \tag{B.4}$$

Then the bilinear map of (h, k) to functions on the disk given by

$$T(h,k)(z) = \left(\sum_{\kappa\in\mathcal{A}} h(\kappa) \frac{(1-|\kappa|^2)^{1+s}}{|1-\overline{\kappa}z|^{2+s}}\right)\left(\sum_{\gamma\in\mathcal{B}} k(\gamma) \frac{(1-|\gamma|^2)^{1+s}}{|1-\overline{\gamma}z|^{1+s}}\right)$$

is bounded from $\ell^2(\mathcal{A}) \times \ell^2(\mathcal{B})$ to $L^2(\mathbb{D})$.

PROOF. We will verify the hypotheses of the Theorem B.8. The kernel function here is

$$H(z,\kappa,\gamma) = \frac{(1-|\kappa|^2)^{1+s}}{|1-\overline{\kappa}z|^{2+s}} \frac{(1-|\gamma|^2)^{1+s}}{|1-\overline{\gamma}z|^{1+s}}, \quad z \in \mathbb{D}, \kappa \in \mathcal{A}, \gamma \in \mathcal{B},$$

with Lebesgue measure on $\mathbb{D}$, and counting measure on $\mathcal{A}$ and $\mathcal{B}$. We will take as Schur functions

$$h(z) = (1-|z|^2)^{-1/4},\ k(\kappa) = (1-|\kappa|^2)^{1/4} \text{ and } m(\gamma) = (1-|\gamma|^2)^{\varepsilon/2},$$

on $\mathbb{D}$, $\mathcal{A}$ and $\mathcal{B}$ respectively, where $\varepsilon = \varepsilon(\alpha, s) > 0$ will be chosen sufficiently small later. We must then verify

$$\sum_{\kappa\in\mathcal{A}}\sum_{\gamma\in\mathcal{B}} \frac{(1-|\kappa|^2)^{3/2+s}}{|1-\overline{\kappa}z|^{2+s}} \frac{(1-|\gamma|^2)^{1+\varepsilon+s}}{|1-\overline{\gamma}z|^{1+s}} \leq A^2(1-|z|^2)^{-1/2}, \tag{B.5}$$

for $z \in \mathbb{D}$, and

$$\int_{\mathbb{D}} \frac{(1-|\kappa|^2)^{1+s}}{|1-\overline{\kappa}z|^{2+s}} \frac{(1-|\gamma|^2)^{1+s}}{|1-\overline{\gamma}z|^{1+s}} (1-|z|^2)^{-1/2} dA \leq B^2(1-|\kappa|^2)^{1/2}(1-|\gamma|^2)^{\varepsilon}, \tag{B.6}$$

for $\kappa \in \mathcal{A}$ and $\gamma \in \mathcal{B}$.

To prove (B.5) we write

$$\sum_{\kappa\in\mathcal{A}}\sum_{\gamma\in\mathcal{B}} \frac{(1-|\kappa|^2)^{3/2+s}}{|1-\overline{\kappa}z|^{2+s}} \frac{(1-|\gamma|^2)^{1+\varepsilon+s}}{|1-\overline{\gamma}z|^{1+s}} =$$

$$\left(\sum_{\kappa\in\mathcal{A}} \frac{(1-|\kappa|^2)^{3/2+s}}{|1-\overline{\kappa}z|^{2+s}}\right)\left(\sum_{\gamma\in\mathcal{B}} \frac{(1-|\gamma|^2)^{1+\varepsilon+s}}{|1-\overline{\gamma}z|^{1+s}}\right).$$

Using (B.1) we then obtain

$$\sum_{\kappa\in\mathcal{A}} \frac{(1-|\kappa|^2)^{3/2+s}}{|1-\overline{\kappa}z|^{2+s}} \lesssim \int_{\mathbb{D}} \frac{(1-|w|^2)^{-1/2+s}}{|1-\overline{w}z|^{2+s}} dA \lesssim (1-|z|^2)^{-1/2}$$

and

$$\sum_{\gamma\in\mathcal{B}}\frac{(1-|\gamma|^2)^{1+\varepsilon+s}}{|1-\overline{\gamma}z|^{1+s}}\lesssim\int_{\zeta\in V_G}\frac{(1-|\zeta|^2)^{-1+\varepsilon+s}}{|1-\overline{\zeta}z|^{1+s}}dA\lesssim 1,$$

which yields (B.5).

We now prove (B.6) We will make repeated use of (B.4) as well as its consequence via the triangle inequality: $\forall\kappa\in\mathcal{A}$, $\gamma\in\mathcal{B}$ $(1-|\kappa|^2)\lesssim|\kappa-\gamma|$. We set $\kappa^*=\kappa/|\kappa|$, $\gamma^*=\gamma/|\gamma|$.

$$\begin{aligned}
&\int_{\mathbb{D}}\frac{(1-|\kappa|^2)^{1+s}}{|1-\overline{\kappa}z|^{2+s}}\frac{(1-|\gamma|^2)^{1+s}}{|1-\overline{\gamma}z|^{1+s}}(1-|z|^2)^{-1/2}dA\\
&=\int\limits_{|z-\gamma^*|\leq 1-|\gamma|^2}+\int\limits_{1-|\gamma|^2\leq|z-\gamma^*|\leq\frac{1}{2}|\kappa-\gamma|}\\
&+\int\limits_{|z-\kappa^*|\leq 1-|\kappa|^2}+\int\limits_{1-|\kappa|^2\leq|z-\kappa^*|\leq\frac{1}{2}|\kappa-\gamma|}+\int\limits_{|z-\gamma^*|,|z-\kappa^*|\geq|\kappa-\gamma|}\cdots dA\\
&=I+II+III+IV+V.
\end{aligned}$$

We have

$$\begin{aligned}
I&\approx\frac{(1-|\kappa|^2)^{1+s}}{|\kappa-\gamma|^{2+s}}\int_{|z-\gamma^*|\leq 1-|\gamma|^2}(1-|z|^2)^{-1/2}dA\\
&\approx\frac{(1-|\kappa|^2)^{1+s}(1-|\gamma|^2)^{3/2}}{|\kappa-\gamma|^{2+s}}\lesssim(1-|\kappa|^2)^{1/2}(1-|\gamma|^2)^{3(1-\alpha)/2}.
\end{aligned}$$

Similarly we have

$$\begin{aligned}
II&\approx\frac{(1-|\kappa|^2)^{1+s}(1-|\gamma|^2)^{1+s}}{|\kappa-\gamma|^{2+s}}\int\limits_{1-|\gamma|^2\leq|z-\gamma^*|\leq\frac{1}{2}|\kappa-\gamma|}\frac{(1-|z|^2)^{-1/2}}{|z-\gamma^*|^{1+s}}dA\\
&\approx\frac{(1-|\kappa|^2)^{1+s}(1-|\gamma|^2)^{1+s}}{|\kappa-\gamma|^{2+s}}(1-|\gamma|^2)^{1/2-s}\\
&=\frac{(1-|\kappa|^2)^{1+s}(1-|\gamma|^2)^{3/2}}{|\kappa-\gamma|^{2+s}}\leq C(1-|\kappa|^2)^{1/2}(1-|\gamma|^2)^{3(1-\alpha)/2}.
\end{aligned}$$

Continuing we obtain

$$III\approx\frac{(1-|\kappa|^2)^{1/2}(1-|\gamma|^2)^{1+s}}{|\kappa-\gamma|^{1+s}}\leq C(1-|\kappa|^2)^{1/2}(1-|\gamma|^2)^{(1+s)(1-\alpha)},$$

and similarly,

$$IV\lesssim(1-|\kappa|^2)^{1/2}(1-|\gamma|^2)^{\varepsilon},$$

for some $\varepsilon>0$. Finally

$$\begin{aligned}
V&\approx\int\limits_{|z-\gamma^*|,|z-\kappa^*|\geq|\kappa-\gamma|}\frac{(1-|\kappa|^2)^{1+s}}{|z-\kappa^*|^{2+s}}\frac{(1-|\gamma|^2)^{1+s}}{|z-\gamma^*|^{1+s}}(1-|z|^2)^{-1/2}dA\\
&\approx\frac{(1-|\kappa|^2)^{1+s}(1-|\gamma|^2)^{1+s}}{|\kappa-\gamma|^{3/2+2s}}\\
&\lesssim(1-|\kappa|^2)^{1/2}(1-|\gamma|^2)^{(1+s)(1-\alpha)}.
\end{aligned}$$

□

Bibliography

[1] D. R. Adams and L. I. Hedberg, *Function spaces and potential theory*, Grundlehren der Mathematischen Wissenschaften [Fundamental Principles of Mathematical Sciences], vol. 314, Springer-Verlag, Berlin, 1996. MR1411441

[2] J. Agler, *A disconjugacy theorem for Toeplitz operators*, Amer. J. Math. **112** (1990), no. 1, 1–14, DOI 10.2307/2374849. MR1037599

[3] J. Agler, *Some interpolation theorems of Nevanlinna–Pick type* (1988), preprint.

[4] J. Agler and J. E. McCarthy, *Pick interpolation and Hilbert function spaces*, Graduate Studies in Mathematics, vol. 44, American Mathematical Society, Providence, RI, 2002. MR1882259

[5] J. Agler and M. Stankus, *m-isometric transformations of Hilbert space. III*, Integral Equations Operator Theory **24** (1996), no. 4, 379–421, DOI 10.1007/BF01191619. MR1382018

[6] J. Agler and M. Stankus, *m-isometric transformations of Hilbert space. II*, Integral Equations Operator Theory **23** (1995), no. 1, 1–48, DOI 10.1007/BF01261201. MR1346617

[7] J. Agler and M. Stankus, *m-isometric transformations of Hilbert space. I*, Integral Equations Operator Theory **21** (1995), no. 4, 383–429, DOI 10.1007/BF01222016. MR1321694

[8] P. Ahern, *Exceptional sets for holomorphic Sobolev functions*, Michigan Math. J. **35** (1988), no. 1, 29–41, DOI 10.1307/mmj/1029003680. MR931938

[9] P. R. Ahern and D. N. Clark, *On inner functions with H^p-derivative*, Michigan Math. J. **21** (1974), 115–127. MR0344479

[10] P. Ahern and W. Cohn, *Exceptional sets for Hardy Sobolev functions, $p > 1$*, Indiana Univ. Math. J. **38** (1989), no. 2, 417–453, DOI 10.1512/iumj.1989.38.38020. MR997390

[11] T. Äkkinen and K. Rajala, *Exponential integrability of mappings of finite distortion*, Rev. Mat. Iberoam. **31** (2015), no. 4, 1459–1476, DOI 10.4171/RMI/876. MR3438397

[12] A. Aleman, *A class of integral operators on spaces of analytic functions*, Topics in complex analysis and operator theory, Univ. Málaga, Málaga, 2007, pp. 3–30. MR2394654

[13] A. Aleman, *The multiplication operator on Hilbert spaces of analytic functions*, Habilitationsschrift, Fern Universität, Hagen, 1993.

[14] A. Aleman, *Hilbert spaces of analytic functions between the Hardy and the Dirichlet space*, Proc. Amer. Math. Soc. **115** (1992), no. 1, 97–104, DOI 10.2307/2159570. MR1079693

[15] A. Aleman, M. Hartz, J. E. McCarthy, and S. Richter, *Interpolating sequences in spaces with the complete Pick property*, preprint, available at `arxiv.org/abs/1701.04885`.

[16] A. Aleman and K.-M. Perfekt, *Hankel forms and embedding theorems in weighted Dirichlet spaces*, Int. Math. Res. Not. IMRN **19** (2012), 4435–4448, DOI 10.1093/imrn/rnr195. MR2981715

[17] A. Aleman, S. Richter, and C. Sundberg, *Invariant subspaces for the backward shift on Hilbert spaces of analytic functions with regular norm*, Bergman spaces and related topics in complex analysis, Contemp. Math., vol. 404, Amer. Math. Soc., Providence, RI, 2006, pp. 1–25, DOI 10.1090/conm/404/07631. MR2244001

[18] D. Alpay and H. T. Kaptanoğlu, *Toeplitz operators on Arveson and Dirichlet spaces*, Integral Equations Operator Theory **58** (2007), no. 1, 1–33, DOI 10.1007/s00020-007-1493-1. MR2312443

[19] É. Amar, *Interpolating sequences and Carleson measures in the Hardy-Sobolev spaces of the ball in $\mathbb{C}^n$*, Studia Math. **241** (2018), no. 2, 101–133, DOI 10.4064/sm8405-8-2017. MR3744980

[20] E. Amar, *A Carleson type condition for interpolating sequences in the Hardy spaces of the ball of $\mathbb{C}^n$*, Publ. Mat. **53** (2009), no. 2, 481–488, DOI 10.5565/PUBLMAT_53209_09. MR2543861

[21] E. Amar, *On the Toëplitz corona problem*, Publ. Mat. **47** (2003), no. 2, 489–496, DOI 10.5565/PUBLMAT_47203_11. MR2006496

[22] É. Amar, *On the corona problem*, J. Geom. Anal. **1** (1991), no. 4, 291–305, DOI 10.1007/BF02921307. MR1129344

[23] C.-G. Ambrozie and D. Timotin, *A von Neumann type inequality for certain domains in* $\mathbf{C}^n$, Proc. Amer. Math. Soc. **131** (2003), no. 3, 859–869, DOI 10.1090/S0002-9939-02-06321-9. MR1937424

[24] J. M. Anderson, J. Clunie, and Ch. Pommerenke, *On Bloch functions and normal functions*, J. Reine Angew. Math. **270** (1974), 12–37. MR0361090

[25] J. Anderson, *Extensions, restrictions, and representations of states on* C^**-algebras*, Trans. Amer. Math. Soc. **249** (1979), no. 2, 303–329, DOI 10.2307/1998793. MR525675

[26] M. E. L. Andersson, *On the* H^p *corona problem*, Bull. Sci. Math. **118** (1994), no. 3, 287–306. MR1284781

[27] M. Andersson, *The* H^2 *corona problem and* $\overline{\partial}_b$ *in weakly pseudoconvex domains*, Trans. Amer. Math. Soc. **342** (1994), no. 1, 241–255, DOI 10.2307/2154692. MR1145727

[28] M. Andersson and H. Carlsson, Q_p *spaces in strictly pseudoconvex domains*, J. Anal. Math. **84** (2001), 335–359, DOI 10.1007/BF02788114. MR1849206

[29] M. Andersson and H. Carlsson, *Estimates of solutions of the* H^p *and BMOA corona problem*, Math. Ann. **316** (2000), no. 1, 83–102, DOI 10.1007/s002080050005. MR1735080

[30] M. Andersson and H. Carlsson, *Wolff type estimates and the* H^p *corona problem in strictly pseudoconvex domains*, Ark. Mat. **32** (1994), no. 2, 255–276, DOI 10.1007/BF02559572. MR1318533

[31] V. V. Andreev, *An extremal function for the Chang-Marshall inequality over the Beurling functions*, Proc. Amer. Math. Soc. **133** (2005), no. 7, 2069–2076, DOI 10.1090/S0002-9939-05-07712-9. MR2137873

[32] V. V. Andreev and A. L. Matheson, *Quadratic extremal problems on the Dirichlet space*, Complex Var. Theory Appl. **48** (2003), no. 9, 767–776, DOI 10.1080/0278107031000154885. MR2002238

[33] V. V. Andreev and A. Matheson, *Extremal functions and the Chang-Marshall inequality*, Pacific J. Math. **162** (1994), no. 2, 233–246. MR1251899

[34] J. Arazy and S. D. Fisher, *The uniqueness of the Dirichlet space among Möbius-invariant Hilbert spaces*, Illinois J. Math. **29** (1985), no. 3, 449–462. MR786732

[35] J. Arazy, S. D. Fisher, S. Janson, and J. Peetre, *An identity for reproducing kernels in a planar domain and Hilbert-Schmidt Hankel operators*, J. Reine Angew. Math. **406** (1990), 179–199. MR1048240

[36] J. Arazy, S. D. Fisher, and J. Peetre, *Hankel operators on weighted Bergman spaces*, Amer. J. Math. **110** (1988), no. 6, 989–1053, DOI 10.2307/2374685. MR970119

[37] N. Arcozzi, *Carleson measures for analytic Besov spaces: the upper triangle case*, JIPAM. J. Inequal. Pure Appl. Math. **6** (2005), no. 1, Article 13, 15. MR2122940

[38] N. Arcozzi, D. Blasi, and J. Pau, *Corrigendum on "Interpolating sequences on analytic Besov type spaces" [MR2541368]*, Indiana Univ. Math. J. **59** (2010), no. 6, 1931–1934, DOI 10.1512/iumj.2010.59.4724. MR2919744

[39] N. Arcozzi, P. Mozolyako, K.-M. Perfekt, S. Richter, and G. Sarfatti, *Some Hilbert spaces related with the Dirichlet space*, Concr. Oper. **3** (2016), 94–101, DOI 10.1515/conop-2016-0011. MR3518370

[40] N. Arcozzi, D. Blasi, and J. Pau, *Interpolating sequences on analytic Besov type spaces*, Indiana Univ. Math. J. **58** (2009), no. 3, 1281–1318, DOI 10.1512/iumj.2009.58.3589. MR2541368

[41] N. Arcozzi and R. Rochberg, *Topics in dyadic Dirichlet spaces*, New York J. Math. **10** (2004), 45–67. MR2052364

[42] N. Arcozzi, R. Rochberg, and E. T. Sawyer, *Onto Interpolating Sequences for the Dirichlet Space* (2016), available at `arXiv:1605.02730`.

[43] N. Arcozzi, R. Rochberg, and E. Sawyer, *Two variations on the Drury-Arveson space*, Hilbert spaces of analytic functions, CRM Proc. Lecture Notes, vol. 51, Amer. Math. Soc., Providence, RI, 2010, pp. 41–58. MR2648865

[44] N. Arcozzi, R. Rochberg, and E. Sawyer, *Capacity, Carleson measures, boundary convergence, and exceptional sets*, Perspectives in partial differential equations, harmonic analysis

and applications, Proc. Sympos. Pure Math., vol. 79, Amer. Math. Soc., Providence, RI, 2008, pp. 1–20, DOI 10.1090/pspum/079/2500486. MR2500486

[45] N. Arcozzi, R. Rochberg, and E. Sawyer, *Carleson measures for the Drury-Arveson Hardy space and other Besov-Sobolev spaces on complex balls*, Adv. Math. **218** (2008), no. 4, 1107–1180, DOI 10.1016/j.aim.2008.03.001. MR2419381

[46] N. Arcozzi, R. Rochberg, and E. Sawyer, *The diameter space, a restriction of the Drury-Arveson-Hardy space*, Function spaces, Contemp. Math., vol. 435, Amer. Math. Soc., Providence, RI, 2007, pp. 21–42, DOI 10.1090/conm/435/08363. MR2359415

[47] N. Arcozzi, R. Rochberg, and E. Sawyer, *Carleson measures and interpolating sequences for Besov spaces on complex balls*, Mem. Amer. Math. Soc. **182** (2006), no. 859, vi+163, DOI 10.1090/memo/0859. MR2229732

[48] N. Arcozzi, R. Rochberg, and E. Sawyer, *Carleson measures for analytic Besov spaces*, Rev. Mat. Iberoamericana **18** (2002), no. 2, 443–510, DOI 10.4171/RMI/326. MR1949836

[49] N. Arcozzi, R. Rochberg, E. T. Sawyer, and B. D. Wick, *Potential theory on trees, graphs and Ahlfors-regular metric spaces*, Potential Anal. **41** (2014), no. 2, 317–366, DOI 10.1007/s11118-013-9371-8. MR3232028

[50] N. Arcozzi, R. Rochberg, E. Sawyer, and B. D. Wick, *Distance functions for reproducing kernel Hilbert spaces*, Function spaces in modern analysis, Contemp. Math., vol. 547, Amer. Math. Soc., Providence, RI, 2011, pp. 25–53, DOI 10.1090/conm/547/10805. MR2856478

[51] N. Arcozzi, R. Rochberg, E. T. Sawyer, and B. D. Wick, *The Dirichlet space: a survey*, New York J. Math. **17A** (2011), 45–86. MR2782728

[52] N. Arcozzi, R. Rochberg, E. Sawyer, and B. D. Wick, *Function spaces related to the Dirichlet space*, J. Lond. Math. Soc. (2) **83** (2011), no. 1, 1–18, DOI 10.1112/jlms/jdq053. MR2763941

[53] N. Arcozzi, R. Rochberg, E. Sawyer, and B. D. Wick, *Bilinear forms on the Dirichlet space*, Anal. PDE **3** (2010), no. 1, 21–47, DOI 10.2140/apde.2010.3.21. MR2663410

[54] N. Aronszajn, *Theory of reproducing kernels*, Trans. Amer. Math. Soc. **68** (1950), 337–404, DOI 10.2307/1990404. MR0051437

[55] W. Arveson, *Quotients of standard Hilbert modules*, Trans. Amer. Math. Soc. **359** (2007), no. 12, 6027–6055, DOI 10.1090/S0002-9947-07-04209-2. MR2336315

[56] W. Arveson, *Interpolation problems in nest algebras*, J. Functional Analysis **20** (1975), no. 3, 208–233, DOI 10.1016/0022-1236(75)90041-5. MR0383098

[57] S. Axler, *Interpolation by multipliers of the Dirichlet space*, Quart. J. Math. Oxford Ser. (2) **43** (1992), no. 172, 409–419, DOI 10.1093/qmathj/43.4.409. MR1188383

[58] S. Axler, *The Bergman space, the Bloch space, and commutators of multiplication operators*, Duke Math. J. **53** (1986), no. 2, 315–332, DOI 10.1215/S0012-7094-86-05320-2. MR850538

[59] S. Axler, P. Bourdon, and W. Ramey, *Harmonic function theory*, Graduate Texts in Mathematics, vol. 137, Springer-Verlag, New York, 1992. MR1184139

[60] J. A. Ball, T. T. Trent, and V. Vinnikov, *Interpolation and commutant lifting for multipliers on reproducing kernel Hilbert spaces*, Operator theory and analysis (Amsterdam, 1997), Oper. Theory Adv. Appl., vol. 122, Birkhäuser, Basel, 2001, pp. 89–138.

[61] G. Bao, Z. Lou, R. Qian, and H. Wulan, *On multipliers of Dirichlet type spaces*, Complex Anal. Oper. Theory **9** (2015), no. 8, 1701–1732, DOI 10.1007/s11785-015-0444-0. MR3415136

[62] A. F. Beardon and D. Minda, *The hyperbolic metric and geometric function theory*, Quasiconformal mappings and their applications, Narosa, New Delhi, 2007, pp. 9–56. MR2492498

[63] F. Beatrous, *Boundary estimates for derivatives of harmonic functions*, Studia Math. **98** (1991), no. 1, 53–71, DOI 10.4064/sm-98-1-53-71. MR1110098

[64] F. Beatrous and J. Burbea, *Holomorphic Sobolev spaces on the ball*, Dissertationes Math. (Rozprawy Mat.) **276** (1989), 60. MR1010151

[65] I. Benjamini and Y. Peres, *Random walks on a tree and capacity in the interval* (English, with English and French summaries), Ann. Inst. H. Poincaré Probab. Statist. **28** (1992), no. 4, 557–592. MR1193085

[66] H. Bercovici, C. Foias, and C. Pearcy, *Dual algebras with applications to invariant subspaces and dilation theory*, CBMS Regional Conference Series in Mathematics, vol. 56, Published for the Conference Board of the Mathematical Sciences, Washington, DC; by the American Mathematical Society, Providence, RI, 1985. MR787041

[67] S. Bergman, *The kernel function and conformal mapping*, Second, revised edition, American Mathematical Society, Providence, R.I., 1970. Mathematical Surveys, No. V. MR0507701

[68] A. Beurling, *On two problems concerning linear transformations in Hilbert space*, Acta Math. **81** (1948), 17, DOI 10.1007/BF02395019. MR0027954

[69] A. Beurling, *Ensembles exceptionnels* (French), Acta Math. **72** (1940), 1–13, DOI 10.1007/BF02546325. MR0001370

[70] A. Beurling, *Études sur un problème de majoration*, Doctoral thesis, Fern Universität, Hagen, 1933.

[71] A. Beurling and J. Deny, *Espaces de Dirichlet. I. Le cas élémentaire* (French), Acta Math. **99** (1958), 203–224, DOI 10.1007/BF02392426. MR0098924

[72] C. J. Bishop, *Interpolating sequences for the Dirichlet space and its multipliers* (1994), preprint, available at `https://www.math.stonybrook.edu/~bishop/papers/mult.pdf`.

[73] D. Blasi and J. Pau, *A characterization of Besov-type spaces and applications to Hankel-type operators*, Michigan Math. J. **56** (2008), no. 2, 401–417, DOI 10.1307/mmj/1224783520. MR2492401

[74] S. Bochner, *Über orthogonale Systeme analytischer Funktionen* (German), Math. Z. **14** (1922), no. 1, 180–207, DOI 10.1007/BF01215898. MR1544547

[75] B. Bøe, *An interpolation theorem for Hilbert spaces with Nevanlinna-Pick kernel*, Proc. Amer. Math. Soc. **133** (2005), no. 7, 2077–2081, DOI 10.1090/S0002-9939-05-07722-1. MR2137874

[76] B. Böe, *A norm on the holomorphic Besov space*, Proc. Amer. Math. Soc. **131** (2003), no. 1, 235–241, DOI 10.1090/S0002-9939-02-06529-2. MR1929043

[77] B. Böe, *Interpolating sequences for Besov spaces*, J. Funct. Anal. **192** (2002), no. 2, 319–341, DOI 10.1006/jfan.2002.3905. MR1923404

[78] B. Bøe and A. Nicolau, *Interpolation by functions in the Bloch space*, J. Anal. Math. **94** (2004), 171–194, DOI 10.1007/BF02789046. MR2124459

[79] K. Bogdan, *On the zeros of functions with finite Dirichlet integral*, Kodai Math. J. **19** (1996), no. 1, 7–16, DOI 10.2996/kmj/1138043542. MR1374458

[80] A. Bonami, M. M. Peloso, and F. Symesak, *On Hankel operators on Hardy and Bergman spaces and related questions*, Actes des Rencontres d'Analyse Complexe (Poitiers-Futuroscope, 1999), Atlantique, Poitiers, 2002, pp. 131–153. MR1944202

[81] A. Borichev, *Boundary behavior in Dirichlet-type spaces*, Uppsala University Department of Mathematics Report **3** (1984).

[82] A. Borichev, *A note on Dirichlet-type spaces*, Uppsala University Department of Mathematics Report **11** (1983).

[83] P. S. Bourdon, *Cellular-indecomposable operators and Beurling's theorem*, Michigan Math. J. **33** (1986), no. 2, 187–193, DOI 10.1307/mmj/1029003347. MR837576

[84] J. Bourgain, *On the radial variation of bounded analytic functions on the disc*, Duke Math. J. **69** (1993), no. 3, 671–682, DOI 10.1215/S0012-7094-93-06928-1. MR1208816

[85] L. Brown and W. Cohn, *Some examples of cyclic vectors in the Dirichlet space*, Proc. Amer. Math. Soc. **95** (1985), no. 1, 42–46, DOI 10.2307/2045570. MR796443

[86] L. Brown and A. L. Shields, *Cyclic vectors in the Dirichlet space*, Trans. Amer. Math. Soc. **285** (1984), no. 1, 269–303, DOI 10.2307/1999483. MR748841

[87] J. Burbea, *A Dirichlet norm inequality and some inequalities for reproducing kernel spaces*, Proc. Amer. Math. Soc. **83** (1981), no. 2, 279–285, DOI 10.2307/2043511. MR624914

[88] A. Cantón, J. L. Fernández, D. Pestana, and J. M. Rodríguez, *On harmonic functions on trees*, Potential Anal. **15** (2001), no. 3, 199–244.

[89] L. Carleson, *Interpolations by bounded analytic functions and the corona problem*, Ann. of Math. (2) **76** (1962), 547–559, DOI 10.2307/1970375. MR0141789

[90] L. Carleson, *A representation formula for the Dirichlet integral*, Math. Z. **73** (1960), 190–196, DOI 10.1007/BF01162477. MR0112958

[91] L. Carleson, *An interpolation problem for bounded analytic functions*, Amer. J. Math. **80** (1958), 921–930, DOI 10.2307/2372840. MR0117349

[92] L. Carleson, *On the zeros of functions with bounded Dirichlet integrals*, Math. Z. **56** (1952), 289–295, DOI 10.1007/BF01174755. MR0051298

[93] L. Carleson, *Sets of uniqueness for functions regular in the unit circle*, Acta Math. **87** (1952), 325–345, DOI 10.1007/BF02392289. MR0050011

[94] P. Cartier, *Harmonic analysis on trees*, Harmonic analysis on homogeneous spaces (Proc. Sympos. Pure Math., Vol. XXVI, Williams Coll., Williamstown, Mass., 1972), Amer. Math. Soc., Providence, R.I., 1973, pp. 419–424. MR0338272

[95] C. Cascante and J. Fàbrega, *Bilinear forms on weighted Besov spaces*, J. Math. Soc. Japan **68** (2016), no. 1, 383–403, DOI 10.2969/jmsj/06810383. MR3454563

[96] C. Cascante, J. Fàbrega, and J. M. Ortega, *Bilinear forms on homogeneous Sobolev spaces*, J. Math. Anal. Appl. **457** (2018), no. 1, 722–750, DOI 10.1016/j.jmaa.2017.08.033. MR3702728

[97] C. Cascante and J. M. Ortega, *On a characterization of bilinear forms on the Dirichlet space*, Proc. Amer. Math. Soc. **140** (2012), no. 7, 2429–2440, DOI 10.1090/S0002-9939-2011-11409-6. MR2898705

[98] C. Cascante and J. M. Ortega, *Carleson measures on spaces of Hardy-Sobolev type*, Canad. J. Math. **47** (1995), no. 6, 1177–1200, DOI 10.4153/CJM-1995-060-4. MR1370513

[99] C. Cascante, J. M. Ortega, and I. E. Verbitsky, *Trace inequalities of Sobolev type in the upper triangle case*, Proc. London Math. Soc. (3) **80** (2000), no. 2, 391–414, DOI 10.1112/S0024611500012260. MR1734322

[100] J. G. Caughran, *Zeros of analytic functions with infinitely differentiable boundary values*, Proc. Amer. Math. Soc. **24** (1970), 700–704, DOI 10.2307/2037305. MR0252649

[101] J. G. Caughran, *Two results concerning the zeros of functions with finite Dirichlet integral*, Canad. J. Math. **21** (1969), 312–316, DOI 10.4153/CJM-1969-033-5. MR0236396

[102] G. A. Chacón, G. R. Chacón, and J. Giménez, *Composition operators on the Dirichlet space and related problems*, Bol. Asoc. Mat. Venez. **13** (2006), no. 2, 155–164. MR2302622

[103] I. Chalendar, E. A. Gallardo-Gutiérrez, and J. R. Partington, *Weighted composition operators on the Dirichlet space: boundedness and spectral properties*, Math. Ann. **363** (2015), no. 3-4, 1265–1279, DOI 10.1007/s00208-015-1195-y. MR3412358

[104] N. Chalmoukis, *Onto interpolation for the Dirichlet Space and for $W^{1,2}(\mathbb{D})$* (2018), preprint, available at `arXiv:1807.08193`.

[105] S.-Y. A. Chang and D. E. Marshall, *On a sharp inequality concerning the Dirichlet integral*, Amer. J. Math. **107** (1985), no. 5, 1015–1033, DOI 10.2307/2374345. MR805803

[106] P. Charpentier, *Formules explicites pour les solutions minimales de l'équation $\bar{\partial}u = f$ dans la boule et dans le polydisque de* $\mathbf{C}^n$ (French), Ann. Inst. Fourier (Grenoble) **30** (1980), no. 4, 121–154. MR599627

[107] Y. Chen, *Commuting Toeplitz operators on the Dirichlet space*, J. Math. Anal. Appl. **357** (2009), no. 1, 214–224, DOI 10.1016/j.jmaa.2009.03.068. MR2526821

[108] J. Cima and A. Matheson, *A nonlinear functional on the Dirichlet space*, J. Math. Anal. Appl. **191** (1995), no. 2, 380–401, DOI 10.1006/jmaa.1995.1136. MR1324020

[109] J. A. Cima and W. T. Ross, *The backward shift on the Hardy space*, Mathematical Surveys and Monographs, vol. 79, American Mathematical Society, Providence, RI, 2000. MR1761913

[110] L. A. Coburn, *Sharp Berezin Lipschitz estimates*, Proc. Amer. Math. Soc. **135** (2007), no. 4, 1163–1168, DOI 10.1090/S0002-9939-06-08569-8. MR2262921

[111] W. Cohn, S. H. Ferguson, and R. Rochberg, *Boundedness of higher order Hankel forms, factorization in potential spaces and derivations*, Proc. London Math. Soc. (3) **82** (2001), no. 1, 110–130, DOI 10.1112/S0024611500012727. MR1794259

[112] R. Coifman, P.-L. Lions, Y. Meyer, and S. Semmes, *Compensated compactness and Hardy spaces* (English, with English and French summaries), J. Math. Pures Appl. (9) **72** (1993), no. 3, 247–286. MR1225511

[113] W. S. Cohn and I. E. Verbitsky, *On the trace inequalities for Hardy-Sobolev functions in the unit ball of* $\mathbf{C}^n$, Indiana Univ. Math. J. **43** (1994), no. 4, 1079–1097, DOI 10.1512/iumj.1994.43.43047. MR1322611

[114] W. S. Cohn and I. E. Verbitsky, *Nonlinear potential theory on the ball, with applications to exceptional and boundary interpolation sets*, Michigan Math. J. **42** (1995), no. 1, 79–97, DOI 10.1307/mmj/1029005154. MR1322190

[115] R. R. Coifman and Y. Meyer, *Une généralisation du théorème de Calderón sur l'intégrale de Cauchy* (French), Fourier analysis (Proc. Sem., El Escorial, 1979), Asoc. Mat. Espa nola, vol. 1, Asoc. Mat. Espa nola, Madrid, 1980, pp. 87–116. MR582251

[116] R. R. Coifman and T. Murai, *Commutators on the potential-theoretic energy spaces*, Tohoku Math. J. (2) **40** (1988), no. 3, 397–407, DOI 10.2748/tmj/1178227981. MR957051

[117] R. R. Coifman and R. Rochberg, *Representation theorems for holomorphic and harmonic functions in* L^p, Representation theorems for Hardy spaces, Astérisque, vol. 77, Soc. Math. France, Paris, 1980, pp. 11–66. MR604369

[118] R. R. Coifman, R. Rochberg, and G. Weiss, *Factorization theorems for Hardy spaces in several variables*, Ann. of Math. (2) **103** (1976), no. 3, 611–635, DOI 10.2307/1970954. MR0412721

[119] R. R. Coifman and G. Weiss, *Extensions of Hardy spaces and their use in analysis*, Bull. Amer. Math. Soc. **83** (1977), no. 4, 569–645, DOI 10.1090/S0002-9904-1977-14325-5. MR0447954

[120] J. B. Conway, *Subnormal operators*, Research Notes in Mathematics, vol. 51, Pitman (Advanced Publishing Program), Boston, Mass.-London, 1981. MR634507

[121] Ş. Costea, E. T. Sawyer, and B. D. Wick, *The corona theorem for the Drury-Arveson Hardy space and other holomorphic Besov-Sobolev spaces on the unit ball in* $\mathbb{C}^n$, Anal. PDE **4** (2011), no. 4, 499–550, DOI 10.2140/apde.2011.4.499. MR3077143

[122] Ş. Costea, E. T. Sawyer, and B. D. Wick, *Electronic Supplement to: The corona theorem for the Drury-Arveson Hardy space and other holomorphic Besov-Sobolev spaces on the unit ball in* $\mathbb{C}^n$, Anal. PDE **4** (2011), no. 4, 499–550, available at `https://msp.org/apde/2011/4-4/apde-v4-n4-x01-suppl.pdf`.

[123] Ş. Costea, E. T. Sawyer, and B. D. Wick, *BMO estimates for the* $H^\infty(\mathbb{B}_n)$ *Corona problem*, J. Funct. Anal. **258** (2010), no. 11, 3818–3840, DOI 10.1016/j.jfa.2009.12.015. MR2606875

[124] C. C. Cowen and B. D. MacCluer, *Composition operators on spaces of analytic functions*, Studies in Advanced Mathematics, CRC Press, Boca Raton, FL, 1995. MR1397026

[125] F. Di Biase, *Fatou type theorems*, Progress in Mathematics, vol. 147, Birkhäuser Boston, Inc., Boston, MA, 1998. Maximal functions and approach regions. MR1483892

[126] F. Di Biase, *Review of Twomey, J. B. "The boundary values of a function in the Dirichlet space" [MR2760420]* (2012).

[127] A. E. Djrbashian and F. A. Shamoian, *Topics in the theory of* A^p_α *spaces*, Teubner-Texte zur Mathematik [Teubner Texts in Mathematics], vol. 105, BSB B. G. Teubner Verlagsgesellschaft, Leipzig, 1988. With German, French and Russian summaries. MR1021691

[128] J. Douglas, *Solution of the problem of Plateau*, Trans. Amer. Math. Soc. **33** (1931), no. 1, 263–321, DOI 10.2307/1989472. MR1501590

[129] R. G. Douglas, *Connections of the corona problem with operator theory and complex geometry*, The corona problem, Fields Inst. Commun., vol. 72, Springer, New York, 2014, pp. 47–68, DOI 10.1007/978-1-4939-1255-1_3. MR3329541

[130] R. G. Douglas, S. G. Krantz, E. T. Sawyer, S. Treil, and B. D. Wicks, *A history of the corona problem*, The corona problem, Fields Inst. Commun., vol. 72, Springer, New York, 2014, pp. 1–29, DOI 10.1007/978-1-4939-1255-1_1. MR3329539

[131] J. J. Duistermaat and Y. J. Lee, *Toeplitz operators on the Dirichlet space*, J. Math. Anal. Appl. **300** (2004), no. 1, 54–67, DOI 10.1016/j.jmaa.2004.05.031. MR2100238

[132] P. L. Duren, *Theory of* H^p *spaces*, Pure and Applied Mathematics, Vol. 38, Academic Press, New York-London, 1970. MR0268655

[133] P. Duren and A. Schuster, *Bergman spaces*, Mathematical Surveys and Monographs, vol. 100, American Mathematical Society, Providence, RI, 2004. MR2033762

[134] P. Duren and R. Weir, *The pseudohyperbolic metric and Bergman spaces in the ball*, Trans. Amer. Math. Soc. **359** (2007), no. 1, 63–76, DOI 10.1090/S0002-9947-06-04064-5. MR2247882

[135] D. E. Edmunds and W. D. Evans, *Hardy operators, function spaces and embeddings*, Springer Monographs in Mathematics, Springer-Verlag, Berlin, 2004. MR2091115

[136] O. El-Fallah, K. Kellay, J. Mashreghi, and T. Ransford, *A primer on the Dirichlet space*, Cambridge Tracts in Mathematics, vol. 203, Cambridge University Press, Cambridge, 2014. MR3185375

[137] O. El-Fallah, K. Kellay, J. Mashreghi, and T. Ransford, *A self-contained proof of the strong-type capacitary inequality for the Dirichlet space*, Complex analysis and potential theory, CRM Proc. Lecture Notes, vol. 55, Amer. Math. Soc., Providence, RI, 2012, pp. 1–20. MR2986889

[138] O. El-Fallah, K. Kellay, and T. Ransford, *Invariant subspaces of the Dirichlet space*, Hilbert spaces of analytic functions, CRM Proc. Lecture Notes, vol. 51, Amer. Math. Soc., Providence, RI, 2010, pp. 133–141. MR2648871

[139] O. El-Fallah, K. Kellay, and T. Ransford, *On the Brown-Shields conjecture for cyclicity in the Dirichlet space*, Adv. Math. **222** (2009), no. 6, 2196–2214, DOI 10.1016/j.aim.2009.07.011. MR2562781

[140] O. El-Fallah, K. Kellay, and T. Ransford, *Cyclicity in the Dirichlet space*, Ark. Mat. **44** (2006), no. 1, 61–86, DOI 10.1007/s11512-005-0008-z. MR2237211

[141] O. El-Fallah, Y. Elmadani, and K. Kellay, *Cyclicity and invariant subspaces in Dirichlet spaces*, J. Funct. Anal. **270** (2016), no. 9, 3262–3279, DOI 10.1016/j.jfa.2016.02.027. MR3475457

[142] L. D. Edholm and J. D. McNeal, *The Bergman projection on fat Hartogs triangles: L^p boundedness*, Proc. Amer. Math. Soc. **144** (2016), no. 5, 2185–2196, DOI 10.1090/proc/12878. MR3460177

[143] M. Essén, *Sharp estimates of uniform harmonic majorants in the plane*, Ark. Mat. **25** (1987), no. 1, 15–28, DOI 10.1007/BF02384434. MR918377

[144] X. Fang, *Invariant subspaces of the Dirichlet space and commutative algebra*, J. Reine Angew. Math. **569** (2004), 189–211, DOI 10.1515/crll.2004.025. MR2057238

[145] C. Fefferman, *Harmonic analysis and H^p spaces*, Studies in harmonic analysis (Proc. Conf., DePaul Univ., Chicago, Ill., 1974), Math. Assoc. Amer., Washington, D.C., 1976, pp. 38–75. MAA Stud. Math., Vol. 13. MR0461001

[146] C. Fefferman, *Characterizations of bounded mean oscillation*, Bull. Amer. Math. Soc. **77** (1971), 587–588, DOI 10.1090/S0002-9904-1971-12763-5. MR0280994

[147] H. G. Feichtinger and F. Voigtlaender, *From Frazier-Jawerth characterizations of Besov spaces to wavelets and decomposition spaces*, Functional analysis, harmonic analysis, and image processing: a collection of papers in honor of Björn Jawerth, Contemp. Math., vol. 693, Amer. Math. Soc., Providence, RI, 2017, pp. 185–216. MR3682610

[148] A. Feintuch, *Robust control theory in Hilbert space*, Applied Mathematical Sciences, vol. 130, Springer-Verlag, New York, 1998. MR1482802

[149] S. H. Ferguson and M. T. Lacey, *A characterization of product BMO by commutators*, Acta Math. **189** (2002), no. 2, 143–160, DOI 10.1007/BF02392840. MR1961195

[150] A. Figà-Talamanca and C. Nebbia, *Harmonic analysis and representation theory for groups acting on homogeneous trees*, London Mathematical Society Lecture Note Series, vol. 162, Cambridge University Press, Cambridge, 1991. MR1152801

[151] T. M. Flett, *A high-indices theorem*, Proc. London Math. Soc. (3) **7** (1957), 142–149, DOI 10.1112/plms/s3-7.1.142. MR0086914

[152] C. Foias, A. E. Frazho, I. Gohberg, and M. A. Kaashoek, *Metric constrained interpolation, commutant lifting and systems*, Operator Theory: Advances and Applications, vol. 100, Birkhäuser Verlag, Basel, 1998. MR1635831

[153] G. B. Folland and E. M. Stein, *Estimates for the $\bar{\partial}_b$ complex and analysis on the Heisenberg group*, Comm. Pure Appl. Math. **27** (1974), 429–522, DOI 10.1002/cpa.3160270403. MR0367477

[154] F. Forelli and W. Rudin, *Projections on spaces of holomorphic functions in balls*, Indiana Univ. Math. J. **24** (1974/75), 593–602, DOI 10.1512/iumj.1974.24.24044. MR0357866

[155] A. E. Frazho and W. Bhosri, *An operator perspective on signals and systems*, Operator Theory: Advances and Applications, vol. 204, Birkhäuser Verlag, Basel, 2010. Linear Operators and Linear Systems. MR2584037

[156] M. Frazier, B. Jawerth, and G. Weiss, *Littlewood-Paley theory and the study of function spaces*, CBMS Regional Conference Series in Mathematics, vol. 79, Published for the Conference Board of the Mathematical Sciences, Washington, DC; by the American Mathematical Society, Providence, RI, 1991. MR1107300

[157] O. Frostman, *Sur les produits de Blaschke* (French), Kungl. Fysiografiska Sällskapets i Lund Förhandlingar [Proc. Roy. Physiog. Soc. Lund] **12** (1942), no. 15, 169–182. MR0012127

[158] P. Galanopoulos and J. Pau, *Hankel operators on large weighted Bergman spaces*, Ann. Acad. Sci. Fenn. Math. **37** (2012), no. 2, 635–648, DOI 10.5186/aasfm.2012.3739. MR2987091

[159] S. R. Garcia, J. Mashreghi, and W. T. Ross, *Introduction to model spaces and their operators*, Cambridge Studies in Advanced Mathematics, vol. 148, Cambridge University Press, Cambridge, 2016. MR3526203

[160] E. Gagliardo, *On integral trasformations with positive kernel*, Proc. Amer. Math. Soc. **16** (1965), 429–434, DOI 10.2307/2034667. MR0177314

[161] J. B. Garnett, *Bounded analytic functions*, 1st ed., Graduate Texts in Mathematics, vol. 236, Springer, New York, 2007. MR2261424

[162] É. Ghys, *Poincaré and his disk*, The scientific legacy of Poincaré, Hist. Math., vol. 36, Amer. Math. Soc., Providence, RI, 2010, pp. 17–46. MR2647617

[163] A. M. Gleason, *Function algebras*, Seminars on Analytic Functions, vol. 2, Institute for Advanced Study, 1957.

[164] L. Grafakos, *Modern Fourier analysis*, 3rd ed., Graduate Texts in Mathematics, vol. 250, Springer, New York, 2014. MR3243741

[165] L. Grafakos, *Classical Fourier analysis*, 3rd ed., Graduate Texts in Mathematics, vol. 249, Springer, New York, 2014. MR3243734

[166] L. Grafakos and R. H. Torres, *On multilinear singular integrals of Calderón-Zygmund type*, Proceedings of the 6th International Conference on Harmonic Analysis and Partial Differential Equations (El Escorial, 2000), Publ. Mat. **Vol. Extra** (2002), 57–91, DOI 10.5565/PUBLMAT_Esco02_04. MR1964816

[167] P. R. Halmos, *Ten problems in Hilbert space*, Bull. Amer. Math. Soc. **76** (1970), 887–933, DOI 10.1090/S0002-9904-1970-12502-2. MR0270173

[168] P. R. Halmos, *Shifts on Hilbert spaces*, J. Reine Angew. Math. **208** (1961), 102–112, DOI 10.1515/crll.1961.208.102. MR0152896

[169] P. R. Halmos and V. S. Sunder, *Bounded integral operators on L^2 spaces*, Ergebnisse der Mathematik und ihrer Grenzgebiete [Results in Mathematics and Related Areas], vol. 96, Springer-Verlag, Berlin-New York, 1978. MR517709

[170] D. H. Hamilton, *Quadratic inequalities and interpolation*, J. Analyse Math. **43** (1983/84), 26–50, DOI 10.1007/BF02790177. MR777411

[171] M. Hartz, *On the isomorphism problem for multiplier algebras of Nevanlinna-Pick spaces*, Canad. J. Math. **69** (2017), no. 1, 54–106, DOI 10.4153/CJM-2015-050-6. MR3589854

[172] H. Hedenmalm, *Bloch functions and asymptotic tail variance*, Adv. Math. **313** (2017), 947–990, DOI 10.1016/j.aim.2017.04.016. MR3649242

[173] H. Hedenmalm, *A factorization theorem for square area-integrable analytic functions*, J. Reine Angew. Math. **422** (1991), 45–68, DOI 10.1515/crll.1991.422.45. MR1133317

[174] H. Hedenmalm, B. Korenblum, and K. Zhu, *Theory of Bergman spaces*, Graduate Texts in Mathematics, vol. 199, Springer-Verlag, New York, 2000. MR1758653

[175] H. Hedenmalm and A. Shields, *Invariant subspaces in Banach spaces of analytic functions*, Michigan Math. J. **37** (1990), no. 1, 91–104, DOI 10.1307/mmj/1029004068. MR1042516

[176] O. Helmer, *Divisibility properties of integral functions*, Duke Math. J. **6** (1940), 345–356. MR0001851

[177] H. Helson, *Lectures on invariant subspaces*, Academic Press, New York-London, 1964. MR0171178

[178] K. Hoffman, *Banach spaces of analytic functions*, Prentice-Hall Series in Modern Analysis, Prentice-Hall, Inc., Englewood Cliffs, N. J., 1962. MR0133008

[179] C. Horowitz, *Factorization theorems for functions in the Bergman spaces*, Duke Math. J. **44** (1977), no. 1, 201–213. MR0427650

[180] C. Horowitz, *Zeros of functions in the Bergman spaces*, Duke Math. J. **41** (1974), 693–710. MR0357747

[181] Y. Hu and Y. Shen, *On quasisymmetric homeomorphisms*, Israel J. Math. **191** (2012), no. 1, 209–226, DOI 10.1007/s11856-011-0204-4. MR2970868

[182] R. Howard and A. R. Schep, *Norms of positive operators on L^p-spaces*, Proc. Amer. Math. Soc. **109** (1990), no. 1, 135–146, DOI 10.2307/2048373. MR1000156

[183] S. Janson and J. Peetre, *Paracommutators—boundedness and Schatten-von Neumann properties*, Trans. Amer. Math. Soc. **305** (1988), no. 2, 467–504, DOI 10.2307/2000875. MR924766

[184] S. Janson, J. Peetre, and R. Rochberg, *Hankel forms and the Fock space*, Rev. Mat. Iberoamericana **3** (1987), no. 1, 61–138, DOI 10.4171/RMI/46. MR1008445

[185] F. John and L. Nirenberg, *On functions of bounded mean oscillation*, Comm. Pure Appl. Math. **14** (1961), 415–426, DOI 10.1002/cpa.3160140317. MR0131498

[186] P. W. Jones, *L^∞ estimates for the $\bar\partial$ problem in a half-plane*, Acta Math. **150** (1983), no. 1-2, 137–152, DOI 10.1007/BF02392970. MR697611

[187] M. Jovović and B. MacCluer, *Composition operators on Dirichlet spaces*, Acta Sci. Math. (Szeged) **63** (1997), no. 1-2, 229–247. MR1459789

[188] Th. Kaluza, *Über die Koeffizienten reziproker Potenzreihen* (German), Math. Z. **28** (1928), no. 1, 161–170, DOI 10.1007/BF01181155. MR1544949

[189] N. H. Katz and C. Pereyra, *On the two weights problem for the Hilbert transform*, Rev. Mat. Iberoamericana **13** (1997), no. 1, 211–243, DOI 10.4171/RMI/220. MR1462332

[190] K. Kellay and J. Mashreghi, *On zero sets in the Dirichlet space*, J. Geom. Anal. **22** (2012), no. 4, 1055–1070, DOI 10.1007/s12220-011-9229-5. MR2965362

[191] R. Kerman and E. Sawyer, *Carleson measures and multipliers of Dirichlet-type spaces*, Trans. Amer. Math. Soc. **309** (1988), no. 1, 87–98, DOI 10.2307/2001160. MR957062

[192] J. R. Kinney, *Tangential limits of functions of the class S_α*, Proc. Amer. Math. Soc. **14** (1963), 68–70, DOI 10.2307/2033956. MR0143916

[193] P. Koosis, *Carleson's interpolation theorem deduced from a result of Pick*, Complex analysis, operators, and related topics, Oper. Theory Adv. Appl., vol. 113, Birkhäuser, Basel, 2000, pp. 151–162. MR1771759

[194] P. Koosis, *Introduction to H_p spaces*, London Mathematical Society Lecture Note Series, vol. 40, Cambridge University Press, Cambridge-New York, 1980. With an appendix on Wolff's proof of the corona theorem. MR565451

[195] A. Korányi, M. A. Picardello, and M. H. Taibleson, *Hardy spaces on nonhomogeneous trees*, Symposia Mathematica, Vol. XXIX (Cortona, 1984), Sympos. Math., XXIX, Academic Press, New York, 1987, pp. 205–265. With an appendix by Picardello and Wolfgang Woess. MR951187

[196] S. G. Krantz, *The corona problem in several complex variables*, The corona problem, Fields Inst. Commun., vol. 72, Springer, New York, 2014, pp. 107–126, DOI 10.1007/978-1-4939-1255-1_6. MR3329544

[197] S. G. Krantz, *Geometric analysis of the Bergman kernel and metric*, Graduate Texts in Mathematics, vol. 268, Springer, New York, 2013. MR3114665

[198] S. G. Krantz and S.-Y. Li, *Some remarks on the corona problem on strongly pseudoconvex domains in $\mathbf{C}^n$*, Illinois J. Math. **39** (1995), no. 2, 323–349. MR1316541

[199] S. G. Krantz, S.-Y. Li, and R. Rochberg, *The effect of boundary geometry on Hankel operators belonging to the trace ideals of Bergman spaces*, Integral Equations Operator Theory **28** (1997), no. 2, 196–213, DOI 10.1007/BF01191818. MR1451501

[200] M. Lacey and E. Terwilleger, *Hankel operators in several complex variables and product BMO*, Houston J. Math. **35** (2009), no. 1, 159–183. MR2491875

[201] J. Lech, *Essentially normal multiplication operators on the Dirichlet space*, Michigan Math. J. **42** (1995), no. 1, 127–140, DOI 10.1307/mmj/1029005158. MR1322194

[202] Y. J. Lee and K. Zhu, *Sums of products of Toeplitz and Hankel operators on the Dirichlet space*, Integral Equations Operator Theory **71** (2011), no. 2, 275–302, DOI 10.1007/s00020-011-1901-4. MR2838146

[203] P. Lefèvre, D. Li, H. Queffélec, and L. Rodríguez-Piazza, *Approximation numbers of composition operators on the Dirichlet space*, Ark. Mat. **53** (2015), no. 1, 155–175, DOI 10.1007/s11512-013-0194-z. MR3319618

[204] A. Lerner and F. Nazarov, *Intuitive dyadic calculus: the basics* (2015), preprint, available at `arXiv:1508.05639`.

[205] M. Levi, *Analytic and Geometric Aspects of the Theory of the Dirichlet Space*, Masters thesis, University of Milano, 2015.

[206] H. Li and D. H. Luecking, *BMO on strongly pseudoconvex domains: Hankel operators, duality and $\bar\partial$-estimates*, Trans. Amer. Math. Soc. **346** (1994), no. 2, 661–691, DOI 10.2307/2154865. MR1273537

[207] E. Ligocka, *Estimates in Sobolev norms $\|\cdot\|_p^s$ for harmonic and holomorphic functions and interpolation between Sobolev and Hölder spaces of harmonic functions*, Studia Math. **86** (1987), no. 3, 255–271, DOI 10.4064/sm-86-3-255-271. MR917051

[208] K.-C. Lin, *On the H^p solutions to the corona equation*, Bull. Sci. Math. **118** (1994), no. 3, 271–286. MR1284780

[209] Y. L. Lin, *The interplay between harmonic analysis, function theory and operators*, ProQuest LLC, Ann Arbor, MI, 2009. Thesis (Ph.D.)–Washington University in St. Louis. MR2713390

[210] S. Lindberg, *On the Hardy space theory of compensated compactness quantities*, Arch. Ration. Mech. Anal. **224** (2017), no. 2, 709–742, DOI 10.1007/s00205-017-1087-2. MR3614758

[211] D. H. Luecking, *Characterizations of certain classes of Hankel operators on the Bergman spaces of the unit disk*, J. Funct. Anal. **110** (1992), no. 2, 247–271, DOI 10.1016/0022-1236(92)90034-G. MR1194989

[212] S. Luo, *Reducing subspaces of multiplication operators on the Dirichlet space*, Integral Equations Operator Theory **85** (2016), no. 4, 539–554, DOI 10.1007/s00020-016-2295-0. MR3551231

[213] S. Luo, *On the index of invariant subspaces in the space of weak products of Dirichlet functions*, Complex Anal. Oper. Theory **9** (2015), no. 6, 1311–1323, DOI 10.1007/s11785-014-0419-6. MR3390194

[214] S. Luo and S. Richter, *Hankel operators and invariant subspaces of the Dirichlet space*, J. Lond. Math. Soc. (2) **91** (2015), no. 2, 423–438, DOI 10.1112/jlms/jdv001. MR3355109

[215] S. Luo and J. Xiao, *Toeplitz operator for Dirichlet space through Sobolev multiplier algebra*, Taiwanese J. Math. **22** (2018), no. 2, 383–399, DOI 10.11650/tjm/170805. MR3780724

[216] R. Lyons and Y. Peres, *Probability on trees and networks*, Cambridge Series in Statistical and Probabilistic Mathematics, vol. 42, Cambridge University Press, New York, 2016. MR3616205

[217] P. Malliavin, *Sur l'analyse harmonique de certaines classes de séries de Taylor* (French), Symposia Mathematica, Vol. XXII (Convegno sull'Analisi Armonica e Spazi di Funzioni su Gruppi Localmente Compatti, INDAM, Rome, 1976), Academic Press, London, 1977, pp. 71–91. MR0585689

[218] A. W. Marcus, D. A. Spielman, and N. Srivastava, *Interlacing families II: Mixed characteristic polynomials and the Kadison-Singer problem*, Ann. of Math. (2) **182** (2015), no. 1, 327–350, DOI 10.4007/annals.2015.182.1.8. MR3374963

[219] D. E. Marshall, *A new proof of a sharp inequality concerning the Dirichlet integral*, Ark. Mat. **27** (1989), no. 1, 131–137, DOI 10.1007/BF02386365. MR1004727

[220] D. E. Marshall and C. Sundberg, *Interpolating sequences for the multipliers of the Dirichlet space* (1994), preprint, available at `http://www.math.washington.edu/~marshall/preprints/interp.pdf`.

[221] *Mathematical Sciences Research Institute*, available at `www.msri.org/summer_schools/590`.

[222] J. Mashreghi, T. Ransford, and M. Shabankhah, *Arguments of zero sets in the Dirichlet space*, Hilbert spaces of analytic functions, CRM Proc. Lecture Notes, vol. 51, Amer. Math. Soc., Providence, RI, 2010, pp. 143–148. MR2648872

[223] J. Mashreghi and M. Shabankhah, *Admissible functions for the Dirichlet space*, Studia Math. **198** (2010), no. 2, 147–156, DOI 10.4064/sm198-2-3. MR2640074

[224] J. Mashreghi and M. Shabankhah, *Zero sets and uniqueness sets with one cluster point for the Dirichlet space*, J. Math. Anal. Appl. **357** (2009), no. 2, 498–503, DOI 10.1016/j.jmaa.2009.04.023. MR2557661

[225] A. Matheson and A. R. Pruss, *Properties of extremal functions for some nonlinear functionals on Dirichlet spaces*, Trans. Amer. Math. Soc. **348** (1996), no. 7, 2901–2930, DOI 10.1090/S0002-9947-96-01656-X. MR1357401

[226] V. G. Maz′ya and I. E. Verbitsky, *Capacitary inequalities for fractional integrals, with applications to partial differential equations and Sobolev multipliers*, Ark. Mat. **33** (1995), no. 1, 81–115, DOI 10.1007/BF02559606. MR1340271

[227] V. G. Maz'ya and I. E. Verbitsky, *The Schrödinger operator on the energy space: boundedness and compactness criteria*, Acta Math. **188** (2002), no. 2, 263–302, DOI 10.1007/BF02392684. MR1947894

[228] J. E. McCarthy, *Pick's theorem—what's the big deal?*, Amer. Math. Monthly **110** (2003), no. 1, 36–45, DOI 10.2307/3072342. MR1952746

[229] J. D. McPhail, *A weighted interpolation problem for analytic functions*, Studia Math. **96** (1990), no. 2, 105–116, DOI 10.4064/sm-96-2-105-116. MR1052627

[230] J. Mercer, *Functions of positive and negative type and their connection with the theory of integral equations*, Transactions of the Royal Society A **209** (1909), no. 1, 415–446.

[231] E. H. Moore, *General Analysis*, Mem Amer Phil Soc, I (1935).

[232] C. Muscalu and W. Schlag, *Classical and multilinear harmonic analysis. Vol. II*, Cambridge Studies in Advanced Mathematics, vol. 138, Cambridge University Press, Cambridge, 2013. MR3052499

[233] C. Muscalu and W. Schlag, *Classical and multilinear harmonic analysis. Vol. I*, Cambridge Studies in Advanced Mathematics, vol. 137, Cambridge University Press, Cambridge, 2013. MR3052498

[234] S. Nag and D. Sullivan, *Teichmüller theory and the universal period mapping via quantum calculus and the $H^{1/2}$ space on the circle*, Osaka J. Math. **32** (1995), no. 1, 1–34. MR1323099

[235] A. Nagel, W. Rudin, and J. H. Shapiro, *Tangential boundary behavior of functions in Dirichlet-type spaces*, Ann. of Math. (2) **116** (1982), no. 2, 331–360, DOI 10.2307/2007064. MR672838

[236] K. Naimark and M. Solomyak, *Eigenvalue estimates for the weighted Laplacian on metric trees*, Proc. London Math. Soc. (3) **80** (2000), no. 3, 690–724, DOI 10.1112/S0024611500012272. MR1744781

[237] F. Nazarov, S. Treil, and A. Volberg, *Two weight estimate for the Hilbert transform and corona decomposition for non-doubling measures* (2010), preprint, available at `arXiv:1003.1596`.

[238] F. Nazarov, S. Treil, and A. Volberg, *The Bellman functions and two-weight inequalities for Haar multipliers*, J. Amer. Math. Soc. **12** (1999), no. 4, 909–928, DOI 10.1090/S0894-0347-99-00310-0. MR1685781

[239] Z. Nehari, *On bounded bilinear forms*, Ann. of Math. (2) **65** (1957), 153–162, DOI 10.2307/1969670. MR0082945

[240] Y. A. Neretin, *Combinatorial analogues of the group of diffeomorphisms of the circle*, Izv. Ross. Akad. Nauk Ser. Mat. **56** (1992), no. 5, 1072–1085; English transl., Russian Acad. Sci. Izv. Math. **41** (1993), no. 2, 337–349.

[241] Yu. A. Neretin, *Groups of hierarchomorphisms of trees and related Hilbert spaces*, J. Funct. Anal. **200** (2003), no. 2, 505–535, DOI 10.1016/S0022-1236(02)00146-5. MR1979021

[242] R. Nevanlinna, *Uber beschrankte Functionen, die in gegenbennen Punkten vorgeschrieben Werte annehmen*, Ann. Acad. Sci. Fenn. Ser. A **13** (1919), no. 1.

[243] A. Nicolau and J. Xiao, *Bounded functions in Möbius invariant Dirichlet spaces*, J. Funct. Anal. **150** (1997), no. 2, 383–425, DOI 10.1006/jfan.1997.3114. MR1479545

[244] N. K. Nikolski, *Operators, functions, and systems: an easy reading. Vol. 2*, Mathematical Surveys and Monographs, vol. 93, American Mathematical Society, Providence, RI, 2002. Model operators and systems; Translated from the French by Andreas Hartmann and revised by the author. MR1892647

[245] N. K. Nikolski, *Operators, functions, and systems: an easy reading. Vol. 1*, Mathematical Surveys and Monographs, vol. 92, American Mathematical Society, Providence, RI, 2002. Hardy, Hankel, and Toeplitz; Translated from the French by Andreas Hartmann. MR1864396

[246] N. K. Nikol′skiĭ, *Treatise on the shift operator*, Grundlehren der Mathematischen Wissenschaften [Fundamental Principles of Mathematical Sciences], vol. 273, Springer-Verlag, Berlin, 1986. Spectral function theory; With an appendix by S. V. Hruščev [S. V. Khrushchëv] and V. V. Peller; Translated from the Russian by Jaak Peetre. MR827223

[247] R. F. Olin and J. E. Thomson, *Cellular-indecomposable subnormal operators*, Integral Equations Operator Theory **7** (1984), no. 3, 392–430, DOI 10.1007/BF01208382. MR756764

[248] J. M. Ortega and J. Fàbrega, *Pointwise multipliers and decomposition theorems in analytic Besov spaces*, Math. Z. **235** (2000), no. 1, 53–81, DOI 10.1007/s002090000123. MR1785071

[249] J. M. Ortega and J. Fàbrega, *Pointwise multipliers and corona type decomposition in BMOA* (English, with English and French summaries), Ann. Inst. Fourier (Grenoble) **46** (1996), no. 1, 111–137. MR1385513

[250] N. Øvrelid, *Integral representation formulae for differential forms, and solutions of the $\overline{\partial}$-equation*, Fonctions analytiques de plusieurs variables et analyse complexe (Colloq. Internat. CNRS, No. 208, Paris, 1972), Gauthier-Villars, Paris, 1974, pp. 180–198. "Agora Mathematica", No. 1. MR0481077

[251] P. Pankka, P. Poggi-Corradini, and K. Rajala, *Sharp exponential integrability for traces of monotone Sobolev functions*, Nagoya Math. J. **192** (2008), 137–149, DOI 10.1017/S0027763000026015. MR2477615

[252] J. Pau and J. Á. Peláez, *Schatten classes of integration operators on Dirichlet spaces*, J. Anal. Math. **120** (2013), 255–289, DOI 10.1007/s11854-013-0020-3. MR3095154

[253] J. Pau and J. Á. Peláez, *On the zeros of functions in Dirichlet-type spaces*, Trans. Amer. Math. Soc. **363** (2011), no. 4, 1981–2002, DOI 10.1090/S0002-9947-2010-05108-6. MR2746672

[254] V. I. Paulsen, *An Introduction to the Theory of Reproducing Kernel Hilbert Spaces* (2006), available at `https://www.math.uh.edu/~vern/rkhs.pdf`.

[255] V. I. Paulsen and M. Raghupathi, *An introduction to the theory of reproducing kernel Hilbert spaces*, Cambridge Studies in Advanced Mathematics, vol. 152, Cambridge University Press, Cambridge, 2016. MR3526117

[256] J. Peetre, *New thoughts on Besov spaces*, Mathematics Department, Duke University, Durham, N.C., 1976. Duke University Mathematics Series, No. 1. MR0461123

[257] V. V. Peller, *Hankel operators and their applications*, Springer Monographs in Mathematics, Springer-Verlag, New York, 2003. MR1949210

[258] M. M. Peloso, *Besov spaces, mean oscillation, and generalized Hankel operators*, Pacific J. Math. **161** (1993), no. 1, 155–184. MR1237143

[259] M. M. Peloso, *Möbius invariant spaces on the unit ball*, Michigan Math. J. **39** (1992), no. 3, 509–536, DOI 10.1307/mmj/1029004604. MR1182505

[260] M. C. Pereyra, *Dyadic harmonic analysis and weighted inequalities: the sparse revolution* (2018), available at `arXiv:1812.00850`.

[261] M. C. Pereyra, *Lecture notes on dyadic harmonic analysis*, Second Summer School in Analysis and Mathematical Physics (Cuernavaca, 2000), Contemp. Math., vol. 289, Amer. Math. Soc., Providence, RI, 2001, pp. 1–60, DOI 10.1090/conm/289/04874. MR1864538

[262] G. Pick, *Über die Beschränkungen analytischer Funktionen, welche durch vorgegebene Funktionswerte bewirkt werden* (German), Math. Ann. **77** (1915), no. 1, 7–23, DOI 10.1007/BF01456817. MR1511844

[263] P. Poggi-Corradini and K. Rajala, *An egg-yolk principle and exponential integrability for quasiregular mappings*, J. Lond. Math. Soc. (2) **76** (2007), no. 2, 531–544, DOI 10.1112/jlms/jdm078. MR2363431

[264] Ch. Pommerenke, *Schlichte Funktionen und analytische Funktionen von beschränkter mittlerer Oszillation* (German), Comment. Math. Helv. **52** (1977), no. 4, 591–602, DOI 10.1007/BF02567392. MR0454017

[265] P. Quiggin, *For which reproducing kernel Hilbert spaces is Pick's theorem true?*, Integral Equations Operator Theory **16** (1993), no. 2, 244–266, DOI 10.1007/BF01358955. MR1205001

[266] T. Ransford, *Potential theory in the complex plane*, London Mathematical Society Student Texts, vol. 28, Cambridge University Press, Cambridge, 1995. MR1334766

[267] S. Richter, *A representation theorem for cyclic analytic two-isometries*, Trans. Amer. Math. Soc. **328** (1991), no. 1, 325–349, DOI 10.2307/2001885. MR1013337

[268] S. Richter, *Invariant subspaces of the Dirichlet shift*, J. Reine Angew. Math. **386** (1988), 205–220, DOI 10.1515/crll.1988.386.205. MR936999

[269] S. Richter, W. T. Ross, and C. Sundberg, *Zeros of functions with finite Dirichlet integral*, Proc. Amer. Math. Soc. **132** (2004), no. 8, 2361–2365, DOI 10.1090/S0002-9939-04-07361-7. MR2052414

[270] S. Richter and A. Shields, *Bounded analytic functions in the Dirichlet space*, Math. Z. **198** (1988), no. 2, 151–159, DOI 10.1007/BF01163287. MR939532

[271] S. Richter and C. Sundberg, *Weak products of Dirichlet functions*, J. Funct. Anal. **266** (2014), no. 8, 5270–5299, DOI 10.1016/j.jfa.2014.02.005. MR3177336

[272] S. Richter and C. Sundberg, *Invariant subspaces of the Dirichlet shift and pseudocontinuations*, Trans. Amer. Math. Soc. **341** (1994), no. 2, 863–879, DOI 10.2307/2154587. MR1145733

[273] S. Richter and C. Sundberg, *Multipliers and invariant subspaces in the Dirichlet space*, J. Operator Theory **28** (1992), no. 1, 167–186. MR1259923

[274] S. Richter and C. Sundberg, *A formula for the local Dirichlet integral*, Michigan Math. J. **38** (1991), no. 3, 355–379, DOI 10.1307/mmj/1029004388. MR1116495

[275] S. Richter and J. Sunkes, *Hankel operators, invariant subspaces, and cyclic vectors in the Drury-Arveson space*, Proc. Amer. Math. Soc. **144** (2016), no. 6, 2575–2586, DOI 10.1090/proc/12922. MR3477074

[276] S. Richter and B. D. Wick, *A remark on the multipliers on spaces of weak products of functions*, Concr. Oper. **3** (2016), 25–28, DOI 10.1515/conop-2016-0004. MR3491880

[277] Riesz F. and Riesz M., *Über die Randwerte einer analytischen Funktion*, Quatrième Congrès des Mathématiciens Scandinaves, Stockholm **3** (1916), 27–55.

[278] M. Riesz, *Sur certaines inégalités dans la théorie des fonctions avec quelques remarques sur les géométries non-euclidiennes*, Kungl. Fysiografiska Sällskapets i Lund Förhandlingar **1** (1931), 21 pp.

[279] R. Rochberg, *Complex hyperbolic geometry and Hilbert spaces with complete Pick kernels*, J. Funct. Anal. **276** (2019), no. 5, 1622–1679, DOI 10.1016/j.jfa.2018.08.017. MR3912786

[280] R. Rochberg, *Structure in the spectra of some multiplier algebras*, The corona problem, Fields Inst. Commun., vol. 72, Springer, New York, 2014, pp. 177–200, DOI 10.1007/978-1-4939-1255-1_9. MR3329547

[281] R. Rochberg, *Projected composition operators on the Hardy space*, Indiana Univ. Math. J. **43** (1994), no. 2, 441–458, DOI 10.1512/iumj.1994.43.43018. MR1291524

[282] R. Rochberg, *Decomposition theorems for Bergman spaces and their applications*, Operators and function theory (Lancaster, 1984), NATO Adv. Sci. Inst. Ser. C Math. Phys. Sci., vol. 153, Reidel, Dordrecht, 1985, pp. 225–277. MR810448

[283] R. Rochberg and S. Semmes, *A decomposition theorem for BMO and applications*, J. Funct. Anal. **67** (1986), no. 2, 228–263, DOI 10.1016/0022-1236(86)90038-8. MR845199

[284] R. Rochberg and Z. J. Wu, *A new characterization of Dirichlet type spaces and applications*, Illinois J. Math. **37** (1993), no. 1, 101–122. MR1193132

[285] R. Rochberg and Z. J. Wu, *Toeplitz operators on Dirichlet spaces*, Integral Equations Operator Theory **15** (1992), no. 2, 325–342, DOI 10.1007/BF01204241. MR1147284

[286] M. Rosenblum, *A corona theorem for countably many functions*, Integral Equations Operator Theory **3** (1980), no. 1, 125–137, DOI 10.1007/BF01682874. MR570865

[287] W. T. Ross, *The classical Dirichlet space*, Recent advances in operator-related function theory, Contemp. Math., vol. 393, Amer. Math. Soc., Providence, RI, 2006, pp. 171–197, DOI 10.1090/conm/393/07378. MR2198379

[288] W. Rudin, *Function theory in the unit ball of* $\mathbf{C}^n$, Grundlehren der Mathematischen Wissenschaften [Fundamental Principles of Mathematical Science], vol. 241, Springer-Verlag, New York-Berlin, 1980. MR601594

[289] W. Rudin, *Function theory in polydiscs*, W. A. Benjamin, Inc., New York-Amsterdam, 1969. MR0255841

[290] W. Rudin, *The radial variation of analytic functions*, Duke Math. J. **22** (1955), 235–242. MR0079093

[291] S. Saitoh, *Theory of reproducing kernels and its applications*, Pitman Research Notes in Mathematics Series, vol. 189, Longman Scientific & Technical, Harlow; copublished in the United States with John Wiley & Sons, Inc., New York, 1988. MR983117

[292] S. Saitoh, *Some inequalities for analytic functions with a finite Dirichlet integral on the unit disc*, Math. Ann. **246** (1979/80), no. 1, 69–77, DOI 10.1007/BF01352026. MR554132

[293] D. Sarason, *Sub-Hardy Hilbert spaces in the unit disk*, University of Arkansas Lecture Notes in the Mathematical Sciences, vol. 10, John Wiley & Sons, Inc., New York, 1994. A Wiley-Interscience Publication. MR1289670

[294] E. T. Sawyer, *Function theory: interpolation and corona problems*, Fields Institute Monographs, vol. 25, American Mathematical Society, Providence, RI; Fields Institute for Research in Mathematical Sciences, Toronto, ON, 2009. MR2476953

[295] E. T. Sawyer, *A characterization of two weight norm inequalities for fractional and Poisson integrals*, Trans. Amer. Math. Soc. **308** (1988), no. 2, 533–545, DOI 10.2307/2001090. MR930072

[296] E. T. Sawyer and B. D. Wick, *The corona problem for kernel multiplier algebras*, Integral Equations Operator Theory **86** (2016), no. 4, 495–544, DOI 10.1007/s00020-016-2329-7. MR3578038

[297] E. Schippers and W. Staubach, *A symplectic functional analytic proof of the conformal welding theorem*, Proc. Amer. Math. Soc. **143** (2015), no. 1, 265–278, DOI 10.1090/S0002-9939-2014-12225-8. MR3272752

[298] C. F. Schubert, *The corona theorem as an operator theorem*, Proc. Amer. Math. Soc. **69** (1978), no. 1, 73–76, DOI 10.2307/2043192. MR0482355

[299] J. Schur, *Bemerkungen zur Theorie der beschränkten Bilinearformen mit unendlich vielen Veränderlichen* (German), J. Reine Angew. Math. **140** (1911), 1–28, DOI 10.1515/crll.1911.140.1. MR1580823

[300] A. P. Schuster and K. Seip, *Weak conditions for interpolation in holomorphic spaces*, Publ. Mat. **44** (2000), no. 1, 277–293, DOI 10.5565/PUBLMAT_44100_11. MR1775765

[301] D. Seco, *Smooth sets and two problems in the Dirichlet space*, Doctoral thesis, Universitat Autonoma de Barcelona, 2013.

[302] K. Seip, *Interpolation and sampling in spaces of analytic functions*, University Lecture Series, vol. 33, American Mathematical Society, Providence, RI, 2004. MR2040080

[303] A. Serra, *New examples of non-complete Pick kernels*, Integral Equations Operator Theory **53** (2005), no. 4, 553–572, DOI 10.1007/s00020-005-1363-7. MR2187439

[304] J.-P. Serre, *Trees*, Springer Monographs in Mathematics, Springer-Verlag, Berlin, 2003. Translated from the French original by John Stillwell; Corrected 2nd printing of the 1980 English translation. MR1954121

[305] J. H. Shapiro, *Composition operators and classical function theory*, Universitext: Tracts in Mathematics, Springer-Verlag, New York, 1993. MR1237406

[306] H. S. Shapiro and A. L. Shields, *On the zeros of functions with finite Dirichlet integral and some related function spaces*, Math. Z. **80** (1962), 217–229, DOI 10.1007/BF01162379. MR0145082

[307] H. S. Shapiro and A. L. Shields, *On some interpolation problems for analytic functions*, Amer. J. Math. **83** (1961), 513–532, DOI 10.2307/2372892. MR0133446

[308] S. Shimorin, *Complete Nevanlinna-Pick property of Dirichlet-type spaces*, J. Funct. Anal. **191** (2002), no. 2, 276–296, DOI 10.1006/jfan.2001.3871. MR1911187

[309] S. M. Shimorin, *Reproducing kernels and extremal functions in Dirichlet-type spaces*, Zap. Nauchn. Sem. S.-Peterburg. Otdel. Mat. Inst. Steklov. (POMI) **255** (1998), no. Issled. po Lineĭn. Oper. i Teor. Funkts. 26, 198–220, 254; English transl., J. Math. Sci. (New York) **107** (2001), no. 4, 4108–4124.

[310] V. I. Smirnov, *On boundary values of functions regular in a disk [J. Soc. Phys.-Math. Leningrade* **2** *(1928), no. 2, 22–37]*, Topics in interpolation theory (Leipzig, 1994), Oper. Theory Adv. Appl., vol. 95, Birkhäuser, Basel, 1997, pp. 481–494. Translated from the French. MR1473270

[311] W. Smith and D. A. Stegenga, *Poincaré domains in the plane*, Complex analysis, Joensuu 1987, Lecture Notes in Math., vol. 1351, Springer, Berlin, 1988, pp. 312–327, DOI 10.1007/BFb0081265. MR982095

[312] P. M. Soardi, *Potential theory on infinite networks*, Lecture Notes in Mathematics, vol. 1590, Springer-Verlag, Berlin, 1994. MR1324344

[313] M. Spivak, *A comprehensive introduction to differential geometry. Vol. One*, Published by M. Spivak, Brandeis Univ., Waltham, Mass., 1970. MR0267467

[314] D. A. Stegenga, *Multipliers of the Dirichlet space*, Illinois J. Math. **24** (1980), no. 1, 113–139. MR550655

[315] E. M. Stein, *Singular integrals: the roles of Calderón and Zygmund*, Notices Amer. Math. Soc. **45** (1998), no. 9, 1130–1140. MR1640159

[316] E. M. Stein, *Harmonic analysis: real-variable methods, orthogonality, and oscillatory integrals*, Princeton Mathematical Series, vol. 43, Princeton University Press, Princeton, NJ, 1993. With the assistance of Timothy S. Murphy; Monographs in Harmonic Analysis, III. MR1232192

[317] E. M. Stein, *Singular integrals and differentiability properties of functions*, Princeton Mathematical Series, No. 30, Princeton University Press, Princeton, N.J., 1970. MR0290095

[318] B. A. Taylor and D. L. Williams, *Ideals in rings of analytic functions with smooth boundary values*, Canad. J. Math. **22** (1970), 1266–1283, DOI 10.4153/CJM-1970-143-x. MR0273024

[319] V. A. Tolokonnikov, *Interpolating Blaschke products and ideals of the algebra H^∞* (Russian, with English summary), Zap. Nauchn. Sem. Leningrad. Otdel. Mat. Inst. Steklov. (LOMI) **126** (1983), 196–201. Investigations on linear operators and the theory of functions, XII. MR697439

[320] V. A. Tolokonnikov, *The corona theorem in algebras of bounded analytic functions*, Akad. Nauk SSSR, Inst. Analit. Priborostr. Nauchn. Tekhn. Ob ed., Leningrad (1983), 61–95; English transl., *American Mathematical Society Translations. Series 2. Vol. 149* **149** (1991), viii+220. Thirteen papers in algebra, functional analysis, topology, and probability, translated from the Russian.

[321] V. A. Tolokonnikov, *Estimates in Carleson's corona theorem and finitely generated ideals of the algebra H^∞* (Russian), Funktsional. Anal. i Prilozhen. **14** (1980), no. 4, 85–86. MR595742

[322] S. Treil, *Estimates in the corona theorem and ideals of H^∞: a problem of T. Wolff*, J. Anal. Math. **87** (2002), 481–495, DOI 10.1007/BF02868486. Dedicated to the memory of Thomas H. Wolff. MR1945294

[323] S. R. Treil′, *Angles between co-invariant subspaces, and the operator corona problem. The Szőkefalvi-Nagy problem* (Russian), Dokl. Akad. Nauk SSSR **302** (1988), no. 5, 1063–1068; English transl., Soviet Math. Dokl. **38** (1989), no. 2, 394–399. MR981054

[324] S. Treil and B. D. Wick, *Analytic projections, corona problem and geometry of holomorphic vector bundles*, J. Amer. Math. Soc. **22** (2009), no. 1, 55–76, DOI 10.1090/S0894-0347-08-00611-5. MR2449054

[325] S. Treil and B. D. Wick, *The matrix-valued H^p corona problem in the disk and polydisk*, J. Funct. Anal. **226** (2005), no. 1, 138–172, DOI 10.1016/j.jfa.2005.04.010. MR2158178

[326] T. T. Trent, *A corona theorem for multipliers on Dirichlet space*, Integral Equations Operator Theory **49** (2004), no. 1, 123–139, DOI 10.1007/s00020-002-1196-6. MR2057771

[327] T. T. Trent, *Solutions for the $H^\infty(D^n)$ corona problem belonging to* $\exp(L^{\frac{1}{2n-1}})$, Recent advances in matrix and operator theory, Oper. Theory Adv. Appl., vol. 179, Birkhäuser, Basel, 2008, pp. 309–328, DOI 10.1007/978-3-7643-8539-2_18. MR2397862

[328] T. T. Trent and X. Zhang, *A matricial corona theorem. II*, Proc. Amer. Math. Soc. **135** (2007), no. 9, 2845–2854, DOI 10.1090/S0002-9939-07-08806-5. MR2317961

[329] T. Trent and X. Zhang, *A matricial corona theorem*, Proc. Amer. Math. Soc. **134** (2006), no. 9, 2549–2558, DOI 10.1090/S0002-9939-06-08172-X. MR2213732

[330] H. Triebel, *Theory of function spaces. III*, Monographs in Mathematics, vol. 100, Birkhäuser Verlag, Basel, 2006. MR2250142

[331] H. Triebel, *Theory of function spaces. II*, Monographs in Mathematics, vol. 84, Birkhäuser Verlag, Basel, 1992. MR1163193

[332] H. Triebel, *Theory of function spaces*, Monographs in Mathematics, vol. 78, Birkhäuser Verlag, Basel, 1983. MR781540

[333] J. B. Twomey, *The boundary values of a function in the Dirichlet space*, Math. Proc. R. Ir. Acad. **110A** (2010), no. 2, 149–162, DOI 10.3318/PRIA.2010.110.2.149. MR2760420

[334] J. B. Twomey, *Tangential boundary behaviour of harmonic and holomorphic functions*, J. London Math. Soc. (2) **65** (2002), no. 1, 68–84, DOI 10.1112/S0024610701002873. MR1875136

[335] N. Th. Varopoulos, *BMO functions and the $\overline{\partial}$-equation*, Pacific J. Math. **71** (1977), no. 1, 221–273. MR0508035

[336] I. È. Verbitskiĭ, *Multipliers in spaces with "fractional" norms, and inner functions* (Russian), Sibirsk. Mat. Zh. **26** (1985), no. 2, 51–72, 221. MR788329

[337] I. E. Verbitsky, *Nonlinear potentials and trace inequalities*, The Maz′ya anniversary collection, Vol. 2 (Rostock, 1998), Oper. Theory Adv. Appl., vol. 110, Birkhäuser, Basel, 1999, pp. 323–343. MR1747901

[338] A. Volberg and B. D. Wick, *Bergman-type singular integral operators and the characterization of Carleson measures for Besov-Sobolev spaces and the complex ball*, Amer. J. Math. **134** (2012), no. 4, 949–992, DOI 10.1353/ajm.2012.0028. MR2956255

[339] B. D. Wick, *Internet Analysis Seminar 2010-2011*, available at `internetanalysisseminar.gatech.edu/dirichlet.html`.

[340] Z. Wu, *Function theory and operator theory on the Dirichlet space*, Holomorphic spaces (Berkeley, CA, 1995), Math. Sci. Res. Inst. Publ., vol. 33, Cambridge Univ. Press, Cambridge, 1998, pp. 179–199. MR1630650

[341] Z. Wu, *A class of bilinear forms on Dirichlet type spaces*, J. London Math. Soc. (2) **54** (1996), no. 3, 498–514, DOI 10.1112/jlms/54.3.498. MR1413894

[342] Z. J. Wu, *Hankel and Toeplitz operators on Dirichlet spaces*, Integral Equations Operator Theory **15** (1992), no. 3, 503–525, DOI 10.1007/BF01200333. MR1155718

[343] J. Xiao, *The $\overline{\partial}$-problem for multipliers of the Sobolev space*, Manuscripta Math. **97** (1998), no. 2, 217–232, DOI 10.1007/s002290050098. MR1651405

[344] S. Zaremba, *L'equation biharmonique et une classe remarquable de fonctions fondamentales harmoniques*, Bulletin International de l'Académie des Sciences de Cracovie **3** (1907), 147–196.

[345] R. Zhao and K. Zhu, *Theory of Bergman spaces in the unit ball of $\mathbb{C}^n$* (English, with English and French summaries), Mém. Soc. Math. Fr. (N.S.) **115** (2008), vi+103 pp. (2009). MR2537698

[346] K. Zhu, *Spaces of holomorphic functions in the unit ball*, Graduate Texts in Mathematics, vol. 226, Springer-Verlag, New York, 2005. MR2115155

[347] K. H. Zhu, *Hankel operators on the Bergman space of bounded symmetric domains*, Trans. Amer. Math. Soc. **324** (1991), no. 2, 707–730, DOI 10.2307/2001738. MR1093426

[348] K. H. Zhu, *Operator theory in function spaces*, Monographs and Textbooks in Pure and Applied Mathematics, vol. 139, Marcel Dekker, Inc., New York, 1990. MR1074007

[349] A. Zygmund, *Trigonometric series. Vol. I, II*, 3rd ed., Cambridge Mathematical Library, Cambridge University Press, Cambridge, 2002. With a foreword by Robert A. Fefferman. MR1963498

Index

Selected Published Titles in This Series

239 **Nicola Arcozzi, Richard Rochberg, Eric T. Sawyer, and Brett D. Wick,** The Dirichlet Space and Related Function Spaces, 2019

237 **Dusa McDuff, Mohammad Tehrani, Kenji Fukaya, and Dominic Joyce,** Virtual Fundamental Cycles in Symplectic Topology, 2019

236 **Bernard Host and Bryna Kra,** Nilpotent Structures in Ergodic Theory, 2018

235 **Habib Ammari, Brian Fitzpatrick, Hyeonbae Kang, Matias Ruiz, Sanghyeon Yu, and Hai Zhang,** Mathematical and Computational Methods in Photonics and Phononics, 2018

234 **Vladimir I. Bogachev,** Weak Convergence of Measures, 2018

233 **N. V. Krylov,** Sobolev and Viscosity Solutions for Fully Nonlinear Elliptic and Parabolic Equations, 2018

232 **Dmitry Khavinson and Erik Lundberg,** Linear Holomorphic Partial Differential Equations and Classical Potential Theory, 2018

231 **Eberhard Kaniuth and Anthony To-Ming Lau,** Fourier and Fourier-Stieltjes Algebras on Locally Compact Groups, 2018

230 **Stephen D. Smith,** Applying the Classification of Finite Simple Groups, 2018

229 **Alexander Molev,** Sugawara Operators for Classical Lie Algebras, 2018

228 **Zhenbo Qin,** Hilbert Schemes of Points and Infinite Dimensional Lie Algebras, 2018

227 **Roberto Frigerio,** Bounded Cohomology of Discrete Groups, 2017

226 **Marcelo Aguiar and Swapneel Mahajan,** Topics in Hyperplane Arrangements, 2017

225 **Mario Bonk and Daniel Meyer,** Expanding Thurston Maps, 2017

224 **Ruy Exel,** Partial Dynamical Systems, Fell Bundles and Applications, 2017

223 **Guillaume Aubrun and Stanisław J. Szarek,** Alice and Bob Meet Banach, 2017

222 **Alexandru Buium,** Foundations of Arithmetic Differential Geometry, 2017

221 **Dennis Gaitsgory and Nick Rozenblyum,** A Study in Derived Algebraic Geometry, 2017

220 **A. Shen, V. A. Uspensky, and N. Vereshchagin,** Kolmogorov Complexity and Algorithmic Randomness, 2017

219 **Richard Evan Schwartz,** The Projective Heat Map, 2017

218 **Tushar Das, David Simmons, and Mariusz Urbański,** Geometry and Dynamics in Gromov Hyperbolic Metric Spaces, 2017

217 **Benoit Fresse,** Homotopy of Operads and Grothendieck–Teichmüller Groups, 2017

216 **Frederick W. Gehring, Gaven J. Martin, and Bruce P. Palka,** An Introduction to the Theory of Higher-Dimensional Quasiconformal Mappings, 2017

215 **Robert Bieri and Ralph Strebel,** On Groups of PL-homeomorphisms of the Real Line, 2016

214 **Jared Speck,** Shock Formation in Small-Data Solutions to 3D Quasilinear Wave Equations, 2016

213 **Harold G. Diamond and Wen-Bin Zhang (Cheung Man Ping),** Beurling Generalized Numbers, 2016

212 **Pandelis Dodos and Vassilis Kanellopoulos,** Ramsey Theory for Product Spaces, 2016

211 **Charlotte Hardouin, Jacques Sauloy, and Michael F. Singer,** Galois Theories of Linear Difference Equations: An Introduction, 2016

210 **Jason P. Bell, Dragos Ghioca, and Thomas J. Tucker,** The Dynamical Mordell–Lang Conjecture, 2016

209 **Steve Y. Oudot,** Persistence Theory: From Quiver Representations to Data Analysis, 2015

For a complete list of titles in this series, visit the
AMS Bookstore at **www.ams.org/bookstore/survseries/**.